Journal of Applied Logics – IfCoLog Journal of Logics and their Applications

Volume 9, Number 1

January 2022

Disclaimer

Statements of fact and opinion in the articles in Journal of Applied Logics – IfCoLog Journal of Logics and their Applications (JALs-FLAP) are those of the respective authors and contributors and not of the JALs-FLAP. Neither College Publications nor the JALs-FLAP make any representation, express or implied, in respect of the accuracy of the material in this journal and cannot accept any legal responsibility or liability for any errors or omissions that may be made. The reader should make his/her own evaluation as to the appropriateness or otherwise of any experimental technique described.

ISBN 978-1-84890-389-0
ISSN (E) 2631-9829
ISSN (P) 2631-9810

College Publications
Scientific Director: Dov Gabbay
Managing Director: Jane Spurr

http://www.collegepublications.co.uk

Editorial Board

Editors-in-Chief
Dov M. Gabbay and Jörg Siekmann

Marcello D'Agostino
Natasha Alechina
Sandra Alves
Arnon Avron
Jan Broersen
Martin Caminada
Balder ten Cate
Agata Ciabattoni
Robin Cooper
Luis Farinas del Cerro
Esther David
Didier Dubois
PM Dung
David Fernandez Duque
Jan van Eijck
Marcelo Falappa
Amy Felty
Eduaro Fermé
Melvin Fitting
Michael Gabbay
Murdoch Gabbay
Thomas F. Gordon
Wesley H. Holliday
Sara Kalvala
Shalom Lappin
Beishui Liao
David Makinson
Réka Markovich
George Metcalfe
Claudia Nalon
Valeria de Paiva
Jeff Paris
David Pearce
Pavlos Peppas
Brigitte Pientka
Elaine Pimentel
Henri Prade
David Pym
Ruy de Queiroz
Ram Ramanujam
Chrtian Retoré
Ulrike Sattler
Jörg Siekmann
Marija Slavkovik
Jane Spurr
Kaile Su
Leon van der Torre
Yde Venema
Rineke Verbrugge
Heinrich Wansing
Jef Wijsen
John Woods
Michael Wooldridge
Anna Zamansky

Scope and Submissions

This journal considers submission in all areas of pure and applied logic, including:

- pure logical systems
- proof theory
- constructive logic
- categorical logic
- modal and temporal logic
- model theory
- recursion theory
- type theory
- nominal theory
- nonclassical logics
- nonmonotonic logic
- numerical and uncertainty reasoning
- logic and AI
- foundations of logic programming
- belief change/revision
- systems of knowledge and belief
- logics and semantics of programming
- specification and verification
- agent theory
- databases
- dynamic logic
- quantum logic
- algebraic logic
- logic and cognition
- probabilistic logic
- logic and networks
- neuro-logical systems
- complexity
- argumentation theory
- logic and computation
- logic and language
- logic engineering
- knowledge-based systems
- automated reasoning
- knowledge representation
- logic in hardware and VLSI
- natural language
- concurrent computation
- planning

This journal will also consider papers on the application of logic in other subject areas: philosophy, cognitive science, physics etc. provided they have some formal content.

Submissions should be sent to Jane Spurr (jane@janespurr.net) as a pdf file, preferably compiled in LaTeX using the IFCoLog class file.

CONTENTS

ARTICLES

THE PROCEEDINGS OF THE XIX EBL — 19TH BRAZILIAN LOGIC CONFERENCE, 2019. A PREFACE

ITALA D'OTTAVIANO (UNICAMP), RICARDO SILVESTRE (UFCG) LEANDRO SUGUITANI (UFBA) PETRUCIO VIANA (UFF)
Guest Editors

The XIX EBL, 19th Brazilian Logic Conference, was held on the premises of the Hotel Nord Luxxor Tambaú, João Pessoa, PB, Brazil, during May 6-10, 2019. This venue, as the name indicates, is located in the Tambaú neighborhood, in front of one of the main beaches of the city.

João Pessoa, or Jampa—as the city is affectionately called by its residents—is the capital of Paraíba, a state located in the Northeast region of Brazil. Thinking about the Brazilian Northeast is thinking about beaches, coconut trees and warm weather all year round. In João Pessoa, the scenery is no different and, in addition to these attractions, those who visit Jampa can enjoy its natural beauty, a varied cuisine, rich folklore and precious contact with its friendly residents. All these features made Jampa the perfect place to hold the EBL 2019, allowing its participants to enjoy a rich cultural and academic environment.

The 163 participants from 12 countries held several fruitful discussions and exchanges that contributed to the success of the conference and to make it truly international in scope. The 6 keynote talks, 97 communications, 6 tutorials, and 5 round tables provided ample opportunity for discussion.

The six keynote speakers offered plenary lectures intercrossing critical areas of Logic: Luis Soares Barbosa (U. Minho, Portugal) "What Coalgebra Can Do for You?"; Mario Benevides (UFRJ, Brazil) "What Makes a Logic Dynamic?"; Mirna Dzamonja (UEA, United Kingdon) "Logic as a Modelling Tool"; Eberhard Guhe (Fudan, China) "Defeasible Reasoning in Navya-Nyāya"; Catarina Dutilh Novaes (UvA, Netherlands) "Paradoxes and Structural Rules from a Dialogical Perspective"; Elaine Pimentel (UFRN, Brazil) "Modalities as Prices: a Game Model of

Generous financial support for the conference was provided by CNPq. The conference was officially sponsored and organized by Universidade Federal de Campina Grande (UFCG), Universidade Federal da Paraíba (UFPB), and Sociedade Brasileira de Lógica (SBL).

Intuitionistic Linear Logic with Subexponentials"; Frank Sautter (UFSM, Brazil) "On Teaching Logic for Undergraduate Philosophy Students"; Edward Zalta (Stanford, EUA) "Logic and Existence: How Logic and Metaphysics are Entangled".

The 97 communications were split into nine main sections: Algebraic Approaches; Logic and Computer Science; Logic and Philosophy; Logic Teaching; Mathematics and Computer Science; Metalogic; Modal Logics; Models and Sets; and Proofs and Automated Proving. They are listed in the online Book of Abstracts to be found at `https://ebl2019.ci.ufpb.br/assets/Book_of_Abstracts_EBL_2019.pdf`.

As part of the 19th Brazilian Logic Conference — as it has become already traditional — a Logic School, aimed at advanced graduate students and other people interested in the study of logic and related areas, was organized. The Logic School had six tutorials presented in English or Portuguese, covering the most varied aspects of the relationships between logic, computation, mathematics, and philosophy: Jean-Yves Beziau (UFRJ, Brazil) "Classical Propositional Logic: A Universal Logic Approach"; Juliana Bueno-Soler (Unicamp, Brazil) and Walter Carnielli (Unicamp, Brazil) "Lógica, Probabilidade: Encontros e Desencontros"; Marcelo E. Coniglio (Unicamp, Brazil) "Semânticas não Determinísticas"; Eberhard Guhe (Fudan, China) "Some Highlights of Indian Logic"; Samuel Gomes da Silva (UFBA, Brazil) "Uma Miscelânea de Aplicações de Ultrafiltros em Matemática"; Renata Wassermann (USP, Brazil) "Logics in Artificial Intelligence".

During the event a posthumous tribute was paid to Giovanni Queiroz, Arley Ramos Moreno, Oswaldo Porchat Pereira and Carolina Blasio. There was also a book launching session for the following titles: Paraconsistent Logic: Consistency, Contradiction and Negation, by Walter Carnielli and Marcelo Esteban Coniglio (Springer, 2016); Alfred Tarski: Lectures at Unicamp in 1975, by Leandro Suguitani, Jorge Petrúcio Viana and Itala M. Loffredo D'Ottaviano (Editors) (Unicamp University Press, 2016); Para além das Colunas de Hércules, uma História da Paraconsistência: de Heráclito a Newton da Costa (Beyond the columns of Hercules, a history of paraconsistency: from Heraclitus to Newton da Costa, in Portuguese), by Evandro Luís Gomes and Itala M. Loffredo D'Ottaviano (Unicamp University Press, 2017). It also is worth mentioning that the Brazilian Logic Society, in a deliberative assembly, launched the "Aula Magna Itala Maria Loffredo D'Ottaviano", a lecture to be delivered in every Brazilian Logic Conference preferably by women.

After the end of the conference, a call for papers was launched, aiming at the publication of a special issue of The Journal of Applied Logics: IfCoLog Journal of Logics and their Applications `https://www.collegepublications.co.uk/ifcolog/`, containing the full versions of the communications presented at EBL 2019. All the plenary speakers and contributing authors were invited to submit a full version of their

EBL 2019 talks and communications, as well as papers of general interest within the conference themes: Philosophical and Mathematical Logic, and Applications; History and Philosophy of Logic; Non-classical Logic and Applications; Philosophy of Formal Sciences; Foundations of Computer Science, Physics, and Mathematics; Logic Education (Logic Teaching). A total of 28 full papers were submitted and subjected to a standard double-blind refereeing process. The following 18 articles were selected for publication in this special volume: Induced morphisms between Heyting-valued models, by J. Alvim, A. Cahali and H. Mariano; Coalgebra for the working software engineer, by L. Barbosa; On the order theory for $\mathcal{C}^\infty-$reduced $\mathcal{C}^\infty-$Rings and applications, by J. Berni, R. Figueiredo and H. Mariano; Strong normalization for Np-Systems via Mimp-Graphs, by V. Costa and E. Haeusler; The Generalized Continuum Hypothesis and two parametrized families of hit-and-miss games, by S. G. da Silva; G'3 as the logic of modal 3-valued Heyting algebras, by A. Figallo-Orellano, M. Coniglio, A. Hernandez-Tello and M. Perez-Gaspar; A logical framework to reason about Reo circuits, by E. Grilo, B. Lopes and D. Toledo; Defeasible reasoning in Navya-Nyaya, by E. Guhe; Huge normal proofs in M $\supset$ are redundant, by E. Haeusler; Extensive measurement with unrestricted concatenation and no maximal elements, by G. Kyriazis; An algebraic (set) theory of surreal numbers, I, by H. Mariano and D. Rangel; Horn-geometric axioms to faithfully quadratic rings, by H. Mariano and H. Ribeiro; On superrings of polynomials and algebraically closed multifields, by H. Mariano and K. Roberto; Connecting abstract logics and adjunctions between Institutions and π-Institutions, by H. Mariano, D. Pinto, G. Rios and D. Souza; Some classical modal logics with a necessity/impossibility operator, by C. Mortari; Logicism in the eyes of the author of Tractatus Logico-Philosophicus (and of Philosophical Remarks), by P. Noguez; Proof-search, models, analytic tableaux and counter-models, in huge constructive semantics for for Minimal and Intuitionistic Propositional Logic, by W. Sanz; A note on Tarski's remarks about the non-admissibility of a general theory of semantics, by G. Sarmento.

The editors would like to extend their gratitude and appreciation to the referees whose (necessarily anonymous) efforts helped to select the papers that appear in this volume and to improve their contents. We think it is important to emphasize that the entire refereeing process was carried out during the difficult period in which the world was adapting to the pandemic caused by the Coronavirus (COVID-19). Within this perspective, authors, editors and reviewers made every possible effort to ensure that the accepted articles had the necessary quality for publication in The Journal of Applied Logics: IfCoLog Journal of Logics and their Applications.

Below is some additional information about the Brazilian Logic Conference (EBL):

The Brazilian Logic Conferences. The Brazilian Logic Conference (EBL) is a traditional event of the Brazilian Logic Society (SBL) `http://sbl.org.br`. It has been occurring since 1979. Congregating logicians of different fields and with different backgrounds—from undergraduate students to senior researchers—the meeting is an important moment for the Brazilian and South-American logical community to gather together and discuss recent developments of the field. The areas of Logic covered spread over Foundations and Philosophy of Science, Analytic Philosophy, Mathematics, Computer Science, Informatics, Linguistics and Artificial Intelligence. The goal of the EBL meeting is to encourage the dissemination and discussion of research papers in Logic in a broad sense. It usually has among the participants several invited speakers from different continents. Previous editions of the EBL have attracted researchers from all over Latin America and elsewhere.

Conference Organisation. The following colleagues from Brazil served at the scientific and organizing committees, for whom special thanks are due for their invaluable assistance in organizing, finding funds and granting the high scientific level of the conference:

Scientific Committee. Alexandre Rademaker (IBM Research); Bruno Lopes Vieira (UFF); Carlos Olarte (UFRN); Ciro Russo (UFBA); Cláudia Nalon (UnB); Daniele Nantes Sobrinho (UnB); Edward Hermann Haeusler (PUC-Rio); Gisele Secco (UFSM); Itala Maria Loffredo D'Ottaviano (Unicamp); João Marcos (UFRN); Petrucio Viana (UFF); Leandro Oliva Suguitani (UFBA); Luiz Carlos Pereira (PUC-Rio/UERJ); Marcelo Coniglio (Unicamp); Marcelo Finger (USP); Maurício Ayala-Rincón (UnB); Nastassja Pugliese (USP); Renata Wassermann (USP); Newton C. A. da Costa (UFSC); Valeria de Paiva (Nuance Communications); Walter Carnielli (Unicamp).

Organizing Committee. Ricardo Sousa Silvestre (UFCG, co-chair); Ana Thereza DÃijrmaier (UFPB, co-chair); Cezar A. Mortari (UFSC, co-chair); Garibaldi Sarmento (UFPB); Samuel Gomes da Silva (UFBA); Lucídio Cabral (UFPB); Wagner Sanz (UFG); Tiago Massoni (UFCG); Bruno Petrato Bruck (UFPB); Teobaldo Leite Bulhões Júnior (UFPB); Andrei de Araujo Formiga (UFPB); Pedro Carné (UFCG); Marcio Kléos Freire Pereira (UFMA).

Acknowledgement

We would like to thank all the people who made this special edition containing the Proceedings to the EBL 2019 possible.

Received November 2021

INDUCED MORPHISMS BETWEEN HEYTING-VALUED MODELS

José Goudet Alvim
Institute of Mathematics and Statistics, University of São Paulo, Brazil
jose.alvim@usp.br

Arthur Francisco Schwerz Cahali
Institute of Mathematics and Statistics, University of São Paulo, Brazil
arthur.cahali@usp.br

Hugo Luiz Mariano
Institute of Mathematics and Statistics, University of São Paulo, Brazil
hugomar@ime.usp.br

Abstract

To the best of our knowledge, there are very few results on how Heyting-valued models are affected by the morphisms on the complete Heyting algebras that determine them: the only cases found in the literature are concerning automorphisms of complete Boolean algebras and complete embeddings between them (*i.e.*, injective Boolean algebra homomorphisms that preserve arbitrary suprema and arbitrary infima). In the present work, we consider and explore how more general kinds of morphisms between complete Heyting algebras $\mathbb{H}$ and $\mathbb{H}'$ induce arrows between $V^{(\mathbb{H})}$ and $V^{(\mathbb{H}')}$, and between their corresponding localic topoi $\mathbf{Set}^{(\mathbb{H})}$ ($\simeq \mathbf{Sh}(\mathbb{H})$) and $\mathbf{Set}^{(\mathbb{H}')}$ ($\simeq \mathbf{Sh}(\mathbb{H}')$). More specifically: any *geometric morphism* $f^* : \mathbf{Set}^{(\mathbb{H})} \to \mathbf{Set}^{(\mathbb{H}')}$ (that automatically came from a unique locale morphism $f : \mathbb{H} \to \mathbb{H}'$) can be "lifted" to an arrow $\tilde{f} : V^{(\mathbb{H})} \to V^{(\mathbb{H}')}$. We also provide some semantic preservation results concerning this arrow $\tilde{f} : V^{(\mathbb{H})} \to V^{(\mathbb{H}')}$.

Keywords: Heyting-valued models; localic topoi; geometric morphisms.

We want to express our gratitude to the referees for their careful reading and valuable suggestions that have improved the submitted version.

Introduction

The expression "Heyting-valued model of set theory" has two (related) meanings, both parametrized by a complete Heyting algebra $\mathbb{H}$:

(i) The canonical Heyting-valued models in set theory, $V^{(\mathbb{H})}$, as introduced in the setting of complete boolean algebras in the 1960s by D. Scott, P. Vopěnka and R. M. Solovay in an attempt to help understand the then recently introduced notion of *forcing* in ZF set theory developed by P. Cohen ([9], [10], [1]);

(ii) The (local) "set-like" behavior of categories called *topoi*, particularly in the case of the (localic) topoi of the form $\mathbf{Sh}(\mathbb{H})$ ([4], [2]).

The concept of a Heyting/Boolean-valued model is nowadays a general model-theoretic notion, whose definition is independent of forcing in set theory: it is a generalization of the ordinary Tarskian notion of structure where the truth values of formulas are not limited to "true" and "false", but instead take values in some fixed complete Heyting algebra $\mathbb{H}$. More precisely, an $\mathbb{H}$-valued model M in a first-order language L consists of an underlying set M and an assignment $[\![\varphi]\!]_{\mathbb{H}}$ of an element of $\mathbb{H}$ to each formula φ with parameters in M, satisfying convenient conditions.

The canonical Heyting-valued model in set theory associated to $\mathbb{H}$ is the pair $\left\langle V^{(\mathbb{H})}, [\![\ ]\!]_{\mathbb{H}} \right\rangle$, where both components are recursively defined. Explicitly, $V^{(\mathbb{H})}$ is the proper class $V^{(\mathbb{H})} := \bigcup_{\beta \in \mathrm{On}} V_{\beta}^{(\mathbb{H})}$, where $V_{\beta}^{(\mathbb{H})}$ is the set of all functions f such that $\mathrm{dom}(f) \subseteq V_{\alpha}^{(\mathbb{H})}$, for some $\alpha < \beta$, and $\mathrm{img}(f) \subseteq \mathbb{H}$. Whenever $\mathbb{H}$ is a complete boolean algebra $\left\langle V^{(\mathbb{H})}, [\![\ ]\!]_{\mathbb{H}} \right\rangle$ is a model of ZFC in the sense that for each axiom σ of ZFC, $[\![\sigma]\!]_{\mathbb{H}} = 1_{\mathbb{H}}$; more generaly, if $\mathbb{H}$ is a complete Heyting algebra, then $\left\langle V^{(\mathbb{H})}, [\![\ ]\!]_{\mathbb{H}} \right\rangle$ is a model IZF, the intuitionistic counterpart of ZFC.

On the other hand, $V^{(\mathbb{H})}$ may give rise to a localic topos, $\mathbf{Set}^{(\mathbb{H})}$, that is equivalent to the (Grothendieck) topos $\mathbf{Sh}(\mathbb{H})$ of all sheaves over the locale (= complete Heyting algebra) $\mathbb{H}$ ([1], [2]). The objects of $\mathbf{Set}^{(\mathbb{H})}$ are equivalence classes of members of $V^{(\mathbb{H})}$ and the arrows are (equivalence classes of) members f of $V^{(\mathbb{H})}$ such that "$V^{(\mathbb{H})}$ believes, with probability $1_{\mathbb{H}}$, that f is a function". A general topos encodes an internal (higher-order) intuitionistic logic, given by the "forcing-like" Kripke-Joyal semantics, and some form of (local) set-theory ([2], [4]); a localic Grothendieck topos is guided by a better behaved internal logic and set theory.

All the considerations above concern a fixed complete Heyting algebra $\mathbb{H}$. However, to the best of our knowledge, there are very few results on how Heyting-valued models are affected by the morphisms between their algebras. The only cases found in the literature ([1]) are concerning automorphisms of complete Heyting algebras and complete embeddings (*i.e.*, injective Heyting algebra homomorphisms that preserves arbitrary suprema and arbitrary infima). In the present work, we consider and explore how more general kinds of morphisms between complete Heyting algebras $\mathbb{H}$ and $\mathbb{H}'$ induce arrows between $\mathrm{V}^{(\mathbb{H})}$ and $\mathrm{V}^{(\mathbb{H}')}$, and between their corresponding Heyting topoi $\mathbf{Set}^{(\mathbb{H})}$ ($\simeq \mathbf{Sh}(\mathbb{H})$) and $\mathbf{Set}^{((\mathbb{H}'))}$ ($\simeq \mathbf{Sh}(\mathbb{H}')$). The result is: any *geometric morphism* $f^* : \mathbf{Set}^{(\mathbb{H})} \to \mathbf{Set}^{(\mathbb{H}')}$ (that automatically came from a unique locale morphism $f : \mathbb{H} \to \mathbb{H}'$) can be "lifted" to an arrow $\tilde{f} : \mathrm{V}^{(\mathbb{H})} \to \mathrm{V}^{(\mathbb{H}')}$.

Outline: In **Section 1** we provide the main definitions on sheaves over locales (= complete Heyting algebras), topoi and Heyting-valued models of IZF. **Section 2** is devoted to present the equivalent descriptions of the categories of sheaves over a locale, in particular establishing a connection between the cumulative construction of Heyting valued models and localic topoi. **Section 3** contains the main results of this work: the "lifting" of all geometric morphisms $f^* : \mathbf{Sh}(\mathbb{H}) \to \mathbf{Sh}(\mathbb{H}')$ to arrows $\tilde{f} : \mathrm{V}^{(\mathbb{H})} \to \mathrm{V}^{(\mathbb{H}')}$ and the corresponding semantic preservation results. We end this work in **Section 4** presenting some remarks on possible further developments.

1 Preliminaries

For the reader's convenience, we provide here the main definitions and results on topooi and Heyting valued models of set theory. Our main references for category theory are [11] and [3]; for topos theory [4] and [2] and for Boolean and Heyting valued models [1].

1.1 Topoi and Grothendieck Topoi

If $\langle X, \mathcal{O}(X)\rangle$ is a topological space, then the family of sets of continuous functions $(\mathcal{C}(U,\mathbb{R}))_{U\in\mathcal{O}(X)}$ has the property that, for any open subset U and any open covering $U = \bigcup_{i\in I} V_i$ every family of continuous functions $(f_i \in \mathcal{C}(V_i,\mathbb{R}))_{i\in I}$ that is *compatible* ($f_{i|V_i\cap V_j} = f_{j|V_i\cap V_j}$, $\forall i,j \in I$), has a unique *gluing* $f \in \mathcal{C}(U,\mathbb{R})$ ($f_{|V_i} = f_i$, $\forall i \in I$): This holds since the property of being continuous is a local property; an analogous remark holds for the $\mathcal{C}^\infty$ functions if X is a smooth manifold. Formally, this is captured by the following:

Definition 1.1. *Let $\langle X, \mathcal{O}(X)\rangle$ be a topological space. Regarding the poset of open sets $(\mathcal{O}(X), \subseteq)$ as a category, a presheaf on X is a functor $F : \mathcal{O}(X)^{op} \to$ **Set**. A sheaf on X is a presheaf F such that, for every open $U \in \mathcal{O}(X)$ and every open covering $\{U_i \in \mathcal{O}(X) \mid i \in I\}$ of U, the diagram below is an equalizer:*

$$F(U) \longrightarrow \prod_{i \in I} F(U_i) \rightrightarrows \prod_{\langle i,j \rangle \in I \times I} F(U_i \cap U_j)$$

We denote the category of presheaves on X by $\mathbf{Psh}(X)$ *and the category of sheaves on X by* $\mathbf{Sh}(X)$.

Note that the definition of a sheaf depends only on the lattice of opens, therefore we may define presheaves and sheaves for any *locale* $\langle \mathbb{H}, \leq \rangle$, *i.e.* a complete lattice satisfying the following distributive law:

$$a \wedge \bigvee_{i \in I} c_i = \bigvee_{i \in I} a \wedge c_i.$$

Locales are precisely the complete Heyting algebras, where

$$a \to b = \bigvee \{c \in \mathbb{H} : a \wedge c \leq b\}.$$

It is also possible to define sheaves in more general categories, using Grothendieck topologies.

Definition 1.2. *Let $\mathcal{C}$ be a small category. A Grothendieck topology on $\mathcal{C}$ is a function J which assigns to each object $c \in Obj(\mathcal{C})$ a family $J(c)$ of sieves on c, satisfying:*

1. *Maximal sieve:* $\bigcup_{a \in \mathcal{C}_0} \mathcal{C}(a, c) \in J(c)$*, for all* $c \in Obj(\mathcal{C})$*;*

2. *Pullback stability: given $c \in Obj(\mathcal{C})$, for every $S \in J(c)$ and every arrow $f : a \to c$, the pullback f^*S of the sieve S along f is an element of $J(a)$;*

3. *Transitivity: given $c \in Obj(\mathcal{C})$ and $S \in J(c)$, if R is a sieve on c such that f^*R is a sieve on a for every $f : a \to c$ in S, then $R \in J(c)$.*

We call the pair $(\mathcal{C}, J)$ a (small) site.

Every locale $(\mathbb{H}, \leq)$ gives rise to a Grothendieck topology: if $c \in \mathbb{H}$, then $J(c)$ is the set of all coverings of c that are downward closed. Another important example is the Zariski topology in algebraic geometry.

Definition 1.3. *Let $(\mathcal{C}, J)$ be a site and $F : \mathcal{C}^{op} \to$* **Set** *a presheaf. We say F is a sheaf on $(\mathcal{C}, J)$ if, for all $c \in Obj(\mathcal{C})$, every collection $\{f_i : a_i \to c \mid i \in I\} \in J(c)$ and every F-compatible family $\{s_i \in F(a_i) \mid i \in I\}$, there exists a unique $s \in F(c)$ such that $F(f_i)(s) = s_i$, for all $i \in I$. We denote the category of sheaves on $(\mathcal{C}, J)$ by* $\mathbf{Sh}(\mathcal{C}, J)$.

A Grothendieck topos is a category that is equivalent to the topos of sheaves on a site. Apart from categories of sheaves, the category of presheaves is also an example of a Grothendieck topos.

Some properties of Grothendieck topoi are of particular interest for developing logic in the context of category theory, such as containing a subobject classifier and being Cartesian closed.

Definition 1.4. *Given a category $\mathcal{C}$ with pullbacks, a subobject classifier in $\mathcal{C}$ is a mono $\top : u \rightarrowtail \Omega$ satisfying: for every other mono $m : a \rightarrowtail b$, there exists a unique $\chi_m : b \to \Omega$ such that the following diagram is a pullback:*

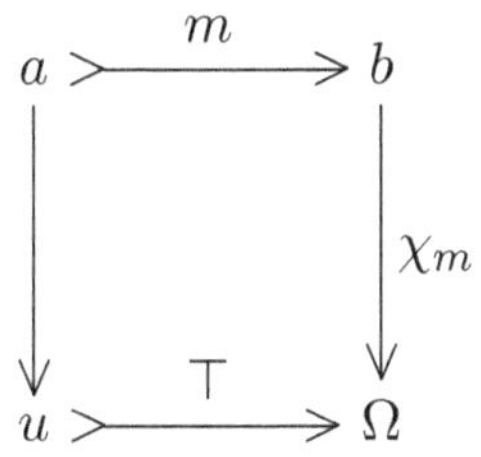

Definition 1.5. *Let $\mathcal{C}$ be a locally small category with binary products. The category $\mathcal{C}$ is called Cartesian closed if, for every $b \in Obj(\mathcal{C})$, the product functor $- \times b$ has a right adjoint (the exponentiation functor $(-)^b$).*

Both these properties are used to define a more general notion of topos: elementary topos.

1.2 Localic Topoi

Definition 1.6. *A topos is said to be localic if it is equivalent to the topos of sheaves on a locale.*

Theorem 1.7. *For a Grothendieck topos $\mathcal{T}$, the following conditions are equivalent:*

1. *$\mathcal{T}$ is a localic topos;*
2. *the subobjects of the terminal object constitute a family of generators of $\mathcal{T}$.*

Theorem 1.8. *For a Grothendieck topos $\mathcal{T}$, the following conditions are equivalent:*

1. *$\mathcal{T}$ is a localic and Boolean topos;*
2. *$\mathcal{T}$ satisfies the axiom of choice;*
3. *$\mathcal{T} \simeq \mathbf{Sh}(\mathbb{B})$, for some Boolean algebra $\mathbb{B}$.*

A continuous function between topological spaces defines a $(\wedge, \bigvee)$-preserving morphism (*i.e.*, the left adjoint of a morphism of locales) between the locales of open sets, and a geometric morphism between the corresponding sheaf topoi:

$$
\begin{array}{ccccc}
X & & \mathcal{O}(X) & & \mathbf{Sh}\,(\mathcal{O}(X)) \\
\Big\downarrow f & \mapsto & \Big\uparrow f^{-1} & \mapsto & \varphi_* \Big\downarrow \Big\uparrow \varphi^* \\
X' & & \mathcal{O}(X') & & \mathbf{Sh}\,(\mathcal{O}(X'))
\end{array}
$$

That is, (φ_*, φ^*) is a pair of functors such that $\varphi^* \dashv \varphi_*$ and φ^* preserves finite limits.

This mapping from the category of topological spaces to the category of topoi and geometric morphisms is not full nor faithful. However, the mapping from the category of locales to the category of topoi and geometric morphisms is fully faithful:

Theorem 1.9. *The mapping below, given by* $\varphi_*(F) = F \circ f^*$, *for every sheaf* F *on* $\mathbb{H}$,

$$
\begin{array}{ccc}
\mathbb{H} & & \mathbf{Sh}\,(\mathbb{H}) \\
f_* \Big\downarrow \Big\uparrow f^* \quad (\wedge, \bigvee) & \mapsto & \varphi_* \Big\downarrow \Big\uparrow \varphi^* \\
\mathbb{H}' & & \mathbf{Sh}\,(\mathbb{H}')
\end{array}
$$

defines a fully faithful functor $\mathbf{Sh} : \mathbf{Loc} \to \mathbf{Topos}_{geo}$.

1.3 Locale-Valued Models

Definition 1.10. *Locale-Valued Model*

We define, for a locale $\mathbb{H}$, *the universe of* $\mathbb{H}$*-names by ordinal recursion. Given an ordinal* α *let*

$$V_\alpha^{(\mathbb{H})} = \left\{ f \in \mathbb{H}^X \mid \exists \beta < \alpha, X \subseteq V_\beta^{(\mathbb{H})} \right\}$$

It is readily seen that $V_\alpha^{(\mathbb{H})} \subset V_{\alpha+1}^{(\mathbb{H})}$ *and that for limit ordinals it is simply the union of the earlier stages. So we let the (proper class)* $V^{(\mathbb{H})}$ *be defined as:*

$$V^{(\mathbb{H})} = \bigcup_{\alpha \in \mathrm{On}} V_\alpha^{(\mathbb{H})}$$

Furthermore, we define for elements of this universe $V^{(\mathbb{H})}$ a function $\varrho(x)$ defined as:

$$\varrho(x) = \min\left\{\alpha \in \mathrm{On} \mid x \in V_{\alpha}^{(\mathbb{H})}\right\}$$

Which is trivially well-founded.

Definition 1.11. *Atomic Formulas' Values*

We endow this class with two[1] binary functions on $\mathbb{H}$, namely $[\![\cdot \in \cdot]\!]$ and $[\![\cdot = \cdot]\!]$ defined by simultaneous recursion on a well-founded relation.

Given a locale $\mathbb{H}$ define first:

$$\langle x, y\rangle \prec \langle u, v\rangle \iff (x = u \wedge y \in dom(v)) \vee (x \in dom(u) \wedge y = v)$$

We will later see that this is a well-founded relation on $V^{(\mathbb{H})} \times V^{(\mathbb{H})}$. By recursion on $\prec$, define for pairs of elements in $V^{(\mathbb{H})}$, the following:

$$\begin{aligned} [\![\cdot \in \cdot]\!] : V^{(\mathbb{H})} \times V^{(\mathbb{H})} &\longrightarrow \mathbb{H} \\ \langle x, y\rangle &\longmapsto \bigvee_{u \in dom(y)} y(u) \wedge [\![x = u]\!] \end{aligned}$$

$$\begin{aligned} [\![\cdot \ni \cdot]\!] : V^{(\mathbb{H})} \times V^{(\mathbb{H})} &\longrightarrow \mathbb{H} \\ \langle x, y\rangle &\longmapsto \bigvee_{v \in dom(x)} x(v) \wedge [\![v = y]\!] \end{aligned}$$

$$\begin{aligned} [\![\cdot = \cdot]\!] : V^{(\mathbb{H})} \times V^{(\mathbb{H})} &\longrightarrow \mathbb{H} \\ \langle x, y\rangle &\longmapsto \bigwedge_{\substack{u \in dom(y) \\ v \in dom(x)}} (y(u) \to [\![x \ni u]\!]) \wedge (x(v) \to [\![v \in y]\!]) \end{aligned}$$

We call these the $\mathbb{H}$-values of the membership, co-membership and, equality, respectively. Where ambiguity may arise, we make a distinction between valuations of different locale-valued models by subscripting $[\![\ ]\!]$ with the relevant locale, or by simply placing a prime symbol over one of the otherwise ambiguous evaluation function: "$[\![\ ]\!]'$".

Proposition 1.12. *The relation $\prec$ is well-founded.*

[1]Technically three.

Proof. Firstly, define $\varrho'(x,y) = \min\{\varrho(x), \varrho(y)\}$. Take any subclass X of $\mathrm{V}^{(\mathbb{H})} \times \mathrm{V}^{(\mathbb{H})}$ and consider its image under ϱ'. For one, it has a minimum, due to the well-orderedness of On, let us call this minimum value α and one pair in X such that it attains value α we name $\langle x, y\rangle$. Without loss of generality, we assume $\varrho(x) = \alpha$.

Suppose now there is some $\langle f, g\rangle$ such that $\langle f, g\rangle \prec \langle x, y\rangle$. By definition:

$$(f \in \mathrm{dom}(x) \wedge g = y) \vee (f = x \wedge g \in \mathrm{dom}(y))$$

Breaking the disjunction we realize that the value of $\langle f, g\rangle$ under ϱ' must be no more than α:

$$\dfrac{\dfrac{f \in \mathrm{dom}(x) \wedge g = y}{\varrho'(f,g) \leq \varrho'(x,y)} \qquad \dfrac{f = x \wedge g \in \mathrm{dom}(y)}{\varrho'(f,g) \leq \varrho'(x,y)}}{\varrho'(f,g) \leq \varrho'(x,y)}$$

But since α is minimal, it follows that they must be the same. Therefore, f can't be in the domain of x, for then it would have a smaller rank than α. One is forced to conclude that, under this assumption, $\langle f, g\rangle \prec \langle x, y\rangle \to g \in \mathrm{dom}(y)$, and since it is necessary that $f = x$. Consequently, because the relation $(\cdot \in \mathrm{dom}(\cdot))$ is well-founded, one is forced to concede that if a descending $\prec$-chain does not stabilize, so too will a descending $(\cdot \in \mathrm{dom}(\cdot))$-chain, a contradiction. □

We are thus entitled to make the definition of those $\mathbb{H}$-values as we previously claimed. The definitions of $\mathbb{H}$-values are then extended to the class of the language of $\mathcal{L}_\in$-formulas enriched with constant symbols for each member of the class $\mathrm{V}^{(\mathbb{H})}$:

Definition 1.13. *$\mathbb{H}$-valuation of Formulas*

We define the value of a ${\mathcal{L}_\in}^{\mathbb{H}}$-formula φ which is the language $\mathcal{L}_\in$ extended by constant symbols for each element of $V^{(\mathbb{H})}$ the valuation $[\![\varphi]\!]$ inductively on the complexity of φ.

$$[\![\varphi]\!] : (\mathrm{Vars}_{{\mathcal{L}_\in}^{\mathbb{H}}} \to V^{(\mathbb{H})}) \to \mathbb{H}$$

For an atomic formula φ involving only constants symbols as terms, $[\![\varphi]\!]$ is simply the value of the corresponding function defined earlier, that is if $\varphi \equiv aRb$ then, its valuation is the constant function:

$$[\![\varphi]\!] = [\![aRb]\!]$$

For atomic formulas with free variables, the valuation is a function of the language's variable symbols. Given a function that assigns values to its free variables, it yields the value of the corresponding valuation, i.e. *if* $\varphi \equiv x_i R x_j$ *for variable symbols* x_i, x_j.

$$\llbracket \varphi \rrbracket : \left(\mathrm{Vars}_{\mathcal{L}_\in^{\mathbb{H}}} \to V^{(\mathbb{H})}\right) \to \mathbb{H}$$
$$v \mapsto \llbracket v(x_i) R v(x_j) \rrbracket$$

If φ *has mixed constants and variable symbols, the definition is analogous but fixing the constants.*

When φ *has free variables, we often write* $\llbracket \varphi(x, y \cdots) \rrbracket$ *to denote the function on its free variables, rather than writing* $\llbracket \varphi \rrbracket (\cdots)$. *For complex formulas, we define, for negation:*

$$\llbracket \neg\varphi \rrbracket = \llbracket \varphi \rrbracket \to \bot$$

For a binary connective $*$ *among* $\to$, $\wedge$, $\vee$*:*

$$\llbracket \varphi * \psi \rrbracket = \llbracket \varphi \rrbracket * \llbracket \psi \rrbracket$$

Given x_i *a variable symbol and* $x \in V^{(\mathbb{H})}$ *and a function*

$$v : \mathrm{Vars}_{\mathcal{L}_\in^{\mathbb{H}}} \to V^{(\mathbb{H})}$$

Define now, the function

$$v_{[x_i|x]}(s) = \begin{cases} v(s), \text{ if } s \neq x_i \\ x, \text{ otherwise} \end{cases}$$

For quantifiers:

$$\llbracket \forall x_i : \varphi \rrbracket : \left(\mathrm{Vars}_{\mathcal{L}_\in^{\mathbb{H}}} \to V^{(\mathbb{H})}\right) \to \mathbb{H}$$
$$v \mapsto \bigwedge_{x \in V^{(\mathbb{H})}} \llbracket \varphi \rrbracket \left(v_{[x_i|x]}\right)$$

And dually,

$$\llbracket \exists x_i : \varphi \rrbracket : \left(\mathrm{Vars}_{\mathcal{L}_\in^{\mathbb{H}}} \to V^{(\mathbb{H})}\right) \to \mathbb{H}$$
$$v \mapsto \bigvee_{x \in V^{(\mathbb{H})}} \llbracket \varphi \rrbracket \left(v_{[x_i|x]}\right)$$

If two functions $f, g : \mathrm{Vars}_{\mathcal{L}_\in^{\mathbb{H}}} \to V^{(\mathbb{H})}$ *coincide on* free_φ*, then* $[\![\varphi]\!](f) = [\![\varphi]\!](g)$*, then the sentences are constant functions — and we will often omit the fact that these valuations are indeed functions of the values we assign to the variable symbols of the formulas and concern ourselves with sentences, which correspond to values in* $\mathbb{H}$*.*

We won't distinguish between the constant symbols corresponding to elements of $V^{(\mathbb{H})}$ *and the elements they correspond to.*

We simply state the following results without proof — as they can be straightforwardly adapted from those on [1].

Theorem 1.14. *Properties of Formula valuation*

1. $[\![x = x]\!] = 1$.
2. $\forall x \in dom(y) : y(x) \leq [\![x \in y]\!]$.
3. $[\![x = y]\!] = [\![y = x]\!]$.
4. $[\![x \in y]\!] = [\![y \ni x]\!]$.
5. $[\![x = y]\!] \wedge [\![y = z]\!] \leq [\![x = z]\!]$.
6. $[\![x = y]\!] \wedge [\![y \in z]\!] \leq [\![x \in z]\!]$.
7. $[\![x \in y]\!] \wedge [\![y = z]\!] \leq [\![x \in z]\!]$.
8. $[\![x = y]\!] \wedge x(u) \leq [\![u \in y]\!]$.
9. $[\![x = y]\!] \wedge [\![\varphi(x)]\!] = [\![y = x]\!] \wedge [\![\varphi(y)]\!]$.
10. $[\![\exists u \in x : \varphi(u)]\!] = \bigvee_{u \in dom(x)} x(u) \wedge [\![\varphi(u)]\!]$
11. $[\![\forall u \in x : \varphi(u)]\!] = \bigwedge_{u \in dom(x)} x(u) \to [\![\varphi(u)]\!]$

Where $Qu \in x : \varphi(u)$ *is the usual shorthand for either* $[\forall u : u \in x \to \varphi(u)]$ *or* $[\exists u : u \in x \wedge \varphi(u)]$ *for* Q *standing for "*$\forall$*" or "*$\exists$*".*

We then must define a "localic semantic", or a notion of truth for that structure so that we may claim that it actually models some form of set theory.

Definition 1.15. $\mathbb{H}$*-Semantic /* $\mathbb{H}$*-Validity / Localic Semantic*
We define a Tarskian-like $\models$ *for each* $\mathbb{H}$ *to say:*

$$V^{(\mathbb{H})} \models \varphi \iff [\![\varphi]\!] = \top$$

Since $[\![\varphi]\!]$ is a function in disguise, to properly make this comparison, it either must be constant or we must convert $\top$ to the constantly tautological function. The latter allows us to interpret formulas with free variables, and will assign them truth if they are always true under any valuation.

Hence, we extend the notion for, given $\sigma : \mathrm{Vars}_{\mathcal{L}_\in^{\mathbb{H}}} \to V^{(\mathbb{H})}$,

$$V^{(\mathbb{H})} \models_\sigma \varphi \iff [\![\varphi]\!](\sigma) = \top$$

Proposition 1.16. *Some properties of the localic $\models$*

1. $V^{(\mathbb{H})} \models_\sigma \varphi$ *and* $V^{(\mathbb{H})} \models_\sigma \psi \iff V^{(\mathbb{H})} \models_\sigma \varphi \wedge \psi$
2. $V^{(\mathbb{H})} \models_\sigma \forall x : \varphi(x) \iff$ *for all* X, $V^{(\mathbb{H})} \models_{\sigma[x|X]} \varphi(x)$
3. *For some* X, $V^{(\mathbb{H})} \models_{\sigma[x|X]} \varphi(x) \Rightarrow V^{(\mathbb{H})} \models_\sigma \exists x : \varphi(x)$
4. $V^{(\mathbb{H})} \models_\sigma \neg\varphi \Rightarrow V^{(\mathbb{H})} \not\models_\sigma \varphi$
5. $V^{(\mathbb{H})} \models_\sigma \varphi$ *or* $V^{(\mathbb{H})} \models_\sigma \psi \Rightarrow V^{(\mathbb{H})} \models_\sigma \varphi \vee \psi$
6. $V^{(\mathbb{H})} \models_\sigma \psi$ *or* $V^{(\mathbb{H})} \models_\sigma \neg\varphi \Rightarrow V^{(\mathbb{H})} \models_\sigma \varphi \to \psi$

Furthermore, modus ponens; generalization; instances of intuitionistic tautology (or classical tautologies if $\mathbb{H}$ is Boolean) and the intuitionistic first order logic axioms are all valid under the eyes of our $\mathbb{H}$-validity.

In fact, $\mathbb{H}$-$\models$ is sound with respect to $\vdash$.

Remark. *The difference between the Tarskian and this Localic semantic is that some equivalences that hold for Tarski's do not hold in Localic semantic (in the nontrivial cases,* i.e., *$\mathbb{H} \neq \{0\}$ or $\mathbb{H} \neq 2$). Also, there is very little reason – a priori – to expect there to be a witness to an existential formula, as it is the arbitrary supremum of values of other formulas.*

The supremum in existential formulas may be attained by witnesses, but this relies on the additional property of $\mathbb{H}$ being Boolean, otherwise only a weaker statement holds.

There is, in fact, a canonical representation of elements of our universe V and the many universes $\mathrm{V}^{(\mathbb{H})}$. For the case $\mathbb{H} = 2 = \{0, 1\}$, this canonical representation establishes a sort of model equivalence, has a good left inverse and is onto modulus $\mathrm{V}^{(\mathbb{H})}$-equality with value $\top$.

Definition 1.17. *Immersion of V in $V^{(\mathbb{H})}$*

Define, $\in$-recursively:

$$\hat{x} = \{\langle \hat{y}, \top \rangle \mid y \in x\}$$

Also, for $x \in V^{(2)}$ define $(\cdot \in dom(\cdot))$-recursively:

$$\check{x} = \{\check{y} \mid y \in dom(x) \wedge x(y) = \top\}$$

Proposition 1.18. *The following hold:*

1. $\forall x, y : x \in y \leftrightarrow [\![\hat{x} \in \hat{y}]\!] = 1$
2. $\forall x', y' \in V^{(2)} : \check{x'} \in \check{y'} \leftrightarrow [\![x' \in y']\!] = 1$
3. $\forall x, y : x = y \leftrightarrow [\![\hat{x} = \hat{y}]\!] = 1$
4. $\forall x', y' \in V^{(2)} : \check{x'} = \check{y'} \leftrightarrow [\![x' = y']\!] = 1$
5. $\forall x : \check{\hat{x}} = x$
6. $\forall x' \in V^{(2)} : \left[\!\!\left[\hat{\check{x'}} = x'\right]\!\!\right] = 1$

Corollary 1.19. *Suppose $\mathbb{H} \neq \{0\}$.*

1. $V \models_\sigma \varphi \iff V^{(2)} \models_{\hat{\_}\circ\sigma} \varphi$
2. $\varphi \in \Sigma_0 \Rightarrow \left[V \models_\sigma \varphi \iff V^{(\mathbb{H})} \models_{\hat{\_}\circ\sigma} \varphi\right]$
3. $\varphi \in \Sigma_1 \Rightarrow \left[V \models_\sigma \varphi \Rightarrow V^{(\mathbb{H})} \models_{\hat{\_}\circ\sigma} \varphi\right]$

Theorem 1.20. *$V^{(\mathbb{H})}$ are models of Intuitionistic Set Theory. Furthermore, if $\mathbb{H}$ is Boolean, it validates classical set theory and the Axiom of Choice (provided the base universe already did).*

This is to say, for all φ axioms of the appropriate theory:

$$V^{(\mathbb{H})} \models \varphi$$

Again we provice no proof since this result is well established ([1]).

2 $\mathbf{V}^{(\mathbb{H})}$ and Equivalent Descriptions of $\mathbf{Sh}(\mathbb{H})$

In this section, that is based on [1] and [4], we present some equivalent descriptions of the category of sheaves of a complete Heyting algebra $\mathbb{H}$, $\mathbf{Sh}(\mathbb{H}) \simeq \mathbb{H}\text{-}\mathbf{Set} \simeq \mathbf{Set}^{(\mathbb{H})}$. This is not only for the reader's convenience, but also because we will later need a detailed description of the equivalence $\mathbb{H}\text{-}\mathbf{Set} \simeq \mathbf{Set}^{(\mathbb{H})}$, which is only sketched in the appendix of [1]. We start by providing the definitions of these categories.

Definition 2.1. *Consider the equivalence relation in $V^{(\mathbb{H})}$ given by $f \equiv g$ if, and only if, $[\![f = g]\!] = 1$. The category $\mathbf{Set}^{(\mathbb{H})}$ is defined as:*

$$Obj\left(\mathbf{Set}^{(\mathbb{H})}\right) := {V^{(\mathbb{H})}}/{\equiv}$$

$$\mathbf{Set}^{(\mathbb{H})}([x],[y]) := \left\{[\phi] \in \mathbf{Set}^{(\mathbb{H})} \;\middle|\; [\![fun(\phi : x \to y)]\!] = 1\right\}$$

The arrows do not depend on the choice of representative of the equivalence classes $[x]$ and $[y]$. The composition and identity are defined as in $\mathbf{Set}^{(\mathbb{H})}$ but with the quotient being taken.

Definition 2.2. *An $\mathbb{H}$-set is a pair $\langle X, \delta\rangle$ such that X is a set and $\delta : X \times X \to \mathbb{H}$ satisfies, for every $x, y, z \in X$,*

1. $\delta(x,y) = \delta(y,x)$;
2. $\delta(x,y) \wedge \delta(y,z) \leq \delta(x,z)$.

Definition 2.3. *A morphism $\phi : \langle X, \delta\rangle \to \langle X', \delta'\rangle$ of $\mathbb{H}$-sets is a function $\phi : X \times X' \to \mathbb{H}$ such that, for all $x, y \in X$ e $x', y' \in X'$:*

1. $\delta'(x',y') \wedge \phi(x,y') \leq \phi(x,x')$;
2. $\delta(x,y) \wedge \phi(x,y') \leq \phi(y,y')$;
3. $\phi(x,x') \wedge \phi(x,y') \leq \delta'(x',y')$;
4. $\bigvee_{z' \in X'} \phi(x,z') = \delta(x,x)$.

A morphism of $\mathbb{H}$-sets, then, can be understood as an $\mathbb{H}$-valued functional relation.

Given morphisms $\phi : \langle X, \delta\rangle \to \langle X', \delta'\rangle$ and $\psi : \langle X', \delta'\rangle \to \langle X'', \delta''\rangle$ of $\mathbb{H}$-sets, their composition $\psi \circ \phi$ is given by:

$$(\psi \circ \phi)(x, x'') = \bigvee_{x' \in X'} \phi(x,x') \wedge \psi(x',x'')$$

for all $x \in X, x'' \in X''$. The identity morphism $id_{\langle X,\delta\rangle}$ is the function such that:

$$id_{\langle X,\delta\rangle}(x,y) = \delta(x,y), \text{ for all } x, y \in X$$

Thus, we can define the category $\mathbb{H}$-**Set**, of $\mathbb{H}$-sets and their morphisms. One result on $\mathbb{H}$-sets morphisms will be particularly useful later on:

Proposition 2.4. *Given morphisms $\phi, \psi : \langle X,\delta\rangle \to \langle X',\delta'\rangle$ of $\mathbb{H}$-sets, the following conditions are equivalent:*

1. $\phi = \psi$;
2. $\phi(x,x') \leq \psi(x,x')$*, for all* $x \in X$ *and* $x' \in X'$.

Definition 2.5. *A singleton of an $\mathbb{H}$-set $\langle X,\delta\rangle$ is a mapping $\sigma : X \to H$ such that, for every $x, y \in X$,*

1. $\sigma(x) \wedge \sigma(y) \leq \delta(x,y)$;
2. $\sigma(x) \wedge \delta(x,y) \leq \sigma(y)$.

Note that, given $x \in X$, the function $\sigma_x : X \to H$ such that $\sigma_x(y) = \delta(x,y)$, for all $y \in H$, defines a singleton.

Definition 2.6. *Consider $\sigma(X)$ the collection of singletons of an $\mathbb{H}$-set $\langle X,\delta\rangle$. $\langle X,\delta\rangle$ is said to be complete if the function $\Upsilon : X \to \sigma(X)$, given by $\Upsilon(x) = \sigma_x$, for all $x \in X$, is bijective. We denote the full subcategory of complete $\mathbb{H}$-sets by* **c**$\mathbb{H}$-**Set**.

There is also an alternative description of complete $\mathbb{H}$-sets:

Proposition 2.7. **c**$\mathbb{H}$-**Set** *is isomorphic to the category whose objects are complete $\mathbb{H}$-sets and arrows are functions $f : X \to X'$ such that:*

1. $\delta(x,y) \leq \delta'(f(x), f(y))$;
2. $\delta(x,x) = \delta'(f(x), f(x))$;

for all $x, y \in X$. The composition is given by usual function composition, and the identity arrow is the identity function.

Theorem 2.8. *Let $\langle X,\delta\rangle$ be an $\mathbb{H}$-set. Define the $\mathbb{H}$-set $\langle \sigma(X), \sigma(\delta)\rangle$ where*

$$\sigma(\delta)(\rho,\tau) = \bigvee_{x\in X} \rho(x) \wedge \tau(x), \text{ for all } (\rho,\tau) \in \sigma(X) \times \sigma(X)$$

Then, $\langle \sigma(X), \sigma(\delta)\rangle$ is complete and isomorphic to $\langle X,\delta\rangle$.

The inverse isomorphisms $\phi : \langle X, \delta \rangle \to \langle \sigma(X), \sigma(\delta) \rangle$ e $\psi : \langle (\sigma(X), \sigma(\delta)) \rangle \to \langle X, \delta \rangle$ are given by:

$$\phi(x, \rho) = \rho(x), \text{ for all } (x, \rho) \in X \times \sigma(X)$$
$$\psi(\rho, x) = \rho(x), \text{ for all } (\rho, x) \in \sigma(X) \times X$$

Corollary 2.9. *There is an equivalence of categories:* $\mathbb{H}\text{-}\mathbf{Set} \simeq \mathbf{c}\mathbb{H}\text{-}\mathbf{Set}$.

We can thereby define the functor $\Gamma : \mathbf{Sh}(\mathbb{H}) \to \mathbf{c}\mathbb{H}\text{-}\mathbf{Set}$ by:

$$\Gamma(F) = \langle X_F, \delta_F \rangle, \text{ for every sheaf } F \text{ on } \mathbb{H}, \text{ where } X_F := \coprod_{a \in \mathbb{H}} F(a) \text{ and}$$

$$\delta_F \text{ is given by } \langle (s, b), (t, c) \rangle \mapsto \bigvee \left\{ d \leq b \wedge c \mid s\!\upharpoonright_d^b = t\!\upharpoonright_d^c \right\}$$

$\Gamma(\eta) : \Gamma(F) \to \Gamma(G)$, for every natural transformation $\eta : F \Rightarrow G$
in $\mathbf{Sh}(\mathbb{H})$, where $\Gamma(\eta)(s, b) = (\eta_b(s), b)$, for all $(s, b) \in \Gamma(F)$

Theorem 2.10. *The functor* $\Gamma : \mathbf{Sh}(\mathbb{H}) \to \mathbf{c}\mathbb{H}\text{-}\mathbf{Set}$ *defined above is fully faithful, and for all complete* $\mathbb{H}$*-set* $\langle X, \delta \rangle$ *there exists a sheaf* F *on* $\mathbb{H}$ *such that* $\langle X, \delta \rangle \cong \Gamma(F)$. *Therefore,* Γ *defines an equivalence of categories* $\mathbf{Sh}(\mathbb{H}) \simeq \mathbf{c}\mathbb{H}\text{-}\mathbf{Set}$.

Finally, to show the equivalence between $\mathbb{H}\text{-}\mathbf{Set}$ and $\mathbf{Set}^{(\mathbb{H})}$[2], we will need two constructions on $V^{(\mathbb{H})}$:

Let $\langle X, \delta \rangle$ be an $\mathbb{H}$-set. For each $x \in X$, define $\dot{x} \in V^{(\mathbb{H})}$ as:

$$\text{dom}(\dot{x}) := \{\hat{z} \mid z \in X\} \quad \text{and} \quad \dot{x}(\hat{z}) := \delta(x, z), \text{ for all } z \in X$$

Then, define $X^\dagger \in V^{(\mathbb{H})}$ as

$$\text{dom}\left(X^\dagger\right) := \{\dot{x} \mid x \in X\} \quad \text{and} \quad X^\dagger(\dot{x}) := \delta(x, x), \text{ for all } x \in X$$

Similarly, given a morphism $\phi : \langle X, \delta \rangle \to \langle X', \delta' \rangle$ of $\mathbb{H}$-sets, we may consider $\varphi^\dagger \in V^{(\mathbb{H})}$ given by:

$$\text{dom}\left(\phi^\dagger\right) := \left\{ \langle \dot{x}, \dot{x}' \rangle^{(\mathbb{H})} \mid x \in X, x' \in X' \right\}$$
$$\phi^\dagger \left(\langle \dot{x}, \dot{x}' \rangle^{(\mathbb{H})} \right) := \phi(x, x'), \text{ for all } x \in X, x' \in X'$$

Since $V^{(\mathbb{H})} \models \text{fun}\left(\phi^\dagger\right)$, we may define a functor $\Phi : \mathbb{H}\text{-}\mathbf{Set} \to \mathbf{Set}^{(\mathbb{H})}$ by taking $\Phi(X, \delta) = \left[X^\dagger\right]$, for every $\mathbb{H}$-set $\langle X, \delta \rangle$, and $\Phi(\phi) = \phi^\dagger$, for every arrow $\phi \in \text{Arr}(\mathbb{H}\text{-}\mathbf{Set})$.

---

[2]Here we follow [1], but we provide a more complete and accurate description.

On the other hand, given $u \in \mathrm{V}^{(\mathbb{H})}$, define $X_u := \mathrm{dom}\,(u)$ and $\delta_u : X_u \times X_u \to \mathbb{H}$ as

$$\delta_u(x,y) := [\![x \in u]\!] \wedge [\![x = y]\!] \wedge [\![y \in u]\!]\,, \text{ for all } x,y \in X_u$$

Observe, however, that $[\![u = u']\!] = 1$ does not imply $X_u = \mathrm{dom}\,(u) = \mathrm{dom}\,(u') = X_{u'}$, and that we may not define an $\mathbb{H}$-set using $[\mathrm{dom}\,(u)]$ since this class is not a set (later we will show that $\left\{u' \in \mathrm{V}^{(H)} \mid [\![u = u']\!] = 1\right\}$ is a proper class). In that case, we will use Scott's trick to define a functor $\Psi : \mathbf{Set}^{(\mathbb{H})} \to \mathbb{H}\text{-}\mathbf{Set}$.

Firstly, if $[\![u = u']\!] = 1$, then $\langle X_u, \delta_u\rangle \cong \langle X_{u'}, \delta_{u'}\rangle$. Indeed, define $\lambda_{u,u'} : \langle X_u, \delta_u\rangle \to \langle X_{u'}, \delta_{u'}\rangle$ such that

$$\lambda_{u,u'}(x,x') := [\![x \in u]\!] \wedge [\![x = x']\!] \wedge [\![x' \in u']\!]\,, \text{ for all } x \in \mathrm{dom}\,(u)\,, x' \in \mathrm{dom}\,(u')$$

We verify this is a morphism of $\mathbb{H}$-sets. Let $x, y \in X_u$ and $x', y' \in X_{u'}$.

1. $\delta_{u'}(x',y') \wedge \lambda_{u,u'}(x,y') \leq \lambda_{u,u'}(x,x')$. Indeed,

$$\begin{aligned}
&\delta_{u'}(x',y') \wedge \lambda_{u,u'}(x,y') = \\
&\qquad = [\![x' \in u']\!] \wedge [\![x' = y']\!] \wedge [\![y' \in u']\!] \wedge [\![x \in u]\!] \wedge [\![x = y']\!] \wedge [\![y' \in u']\!] \\
&\qquad \leq [\![x' \in u']\!] \wedge [\![x' = y']\!] \wedge [\![x \in u]\!] \wedge [\![x = y']\!] \\
&\qquad \leq [\![x \in u]\!] \wedge [\![x = x']\!] \wedge [\![x' \in u']\!] \\
&\qquad = \lambda_{u,u'}(x,x')
\end{aligned}$$

2. $\delta_u(x,y) \wedge \lambda_{u,u'}(x,y') \leq \lambda_{u,u'}(y,y')$. Indeed,

$$\begin{aligned}
&\delta_u(x,y) \wedge \lambda_{u,u'}(x,y') = \\
&\qquad = [\![x \in u]\!] \wedge [\![x = y]\!] \wedge [\![y \in u]\!] \wedge [\![x \in u]\!] \wedge [\![x = y']\!] \wedge [\![y' \in u']\!] \\
&\qquad \leq [\![x = y]\!] \wedge [\![y \in u]\!] \wedge [\![x = y']\!] \wedge [\![y' \in u']\!] \\
&\qquad \leq [\![y \in u]\!] \wedge [\![y = y']\!] \wedge [\![y' \in u']\!] \\
&\qquad = \lambda_{u,u'}(y,y')
\end{aligned}$$

3. $\lambda_{u,u'}(x,x') \wedge \lambda_{u,u'}(x,y') \leq \delta_{u'}(x',y')$. Indeed,

$$\begin{aligned}
&\lambda_{u,u'}(x,x') \wedge \lambda_{u,u'}(x,y') = \\
&\qquad = [\![x \in u]\!] \wedge [\![x = x']\!] \wedge [\![x' \in u']\!] \wedge [\![x \in u]\!] \wedge [\![x = y']\!] \wedge [\![y' \in u']\!] \\
&\qquad \leq [\![x = x']\!] \wedge [\![x' \in u']\!] \wedge [\![x = y']\!] \wedge [\![y' \in u']\!] \\
&\qquad \leq [\![x' \in u']\!] \wedge [\![x' = y']\!] \wedge [\![y' \in u']\!] \\
&\qquad = \delta_{u'}(x',y')
\end{aligned}$$

4. $\bigvee_{z' \in X_{u'}} \lambda_{u,u'}(x, z') = \delta_u(x, x)$. Indeed, using that $\delta_u(x, x) = [\![x \in u]\!]$,

- on one hand, for every $z' \in X_{u'}$,

$$\lambda_{u,u'}(x, z') = [\![x \in u]\!] \wedge [\![x = z']\!] \wedge [\![z' \in u']\!] \leq [\![x \in u]\!]$$

  Therefore, $\bigvee_{z' \in X_{u'}} \lambda_{u,u'}(x, z') \leq \delta_u(x, x)$;

- on the other hand, for every $z \in X_{u'}$,

$$u'(z') \wedge [\![z' = x]\!] = [\![x = z']\!] \wedge u'(z') \wedge [\![z' = z']\!] \leq$$
$$\leq [\![x = z']\!] \wedge \left(\bigvee_{t' \in X_{u'}} u'(t') \wedge [\![t' = z']\!] \right) = [\![x = z']\!] \wedge [\![z' \in u']\!]$$

  Thus,

$$[\![x \in u']\!] = \bigvee_{z' \in X_{u'}} u'(z') \wedge [\![z' = x]\!] \leq \bigvee_{z' \in X_{u'}} [\![x = z']\!] \wedge [\![z' \in u']\!]$$

  But observe that $[\![u = u']\!] = 1$ implies $[\![x \in u]\!] = [\![x \in u']\!]$, so:

$$\begin{aligned} [\![x \in u]\!] &= [\![x \in u]\!] \wedge [\![x \in u']\!] \\ &\leq [\![x \in u]\!] \wedge \left(\bigvee_{z' \in X_{u'}} [\![x = z']\!] \wedge [\![z' \in u']\!] \right) \\ &= \bigvee_{z' \in X_{u'}} [\![x \in u]\!] \wedge [\![x = z']\!] \wedge [\![z' \in u']\!] \end{aligned}$$

  That is, $\delta_u(x, x) \leq \bigvee_{z' \in X_{u'}} \lambda_{u,u'}(x, z')$.

Finally, we verify that $\lambda_{u,u'}$ is an isomorphism, with inverse morphism $\lambda_{u,u'}^{-1} = \lambda_{u',u} : \langle X_{u'}, \delta_{u'} \rangle \to \langle X_u, \delta_u \rangle$. For all $x, y \in X_u$,

$$(\lambda_{u',u} \circ \lambda_{u,u'})(x, y) = \bigvee_{x' \in X_{u'}} \lambda_{u,u'}(x, x') \wedge \lambda_{u',u}(x', y) =$$
$$= \bigvee_{x' \in X_{u'}} [\![x \in u]\!] \wedge [\![x = x']\!] \wedge [\![x' \in u']\!] \wedge [\![y \in u]\!] \wedge [\![y = x']\!] \wedge [\![x' \in u']\!] \leq$$
$$\leq [\![x \in u]\!] \wedge [\![x = y]\!] \wedge [\![y \in u]\!] = \delta_u(x, y)$$

Therefore, using Proposition 2.4, we conclude that $\lambda_{u',u} \circ \lambda_{u,u'} = id_{\langle X,\delta\rangle}$. Analogously, it can be verified that $\lambda_{u,u'} \circ \lambda_{u',u} = id_{\langle X',\delta'\rangle}$.

Now, for each $[u] \in \mathbf{Set}^{(\mathbb{H})}$, let $I^{[u]}$ be the category given by:

$$\mathrm{Obj}\left(I^{[u]}\right) := [u]_m \qquad\qquad \mathrm{Arr}\left(I^{[u]}\right) := [u]_m \times [u]_m$$

where $[u]_m$ is the equivalence class of the elements with minimum rank. Consider the functor $F^{[u]} : I^{[u]} \to \mathbb{H}\text{-}\mathbf{Set}$ such that

$$F^{[u]}(u') := \langle X_u, \delta_u \rangle \,, \text{ for all } u' \in [u]_m$$

$$F^{[u]}(u', u'') := \lambda_{u',u''} : \langle X_{u'}, \delta_{u'} \rangle \to \langle X_{u''}, \delta_{u''} \rangle \,, \text{ for all } u', u'' \in [u]_m$$

At last, we may define the functor $\Psi : \mathbf{Set}^{(\mathbb{H})} \to \mathbb{H}\text{-}\mathbf{Set}$ as $\Psi([u]) = \lim\limits_{u' \in [u]_m} F^{[u]}(u')$.

This functor can also be described more explicitly. The product of a family of $\mathbb{H}$-sets $\{\langle X_i, \delta_i \rangle \mid i \in I\}$ is given by $\langle P, \delta \rangle$, where the set is simply the Cartesian product $P = \prod\limits_{i \in I} X_i$ and $\delta : P \times P \to \mathbb{H}$ is given by:

$$\delta\left(\langle x_i \rangle_{i \in I}, \langle x'_i \rangle_{i \in I}\right) = \bigwedge_{i \in I} \delta(x, x')$$

The projections $\pi_j : P \times X_j \to \mathbb{H}$ are given by

$$\pi_j\left(\langle x_i \rangle_{i \in I}, x'_j\right) = \delta_j(x_j, x'_j)$$

for each $j \in I$ (see [4], exercise 2.13.15). The equalizer of two morphisms $\phi, \psi : \langle X, \delta \rangle \to \langle X', \delta' \rangle$ of $\mathbb{H}$-sets is $\langle X, \tau \rangle$, where

$$\tau(x, y) = \bigvee_{x' \in X'} \phi(x, x') \wedge \psi(y, x')$$

(see [4], exercise 2.13.16).

We can then use the construction of limits by products and equalizers (see [3], Theorem 2.8.1), denoting $\Psi([u])$ by $\lim F^{[u]}$:

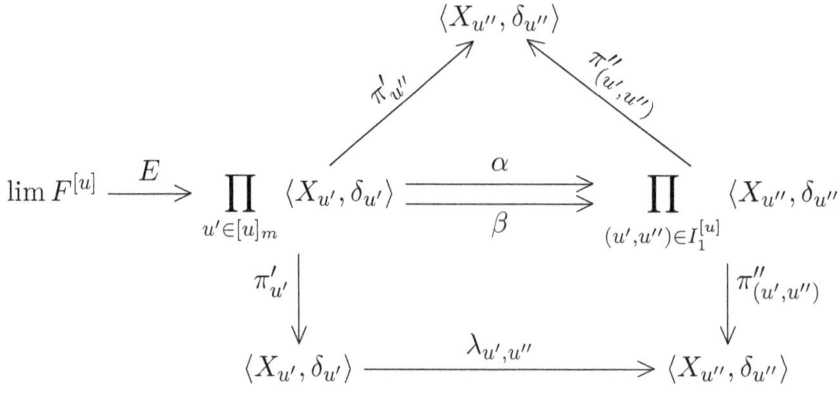

where π', π'' are the projections (of the corresponding products) and $\langle \lim F, E \rangle$ is the equalizer of α and β, which are the morphisms that make the diagram commute. That is,

$$\pi''_{(u',u'')} \circ \alpha = \pi'_{u''} \qquad\qquad \pi''_{(u',u'')} \circ \beta = \lambda_{u',u''} \circ \pi'_{u'}$$

We can proceed similarly for the arrows of the category. For each $f \in \mathrm{V}^{(\mathbb{H})}$ such that $[\![\mathrm{fun}\,(f : u \to v)]\!] = 1$, define $\lambda_f : \langle X_u, \delta_u \rangle \to \langle X_v, \delta_v \rangle$ as:

$$\lambda_f(x, y) = [\![x \in u]\!] \wedge [\![(x, y) \in f]\!] \wedge [\![y \in v]\!]$$

Now, given $f' \in \mathrm{V}^{(\mathbb{H})}$ such that $[\![\mathrm{fun}\,(f' : u' \to v')]\!] = 1$ and $[\![f = f']\!] = 1$ (which already implies $u \equiv u'$ and $v \equiv v'$), we obtain the following commutative diagram:

$$\begin{array}{ccc}
\langle X_u, \delta_u \rangle & \overset{\lambda_{u,u'}}{\rightarrowtail\!\!\!\twoheadrightarrow} & \langle X_{u'}, \delta_{u'} \rangle \\
\Big\downarrow \lambda_f & & \Big\downarrow \lambda_{f'} \\
\langle X_v, \delta_v \rangle & \overset{\lambda_{v,v'}}{\rightarrowtail\!\!\!\twoheadrightarrow} & \langle X_{v'}, \delta_{v'} \rangle
\end{array}$$

Thus, we may define an arrow $\Psi([f]) : \lim\limits_{u' \in [u]_m} F(u') \to \lim\limits_{v' \in [v]_m} F(v')$.

Theorem 2.11. *The functors Φ, Ψ constructed above define an equivalence of categories: $\mathbb{H}$-**Set** $\simeq$ $\mathbf{Set}^{(\mathbb{H})}$.*

3 Induced morphisms in Heyting valued models

Previously, we saw (see Definition 1.17) an injection $V \to \mathrm{V}^{(\mathbb{B})}$ given by $\hat{\cdot}$ which preserves the truth values of Σ_1 formulas (see Corollary 1.19). Currently, it is known that if $\phi : \mathbb{A} \to \mathbb{B}$ is a complete (that is, preserving arbitrary suprema and infima) and injective morphism of Heyting algebras, we can define a map $\tilde{\phi} : \mathrm{V}^{(\mathbb{A})} \to \mathrm{V}^{(\mathbb{B})}$ that is injective and such that: for all $x, y \in \mathrm{V}^{(\mathbb{A})}$,

$$\phi\,[\![x = y]\!]_{\mathbb{A}} = [\![\tilde{\phi}(x) = \tilde{\phi}(y)]\!]_{\mathbb{B}}$$
$$\phi\,[\![x \in y]\!]_{\mathbb{A}} = [\![\tilde{\phi}(x) \in \tilde{\phi}(y)]\!]_{\mathbb{B}}$$

For Δ_0 formulas, the equality, trivially, still holds. One gets the following inequality for any Σ_1 formula ψ:

$$\phi\,[\![\psi(x_1, \cdots, x_n)]\!]_{\mathbb{A}} \leq [\![\psi(\tilde{\phi}(x_1), \cdots, \tilde{\phi}(x_n))]\!]_{\mathbb{B}}$$

It is relatively straightforward to relax these conditions to injective functions that preserve only arbitrary suprema and finite infima[3] to obtain the useful inequalities:

$$\phi\,[\![x = y]\!]_{\mathbb{A}} \leq [\![\tilde{\phi}(x) = \tilde{\phi}(y)]\!]_{\mathbb{B}}$$
$$\phi\,[\![x \in y]\!]_{\mathbb{A}} \leq [\![\tilde{\phi}(x) \in \tilde{\phi}(y)]\!]_{\mathbb{B}}$$

And we still have the inequality for Σ_1 formulas.

Our efforts were in providing a possible generalization of this construction for non-injective maps that preserve arbitrary suprema and finite infima. In this section, we focus on that and on some difficulties we faced in the process.

The reason for our search is that it is taken as a fact that the category of Heyting/Boolean valued models is related to other categories endowed with these morphisms. Despite us having a *horizontal* connection between models and topoi

$$\mathbb{H} \rightsquigarrow \mathrm{V}^{(\mathbb{H})} \rightsquigarrow \mathbf{Set}^{(\mathbb{H})} \simeq \mathbb{H}\text{-}\mathbf{Set}$$

the vertical connections between arrows from $\mathbb{H} \to \mathbb{H}'$, $\mathrm{V}^{(\mathbb{H})} \to \mathrm{V}^{(\mathbb{H}')}$, etc. does not seem to have been widely explored in the literature. The only studied cases were automorphisms of complete Boolean algebra, complete monomorphisms between complete Boolean algebras (see exercise 3.12 in [1]) and retractions associated to those morphisms (see chapter 3 of [7]).

This constitutes one of our main motivations to study if (and how) we could induce arrows between models from more general arrows between complete Heyting algebras. The other one is purely categorical: can geometric morphisms between localic topoi be lifted to morphisms between their associated Heyting value models?

Here we present the main results of our work: we consider and explore how more general kinds of morphisms between complete Heyting algebras $\mathbb{H}$ and $\mathbb{H}'$ induce arrows between $V^{\mathbb{H}}$ and $V^{\mathbb{H}'}$, and between their corresponding Heyting topoi $\mathbf{Set}^{(\mathbb{H})}(\simeq \mathbf{Sh}\,(\mathbb{H}))$ and $\mathbf{Set}^{(\mathbb{H}')}(\simeq \mathbf{Sh}\,(\mathbb{H}'))$.

In the remainder of the section, $\mathbb{H}$ and $\mathbb{H}'$ will denote complete Heyting algebras, and $f : \mathbb{H} \to \mathbb{H}'$ shall be a locale morphism (notation: $f \in \mathbf{Loc}(\mathbb{H}, \mathbb{H}')$), *i.e.*, f is a function that preserves arbitrary suprema and finite infima.

3.1 Induced morphisms

Definition 3.1. (First proposal) *We recursively define a family*

$$\left\{\tilde{f}_\alpha : V_\alpha^{(\mathbb{H})} \rightharpoonup V_\alpha^{(\mathbb{H}')} \;\middle|\; \alpha \in \mathrm{On}\right\}$$

[3]Geometric morphisms or Locale morphisms, which are related to Topoi and Sheaves over Locales.

where $\rightharpoonup$ *indicates that* $\tilde{f}_\alpha$ *are "semi-functions". That is, for all* $x \in V_\alpha^{(\mathbb{H})}$, *there exists* $x' \in V^{(\mathbb{H}')}$ *such that* $\langle x, x' \rangle \in \tilde{f}_\alpha$ *in the following way: for every* $\alpha \in \mathrm{On}$ *and every* $\langle x, x' \rangle \in V_\alpha^{(\mathbb{H})} \times V^{(\mathbb{H}')}$, $\langle x, x' \rangle \in \tilde{f}_\alpha$ *if, and only if, there exists a surjection* $\varepsilon : dom(x) \twoheadrightarrow dom(x')$ *such that* $\langle u, \varepsilon(u) \rangle \in \tilde{f}_{\varrho(x)}$ *for all* $u \in dom(x)$, *and the following diagram commutes:*

$$\begin{array}{ccc} dom(x) & \xrightarrow{\;x\;} & \mathbb{H} \\ \varepsilon \big\downarrow & & \big\downarrow f \\ dom(x') & \xrightarrow{\;x'\;} & \mathbb{H}' \end{array}$$

Under these conditions, we say that ε *witnesses* $\langle x, x' \rangle \in \tilde{f}_\alpha$. *Note that, if we suppose (by induction) that* $\tilde{f}_\beta$ *is defined for every* $\beta < \alpha$, *then the semi-function* $\tilde{f}_{\varrho(x)} : V_{\varrho(x)}^{(\mathbb{H})} \rightharpoonup V_{\varrho(x)}^{(\mathbb{H}')}$ *is defined, therefore* $dom(x') \subseteq V_{\varrho(x)}^{(\mathbb{H}')}$ *and* $x' \in V_\alpha^{(\mathbb{H}')}$.

Thus, we define $\tilde{f} := \bigcup_{\alpha \in \mathrm{On}} \tilde{f}_\alpha$ *and*

$$dom\left(\tilde{f}\right) = \bigcup_{\alpha \in \mathrm{On}} dom\left(\tilde{f}_\alpha\right) = \bigcup_{\alpha \in \mathrm{On}} V_\alpha^{(\mathbb{H})} = V^{(\mathbb{H})}$$

so that $\tilde{f}$ *is also a semi-function* $\tilde{f} : V^{(\mathbb{H})} \rightharpoonup V^{(\mathbb{H}')}$.

Proposition 3.2. *If* f *is injective, then, for all* $\alpha \in \mathrm{On}$, $\tilde{f}_\alpha$ *is an injective function.*

Proof. By induction. Suppose that $\tilde{f}_\beta$ is an injective function for all $\beta < \alpha$ and let $\langle x, x' \rangle \in \tilde{f}_\alpha$. Then, $\varepsilon = \tilde{f}_{\varrho(x)}\restriction : \mathrm{dom}(x) \twoheadrightarrow \mathrm{dom}(x')$ is a bijection, because it's surjective by definition, and, since $\langle u, \varepsilon(u) \rangle \in \tilde{f}_{\varrho(x)}$ for all $u \in \mathrm{dom}(x)$, the induction hypothesis implies that $\varepsilon = \tilde{f}_{\varrho(x)}\restriction$ is injective. Therefore, using the commutative diagram from the definition, x' is uniquely determined by $x' = f \circ x \circ \varepsilon^{-1}$, that is, $\tilde{f}_\alpha$ is a function. Besides, if $x \neq y$ (in $\mathrm{V}_\alpha^{(\mathbb{H})}$), then f being injective implies

$$\tilde{f}_\alpha(x) = f \circ x \circ \varepsilon^{-1} \neq f \circ y \circ \varepsilon^{-1} = \tilde{f}_\alpha(y)$$

so that $\tilde{f}_\alpha$ is also injective. $\square$

Hence, this function covers the result stated in [1], exercise 3.12.

Remark. *Naive attempts to extend this initial proposal are fated to fail, for, in the absence of injectivity, the defined relation is not a function.*

In fact, note that Definition 3.1 presents a serious problem: it does not guarantee that $dom\left(\tilde{f}\right)_\alpha = V_\alpha^{(\mathbb{H})}$*, for each* $\alpha \in \mathrm{On}$*. For example, consider the Boolean algebras* $\mathbf{2} = \{0,1\}$ *and* $\mathbf{4} = \{0, a, \neg a, 1\}$ *(with* $0 \neq a \neq 1$*) and the function* $f : \mathbf{4} \to \mathbf{2}$ *given by* $f(a) = 0$ *and* $f(\neg a) = 1$*. Firstly,*

$$V_0^{(\mathbf{4})} = \emptyset \qquad\qquad V_0^{(\mathbf{2})} = \emptyset$$
$$V_1^{(\mathbf{4})} = \{\emptyset\} \qquad\qquad V_1^{(\mathbf{2})} = \{\emptyset\}$$
$$V_2^{(\mathbf{4})} = \{\{(\emptyset,0)\}, \{(\emptyset,1)\}, \{(\emptyset,a)\}, \{(\emptyset,\neg a)\}\} \qquad\qquad V_2^{(\mathbf{2})} = \{\{(\emptyset,0)\}, \{(\emptyset,1)\}\}$$

Let $x := \{(\{(\emptyset,0)\},0)\}, (\{(\emptyset,a)\},1)\}$*,* $u := \{(\emptyset,0)\}$ *and* $v := \{(\emptyset,1)\}$*. It can be easily verified that* x *is such that* $\langle u, x\rangle, \langle v, x\rangle \in \tilde{f}_2$*.*

We'd like to show that x *makes the relation pathological. Thus, consider* $x' \in V^{(\mathbf{2})}$ *and suppose there exists some* $\varepsilon : dom(x) \twoheadrightarrow dom(x')$ *such that* $\langle u, \varepsilon(u)\rangle, \langle v, \varepsilon(v)\rangle \in \tilde{f}_2$*, i.e.* $\varepsilon(u) = \varepsilon(v) = \{(\emptyset, 0)\}$*. But in this case, we cannot guarantee the diagram in the definition commutes, since we would have:*

$$0 = \varphi(x(u)) = x'(\varepsilon(u)) = x'(\varepsilon(v)) = \varphi(x(v)) = 1$$

Therefore, there exists no such $x' \in V^{(\mathbf{2})}$ *for which* $\langle x, x'\rangle \in \tilde{f}_3$*.*

To deal with this issue, we add more elements to the image of the semi-function, closing it by the equivalence relation $\equiv$ (another option would be to close the images only for the equivalent members with minimum rank).

Definition 3.3. *Generalized Connection between* $V^{(\mathbb{H})}$*s*

Let $f \in \mathbf{Loc}(\mathbb{H}, \mathbb{H}')$*. Define the following compatible family of relations by ordinal recursion:*

$$x \,\tilde{f}_\alpha\, y \iff \exists(\varepsilon : dom(x) \twoheadrightarrow dom(y)) : (y \circ \varepsilon = f \circ x) \wedge$$
$$\forall u \in dom(x) : \exists v \in V^{(\mathbb{H}')} : \exists \beta < \alpha : (u \,\tilde{f}_\beta\, v) \wedge [\![v = \varepsilon(u)]\!] = 1$$

$$\tilde{f} = \bigcup_{\alpha \in \mathrm{On}} \tilde{f}_\alpha$$

This definition in particular was used because of the following: the requirement of the existence of a surjective function is due to our need that every object that is related to x has its (domain's) elements determined by elements of (the domain of) x. This is true in the injective case, where the function is the witness of this existential.

In the non-injective case, ε "essentially"[4] is going to be a "piece" or "fragment" of $\tilde{f}$ that happens to be a function and behaves *similarly* to how $\tilde{f}$ would if f was injective.

We demand that $y \circ \varepsilon = f \circ x$, to extend the original idea of the construction by injective morphisms to more general ones: $\tilde{f}(x)(\tilde{f}(u)) = (f \circ x)(u)$, $u \in dom(x)$. If y is related to x, then there is a function fragment of $\tilde{f}$ which makes the above commute.

It is, however, not enough to ask only this, since one such y could be chosen *ad hoc* without the members of its domain being related to the members of x's. There is no hope for us to attain the imposed conditions of inequalities of atomic formulas, which depend recursively on the domains of the involved objects, if we do not impose some similarly recursive demands on the relation.

Thus, the final condition says that for every member u of x's domain, there was some u in some previous step which to which it was related. Surely $\tilde{f}$ is only very rarely a function, but after taking the quotient by $[\![\cdot = \cdot]\!]$-equivalence it is a function.

Were we to remove $[\![v = \varepsilon(u)]\!] = 1$, and simply require that $u \ \tilde{f}_\beta \ \varepsilon(u)$, the definition would coincide for injective functions, but in general the domain of $\tilde{f}$ as a relation would not be total, *i.e.* it wouldn't be all of $V^{(\mathbb{H})}$.

Theorem 3.4. *$\tilde{f}$'s domain is total*

This is: for all morphism $f : \mathbb{H} \to \mathbb{H}'$,

$$\forall x \in V^{(\mathbb{H})} : \exists x' \in V^{(\mathbb{H}')} : x \ \tilde{f} \ x'$$

Proof. The proof follows from the 2 facts below:

Fact 3.5. *Suppose that $\forall u \in dom(x) : \forall \kappa \in \mathrm{On} : \exists (u' : \kappa \rightarrowtail V^{(\mathbb{H}')}) : \forall \alpha < \kappa : u \ \tilde{f} \ u'_\alpha$. In this case, it is trivial to see that there is an $X' \subset V^{(\mathbb{H}')}$ such that there is a bijection $\varepsilon : dom(x) \to dom(x') = X'$ such that $\forall u \in dom(x) : u \ \tilde{f} \ \varepsilon(u)$.*

Thus, let $x' = f \circ x \circ \varepsilon^{-1}$. It is evident that $x \ \tilde{f} \ x'$. Therefore, if there exists a proper class of elements to the right of every member of the domain of x, then there is some x' such that $x \ \tilde{f} \ x'$.

Fact 3.6. *Suppose that $x \neq \emptyset$ and that $\exists x' \in V^{(\mathbb{H}')} : x \ \tilde{f} \ x'$. Let $u \in dom(x)$, $u \in dom(x')$ such that $u \ \tilde{f} \ u'$ and consider $\varepsilon : dom(x) \twoheadrightarrow dom(x')$ witnessing $x \ \tilde{f} \ x'$.*

Trivially,

$$\exists \alpha \in \mathrm{On} : \forall \xi > \alpha : \forall t' := \left[u' \cup \left\{\left\langle \hat{\xi}, 0 \right\rangle\right\}\right] \to [\![t' = u']\!] = 1$$

[4]Up to $[\![\cdot = \cdot]\!] = 1$-equivalence.

Simply because we are adding some object which was not in the domain of u' whose value under t will be 0*, and in the equality, the* 0 *will be in the antecedent of the implication.*

So, for each ordinal bigger than α, we obtain a different set which is equal to u' with "probability" 1*, and thus, x must have a proper class of elements y such that $x\ \tilde{f}\ y$.*

Joining the previous results, we have:

$$\forall x \in \mathrm{V}^{(\mathbb{H})} : x \neq \emptyset \to ([\forall u \in \mathrm{dom}\,(x) : \exists u' : u\ \tilde{f}\ u'] \to \exists x' : x\ \tilde{f}\ x')$$

As the consequent is true when $x = \emptyset$ — for $\emptyset\ \tilde{f}\ \emptyset$ — we have:

$$\forall x \in \mathrm{V}^{(\mathbb{H})} : [\forall u \in \mathrm{dom}\,(x) : \exists u' : u\ \tilde{f}\ u'] \to \exists x' : x\ \tilde{f}\ x'$$

By regularity:

$$\forall x \in \mathrm{V}^{(\mathbb{H})} : \exists x' : x\ \tilde{f}\ x'$$

□

Alternatively, we could assume in the Definition 3.1 that $\langle x, x'\rangle \in \tilde{f}_\alpha$ if, and only if, there exists a witness $\varepsilon : \mathrm{dom}\,(x) \twoheadrightarrow \mathrm{dom}\,(x')$ as in the original definition, and, for all $u \in \mathrm{dom}\,(x)$, there exists $u' \in \mathrm{V}^{(\mathbb{H}')}$ such that $\langle u, u'\rangle \in \tilde{f}_{\varrho(x)}$ and $[\![\varepsilon(u) = u']\!]' = 1'$.

We choose the first condition mentioned for our final definition, however all are equivalent in the quotient by $\equiv$.

Definition 3.1. (Amended version) *Adding to the original definition, we also assume that if $\langle x, x'\rangle \in \tilde{f}_\alpha$ (that is, there exists a witness for that) and $[\![x' = y']\!]' = 1'$, then $\langle x, y'\rangle \in \tilde{f}_\alpha$.*

Remark. *We observe that $V^{(\mathbb{H})}$ is a proper class (for $\mathbb{H} \neq \{0\}$), since there exists an injection $V \rightarrowtail V^{(\mathbb{H})}$. It can also be shown that, for all $x \in V^{(\mathbb{H})}$, $\{y \in V^{(\mathbb{H})} \mid [\![x = y]\!] = 1\}$ is a proper class. Indeed, for all $\Sigma \subseteq V^{(\mathbb{H})}$ such that $\Sigma \cap dom(x) = \emptyset$, we may define $y_\Sigma : dom(x) \cup \Sigma \to \mathbb{H}$ as:*

$$y_\Sigma(u) = \begin{cases} x(u) & , \text{if } u \in dom(x) \\ 0 & , \text{if } u \in \Sigma \end{cases}$$

so that $[\![x = y_\Sigma]\!] = 1$.

Finally, we show that $\tilde{f}_\alpha$ does actually have the desired domain; the argument is similar to the one used above to prove the injective case.

Proposition 3.7. *For all $\alpha \in \mathrm{On}$, $dom\left(\tilde{f}_\alpha\right) = V_\alpha^{(\mathbb{H})}$ (and its image is closed by $\equiv$).*

Proof. By induction: suppose that, for all $\beta \in \mathrm{On}$, with $\beta < \alpha$, $\mathrm{dom}\left(\tilde{f}_\beta\right) = \mathrm{V}_\beta^{(\mathbb{H})}$. Let $x \in \mathrm{V}_\alpha^{(\mathbb{H})}$, and notice that, for all $u \in \mathrm{dom}\,(x) \subseteq \mathrm{V}_{\varrho(x)}^{(\mathbb{H})}$, the image of $\tilde{f}_{\varrho(x)}(\{u\})$ is a proper class, since it is non-empty and closed by $\equiv$ (by definition). Therefore, using the axiom of replacement, we may define an injection $\varepsilon : \mathrm{dom}\,(x) \rightarrowtail \mathrm{V}^{(\mathbb{H}')}$ satisfying $\langle u, \varepsilon(u)\rangle \in \tilde{f}_{\varrho(x)}$, which we may restrict to a bijection $\tau : \mathrm{dom}\,(x) \to \varepsilon(\mathrm{dom}\,(x))$. Hence, just take $x' \in \mathrm{V}^{(\mathbb{H}')}$ as $x' : \varepsilon(\mathrm{dom}\,(x)) \to \mathbb{H}'$ given by $x' := f \circ x \circ \tau^{-1}$, so that $\langle x, x'\rangle \in \tilde{f}_\alpha$, and thus $\mathrm{dom}\left(\tilde{f}_\alpha\right) = \mathrm{V}_\alpha^{(\mathbb{H})}$. □

Note that with this new definition, for all $x \in \mathrm{V}_\alpha^{(\mathbb{H})}$ there exists $x' \in \mathrm{V}^{(\mathbb{H}')}$ such that $\langle x, x'\rangle \in \tilde{f}_\alpha$ is witnessed by a bijection $\tau : \mathrm{dom}\,(x) \to \mathrm{dom}\,(x')$ (not only a surjection). In fact, we could have assumed the existence of a bijective witness in the definition of $\tilde{f}_\alpha$, and again that would be equivalent to the other possible definitions in the quotient by "$[\![\cdot = \cdot]\!] = 1$" equivalence relation.

3.2 Semantical preservation results

Theorem 3.8. *For all $\langle x, x'\rangle, \langle y, y'\rangle, \langle z, z'\rangle \in \tilde{f}$,*

$$f\left([\![y \in x]\!]\right) \leq' [\![y' \in x']\!]' \qquad \text{and} \qquad f\left([\![x = z]\!]\right) \leq' [\![x' = z']\!]'$$

Proof. The proof is by induction on the well-founded relation

$$\langle u, x\rangle \prec \langle v, y\rangle \iff (u = v \text{ and } x \in \mathrm{dom}\,(y)) \text{ or } (u \in \mathrm{dom}\,(v) \text{ and } x = y)$$

Let $\varepsilon : \mathrm{dom}\,(x) \twoheadrightarrow \mathrm{dom}\,(x')$ satisfying the conditions of 3.1. Then:

$$\begin{aligned}
&f\left([\![y \in x]\!]\right) = \\
&= f\left(\bigvee_{u \in \mathrm{dom}(x)} x(u) \wedge [\![u = y]\!]\right) && \text{(by definition)}\\
&= \bigvee_{u \in \mathrm{dom}(x)}{}' f(x(u)) \wedge' f\left([\![u = y]\!]\right) && \left(\text{since } f \text{ preserves } \wedge, \bigvee\right)\\
&\leq \bigvee_{u \in \mathrm{dom}(x)}{}' f(x(u)) \wedge' [\![\varepsilon(u) = y']\!] && \left(\text{induction hypothesis } \langle u, \varepsilon(u)\rangle \in \tilde{f}\right)\\
&= \bigvee_{u \in \mathrm{dom}(x)}{}' x'(\varepsilon(u)) \wedge' [\![\varepsilon(u) = y']\!] && \left(\text{using that } \langle x, x'\rangle, \langle u, \varepsilon(u)\rangle \in \tilde{f}\right)
\end{aligned}$$

$$= \bigvee_{u' \in \mathrm{dom}(x')}^{\prime} x'(u') \wedge' [\![u' = y']\!] \qquad \text{(since } \varepsilon \text{ is surjective)}$$

$$= [\![y' \in x']\!]' \qquad \text{(by definition)}$$

Now, since $\mathbb{H}$ is a Heyting algebra, note that the fact that f preserves meets implies that f is increasing, and also implies that $f(a \to b) \leq f(a) \to f(b)$, for all $a, b \in \mathbb{H}$. With that, let $\tau : \mathrm{dom}\,(z) \twoheadrightarrow \mathrm{dom}\,(z')$ satisfying the conditions of 3.1. Then:

$$f\left([\![x = z]\!]\right) =$$

$$= f\left(\bigwedge_{\substack{u \in \mathrm{dom}(x) \\ v \in \mathrm{dom}(z)}} (x(u) \to [\![u \in z]\!]) \wedge (z(v) \to [\![v \in x]\!]) \right) \qquad \text{(by definition)}$$

$$\leq \bigwedge_{\substack{u \in \mathrm{dom}(x) \\ v \in \mathrm{dom}(z)}}^{\prime} f\left(x(u) \to [\![u \in z]\!]\right) \wedge' f\left(z(v) \to [\![v \in x]\!]\right) \qquad (f \text{ is increasing})$$

$$\leq \bigwedge_{\substack{u \in \mathrm{dom}(x) \\ v \in \mathrm{dom}(z)}}^{\prime} \left(f(x(u)) \to' f\left([\![u \in z]\!]\right)\right) \wedge' \left(f(z(v)) \to' f\left([\![v \in x]\!]\right)\right) \qquad \text{(comments above)}$$

$$\leq \bigwedge_{\substack{u \in \mathrm{dom}(x) \\ v \in \mathrm{dom}(z)}}^{\prime} \left(f(x(u)) \to' [\![\varepsilon(u) \in z']\!]'\right) \wedge' \left(f(z(v)) \to' [\![\tau(v) \in x']\!]'\right) \qquad \text{(comments below)}$$

$$= \bigwedge_{\substack{u \in \mathrm{dom}(x) \\ v \in \mathrm{dom}(z)}}^{\prime} \left(x'(\varepsilon(u)) \to' [\![\varepsilon(u) \in z']\!]'\right) \wedge' \left(z'(\tau(v)) \to' [\![\tau(v) \in x']\!]'\right) \qquad \text{(they are elements of } \tilde{f}\text{)}$$

$$= \bigwedge_{\substack{u' \in \mathrm{dom}(x') \\ v' \in \mathrm{dom}(z')}}^{\prime} \left(x'(u') \to' [\![u' \in z']\!]'\right) \wedge' \left(z'(v') \to' [\![v' \in x']\!]'\right) \qquad (\varepsilon, \tau \text{ are surjections})$$

$$= [\![x' = z']\!]' \qquad \text{(by definition)}$$

In the fourth step, we use that the implication is increasing in the second coordinate, and that by the induction hypothesis we have:

$$f\left(\llbracket u\in z\rrbracket\right)\leq\llbracket\varepsilon(u)\in z'\rrbracket'\quad\text{and}\quad f\left(\llbracket v\in x\rrbracket\right)\leq\llbracket\tau(v)\in x'\rrbracket'$$

□

This result easily extends to positive formulas (with only $\wedge,\vee$) with bounded quantifiers (of the form $\exists u\in x$ and $\forall u\in x$), using that corollary 1.18 from [1] gives us that:

$$\llbracket\exists u\in x\ \varphi(u)\rrbracket=\bigvee_{u\in\mathrm{dom}(x)}x(u)\wedge\llbracket\varphi(u)\rrbracket$$

$$\llbracket\forall u\in x\ \varphi(u)\rrbracket=\bigwedge_{u\in\mathrm{dom}(x)}x(u)\to\llbracket\varphi(u)\rrbracket$$

Corollary 3.9. *Let φ be a positive formula with bounded quantifiers. Then, for all $\langle a_1,a_1'\rangle,...,\langle a_n,a_n'\rangle\in\tilde{f}$, we have:*

$$f\left(\llbracket\varphi(a_1,...,a_n)\rrbracket\right)\leq'\llbracket\varphi(a_1',...,a_n')\rrbracket'$$

Proof. By induction in the complexity of the formula. The initial case, for atomic sentences, was shown in the previous theorem, and the cases with $\wedge$ and $\vee$ are immediate from the fact that f preserves finite meets and joins. For quantifiers, the proof is similar to last theorem's proof. Let $\langle x,x'\rangle\in\tilde{f}$ with witness $\varepsilon:\mathrm{dom}(x)\twoheadrightarrow\mathrm{dom}(x')$. Then:

$$\begin{aligned}
&f\left(\llbracket\exists u\in x\ \varphi(u)\rrbracket\right)=\\
&=f\left(\bigvee_{u\in\mathrm{dom}(x)}x(u)\wedge\llbracket\varphi(u)\rrbracket\right) && \text{(by definition)}\\
&=\bigvee_{u\in\mathrm{dom}(x)}{}' f(x(u))\wedge' f\left(\llbracket\varphi(u)\rrbracket\right) && \left(\text{since } f \text{ preserves } \wedge,\bigvee\right)\\
&\leq\bigvee_{u\in\mathrm{dom}(x)}{}' f(x(u))\wedge'\llbracket\varphi(\varepsilon(u))\rrbracket && \left(\text{assumption and } \langle u,\varepsilon(u)\rangle\in\tilde{f}\right)\\
&=\bigvee_{u\in\mathrm{dom}(x)}{}' x'(\varepsilon(u))\wedge'\llbracket\varphi(\varepsilon(u))\rrbracket && \left(\text{using } \langle x,x'\rangle,\langle u,\varepsilon(u)\rangle\in\tilde{f}\right)\\
&=\bigvee_{u'\in\mathrm{dom}(x')}{}' x'(u')\wedge'\llbracket\varphi(u')\rrbracket && \text{(since } \varepsilon \text{ is surjective)}
\end{aligned}$$

$$= [\![\exists u' \in x' \; \varphi(u')]\!]' \qquad \text{(by definition)}$$

Similarly, we have

$$\begin{aligned}
& f\left([\![\forall u \in x \; \varphi(u)]\!]\right) = \\
&= f\left(\bigwedge_{u \in \mathrm{dom}(x)} x(u) \to [\![\varphi(u)]\!]\right) && \text{(by definition)} \\
&\leq \bigwedge'_{u \in \mathrm{dom}(x)} f\left(x(u) \to [\![\varphi(u)]\!]\right) && \text{(since } f \text{ is increasing)} \\
&\leq \bigwedge'_{u \in \mathrm{dom}(x)} f(x(u)) \to' f\left([\![\varphi(u)]\!]\right) && \text{(since } f(a \to b) \leq f(a) \to f(b)) \\
&\leq \bigwedge'_{u \in \mathrm{dom}(x)} f(x(u)) \to' [\![\varphi(\varepsilon(u))]\!]' && \text{(induction hypothesis)} \\
&= \bigwedge'_{u \in \mathrm{dom}(x)} x'(\varepsilon(u)) \to' [\![\varphi(\varepsilon(u))]\!]' && \text{(using that } \langle x, x' \rangle \in \tilde{f}) \\
&= \bigwedge'_{u' \in \mathrm{dom}(x')} x'(u') \to' [\![\varphi(u')]\!]' && \text{(since } \varepsilon \text{ is surjective)} \\
&= [\![\forall u' \in x' \; \varphi(u')]\!]' && \text{(by definition)}
\end{aligned}$$

□

3.3 Functorial properties

Another consequence of the previous theorem is that, if $[\![x = z]\!] = 1_H$, then, since $1_H = \bigwedge \emptyset$, we obtain:

$$f\left([\![x = z]\!]\right) = f(1_H) = 1_{H'} \leq [\![x' = z']\!]'$$

that is, $[\![x' = z']\!]' = 1_{H'}$. Therefore, when we take the quotient by $\equiv$, the semi-function $\tilde{f}$ defines an object mapping $\overline{f} : \mathbf{Set}^{(\mathbb{H})} \to \mathbf{Set}^{(\mathbb{H}')}$.

Proposition 3.10.

1. $\overline{id_{\mathbb{H}}} = id_{\mathbf{Set}^{(\mathbb{H})}} : \mathbf{Set}^{(\mathbb{H})} \to \mathbf{Set}^{(\mathbb{H})}$;
2. *if $f' : \mathbb{H}' \to \mathbb{H}''$ preserves finite meets and arbitrary joins, then $\overline{f'} \circ \overline{f} = \overline{f' \circ f} : \mathbf{Set}^{(\mathbb{H})} \to \mathbf{Set}^{(\mathbb{H}'')}$.*

Proof.

1. We show that, for all $\alpha \in \mathrm{On}$, if $\langle x, y'\rangle \in (\widetilde{id_{\mathbb{H}}})_\alpha$, then $[\![x = y']\!] = 1$. Suppose, inductively, that this is the case for all $\beta < \alpha$, and let $\varepsilon : \mathrm{dom}\,(x) \twoheadrightarrow \mathrm{dom}\,(x')$ (with $[\![x' = y']\!] = 1$) witness $\langle x, y'\rangle \in (\widetilde{id_{\mathbb{H}}})_\alpha$. Then, $x' \circ \varepsilon = id_{\mathbb{H}} \circ x = x$, and for all $u \in \mathrm{dom}\,(x)$, $\langle u, \varepsilon(x)\rangle \in (\widetilde{id_{\mathbb{H}}})_{\varrho(x)}$, thereby $[\![\varepsilon(u) = u]\!] = 1$ (using the induction hypothesis). Thus, for all $u \in \mathrm{dom}\,(x)$, since ε is surjective we have:

$$x(u) = x'(\varepsilon(u)) = x'(\varepsilon(u)) \wedge [\![\varepsilon(u) = u]\!] \leq \bigvee_{w' \in \mathrm{dom}(x')} x'(w') \wedge [\![w' = u]\!] = [\![u \in x']\!]$$

Similarly, for all $v' \in \mathrm{dom}\,(x')$, there exists $v \in \mathrm{dom}\,(x)$ such that $\varepsilon(v) = v'$, and

$$x'(v') = x'(\varepsilon(v)) = x(v) = x(v) \wedge [\![\varepsilon(v) = v]\!] \leq \bigvee_{w \in \mathrm{dom}(x)} x(w) \wedge [\![w = v']\!] = [\![v' \in x]\!]$$

Observe that $x(u) \leq [\![u \in x']\!]$ if, and only if, $1 \leq x(u) \to [\![u \in x']\!]$, for all $u \in \mathrm{dom}\,(x)$; that is, $\bigwedge_{u \in \mathrm{dom}(x)} x(u) \to [\![u \in x']\!] = 1$. Analogously, $x'(v') \leq [\![v' \in x']\!]$ for all $v' \in \mathrm{dom}\,(x')$ is equivalent to $\bigwedge_{v' \in \mathrm{dom}(x')} x'(v') \to [\![v' \in x]\!] = 1$. Therefore:

$$[\![x = x']\!] = \bigwedge_{u \in \mathrm{dom}(x)} (x(u) \to [\![u \in x']\!]) \wedge \bigwedge_{v' \in \mathrm{dom}(x')} (x'(v') \to [\![v' \in x]\!]) = 1$$

Finally, since $[\![x = x']\!] = 1 = [\![x' = y']\!]$, we may conclude that $[\![x = y']\!] = 1$, as desired.

Now, by the definition of $\equiv$, $[\![x = y']\!] = 1$ if, and only if, $[x] = [y']$. As a result, taking the quotient, $\langle [x], [y']\rangle \in \overline{id_{\mathbb{H}}}$ if, and only if, $[x] = [y']$, hence $\overline{id_{\mathbb{H}}}$ is the identity in $\mathbf{Set}^{(\mathbb{H})}$.

2. Let $\langle [x], [z'']\rangle \in \overline{f'} \circ \overline{f}$, *i.e.*, there exists $[y'] \in \mathbf{Set}^{(\mathbb{H})}$ such that $\langle [x], [y']\rangle \in \overline{f}$ and $\langle [y'], [z'']\rangle \in \overline{f'}$. Consider $\varepsilon : \mathrm{dom}\,(x) \twoheadrightarrow \mathrm{dom}\,(x')$, with $x' \in [y']$, a witness of $\langle [x], [y']\rangle \in \overline{f}$, and $\varepsilon' : \mathrm{dom}\,(x') \twoheadrightarrow \mathrm{dom}\,(y'')$, with $y'' \in [z'']$, a witness of $\langle [y'], [z'']\rangle = \langle [x'], [z'']\rangle \in \overline{f'}$. Then, $\varepsilon' \circ \varepsilon : \mathrm{dom}\,(x) \to \mathrm{dom}\,(y'')$ witnesses $\langle [x], [z'']\rangle \in \overline{f' \circ f}$. That is, we have shown that $\overline{f'} \circ \overline{f} \subseteq \overline{f' \circ f}$, and since both are functions, we obtain $\overline{f'} \circ \overline{f} = \overline{f' \circ f}$.

□

At last, using that $f\left([\![y \in x]\!]\right) \leq' [\![y' \in x']\!]'$, it can be shown that, if we have $[\![\text{fun}\,(h : x \to y)]\!] = 1$, then $\left[\!\!\left[\text{fun}\left(\tilde{f}(h) : \tilde{f}(x) \to \tilde{f}(y)\right)\right]\!\!\right] = 1'$, and:

$$\tilde{f}(id_x) = id_{\tilde{f}(x)} : \tilde{f}(x) \to \tilde{f}(x) \quad \text{and} \quad \left[\!\!\left[\text{fun}\left(\tilde{f}(id_x) : \tilde{f}(x) \to \tilde{f}(x)\right)\right]\!\!\right]' = 1'$$

Besides, if $[\![\text{fun}\,(g : y \to z)]\!] = 1$, then:

$$\tilde{f}(g \circ h) = \tilde{f}(g) \circ \tilde{f}(h) : \tilde{f}(x) \to \tilde{f}(z) \quad \text{and} \quad \left[\!\!\left[\text{fun}\left(\tilde{f}(g \circ h) : \tilde{f}(x) \to \tilde{f}(z)\right)\right]\!\!\right]' = 1'$$

That is, by taking the quotient, $\overline{f} : \mathbf{Set}^{(\mathbb{H})} \to \mathbf{Set}^{(\mathbb{H}')}$ actually defines a functor.

As we saw in the first section, a $(\wedge, \bigvee)$-preserving function between Heyting algebras induces a functor between the corresponding sheaf topos which preserves finite limits and arbitrary colimits (the left adjoint of a geometric morphism). More precisely, using the natural equivalences $\mathbb{H}$-$\mathbf{Set} \simeq \mathbf{Sh}\,(\mathbb{H})$ and $\mathbb{H}'$-$\mathbf{Set} \simeq \mathbf{Sh}\,(\mathbb{H}')$, such a function $f : \mathbb{H} \to \mathbb{H}'$ gives rise to a functor $\varphi_f : \mathbb{H}$-$\mathbf{Set} \to \mathbb{H}'$-$\mathbf{Set}$ given by:

$$\begin{array}{ccc} \langle X, \delta \rangle & & \langle X, f \circ \delta \rangle \\ \phi \Big\downarrow & \longmapsto & \Big\downarrow f \circ \phi \\ \langle Y, \tau \rangle & & \langle Y, f \circ \tau \rangle \end{array}$$

where $f \circ \phi : X \times Y \to \mathbb{H}'$. Thus, we investigate how $\tilde{f}$ may induce a morphism of $\mathbb{H}'$-sets.

Proposition 3.11. *Let $\langle x, x' \rangle \in \tilde{f}$ with $\varepsilon : dom\,(x) \twoheadrightarrow dom\,(x')$ as witness. Consider the function $\varepsilon^{H'} : dom\,(x) \times dom\,(x') \to \mathbb{H}'$ given by:*

$$\varepsilon^{H'}(u, v') := f\left([\![u \in x]\!]\right) \wedge' [\![\varepsilon(u) = v']\!]' \wedge' [\![v' \in x']\!]'$$

for all $\langle u, v' \rangle \in dom\,(x) \times dom\,(x')$. Then, $\varepsilon^{H'}$ defines a morphism of $\mathbb{H}$-sets $\varepsilon^{H'}$: $(dom\,(x)\,, f \circ \delta_x) \to (dom\,(x')\,, \delta_{x'})$ which does not depend on the choice of witness, where

$$\delta_x(u, v) := [\![u \in x]\!] \wedge [\![u = v]\!]\,, \ \textit{for all } u, v \in dom\,(x)$$
$$\delta_{x'}(u', v') := [\![u' \in x']\!]' \wedge' [\![u' = v']\!]'\,, \ \textit{for all } u', v' \in dom\,(x')$$

Note that δ_x and $\delta_{x'}$ are exactly the ones used in the equivalence $\mathbb{H}$-$\mathbf{Set} \simeq \mathbf{Set}^{(\mathbb{H})}$ because, since $[\![u \in x]\!] \wedge [\![u = v]\!] \leq [\![v \in x]\!]$, we have:

$$[\![u \in x]\!] \wedge [\![u = v]\!] \wedge [\![v \in x]\!] = [\![u \in x]\!] \wedge [\![u = v]\!]$$

Proof. We verify the four conditions that define a morphism of $\mathbb{H}$-sets. Let $u, v \in \operatorname{dom}(x)$ and $u', v' \in \operatorname{dom}(x')$.

1. $\delta_{x'}(u', v') \wedge' \varepsilon^{H'}(u, v') \leq \varepsilon^{H'}(u, u')$. Indeed,

$$\begin{aligned}
&\delta_{x'}(u', v') \wedge' \varepsilon^{H'}(u, v') = \\
&\quad = [\![u' \in x']\!]' \wedge' [\![u' = v']\!]' \wedge' f([\![u \in x]\!]) \wedge' [\![\varepsilon(u) = v']\!]' \wedge' [\![v' \in x']\!]' \\
&\quad \leq f([\![u \in x]\!]) \wedge' [\![\varepsilon(u) = u']\!]' \wedge' [\![u' \in x']\!]' \\
&\quad = \varepsilon^{H'}(u, u')
\end{aligned}$$

2. $(f \circ \delta_x)(u, v) \wedge' \varepsilon^{H'}(u, v') \leq \varepsilon^{H'}(v, v')$. Indeed,

$$\begin{aligned}
&(f \circ \delta_x)(u, v) \wedge' \varepsilon^{H'}(u, v') = \\
&\quad = f([\![u \in x]\!] \wedge [\![u = v]\!]) \wedge' f([\![u \in x]\!]) \wedge' [\![\varepsilon(u) = v']\!]' \wedge' [\![v' \in x']\!]' \\
&\quad \leq f([\![v \in x]\!]) \wedge' [\![\varepsilon(u) = \varepsilon(v)]\!]' \wedge' [\![\varepsilon(u) = v']\!]' \wedge' [\![v' \in x']\!]' \\
&\quad \leq f([\![v \in x]\!]) \wedge' [\![\varepsilon(v) \in v']\!]' \wedge' [\![v' \in x']\!]' \\
&\quad = \varepsilon^{H'}(v, v')
\end{aligned}$$

3. $\varepsilon^{H'}(u, u') \wedge \varepsilon^{H'}(u, v') \leq \delta_{x'}(u', v')$. Indeed,

$$\begin{aligned}
&\varepsilon^{H'}(u, u') \wedge \varepsilon^{H'}(u, v') = \\
&\quad = f([\![u \in x]\!]) \wedge' [\![\varepsilon(u) = u']\!]' \wedge' [\![u' \in x']\!]' \wedge' f([\![u \in x]\!]) \wedge' \\
&\qquad\qquad \wedge' [\![\varepsilon(u) = v']\!]' \wedge' [\![v' \in x']\!]' \\
&\quad \leq f([\![u \in x]\!]) \wedge' [\![u' = v']\!]' \wedge' [\![u' \in x']\!]' \wedge' [\![v' \in x']\!]' \\
&\quad \leq [\![v' \in x']\!]' \wedge' [\![u' = v']\!]' \\
&\quad = \delta_{x'}(u', v')
\end{aligned}$$

4. $\bigvee'_{w' \in \operatorname{dom}(x')} \varepsilon^{H'}(u, w') = (f \circ \delta_x)(u, u)$. Indeed,

   - for all $w' \in \operatorname{dom}(x')$, using that f preserves 1, we have:

$$\begin{aligned}
\varepsilon^{H'}(u, w') &= f([\![u \in x]\!]) \wedge' [\![\varepsilon(u) = w']\!]' \wedge' [\![w' \in x']\!]' \\
&\leq f([\![u \in x]\!]) \wedge' 1_{H'} = f([\![u \in x]\!]) \wedge' f(1_H) \\
&= f([\![u \in x]\!]) \wedge' f([\![u \in u]\!]) = f([\![u \in x]\!] \wedge [\![u = u]\!])
\end{aligned}$$

$$= (f \circ \delta_x)(u, u)$$

thus, $\bigvee'_{w' \in \mathrm{dom}(x')} \varepsilon^{H'}(u, w') \leq (f \circ \delta_x)(u, u)$;

- on the other hand, since $\varepsilon(u) \in \mathrm{dom}(x')$,

$$\begin{aligned}\bigvee'_{w' \in \mathrm{dom}(x')} \varepsilon^{H'}(u, w') \geq \varepsilon^{H'} \langle u, \varepsilon(u) \rangle = \\ &= f([\![u \in x]\!]) \wedge' [\![\varepsilon(u) = \varepsilon(u)]\!]' \wedge' [\![\varepsilon(u) \in x']\!]' \\ &= f([\![u \in x]\!]) \wedge' 1_{H'} \wedge' [\![\varepsilon(u) \in x']\!]' \\ &= f([\![u \in x]\!]) \wedge' [\![\varepsilon(u) \in x']\!]' \\ &\geq f([\![u \in x]\!]) \wedge' f([\![u \in x]\!]) = f([\![u \in x]\!]) \\ &= f([\![u \in x]\!] \wedge 1_H) = f([\![u \in x]\!] \wedge [\![u = u]\!]) \\ &= (f \circ \delta_x)(u, u)\end{aligned}$$

Therefore, $\bigvee'_{w' \in \mathrm{dom}(x')} \varepsilon^{H'}(u, w') \geq (f \circ \delta_x)(u, u)$.

Finally, note that this result does not depend on the choice of witness: let $\langle u, v' \rangle \in \mathrm{dom}(x) \times \mathrm{dom}(x')$ and $\tau : \mathrm{dom}(x) \twoheadrightarrow \mathrm{dom}(x')$ be a witness of $\langle x, x' \rangle \in \tilde{f}$. Then, since $u \in \mathrm{dom}(x)$, we have $\langle u, \varepsilon(u) \rangle \in \tilde{f}$ e $\langle u, \tau(u) \rangle \in \tilde{f}$; and since $1 = [\![u = u]\!]$, the previous theorem gives us:

$$1' = f(1) = f([\![u = u]\!]) \leq [\![\tau(u) = \varepsilon(u)]\!]'$$

Thus,

$$\begin{aligned}\varepsilon^{H'}(u, v') &= f([\![u \in x]\!]) \wedge' [\![\varepsilon(u) = v']\!]' \wedge' [\![v' \in x']\!]' = \\ &= f([\![u \in x]\!]) \wedge' 1_{H'} \wedge' [\![\varepsilon(u) = v']\!]' \wedge' [\![v' \in x']\!]' \\ &= f([\![u \in x]\!]) \wedge' [\![\tau(u) = \varepsilon(u)]\!]' \wedge' [\![\varepsilon(u) = v']\!]' \wedge' [\![v' \in x']\!]' \\ &\leq f([\![u \in x]\!]) \wedge' [\![\tau(u) = v']\!]' \wedge' [\![v' \in x']\!]' \\ &= \tau^{H'}(u, v')\end{aligned}$$

thereby $\varepsilon^{H'}(u, v') \leq \tau^{H'}(u, v')$. The proof that $\tau^{H'}(u, v') \leq \varepsilon^{H'}(u, v')$ is analogous. □

Remark. *The idea now would be to show that such morphisms $\varepsilon^{H'}$ are isomorphisms, which could be used to build a natural isomorphism between $\overline{f}$ and φ_f. To show that, a possibility would be to use the characterization of monomorphisms and epimorphisms in* $\mathbb{H}$-**Set** *(see [4], Propositions 2.8.8 and 2.8.7), that is, to show that for all $u, v \in dom(x)$ and $u' \in dom(x')$:*

- $\varepsilon^{H'}(u, u') \wedge \varepsilon^{H'}(v, u') \leq (f \circ \delta_x)(u, v)$ *(which is equivalent to $\varepsilon^{H'}$ being monic);*
- $\bigvee_{w \in dom(x)} \varepsilon^{H'}(w, u') = \delta_{x'}(u', u')$ *(which is equivalent to $\varepsilon^{H'}$ being epic);*

and, since $\mathbb{H}$-**Set** *is a topos, $\varepsilon^{H'}$ would be an isomorphism.*

Now, expanding the definitions,

$$\begin{aligned}
&\varepsilon^{H'}(u, u') \wedge \varepsilon^{H'}(v, u') = \\
&= f([\![u \in x]\!]) \wedge' [\![\varepsilon(u) = u']\!]' \wedge' [\![u' \in x']\!]' \wedge' f([\![v \in x]\!]) \wedge' [\![\varepsilon(v) = u']\!]' \wedge' \\
&\qquad\qquad \wedge' [\![u' \in x']\!]' \\
&\leq f([\![u \in x]\!]) \wedge' [\![\varepsilon(u) = u']\!]' \wedge' [\![\varepsilon(v) = u']\!]' \\
&\leq f([\![u \in x]\!]) \wedge' [\![\varepsilon(u) = \varepsilon(v)]\!]'
\end{aligned}$$

and we want to show that $\leq f([\![u \in x]\!] \wedge [\![u = v]\!]) = (f \circ \delta_x)(u.v)$

$$\begin{aligned}
&\bigvee_{w \in dom(x)} \varepsilon^{H'}(w, u') = \\
&\qquad = \bigvee_{w \in dom(x)} f([\![w \in x]\!]) \wedge' [\![\varepsilon(w) = u']\!]' \wedge' [\![u' \in x']\!]' \\
&\qquad = \bigvee_{w \in dom(x)} f([\![w \in x]\!]) \wedge' [\![\varepsilon(w) = \varepsilon(t)]\!]' \wedge' [\![\varepsilon(t) \in x']\!]' \qquad (\varepsilon \text{ is surjective}) \\
&\qquad \geq f([\![t \in x]\!]) \wedge' [\![\varepsilon(t) = \varepsilon(t)]\!]' \wedge' [\![\varepsilon(t) \in x']\!]'
\end{aligned}$$

and we want to show that $= [\![u' \in x']\!]' = \delta_{x'}(u', u')$

(the other inequality for the epimorphism condition is trivial, because of the meet's properties).

These inequalities can be achieved whenever f preserves meets and preserves strictly the values of atomic formulas that is: if $\langle x, x' \rangle, \langle y, y' \rangle, \langle z, z' \rangle \in \tilde{f}$, then

$$f([\![x \in y]\!]) = [\![x' \in y']\!]' \qquad\qquad f([\![x = z]\!]) = [\![x' = z'']\!]'$$

Therefore, observing the proof of the aforementioned theorem, note that we may obtain these inequalities (and, thus, that $\varepsilon^{\mathbb{H}'}$ is iso) at least in the case that $f : \mathbb{H} \to$

$\mathbb{H}'$ preserves (strictly) the implication and both arbitrary meets and joins. With that hypothesis, we could also adapt the corollary to the theorem to obtain the strict preservation of $\mathbb{H}$-values of all formulas with bounded quantifiers.

4 Final Remarks and Future Works

Remark. *The categorical and semantical correspondences between local set theories (= topoi, see [2]) and cumulative (constructions in) set theories has been studied since the late 1970s : [6], [8], [5], [13], [14]. It will be interesting to determine in what level this semantical correspondence is compatible with the change of basis given by a locale morphism $f : \mathbb{H} \to \mathbb{H}'$.*
*Possible extensions of this correspondence to other kinds of categories associated to other complete lattices (eventually endowed with additional structure [LT15]) could give us a clue of what are the "right semantical notions" of the less structured side of the correspondence (*i.e.*, the cumulative construction), since the notion of $\mathbb{H}$-set can be extended to more general algebras ([12]).*

Remark. *In a different direction, another aspect that could be analysed is if the "lifting property" through $V^{(\mathbb{H})} \twoheadrightarrow \mathbf{Sh}(\mathbb{H})$ also holds for other natural topoi morphisms, such as the* logical functors. *Since logical functors and (the left part of) geometric morphism coincide only trivially (*i.e. *iff when both are equivalences of categories), this will be in fact a new direction to pursue.*

Note that the "conceptual orthogonality" between the two kind of functors between $\mathbf{Sh}(\mathbb{H})$ and $\mathbf{Sh}(\mathbb{H}')$ occurs already in the algebraic level for arrows $\mathbb{H} \to \mathbb{H}'$. More precisely, given a non-trivial complete Boolean algebra $(\mathbb{B}, \leq)$ and the unique injective morphism $i : \mathbf{2} \hookrightarrow \mathbb{B}$ (where $\mathbf{2} = \{0, 1\}$), we get three kinds of morphisms $\mathbb{B} \to \mathbf{2}$:

- *$l : \mathbb{B} \to \mathbf{2}$ is the left adjoint of i (given by $l(x) = 0 \Leftrightarrow x = 0$): it preserves only the suprema;*
- *$r : \mathbb{B} \to \mathbf{2}$ is the right adjoint of i (given by $r(x) = 1 \Leftrightarrow x = 1$): it preserves only the infima;*
- *$U : \mathbb{B} \to \mathbf{2}$ is the quotient by an ultrafilter U, that preserves 0, 1, negation, implication, finite sups and finite infs.*

On the other hand, note that a logical functor $\mathbf{Sh}(\mathbb{H}) \to \mathbf{Sh}(\mathbb{H}')$ induces a Heyting algebra morphism $\mathbb{H} \to \mathbb{H}'$ (since $\mathbb{H} \cong Subobj(\mathbf{1})$). Therefore, we would expect to be able to establish a correspondence between other kind of morphisms $\mathbb{H} \to \mathbb{H}'$ and the logical functors $\mathbf{Sh}(\mathbb{H}) \to \mathbf{Sh}(\mathbb{H}')$ and ask how they are related to some alternative notion of induced arrow $V^{(\mathbb{H})} \to V^{(\mathbb{H}')}$.

In particular, it seems be natural to consider the connections between the various "forcing relations" (according the previous remark), classical and intuitionistic, related to the canonical morphisms between complete Boolean algebras associated to a complete Heyting algebra Reg$(\mathbb{H}) \hookrightarrow \mathbb{H}$ *and* $\mathbb{H} \twoheadrightarrow \dfrac{\mathbb{H}}{\langle x \leftrightarrow \neg\neg x \rangle}$.

References

[1] John L. Bell. *Set theory: Boolean-valued models and independence proofs.* 3rd ed. Oxfor Logic Guides, vol. 47. Oxford, United Kingdom: Clarendon Press, 2005.

[2] John L. Bell. *Toposes and local set theories: an introduction.* Oxford Logic Guides, vol. 14. Oxford, United Kingdom: Clarendon Press, 1988.

[3] Francis Borceux. *Handbook of Categorical Algebra.* Vol. 1. Encyclopedia of mathematics and its applications, vol. 50. Cambridge, United Kingdom: Cambridge University Press, 2008.

[4] Francis Borceux. *Handbook of Categorical Algebra.* Vol. 3. Encyclopedia of mathematics and its applications, vol. 52. Cambridge, United Kingdom: Cambridge University Press, 2008.

[5] Andreas R. Blass and Andre Scedrov. "Complete topoi representing models of set theory". In: *Annals of Pure and Applied Logic* 57 (1992), pp. 1–26. doi: https://doi.org/10.1016/0168-0072(92)90059-9.

[6] Michael P. Fourman. "Sheaf models for set theory". In: *Annals of Pure and Applied Logic* 19 (1980), pp. 91–101. doi: https://doi.org/10.1016/0022-4049(80)90096-1.

[7] Fiorella Guichardaz. "Limits of Boolean algebras and Boolean valued models". MA thesis. Universitá degli studi di Torino, 2013.

[8] Susumu Hayashi. "On set theories in toposes". In: *Logic Symposia Hakone 1979, 1980.* Ed. by Gert H. Müller, Gaisi Takeuti, and Tosiyuki Tugué. Lecture Notes in Mathematics, vol. 891. Heidelberg, Germany: Springer-Verlag, 1981. Chap. 2, pp. 23–29.

[9] Thomas J. Jech. *Set Theory.: The third millennium edition, revised and expanded.* Springer Monographs in Mathematics. Heidelberg, Germany: Springer-Verlag, 2003.

[10] Kenneth Kunen. *Set Theory.* Studies in Logic, vol. 34. London, United Kingdom: College Publications, 2011.

[11] Saunders Mac Lane. *Categories for the Working Mathematician.* 2nd ed. Graduate texts in mathematics, vol. 5. New York, United States: Springer-Verlag, 1998.

[LT15] Benedikt Löwe and Sourav Tarafder. "Generalized algebra-valued models of set theory". In: *The Review of Symbolic Logic* 8 (2015), pp. 192–205. doi: https://doi.org/10.1017/S175502031400046X.

[12] Caio A. Mendes. "Sheaf-like categories and applications". In private communication. 2019.

[13] Michael A. Shulman. "Stack semantics and the comparison of material and structural set theories". In: *arXiv e-prints* (2010). arXiv: 1004.3802v1.

[14] Keita Yamamoto. "Toposes from Forcing for Intuitionistic ZF with Atoms". In: *arXiv e-prints* (2018). arXiv: 1702.03399v2.

 Received 4 December 2019

Coalgebra
for the working software engineer

Luís Soares Barbosa*
INL - International Iberian Nanotechnology Laboratory
High-Assurance Software Lab - INESC TEC
Universidade do Minho
Braga, Portugal
`lsb@di.uminho.pt`

Abstract

Often referred to as 'the mathematics of dynamical, state-based systems', Coalgebra claims to provide a compositional and uniform framework to specify, analyse and reason about state and behaviour in computing. This paper addresses this claim by discussing why Coalgebra matters for the design of models and logics for computational phenomena. To a great extent, in this domain one is interested in properties that are preserved along the system's evolution, the so-called 'business rules' or system's invariants, as well as in liveness requirements, stating that e.g. some desirable outcome will be eventually produced. Both classes are examples of modal assertions, i.e. properties that are to be interpreted across a transition system capturing the system's dynamics. The relevance of modal reasoning in computing is witnessed by the fact that most university syllabi in the area include some incursion into modal logic, in particular in its temporal variants. The novelty is that, as it happens with the notions of transition, behaviour, or observational equivalence, modalities in Coalgebra acquire a shape . That is, they become parametric on whatever type of behaviour, and corresponding coinduction scheme, seems appropriate for addressing the problem at hand. In this context, the paper revisits Coalgebra from a computational perspective, focussing on three topics central to software design: how systems are *modelled*, how models are *composed*, and finally, how *properties* of their behaviours can be expressed and verified.

*Supported by the ERDF – European Regional Development Fund through the Operational Programme for Competitiveness and Internationalisation — COMPETE 2020 Programme and by National Funds through the Portuguese funding agency, FCT, within project POCI-01-0145-FEDER-030947.

1 Introduction

1.1 Coalgebra ...

To define an (inductive) data structure, as typically taught in a first undergraduate course on programming, one essentially specifies its 'assembly process'. For example, one builds a sequence in a data domain D, either by taking an empty list or by adjoining a fresh element to an existing sequence. Thus, declaring a sequence data type yields a function $\zeta : \mathbf{1} + D \times U \longrightarrow U$, where U stands for the data type being defined. The structured domain of function ζ captures a signature of *constructors* ($nil : \mathbf{1} \longrightarrow U$, $cons : D \times U \longrightarrow U$), composed additively (*i.e.* $\zeta = [nil, cons]$). The whole procedure resembles the way in which an algebraic structure is defined.

Reversing an 'assembly process' swaps structure from the domain to the codomain of the arrow, which now captures the result of a 'decomposition' or 'observation' process. In the example at hand this is performed by the familiar *head* and *tail selectors* joined together into

$$\alpha : U \longrightarrow \mathbf{1} + D \times U \tag{1}$$

where $\alpha = \underline{*} \lhd empty? \rhd \langle head, tail \rangle$ either returns a token $*$, when observing an empty sequence, or its decomposition in the top element and the remaining tail[1].

This reversal of perspective also leads to a different understanding of what U may stand for. The product $D \times U$ captures the fact that both the head and the tail of a sequence are selected (or *observed*) simultaneously. In fact, once one is no longer focused on how to construct U, but simply on what can be observed of it, finiteness is no longer required: both finite or infinite sequences can be observed through the process above. Therefore, U can be more accurately thought of as a *state space* of a machine generating a finite or infinite sequence of values of type D. Elements of U, in this example, can no longer be distinguished by construction, but should rather be identified when generating the same sequence. That is to say, when it becomes impossible to distinguish them through the observations allowed by the 'shape' structuring the codomain of α.

Function (1) is an example of a *coalgebra*. Its ingredients are: a carrier U (intuitively the state space of a machine), the *shape* of allowed observations, technically a functor $\mathcal{F}(X) = \mathbf{1} + D \times X$, and the observation *dynamics* given by function α, *i.e.* the machine itself. Formally, a $\mathcal{F}$-coalgebra is a pair $\langle U, \alpha \rangle$ consisting of an object U and a map $\alpha : U \longrightarrow \mathcal{F}\ U$. The latter maps states to structured collections of successor states. By varying $\mathcal{F}$, *i.e.* the shape of the underlying transitions, one may

[1]Notation $\underline{e}$ denotes the constant function $\lambda\ x.e$; the conditional 'if b then a else c' is written $a \lhd b \rhd c$.

capture a large class of semantic structures used to model computational phenomena as (more or less complex) transition systems. Going even further, $\mathcal{F}$ is not restricted to be an endofunctor in *Set*, the category of sets and functions. For example, as we will see later, the category of topological spaces emerges as the natural host for coalgebras modelling continuous systems. The study of the common properties of all these systems is the subject of *Universal Coalgebra*, as developed systematically by a number of authors from the pioneering work of J. J. M. M. Rutten [61].

A morphism between two $\mathcal{F}$-coalgebras, $\langle U, \alpha \rangle$ and $\langle V, \beta \rangle$, is a map h between carriers U and V which preserves the dynamics, *i.e.* such that $\beta \cdot h = \mathcal{F}\ h \cdot \alpha$. As one would expect, $\mathcal{F}$-coalgebras and their morphisms form a category $C_{\mathcal{F}}$ where both composition and identities are inherited from the host category C. Along this paper, C will always be *Set*, but in a few explicitly mentioned cases.

This sets Coalgebra as a suitable mathematical framework for the study of dynamical systems in both a *compositional* and *uniform* way. The qualifier *uniform* requires some extra explanation: coalgebraic concepts (*i.e.* models, constructions, logics, and proof principles) are parametric on, or *typed* by, the functor that characterises the underlying transition structure. The point is that, in Mathematics as in Software Engineering, going parametric allows us to focus on the abstract structure of a problem such that, on solving it, what we actually solve is a whole class of problems. The obvious limits of human reasoning make such an economy of resources the hallmark of rational thinking. And so we are back to Engineering.

This paper aims at introducing Coalgebra as a (conceptual) tool for the working software engineer. The title is borrowed from Saunders Mac Lane's famous book *Categories for the working mathematician* first published in 1971. Category theory is the study of mathematical structures focussed on the ways they interact rather than on what they pretend to be. Roughly speaking, categories deal with (typed) *arrows* and their composition, in the same sense that sets deal with *elements*, their aggregation and membership. The theory uncovers universal properties, through which whole families of arrows can be factored out in essentially unique ways, characterises constructions uniformly applicable to structures and their transformations, and unveils dual universes by simply reversing arrows.

Coalgebras are arrows in a category. Their theory brings to scene a mathematical space in which key ingredients of computational systems find their place: state, behaviour, observation, interaction. Objects, automata, state-based components, services, processes are part of our vocabulary to talk about systems which compute by reacting to contextual stimuli received along their overall computation. Typically, reactive systems rely on the cooperation of distributed, heterogeneous, often anonymous components organised into open software architectures prepared to survive in loosely-coupled environments and adapt to changing application requirements. In a sense, the object of Software Engineering is nothing more than the (emergent) behaviour of computing systems, for which Coalgebra provides a suitable foundation. As Robin Milner put it in his Turing Award Lecture [50],

> *From being a prescription for how to do something – in Turing's terms a 'list of instructions', software becomes much more akin to a description of behaviour, not only programmed on a computer, but also occurring by hap or design inside or outside it.*

Indeed, the origins of Coalgebra, in its applications to Computer Science, may be traced back to Peter Aczel's attempt [1] to characterise bisimulation and providing a precise semantics to Milner's calculus of communicating systems.

1.2 ... for the working software engineer

This paper is not a systematic presentation of coalgebra theory, let alone a tutorial. My aim is much humbler: to make a case for Coalgebra in relation to three main topics unavoidable in any roadmap to Software Engineering – systems' *models*, *architectures*, and *properties*. Each of them will give me the opportunity to introduce a number of concepts and constructions in Coalgebra, as well as to provide a brief illustration based on current research developed by my research team.

Models. Models are pervasive in the engineering practice, and the software domain is not an exception. Irrespective of the myriad of (textual, diagrammatic, formal, etc.) notations used in practice, models should always be understood in the sense they are in *e.g.* school physics problem-solving. There, once a problem is understood, a mathematical model is built as an appropriate abstraction, on top of which one reasons about the behaviour of the system until a 'solution' is found. We will discuss how several variants of transition systems can be modelled coalgebraically. The characterisation of systems' behaviour, and the definition of suitable notions of equivalence for state-based systems will also be addressed. As an illustration, we will revisit recent results on modelling hybrid automata as coalgebras.

Architectures. Software architecture emerged as a proper discipline within Software Engineering from the need to explicitly consider, in the development of increasingly larger and more complex systems, their overall structure, organisation, and emergent behaviour. As a model, an architecture acts as an abstraction of a system that suppresses details of its constituents, except for those which affect the ways they use, are used by, relate to, or interact with other components. This topic will be illustrated by revisiting an architectural calculus of state-based software components framed as generalised Mealy machines, in which the strict deterministic discipline is relaxed to capture more complex behavioural patterns. In particular, we will mention the interplay between the two basic modes in which software can be composed: sequentially and concurrently, *i.e.* along a temporal or a spatial dimension, respectively. This leads to particular instances of what is known in Mathematics as an *interchange law*. Again, some aspects will be instantiated for the less regular case of components which, like sensors in a network, exhibit forms of continuous evolution.

Properties. A plethora of logics is used in Software Engineering to support the specification of systems' requirements and properties, as well as to verify whether, or to what extent, they are enforced in specific implementations. Broadly speaking, the logics of dynamical systems are *modal*, *i.e.* they provide operators which qualify formulas as holding in a certain *mode*. In mediaeval Scholastics such modes represented the strength of assertion (e.g. 'necessity' or 'possibility'). In temporal reasoning they can refer to a future or past instant, or a collection thereof. Similarly, one may express epistemic states (e.g. 'as everyone knows'), deontic obligations (e.g. 'when legally entitled'), or spatial states (e.g. 'in every point of a surface'). Regarding dynamical systems as transformations of state spaces according to specific transition shapes, *i.e.* as coalgebras for particular functors, such modes refer to particular configurations of successor states as defined, or induced, by the coalgebra dynamics. Again, Coalgebra provides a *uniform* characterisation by letting functor $\mathcal{F}$ induce 'canonical' notions of modality and the corresponding logic. General questions in modal logic, such as the trade-off between expressiveness and computational tractability, or the relationship between logical equivalence and bisimilarity, can be addressed at this (appropriate) level of abstraction. We will revisit modal logic from a coalgebraic perspective and illustrate this discussion mentioning a logic to express properties of n-layered, hierarchical transition systems.

Paper structure. Models, architectures and properties are revisited, from a coalgebraic viewpoint, in the following sections. We will try to substantiate the claim that Coalgebra is the right mathematics to model and reason about state-based systems. On the other hand, we will argue that the coalgebraic approach is generic and compositional: constructions, techniques and tools apply to a large class of application areas and can be combined in a modular way. Finally, section 5 concludes with a brief discussion of current research directions and of what the future might bring for this area.

2 Models

2.1 State and behaviour

Although information technology became ubiquitous in modern life long before a solid scientific methodology, let alone formal foundations, has been put forward, the ultimate goal of a software engineering discipline is the development of methods, techniques and tools for formal – and preferably automatic – analysis and verification of computational systems. Analysis and verification are usually performed on suitable abstractions of the real systems, rather than on the systems themselves. Coalgebra provides a framework to build such abstractions, or *models*, as state-based transition systems parametric on a transition *shape*, or *type*, given by an endofunctor $\mathcal{F}$ in a host category. The choice of $\mathcal{F}$ determines not only the expressivity of the model, but also a canonical notion of *behaviour* and *observational equivalence*.

Consider, for example, an elementary model of an object whose internal state is observable through an *attribute* $\mathsf{at} : U \longrightarrow B$ and may evolve by reacting to external stimuli through a *method*[2] $\overline{\mathsf{m}} : U \longrightarrow U^A$. This defines a coalgebra

$$p \mathrel{\widehat{=}} \langle \overline{\mathsf{m}}, \mathsf{at} \rangle : U \longrightarrow U^A \times B$$

for the functor $\mathcal{F}\,X = X^A \times B$, known in the literature as a Moore machine. A bit of syntactic sugar recovers the usual transitional notation:

$$u \xrightarrow{a}_p u' \Leftrightarrow \overline{\mathsf{m}}\, u\, a = u' \quad \text{and} \quad u \downarrow_p b \Leftrightarrow \mathsf{at}\, u = b$$

The notion of a coalgebra *morphism* $h : p \longrightarrow p'$ boils down to the following commuting diagram

$$\begin{array}{ccc} U & \xrightarrow{p} & U^A \times B \\ {\scriptstyle h}\downarrow & & \downarrow{\scriptstyle h^A \times id} \\ V & \xrightarrow{p'} & V^A \times B \end{array}$$

or, avoiding exponentials,

$$\begin{array}{ccc} U & \xrightarrow{\mathsf{at}} & B \\ {\scriptstyle h}\downarrow & & \downarrow{\scriptstyle id} \\ V & \xrightarrow{\mathsf{at}'} & B \end{array} \qquad \begin{array}{ccc} U \times A & \xrightarrow{\mathsf{m}} & U \\ {\scriptstyle h \times id}\downarrow & & \downarrow{\scriptstyle h} \\ V \times A & \xrightarrow{\mathsf{m}'} & V \end{array}$$

because

$$\begin{array}{cl} & \langle \overline{\mathsf{m}}', \mathsf{at}' \rangle \cdot h \;=\; (h^A \times id) \cdot \langle \overline{\mathsf{m}}, \mathsf{at} \rangle \\ \Leftrightarrow & \qquad \{ \text{ products} \} \\ & \langle \overline{\mathsf{m}}' \cdot h, \mathsf{at}' \cdot h \rangle \;=\; \langle h^A \cdot \overline{\mathsf{m}}, \mathsf{at} \rangle \\ \Leftrightarrow & \qquad \{ \text{ structural equality} \} \end{array}$$

[2]Notation $\overline{f}$ stands for the curried version of a function f.

$$
\begin{array}{ll}
& \overline{\mathsf{m}}' \cdot h = h^A \cdot \overline{\mathsf{m}} \ \wedge\ \mathsf{at}' \cdot h = \mathsf{at} \\
\Leftrightarrow & \qquad \{\text{ exponentials}\} \\
& \overline{\mathsf{m}' \cdot (h \times id)} = \overline{h \cdot \mathsf{m}} \ \wedge\ \mathsf{at}' \cdot h = \mathsf{at} \\
\Leftrightarrow & \qquad \{\text{ curry is a bijection}\} \\
& \mathsf{m}' \cdot (h \times id) = h \cdot \mathsf{m} \ \wedge\ \mathsf{at}' \cdot h = \mathsf{at}
\end{array}
$$

The behaviour of p, denoted in the sequel by $[\![(p)]\!]$, at a state $u \in U$, is revealed by successive observations (or experiments) triggered by the input of different sequences $s = [a_0, a_1, \ldots]$ in A^*:

$$
\mathsf{at}\ u,\ \mathsf{at}\ (\overline{\mathsf{m}}\ u\, a_0),\ \mathsf{at}\ (\overline{\mathsf{m}}\ (\overline{\mathsf{m}}\ u\, a_0)\, a_1), \ldots
$$

which entails the following recursive definition of $[\![(p)]\!]$:

$$
[\![(p)]\!] u\, \underline{\mathsf{nil}} \mathrel{\widehat{=}} \mathsf{at}\, u \quad \text{and} \quad [\![(p)]\!]\, u\, (cons\, \langle a, t\rangle) \mathrel{\widehat{=}} [\![(p)]\!]\, (\mathsf{m}\, \langle u, a\rangle)\, t\ .
$$

Therefore, behaviours are elements of B^{A^*}, and can be thought of as rooted trees whose branches are labelled by sequences of inputs in A and leaves by values in B. Moreover, they organise themselves into a Moore machine over B^{A^*},

$$
\omega_{\mathcal{F}} \mathrel{\widehat{=}} \langle \overline{\mathsf{m}}_\omega, \mathsf{at}_\omega\rangle : B^{A^*} \longrightarrow (B^{A^*})^A \times B\ .
$$

where

$$
\begin{array}{rcll}
\mathsf{at}_\omega\, f & \widehat{=} & f\ nil & \textit{i.e.}\ \text{the value of the attribute before any input} \\
\overline{\mathsf{m}}_\omega\, f\, a & \widehat{=} & \lambda\, s\, .\ f(cons\langle a, s\rangle) & \textit{i.e.}\ \text{input determines subsequent evolution}
\end{array}
$$

The coalgebra $\omega_{\mathcal{F}}$ whose states are the $\mathcal{F}$-behaviours themselves plays a specific role: it is *final* among all $\mathcal{F}$-coalgebras. Actually, for any $p = \langle \overline{\mathsf{m}}, \mathsf{at}\rangle$, $[\![(p)]\!]$ is the unique morphism $[\![(p)]\!] : p \longrightarrow \omega_{\mathcal{F}}$. Note that

$$
\begin{array}{ll}
& \mathsf{at}_\omega \cdot [\![(p)]\!] = \mathsf{at} \\
\Leftrightarrow & \qquad \{\text{ introduction of variables }\} \\
& \mathsf{at}_\omega \langle [\![(p)]\!]\, u\rangle = \mathsf{at}\, u \\
\Leftrightarrow & \qquad \{\text{ definition of } \mathsf{at}_\omega \} \\
& \langle [\![(p)]\!]\, u\rangle\, nil = \mathsf{at}\, u \\
\Leftrightarrow & \qquad \{\text{ definition of } [\![(p)]\!] \} \\
& true
\end{array}
$$

$$
\begin{array}{ll}
& \mathsf{m}_\omega \cdot ([\![(p)]\!] \times id) = [\![(p)]\!] \cdot \mathsf{m} \\
\Leftrightarrow & \qquad \{\text{ introduction of variables and application }\} \\
& \mathsf{m}_\omega \langle [\![(p)]\!]\, u, a\rangle = [\![(p)]\!] (\mathsf{m}\, \langle u, a\rangle) \\
\Leftrightarrow & \qquad \{\text{ definition of } \mathsf{m}_\omega \} \\
& \lambda\, s\, .\ [\![(p)]\!]\, u\, (cons\, \langle a, s\rangle) = [\![(p)]\!] (\mathsf{m}\, \langle u, a\rangle) \\
\Leftrightarrow & \qquad \{\text{ introduction of variables and application }\} \\
& [\![(p)]\!]\, u\, (cons\, \langle a, t\rangle) = [\![(p)]\!]\, (\mathsf{m}\, \langle u, a\rangle)\, t \\
\Leftrightarrow & \qquad \{\text{ definition of } [\![(p)]\!] \} \\
& true
\end{array}
$$

with uniqueness being easily established. In general, denoting by $\omega_{\mathcal{F}} : \Omega_{\mathcal{F}} \longrightarrow \mathcal{F}(\Omega_{\mathcal{F}})$ the final coalgebra for a functor $\mathcal{F}$, finality can be expressed as a universal property by the following equivalence:

$$k = [\![p]\!] \;\Leftrightarrow\; \omega_{\mathcal{F}} \cdot k = \mathcal{F}(k) \cdot p \tag{2}$$

Finality is a powerful tool. For example, the assertion that any two states from coalgebras p and q connected by an arbitrary morphism $h : p \longrightarrow q$ generate the same behaviour, *i.e.* $[\![p]\!] = [\![q]\!] \cdot h$, is a direct consequence of *uniqueness* (the right to left implication in equivalence (2)) as depicted in the diagram below[3].

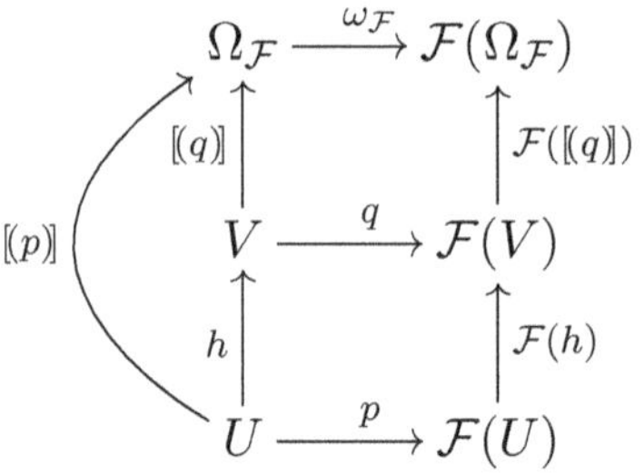

Similarly, *existence* (the dual, left to right implication) provides a *definition* principle for operators over behaviours. Each of those has its source equipped with coalgebra structure p specifying the 'one-step' dynamics. Then $[\![p]\!]$ gives the rest: the operator becomes defined by specifying its output under all different observers as recorded in functor $\mathcal{F}$.

An important observation is that the dynamics of the final coalgebra is an isomorphism. Isomorphisms being self-dual, this also entails that the initial algebra of a functor is an isomorphism as well, which was the original statement of this result known as Lambek's lemma. The proof relies on the universal property and, in this sense, is illustrative of a proof by coinduction presented equationally.

One starts by assuming the existence of an inverse $\alpha_{\mathcal{F}}$ to $\omega_{\mathcal{F}}$, which entails $\alpha_{\mathcal{F}} \cdot \omega_{\mathcal{F}} = id_{\Omega_{\mathcal{F}}}$ and $\omega_{\mathcal{F}} \cdot \alpha_{\mathcal{F}} = id_{\mathcal{F}(\Omega_{\mathcal{F}})}$. Then, one of these requirements is used to conjecture a definition for $\alpha_{\mathcal{F}}$ (an engineer would say an 'implementation' ...). Note the use of the fact that $[\![\omega_{\mathcal{F}}]\!] = id_{\Omega_{\mathcal{F}}}$, entailing a 'reflection' law, to introduce, rather than eliminate, the behaviour morphism in the calculation. Finally, one checks the validity of the conjecture above by verifying with it the remaining requirement.

[3]The diagram captures *a fusion* property useful in behaviour reasoning.

Putting both arguments side by side, the proof goes as follows:

$$
\begin{array}{cl}
 & \alpha_{\mathcal{F}} \cdot \omega_{\mathcal{F}} = id_{\Omega_{\mathcal{F}}} \\
\Leftrightarrow & \quad \{ \text{ reflection } \} \\
 & \alpha_{\mathcal{F}} \cdot \omega_{\mathcal{F}} = [\!(\omega_{\mathcal{F}})\!] \\
\Leftrightarrow & \quad \{ \text{ universality } \} \\
 & \omega_{\mathcal{F}} \cdot \alpha_{\mathcal{F}} \cdot \omega_{\mathcal{F}} = \mathcal{F}(\alpha_{\mathcal{F}} \cdot \omega_{\mathcal{F}}) \cdot \omega_{\mathcal{F}} \\
\Leftrightarrow & \quad \{ \text{ } \mathcal{F} \text{ preserves composition } \} \\
 & \omega_{\mathcal{F}} \cdot \alpha_{\mathcal{F}} \cdot \omega_{\mathcal{F}} = \mathcal{F}(\alpha_{\mathcal{F}}) \cdot \mathcal{F}(\omega_{\mathcal{F}}) \cdot \omega_{\mathcal{F}} \\
\Leftarrow & \quad \{ \text{ cancel } \omega_{\mathcal{F}}\text{; universality } \} \\
 & \alpha_{\mathcal{F}} = [\!(\mathcal{F}(\omega_{\mathcal{F}}))\!]
\end{array}
$$

$$
\begin{array}{cl}
 & \omega_{\mathcal{F}} \cdot \alpha_{\mathcal{F}} \\
= & \quad \{ \text{ replace by derived conjecture } \} \\
 & \omega_{\mathcal{T}} \cdot [\!(\mathcal{F}(\omega_{\mathcal{F}}))\!] \\
= & \quad \{ \text{ } [\!(\mathcal{F}(\omega_{\mathcal{F}}))\!] \text{ is a morphism } \} \\
 & \mathcal{F}([\!(\mathcal{F}(\omega_{\mathcal{F}}))\!]) \cdot \mathcal{F}(\omega_{\mathcal{F}}) \\
= & \quad \{ \text{ } \mathcal{F} \text{ preserves composition } \} \\
 & \mathcal{F}([\!(\mathcal{F}(\omega_{\mathcal{F}}))\!] \cdot \omega_{\mathcal{F}}) \\
= & \quad \{ \text{ just proved } \} \\
 & \mathcal{F}(id_{\Omega_{\mathcal{F}}}) \\
= & \quad \{ \text{ } \mathcal{F} \text{ preserves identities } \} \\
 & id_{\mathcal{F}(id_{\Omega_{\mathcal{F}}})}
\end{array}
$$

Lambek's lemma characterises both initial algebras and final coalgebras for a functor $\mathcal{F}$ as fixed points of equation $X = \mathcal{F}(X)$. The terminology comes from an analogy with what happens in a partial order $\langle P, \leq \rangle$ seen as a category. A functor is then just a monotone function, and therefore a coalgebra is an element x of P such that $x \leq \mathcal{F}(x)$. The final coalgebra is, then, an element $m \leq \mathcal{F}(m)$ such that, for all $x \in P$, $x \leq \mathcal{F}(x) \Rightarrow x \leq m$, which, by Tarski's theorem, is the greatest fixpoint of $\mathcal{F}$ with respect to $\leq$.

Whenever final coalgebras exist, which is the case for every bounded *Set* endofunctor, they provide a canonical, often intuitive interpretation of behaviour. Even when this is not the case, behaviours can be approximated by an ordinal indexed sequence of objects such that each element b_α encodes behaviour that can be generated (or exhibited) in α steps.

As mentioned before, varying the functor $\mathcal{F}$ one obtains different models, with tuned notions of morphism and behaviour. For example, making $B = \mathbf{2}$ in $\mathcal{F}$ characterises deterministic automata on the alphabet A, whose behaviours are identified with the recognised languages. Actually, the state space of $\omega_{\mathcal{F}}$ becomes $\mathbf{2}^{A^*}$, *i.e.* each state is a subset of A^*, and its dynamics is given by $\langle \overline{\mathsf{m}}_\omega, \mathsf{at}_\omega \rangle : \mathbf{2}^{A^*} \longrightarrow (\mathbf{2}^{A^*})^A \times \mathbf{2}$, where

$$\mathsf{at}_\omega\, s \;=\; \mathsf{nil} \in s \quad \text{and} \quad \overline{\mathsf{m}}_\omega\, s \;=\; \lambda\, a\, .\; \{cons\langle a, x\rangle \mid x \in s\}\ .$$

Variants of Moore machines can be obtained by specifying a particular behavioural effect $\mathcal{T}$:

$$p : U \longrightarrow \mathcal{T}(U)^A \times B$$

thus enforcing a particular branching structure upon p. For example $\mathcal{T}(X) = X + \mathbf{1}$

makes the automata partial, whereas $\mathcal{T} = \mathcal{P}$, for $\mathcal{P}-$ the finite, covariant powerset functor, introduces non-determinism.

Another classical distinction concerns whether the 'output' B depends on the 'input' A. For example, a generic Mealy machine would be specified by a coalgebra

$$p : U \longrightarrow \mathcal{T}(U \times B)^A \ .$$

Both Moore and Mealy machines are examples of what are usually called *reactive* transition systems, due to the explicit presence of an 'input' universe. A coalgebra for $\mathcal{F}(X) = \mathcal{T}(B \times X)$, on the other hand, stands for a so-called *generative* model, as values are produced, rather than consumed, on transitions. For example, processes in a process algebra are typically modelled as a (the final) coalgebra for $\mathcal{F}(X) = \mathcal{P}(B \times X)$. Probabilistic automata are based on the distribution functor $\mathcal{D}(X) = \{\mu : X \longrightarrow \mathbb{R}_{\geq 0} \mid \sum_{x \in X} \mu x = 1\}$. The large collection of variants of automata capturing some form of probabilistic evolution was systematically studied by Ana Sokolova [66] in a coalgebraic setting. Examples of a reactive probabilistic automata and a stratified one, in which Markovian and regular transitions may alternate, are depicted in diagrams (a) and (b) below.

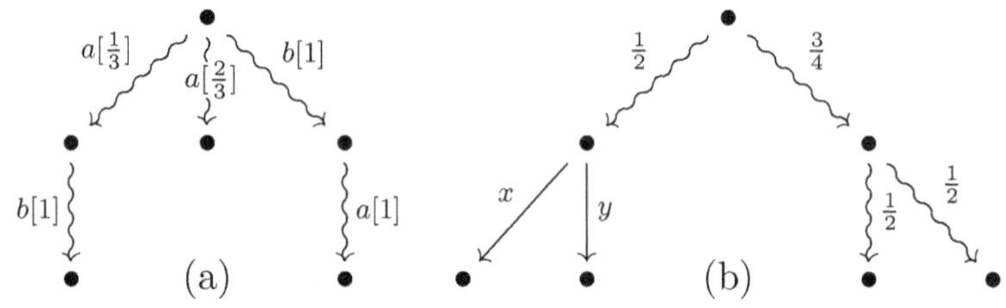

The relevant functors are, respectively, $\mathcal{F}(X) = (\mathcal{D}(X) + \mathbf{1})^A$ and $\mathcal{F}(X) = \mathcal{D}(X) + (B \times S) + \mathbf{1}$. More complex transitions come from combining different effects. For example a Segala probabilistic automata is a coalgebra $p : U \longrightarrow \mathcal{P}(B \times \mathcal{D}(U))$. Note than more than one transition may be chosen non-deterministically from a given state, but once the choice is made outcomes with different probabilities are possible. Specified in a common coalgebraic setting, all such variants can be analysed and their expressivity compared through the identification of suitable natural transformations between the 'shape' functors; moreover, one typically obtains more general results and shorter proofs. Later, in subsection 2.3, recent work in my group on a similar exercise for hybrid automata will be commented. First, however, an essential ingredient for a modelling discipline is still missing: a notion of model equivalence.

2.2 Equivalences

The comparison, replacement and reuse of models entails the need for suitable notions of equivalence. In a coalgebraic setting, this is *observational*: two states $u \in U$ and $v \in V$ in $\mathcal{F}$-coalgebras $p : U \longrightarrow \mathcal{F}(U)$ and $q : V \longrightarrow \mathcal{F}(V)$, are identified if they cannot be distinguished by observations as allowed by $\mathcal{F}$. Actually, in this case they generate the same behaviour. For example, equivalent states in a Moore machine $p : U \longrightarrow U^A \times \mathbf{2}$, *i.e.* a deterministic automaton, do recognise the same language.

Whenever $\mathcal{F}$ admits a final coalgebra $\omega_{\mathcal{F}}$, the notion of *observational equivalence*, represented in the sequel as $\equiv_{\mathcal{F}}$, can be made precise is the obvious way:

$$u \equiv_{\mathcal{F}} v \Leftrightarrow [\![p]\!]u = [\![q]\!]v \tag{3}$$

If that is not case, the definition can be generalised by requiring the existence of a coalgebra $\xi : S \longrightarrow \mathcal{F}(S)$ and a (epic) cospan $p \xrightarrow{r_1} \xi \xleftarrow{r_2} q$ in $C_{\mathcal{F}}$ (or equivalently a epic cospan in the host category C whose legs lift to $\mathcal{F}$-coalgebra morphisms, as depicted in diagram (a) below) such that $r_1 u = r_2 v$.

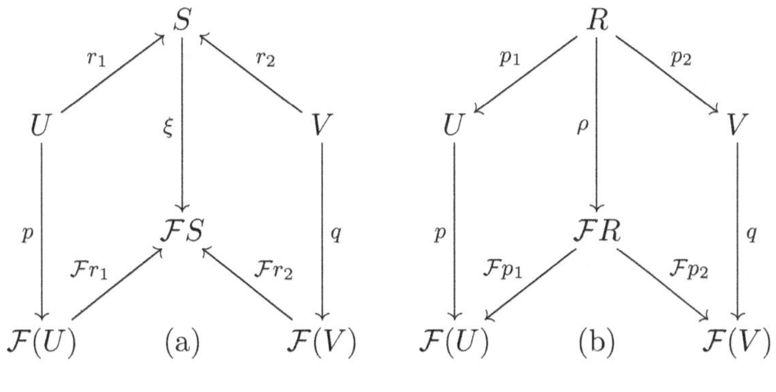

It is worthwhile to stress that the way $\xi : S \longrightarrow \mathcal{F}(S)$ is defined is *dual* to the one used in Algebra (the other, perhaps more familiar, half of the universe) to give a congruence. Indeed, a congruence is an equivalence relation compatible with the constructors in the algebra signature, captured by a functor $\mathcal{G}$. This means that there exists an algebra $\zeta : \mathcal{G}(A) \longrightarrow A$ and a (monic) span $a \xleftarrow{p_1} \zeta \xrightarrow{p_2} b$ to $\mathcal{G}$-algebras a and b. Not surprisingly, thus, ξ is called a *cocongruence*: congruent terms in Algebra have cocongruent behaviours as a counterpart in Coalgebra.

Interestingly enough, indistinguishability by observation is often given in terms of *bisimilarity*, *i.e.* the existence of a bisimulation containing the pair of states under consideration. A bisimulation is defined as the *analogue*, rather than the *dual*, to a compatible relation in Algebra, *i.e.* as a (monic) span $p \xleftarrow{p_1} \rho \xrightarrow{p_2} q$ in $C_{\mathcal{F}}$ as depicted in (b). In *Set*, R is a relation in $U \times V$ whose projections p_1, p_2 lift to coalgebra morphisms, which is the original definition of bisimulation given by Aczel and Mendler [2]. We write $u \sim_{\mathcal{F}} v$ if there exists a bisimulation R such that $u = p_1 t$ and $v = p_2 t$ for a $t \in R$, and say that u and v are bisimilar states.

$$\begin{array}{ccccc} U & \xleftarrow{p_1} & G_h & \xrightarrow{p_2} & V \\ {\scriptstyle p}\downarrow & & {\scriptstyle \gamma}\downarrow & & \downarrow{\scriptstyle q} \\ \mathcal{F}(U) & \xleftarrow{\mathcal{F}(p_1)} & \mathcal{F}(G_h) & \xrightarrow{\mathcal{F}(p_2)} & \mathcal{F}(V) \end{array}$$

A rather obvious example of a bisimulation is provided by the graph of any coalgebra morphism. Indeed, let $h : p \longrightarrow q$ and $G_h = \{\langle u, hu \rangle \mid u \in U\}$, as usual. Taking, in the diagram on the right, $\gamma = \mathcal{F}(p_1)^\circ \cdot p \cdot p_1$, both squares commute because

$$\begin{array}{cl} & p \cdot p_1 = \mathcal{F}(p_1) \cdot \gamma \\ \Leftrightarrow & \quad \{ \text{ definition of } \gamma \} \\ & p \cdot p_1 = \mathcal{F}(p_1) \cdot \mathcal{F}(p_1)^\circ \cdot p \cdot p_1 \\ \Leftrightarrow & \quad \{ \text{ converse } \} \\ & p \cdot p_1 = p \cdot p_1 \end{array}$$

$$\begin{array}{cl} & q \cdot p_2 = \mathcal{F}(p_2) \cdot \gamma \\ \Leftrightarrow & \quad \{ \text{ definition of } \gamma \} \\ & q \cdot p_2 = \mathcal{F}(p_2) \cdot \mathcal{F}(p_1)^\circ \cdot p \cdot p_1 \\ \Leftrightarrow & \quad \{ h = p_2 \cdot p_1^\circ, \text{ functors } \} \\ & q \cdot p_2 = \mathcal{F}(h) \cdot p \cdot p_1 \\ \Leftrightarrow & \quad \{ h \text{ is a morphism } \} \\ & q \cdot p_2 = q \cdot h \cdot p_1 \\ \Leftrightarrow & \quad \{ h = p_2 \cdot p_1^\circ \} \\ & q \cdot p_2 = q \cdot p_2 \end{array}$$

Conversely, whenever a graph G_h is a bisimulation, then h is a coalgebra morphism. Since p_1 is bijective, so is its converse p_1°. Thus, composition $h = p_2 \cdot p_1^\circ$ is a morphism.

There is an alternative definition of bisimulation which is closer to the intuitive interpretation as a binary relation over states which is closed for the coalgebra dynamics. It reads: R is a bisimulation if

$$\langle u, v \rangle \in R \;\Rightarrow\; \langle pu, qv \rangle \in \overline{\mathcal{F}}(R) \qquad (4)$$

where $\overline{\mathcal{F}}(R)$ is the so-called a *relation lifting* of R through functor $\mathcal{F}$. This can be defined inductively for a wide class of functors, including all mentioned up to now in this paper, but a more general definition, applicable to any *Set* endofunctor, can be given as the image of $\mathcal{F}(R)$ under the split $\langle \mathcal{F}(p_1), \mathcal{F}(p_2) \rangle$, where p_1 and p_2 are, as before, the projections of R onto U and V, respectively; thus,

$$\overline{\mathcal{F}}(R) \;=\; \{\langle \mathcal{F}(p_1)\, t, \mathcal{F}(p_2)\, t \rangle \mid t \in \mathcal{F}(R)\} \;.$$

For example, applying (4) to the functor used above to specify Moore machines, leads to the following definition of bisimulation:

$$\langle u, v\rangle \in R \quad \Rightarrow \quad \mathsf{at}_p\ u = \mathsf{at}_q\ v \quad \text{and} \quad \langle \overline{\mathsf{m}}_p\ u\ a, \overline{\mathsf{m}}_q\ v\ a\rangle \in R, \quad \text{for all } a \in A\ .$$

This means that all states related by R support identical observations and enforce that their successor states are also related by R.

The two definitions of bisimulation discussed here are indeed equivalent. This is shown in [41] taking $\overline{\mathcal{F}}$ as a functor in a category whose objects are relations and morphisms the corresponding spans[4]. The Aczel-Mendler definition has a wider application for functors in arbitrary categories. The one based on relation lifting, on the other hand, is closer to the intuitive notion in Process Algebra [58] that a bisimulation is a closed relation.

Whatever definition one uses, the fact is that in Coalgebra bisimulation, just as behaviour, acquires a shape given by $\mathcal{F}$. Moreover, all folklore results from Process Algebra hold for coalgebraic bisimulations. In particular, the set of bisimulations linking two coalgebras forms a complete lattice for relation inclusion with joins given by unions. The largest bisimulation in this lattice is the bisimilarity relation denoted by $\sim_{\mathcal{F}}$. This is actually the greatest fixed point of a map $R \mapsto \{\langle u, v\rangle \mid \langle pu, qv\rangle \in \overline{\mathcal{F}}(R)\}$, from a direct application of the Knaster–Tarski theorem, based on $\overline{\mathcal{F}} : \mathcal{P}(U \times V) \longrightarrow \mathcal{P}(\mathcal{F}(U) \times \mathcal{F}(V))$ being monotone. Bisimulations are closed for union and converse, but not necessarily for relational composition.

Bisimilarity, however, is strictly weaker than observation equivalence. Actually, the choice of a concept that is an analogue of, rather than a dual to, a congruence is largely motivated by historical reasons [62]. Moreover, unlike $\equiv$, bisimulations may be constructed iteratively, and therefore is amenable to automation. Quite efficient algorithms for checking bisimilarity are indeed available. For most functors of interest in current applications to Software Engineering, both notions coincide. It is instructive, however, to take a while to understand what is indeed required from the 'shape' functor in order to guarantee such a coincidence.

First of all notice it is not difficult to see that $\sim\ \subseteq \equiv$. Consider again $\mathcal{F}$-coalgebras $p : U \longrightarrow \mathcal{F}(U)$ and $q : V \longrightarrow \mathcal{F}(V)$. Now form the pushout (S, r_1, r_2) of a bisimulation R and its projections as depicted in the following diagram:

[4]Regarding $\mathcal{F}$ as a relator in the category of sets and relations leads to a very compact proof [9] of this result.

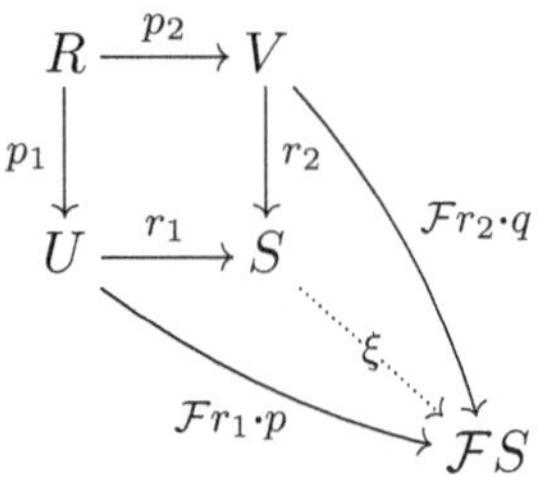

Then, arrows $\mathcal{F}(r_1) \cdot p : U \longrightarrow \mathcal{F}(S)$ and $\mathcal{F}(r_2) \cdot q : V \longrightarrow \mathcal{F}(S)$ determine a unique coalgebra ξ such that $\mathcal{F}(r_1) \cdot p = \xi \cdot r_1$ and $\mathcal{F}(r_2) \cdot q = \xi \cdot r_2$ as required. Suppose now that $u \equiv v$, *i.e.* that there is an epic cospan $p \xrightarrow{r_1} \xi \xleftarrow{r_2} q$ in $C_{\mathcal{F}}$ as depicted in the fore square of the cube in the diagram below.

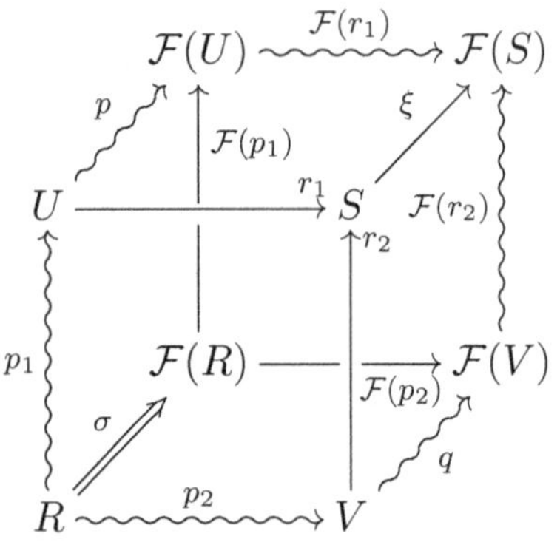

Form its pullback, with $R = \{\langle u, v\rangle \in U \times V \mid r_1 u = r_2 v\}$. The lifting of this square through $\mathcal{F}$ is represented in the back square of the cube.

Coalgebras p, q and ξ link both squares. Observe that the following two paths from R to $\mathcal{F}(S)$, depicted with curly arrows in the cube, coincide:

$$
\begin{array}{ll}
 & \mathcal{F}(r_1) \cdot p \cdot p_1 \\
= & \quad \{ \; r_1 : p \longrightarrow \xi \text{ is a coalgebra morphism } \} \\
 & \xi \cdot r_1 \cdot p_1 \\
= & \quad \{ \; \text{the pullback square commutes } \} \\
 & \xi \cdot r_2 \cdot p_2 \\
= & \quad \{ \; r_2 : q \longrightarrow \xi \text{ is a coalgebra morphism } \} \\
 & \mathcal{F}(r_2) \cdot q \cdot p_2
\end{array}
$$

If $\mathcal{F}(R)$, together with $\mathcal{F}(p_1)$ and $\mathcal{F}(p_2)$, is also a pullback, then, given the equality just proved, there exists a coalgebra $\sigma : R \longrightarrow \mathcal{F}(R)$ and coalgebra morphisms

$p_1 : \sigma \longrightarrow p$ and $p_2 : \sigma \longrightarrow q$. This means that R is a bisimulation. Notice there is no need for σ to be unique, therefore all one has to require from functor $\mathcal{F}$ is that it preserves *weak* pullbacks.

Under this apparently weird condition, bisimilarity and observational equivalence coincide. That is to say, two states generate the same behaviour if and only if they are bisimilar. Therefore, every bisimulation over the final coalgebra is a coreflexive, *i.e.* a subset of the identity relation. Furthermore this condition guarantees that the relational composition of two bisimulations is still a bisimulation, as one is used to from the (well-behaved) domain of Process Algebra.

Most *Set* endofunctors useful for the software engineer, which do indeed preserve weak pullbacks, belong to the class of extended polynomial functors

$$\mathcal{F} \ni Id \mid K \mid Id^K \mid \mathcal{P} \mid \mathcal{F} \times \mathcal{F} \mid \mathcal{F} + \mathcal{F} \mid \mathcal{F} \cdot \mathcal{F}$$

where K is a set, and $\mathcal{P}$ is the finite, covariant powerset functor. The distribution functor $\mathcal{D}$, mentioned above, is also often considered, as well as the star functor and other solutions of datatype equations.

Bisimilarity provides a technique for coinductive proofs, *i.e.* a sound tool to establish observational equivalence, which is complete for the class of functors preserving weak pullbacks. To establish equality of the behaviour generated by two state values it is enough to build a bisimulation containing them. This corresponds to the following procedure: i) iteratively strengthen the statement to be proved (from equality $u = v$ to a larger set containing the pair $\langle u, v\rangle$), and then ii) ensure that such a set is closed for the coalgebra dynamics (*i.e.* it forms a bisimulation). Actually what is going on underneath is an *unfolding* process which, typically, does not terminate, but reveals longer and longer prefixes of the result: every element in the result gets uniquely determined along this process. Inductive reasoning requires that, by repeatedly unfolding the definition, arguments become *smaller*, *i.e.* closer to the elementary constructors of the algebra. In Coalgebra our attention shifts from argument's structural shrinking to the progressive construction of the behaviour which becomes richer in informational contents.

2.3 Illustration: Hybrid automata

Hybrid automata were proposed more than two decades ago as a family of models capturing the interaction of discrete (computational) systems with continuous (physical) processes. Essentially, they are finite state machines with a finite set of continuous variables whose values are typically described by a set of ordinary differential equations. Since the publication of T. Henzinger seminal paper [36] in 1996, several different characterisations emerged independently. They were often driven by applications, seeking to capture a specific feature or property of the system to be modelled.

As has happened before, for example in the case of probabilistic automata [66], Coalgebra helps to organise the landscape by characterising hybrid automata, and associated notions of bisimulation, in a uniform way, parametric on the concrete functor expressing the specific variant of interest. The coalgebraic perspective promotes a 'black-box' view where discrete transitions are kept internal to the automaton and continuous evolutions make up the external, observable behaviour. This is in contrast with the traditional representation in which both discrete steps and continuous evolutions are joined in the same transition relation.Therefore, the general shape for these models are coalgebras typed as

$$p : U \longrightarrow \mathcal{G}(U) \times \mathcal{H}(O) \tag{5}$$

where $\mathcal{H}$ captures the continuous evolution of a quantity O over time. Functor $\mathcal{H}$ was introduced in a recent paper [55] as an endofunctor in the category Top of topological spaces and continuous functions. In broad terms, working in Top is motivated by the key role that continuity plays in this setting, and by the possibility to handle, within the coalgebraic framework, classical properties of dynamical systems. For example, a notion of *robustness* (a system is robust if small changes in the input lead to very similar evolutions) can be addressed by varying the topology on the space of inputs. The topic, however, will not be pursued in detail here.

The functor $\mathcal{H}$ is defined as

$$\mathcal{H}(X) \ \widehat{=}\ \{ \langle f, d \rangle \in X^{\mathsf{T}} \times \mathsf{D} \mid f \cdot \curlywedge_d = f \} \quad \text{and} \quad \mathcal{H}(h) \ \widehat{=}\ h^{\mathsf{T}} \times id \tag{6}$$

where T abbreviates $\mathbb{R}_{\geq 0}$, $\mathsf{D} = [0, \infty]$ is the one-point compactification of $\mathbb{R}_{\geq 0}$ and $h^{\mathsf{T}} f = h \cdot f$. Condition $f \cdot \curlywedge_d = f$, for $\curlywedge_d \ \widehat{=}\ id \lhd (\leq_d) \rhd \underline{d}$, means that f becomes constant after time instant d.

To illustrate this model, consider a bouncing ball dropped at some positive height and with no initial velocity. Due to the gravitational pull, it will fall into the ground but then bounce back up, losing, of course, part of its kinetic energy in the process.

This can be seen as a hybrid component whose (continuous) observable behaviour is the evolution of its spacial position (P), whereas the internal memory records the initial velocity (V) and position updated at each bounce:

$$b : V \times P \longrightarrow (V \times P) \times \mathcal{H}(P)$$

The discrete behaviour $b_d : V \times P \rightarrow V \times P$ (which updates the discrete state, *i.e.* the initial velocity and position pair) is computed by multiplying the current velocity by the coefficient of restitution to obtain the new initial velocity for the next bounce and updating position to 0. Formally,

$$b_d\, \langle v, p\rangle \mathrel{\widehat{=}} \langle vel_g \langle v, zpos_g \langle v, p\rangle\rangle \times -0.5,\, 0\rangle$$

where 0.5 is the coefficient of restituion, and current velocity is computed as $vel_a \langle v, t\rangle \mathrel{\widehat{=}} v - at$. Function $zpos_a\ (v,p) \mathrel{\widehat{=}} \frac{\sqrt{2ap+v^2}+v}{a}$ returns the time needed to reach the ground, given a positive height and a current velocity. On the one hand, the continuous part $b_c : V \times P \to \mathcal{H}P$ is computed by

$$b_c \mathrel{\widehat{=}} \langle pos_g(v,p), zpos_g\rangle\ ,$$

where $pos_a : V \times P \longrightarrow P^T$ is given by $pos_a\,\langle v,p\rangle \mathrel{\widehat{=}} \lambda t.\ (p + vt - \frac{1}{2}at^2)$, and g is the gravitational constant. Putting both components together

$$b \mathrel{\widehat{=}} \langle b_d, b_c\rangle\ .$$

The behaviour of a $(- \times \mathcal{H}(O))$-coalgebra p at a state $u \in U$, is a function that computes a stream of (observable) continuous evolutions generated by p from state u. Actually, the functor $- \times \mathcal{H}(O)$ has a final coalgebra, *i.e.* the following diagram commutes uniquely

$$\begin{array}{ccc} \mathcal{H}(O)^\omega & \xrightarrow{\ \omega\ } & \mathcal{H}(O)^\omega \times \mathcal{H}(O) \\ {\scriptstyle [\![(p)]\!]}\uparrow & & \uparrow{\scriptstyle [\![(p)]\!] \times id} \\ U & \xrightarrow{\ p\ } & U \times \mathcal{H}(O) \end{array}$$

where X^ω denotes the set of streams over X, and $\omega \mathrel{\widehat{=}} \langle tail, head\rangle$ is the dynamics of the final coalgebra.

For the bouncing ball, assuming the initial velocity and position pair is $\langle 0, 5\rangle$, the plot below depicts the first three elements of the generated stream.

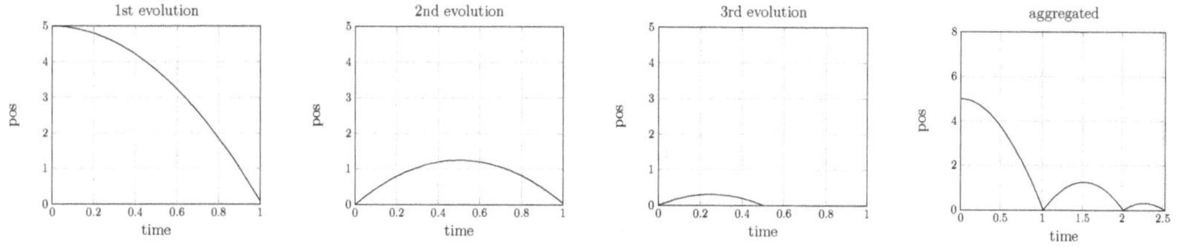

Varying $\mathcal{G}$ in (5), one is able to capture different variants of hybrid automata and compute the corresponding notions of behaviour and bisimulation. For example, instantiating $\mathcal{G}$ to $\mathcal{P}$ or $\mathcal{D}$, leads to non-deterministic or reactive Markov hybrid automata. More complex variants, already studied in the literature [28], combine non-determinism with probabilities, in the sense that, at each transition, a distribution function over states is non-deterministically chosen. They come up as coalgebras $p : U \longrightarrow \mathcal{PD}(U) \times \mathcal{H}(O)$. Another interesting case makes $\mathcal{G}(U) = K^U$, for K a set of weights, thus associating costs to discrete transitions. New types of hybrid automata can also be studied in this setting. For example, $\mathcal{G} = \Delta$, for Δ the diagonal functor, gives rise to arrows of type $p : U \longrightarrow \Delta U \times \mathcal{H}(O)$, explored in [54]. These correspond to deterministic hybrid automata able to replicate themselves at each discrete transition to capture, for example, cellular replication when an organism reaches a specific saturation.

3 Architectures

3.1 Composition and refinement

Coalgebra provides a uniform framework for modelling state-based systems. The architectural problem in Software Engineering addresses the ways in which such systems can be composed. Composition has a 'geometrical' flavour: components have boundaries (*i.e.* interfaces) and organise themselves in two dimensions, *temporal* and *spatial*, as in a Cartesian plane, as depicted below. The boundary shared by vertically composed components represents handling control from one, which is terminating a particular execution thread, to another which is launching a new one. Dually, horizontal composition corresponds to concurrent evolutions being juxtaposed or eventually interacting through exchange of values or a common involvement in shared actions.

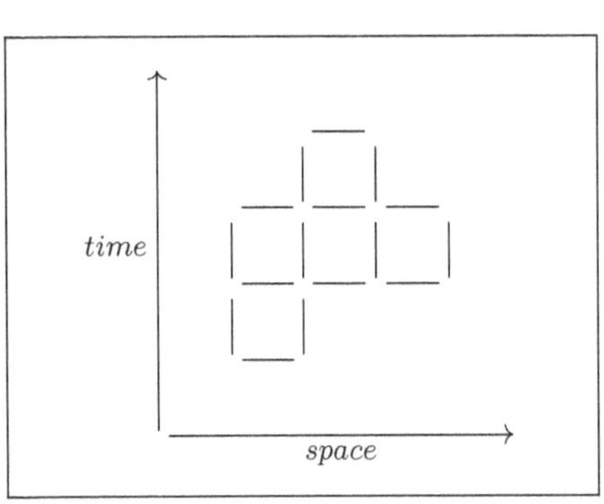

The ways in which these two basic forms of composition are defined certainly varies. For example, if components are terminating imperative programs, vertical interfaces are often realised through shared, global state variables. On the other hand, horizontal boundaries may represent the history of recorded interactions. In components modelled coalgebraically [7, 10], vertical composition is a form of pipelining (interfacing through shared output-input types), whereas horizontal composition is realised by some form of parallel paired evolution.

At this abstract level, architectural calculi build on these two forms of composition, hereby represented by ; and $\Box$, respectively. Associativity of both operators is clearly expected, as the 'geometry' is invariant for the ways parentheses are placed. Commutativity of $\Box$ conforms to the intuition that parallel composition is unordered, but this conceptual assumption may be differently interpreted by different tensors. A similar observation applies to idempotency: it would be expected when $\Box$ encodes a choice between alternative components, but less so for a synchronous product. The interaction between the temporal and the spatial dimensions is captured by distribution, which follows the same pattern of an *interchange* law in category theory,

$$(p \Box q) ; (p' \Box q') \;=\; (p ; p') \Box (q ; q') \;. \tag{7}$$

The law is rather familiar. For example, taking $\Box$ as parallel composition in a process algebra and ; as action prefixing, it introduces interleaving on moving from the right to the left hand side.

In practice, software architectures are described through a myriad of concrete languages and formalisms, often of a graphical nature which may allow for rather flexible coordination patterns. Coalgebra has been used to provide semantics to some of these formalisms, from classical process algebra [11] and the π-calculus [51] to statecharts [29] and different UML diagrams. In each context, composition operators to build new (coalgebraic models of) components from old, like variants of $\Box$ and ;, are specified and their properties studied. A concrete component calculus will be reviewed below to illustrate a possible application of Coalgebra to architectural design. Before that, however, we would like to address two main issues on this discussion.

The first one concerns *composition* of components modelled coalgebraically. To make things concrete, and already anticipating the illustration section, consider generalised Mealy machines

$$p : U \times I \longrightarrow \mathcal{T}(U \times O) \tag{8}$$

which can be seen as coalgebras for $\mathcal{F}(X) = \mathcal{T}(X \times O)^I$. This provides an elementary model of state-based software components characterised by

- an internal state space,
- input and output observation universes to ensure the flow of data,
- the possibility of interaction with other components during the overall computation,

- a behavioural effect which specifies their branching structure.

We have already seen some possible variants for $\mathcal{T}$. However, if one wants to compose this sort of components and define a calculus to reason about their interconnection, the first step is to enforce extra structure upon $\mathcal{T}$, namely that of a strong *monad*. Reacting to an input i, the coalgebra will not simply produce an output and a continuation state, but a $\mathcal{T}$-structure of such pairs. The monadic structure provides tools to handle the 'nesting' of such computations. Unit (η) and multiplication (μ) correspond, respectively, to a value embedding and a 'flatten' operation to reduce nested behavioural effects. The latter is the key element to define effect-aware composition, which, as discussed below, is based on composition in the Kleisli category for $\mathcal{T}$. Recall, for future reference, the definition: the composition of two monadic arrows $m : I \longrightarrow \mathcal{T}(Z)$ and $n : Z \longrightarrow \mathcal{T}(O)$, is given by $n \bullet m \;\hat{=}\; \mu \cdot \mathcal{T}(n) \cdot m$. Therefore, in this setting, the abstract operator ; builds on composition in the universe of $\mathcal{T}$-computations, *i.e.* in the corresponding Kleisli category, whereas $\square$ will be a tensor in this category. Strength, either in its right ($\tau_r : U \times \mathcal{T}(V) \longrightarrow \mathcal{T}(U \times V)$) or left ($\tau_l$) version, handles context information. A sort of distributive law $\delta : \mathcal{T}(U) \times \mathcal{T}(V) \longrightarrow \mathcal{T}(U \times V)$ is obtained by composing the right and left strengths. Whenever the order in which this composition is performed does not matter, the monad is said to be commutative. As one may guess this will impact on commutativity of tensors $\square$ connecting such models.

The second comment which is in order concerns the interpretation of the equality symbol in equation (7). In a coalgebraic setting the obvious choice is observational equality for the relevant functor $\mathcal{T}$. In engineering practice, however, one is sometimes interested in establishing weaker relationships. For example, (7) may be presented as an inequation to convey the intuition that a sequential computation is a special case of a parallel one, as in concurrent Kleene algebra [37].

This opens an important issue in architectural design – *refinement* – that we will briefly review from two different perspectives.

In data refinement, there is a 'recipe' to identify a refinement situation: look for an abstraction function to witness it. In other words: look for a morphism in the relevant category from the 'concrete' to the 'abstract' model such that the latter can be recovered from the former up to a suitable notion of equivalence, though typically not in a unique way. In a coalgebraic framework, however, some extra care is in order. The reason is obvious: coalgebra morphisms entail bisimilarity. Therefore one has to look for a somewhat *weaker* notion of a morphism between coalgebras.

The first approach to be mentioned here [68] works for extended polynomial endofunctors in *Set* and resorts to the notion of a preorder $\leq$ on a functor $\mathcal{T}$. This is itself a functor which makes the following diagram commute:

$$\begin{array}{ccc} & PreOrd & \\ {\leq_{\mathcal{T}}} \nearrow & \downarrow & \\ Set \xrightarrow{\mathcal{T}} & Set & \end{array} \quad \textit{i.e.} \quad \begin{array}{ccc} & \langle \mathcal{T}(V), \leq_{\mathcal{T}(V)} \rangle & \\ \nearrow & \downarrow & \\ V \longmapsto & \mathcal{T}(V) & \end{array}$$

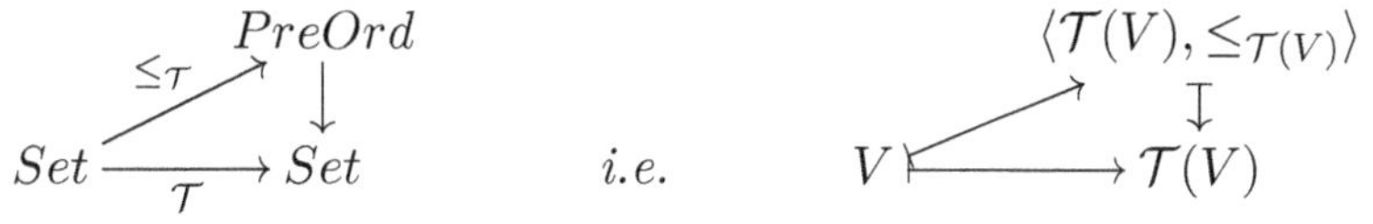

This means that for any function $h : V \longrightarrow U$, $\mathcal{T}(h)$ preserves the order, *i.e.* in a pointfree formulation, $\mathcal{T}(h) \cdot \leq_{\mathcal{T}(V)} \subseteq \leq_{\mathcal{T}(U)} \cdot \mathcal{T}(h)$.

Given two $\mathcal{T}$-coalgebras $\beta : V \longrightarrow \mathcal{T}(V)$ and $\alpha : U \longrightarrow \mathcal{T}(U)$, one may now define, with respect to a preorder $\leq$, the notions of a *preserving* and a *reflecting* morphism as a function h from V to U such that

$$\mathcal{T}(h) \cdot \beta \leq \alpha \cdot h \quad \text{and} \quad \alpha \cdot h \leq \mathcal{T}(h) \cdot \beta \,,$$

respectively. The notation $\dot{\leq}$ is used for the pointwise lifting of the preorder $\leq$ to the functional level, *i.e.* $f \dot{\leq} g \Leftrightarrow \forall_x . \ f \ x \leq g \ x$,or, equivalently, $f \dot{\leq} g \Leftrightarrow f \subseteq \leq \cdot g$. The names chosen for these morphisms come from the fact that indeed they respectively preserve and reflect state transitions induced by coalgebras, *i.e.*

$$v \longrightarrow_\beta v' \ \Rightarrow \ h \, v \longrightarrow_\alpha h \, v' \quad \text{and} \quad h \, v' \longrightarrow_\alpha u' \Rightarrow \exists_{v' \in V} . \ v \longrightarrow_\beta v' \wedge u' = h \, v'$$

where $u' \longrightarrow_\alpha u \Leftrightarrow u' \in_{\mathcal{T}} \alpha \, u$ is an instance of datatype membership [38], defined inductively for the class of relevant functors [8] and verifying

$$h \cdot \in_{\mathcal{T}} \ = \ \in_{\mathcal{T}} \cdot \mathcal{T} h \qquad (9)$$

for any function h.

A *refinement preorder* is a preorder $\leq$ on an endofunctor $\mathcal{T}$ satisfying the following compatibility condition with the membership relation: for all $x \in X$ and $x_1, x_2 \in \mathcal{T}(X)$,

$$x \in_{\mathcal{T}} x_1 \wedge x_1 \leq x_2 \ \Rightarrow \ x \in_{\mathcal{T}} x_2$$

or, again in a pointfree formulation,

$$\in_{\mathcal{T}} \cdot \leq \ \subseteq \ \in_{\mathcal{T}} . \qquad (10)$$

It is easy to see that reflecting morphisms form a category and similarly for the dual case. The point, however, is that the exact meaning of a refinement assertion $p \leq q$ above depends on the concrete refinement preorder adopted. But what do we know about such preorders?

Condition (10) is equivalent to $\leq \ \subseteq \ \in_{\mathcal{T}} \backslash \in_{\mathcal{T}}$ by direct application of the Galois connection which defines relational *division*, *i.e.* $R \cdot X \subseteq S \Leftrightarrow X \subseteq R \backslash S$. Clearly,

this provides an upper bound for refinement preorders, the lower bound being the identity. Note that $\in_{\mathcal{T}} \setminus \in_{\mathcal{T}}$ corresponds to the lifting of $\in_{\mathcal{T}}$ to (structural) inclusion, *i.e.* $x\ (\in_{\mathcal{T}} \setminus \in_{\mathcal{T}})\ y \Leftrightarrow \forall_{e \in_{\mathcal{T}} x} \cdot\ e \in_{\mathcal{T}} y$. Different refinement preorders have been studied [8] and will not be detailed here. In broad terms, refinement based on *preserving* morphisms generalises the usual axis of *non-determinism reduction* in a functorial way. On the other hand, *reflecting* morphisms witness a similar functorial generalisation of *definition increase*.

The second approach, developed along a series of papers by I. Hasuo [31, 35, 69, 70], plays a similar game but in a different category. It applies to coalgebras $U \to \mathcal{T}(\mathcal{F}(U))$ where $\mathcal{T}$ is a monad in *Set*, capturing the branching behavioural effect, $\mathcal{F}$ is a functor which determines the linear-time behaviour, and a distributive law, *i.e.* a natural transformation $\lambda : \mathcal{F}\mathcal{T} \Longrightarrow \mathcal{T}\mathcal{F}$ is assumed to hold. Thus, a $\mathcal{T}\mathcal{F}$-coalgebra in *Set* corresponds to a $\overline{\mathcal{F}}$-coalgebra in the Kleisli category $Kleisli(\mathcal{T})$ for monad $\mathcal{T}$. Functor $\overline{\mathcal{F}}$ is the canonical lifting of $\mathcal{F}$ to $Kleisli(\mathcal{T})$ which coincides with $\mathcal{F}$ on objects and maps an arrow $h : U \leftrightarrow V$ to $\overline{\mathcal{F}}(h) = \lambda_V \cdot \mathcal{F}(h) : \mathcal{F}(U) \leftrightarrow \mathcal{F}(V)$. Notice that notation $U \leftrightarrow V$ stands for an arrow in $Kleisli(\mathcal{T})$, *i.e.* a *Set* function $U \to \mathcal{T}(V)$.

Once this setting is defined, all one has to do is to play the coalgebraic game as usual. In particular, *reflecting* and *preserving* morphisms as introduced above, emerge now as lax and oplax morphisms, renamed in this context to *forward* and *backward* simulations. However, rather than defining what we have called above refinement preorders, essentially based on the functor structure, the novelty of this approach is to build on the fact that, for the class of functors considered, the homsets in the Kleisli catagory are dcpo$_\perp$-enriched, therefore carrying a notion of order. This means that the set of arrows, say from U to V in $Kleisli(\mathcal{T})$ forms a dcpo with a minimum element $\perp$. The crucial observation is that in such circumstances an initial $\mathcal{F}$-algebra in *Set* yields a final $\overline{\mathcal{F}}$-coalgebra in $Kleisli(\mathcal{T})$. Actually, this comes from Smyth and Plotkin's classical work on limit-colimit coincidence [65].

The basic result is as follows: An initial $\mathcal{F}$-algebra $\zeta : \mathcal{F}(W) \longrightarrow W$ in *Set* lifts to an initial $\overline{\mathcal{F}}$-algebra in $Kleisli(\mathcal{T})$, $\eta_W \cdot \zeta$, which coincides with the final $\overline{\mathcal{F}}$-coalgebra. Its dynamics is given by

$$\omega = \eta_{\mathcal{F}(W)} \cdot \zeta^\circ : W \leftrightarrow \mathcal{F}(W)$$

in $Kleisli(\mathcal{T})$.

Coinduction in the Kleisli category works as expected, entailing a unique behaviour map from any other $\overline{\mathcal{F}}$-coalgebra $p : U \leftrightarrow \overline{\mathcal{F}}(U)$ as depicted in the diagram below.

$$\begin{array}{ccc} W & \overset{\omega}{\longrightarrow} & \overline{\mathcal{F}}(W) \\ {\scriptstyle tr_p}\uparrow & & \uparrow{\scriptstyle \overline{\mathcal{F}}(tr_p)=\lambda\cdot\mathcal{F}(tr_p)} \\ U & \overset{p}{\longrightarrow} & \overline{\mathcal{F}}(U) \end{array}$$

This behaviour map in the the Kleisli category, denoted by tr_p, corresponds to the *trace semantics* of the original coalgebra in Set. Just to build up intuition, let us compute tr_p for a non-deterministic automaton, *i.e.* a coalgebra $p : U \longrightarrow \mathcal{P}(\mathbf{1} + A \times U)$.

Notice that $\mathcal{T} = \mathcal{P}$ and $\mathcal{F} = \mathbf{1} + A \times -$. The carrier of the final $\overline{\mathcal{F}}$-coalgebra is $W = A^*$, as $[nil, cons] : \mathbf{1} + A \times A^* \longrightarrow A^*$ is the initial $\mathcal{F}$-algebra in Set. Therefore, $tr_p : U \longrightarrow \mathcal{P}(A^*)$ is such that

$$\begin{cases} nil \in tr_p u & \Leftarrow \ u = \text{ß}_1 * \\ cons\langle a, s\rangle \in tr_p u & \Leftarrow \ u = \text{ß}_2\langle a, u'\rangle \text{ and } s \in tr_p u' \end{cases}$$

where $\text{ß}_1, \text{ß}_2$ are the coproduct injections, which corresponds to the language accepted by the automaton p.

This example drives us in the right direction: the behaviour of a coalgebra in the Kleisli of the monad capturing the intended behavioural effect gives its *trace semantics*. Forward and backward simulations computed in exactly the same setting, as indicated above, entail notions of refinement which are sound with respect to trace inclusion (and even complete for a combination of both kinds of simulation). The point to stress, however, is that, just as the genericity of Coalgebra makes bisimulation acquire the shape of the relevant functor, it does the same to trace semantics. Similarly, different notions of simulation, *e.g.* for probabilistic and weighted coalgebras, have been extensively studied [69].

This construction of trace semantics, of which reference [35] gives a detailed account, is limited to the family of functors mentioned above. For example, it does not apply to $\mathcal{T} = \mathcal{D}$ which induces a trivial order in the Kleisli homsets; the subdistribution functor

$$\mathcal{D}_{\leq}(X) \;=\; \{\mu : X \longrightarrow \mathbb{R}_{\geq 0} \mid \sum_{x \in X} \mu x \leq 1\}$$

can be used instead – the software engineer may think of what is missing to 1, in each transition, as the probability of some sort of 'systemic' failure, such as deadlock, to occur. On the other hand, although the finite powerset monad can serve as $\mathcal{T}$ in several contexts, it cannot, for example, in combination with $\mathcal{F}(X) = A \times X$, because the initial algebra is then the empty set, thus yielding a trivial trace.

A recent, alternative path [42] to compute coalgebraic trace semantics based on *determinisation*, rather that on order enrichment, seems particularly fruitful.

The idea is borrowed from automata theory where determinisation refers to the algorithmic construction of the deterministic equivalent to a non-deterministic automata. The latter often provides a smaller representation of the problem at hand but its processing is computationally harder. A similar process converts a partial into a total automaton. Both of them have a common shape: more transitions are added but the behaviour of the non-deterministic or the partial automata is given in terms of the deterministic, total case. The coalgebraic generalization is based on a different decomposition, studying coalgebras of type $U \to \mathcal{F}(\mathcal{T}(U))$, rather than $U \to \mathcal{T}(\mathcal{F}(U))$. It puts new conditions on the functors of interest and lifts the constructions not to the Kleisli, but to the Eilenberg-Moore category of the behavioural effect monad. Interestingly enough, it captures cases that fail to have $\mathrm{dcpo}_\perp$-enriched Kleisli homsets . One such example concerns coalgebras

$$p : U \longrightarrow \mathcal{M}(\mathbf{1} + A \times U)$$

where $\mathcal{M}(X) = \mathbb{N}^X$ is the multiset monad, and corresponds to a quite general form of weighted transition systems.

3.2 Illustration: Variants of a component calculus

My first contact with Coalgebra, in the context of my own doctoral studies, focused on the development of a calculus for software components modelled as monadic Mealy machines [7], typed as (8) above. A component model in such a setting is a pointed coalgebra

$$p \mathrel{\widehat{=}} \langle u_p \in U_p, \overline{a}_p : U_p \longrightarrow \mathcal{T}(U_p \times O)^I \rangle \tag{11}$$

where u_p is the initial state and the coalgebra dynamics is captured by currying a state-transition function $a_p : U_p \times I \longrightarrow \mathcal{T}\ (U_p \times O)$.

The basic architectural operator is *pipeline* – a form of sequential composition which amounts to the Kleisli composition for monad $\mathcal{T}$ of its arguments suitably extended to each other's state space. It is worthwhile to detail the construction. Given $a_p : U_p \times I \longrightarrow \mathcal{T}(U_p \times O)$, its (left) state extension to X is computed as

$$\begin{aligned} a_{X|p} \mathrel{\widehat{=}} (X \times U_p) \times I &\xrightarrow{a^\circ} X \times (U_p \times I) \xrightarrow{id \times a_p} X \times \mathcal{T}(U_p \times O) \\ &\xrightarrow{\tau_r} \mathcal{T}(X \times (U_p \times O)) \xrightarrow{\mathcal{T}(a)} \mathcal{T}((X \times U_p) \times O) \end{aligned}$$

where a is the associativity natural isomorphism and τ_r the right strength for $\mathcal{T}$. The right state extension, $p|X$, is defined similarly. Therefore, for components with dynamics $a_p : U_p \times I \longrightarrow \mathcal{T}(U_p \times Z)$ and $a_q : U_p \times Z \longrightarrow \mathcal{T}(U_p \times O)$, their pipeline is defined as

$$p\,;q \mathrel{\widehat{=}} \langle\langle u_p, u_q\rangle, \overline{a}_{p;q}\rangle \quad \text{with} \quad a_{p;q} \mathrel{\widehat{=}} a_{U_p|q} \bullet a_{p|U_q}\ . \tag{12}$$

Having defined generic components as (pointed) coalgebras, one may wonder how do they get composed and what kind of calculus emerges from this framework. Actually, interfaces are sets representing the input and output range of a component. Consequently, components are arrows between interfaces and arrows between components are arrows between arrows. Formally, this leads to the notion of a *bicategory*[5] to structure our reasoning universe. We take interfaces (*i.e.* sets modelling observation universes of components) as *objects* of a bicategory Cp, whose *arrows* are pointed coalgebras. For each pair $\langle I, O \rangle$ of interface objects, a (hom-)category $Cp(I, O)$ is defined, whose arrows $h : \langle u_p, \overline{a}_p \rangle \longrightarrow \langle u_q, \overline{a}_q \rangle$ satisfy the expected *morphism* and *initial state preservation* conditions:

$$\overline{a}_q \cdot h \;=\; \mathcal{T}(h \times O)^I \cdot \overline{a}_p \quad \text{and} \quad h\,(u_p) \;=\; u_q\,. \tag{13}$$

Composition is inherited from Set and the identity $1_p : p \longrightarrow p$ on component p is defined as the identity id_{U_p} on its carrier. Next, for each triple of objects (I, K, O), a composition law is given by a functor

$$;_{I,K,O} : Cp(I, K) \times Cp(K, O) \longrightarrow Cp(I, O)$$

whose action on objects p and q was given above.
The action of ; on 2-cells reduces to $h\,;k = h \times k$. Finally, for each object K, an identity law is given by a functor

$$copy_K : \mathbf{1} \longrightarrow Cp(K, K)$$

whose action on objects is the constant component

$$\langle * \in \mathbf{1}, \overline{a}_{copy_K} \rangle$$

with $a_{copy_K} \;\widehat{=}\; \eta_{\mathbf{1} \times K}$. Similarly, the action on morphisms is the identity on $\mathbf{1}$.

The fact that, for each strong monad $\mathcal{T}$, components form a bicategory amounts not only to a standard definition of the two basic combinators ; and $copy_K$ of a component calculus, but also to setting up its basic laws. Recall that the graph of a morphism is a bisimulation. Therefore, the existence of an initial state-preserving

[5]Basically a *bicategory* [13] is a category in which a notion of arrows between arrows is additionally considered. This means that the the space of morphisms between any given pair of objects, usually referred to as a (hom-)set, acquires itself the structure of a category. Therefore the standard arrow composition and unit laws become functorial, since they transform both objects and arrows of each hom-set in a uniform way. A typical example is Cat itself: the category whose objects are small categories, arrows are functors and arrows between arrows, or 2-cells as they are often called, correspond to natural transformations.

morphism between two components makes them bisimilar, leading to the following laws, for appropriately typed components p, q and r:

$$copy_I \,;\, p \sim p \sim p \,;\, copy_O \quad \text{and} \quad (p\,;\,q)\,;\,r \sim p\,;\,(q\,;\,r)$$

The dynamics of a component specification is essentially 'one step': it describes immediate reactions to possible state/input configurations. Its temporal extension becomes the component's *behaviour*. Formally, the behaviour $[\![(p)]\!]$ of a component p is computed by *coinductive extension*, *i.e.* $[\![(p)]\!] = [\![(\overline{a}_p)]\!]\ u_p$. Behaviours organise themselves in a category Bh, whose objects are sets and arrows $b : I \longrightarrow O$ are elements of the carrier of the final coalgebra $\omega_{I,O}$ for functor $\mathcal{T}(\mathsf{Id} \times O)^I$. Thus, composition in Bh is given by a family of combinators, for each I, K and O, ${;_{I,K,O}}^{Bh} : Bh(I,K) \times Bh(K,O) \longrightarrow Bh(I,O)$, such that ${;_{I,K,O}}^{Bh} \,\widehat{=}\, [\![(\omega_{I,K} ; \omega_{K,O})]\!]$. On the other hand, identities are given by ${copy_K}^{Bh} : \mathbf{1} \longrightarrow Bh(K,K)$ and ${copy_K}^{Bh} \,\widehat{=}\, [\![(\overline{a}_{copy_K})]\!]\ *$, *i.e.* the behaviour of component $copy_K$, for each K.

The basic observation is that the structure of Bh mirrors whatever structure Cp possesses. In fact, the former is isomorphic to a sub-(bi)category of the latter, whose arrows are components defined over the corresponding final coalgebra. Alternatively, we may think of Bh as constructed by quotienting Cp by the greatest bisimulation. However, as final coalgebras are fully abstract with respect to bisimulation, the bicategorical structure collapses. Moreover, as discussed in [7], some tensors in Cp become universal constructions in Bh, for particular instances of $\mathcal{T}$. This also explains why properties holding in Cp up to bisimulation, do hold 'on the nose' in the behaviour category. For example, the ; laws above may be rephrased as

$$copy_I \,;\, b = b = b \,;\, copy_O \quad \text{and} \quad (b\,;\,c)\,;\,d = b\,;\,(c\,;\,d)$$

for suitably typed behaviours b, c and d, in Bh. It is easy to check that Bh is a category and $[\![(\ )]\!]$ is a 2-functor from Cp to Bh. Indeed,

$$\begin{aligned} b\,;\,copy_O &= [\![(\omega_{I,O}\,;\,copy_O)]\!]\ \langle b, *\rangle = [\![(\omega_{I,O})]\!]\ b = b \\ (b\,;\,c)\,;\,d &= [\![((\omega_{I,K}\,;\,\omega_{K,L})\,;\,\omega_{L,O})]\!]\ \langle\langle b, c\rangle, d\rangle = \\ &= [\![(\omega_{I,K}\,;\,(\omega_{K,L}\,;\,\omega_{L,O}))]\!]\ \langle b, \langle c, d\rangle\rangle = b\,;\,(c\,;\,d) \end{aligned}$$

On the other hand, note that $[\![({copy_K}^{Cp})]\!] = {copy_K}^{Bh}$ and

$$\begin{aligned} [\![((p\,;^{Cp} q))]\!] &= [\![(\overline{a}_{p;q})]\!](\langle u_p, u_q\rangle) \\ &= [\![(\omega_{I,K}\,;\,\omega_{K,O})]\!] \cdot ([\![(\overline{a}_p)]\!] \times [\![(\overline{a}_q)]\!])\ \langle u_p, u_q\rangle \\ &= ;^{Bh} \cdot ([\![(\overline{a}_p)]\!] \times [\![(\overline{a}_q)]\!])\ \langle u_p, u_q\rangle \\ &= ;^{Bh}\ \langle [\![(\overline{a}_p)]\!]\ u_p, [\![(\overline{a}_q)]\!]\ u_q\rangle \\ &= [\![(p)]\!]\ ;^{Bh}\ [\![(q)]\!] \end{aligned}$$

Some detail put above in describing the structure of Cp and Bh aims at emphasising an important aspect from the architectural point of view: behaviour descriptions are compositional in a sense that is compatible with composition at the (state based) component level. Such compatibility comes exactly from the bicategorial structure. Or, as put by I. Hasuo, C. Heunen, B. Jacobs and A. Sokolova [33], as a manifestation of the *microcosm* principle which states that the same algebraic structure is carried by a category and by one of its objects which assumes a prototypical role. Examples abound in the literature on 'categorification' [4], a typical one is that of a monoid object inside a monoidal category. It is interesting that the same kind of phenomena arises in our context.

A whole component calculus, parametric on a behaviour monad $\mathcal{T}$, can be developed on Cp. The relevant structure lifts naturally to Bh defining a particular (typed) 'process' algebra. We will not go into detail here, but to mention the basic ingredients considered in all approaches documented in the literature [7, 10, 33, 32].

The first one is the representation of functions in Cp: A function $f : A \longrightarrow B$ is lifted to a component $\ulcorner f \urcorner \;\widehat{=}\; \langle * \in \mathbf{1}, \overline{a}_{\ulcorner f \urcorner} \rangle$ over $\mathbf{1}$ whose action is given by the currying of

$$a_{\ulcorner f \urcorner} \;\widehat{=}\; \mathbf{1} \times A \xrightarrow{id \times f} \mathbf{1} \times B \xrightarrow{\eta_{(\mathbf{1} \times B)}} \mathcal{T}(\mathbf{1} \times B)$$

Up to bisimulation, function lifting is functorial, that is, for $g : I \longrightarrow K$ and $f : K \longrightarrow O$ functions, one has

$$\ulcorner f \cdot g \urcorner \sim \ulcorner g \urcorner \,;\, \ulcorner f \urcorner \quad \text{and} \quad \ulcorner id_I \urcorner \sim copy_I$$

Actually, lifting canonical Set arrows to Cp is a simple way to explore the structure of Cp itself. For instance, $?_I : \emptyset \longrightarrow I$ keeps its naturality as, for any $p : I \longrightarrow O$, the corresponding diagram below commutes up to bisimulation, because both $\ulcorner ?_I \urcorner$ and $\ulcorner ?_O \urcorner$ are the *inert* components: the absence of input makes reaction impossible. Formally,

$$\ulcorner ?_I \urcorner \,;\, p \sim \ulcorner ?_O \urcorner \tag{14}$$

Naturality is lost, however, in the lifting of $!_I : I \longrightarrow \mathbf{1}$: the diagram fails to commute for non trivial $\mathcal{T}$ (*e.g.* the finite powerset monad).

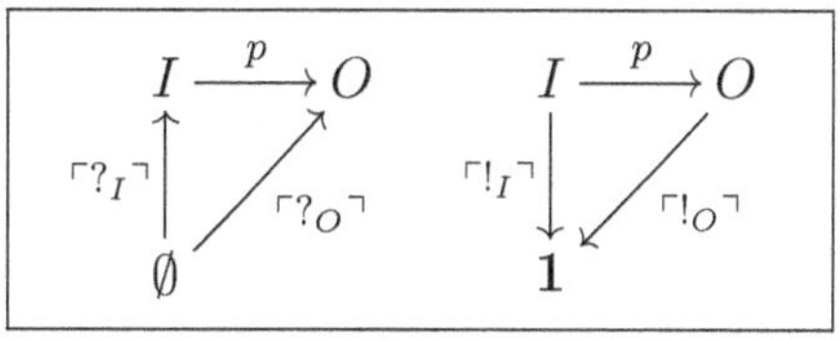

Components over $\mathbf{1}$ defined from identities and structural properties of the underlying category are called *wires*. Typical examples, include the liftings of canonical isomorphisms – *e.g.* associativity, a, or commutativity, s – which leads to bisimilarity up to an isomorphic rearranging of the interface, as well as liftings of embeddings, projections, codiagonals and diagonals, the latter used to *merge* input and *replicate* output types, as in, for example, $\ulcorner\nabla\urcorner\,;p\,;\ulcorner\Delta\urcorner : I+I \longrightarrow O\times O$.

The pre- and post-composition of a component with Cp-lifted functions can be encapsulated into a unique combinator, called *wrapping*, which is reminiscent of the *renaming* connective found in process calculi. It is defined as functor $-[f,g] : Cp(I,O) \longrightarrow Cp(I',O')$, for f and g suitably typed, which is the identity on morphisms and maps component $\langle u_p, \overline{a}_p\rangle$ into $\langle u_p, \overline{a}_{p[f,g]}\rangle$, where

$$a_{p[f,g]} \mathrel{\widehat{=}} U_p \times I' \xrightarrow{id\times f} U_p \times I \xrightarrow{a_p} \mathcal{T}(U_p \times O) \xrightarrow{\mathcal{T}(id\times g)} \mathcal{T}(U_p\times O')$$

Typical properties are, as one could expect,

$$p[f,g] \sim \ulcorner f\urcorner\,;p\,;\ulcorner g\urcorner \quad\text{and}\quad (p[f,g])[f',g'] \sim p[f\cdot f', g'\cdot g]$$

Components can be aggregated in a number of ways, besides the 'pipeline' composition discussed above. Several tensors have been introduced in the literature [7, 33, 52] corresponding to *choice*, *parallel* and *concurrent* composition. We will briefly detail the first one which provides a form of additive composition defined as a lax functor $\boxplus : Cp\times Cp \longrightarrow Cp$. It consists of an action on objects given by $I\boxplus J = I+J$ and a family of functors $\boxplus_{I,O,J,R} : Cp(I,O)\times Cp(J,R) \longrightarrow Cp(I+J, O+R)$ yielding

$$p\boxplus q \mathrel{\widehat{=}} \langle\langle u_p,u_q\rangle \in U_p\times U_q, \overline{a}_{p\boxplus q}\rangle \quad\text{and}\quad a_{p\boxplus q} \mathrel{\widehat{=}} dr^\circ \bullet (a_{p|U_q} + a_{U_p|q}) \bullet dr^\circ\ ,$$

where dr is the right distributivity isomorphism, and mapping pairs of arrows $\langle h_1,h_2\rangle$ into $h_1\times h_2$. When interacting with $p\boxplus q$, the environment chooses either to input a value of type I or one of type J, triggering the corresponding component, p or q, respectively. The following laws arise from the fact that $\boxplus$ is a lax functor in Cp:

$$(p\boxplus p')\,;(q\boxplus q') \sim (p\,;q)\boxplus(p'\,;q')$$
$$copy_{K\boxplus K'} \sim copy_K \boxplus copy_{K'}$$
$$\ulcorner f\urcorner \boxplus \ulcorner g\urcorner \sim \ulcorner f+g\urcorner\ .$$

Moreover, up to isomorphic wiring, $\boxplus$ is a symmetric tensor product in each hom-category, with $nil = \ulcorner id_\emptyset\urcorner$ as unit, *i.e.*

$$(p\boxplus q)\boxplus r \sim p\boxplus(q\boxplus r)$$
$$nil\boxplus p \sim p \text{ and } p\boxplus nil \sim p$$
$$p\boxplus q \sim q\boxplus p\ .$$

The construction of architectural calculi based on generalised Mealy machines initiated in [6], was furhter developed by a several authors. V. Miraldo and J. N. Oliveira [52] focused on lifting the whole calculus to the Kleisli category of the relevant behaviour monad $\mathcal{T}$, under the *motto* 'keep definition, change category'. Some of these categories are paradigmatic universes for dealing, namely, with non-determinism and probabilistic evolution. In the first case the calculus is 'instantiated' in the category of sets and binary relations, *i.e.* the Kleisli for the finite powerset functor. In the second, in the category of (sub-)stochastic matrices, the Kleisli of the (sub-)distribution functor. In both cases, calculation takes advantage of a well-studied universe of *typed* relations and matrices, respectively [57]. The programme is not straightforward, namely in what concerns the lifting of theories (*e.g.* of behavioural equivalence), further than just the definition of combinators and the preservation of monadic strength on moving from the original to the Kleisli category. Again, a somehow heavy requirement is found here: monad $\mathcal{T}$ should induce a $\mathrm{dcpo}_\perp$-enriched Kleisli category (as pointed out above when discussing trace semantics for coalgebras) and should itself be symmetric monoidal, *i.e.* commutative.

These two requirements appear again in the approach developed by B. Jacobs, I. Hasuo, C. Heunen and A. Sokolova, in a series of papers [33, 32, 34]. The whole work is done in the context of a symmetric monoidal category C, equipped with coproducts $+$ and $\emptyset$ over which the tensor $\otimes$ distributes. The behaviour monad $\mathcal{T}$ is assumed to be commutative, with a distributive law $\delta : \mathcal{T}(U) \times \mathcal{T}(V) \longrightarrow \mathcal{T}(U \times V)$. Instead of building on a bicategorial structure as before, components are taken as objects in a category with fixed input/output universes. This entails the need for the introduction of an *indexing* mechanism, similar to the one underlying relabelling in process algebra. Actually, the calculus works directly with arrows $U \times I \to \mathcal{T}(U \times O)$ which lift to coalgebras if C is Cartesian closed. This is not assumed in general, leading to a very general setting; for example state extension discussed above arises simply as an action of the monoidal category on a category of components. A very interesting connection links components in C to Freyd categories [60], which further correspond to J. Hughes' notion of an *arrow* [39], a construction which, like that of a monad, is used to model structured computations in functional programming.

But what is really new in this approach is the introduction of a *trace* operator in the architectural calculus, which provides a formalisation of the notion of a 'loop' in a diagram of components. Semantically, this brings to scene a feedback construction with respect to the additive structure of C, embodying a form of *iteration*. Mathematically, the operator is a trace in the sense of A. Joyal, R. Street and D. Verity [43].

The development of these ideas can be summed up as follows: whenever C has countable coproducts and the behaviour monad $\mathcal{T}$ is commutative and, as before, induces a Kleisli category whose homsets are dcpo$_\perp$ enriched, $Kleisli(\mathcal{T})$ is traced monoidal with respect to coproduct as a monoidal structure. This means that, to each arrow $h : I+Z \longrightarrow \mathcal{T}(O+Z)$ corresponds a traced arrow $Tr^{Kl}(h) : I \longrightarrow \mathcal{T}(O)$, where operator $Tr^{Kl}(\ )$ satisfies the canonical properties for a trace [43]. This lifts to a trace operator in the category of components which also obeys those properties, although only up to isomorphism. Thus, given a component $p = \langle u_p \in U_p, \overline{a}_p\rangle : I + Z \longrightarrow O + Z$ the trace operator builds a new one where output in Z is fed back to p:

$$Tr(p) \mathrel{\widehat{=}} \langle u_p \in U_p, \overline{a}_{Tr(p)}\rangle : I \longrightarrow O \quad \text{with} \quad a_{Tr(p)} \mathrel{\widehat{=}} Tr^{Kl}(\mathcal{T}(dr^\circ) \bullet a_p \bullet dr)$$

Note that $\mathcal{T}(dr^\circ) \bullet a_p \bullet dr$ is typed as $U_p \times I + U_p \times Z \to \mathcal{T}(U_p \times O + U_p \times Z)$ in $Kleisli(\mathcal{T})$.

The requirements on the behaviour monad mentioned above seem to be recurrent when aiming at a richer structure for the development of architectural calculi. They are not met, however, by the functor $\mathcal{H}$ introduced in section 2.3 and intended to capture continuous behaviour.

The functor $\mathcal{H}$, however, extends to a strong monad (both in Set and Top) which means that most of the calculus discussed above can be developed to address components with continuous evolutions both in their output and state space, *i.e.* built as coalgebras for $\mathcal{F}(X) = \mathcal{H}(U \times O)^I$. Basically, all the calculus is kept, but for an additive trace and the interchange law with respect to a tensor capturing parallel evolution. There is, however, a notion of iteration, which, under some circumstances [55], induces a fixed point to give semantics to infinite loops.

In any case, it seems that $\mathcal{H}$-based coalgebras will play a relevant role in generalising the component calculus to the continuous domain, to reason *e.g.* about sensor networks and IoT configurations. Therefore, I'll close this section introducing the associated monadic structure. Recall the definition of $\mathcal{H}$ in (6), section 2.3. The monad structure adds a multiplication, $\mu : \mathcal{H} \cdot \mathcal{H} \Longrightarrow \mathcal{H}$ and its unit $\eta : Id \Longrightarrow \mathcal{H}$. The latter produces trivial evolutions with duration 0. Formally, $\eta_X\ x \mathrel{\widehat{=}} \langle \underline{x},\ 0\rangle$. Multiplication is a bit more complex. Let $\langle f, d\rangle \in (X^T \times D)^T \times D$. Then, the 'flattened' system, $\mu_X\ \langle f, d\rangle$, will return, at each instant t_i, the value $(f\ t_i)0$ until, and if, d is reachead. After that, if $d \neq \infty$, it will evolve according to $fd = \langle g, e\rangle$ for the remaining duration $e - d$. Formally,

$$\mu_X\ \langle f, d\rangle \mathrel{\widehat{=}} \begin{cases} \langle \theta_X \cdot f, d\rangle \ \ +\!\!+\ \ (f\ d) & \text{if } d \neq \infty \\ \langle \theta_X \cdot f,\ \infty\rangle & \text{otherwise} \end{cases}$$

where $\theta : \mathcal{H} \Longrightarrow Id$ is given by $\theta\,\langle f, d\rangle \mathrel{\widehat{=}} f\,0$ and $\langle f, d\rangle \ +\!\!+\ \langle g, e\rangle \mathrel{\widehat{=}} \langle f \ +\!\!+_d\ g, d+e\rangle$ with $f \ +\!\!+_d\ g \mathrel{\widehat{=}} f \lhd (\leq_d) \rhd g\,(\_ - d)$. Note that θ is an Eilenberg-Moore $\mathcal{H}$-algebra: indeed, $\theta \cdot \eta = id$ and $\theta \cdot \mu = \theta \cdot \mathcal{H}\theta$.

It is worthwhile to see what composition means in $Kleisli(\mathcal{H})$. Let $c_1 : I \nrightarrow K$, $c_2 : K \nrightarrow O$, and assume $c_i = \langle f_i, d_i \rangle$, for $i = 1, 2$. Thus[6],

$$
\begin{array}{cl}
& (\mu \cdot \mathcal{H}c_2 \cdot c_1)\ x \\
= & \quad \{\ c_1 = \langle f_1, d_1 \rangle, \text{ and let } d = d_1\ x\ \} \\
& (\mu \cdot \mathcal{H}c_2)\ \langle f_1\ x, d \rangle \\
= & \quad \{ \text{ definition of } \mathcal{H}\ \} \\
& \mu\ \langle c_2 \cdot (f_1\ x), d \rangle \\
= & \quad \{ \text{ definition of } \mu \} \\
& \langle \theta \cdot c_2 \cdot (f_1\ x), d \rangle\ \ +\!\!+\ \ (c_2 \cdot (f_1\ x))\ d \\
= & \quad \{ \text{ definition of } +\!\!+\ \} \\
& \langle\ (\theta \cdot c_2 \cdot (f_1\ x))\ \ +\!\!+_d\ f_2\ ((f_1\ x)\ d),\ \ d + \pi_2(c_2((f_1\ x)\ d))\ \rangle \\
= & \quad \{ \text{ definition of } +\!\!+_d\ \} \\
& \langle\ \theta \cdot c_2\ ((f_1\ x)\ \_\ )\ \lhd\ (\leq_d)\ \rhd\ f_2\ ((f_1\ x)\ d)\ (\_ - d),\ \ d + \pi_2(c_2((f_1\ x)\ d))\ \rangle
\end{array}
$$

Two different cases must be considered. In the first one suppose that c_2 in $(c_2 \bullet c_1)$ is *pre-dynamical* in the standard sense that $\theta \cdot c_2 = id$ (or, at least, an inclusion). In this case composition yields *sequencing*: for the duration of $c_1 x$, $(c_2 \bullet c_1)x$ evolves first according to c_1, and then, on its termination, according to c_2, which receives as input the endpoint of $f_1 x$. Otherwise it yields a form of *modulation*: the second component acts upon the first one.

This behaviour may be illustrated through the following example. Suppose the temperature of a room is to be regulated as follows: start at 10 °C, seek to reach and maintain 20 °C, but in no case surpass 20.5 °C. The system is realised by three elementary components that have to work together: component c_1 to raise the temperature to 20 °C, component c_2 to maintain a given temperature, and, finally, c_3 to ensure the temperature never goes over 20.5 °C. Formally,

$$
\begin{aligned}
c_1\ x &= \langle\ (x + \_\ ),\ 20 \ominus x\ \rangle \\
c_2\ x &= \langle\ x + (\sin\ \_\ ),\ \infty\ \rangle \\
c_3\ x &= \langle\ \underline{x}\ \lhd\ (x \leq 20.5)\ \rhd\ \underline{20.5}\ , 0\ \rangle
\end{aligned}
$$

where $\ominus$ is truncated subtraction (i.e. $x \ominus y$ is $x - y$ if $x > y$ or 0 otherwise). Composing $c_2 \bullet c_1$ yields a component which reads the current temperature, raises it to 20 °C, and then keeps it stable, as exemplified by the plot below (left). If,

[6] I'll omit the simpler, but similar case dealing with infinite durations.

however, temperatures over 20.5 °C occur, composition $c_3 \bullet (c_2 \bullet c_1)$ puts the system back into the right track as illustrated by the plot in the right.

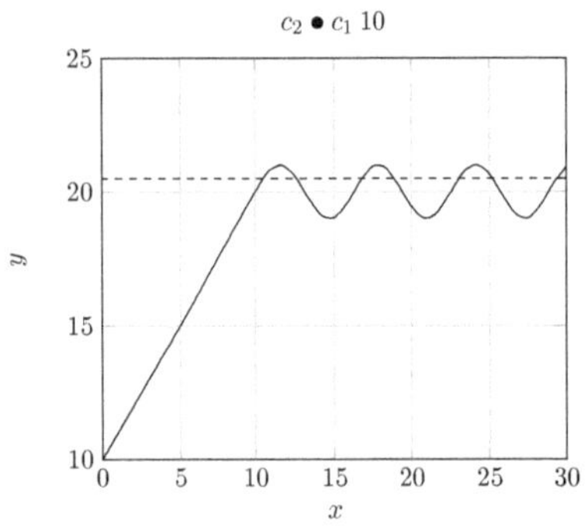

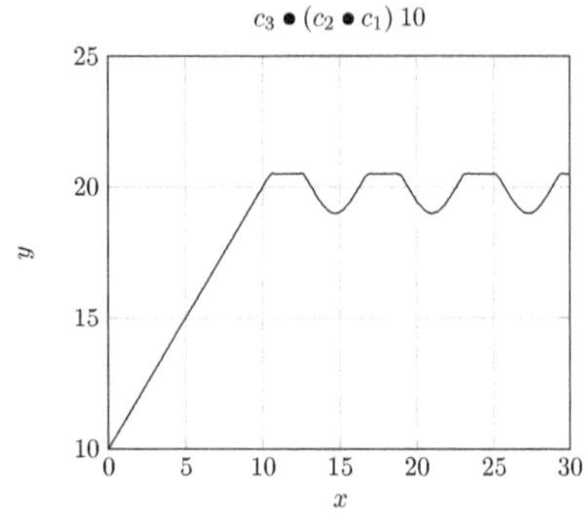

Clearly, c_3 can be regarded as a supervisor system that, for the sake of efficiency, only acts when temperatures exceed the threshold, using just enough power to keep the temperature below the limit. Actually, note that c_3 is able to play a supervisory role precisely because it is not pre-dynamical.

Recent research [55] unveiled most of the structure of an $\mathcal{H}$-based architectural calculus. In particular, $\mathcal{H}$ is shown to be strong in *Top*, its Kleisli category to inherit colimits from *Top*, as expected, and moreover to preserve pullbacks (a little bit harder to prove). A concrete description of the final coalgebra can be done, as well as the systematic definition of old and new combinators for $\mathcal{H}$-based components.

4 Properties

4.1 From invariants to modalities

Requirements, architectural properties, interface specifications, business rules, etc. are common designations for (different kinds of) properties recurring in the practice of Software Engineering. To be unambiguously stated, compared, and even verified against the models of interest, they need to be expressed in suitable logics, preferably equipped with some form of mechanically supported verification framework.

To a great extent, in software design one is interested in properties that are preserved along the system's evolution, the so-called 'business rules', as well as in 'future warranties', stating that *e.g.* some desirable outcome will be eventually produced. Both classes are examples of *modal* assertions, *i.e.* properties that are to be interpreted across a transition system capturing the software dynamics. The relevance of modal reasoning in computing is witnessed by the fact that most university syllabi in the area include some incursion into modal logic, in particular in its temporal variants.

The novelty is that, as it happens with the notions of transition, behaviour, observational equivalence or refinement, modalities in Coalgebra also acquire a *shape*. That is, their definitions become parametric on whatever type of behaviour seems appropriate for addressing the problem at hand.

Let me start with the notion of an *invariant* – a predicate[7] which is supposed to hold in all states of a system, thus configuring what is classically called a *safety* property. If the system dynamics is described by a coalgebra $\alpha : U \longrightarrow \mathcal{F}(U)$, a predicate ϕ over state space U is an invariant if it holds on the 'current' state and on its 'successor' states, which are of course obtained by execution of α. This entails the need to lift ϕ from U to $\mathcal{F}(U)$.

In a relational setting, *i.e.* regarding predicate ϕ as a coreflexive relation and $\mathcal{F}$ as a relator[8], the informal definition of an invariant can be captured by the statement

$$\forall_{u \in U} \,.\; u\, \phi\, u \;\Rightarrow\; (\alpha\, u)\, \mathcal{F}(\phi)\, (\alpha\, u)$$

which, by eliminating variables, is equivalent to

$$\phi \;\subseteq\; \alpha^{\circ} \cdot \mathcal{F}(\phi) \cdot \alpha \tag{15}$$

Clearly, just as bisimulations are preserved by the coalgebra transitions, so are invariants. But what is more, the right hand side of expression (15) defines a 'box' modality over the transition system entailed by coalgebra α:

$$\Box\phi \;\widehat{=}\; \alpha^{\circ} \cdot \mathcal{F}(\phi) \cdot \alpha \tag{16}$$

which rewrites (15) as: ϕ is invariant whenever $\phi \subseteq \Box\phi$. The crucial observation, however, is that the modal operator $\Box$ is parametric on the coalgebra α and, of course, on the functor $\mathcal{F}$. Moreover, invariants induce invariants because, $\Box$ being monotonic,

$$\phi \subseteq \Box\phi \;\Rightarrow\; \Box\phi \subseteq \Box\Box\phi \,.$$

It is instructive to unfold the definition of the $\Box$ modality for specific cases. Taking $\mathcal{F}(X) = \mathcal{P}(X)$, for example, one gets

$$\Box\phi \;=\; \{u \in U \mid (\alpha\, u)\; \mathcal{P}(\phi)\; (\alpha\, u)\}$$

[7]In the sequel we will resort, with no change of notation, to two equivalent representations of a predicate over a set X: as a subset of X or as a coreflexive binary relation, *i.e.* a subset of the identity over X. Thus, $x \in \phi$ iff $x\; \phi\; x$.

[8]I've already mentioned relators in footnote 5. The concept of a *relator* [27] extends that of a functor to relations: $\mathcal{F}(R)$ is a relation from $\mathcal{F}(U)$ to $\mathcal{F}(V)$ provided R is a relation from U to V. Relators are monotone and commute with composition, converse and the identity.

which, regarding predicates as sets, takes the more familiar form

$$\Box\phi \;= \{u \in U \mid \alpha\ u \subseteq \phi\}$$

which corresponds to the standard interpretation of the $\Box$ modality in Kripke semantics. As another example consider the functor $\mathcal{F}(X) = \mathbf{1} + X$. Clearly,

$$\Box\phi \;=\; \{u \in U \mid \alpha\ u = \iota_2\ u' \;\Rightarrow\; u' \in \phi\}\ .$$

The whole construction of a modal logic relative to a coalgebra α can be pursued along similar lines. Such a programme is often referred to as the *temporal logic of coalgebras* [41]. Actually, not only a diamond modality is defined, as usual, by duality, $\diamond\phi \;\widehat{=} \neg\Box\neg\phi$, but 'temporal extensions' of these modalities can be obtained as fixed points. Consider, for example, the definition of $\boxdot\phi$, the *henceforth* ϕ operator which extends the validity of ϕ over all states computed by successive application of α:

$$\boxdot\phi(x) \;\;\widehat{=}\;\; \exists_{\psi}\ .\ \psi \text{ is invariant } \;\wedge\; \psi \subseteq \phi \;\wedge\; \psi\ x$$

Regarding predicates ϕ and ψ as coreflexives and making explicit the *supremum* implicit in the existential quantification one gets,

$$\begin{aligned}
\boxdot\phi \;&=\; \bigcup\{\psi \mid \psi \subseteq \Box\psi \;\wedge\; \psi \subseteq \phi\}\\
&=\qquad \{\ \cap\text{-universal}\}\\
&\qquad \bigcup\{\psi \mid \psi \subseteq \Box\psi\ \cap\ \phi\}\\
&=\qquad \{\ \text{intersection of coreflexives is relational composition}\ \}\\
&\qquad \bigcup\{\psi \mid \psi \subseteq \phi \cdot \Box\psi\}
\end{aligned}$$

which leads to a greatest (post)fixed point definition:

$$\boxdot\phi \;\;=\;\; \nu\psi\ (\phi \cdot \Box\psi)$$

4.2 Coalgebraic logic

The modalities induced by a coalgebra α and considered so far are relative to the 'global' dynamics of α. Depending on applications, however, one may be interested in other types of modalities. For example, suppose α is a coalgebra for functor $\mathcal{F}(X) = A \times X \times X$. Then, it may be relevant to have modalities to take care of just the right or the left successors.

For another, popular example consider $\mathcal{F}(X) = \mathcal{P}X^A$, the 'shape' of a non-deterministic transition system. In this case one may be interested in one 'box' operator per each action $a \in A$ dealing only with transitions labelled by a. Thus, predicate ϕ over U has to be lifted in a specific way to $\mathcal{P}(U)^A$, for each $a \in A$. The corresponding modality will build on such 'user-defined' lifting.

To proceed, a more general notion of predicate lifting is in order. Fortunately, the definition is straightforward [47]: A predicate lifting is simply a natural transformation $\gamma : \mathbf{2}^- \Longrightarrow \mathbf{2}^{\mathcal{F}(-)}$, where $\mathbf{2}^-$ is the contravariant powerset functor. Then, a modality $\Box$, with respect to a coalgebra $\alpha : U \longrightarrow \mathcal{F}(U)$ and a predicate lifting γ, is defined as

$$\Box \ \widehat{=}\ \mathbf{2}^U \xrightarrow{\ \gamma_U\ } \mathbf{2}^{\mathcal{F}(U)} \xrightarrow{\ \alpha^{-1}\ } \mathbf{2}^U$$

where f^{-1} denotes the inverse image of function f, *i.e.* $f^{-1}\,Z = \{u \in U \mid f\ u \in Z\}$[9]. Thus,

$$\Box\phi = \{u \in U \mid \alpha\ u \in \gamma_U \phi\}\ . \tag{17}$$

For the example above, one specifies a family $\{\gamma^a : \mathbf{2}^- \Longrightarrow \mathbf{2}^{\mathcal{P}(-)^A} \mid a \in A\}$ of predicate liftings

$$\gamma^a_U\ \phi \ \widehat{=}\ \{s \in \mathcal{P}(U)^A \mid s\ a \subseteq \phi\}$$

which induces a corresponding family of $\Box$-like modalities

$$[a]\phi = \{u \in U \mid (\alpha\ u)\ a \subseteq \phi\}\ .$$

As one would expect, those are exactly the indexed modalities of Hennessy–Milner logic [67].

The 'global' modality given by equation (16) in the previous section, can be framed in this more general setting by defining the predicate lifting $\gamma_X\ \phi \ \widehat{=}\ \{s \in \mathcal{F}(X) \mid s\ \mathcal{F}(\phi)\ s\}$.

For the general case one may proceed as follows. As a first step define a signature Σ of modal operators $\circledast : X^n \longrightarrow X$, each one with its arity. Then, the syntax of the logic is given by the set of formulas

$$\varphi \ni p \mid \varphi \wedge \varphi \mid \neg\varphi \mid \circledast(\varphi, \cdots, \varphi)$$

for $p \in Prop$, a countable set of propositional variables.

A model M for the logic consists of a coalgebra $\alpha : U \longrightarrow \mathcal{F}(U)$, a valuation $V : Prop \longrightarrow \mathcal{P}(U)$, and, for each n-ary modal symbol $\circledast$, an n-ary predicate lifting $\gamma^{\circledast} : (\mathbf{2}^-)^n \Longrightarrow \mathbf{2}^{\mathcal{F}(-)}$. Formulas are interpreted over a model inductively: Forgetting

[9] Equivalently, regarding set Z as a function from U to the two element set $\mathbf{2}$, $f^{-1} = \mathbf{2}^f$, with $\mathbf{2}^f\ Z = Z \cdot f$.

the modal operators for a while, the result is the standard interpretation over the Boolean algebra $\mathcal{P}(U)$. For example, $[\![p]\!] = V(p)$ and $[\![\varphi_1 \wedge \varphi_2]\!] = [\![\varphi_1]\!] \cap [\![\varphi_2]\!]$, as expected. Each n-ary modal operator, on the other hand, is interpreted as

$$[\![*(\varphi_1, \cdots, \varphi_n)]\!] \;=\; \alpha^{-1}(\gamma_U^{*}([\![\varphi_1]\!], \cdots, [\![\varphi_n]\!]))$$

as in equation (17).

As a final example suppose α is a coalgebra over U for the multiset functor $\mathcal{M}(X) = \mathbb{N}^X$, typically used to capture weighted transition systems. A modal operator $*_N$ could be defined to deal with those successor states that are reachable with a cost (measuring *e.g.* resources or time units used) limited to N. Thus,

$$[\![*_N\ \varphi]\!] \;=\; \{u \in U \mid \forall_{u' \in U} \,.\, (\alpha\ u)\ u' \leq N \Rightarrow u' \in [\![\varphi]\!]\}\ .$$

The corresponding predicate lifting is $\gamma_X^{*_N}\ \phi \widehat{=} \{s \in \mathbb{N}^X \mid \forall_{x \in X} \,.\, s\ x \leq N \Rightarrow x \in \phi\}$. This is similar to what is called in modal logic a *graded* modality, although this qualifier originally refers to a restriction on the cardinality of outgoing transitions from a state, rather than on their weights. Further examples, most useful in software design, are obtained with coalgebras for the distribution monad, for example, to address transitions with a some type of bound on the probability of occurrence.

Of course, the satisfaction relation $\models_M$ for a model M pops out easily. For example,

$$u \models_M *_N\varphi \;\Leftrightarrow\; \forall_{u' \in U} \,.\, (\alpha\ u)\ u' \leq N \Rightarrow u' \in [\![\varphi]\!]\ .$$

The crucial point is the assignment of a specific predicate lifting to each modal operator in Σ. There is no restriction on how such a lifting is defined but the naturality requirement: This is what ensures that the meaning of the operator will not depend on the state space of the particular coalgebra in a possible model. From the working software engineer perspective, this provides the freedom to define the most suitable logic for the problem at hand.

Such a freedom has an obvious drawback: The definition of the logic along the strategy outlined above is not fully parametric on the functor $\mathcal{F}$. It requires the definition of a set of predicate liftings, one for each modal operator, to give the way in which, for each case, a property over the state space is lifted to an $\mathcal{F}$-structured collection of states. The approach sketched above, however, is the most popular in coalgebraic logic [59]. Actually, it can be formulated in a more abstract setting [48] by first extending the signature Σ to an endofunctor in the category of Boolean algebras, and then interpreting the propositional logic, extended with the operators in Σ, as an algebra for such a functor, *i.e.*

$$\Sigma(\mathbf{2}^U) \xrightarrow{\quad \gamma_U \quad} \mathbf{2}^{\mathcal{F}(U)} \xrightarrow{\quad \alpha^{-1} \quad} \mathbf{2}^U\ .$$

Actually, moving from the powerset Boolean algebra to an arbitrary one is possible because, on extending the propositional calculus, one may always identify propositionally equivalent formulas and equip the corresponding quotient with a Boolean algebra structure.

An alternative approach, historically the first to be proposed, builds on L. Moss's original idea [53] of considering functor $\mathcal{F}$ himself as a syntax constructor, therefore leading to a logic which is fully parametric on the functor encoding the system's behaviour. This framework is slightly less general, in the sense that $\mathcal{F}$ is required to preserve weak pullbacks. The main disadvantage, however, from the point of view of Software Engineering applications, is the cumbersome, unintuitive syntax it entails.

In both approaches, however, coalgebraic logic emerges as a powerful, generic theory [25], rather than a way to put together a number of curious examples. The framework is parametric, as discussed, and compositional – a most relevant feature in Computer Science which often requires non trivial combinations of logics. But the hallmark of coalgebraic logic resides in the way most properties one expects to discuss in Logic can be formulated and analysed in this abstract, parametric setting.

A typical example, and most relevant from the applications point of view, concerns the so-called Hennessy–Milner theorem. A modal logic has the Hennessy–Milner property whenever the induced logical equivalence distinguishes between non-bisimilar states and only those. The same applies, in general, to coalgebraic logics. In modal logic, the 'only if' part of the theorem (*i.e.* that logical equivalence entails bisimilarity) requires the underlying Kripke frame to be finitely branching. This is mirrored in the coalgebraic setting through the separability condition which basically says that the logic allows enough predicate liftings to distinguish between all elements in $\mathcal{F}(U)$. The definition of suitable Hilbert calculi as well as the study of expressivity, soundness, completeness and decidability can also be carried out in the abstract setting [59, 47]. In this sense, going coalgebraic seems the right way to do modal logic.

4.3 Illustration: Reasoning about hierarchical designs

Hierarchical transition systems are a popular mathematical structure to represent state-based software applications in which different layers of abstraction are captured by interrelated state machines. The decomposition of high-level states into inner sub-states, and of their transitions into inner sub-transitions, is a common refinement procedure adopted in a number of specification formalisms. The diagram below depicts a high level behavioural model of a strongbox controller in the form of a transition system with three states.

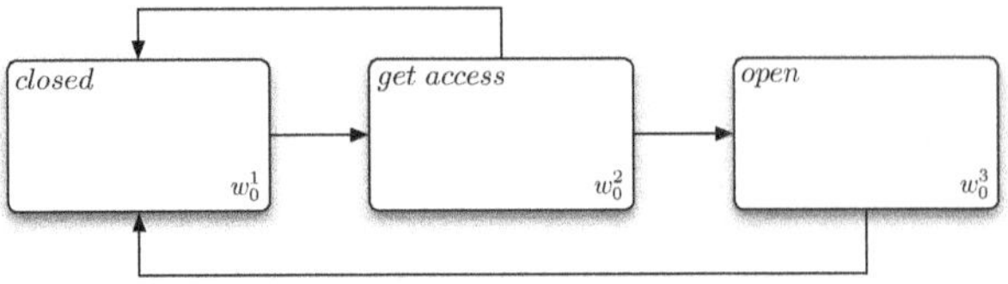

The strongbox can be open, closed, or going through an authentication process. The model can be formalised in some sort of modal logic, so that state transitions can be expressed, possibly combined with hybrid features to refer to specific, individual states. The qualifier *hybrid* [15] refers to an extension of modal languages with symbols, called *nominals*, which explicitly refer to individual states in the underlying Kripke frame[10]. A satisfaction operator $@_i\varphi$ stands for φ holding in the state named by nominal i.

For example, in propositional hybrid logic [23] and assuming

$$\text{Nom} = \{closed, get\ access, open\}$$

as a set of nominals, a number of properties of the the diagram above can be expressed, *e.g.*

- the state *get access* is accessible from the state *closed*: $@_{closed}\Diamond get\ access$,
- or the state *open* is not directly accessible from *closed*: $\Diamond open \rightarrow \neg closed$.

This high level vision of the strongbox controller can be refined by decomposing not only its internal states, but also its transitions. Thus, each 'high-level' state gives rise to a new, local transition system, and each 'upper-level' transition is decomposed into a number of 'intrusive' transitions from sub-states of the 'lower-level' transition system corresponding to the refinement of the original source state, to sub-states of the corresponding refinements of original target states. For instance, the (upper) *close* state can be refined into a (inner) transition system with two (sub) states: one, *idle*, representing the system waiting for the order to proceed for the *get access* state, and another one, *blocked*, capturing a system which is unable to proceed with the opening process (e.g. when authorised access for a given user was definitively denied). In this scenario, the upper level transition from *closed* to *get access* can be realised by, at least, one intrusive transition between the *closed* sub-state *idle* and the *get access* sub-state *identification*, in which the user identification is to be checked before proceeding. This refinement step is illustrated in the diagram below

[10]Notice the same adjective was used in the previous sections with a totally different meaning: to refer to software components with both discrete and continuous behaviour. The designation *cyber-physical* is also used in the later case with a similar meaning.

(left). Still the specifier may go even further. For example, she may like to refine the *get access* sub-state *authorisation* into the more fine-grained transition structure depicted on the right hand side of the diagram. This third-level view includes a sub-state corresponding to each one of the possible three attempts of password validation, as well as an auxiliary state to represent authentication success.

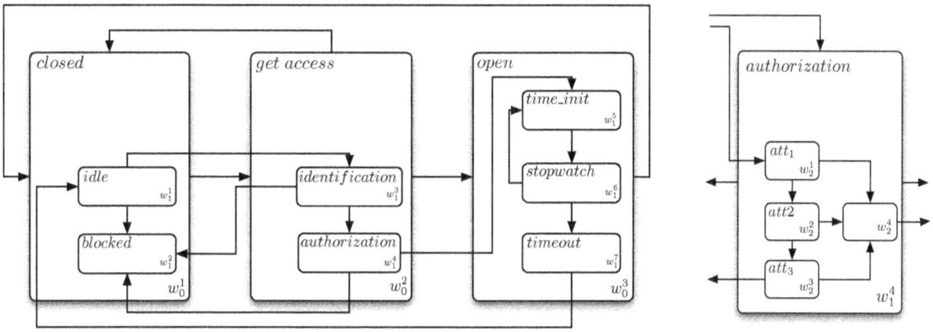

Such a hierarchical way to design a system is quite natural and somehow inherent to well-known design formalisms such as D. Harel's statecharts [29] and the subsequent UML hierarchical state machines, among others. This sort of systems have been studied in my own research group in the context of reconfigurable software architectures [49]. In particular, a hierarchical hybrid logic was proposed to express (and reason about) requirements that typically involve transitions between designated states in different local transition systems.

The whole programme can actually be carried out in a coalgebraic setting. The first observation is that a measure of maturity of coalgebraic logic is its ability to incorporate extensions which are already classical in the modal logic literature. We have briefly mentioned how new 'temporal' operators can be defined through fixed points, as in the modal μ-calculus [46]. Hybrid logic is also easily accommodated in a way which is quite similar to what is done in modal logic with classical Kripke semantics. In particular, the set of formulas is extended with

$$\varphi \ni \cdots \mid i \mid @_i\, \varphi$$

for $i \in Nom$, a set of nominals. The original valuation is extended to $V : Prop \cup Nom \longrightarrow \mathcal{P}(U)$ with the restriction that $V\ i$ is a singleton for each nominal i, *i.e.* a nominal identifies a unique state in the state space. The interpretation of the hybrid operators is the classical one: $[\![@_i\varphi]\!] = \{u \in U \mid V\ i \in [\![\varphi]\!]\}$ and $[\![i]\!] = V\ i$. The only aspect one needs to take into account is the interplay between the satisfaction operators and the modalities induced (or built over) the coalgebra. For example, one has to specify that a formula like $@_i\ \varphi$ must be valid either in the whole model or nowhere. In an Hilbert calculus this can be achieved through an extra axiom, for each modal operator $\circledast$:

$$@_i\varphi \ \Rightarrow\ (\circledast(\varphi_1, \cdots, \varphi_k)\ \Leftrightarrow\ \circledast(\varphi_1 \wedge @_i\varphi, \cdots, \varphi_k \wedge @_i\varphi)$$

capturing the intended validity of $@_i\varphi$ irrespective to the interpretation of each φ_j.

In the definition of a model for this logic the family of accessibility relations considered in [49] is replaced by a family of coalgebras for the same endofunctor, each of which captures the dynamics of the appropriate layer.

Signatures are n-families of disjoint, possible empty, sets of symbols

$$\Delta^n = (\mathrm{Prop}_k, \mathrm{Nom}_k)_{k\in\{0,\cdots,n\}}\ .$$

For example, to specify the strongbox above, one considers a signature Δ^2 for the three layers presented. 0-level symbols consist of the set of nominals

$$\mathrm{Nom}_0 = \{closed_0, get\_access_0, open_0\}$$

and a set of propositions Prop_0. The 1-level signature introduces a set of nominals

$$\begin{aligned}\mathrm{Nom}_1 = \{&idle_1, blocked_1, identification_1, authorization_1, time\_init_1,\\ &stopwatch_1, time\_out_1\}\end{aligned}$$

and, again, a set of propositions Prop_1. Level 3, finally, introduces $att1_2$, $att2_2$ and $att3_2$ in Nom_2. The set of formulas $Fm(\Delta^n)$ is the n-family recursively defined, for each k, by

$$\varphi_0 \ni i_0 \mid p_0 \mid \neg\varphi_0 \mid \varphi_0 \wedge \varphi_0 \mid @_{i_0}\varphi_0 \mid \Box_0\varphi_0$$

$$\varphi_0^b \ni i_0 \mid p_0 \mid @_{i_0}\varphi_0 \mid \Box_0\varphi_0$$

where superscript b qualifies the basic formulas, and

$$\varphi_k \ni \varphi_{k-1}^b \mid i_k \mid p_k \mid \neg\varphi_k \mid \varphi_k \wedge \varphi_k \mid @_{i_k}\varphi_k \mid \diamond_k\, \varphi_k$$

where for any $k \in \{1, \ldots, n\}$, the basic formulas are defined by

$$\varphi_{k-1}^b \ni i_{k-1} \mid p_{k-1} \mid \varphi_{k-2}^b \mid @_{i_{k-1}}\varphi_{k-1} \mid \Box_{k-1}\varphi_{k-1}$$

for $k \in \{2, \cdots, n\}$, $p_k \in \mathrm{Prop}_k$ and $i_k \in \mathrm{Nom}_k$.

This language is able to express quite different properties. For instance, inner-outer relations between named states, e.g. $@_{idle_1}closed_0$ or $@_{att1_2}open_0$, as well as a variety of transitions. Those include, for example, the layered transition $@_{get\_access_0} \diamond_0 open_0$, the 0-internal transition $@_{identification_1} \diamond_1 authorisation_1$ or intrusive transitions like $@_{idle_1} \diamond_1 authorisation_1$ and $get\_access_0 \to \diamond_1 open_0$.

The definition of a model is parameterised by a family of coalgebras defined for the same functor, *i.e.* exhibiting the same type of behavioural effect. A n-layered model $M \in \mathrm{Mod}^n(\Delta^n)$ is a tuple

$$M = \langle W^n, D^n, \alpha^n, V^n \rangle$$

where $W^n = (W_k)_{k\in\{0,\cdots,n\}}$ is a family of disjoint sets of states, and $D^n \subseteq W_0 \times \cdots \times W_n$ is a definition predicate that singles out the chains of states across the n levels which are considered meaningful 'global' states. Denoting by D_k the k-restriction $D^n|_k$ to the first $k+1$ columns, for each $k \in \{0, \cdots, n\}$, it is the case that

$$W_k = \{v_k | D_k\langle w_0, \cdots, w_{k-1}, v_k\rangle, \text{ for some } w_0, \cdots, w_{k-1} \text{ st } D_{k-1}\langle w_0, \cdots, w_{k-1}\rangle\} \ .$$

Then, comes the 'dynamics': $\alpha^n = (\alpha_k : D_k \longrightarrow \mathcal{F}(D_k))_{k\in\{0,\cdots,n\}}$ is a family of $\mathcal{F}$-coalgebras specifying the system's evolution at each level in the hierarchy. Finally, $V^n = (V_k^{\mathrm{Prop}}, V_k^{\mathrm{Nom}})_{k\in\{0,\cdots,n\}}$ is a family of pairs of valuations defined as one could expect:

- $V_k^{\mathrm{Prop}} : \mathrm{Prop}_k \to \mathcal{P}(D_k)$, and
- $V_k^{\mathrm{Nom}} : \mathrm{Nom}_k \to W_k$.

The satisfaction relation takes a similar shape as a family of relations

$$\models^n \ = \ (\models_k)_{k\in\{0,\cdots,n\}}$$

defined, for each $w_r \in W^r$, $r \in \{0, \cdots, k\}$, $k \leq n$, such that $D_k\langle w_0, \cdots w_k\rangle$. The case of interest in the context of this paper is the one for modalities, *i.e.* $M_k, w_0, \cdots, w_k \models_k \Box_k \varphi_k$ iff

$$\forall_{v_0\in W_0,\cdots,v_k\in W_k} . \ \langle v_0, \cdots, v_k\rangle \in \alpha_k\langle w_0, \cdots, w_k\rangle \text{ implies } M, v_0, \cdots, v_k \models_k \varphi_k \ .$$

The hybrid part is given by

- $M_k, w_0, \cdots, w_k \models_k i_k$ iff $w_k = V_k^{\mathrm{Nom}}(i_k)$ and $D_k\langle w_0, \cdots, w_{k-1}, V_k^{\mathrm{Nom}}(i_k)\rangle$,
- $M_k, w_0, \cdots, w_k \models_k @_{i_k}\varphi_k$ iff $M_k, w_0, \cdots w_{k-1}, V_k^{\mathrm{Nom}}(i_k) \models_k \varphi_k$ and $D_k\langle w_0, \cdots w_{k-1}, V_k^{\mathrm{Nom}}(i_k)\rangle$.

The Boolean part, finally, is defined as usual, just taking care of the definability interdependence captured by D^n. Thus,

- $M_k, w_0, \cdots, w_k \models_k \varphi^b_{k-1}$ iff $M_{k-1}, w_0, \cdots, w_{k-1} \models_{k-1} \varphi^b_{k-1}$,
- $M_k, w_0, \cdots, w_k \models_k p_k$ iff $\langle w_0, \cdots, w_k \rangle \in V_k^{\text{Prop}}(p_k)$,
- $M_k, w_0, \cdots, w_k \models_k \varphi_k \wedge \varphi'_k$ iff $M_k, w_0, \cdots, w_k \models_k \varphi_k$ and $M_k, w_0, \cdots, w_k \models_k \varphi'_k$,
- $M_k, w_0, \cdots, w_k \models_k \neg\varphi_k$ iff it is false that $M_k, w_0, \cdots, w_k \models_k \varphi_k$.

The resulting logic is quite expressive. Notions of n-layered bisimilarity and refinement can be introduced [49] along the lines already discussed in this paper, and a Hennessy–Milner theorem proved.

A specific, particularly well-behaved class of layered models, is called *hierarchical*: it requires that the restriction of a coalgebra α_k to the state space of α_{k-1} coincides with the latter. This ensures that the elements in the family of coalgebras are compatible.

The example sketched here is clearly an hierarchical model. Examples of non-hierarchical layered models can be achieved by removing some 0-transitions depicted in the diagram above (e.g. the one linking the named states $closed_0$ and get_access_0). This *hierarchical* condition can be expressed as a naturality condition as follows. Define $\pi_k : D_k \longrightarrow D_{k-1}$ by $\pi_k \langle w_0, \cdots, w_{k-1}, w_k \rangle \mathrel{\widehat{=}} \langle w_0, \cdots, w_{k-1} \rangle$. Then, the model is hierarchical if, for all k, the following diagram commutes[11].

$$\begin{array}{ccc} D_k & \xrightarrow{\alpha_k} & \mathcal{F}(D_k) \\ \downarrow{\scriptstyle \pi_k} & & \downarrow{\scriptstyle \mathcal{F}(\pi_k)} \\ D_{k-1} & \xrightarrow{\alpha_{k-1}} & \mathcal{F}(D_{k-1}) \end{array}$$

For $\mathcal{F} = \mathcal{P}$ this means, for example, that the transitions depicted in the diagram

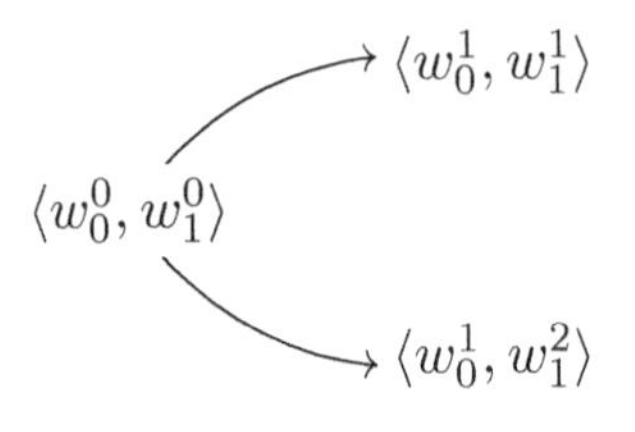

exist at level 1 iff a transition $w_0^0 \longrightarrow w_0^1$ exists at level 0.

[11]This basically means that the family α^n of coalgebras in a model of a hierarchical system, can be regarded as a coalgebra in the category of pre-sheaves $[\overline{n}, Set]$, where $\overline{n}$ is the total order corresponding to the initial n-segment of natural numbers. Such a coalgebra is, of course, a natural transformation.

As another example consider $\mathcal{F} = \mathcal{D}$. In an hierarchical (probabilistic) system the 1-level transitions in the left of the diagram below exist if the 0-level transitions depicted on the right exist as well.

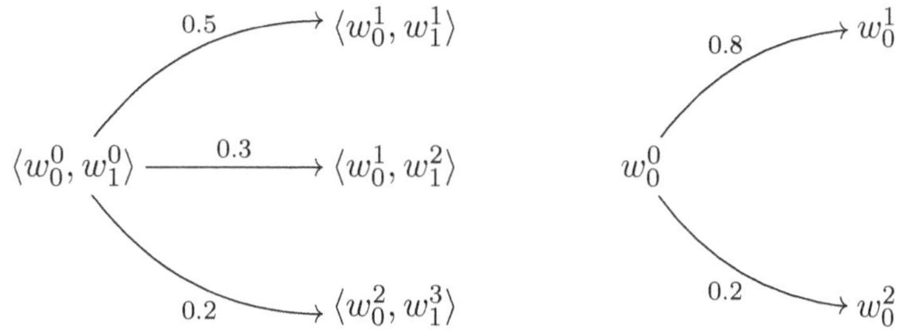

5 Concluding

This paper revisited a few themes in elementary, *i.e.* *Set*-based, Coalgebra in connection with what may be regarded as the kernel activities of a software engineer: *modelling* complex systems, *architecting* their composition and *reasoning* about their behaviour. Models, architectures and properties were therefore the buzzwords chosen to guide this exercise.

As a design discipline, Software Engineering is currently challenged by continuous technological evolution towards very large, heterogeneous, highly dynamic computing systems, which require innovative approaches to master their complexity. Systems whose behaviour cannot be simply characterised in terms of a relation between input and output data, but expresses a continued interaction with their external (computational or physical) and internal (sub-systems) contexts. In this sense, they can be classified as *reactive*, to use a term coined by A. Pnueli and D. Harel [30] in the 1980s. Furthermore, concurrent composition is the norm, rather than the exception.

Developing such systems correctly is very difficult, because it involves not only mastering the complexity of building and deploying large applications on time and within budget, but also managing an open-ended structure of autonomous components, typically distributed, often organised in loosely coupled configurations, and highly heterogeneous. Additional difficulties arise with the need to take into account a plethora of issues such as real-time responsiveness, dynamic reconfiguration, QoS-awareness, self-adaptability, security, dependability, under-specification of third-party components, among many others.

Unfortunately, software technology is still pre-scientific in its lack of sound mathematical foundations to provide an effective basis to predict and certify computationally generated behaviour. In a sense, compared to other Engineering disciplines, we are just living our 17th century, seeking for the right foundations, methods and calculi to move from *ad hoc* to systematic and accountable engineering practices.

My purpose was to explore one of those mathematical frameworks which has the potential to address a large class of computational systems. Indeed, we would suggest Coalgebra as a, probably *the*, mathematics for dynamical, state-based systems.

The essence of the coalgebraic method boils down to a very basic observation: that from a suitable characterisation of the *type* of a system's dynamics, canonical notions of behaviour, observational reasoning (equational and inequational) and modality can be derived in a uniform (*i.e.* parametric) way.

This setting may sound familiar to the working software engineer: the object of her practice, if not of her study, is precisely the ubiquity of the computing phenomena along and across universes of typed arrows. Arrows may stand for functions, algorithms, services or components, programs fulfilling a specification contract, relationships in a UML diagram, processes through mobile ambients, evolutions in a sensor network, links in a software architectural description, circuits coordinating loosely coupled agents, or whatever structures our domain. The type of a coalgebra, an endofunctor, is itself an arrow, and so is, moreover, the coalgebra itself.

In a brief historical overview of the trends and results predating the emergence of Coalgebra in Computer Science, B. Jacobs [41], referring to the work of Arbib, Manes, Goguen, Adamek and others in the late 1970s, on categorical approaches to systems theory, comments: *Their aim was to place sequential machines and control systems in a unified framework* which *(...) led to general notions of state, behaviour, reachability, observability, and realisation of behaviour.* Jacobs remarks, however, that the reason why Coalgebra did not emerge directly from this work was *probably because the setting of modules and vector spaces from which this work arose provided too little categorical infrastructure (especially: no cartesian closure).* The quick expansion of Coalgebra, its techniques and applications, and the capacity shown to capture in a uniform, parametric way a myriad of state-based systems, as well as its mathematical elegance, offers evidence we may be on the right track.

The development of Coalgebra and its application to Computer Science stemmed from different sources, from P. Aczel's non well-founded set theory, accommodating infinitely descending $\in$-chains, to the study of infinite data types and the theory of behavioural specification [40] and satisfaction [14]. Still, it remains an area of active research, with a growing impact not only on the foundations of computing semantics, but also on very concrete programming techniques. Actually, there is a growing interest on the potential of coalgebraic techniques in algorithm understanding and derivation, often based on rediscovering and generalising specific algorithms, for example from automata theory [21, 17, 22, 26]. Unsurprisingly, going generic in the theory often leads to efficient computational solutions. Striking developments on coinductive proof methods, notably the recent work on up-to techniques [18, 19], go in a similar direction.

Without trying to be exhaustive, we would still like to mention a few other current research directions which will certainly have an impact in the coming decade. The first concerns the combination of algebraic and coalgebraic techniques and the discovery of compatible patterns described by distributive laws [45], which, as shown in D. Turi and J. J. M. M. Rutten landmark work, in the late 1990's, correspond to specification formats in operational semantics. The impact of such laws in several constructions, for example in the formulation of trace semantics, as mentioned above, but also in combining monadic and comonadic effects [12] and logic, suggests we are dealing with some sort of very fundamental structures.

Another direction addresses the challenge of quantitative (weighted, probabilistic, continuous) reasoning, once again driven by the broadening spectrum of Software Engineering problems. This is not only pushing the development of Coalgebra within categories different from *Set* [16, 44, 55], but also leading a lot of results on behavioural *metrics* as an alternative to equivalences [24, 3, 5]. Actually, in the context of *e.g.* probabilistic or hybrid systems, working with equivalences entailing the need for exact matching of real numbers is unrealistic. Metrics, on the other hand, can measure how close two systems are and conclude whether they should be taken as equivalent.

But the impact of Coalgebra can also be recognised at a more 'syntactical' level. The work of A. Silva, a former student of this University, on the derivation of specification languages from the functor typing the coalgebra dynamics [64, 63] should be mentioned here. In a more general setting, the points of contact between Coalgebra and current research on graphical languages in which diagrams, syntax and interpretations, are generated as arrows in special families of monoidal categories [71, 20], seem most promising, with applications ranging from the 're-interpretation' of classical control theory to the design of diagrammatic languages to express, *e.g.*, software architectures.

I'm not sure whether this paper was able to raise the interest of the working software engineer in Coalgebra, or, on the other hand, that of the logician who may find in Computer Science a huge domain for the fruitful application of her methods and tools. In 1967, Anthony Oettinger [56], speaking as President of the ACM, recognised that the expression Software Engineering

> *seems strange to classical engineers and to classical mathematicians alike, because, you know, why would a mathematician think of engineering with symbols and, by the same token, why would somebody who thinks of engineering in terms of things we do with pieces of metal or transistors, think of an operation that takes place on paper with pencils and erasers as engineering.*

Almost 50 years later, there is still a need to push back this discipline to where it actually belongs. Fortunately, to continue with A. Oettinger's speech, *there is no question but that the study of symbol systems, of effective algorithms, of efficient algorithms, of the structure of algorithms, is a mathematical discipline.* And, moreover, *there is, in this realm, enough elegance to attract anybody who wants a challenge.*

Doing Software Engineering in lighter, more informal ways, brings to my mind a quotation attributed to Vlad Patryshev in a slightly different context: *It's like talking about electricity without using calculus. Good enough to replace a fuse, not enough to design an amplifier.*

Acknowledgements. Sections 2.3, 3.2 and 4.3 illustrate, through three concrete applications, the potential Coalgebra may have for the working software engineer with respect to each of the topics chosen for this paper: *models*, *architectures* and *properties*. These applications come from current research along which I had the privilege of collaborating with a number of colleagues. In particular, the coalgebraic treatment of hybrid systems was developed by Renato Neves, who introduced the $\mathcal{H}$ monad and a number of exciting results still emerging at the time of writing. The remaining 'illustrations' are also in debt to ongoing collaboration with José Nuno Oliveira, on formal approaches to software architecture, Sun Meng, on coalgebraic refinement, and both Alexandre Madeira and Manuel A. Martins, on hybrid logics for reconfigurable systems.

References

[1] P. Aczel. *Non-Well-Founded Sets.* CSLI Lecture Notes (14), Stanford, 1988.

[2] P. Aczel and N. Mendler. A final coalgebra theorem. In D. Pitt, D. Rydeheard, P. Dybjer, A. Pitts, and A. Poigne, editors, *Proc. Category Theory and Computer Science*, pages 357–365. Springer Lect. Notes Comp. Sci. (389), 1988.

[3] G. Bacci, G. Bacci, K. G. Larsen, and R. Mardare. Computing behavioral distances, compositionally. In Krishnendu Chatterjee and Jirí Sgall, editors, *Mathematical Foundations of Computer Science 2013 - 38th International Symposium, MFCS 2013,*

Klosterneuburg, Austria, August 26-30, 2013., pages 74–85. Springer Lect. Notes Comp. Sci. (8087), 2013.

[4] J. Baez and J. Dolan. Categorification. In Ezra Getzler and Mikhail Kapranov, editors, *Higher Category Theory*, Contemp. Math. 230, pages 1–36. American Mathematical Society, 1998.

[5] P. Baldan, F. Bonchi, H. Kerstan, and B. König. Behavioral metrics via functor lifting. In Venkatesh Raman and S. P. Suresh, editors, *34th International Conference on Foundation of Software Technology and Theoretical Computer Science, FSTTCS 2014, December 15-17, 2014, New Delhi, India*, volume 29 of *LIPIcs*, pages 403–415. Schloss Dagstuhl - Leibniz-Zentrum fuer Informatik, 2014.

[6] L. S. Barbosa. *Components as Coalgebras.* PhD thesis, DI, Universidade do Minho, 2001.

[7] L. S. Barbosa. Towards a calculus of state-based software components. *"Jour. Universal Comp. Sci."*, 9(8):891–909, 2003.

[8] L. S. Barbosa and J. N. Oliveira. Transposing partial components: an exercise on coalgebraic refinement. *Theor. Comp. Sci.*, 365(1-2):2–22, 2006.

[9] L. S. Barbosa, J. N. Oliveira, and A. M. Silva. Calculating invariants as coreflexive bisimulations. In J. Meseguer and G. Rosu, editors, *Algebraic Methodology and Software Technology, 12th International Conference, AMAST 2008, Urbana, IL, USA, July 28-31, 2008, Proceedings*, pages 83–99. Springer Lect. Notes Comp. Sci. (5140), 2008.

[10] L. S. Barbosa, M. Sun, B. K. Aichernig, and N. Rodrigues. On the semantics of componentware: a coalgebraic perspective. In Jifeng He and Zhiming Liu, editors, *Mathematical Frameworks for Component Software: Models for Analysis and Synthesis*, Series on Component-Based Software Development, pages 69–117. World Scientific, 2006.

[11] J. Beaten and W. Weijland. *Process Algebra.* Cambridge University Press, 1990.

[12] M. Behrisch, S. Kerkhoff, and J. Power. Category theoretic understandings of universal algebra and its dual: Monads and lawvere theories, comonads and what? *Electr. Notes Theor. Comput. Sci.*, 286:5–16, 2012.

[13] J. Benabou. Introduction to bicategories. *Springer Lect. Notes Maths. (47)*, pages 1–77, 1967.

[14] M. Bidoit and R. Hennicker. Behavioural theories and the proof of behavioural properties. *Theor. Comput. Sci.*, 165(1):3–55, 1996.

[15] P. Blackburn. Representation, reasoning, and relational structures: a hybrid logic manifesto. *Logic Journal of IGPL*, 8(3):339–365, 2000.

[16] F. Bonchi, M. M. Bonsangue, M. Boreale, J. J. M. M. Rutten, and A. Silva. A coalgebraic perspective on linear weighted automata. *Inf. Comput.*, 211:77–105, 2012.

[17] F. Bonchi, M. M. Bonsangue, H. H. Hansen, P. Panangaden, J. J. M. M. Rutten, and A. Silva. Algebra-coalgebra duality in Brzozowski's minimization algorithm. *ACM Trans. Comput. Log.*, 15(1):3, 2014.

[18] F. Bonchi, D. Petrisan, D. Pous, and J. Rot. Coinduction up-to in a fibrational setting.

In Thomas A. Henzinger and Dale Miller, editors, *Joint Meeting of the Twenty-Third EACSL Annual Conference on Computer Science Logic (CSL) and the Twenty-Ninth Annual ACM/IEEE Symposium on Logic in Computer Science (LICS), CSL-LICS '14, Vienna, Austria, July 14 - 18, 2014*, pages 20:1–20:9. ACM, 2014.

[19] F. Bonchi, D. Petrisan, D. Pous, and J. Rot. Lax bialgebras and up-to techniques for weak bisimulations. In Luca Aceto and David de Frutos-Escrig, editors, *26th International Conference on Concurrency Theory, Madrid, September 1.4, 2015*, volume 42 of *LIPIcs*, pages 240–253. Schloss Dagstuhl - Leibniz-Zentrum fuer Informatik, 2015.

[20] F. Bonchi, P. Sobocinski, and F. Zanasi. Full abstraction for signal flow graphs. In Sriram K. Rajamani and David Walker, editors, *Proceedings of the 42nd Annual ACM SIGPLAN-SIGACT Symposium on Principles of Programming Languages, POPL 2015, Mumbai, India, January 15-17, 2015*, pages 515–526. ACM, 2015.

[21] Filippo Bonchi and Damien Pous. Checking NFA equivalence with bisimulations up to congruence. In Roberto Giacobazzi and Radhia Cousot, editors, *The 40th Annual ACM SIGPLAN-SIGACT Symposium on Principles of Programming Languages, POPL '13, Rome, Italy - January 23 - 25, 2013*, pages 457–468. ACM, 2013.

[22] Filippo Bonchi and Damien Pous. Hacking nondeterminism with induction and coinduction. *Commun. ACM*, 58(2):87–95, 2015.

[23] T. Brauner. *Hybrid Logic and its Proof-Theory*. Applied Logic Series. Springer, 2010.

[24] F. van Breugel and J. Worrell. Approximating and computing behavioural distances in probabilistic transition systems. *Theor. Comput. Sci.*, 360(1-3):373–385, 2006.

[25] C. Cîrstea, A. Kurz, D. Pattinson, L. Schröder, and Y. Venema. Modal logics are coalgebraic. *Comput. J.*, 54(1):31–41, 2011.

[26] N. Foster, D. Kozen, M. Milano, A. Silva, and L. Thompson. A coalgebraic decision procedure for netkat. In Sriram K. Rajamani and David Walker, editors, *Proceedings of the 42nd Annual ACM SIGPLAN-SIGACT Symposium on Principles of Programming Languages, POPL 2015, Mumbai, India, January 15-17, 2015*, pages 343–355. ACM, 2015.

[27] P. J. Freyd and A. Ščedrov. *Categories, Allegories*, volume 39 of *Mathematical Library*. North-Holland, 1990.

[28] E. Haghverdi, P. Tabuada, and G. J. Pappas. Bisimulation relations for dynamical, control, and hybrid systems. *Theoretical Computer Science*, 342(2-3):229–261, 2005.

[29] D. Harel. Statecharts: A visual formalism for complex systems. *Sci. Comput. Program.*, 8(3):231–274, 1987.

[30] D. Harel and A. Pnueli. On the development of reactive systems. In *Logics and Models of Concurrent Systems*, volume 13 of *NATO Adv. Sci. Inst. Ser. F Comput. Systems Sci.*, pages 477–498. Springer-Verlag, 1985.

[31] I. Hasuo. Generic forward and backward simulations. In Christel Baier and Holger Hermanns, editors, *CONCUR 2006 - Concurrency Theory, 17th International Conference, CONCUR 2006, Bonn, Germany, August 27-30, 2006, Proceedings*, volume 4137 of *Lecture Notes in Computer Science*, pages 406–420. Springer, 2006.

[32] I. Hasuo. The microcosm principle and compositionality of GSOS-based component calculi. In Andrea Corradini, Bartek Klin, and Corina Cîrstea, editors, *Algebra and Coalgebra in Computer Science - 4th International Conference, CALCO 2011, Winchester, UK, August 30 - September 2, 2011.*, pages 222–236. Springer Lect. Notes Comp. Sci. (6859), 2011.

[33] I. Hasuo, C. Heunen, B. Jacobs, and A. Sokolova. Coalgebraic components in a many-sorted microcosm. In Alexander Kurz, Marina Lenisa, and Andrzej Tarlecki, editors, *Algebra and Coalgebra in Computer Science, Third International Conference, CALCO 2009, Udine, Italy, September 7-10, 2009. Proceedings*, pages 64–80. Springer Lect. Notes Comp. Sci. (5728), 2009.

[34] I. Hasuo and B. Jacobs. Traces for coalgebraic components. *Mathematical Structures in Computer Science*, 21(2):267–320, 2011.

[35] I. Hasuo, B. Jacobs, and A. Sokolova. Generic trace semantics via coinduction. *Logical Methods in Computer Science*, 3(4), 2007.

[36] T. A. Henzinger. The theory of hybrid automata. In *Proceedings, 11th Annual IEEE Symposium on Logic in Computer Science, New Brunswick, New Jersey, USA, July 27-30, 1996*, pages 278–292. IEEE Computer Society, 1996.

[37] C. A. R. Hoare, S. van Staden, B. Möller, G. Struth, J. Villard, H. Zhu, and P. W. O'Hearn. Developments in concurrent kleene algebra. In Peter Höfner, Peter Jipsen, Wolfram Kahl, and Martin Eric Müller, editors, *Relational and Algebraic Methods in Computer Science - 14th International Conference, RAMiCS 2014, Marienstatt, Germany, April 28-May 1, 2014.*, pages 1–18. Springer Lect. Notes Comp. Sci. (8428), 2014.

[38] P. F. Hoogendijk. *A generic theory of datatypes.* PhD thesis, Department of Computing Science, Eindhoven University of Technology, 1996.

[39] J. Hughes. Generalising monads to arrows. *Sci. Comput. Program.*, 37(1-3):67–111, 2000.

[40] U. L. Hupbach and H. Reichel. On behavioural equivalence of data types. *Elektronische Informationsverarbeitung und Kybernetik*, 19(6):297–305, 1983.

[41] B. Jacobs. *Introduction to Coalgebra. Towards Mathematics of States and Observations.* Cambridge University Press (to appear), 2012. Draft copy: Version 2.0, 2012. Institute for Computing and Information Sciences, Radboud University Nijmegen.

[42] B. Jacobs, A. Silva, and A. Sokolova. Trace semantics via determinization. *J. Comput. Syst. Sci.*, 81(5):859–879, 2015.

[43] A. Joyal, R. Street, and D. Verity. Traced monoidal categories. *Math. Proc. Camb. Phil. Soc.*, 119:447–468, 1996.

[44] H. Kerstan and B. König. Coalgebraic trace semantics for continuous probabilistic transition systems. *Logical Methods in Computer Science*, 9(4), 2013.

[45] B. Klin. Bialgebras for structural operational semantics: An introduction. *Theor. Comput. Sci.*, 412(38):5043–5069, 2011.

[46] D. Kozen. Results on the propositional mu-calculus. *Theor. Comput. Sci.*, 27:333–354,

1983.

[47] C. Kupke and D. Pattinson. Coalgebraic semantics of modal logics: An overview. *Theor. Comput. Sci.*, 412(38):5070–5094, 2011.

[48] A. Kurz and R. L. Leal. Modalities in the stone age: A comparison of coalgebraic logics. *Theor. Comput. Sci.*, 430:88–116, 2012.

[49] A. Madeira, M. A. Martins, and L. S. Barbosa. A logic for n-dimensional hierarchical refinement. In John Derrick, Eerke A. Boiten, and Steve Reeves, editors, *Proceedings 17th International Workshop on Refinement, Refine@FM 2015, Oslo, Norway, 22nd June 2015*, volume 209 of *EPTCS*, pages 40–56, 2016.

[50] R. Milner. Elements of interaction (Turing Award Lecture). *Communications of the ACM*, 36(1):78–89, 1993.

[51] R. Milner. *Communicating and Mobile Processes: the π-Calculus.* Cambridge University Press, 1999.

[52] V. C. Miraldo and J. N. Oliveira. 'keep definition, change category' – a practical approach to state-based system calculi. *Journal of Logical and Algebraic Methods in Programming*, 85(4):449–474, 2016.

[53] L. Moss. Coalgebraic logic. *Ann. Pure & Appl. Logic*, 96(1-3):277–317, 1999.

[54] R. Neves and L. S Barbosa. Hybrid automata as coalgebras. In Augusto Sampaio and Farn Wang, editors, *Theoretical Aspects of Computing - ICTAC 2016 - 13th International Colloquium, Taiwan, Proceedings*, pages 385–402. Springer Lect. Notes Comp. Sci. (9965), 2016.

[55] R. Neves, L. S. Barbosa, D. Hofmann, and M. A. Martins. Continuity as a computational effect. *J. Log. Algebr. Meth. Program.*, 85(5):1057–1085, 2016.

[56] A. G. Oettinger. The hardware-software complementarity. *Commun. ACM*, 10(10):604–606, 1967.

[57] J. N. Oliveira. Towards a linear algebra of programming. *Formal Aspects of Computing*, 24(4-6):433–458, 2012.

[58] D. Park. Concurrency and automata on infinite sequences. In P. Deussen, editor, *Proc. Conf. on Theoretical Computer Science*, pages 167–183. Springer Lect. Notes Comp. Sci. (104), 1981.

[59] D. Pattinson. Coalgebraic modal logic: soundness, completeness and decidability of local consequence. *Theor. Comput. Sci.*, 309(1-3):177–193, 2003.

[60] E. P. Robinson. Variations on algebra: monadicity and generalisations of equational theories. *Formal Asp. Comput.*, 13(3-5):308–326, 2002.

[61] J. J. M. M. Rutten. Universal coalgebra: A theory of systems. *Theor. Comput. Sci.*, 249(1):3–80, 2000. (Revised version of CWI Techn. Rep. CS-R9652, 1996).

[62] D. Sangiorgi. On the origins of bisimulation and coinduction. *ACM Trans. Program. Lang. Syst.*, 31(4), 2009.

[63] A. Silva, F. Bonchi, M. Bonsangue, and J. J. M. M. Rutten. Quantitative Kleene coalgebras. *Inf. Comput.*, 209(5):822–849, 2011.

[64] A. Silva, M. M. Bonsangue, and J. J. M. M. Rutten. Non-deterministic Kleene coalge-

bras. *Logical Methods in Computer Science*, 6(3), 2010.

[65] M. Smyth and G. Plotkin. The category theoretic solution of recursive domain equations. *SIAM Journ. Comput.*, 4(11):761–783, 1982.

[66] A. Sokolova. Probabilistic systems coalgebraically: A survey. *Theor. Comput. Sci.*, 412(38):5095–5110, 2011.

[67] C. Stirling. Modal and temporal logics for processes. *Springer Lect. Notes Comp. Sci. (715)*, pages 149–237, 1995.

[68] M. Sun and L. S. Barbosa. Components as coalgebras: The refinement dimension. *Theor. Comp. Sci.*, 351:276–294, 2006.

[69] N. Urabe and I. Hasuo. Generic forward and backward simulations III: quantitative simulations by matrices. In Paolo Baldan and Daniele Gorla, editors, *CONCUR 2014 - Concurrency Theory - 25th International Conference, CONCUR 2014, Rome, Italy, September 2-5, 2014.*, pages 451–466. Springer Lect. Notes Comp. Sci. (8704), 2014.

[70] N. Urabe and I. Hasuo. Coalgebraic infinite traces and kleisli simulations. In Lawrence S. Moss and Pawel Sobocinski, editors, *6th Conference on Algebra and Coalgebra in Computer Science, CALCO 2015, June 24-26, 2015, Nijmegen, The Netherlands*, volume 35 of *LIPIcs*, pages 320–335. Schloss Dagstuhl - Leibniz-Zentrum fuer Informatik, 2015.

[71] Fabio Zanasi. *Interacting Hopf Algebras- the Theory of Linear Systems.* PhD thesis, École normale supérieure de Lyon, France, 2015.

Received 02 March 2020

On the Order Theory of $\mathscr{C}^\infty$–Reduced $\mathscr{C}^\infty$–Rings and Applications

Jean Cerqueira Berni*
Institute of Mathematics and Statistics of the University of São Paulo – IME-USP
jeancb@ime.usp.br

Rodrigo Figueiredo
Federal Institute of Education, Science and Technology of Minas Gerais – IFMG
rodrigo.f@ifmg.edu.br

Hugo Luiz Mariano
Institute of Mathematics and Statistics of the University of São Paulo – IME-USP
hugomar@ime.usp.br

Abstract

In the present work we carry on the study of the order theory for ($\mathscr{C}^\infty$-reduced) $\mathscr{C}^\infty$-rings initiated in [16] (see also [4]). In particular, we apply some results of the order theory of $\mathscr{C}^\infty$-fields (*e.g.* every such field is real closed) to present another approach to the order theory of general $\mathscr{C}^\infty$-rings: "smooth real spectra" (see [5]). This suggests that a model-theoretic investigation of the class of $\mathscr{C}^\infty$-fields could be interesting and also useful to provide the first steps towards the development of the "Real Algebraic Geometry" of $\mathscr{C}^\infty$-rings.

Keywords: order theory, $\mathscr{C}^\infty$-rings, real spectrum

The authors wish to express their thanks to the anonymous referees for their several helpful comments concerning the structure of this paper as well as for its very detailed revision.

*Sponsored by Coordenação de Aperfeiçoamento do Pessoal de Ensino Superior – CAPES – grant number 88887.467027/2019-00

Introduction

Given a smooth manifold, M, the set $\mathscr{C}^\infty(M,\mathbb{R})$ supports a far richer structure than just of an $\mathbb{R}$-algebra: it interprets not only the symbols for real polynomial functions but for all smooth real functions $\mathbb{R}^n \to \mathbb{R}$, $n \in \mathbb{N}$. Thus, $\mathscr{C}^\infty(M,\mathbb{R})$ is a natural instance of the algebraic structure called *$\mathscr{C}^\infty$-ring.*

It was not until the decades of 1970's and 1980's that a study of the abstract (algebraic) theory of $\mathscr{C}^\infty$-rings was made, mainly in order to construct topos models for "Synthetic Differential Geometry" ([12]). The interest in $\mathscr{C}^\infty$-rings gained strength in recent years, mainly motivated by the differential version of 'Derived Algebraic Geometry" (see [11]).

In this paper we address the study of the order theory of $\mathscr{C}^\infty$−reduced $\mathscr{C}^\infty$−rings, presenting a useful characterization of the "natural order" of a $\mathscr{C}^\infty$−ring, introduced by Moerdijk and Reyes in [16]: given any $\mathscr{C}^\infty$-ring $\mathfrak{A} = (A,\Phi)$, this canonical strict partial order $\prec$ is given by:

$$(a \prec b) \iff (\exists u \in A^\times)(b - a = u^2).$$

Since this natural binary relation given on a generic $\mathscr{C}^\infty$-ring involves invertible elements, we should first analyze these elements of a $\mathscr{C}^\infty$-ring. In order to do so, we shall restrict ourselves to the case of the *$\mathscr{C}^\infty$-reduced* $\mathscr{C}^\infty$-rings. This is carried out in two steps: first proving the results for finitely generated $\mathscr{C}^\infty$-rings and then proving them for arbitrary ones.

Since any $\mathscr{C}^\infty$-ring can be expressed as the quotient of a free object – $\mathscr{C}^\infty(\mathbb{R}^E)$, for some set E – by some (ring-theoretic) ideal, it is appropriate to characterize the equality between their elements by making use of these ring-theoretic ideals. We show that in this context the canonical strict partial order of a generic $\mathscr{C}^\infty$-ring, say $\mathscr{C}^\infty(\mathbb{R}^E)/I$, can be characterized by properties concerning filters of zerosets of functions in I.

In [16], Moerdijk and Reyes prove that every $\mathscr{C}^\infty$-field is real closed (cf. **Theorem 2.10**). This suggests that the class of $\mathscr{C}^\infty$-fields is "well behaved" with respect to its model theory.

We apply, in particular, some results on the order theory of $\mathscr{C}^\infty$-fields – *e.g.*, every such field is real closed (cf. **Theorem 2.10** in [16]). – to present another approach to the order theory of general $\mathscr{C}^\infty$-rings, introducing the so-called "smooth real spectra" (see [5]). This suggests that a model-theoretic study of the class of $\mathscr{C}^\infty$-fields could be interesting and also useful to provide the first steps towards the development of the "Real Algebraic Geometry" of $\mathscr{C}^\infty$-rings in the vein of [19].

Overview of the paper: In the first section we present some preliminary notions and results that are used (implicitly or explicitly) throughout the paper, such as the basic concept of $\mathscr{C}^\infty$–ring and some features of the category of $\mathscr{C}^\infty$–rings, $\mathscr{C}^\infty$–fields, $\mathscr{C}^\infty$–rings of fractions, $\mathscr{C}^\infty$–radical ideals, $\mathscr{C}^\infty$–reduced $\mathscr{C}^\infty$-rings and some facts about the smooth Zariski spectrum. **Section 2** is devoted to present some results that (dually) connect subsets of $\mathbb{R}^E$ to quotients of $\mathscr{C}^\infty(\mathbb{R}^E)$: we present the characterizations of equalities and inequalities between elements of $\mathscr{C}^\infty$–reduced $\mathscr{C}^\infty$–rings, *i.e.*, $\mathscr{C}^\infty$–rings of the form $A = \frac{\mathscr{C}^\infty(\mathbb{R}^E)}{I}$ with $\sqrt[\infty]{I} = I$, by means of the filter of zerosets of functions of I, and we use a Galois connection between filters of zerosets of $\mathbb{R}^E$ and ideals of $\mathscr{C}^\infty(\mathbb{R}^E)$ to show that there are bijections between the set of maximal filters on $\mathbb{R}^E$ and the set of maximal ideals of $\mathscr{C}^\infty(\mathbb{R}^E)$. In **Section 3** we develop a detailed study of the natural strict partial ordering $\prec$ (introduced first by Moerdijk and Reyes) defined on a non-trivial $\mathscr{C}^\infty$-reduced $\mathscr{C}^\infty$-ring, with the aid of the results established in the previous sections (see **Theorem 3.7**). **Section 4** presents some interesting results on $\mathscr{C}^\infty$–fields based on the results from **Section 3**: for instance, every $\mathscr{C}^\infty$-field has $\prec$ as its unique (strict) total ordering compatible with the operations $+$ and $\cdot$, thus being a Euclidean field, cf. **Theorem 4.4** , (in fact, it is a real closed field); this is useful to analyse the concept of "real $\mathscr{C}^\infty$–spectrum of a $\mathscr{C}^\infty$–ring", which seems to be the suitable notion to deal with a smooth version of Real Algebraic Geometry. Finally, **Section 5** brings some concluding remarks, pointing some possible applications of the order structure of a $\mathscr{C}^\infty$–reduced $\mathscr{C}^\infty$–ring to its model theory.

1 Preliminaries

In this section we present the ingredients of the theory of $\mathscr{C}^\infty$-rings needed in the sequel of this work for the reader's convenience: we present the class of $\mathscr{C}^\infty$-rings as the class of models of an algebraic theory, and we describe the main notions of "Smooth Commutative Algebra of $\mathscr{C}^\infty$-rings": smooth rings of fractions, $\mathscr{C}^\infty$-radicals, $\mathscr{C}^\infty$-saturation and the smooth Zariski spectra. The main references used here are [16], [17], [3], [4].

1.1 On the Algebraic Theory of $\mathscr{C}^\infty$-Rings

In order to formulate and study the concept of $\mathscr{C}^\infty$–ring, we use a first order language $\mathscr{L}$ with a denumerable set of variables ($\mathbf{Var}(\mathscr{L}) = \{x_1, x_2, \cdots, x_n, \cdots\}$), whose nonlogical symbols are the symbols of all $\mathscr{C}^\infty$–functions from $\mathbb{R}^m$ to $\mathbb{R}^n$, with $m, n \in \mathbb{N}$, *i.e.*, the non-logical symbols consist only of function symbols, described as follows.

For each $n \in \mathbb{N}$, we have the n–ary **function symbols** of the set $\mathscr{C}^\infty(\mathbb{R}^n, \mathbb{R})$, *i.e.*, $\mathscr{F}_{(n)} = \{f^{(n)} | f \in \mathscr{C}^\infty(\mathbb{R}^n, \mathbb{R})\}$. Thus, the set of function symbols of our language is given by:

$$\mathscr{F} = \bigcup_{n \in \mathbb{N}} \mathscr{F}_{(n)} = \bigcup_{n \in \mathbb{N}} \mathscr{C}^{\infty}(\mathbb{R}^n)$$

Note that our set of constants is $\mathbb{R}$, since it can be identified with the set of all $0-$ary function symbols, *i.e.*, $\mathbf{Const}(\mathscr{L}) = \mathscr{F}_{(0)} = \mathscr{C}^{\infty}(\mathbb{R}^0) \cong \mathscr{C}^{\infty}(\{*\}) \cong \mathbb{R}$.

The terms of this language are defined in the usual way as the smallest set which comprises the individual variables, constant symbols and $n-$ary function symbols followed by n terms ($n \in \mathbb{N}$).

Before we proceed, we give the following:

Definition 1.1. *A $\mathscr{C}^{\infty}-$**structure** on a set A is a pair $\mathfrak{A} = (A, \Phi)$, where:*

$$\begin{array}{rccc} \Phi : & \bigcup_{n \in \mathbb{N}} \mathscr{C}^{\infty}(\mathbb{R}^n, \mathbb{R}) & \to & \bigcup_{n \in \mathbb{N}} \mathrm{Func}\,(A^n; A) \\ & (f : \mathbb{R}^n \stackrel{\mathscr{C}^{\infty}}{\to} \mathbb{R}) & \mapsto & \Phi(f) := (f^A : A^n \to A) \end{array},$$

*that is, Φ interprets the **symbols** of all smooth real functions of n variables as $n-$ary functions on A. Given two $\mathscr{C}^{\infty}$-structures, $\mathfrak{A} = (A, \Phi)$ and $\mathfrak{B} = (B, \Psi)$, a $\mathscr{C}^{\infty}-$**structure homomorphism** is a function $\varphi : A \to B$ such that for any $n \in \mathbb{N}$ and any $f : \mathbb{R}^n \stackrel{\mathscr{C}^{\infty}}{\to} \mathbb{R}$ the following diagram commutes:*

$$\begin{array}{ccc} A^n & \xrightarrow{\varphi^{(n)}} & B^n \\ {\scriptstyle \Phi(f)}\downarrow & & \downarrow{\scriptstyle \Psi(f)} \\ A & \xrightarrow{\varphi} & B \end{array}$$

i.e., $\Psi(f) \circ \varphi^{(n)} = \varphi \circ \Phi(f)$. The class of $\mathscr{C}^{\infty}-$structures and their morphisms compose a category that we denote by $\mathscr{C}^{\infty}\mathbf{Str}$.

We call a $\mathscr{C}^{\infty}-$structure $\mathfrak{A} = (A, \Phi)$ a $\mathscr{C}^{\infty}-$**ring** if it preserves projections and all equations between smooth functions. Formally, we have the following:

Definition 1.2. *Let $\mathfrak{A} = (A, \Phi)$ be a $\mathscr{C}^{\infty}-$structure. We say that $\mathfrak{A}$ (or, when there is no danger of confusion, A) is a $\mathscr{C}^{\infty}-$**ring** if the following is true:*

- *Given any $n, k \in \mathbb{N}$ and any projection $p_k : \mathbb{R}^n \to \mathbb{R}$, we have:*

$$\mathfrak{A} \models (\forall x_1) \cdots (\forall x_n)(p_k(x_1, \cdots, x_n) = x_k)$$

- *For every $f, g_1, \cdots g_n \in \mathscr{C}^{\infty}(\mathbb{R}^m, \mathbb{R})$ with $m, n \in \mathbb{N}$, and every $h \in \mathscr{C}^{\infty}(\mathbb{R}^n, \mathbb{R})$ such that $f = h \circ (g_1, \cdots, g_n)$, one has:*

$$\mathfrak{A} \models (\forall x_1) \cdots (\forall x_m)(f(x_1, \cdots, x_m) = h(g(x_1, \cdots, x_m), \cdots, g_n(x_1, \cdots, x_m)))$$

*Given two $\mathscr{C}^{\infty}$–rings, $\mathfrak{A} = (A, \Phi)$ and $\mathfrak{B} = (B, \Psi)$, a $\mathscr{C}^{\infty}$-**homomorphism** is just a $\mathscr{C}^{\infty}$-structure homomorphism between these $\mathscr{C}^{\infty}$–rings. The category of all $\mathscr{C}^{\infty}$–rings and $\mathscr{C}^{\infty}$–ring homomorphisms make up a full subcategory of $\mathscr{C}^{\infty}$**Str**, that we denote by $\mathscr{C}^{\infty}$**Rng**.*

Remark 1.3 (cf. **Sections 2, 3** and **4** of [3]). *Since $\mathscr{C}^{\infty}$**Rng** is a "variety of algebras" (it is a class of $\mathscr{C}^{\infty}$–structures which satisfy a given set of equations), it is closed under substructures, homomorphic images and products, by **Birkhoff's HSP Theorem**. Moreover:*

- *$\mathscr{C}^{\infty}$**Rng** is a concrete category and the forgetful functor, $U : \mathscr{C}^{\infty}\mathbf{Rng} \to \mathbf{Set}$ creates directed inductive colimits. Since $\mathscr{C}^{\infty}$**Rng** is a variety of algebras, it has all (small) limits and (small) colimits. In particular, it has binary coproducts, that is, given any two $\mathscr{C}^{\infty}$–rings A and B, we have their coproduct $A \stackrel{\iota_A}{\to} A \otimes_{\infty} B \stackrel{\iota_B}{\leftarrow} B$ again in $\mathscr{C}^{\infty}$**Rng**;*
- *Each set X freely generates a C^{∞}-ring, $L(X)$, as follows:*

- for any finite set X' with $\sharp X' = n$ we have $L(X') = \mathscr{C}^{\infty}(\mathbb{R}^{X'}) \cong \mathscr{C}^{\infty}(\mathbb{R}^n, \mathbb{R})$, which is the free C^{∞}-ring on n generators, $n \in \mathbb{N}$;
- for a general set, X, we take $L(X) = \mathscr{C}^{\infty}(\mathbb{R}^X) := \varinjlim_{X' \subseteq_{fin} X} \mathscr{C}^{\infty}(\mathbb{R}^{X'})$, with transition maps induced by restriction from $\mathbb{R}^{X''}$ to $\mathbb{R}^{X'}$ for $X' \subseteq X'' \subseteq_{fin} X$;

- *Given any $\mathscr{C}^{\infty}$–ring A and a set, X, we can freely adjoin the set X of variables to A with the following construction: $A\{X\} := A \otimes_{\infty} L(X)$. The elements of $A\{X\}$ are usually called $\mathscr{C}^{\infty}$–polynomials;*
- *The congruences of $\mathscr{C}^{\infty}$–rings are classified by their "ring-theoretical" ideals;*
- *Every $\mathscr{C}^{\infty}$–ring is the homomorphic image of some free $\mathscr{C}^{\infty}$–ring determined by some set, being isomorphic to the quotient of a free $\mathscr{C}^{\infty}$–ring by some ideal.*

Moreover, since $\mathscr{C}^{\infty}$**Rng** is a variety of algebras, the **Fundamental Theorem of Homomorphism** holds (**Theorem 3** of [3]). We register this result here, for the benefit of the reader:

Fact 1.4. (Fundamental Theorem of the $\mathscr{C}^{\infty}$–Homomorphism) *Let (A, Φ) be a $\mathscr{C}^{\infty}$–ring and $R \subseteq A \times A$ be a $\mathscr{C}^{\infty}$–congruence. For every $\mathscr{C}^{\infty}$–ring (B, Ψ) and for every $\mathscr{C}^{\infty}$–homomorphism $\varphi : (A, \Phi) \to (B, \Psi)$ such that $R \subseteq \ker(\varphi)$, that is, such that:*

$$(a, a') \in R \Rightarrow \varphi(a) = \varphi(a'),$$

there is a unique $\mathscr{C}^{\infty}$–homomorphism:

$$\widetilde{\varphi} : \left(\frac{A}{R}, \overline{\Phi}\right) \to (B, \Psi)$$

such that the following diagram commutes:

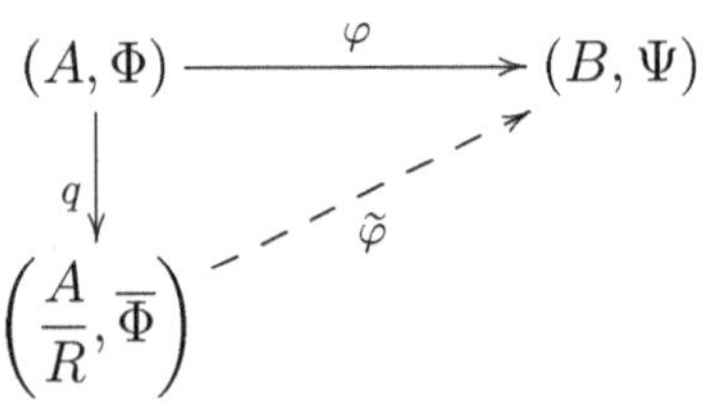

that is, such that $\widetilde{\varphi} \circ q = \varphi$, where $\overline{\Phi}$ is the canonical $\mathscr{C}^{\infty}-$structure induced on the quotient $\frac{A}{R}$.

Within the category of $\mathscr{C}^{\infty}-$rings we can perform a construction that is similar to the "ring of fractions" in Commutative Algebra, as well as define a suitable notion of "radical ideal". We analyze these concepts in the following section.

1.2 On $\mathscr{C}^{\infty}$-Rings of Fractions and $\mathscr{C}^{\infty}$-Radical Ideals

In order to extend the notion of the ring of fractions to the category $\mathscr{C}^{\infty}$**Rng**, we make use of the universal property a ring of fractions must satisfy in **Ring**- except that we must deal with $\mathscr{C}^{\infty}-$rings and $\mathscr{C}^{\infty}-$homomorphisms instead of rings and homomorphisms of rings.

Definition 1.5. *Let $\mathfrak{A} = (A, \Phi)$ be a $\mathscr{C}^{\infty}-$ring and $S \subseteq A$ be a subset. The $\mathscr{C}^{\infty}-$**ring of fractions** of A with respect to S is a $\mathscr{C}^{\infty}-$ring $A\{S^{-1}\}$, together with a $\mathscr{C}^{\infty}-$homomorphism $\eta_S : A \to A\{S^{-1}\}$ satisfying the following properties:*

(1) $(\forall s \in S)(\eta_S(s) \in (A\{S^{-1}\})^{\times})$

(2) If $\varphi : A \to B$ is any $\mathscr{C}^{\infty}-$homomorphism such that for every $s \in S$ we have $\varphi(s) \in B^{\times}$, then there is a unique $\mathscr{C}^{\infty}-$homomorphism $\widetilde{\varphi} : A\{S^{-1}\} \to B$ such that the following triangle commutes:

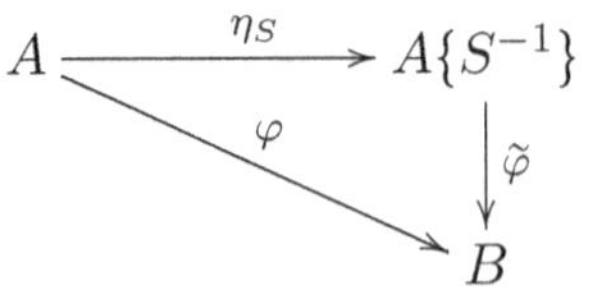

By this universal property, the $\mathscr{C}^{\infty}-$ring of fractions is unique, up to (unique) isomorphisms.

The existence of smooth rings of fractions can be guaranteed by a combination of constructions:

• first consider the addition of $\sharp S$-variables to the $\mathscr{C}^\infty$-ring A:

$$A\{x_s|s \in S\} := A \otimes_\infty \mathscr{C}^\infty(\mathbb{R}^S),$$

and let $j_S : A \to A\{x_s|s \in S\}$ be the (left) canonical morphism;

• now consider the ideal $\langle\{x_s\cdot\iota_A(s)-1|s \in S\}\rangle$ of A generated by $\{x_s\cdot\iota_A(s)-1|s \in S\}$, and take the quotient:

$$A\{x_s|s \in S\} \stackrel{q_S}{\twoheadrightarrow} \frac{A\{x_s|s \in S\}}{\langle\{x_s \cdot \iota_A(s) - 1|s \in S\}\rangle}.$$

Finally, define:

$$A\{S^{-1}\} := \frac{A\{x_s|s \in S\}}{\langle\{x_s \cdot \iota_A(s) - 1|s \in S\}\rangle};$$

and

$$\eta_S := q_S \circ j_S : A \to A\{S^{-1}\}.$$

It is not difficult to see that such a construction satisfies the required universal property.

Example 1.6. *Let $\varphi \in \mathscr{C}^\infty(\mathbb{R}^n)$ and consider the (closed) subset $Z(\varphi) = \{\vec{x} \in \mathbb{R}^n : \varphi(\vec{x}) = 0\} \subseteq \mathbb{R}^n$. Then $\mathscr{C}^\infty(\mathbb{R}^n)\{\varphi^{-1}\} \cong \mathscr{C}^\infty(\mathbb{R}^{n+1})/\langle\{y \cdot \varphi - 1\}\rangle \cong \mathscr{C}^\infty(\mathbb{R}^n \backslash Z(\varphi))$ together with the restriction map $\mathscr{C}^\infty(\mathbb{R}^n) \to \mathscr{C}^\infty(\mathbb{R}^n\backslash Z(\varphi))$ is a $\mathscr{C}^\infty$-homomorphism that satisfies the universal property of $\eta_{\{\varphi\}}$.*

Now we analyze the concept of the "$\mathscr{C}^\infty$–radical ideal" in the theory of $\mathscr{C}^\infty$–rings, which plays a similar role to the one played by radical ideals in Commutative Algebra. This concept was first presented by I. Moerdijk and G. Reyes in [16] in 1986, and explored in more detail in [17].

Unlike many notions in the branch of Smooth Rings such as $\mathscr{C}^\infty$–fields ($\mathscr{C}^\infty$-rings whose underlying rings are fields), $\mathscr{C}^\infty$–domains ($\mathscr{C}^\infty$-rings whose underlying rings are domains) and local $\mathscr{C}^\infty$–rings ($\mathscr{C}^\infty$-rings whose underlying rings are local rings), the concept of a $\mathscr{C}^\infty$–radical of an ideal cannot be brought from Commutative Algebra via the forgetful functor. This happens because when we take the localization of a $\mathscr{C}^\infty$-ring by an arbitrary prime ideal, it is not always true that we get a local $\mathscr{C}^\infty$–ring (see **Example 1.2** of [17]). In order to get a local $\mathscr{C}^\infty$–ring we must require an extra condition, that we are going to see later on.

Recall, from Commutative Algebra, that the radical of an ideal I of a commutative unital ring R is given by:

$$\sqrt{I} = \{x \in R | (\exists n \in \mathbb{N})(x^n \in I)\}.$$

There are several characterizations of this concept, among which we highlight the following ones:

$$\sqrt{I} = \bigcap \{\mathfrak{p} \in \mathrm{Spec}\,(R) | I \subseteq \mathfrak{p}\} = \left\{ x \in R | \left(\frac{R}{I}\right)[(x+I)^{-1}] \cong 0 \right\}.$$

The latter equality is the one which motivates our next definition.

Definition 1.7. (cf. p. 329 of [16]) *Let A be a $\mathscr{C}^\infty$−ring and let $I \subseteq A$ be a proper ideal. The $\mathscr{C}^\infty$−**radical of** I is given by:*

$$\sqrt[\infty]{I} := \left\{ a \in A | \left(\frac{A}{I}\right)\{(a+I)^{-1}\} \cong 0 \right\}$$

Definition 1.8. (cf. **Definition 2.1.5** of [2]) *Given a $\mathscr{C}^\infty$-ring A and a subset $S \subseteq A$, we define the $\mathscr{C}^\infty$−**saturation** of S by:*

$$S^{\infty-\mathrm{sat}} := \{a \in A \mid \eta_S(a) \in A^\times\}.$$

Example 1.9. *Given $\varphi \in \mathscr{C}^\infty(\mathbb{R}^n)$, we have $\{\varphi\}^{\infty-\mathrm{sat}} = \{\psi \in \mathscr{C}^\infty(\mathbb{R}^n) \mid Z(\psi) \subseteq Z(\varphi)\}$.*

The concept of $\mathscr{C}^\infty$−saturation is similar to the ordinary (ring-theoretic) concept of saturation in many aspects (for a detailed account of this concept, see **Section 3.1** of [4]). In particular, we use it to give a characterization of the elements of $\mathscr{C}^\infty$−radical ideals.

Proposition 1.10. [**Proposition 3.48** of [4]] *Let A be a $\mathscr{C}^\infty$−ring and let $I \subseteq A$ be any ideal. We have the following equalities:*

$$\sqrt[\infty]{I} = \{a \in A | (\exists b \in I) \& (\eta_a(b) \in (A\{a^{-1}\})^\times)\} = \{a \in A | I \cap \{a\}^{\infty-\mathrm{sat}} \neq \varnothing\}$$

where $\eta_a : A \to A\{a^{-1}\}$ is the $\mathscr{C}^\infty$−homomorphism of fractions with respect to $\{a\}$.

In ordinary Commutative Algebra, given an element x of a ring R, we say that x is a nilpotent infinitesimal if and only if there is some $n \in \mathbb{N}$ such that $x^n = 0$. Let A be a $\mathscr{C}^\infty$−ring and $a \in A$. D. Borisov and K. Kremnizer in [6] call a an ∞−infinitesimal if, and only if $A\{a^{-1}\} \cong 0$. The next definition describes the notion of a $\mathscr{C}^\infty$−ring being free of ∞−infinitesimals - which is analogous to the notion of "reducedness", of a commutative ring.

Definition 1.11. *A $\mathscr{C}^\infty$–ring A is* $\mathscr{C}^\infty$**–reduced** *if, and only if,* $\sqrt[\infty]{(0)} = (0)$.

Example 1.12. *The simplest example of $\mathscr{C}^\infty$–reduced $\mathscr{C}^\infty$–rings is the free $\mathscr{C}^\infty$–ring on any set of generators E (cf.* **Proposition 4.47** *of [4]).*

Next we register some useful results on $\mathscr{C}^\infty$–radical ideals and $\mathscr{C}^\infty$–reduced $\mathscr{C}^\infty$–rings.

Proposition 1.13 (Proposition 4.33, [4]**).** *Let A', B' be two $\mathscr{C}^\infty$–rings and $\jmath : A' \to B'$ be a monomorphism. If B' is $\mathscr{C}^\infty$–reduced, then A' is also $\mathscr{C}^\infty$–reduced.*

Proposition 1.14. *Let A be a $\mathscr{C}^\infty$–ring. We have:*

(a) An ideal $J \subseteq A$ is a $\mathscr{C}^\infty$–radical ideal if, and only if, $\dfrac{A}{J}$ is a $\mathscr{C}^\infty$–reduced $\mathscr{C}^\infty$–ring.

(b) A proper prime ideal $\mathfrak{p} \subseteq A$ is $\mathscr{C}^\infty$–radical if, and only if, $\dfrac{A}{\mathfrak{p}}$ is a $\mathscr{C}^\infty$–reduced $\mathscr{C}^\infty$–domain.

Proof. See **Corollary 4.31** of [4]. □

Next we present some properties of $\mathscr{C}^\infty$–radical ideals of a $\mathscr{C}^\infty$–ring A regarding some "operations" such as the intersection, the directed union and the preimage by a $\mathscr{C}^\infty$–homomorphism of $\mathscr{C}^\infty$–radical ideals. To simplify the notation, given a $\mathscr{C}^\infty$–ring A, we denote by $\mathfrak{I}^\infty_A$ the set of all its $\mathscr{C}^\infty$–radical ideals. The proofs of the results given in the next proposition can be found in [4].

Proposition 1.15 (Proposition 4.42 of [4]**).** *The following results hold:*

(a) Suppose that $(\forall \alpha \in \Lambda)(I_\alpha \in \mathfrak{I}^\infty_A)$. Then $\bigcap_{\alpha\in\Lambda} I_\alpha \in \mathfrak{I}^\infty_A$, that is, if $(\forall \alpha \in \Lambda)(I_\alpha \in \mathfrak{I}^\infty_A)$, then:

$$\sqrt[\infty]{\bigcap_{\alpha\in\Lambda} I_\alpha} = \bigcap_{\alpha\in\Lambda} I_\alpha = \bigcap_{\alpha\in\Lambda} \sqrt[\infty]{I_\alpha}$$

(b) Let $\{I_\alpha | \alpha \in \Sigma\}$ an upwards directed family of elements of $\mathfrak{I}^\infty_A$. Then $\bigcup_{\alpha\in\Sigma} I_\alpha \in \mathfrak{I}^\infty_A$.

(c) Let A, B be $\mathscr{C}^\infty$–rings, $f : A \to B$ a $\mathscr{C}^\infty$–homomorphism and $J \subseteq B$ any ideal. Then:

$$\sqrt[\infty]{f^{\dashv}[J]} \subseteq f^{\dashv}[\sqrt[\infty]{J}].$$

(d) *Let A, B be $\mathscr{C}^{\infty}-$rings, $f : A \to B$ be a $\mathscr{C}^{\infty}-$homomorphism and $J \subseteq B$ be a $\mathscr{C}^{\infty}-$radical ideal. Then $f^{\dashv}[J]$ is a $\mathscr{C}^{\infty}$-radical ideal of A.*

(e) *Given any two $\mathscr{C}^{\infty}-$radical ideals of a $\mathscr{C}^{\infty}-$ring A, $I, J \in \mathfrak{I}^{\infty}_{A}$, we have:*

$$\sqrt[\infty]{I \cdot J} = \sqrt[\infty]{I \cap J}$$

For a more comprehensive account of $\mathscr{C}^{\infty}-$radical ideals and $\mathscr{C}^{\infty}-$reduced $\mathscr{C}^{\infty}$-rings, we refer the reader to sections **3** and **4** of [4].

1.3 On the Smooth Zariski Spectrum

Recall that the spectrum of a commutative unital ring R consists of all prime ideals of R, together with a spectral topology - given by its "distinguished basic sets", its Zariski topology. Recall, also, that in ordinary Commutative Algebra, every prime ideal is radical - and that the $\mathscr{C}^{\infty}-$version of this implication is false in the context of Smooth Commutative Algebra (not every prime ideal of a $\mathscr{C}^{\infty}-$ring is $\mathscr{C}^{\infty}-$radical). At this point it is natural to look for a $\mathscr{C}^{\infty}-$analog of the Zariski spectrum of a commutative unital ring. Keeping in mind the definitions of the previous subsection, we give the following definition, that can be found in [17]:

Definition 1.16. *For a $\mathscr{C}^{\infty}-$ring A, we define:*

$$\mathrm{Spec}^{\infty}(A) = \{\mathfrak{p} \in \mathrm{Spec}(A) | \mathfrak{p} \ is \mathscr{C}^{\infty} - radical\}$$

equipped with the smooth Zariski topology defined by its basic open sets:

$$D^{\infty}(a) = \{\mathfrak{p} \in \mathrm{Spec}^{\infty}(A) | a \notin \mathfrak{p}\}$$

Proposition 1.17. *For every $\mathscr{C}^{\infty}-$ring A, $\mathrm{Spec}^{\infty}(A)$ is a spectral space. Given two $\mathscr{C}^{\infty}-$rings A, A' and a $\mathscr{C}^{\infty}-$homomorphism $f : A \to A'$, The function:*

$$\begin{array}{cccc} f^{*}: & \mathrm{Spec}^{\infty}(A') & \to & \mathrm{Spec}^{\infty}(A) \\ & \mathfrak{p} & \mapsto & f^{\dashv}[\mathfrak{p}] \end{array}$$

is a spectral map.

Proof. For a detailed proof that $\mathrm{Spec}^{\infty}(A)$ is a spectral space for any $\mathscr{C}^{\infty}-$ring A we refer the reader to **Theorem 5.16** of [4]. For a detailed proof that for any $\mathscr{C}^{\infty}-$rings A, A' and any $\mathscr{C}^{\infty}-$homomorphism $f : A \to A'$, the map f^{*} is spectral, we refer the reader to **Proposition 5.19** of [4]. □

Theorem 1.18. *(**Separation Theorems**, [5]) Let A be a $\mathscr{C}^\infty$–ring, $S \subseteq A$ be a subset of A and I be an ideal of A. Denote by $\langle S \rangle$ the multiplicative submonoid of A generated by S. We have:*

(a) If I is a $\mathscr{C}^\infty$–radical ideal, then:

$$I \cap \langle S \rangle = \varnothing \iff I \cap S^{\infty-\mathrm{sat}} = \varnothing$$

(b) If $S \subseteq A$ is a $\mathscr{C}^\infty$-saturated subset, then:

$$I \cap S = \varnothing \iff \sqrt[\infty]{I} \cap S = \varnothing$$

(c) If $\mathfrak{p} \in \mathrm{Spec}^\infty(A)$, then $A \backslash \mathfrak{p} = (A \backslash \mathfrak{p})^{\infty-\mathrm{sat}}$

(d) If $S \subseteq A$ is a $\mathscr{C}^\infty$-saturated subset, then:

$$I \cap S = \varnothing \iff (\exists \mathfrak{p} \in \mathrm{Spec}^\infty(A))((I \subseteq \mathfrak{p})\&(\mathfrak{p} \cap S = \varnothing)).$$

(e) $\sqrt[\infty]{I} = \bigcap\{\mathfrak{p} \in \mathrm{Spec}^\infty(A) | I \subseteq \mathfrak{p}\}$

Proof. See **Theorem 4.49** of [4]. □

A more detailed account of the smooth Zariski spectrum containing detailed proofs can be found in **Section 5.1** of [4].

2 On Smooth Spaces and Smooth Algebras

Every (finite dimensional) smooth manifold M can be embedded as a closed subspace of some $\mathbb{R}^n$ (**Whitney's Theorem**) and determines a $\mathscr{C}^\infty$-ring, $\mathscr{C}^\infty(M)$. This mapping, $M \mapsto \mathscr{C}^\infty(M)$, extends to a full and faithful contravariant functor into the category of $\mathscr{C}^\infty$-rings (see for instance **Theorem 2.8** of [18]). In this section we present some results that (dually) connects subsets of $\mathbb{R}^E$ and quotients of $\mathscr{C}^\infty(\mathbb{R}^E)$. More precisely, we present some very useful characterizations of equalities and inequalities between elements of $\mathscr{C}^\infty$–reduced $\mathscr{C}^\infty$–rings, *i.e.*, $\mathscr{C}^\infty$–rings of the form $A = \frac{\mathscr{C}^\infty(\mathbb{R}^E)}{I}$ with $\sqrt[\infty]{I} = I$, by means of the filter of zerosets of functions of I.

2.1 The Finitely Generated Case

We start by recalling an important fact about the relation between closed subsets of $\mathbb{R}^n$ and zerosets of $\mathscr{C}^\infty$-functions.

Fact 2.1. *(essentially* **Lemma 1.4** *of [18]) For each open subset $U \subseteq \mathbb{R}^n$ there is a smooth function $\chi : \mathbb{R}^n \to \mathbb{R}$ such that:*

- $(\forall x \in \mathbb{R}^n)(\chi(x) \geqslant 0)$
- $(\forall x \in \mathbb{R}^n)((\chi(x) = 0) \iff (x \notin U))$.

Definition 2.2 **(smooth function).** *Let $X \subseteq \mathbb{R}^n$. A function $f : X \to \mathbb{R}$ defined over X is* smooth *if there is an open subset $U \subseteq \mathbb{R}^n$ such that $X \subseteq U$ and a $\mathscr{C}^\infty$-extension of f, $\widetilde{f} : U \to \mathbb{R}$, such that $\widetilde{f} \upharpoonright_X = f$.*

Fact 2.3 **(Smooth Tietze Theorem).** *Let $F \subseteq \mathbb{R}^n$ be a* closed *set and let $f \in \mathscr{C}^\infty(F)$. Then there is a smooth function $\widetilde{f} \in \mathscr{C}^\infty(\mathbb{R}^n)$ such that $\widetilde{f} \upharpoonright_F = f$. Moreover:*

- *If $(\forall x \in F)(f(x) \neq 0)$, then we can choose a $\mathscr{C}^\infty$-extension $\widetilde{f}$ of f and an open subset $U \subseteq \mathbb{R}^n$ such that $F \subseteq U$ and $(\forall x \in U)(\widetilde{f}(x) \neq 0)$.*
- *If $(\forall x \in F)(f(x) > 0)$, then we can choose a $\mathscr{C}^\infty$-extension $\widetilde{f}$ of f such that $(\forall x \in \mathbb{R}^n)(\widetilde{f}(x) > 0)$.*

Proposition 2.4. *Let $A = \frac{\mathscr{C}^\infty(\mathbb{R}^n)}{I}$ be a $\mathscr{C}^\infty-$reduced $\mathscr{C}^\infty-$ring, so $\sqrt[\infty]{I} = I$. Given $f, g \in \mathscr{C}^\infty(\mathbb{R}^n)$, we have:*

$$(q_I(f) = f + I = g + I = q_I(g)) \iff (\exists \varphi \in I)(\forall x \in Z(\varphi))(f(x) = g(x)).$$

Proof. Suppose $q_I(f) = f + I = g + I = q_I(g)$, so $g - f \in I$. It suffices to take $\varphi = g - f$, so:

$$(\forall x \in Z(\varphi))(g(x) - f(x) = 0)$$

and

$$(\forall x \in Z(\varphi))(f(x) = g(x))$$

Conversely, if there is some $\varphi \in I$ such that $(\forall x \in Z(\varphi))(f(x) = g(x))$, then $Z(\varphi) \subseteq Z(g-f)$ and $\varphi \upharpoonright_{\mathbb{R}^n \setminus Z(g-f)} \in \mathscr{C}^\infty(\mathbb{R}^n \setminus Z(g-f))^\times \cong \mathscr{C}^\infty(\mathbb{R}^n)\{(g-f)^{-1}\}^\times$ (see **Example 1.6**). It follows that $g - f \in \sqrt[\infty]{I} \subseteq I$ and $f + I = g + I$. □

Now we characterize the invertible elements of a $\mathscr{C}^\infty$-reduced $\mathscr{C}^\infty$-ring.

Proposition 2.5. *Let $A = \frac{\mathscr{C}^\infty(\mathbb{R}^n)}{I}$ be a $\mathscr{C}^\infty$–reduced finitely generated $\mathscr{C}^\infty$–ring, so $\sqrt[\infty]{I} = I$. Given $f \in \mathscr{C}^\infty(\mathbb{R}^n)$ we have:*

$$\left(q_I(f) = (f + I) \in \left(\frac{\mathscr{C}^\infty(\mathbb{R}^n)}{I} \right)^\times \right) \iff (\exists \varphi \in I)(\forall x \in Z(\varphi))(f(x) \neq 0).$$

Proof. Suppose, first, that $(f + I) \in \left(\frac{\mathscr{C}^\infty(\mathbb{R}^n)}{I}\right)^\times$, so there is some $h + I \in \frac{\mathscr{C}^\infty(\mathbb{R}^n)}{I}$ such that:

$$(f + I) \cdot (h + I) = 1 + I$$

$$q_I(f) \cdot q_I(h) = q_I(1)$$

$$f \cdot h - 1 \in \ker(q_I)$$

$$\varphi = f \cdot h - 1 \in I,$$

One has:

$$(\forall x \in Z(\varphi))(f(x) \cdot h(x) = 1 \neq 0)$$

and thus:

$$(\forall x \in Z(\varphi))(f(x) = \frac{1}{h(x)} \neq 0)$$

Conversely, suppose that $f \in \mathscr{C}^\infty(\mathbb{R}^n)$ is such that there is some $\varphi \in I$ with $(\forall x \in Z(\varphi))(f(x) \neq 0)$. Since f is a continuous function, there is an open subset $U \subseteq \mathbb{R}^n$ such that $Z(\varphi) \subseteq U$ and $(\forall x \in U)(f(x) \neq 0)$.

We define:

$$\begin{array}{rccc} g: & U \subseteq \mathbb{R}^n & \to & \mathbb{R} \\ & x & \mapsto & \frac{1}{f(x)} \end{array}$$

Thus $g \upharpoonright_{Z(\varphi)}$ is smooth on $Z(\varphi)$ and by **Smooth Tietze's Theorem** (**Fact 2.3**), one is able to construct a $\mathscr{C}^\infty$-function $\widetilde{g} : \mathbb{R}^n \to \mathbb{R}$ such that $\widetilde{g} \upharpoonright_{Z(\varphi)} = g \upharpoonright_{Z(\varphi)}$.

Since we have:

$$(\forall x \in Z(\varphi))(f(x) \cdot \widetilde{g}(x) - 1 = 0)$$

it follows, by **Proposition 2.4**, that $f \cdot \widetilde{g} - 1 \in I$, so:

$$(f + I) \cdot (\widetilde{g} + I) = 1 + I$$

and $f + I \in \left(\frac{\mathscr{C}^\infty(\mathbb{R}^n)}{I}\right)^\times$. □

Combining the previous proposition with **Proposition 2.4**, we obtain the following:

Corollary 2.6. *Let $A = \frac{\mathscr{C}^\infty(\mathbb{R}^n)}{I}$ be a finitely generated $\mathscr{C}^\infty$−reduced $\mathscr{C}^\infty$−ring, so $\sqrt[\infty]{I} = I$. Given $f \in \mathscr{C}^\infty(\mathbb{R}^n)$, are equivalent:*

1. $(\exists u \in \mathscr{C}^\infty(\mathbb{R}^n))(((f + I) = (u^2 + I))\&(u + I \in \left(\frac{\mathscr{C}^\infty(\mathbb{R}^n)}{I}\right)^\times))$
2. $(\exists u \in \mathscr{C}^\infty(\mathbb{R}^n))(\exists \psi \in I)(\forall x \in Z(\psi))(f(x) = u^2(x) \neq 0)$
3. $(\exists \psi \in I)(\forall x \in Z(\psi))(f(x) > 0)$

2.2 The General Case

We know that every closed subset of $\mathbb{R}^n$ is the zeroset of some smooth function $f : \mathbb{R}^n \to \mathbb{R}$ (see **Fact 2.1**). We now expand the notion of zero set to $\mathbb{R}^E$, where E is not necessarily a finite set.

Definition 2.7. *Let E be any set. Consider $\mathbb{R}^E = \{v : E \to \mathbb{R} | v \text{ is a function}\}$ and denote $\mathscr{F}(\mathbb{R}^E) := \{f : \mathbb{R}^E \to \mathbb{R} | f \text{ is a function}\}$. For every $D \subseteq E$, we have the canonical projection:*

$$\begin{array}{rccc} \pi_{ED} : & \mathbb{R}^E & \to & \mathbb{R}^D \\ & v & \mapsto & v \restriction_D : D \to \mathbb{R} \end{array}$$

and this induces a function:

$$\begin{array}{rccc} \mu_{DE} : & \mathscr{F}(\mathbb{R}^D) & \to & \mathscr{F}(\mathbb{R}^E) \\ & f & \mapsto & \mu_{DE}(f) := f \circ \pi_{ED} \end{array}$$

Definition 2.8. *Let E be any set. Define:*

$$\mathscr{C}^\infty(\mathbb{R}^E) := \{f \in \mathscr{F}(\mathbb{R}^E) \mid (\exists E' \subseteq_{\text{fin}} E)(\exists f' \in \mathscr{C}^\infty(\mathbb{R}^{E'}))(f = f' \circ \pi_{EE'})\}.$$

Thus, by a smooth function on $\mathbb{R}^E$ we mean a function $f : \mathbb{R}^E \to \mathbb{R}$ that factors through some projection $\pi_{EE'} : \mathbb{R}^E \to \mathbb{R}^{E'}$ and a smooth function $\widetilde{f} \in \mathscr{C}^\infty(\mathbb{R}^{E'})$, for some $E' \subseteq_{\text{fin}} E$. I.e., given $f \in \mathscr{F}(\mathbb{R}^E)$ we have:

$$f \in \mathscr{C}^\infty(\mathbb{R}^E) \iff (\exists E' \subseteq_{\text{fin}} E)(\exists \widetilde{f} \in \mathscr{C}^\infty(\mathbb{R}^{E'})(f = \widetilde{f} \circ \pi_{EE'}).$$

It is not hard to see that for every $E', E'' \subseteq_{\text{fin}} E$ with $E' \subseteq E''$, the following diagram commutes:

$$\begin{array}{ccc} & \mathscr{C}^\infty(\mathbb{R}^E) & \\ {\scriptstyle \mu_{E'E}} \nearrow & & \nwarrow {\scriptstyle \mu_{E''E}} \\ \mathscr{C}^\infty(\mathbb{R}^{E'}) & \xrightarrow[\mu_{E'E''}]{} & \mathscr{C}^\infty(\mathbb{R}^{E''}) \end{array}$$

Moreover, notice that $\mathscr{C}^\infty(\mathbb{R}^E) \cong \varinjlim_{E' \subseteq_{\text{fin}} E} \mathscr{C}^\infty(\mathbb{R}^{E'})$, where for every $E', E'' \subseteq_{\text{fin}} E$ such that $E' \subseteq E''$, the following triangle commutes:

$$\begin{array}{ccc} & \varinjlim_{E' \subseteq_{\text{fin}} E} \mathscr{C}^\infty(\mathbb{R}^{E'}) & \\ {\scriptstyle \ell_{E'}} \nearrow & & \nwarrow {\scriptstyle \ell_{E''}} \\ \mathscr{C}^\infty(\mathbb{R}^{E'}) & \xrightarrow[\mu_{E'E''}]{} & \mathscr{C}^\infty(\mathbb{R}^{E''}) \end{array}$$

where the morphisms $\ell_{E'}, \ell_{E''}$ are defined as in **Section 3** of [3].

Definition 2.9. *Let E be any set. A subset $X \subseteq \mathbb{R}^E$ is a* ***zeroset*** *if there is some $\varphi \in \mathscr{C}^\infty(\mathbb{R}^E)$ such that $X = Z(\varphi)$, where*

$$Z(\varphi) := \varphi^{\dashv}[\{0\}] = \{\vec{x} \in \mathbb{R}^E : \varphi(\vec{x}) = 0\}$$

The set $\mathscr{Z}_E := \{Z(\varphi) \in \wp(\mathbb{R}^E) : \varphi \in \mathscr{C}^\infty(\mathbb{R}^E)\}$ denotes the set of all zerosets in $\mathbb{R}^E$.

Remark 2.10. • *Let E be an arbitrary set and $\varphi \in \mathscr{C}^\infty(\mathbb{R}^E)$. Select $E' \subseteq_{\text{fin}} E$ and $\varphi' \in \mathscr{C}^\infty(\mathbb{R}^{E'})$ such that $\varphi = \varphi' \circ \pi_{EE'}$. Then $Z(\varphi) = \pi_{EE'}^{\dashv}[Z(\varphi')]$.*

• *If E is a* finite *set, then by* **Fact 2.1**, *$\mathscr{Z}_E = \text{Closed}\,(\mathbb{R}^E) \subseteq \wp(\mathbb{R}^E)$ thus it is stable under* finite *unions and* arbitrary *intersections; in particular, $\varnothing = \bigcup \varnothing$ and $\mathbb{R}^E = \bigcap \varnothing$ are in $\mathscr{Z}_E$.*

• *In general, for an* arbitrary *set E, the subset $\mathscr{Z}_E \subseteq \wp(\mathbb{R}^E)$ is stable just under* finite *unions and* finite *intersections.*

Definition 2.11. *If $I \subseteq \mathscr{C}^\infty(\mathbb{R}^E)$ is an ideal, then $I' = \mu_{E'E}^{\dashv}[I]$ is an ideal of $\mathscr{C}^\infty(\mathbb{R}^{E'})$. We define:*

$$\widehat{I} = \{F \subseteq \mathbb{R}^E \mid (\exists E' \subseteq_{\text{fin}} E)(\exists f \in I' = \mu_{E'E}^{\dashv}[I])(F = \pi_{EE'}^{\dashv}[Z(f)])\}.$$

Proposition 2.12. *Let $A = \frac{\mathscr{C}^\infty(\mathbb{R}^E)}{I}$ be a $\mathscr{C}^\infty$–reduced $\mathscr{C}^\infty$–ring, so $\sqrt[\infty]{I} = I$. Given $f, g \in \mathscr{C}^\infty(\mathbb{R}^E)$, we have:*

$$(q_I(f) = f + I = g + I = q_I(g)) \iff (\exists \varphi \in I)(\forall x \in Z(\varphi))(f(x) = g(x)).$$

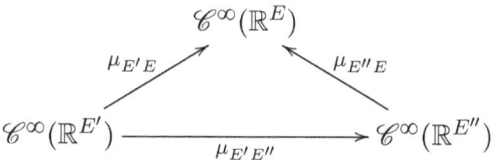

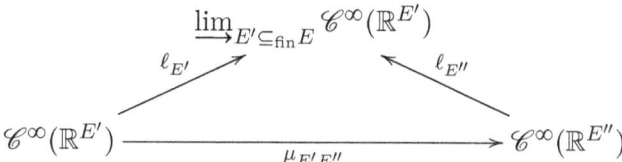

Proof. Given $f, g \in \mathscr{C}^\infty(\mathbb{R}^E)$ such that $q_I(f) = f + I = g + I = q_I(g)$, by definition there are finite subsets $E_f, E_g \subseteq E$ and $\widehat{f} \in \mathscr{C}^\infty(\mathbb{R}^{E_f}), \widehat{g} \in \mathscr{C}^\infty(\mathbb{R}^{E_g})$ such that $f = \mu_{E_f}(\widehat{f}) = \widehat{f} \circ \pi_{E_f} \in \mathscr{C}^\infty(\mathbb{R}^{E_f})$ and $g = \mu_{E_g}(\widehat{g}) = \widehat{g} \circ \pi_{E_g} \in \mathscr{C}^\infty(\mathbb{R}^{E_g})$. Then $E_f \cup E_g \subseteq_{\text{fin}} E$. Let $\widetilde{f} = \mu_{E_f, E_f \cup E_g}(\widehat{f}) \in \mathscr{C}^\infty(\mathbb{R}^{E_f \cup E_g})$ and $\widetilde{g} = \mu_{E_g, E_f \cup E_g}(\widehat{g}) \in \mathscr{C}^\infty(\mathbb{R}^{E_f \cup E_g})$. By hypothesis, $f + I = g + I$, so $f - g \in I$ and $\mu_{E_f \cup E_g, E}(\widetilde{f}) - \mu_{E_f \cup E_g, E}(\widetilde{g}) \in I$. We have, thus, $(\widetilde{f} - \widetilde{g}) \in \mu^{\dashv}_{E_f \cup E_g, E}[I] = \sqrt[\infty]{\mu^{\dashv}_{E_f \cup E_g}[I]}$, since I is a $\mathscr{C}^\infty$-radical ideal (see **Proposition 1.15.(d)**). By the finitely generated case (**Proposition 2.4**), since $\widetilde{f}, \widetilde{g} \in \mathscr{C}^\infty(\mathbb{R}^{E_f \cup E_g})$ and $\widetilde{f} + \mu^{\dashv}_{E_f \cup E_g}[I] = \widetilde{g} + \mu^{\dashv}_{E_f \cup E_g}[I]$, it follows that there is some $\widetilde{\varphi} \in \mu^{\dashv}_{E_f \cup E_g}[I]$ such that:

$$(\forall y \in Z(\widetilde{\varphi}))(\widetilde{f}(y) = \widetilde{g}(y)).$$

Taking $\varphi = \mu_{E_f \cup E_g, E}(\widetilde{\varphi}) = \widetilde{\varphi} \circ \pi_{E, E_f \cup E_g} \in I$, we have:

$$(\forall x \in Z(\varphi))(f(x) = \widetilde{f} \circ \pi_{E_f \cup E_g}(x) = \widetilde{g} \circ \pi_{E_f \cup E_g}(x) = g(x)).$$

On the other hand, suppose $f, g \in \mathscr{C}^\infty(\mathbb{R}^E)$ are such that $(\exists \varphi \in I)(\forall x \in Z(\varphi))(f(x) = g(x))$. Thus, for such φ there is a finite $E_\varphi \subseteq E$ and $\widehat{\varphi} \in \mathscr{C}^\infty(\mathbb{R}^{E_\varphi})$ such that $\varphi = \widehat{\varphi} \circ \pi_{E, E_\varphi}$, and there are also some finite $E_f, E_g \subseteq E$ and some $\widehat{f} \in \mathscr{C}^\infty(\mathbb{R}^{E_f}), \widehat{g} \in \mathscr{C}^\infty(\mathbb{R}^{E_g})$ such that $f = \mu_{E_f, E}(\widehat{f})$ and $g = \mu_{E_g, E}(\widehat{g})$. Let $\widetilde{\varphi} = \mu_{E_\varphi, E_\varphi \cup E_f \cup E_g}(\widehat{\varphi}), \widetilde{f} = \mu_{E_f, E_\varphi \cup E_f \cup E_g}(\widehat{f})$ and $\widetilde{g} = \mu_{E_g, E_\varphi \cup E_f \cup E_g}(\widehat{g})$. By the finitely generated case (**Proposition 2.4**), since $\widetilde{f}, \widetilde{g}, \widetilde{\varphi} \in \mathscr{C}^\infty(\mathbb{R}^{E_\varphi \cup E_f \cup E_g})$, $(\forall x \in Z(\widetilde{\varphi}))(\widetilde{f}(x) = \widetilde{g}(x))$ and

$$\sqrt[\infty]{\mu^{\dashv}_{E_\varphi \cup E_f \cup E_g, E}[I]} = \mu^{\dashv}_{E_\varphi \cup E_f \cup E_g, E}[I],$$

it follows that $\widetilde{f} - \widetilde{g} \in \mu^{\dashv}_{E_\varphi \cup E_f \cup E_g, E}[I]$, so $f - g = \mu_{E_\varphi \cup E_f \cup E_g, E}(\widetilde{f} - \widetilde{g}) \in I$, and $f + I = g + I$. □

Proposition 2.13. *Let E be any set and $I \subseteq \mathscr{C}^\infty(\mathbb{R}^E)$ be a $\mathscr{C}^\infty$-radical ideal. We have, for every $f \in \mathscr{C}^\infty(\mathbb{R}^E)$:*

$$\left(f + I \in \left(\frac{\mathscr{C}^\infty(\mathbb{R}^E)}{I}\right)^\times\right) \iff (\exists \varphi \in I)(\forall x \in Z(\varphi))(f(x) \neq 0)$$

Proof. Given $f \in \mathscr{C}^\infty(\mathbb{R}^E)$ such that $q_I(f) = f + I$ is invertible, let $h, \varphi \in \mathscr{C}^\infty(\mathbb{R}^E)$ such that

$$(f \cdot h - 1) = \varphi \in I.$$

As in the proof of previous proposition, we can select $E' \subseteq_{\text{fin}} E$ and $f', h', \varphi' \in \mathscr{C}^\infty(\mathbb{R}^{E'})$ such that $f = \mu_{E'E}(f'), h = \mu_{E'E}(h'), \varphi = \mu_{E'E}(\varphi')$. Then

$$(f' \cdot h' -' 1) = \varphi' \in I' := \mu_{E'E}^{\dashv}[I].$$

Thus

$$(\forall x' \in \mathbb{R}^{E'})(x' \in Z(\varphi') \Rightarrow f'(x') \neq 0).$$

Since $Z(\varphi) = \pi_{EE'}^{\dashv}[Z(\varphi')]$ and $f = f' \circ \pi_{EE'}$, then

$$(\forall x \in \mathbb{R}^E)(x \in Z(\varphi) \Rightarrow f(x) \neq 0).$$

Conversely, let $f, \varphi \in \mathscr{C}^\infty(\mathbb{R}^E)$ such that $\varphi \in I$ and

$$(\forall x \in \mathbb{R}^E)(x \in Z(\varphi) \to f(x) \neq 0).$$

Select $E' \subseteq_{\text{fin}} E$ and $f', \varphi' \in \mathscr{C}^\infty(\mathbb{R}^{E'})$ such that $f = \mu_{E'E}(f'), \varphi = \mu_{E'E}(\varphi')$.
Then $I' := \mu_{E'E}^{\dashv}[I]$ is a $\mathscr{C}^\infty$-radical ideal of $\mathscr{C}^\infty(\mathbb{R}^{E'})$, $\varphi' \in I'$ and

$$(\forall x' \in \mathbb{R}^{E'})(x' \in Z(\varphi') \to f'(x') \neq 0).$$

By the finitely generated case (**Proposition 2.5**), $f' + I' \in (\mathscr{C}^\infty(\mathbb{R}^{E'})/I')^\times$. Let $h' \in \mathscr{C}^\infty(\mathbb{R}^{E'})$ such that

$$(f' + I')(h' + I') = 1 + I' \in \mathscr{C}^\infty(\mathbb{R}^{E'})/I'.$$

Now define $h := \mu_{EE'}(h')$. Then

$$(f + I)(h + I) = 1 + I \in \mathscr{C}^\infty(\mathbb{R}^E)/I$$

□

Proposition 2.14. *Let E be any set. If $I \subseteq \mathscr{C}^\infty(\mathbb{R}^E)$ is an ideal, then:*

$$\widehat{I} := \{X \in \wp(\mathbb{R}^E) \mid (\exists f \in I)(X = Z(f))\} \subseteq \wp(Z(\mathbb{R}^E))$$

is a filter of zerosets in $\mathbb{R}^E$.

Proof. Note first that

$$\widehat{I} = \{F \subseteq \mathbb{R}^E \mid (\exists E' \subseteq_{\text{fin}} E)(\exists f' \in I' = \mu_{E'}^{\dashv}[I])(F = \pi_{EE'}^{\dashv}[Z(f)])\} \subseteq \wp(\mathscr{Z}(\mathbb{R}^E))$$

Note that $\mathbb{R}^E$ is a zeroset: $\mathbb{R}^E = Z(0_E)$, where

$$\begin{array}{rccc} 0_E : & \mathbb{R}^E & \to & \mathbb{R} \\ & x & \mapsto & 0 \end{array}.$$

Note that $0_E \in \mathscr{C}^\infty(\mathbb{R}^E)$: choose any finite $D \subseteq E$ and consider the $\mathscr{C}^\infty$-function

$$\begin{array}{rccc} 0_D : & \mathbb{R}^D & \to & \mathbb{R} \\ & x & \mapsto & 0 \end{array},$$

so $0_E = \mu_{DE}(0_D) \in \mathscr{C}^\infty(\mathbb{R}^E)$.

Given $G_1, G_2 \in \widehat{I} \subseteq \mathbb{R}^E$, let $g_1, g_2 \in I \subseteq \mathscr{C}^\infty(\mathbb{R}^E)$ such that $G_i = Z(g_i)$ for $i = 1, 2$. There are finite $E', E'' \subseteq E$, $f_1 \in \mu_{E'E}^{\dashv}[I] \subseteq \mathscr{C}^\infty(\mathbb{R}^{E'})$, $f_2 \in \mu_{E''E}^{\dashv}[I] \subseteq \mathscr{C}^\infty(\mathbb{R}^{E''})$ such that $g_1 = f_1 \circ \pi_{EE'}, g_2 = f_2 \circ \pi_{EE''}$. Thus $\pi_{EE'}^{\dashv}[Z(f_1)] = G_1$ and $\pi_{EE''}^{\dashv}[Z(f_2)] = G_2$, where $\pi_{EE'} : \mathbb{R}^E \to \mathbb{R}^{E'}$ and $\pi_{EE''} : \mathbb{R}^E \to \mathbb{R}^{E''}$ are the canonical projections (restrictions). Consider:

$$\begin{array}{ccc} & \mathbb{R}^{E'\cup E''} & \\ {}^{\pi_{E'\cup E'',E'}}\swarrow & & \searrow^{\pi_{E'\cup E'',E''}} \\ \mathbb{R}^{E'} & & \mathbb{R}^{E''} \end{array}$$

where:

$$\begin{array}{rccc} \pi_{E'\cup E'',E'} : & \mathbb{R}^{E'\cup E''} & \to & \mathbb{R}^{E'} \\ & v & \mapsto & v\restriction_{E'} : E' \to \mathbb{R} \end{array}$$

and

$$\begin{array}{rccc} \pi_{E'\cup E'',E''} : & \mathbb{R}^{E'\cup E''} & \to & \mathbb{R}^{E''} \\ & v & \mapsto & v\restriction_{E''} : E'' \to \mathbb{R} \end{array}$$

We have the commutative diagram:

$$\begin{array}{ccccc} & & \mathbb{R}^E & & \\ & \swarrow^{\pi_{EE'}} & \downarrow{\scriptstyle \pi_{E,E'\cup E''}} & \searrow^{\pi_{EE''}} & \\ & & \mathbb{R}^{E'\cup E''} & & \\ & {}^{\pi_{E'\cup E'',E'}}\swarrow & & \searrow^{\pi_{E'\cup E'',E''}} & \\ \mathbb{R}^{E'} & & & & \mathbb{R}^{E''} \end{array}$$

Define $\widetilde{f_1} = f_1 \circ \pi_{E'\cup E'',E'} : \mathbb{R}^{E'\cup E''} \to \mathbb{R}$ and $\widetilde{f_2} = f_2 \circ \pi_{E'\cup E'',E''} : \mathbb{R}^{E'\cup E''} \to \mathbb{R}$, so $\widetilde{F_1} = \pi_{E'\cup E'',E'}^{\dashv}[Z(f_1)] = \widetilde{f_1}^{\dashv}[\{0\}] = Z(\widetilde{f_1}) \subseteq \mathbb{R}^{E'\cup E''}$ and
$\widetilde{F_2} = \pi_{E'\cup E'',E''}^{\dashv}[Z(f_2)] = \widetilde{f_2}^{\dashv}[\{0\}] = Z(\widetilde{f_2}) \subseteq \mathbb{R}^{E'\cup E''}$ are zerosets.

Note that $\widetilde{F_1} \cap \widetilde{F_2}$ is also a zeroset, namely $\widetilde{F_1} \cap \widetilde{F_2} = Z(\widetilde{f_1}^2 + \widetilde{f_2}^2)$, with $\widetilde{f_1}^2 + \widetilde{f_2}^2 \in \mu_{E'\cup E'',E}^{\dashv}[I]$. In fact, we have the commutative diagram:

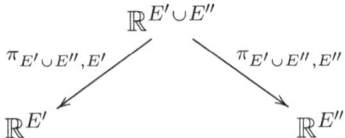

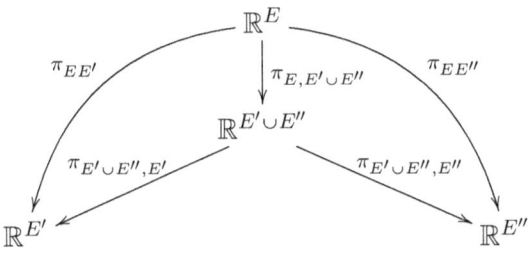

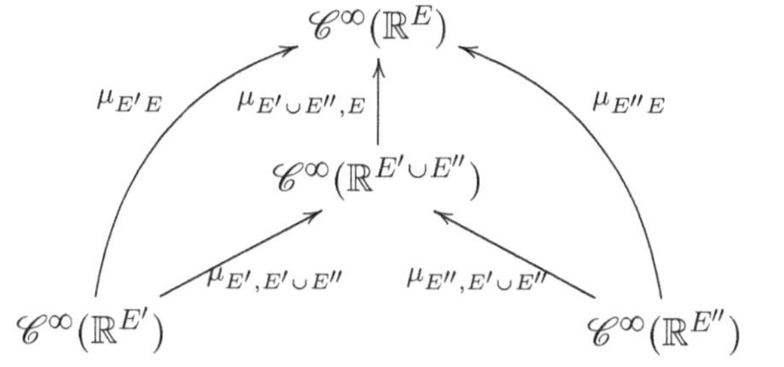

Since the diagram commutes, we have $\mu_{E',E'\cup E''}(f_1) = f_1 \circ \pi_{E'\cup E'',E'} = \widetilde{f_1} \in \mu^{\dashv}_{E'\cup E''}[I]$ and $\mu_{E'',E'\cup E''}(f_2) = f_2 \circ \pi_{E'\cup E'',E''} = \widetilde{f_2} \in \mu^{\dashv}_{E'\cup E'',E}[I]$, so $\widetilde{f_1}^2 + \widetilde{f_2}^2 \in \mu^{\dashv}_{E'\cup E'',E}[I]$.

Then $g_1^2 + g_2^2 \in I$ and

$$\pi^{\dashv}_{E,E'\cup E''}[\widetilde{F_1} \cap \widetilde{F_2}] = \pi^{\dashv}_{E,E'\cup E''}[Z(\widetilde{f_1}^2 + \widetilde{f_2}^2)] = Z(g_1^2 + g_2^2) = G_1 \cap G_2.$$

Let $G \in \widehat{I}$ and $H \in \mathscr{Z}(\mathbb{R}^E)$ such that $G \subseteq H$. Then there are $g \in I, h \in \mathscr{C}^\infty(\mathbb{R}^E)$ such that $G = Z(g), H = Z(h) \in \mathscr{Z}(\mathbb{R}^E)$. Now select $E' \subseteq_{\text{fin}} E$ and $g', h' \in \mathscr{C}^\infty(\mathbb{R}^{E'})$ such that $\mu_{E'E}(g') = g, \mu_{E'E}(h') = h$; thus $\pi^{\dashv}_{EE'}[Z(g')] \subseteq \pi^{\dashv}_{EE'}[Z(h')]$ and $g' \in I' := \mu^{\dashv}_{E'E}[I]$. Let $G' = Z(g'), H' = Z(h') \in \mathscr{Z}(\mathbb{R}^{E'})$, then $G' \subseteq H'$. Since we are dealing with $\mathscr{C}^\infty(\mathbb{R}^{E'})$ with E' finite, the **Smooth Tietze Theorem** (**Fact 2.3**) gives us a smooth function, $\chi_{H'} \in \mathscr{C}^\infty(\mathbb{R}^{E'})$ such that $H' = Z(\chi_{H'})$. We have $Z(h') = H' = G' \cap H' = Z(g') \cap Z(\chi_{H'}) = Z(g'.\chi_{H'})$ and, since $I' = \mu^{\dashv}_{E'E}[I]$ is an ideal, $g'.\chi_{H'} \in I'$.

Since $H = \pi^{\dashv}_{EE'}[H']$, $H' = Z(g'.\chi_{H'})$, $g'.\chi_{H'} \in I'$ and

$$\widehat{I} = \{F \subseteq \mathbb{R}^E \mid (\exists E' \subseteq_{\text{fin}} E)(\exists f' \in I' = \mu^{\dashv}_{E'}[I])(F = \pi^{\dashv}_{EE'}[Z(f)])\} \subseteq \wp(\mathscr{Z}(\mathbb{R}^E)),$$

we have $H \in \widehat{I}$.

□

Proposition 2.15. *Let E be any set. If $\Phi \subseteq \wp(\mathscr{Z}(\mathbb{R}^E))$ is a filter of zerosets in $\mathbb{R}^E$, then:*

$$\breve{\Phi} := \{f \in \mathscr{C}^\infty(\mathbb{R}^E) \mid Z(f) \in \Phi\} \subseteq \mathscr{C}^\infty(\mathbb{R}^E)$$

is an ideal of $\mathscr{C}^\infty(\mathbb{R}^E)$.

Proof. Note first that

$$\check{\Phi} = \{f \in \mathscr{C}^\infty(\mathbb{R}^E) \mid (\exists E' \subseteq_{\text{fin}} E)(\exists f' \in \mathscr{C}^\infty(\mathbb{R}^{E'}))((\mu_{E'E}(f') = f)\&$$
$$\&(\pi^{\dashv}_{EE'}[Z(f')] \in \Phi))\} \subseteq \mathscr{C}^\infty(\mathbb{R}^E).$$

It is easy to see that $0_E \in \check{\Phi}$. In fact, $Z(0_E) = \mathbb{R}^E \in \Phi$.

Given $f \in \check{\Phi} \subseteq \mathscr{C}^\infty(\mathbb{R}^E)$ and $h \in \mathscr{C}^\infty(\mathbb{R}^E)$. Select $E' \subseteq_{\text{fin}} E$ and $f', h' \in \mathscr{C}^\infty(\mathbb{R}^{E'})$ such that $\mu_{E'E}(f') = f, \mu_{E'E}(h') = h$. Then $h \cdot f = \mu_{E'E}(h' \cdot f') \in \mathscr{C}^\infty(\mathbb{R}^E)$ and $Z(h.f) = Z(h) \cup Z(f) \supseteq Z(f) \in \Phi$. Thus $h \cdot f \in \check{\Phi}$.

Let $f, g \in \check{\Phi}$. Select $E' \subseteq_{\text{fin}} E$ and $f', g' \in \mathscr{C}^\infty(\mathbb{R}^{E'})$ such that $\mu_{E'E}(f') = f, \mu_{E'E}(g') = g$. Thus $\pi^{\dashv}_{EE'}[Z(f')], \pi^{\dashv}_{EE'}[Z(g')] \in \Phi$ and $f + g = \mu_{E'E}(f' + g') \in \mathscr{C}^\infty(\mathbb{R}^E)$. Since $Z(f + g) \supseteq Z(f) \cap Z(g) \in \Phi$, we obtain $f + g \in \check{\Phi}$.

□

Proposition 2.16. *Consider the partially ordered sets:*

$$\mathfrak{F} = (\{\Phi \subseteq \wp(\mathscr{Z}(\mathbb{R}^E)) \mid \Phi \textit{ is a filter}\}, \subseteq)$$

and

$$\mathfrak{I} = (\{I \subseteq \mathscr{C}^\infty(\mathbb{R}^E) \mid I \textit{ is an ideal of} \mathscr{C}^\infty(\mathbb{R}^E)\}, \subseteq)$$

The following functions:

$$\begin{array}{rccc} \vee : & \mathfrak{F} & \to & \mathfrak{I} \\ & \Phi & \mapsto & \check{\Phi} \end{array}$$

$$\begin{array}{rccc} \wedge : & \mathfrak{I} & \to & \mathfrak{F} \\ & I & \mapsto & \hat{I} \end{array}$$

form a covariant *Galois connection,* $\wedge \dashv \vee$ *, that is:*

(a) *Given $\Phi_1, \Phi_2 \in \mathfrak{F}$ such that $\Phi_1 \subseteq \Phi_2$, then $\breve{\Phi_1} \subseteq \breve{\Phi_2}$;*

(b) *Given $I_1, I_2 \in \mathfrak{I}$ such that $I_1 \subseteq I_2$ then $\hat{I_1} \subseteq \hat{I_2}$;*

(c) *For every $\Phi \in \mathfrak{F}$ and every $I \in \mathfrak{I}$ we have:*

$$\hat{I} \subseteq \Phi \iff I \subseteq \check{\Phi}$$

Moreover, the mappings $(\vee, \wedge)$ establish a correspondence between:
(1) $\wp(\mathscr{Z}(\mathbb{R}^E))$ and $\mathscr{C}^\infty(\mathbb{R}^E)$;
(2) Proper filters of $(\mathfrak{F}, \subseteq)$ and proper ideals of $(\mathfrak{I}, \subseteq)$.

Proof. Items (a), (b), (c) follows directly from the definitions.

Suppose that $\Phi = \wp(\mathscr{Z}(\mathbb{R}^E))$. Then $\check{\Phi} = \{f \in \mathscr{C}^\infty(\mathbb{R}^E)) \mid f \in \wp(\mathscr{Z}(\mathbb{R}^E))\}$, thus $\check{\Phi} = \mathscr{C}^\infty(\mathbb{R}^E)$.

Suppose that $I = \mathscr{C}^\infty(\mathbb{R}^E)$. Then $\hat{I} = \{Z(f) \in \wp(\mathscr{Z}(\mathbb{R}^E)) \mid f \in I\}$, thus $\hat{I} = \wp(\mathscr{Z}(\mathbb{R}^E))$.

Suppose that Φ is a proper filter. If $f \in \mathscr{C}^\infty(\mathbb{R}^E)$ is such that $Z(f) = \varnothing \notin \Phi$, then $f \in \mathscr{C}^\infty(\mathbb{R}^E)^\times$ and $f \notin \check{\Phi} \subseteq \mathscr{C}^\infty(\mathbb{R}^E)$. Thus $\check{\Phi} \subseteq \mathscr{C}^\infty(\mathbb{R}^E)$ is a proper ideal.

Suppose that I is a proper ideal. So $f \notin I$ whenever $f \in \mathscr{C}^\infty(\mathbb{R}^E)^\times$, i.e. whenever $Z(f) = \varnothing$. Thus $\varnothing \notin \hat{I}$, i. e. $\hat{I}$ is a proper filter.

□

Remark 2.17. *As in any (covariant) Galois connection, we have automatically:*

- $I \subseteq \check{\hat{I}}$; $\Phi \supseteq \hat{\check{\Phi}}$
- $\hat{I} = \hat{\check{\hat{I}}}$; $\check{\Phi} = \check{\hat{\check{\Phi}}}$

The following result gives a more detailed information on these compositions.

Proposition 2.18. *Let $I \subseteq \mathscr{C}^\infty(\mathbb{R}^E)$ be any ideal and $\Phi \subseteq \wp(\mathscr{Z}(\mathbb{R}^E))$ be a filter of zerosets. Then:*

1. $\hat{\check{\Phi}} = \{X \subseteq (\mathbb{R}^E) \mid \exists f \in \mathscr{C}^\infty(\mathbb{R}^E)(X = Z(f), Z(f) \in \Phi)\} = \Phi.$
2. $\check{\hat{I}} = \{g \in \mathscr{C}^\infty(\mathbb{R}^E) \mid \exists f \in \mathscr{C}^\infty(\mathbb{R}^E)(f \in I, Z(g) = Z(f))\} = \sqrt[\infty]{I}.$

Proof. Item (1) and the first equality in item (2) follow directly from the definitions. We will show that

$$\{g \in \mathscr{C}^\infty(\mathbb{R}^E) \mid (\exists f \in \mathscr{C}^\infty(\mathbb{R}^E))(f \in I, Z(g) = Z(f))\} = \sqrt[\infty]{I}$$

Note that: $\sqrt[\infty]{I} = \{g \in \mathscr{C}^\infty(\mathbb{R}^E) \mid (\exists f \in I)((\eta_g(f) \in \mathscr{C}^\infty(\mathbb{R}^E)\{g^{-1}\}^\times)\} =$ $\{g \in \mathscr{C}^\infty(\mathbb{R}^E) \mid (\exists E' \subseteq_{\text{fin}} E)(\exists \widetilde{g} \in \mathscr{C}^\infty(\mathbb{R}^{E'}))(\exists \widetilde{f} \in \mu_{E'E}^{\dashv}[I])(g = \widetilde{g} \circ \pi_{EE'})\&(\eta_{\widetilde{g}}(\widetilde{f}) \in \mathscr{C}^\infty(\mathbb{R}^{E'})\{\widetilde{g}^{-1}\}^\times)\}$

Given $g \in \check{\hat{I}}$, there is some finite $E' \subseteq E$, some $\widetilde{g} \in \mathscr{C}^\infty(\mathbb{R}^{E'})$ with $g = \widetilde{g} \circ \pi_{EE'}$ and some $\widetilde{f} \in \mu_{E'E}^{\dashv}[I]$ such that $\pi_{EE'}^{\dashv}[Z(\widetilde{g})] = Z(g) = Z(f) = \pi_{EE'}^{\dashv}[Z(\widetilde{f})]$. Since $\pi_{EE'} : \mathbb{R}^E \to \mathbb{R}^{E'}$ is surjective, we have $Z(\widetilde{g}) = \pi_{EE'}[\pi_{EE'}^{\dashv}[Z(\widetilde{g})]] = \pi_{EE'}[\pi_{EE'}^{\dashv}[Z(\widetilde{f})]] = Z(\widetilde{f})$. It follows that $\widetilde{f} \restriction_{\mathbb{R}^{E'} \setminus Z(\widetilde{g})} \in \mathscr{C}^\infty(\mathbb{R}^{E'} \setminus Z(\widetilde{g}))^\times$ and, by **Example 1.6**, $\eta_{\widetilde{g}}(\widetilde{f}) \in \mathscr{C}^\infty(\mathbb{R}^{E'})\{\widetilde{g}^{-1}\}^\times$. Since there is $\widetilde{f} \in \mu_{E'E}^{\dashv}[I]$ such that $\eta_{\widetilde{g}}(\widetilde{f}) \in \mathscr{C}^\infty(\mathbb{R}^{E'})\{\widetilde{g}^{-1}\}^\times$, it follows that $g \in \sqrt[\infty]{I}$.

Conversely, given $g \in \sqrt[\infty]{I}$, there is some finite $E' \subseteq_{\text{fin}} E$, some $\widetilde{g} \in \mathscr{C}^\infty(\mathbb{R}^{E'})$ and some $\widetilde{f} \in \mu_{E'E}^{\dashv}[I]$ such that $g = \widetilde{g} \circ \pi_{E'}$ and $\eta_{\widetilde{g}}(\widetilde{f}) \in \mathscr{C}^\infty(\mathbb{R}^{E'})\{\widetilde{g}^{-1}\}^\times$. So $\widetilde{f} \restriction_{\mathbb{R}^{E'} \setminus Z(\widetilde{g})} \in \mathscr{C}^\infty(\mathbb{R}^{E'} \setminus Z(\widetilde{g}))^\times$, $Z(\widetilde{f}) \subseteq Z(\widetilde{g})$ and $\pi_{EE'}^{\dashv}[Z(\widetilde{f})] \subseteq \pi_{EE'}^{\dashv}[Z(\widetilde{g})]$. Since $\pi_{EE'}^{\dashv}[Z(\widetilde{f})] \in \hat{I}$ and $\hat{I}$ is a filter, we have $Z(g) = \pi_{EE'}^{\dashv}[Z(\widetilde{g})] \in \hat{I}$, so $g \in \check{\hat{I}}$. □

Remark 2.19. *The item (2) in the previous proposition ensures that the $\mathscr{C}^\infty-$radical of any ideal of a $\mathscr{C}^\infty-$ring is an ideal.*

Proposition 2.20. *Let E be any set, and consider $A = \mathscr{C}^\infty(\mathbb{R}^E)$. The Galois connection $\wedge \dashv \vee$ establishes bijective correspondences between:*

(a) *The poset of all (proper) filters of zerosets of $\mathbb{R}^E$ and the poset of all (proper) $\mathscr{C}^\infty-$radical ideals of $\mathscr{C}^\infty(\mathbb{R}^E)$, $\mathfrak{I}^\infty = \{I \in \mathfrak{I} \mid \sqrt[\infty]{I} = I\}$;*

(b) *The set of all maximal filters of $(\mathfrak{F}, \subseteq)$ and the set of all maximal ideals of $(\mathfrak{I}, \subseteq)$;*

(c) *The poset of all prime (proper) filters of $(\mathfrak{F}, \subseteq)$ and the poset of all prime (proper) $\mathscr{C}^\infty$-radical ideals of $(\mathfrak{I}, \subseteq)$.*

Proof. We saw in **Proposition 2.16**, that the functions $(\vee, \wedge)$ restricts to maps between proper filters of zerosets of $\mathbb{R}^E$ and proper ideals of $\mathscr{C}^\infty(\mathbb{R}^E)$. Thus the additional parts in items (a) and (c) are automatic.

Ad (a): Let Φ be a filter of zerosets in $\mathbb{R}^E$, then by **Proposition 2.18.(1)** $\Phi = \hat{\check{\Phi}}$. Let I be a $\mathscr{C}^\infty$-radical ideal in $\mathscr{C}^\infty(\mathbb{R}^E)$, then by **Proposition 2.18.(2)** and **Remark 2.17**,

$$\check{\hat{\check{\hat{I}}}} = \check{\hat{I}} = \sqrt[\infty]{I} = I$$

Thus, since $\hat{I} = \hat{\check{\hat{I}}}$ and $\check{\Phi} = \check{\hat{\check{\Phi}}}$, the (increasing) mappings $(\vee, \wedge)$ establish a bijective correspondence between the poset of all filters of zerosets of $\mathbb{R}^E$ and the poset of all $\mathscr{C}^\infty-$radical ideals of $\mathscr{C}^\infty(\mathbb{R}^E)$.

Ad (b): First of all, note that, by a combination of previous results, if I is a proper ideal of $\mathscr{C}^\infty(\mathbb{R}^E)$, then $\sqrt[\infty]{I}$ is also a proper ideal of $\mathscr{C}^\infty(\mathbb{R}^E)$. Thus if I is a (proper) maximal ideal of $\mathscr{C}^\infty(\mathbb{R}^E)$, then $I = \sqrt[\infty]{I}$.

Now, by item (a), the increasing mappings $(\vee, \wedge)$ establishes a bijective correspondence between the poset of all proper filters of zerosets of $\mathbb{R}^E$ and the poset of all proper $\mathscr{C}^\infty-$radical ideals of $\mathscr{C}^\infty(\mathbb{R}^E)$. Thus the mappings $(\vee, \wedge)$ restrict to a pair of inverse bijective correspondence between the set of all maximal filters of zerosets of $\mathbb{R}^E$ and the set of all maximal ideals of $\mathscr{C}^\infty(\mathbb{R}^E)$.

Ad (c): By the bijective correspondence in item (a), it is enough to show that the mappings $(\vee, \wedge)$ restricts to a pair of mappings between the set of all prime filters of zerosets of $\mathbb{R}^E$ and the set of all $\mathscr{C}^\infty$-radical prime ideals of $\mathscr{C}^\infty(\mathbb{R}^E)$.

Let Φ be a prime filter of zerosets of $\mathbb{R}^E$. If $f, g \in \mathscr{C}^\infty(\mathbb{R}^E)$ are such that $f \cdot g \in \check{\Phi}$, then $Z(f \cdot g) = Z(f) \cup Z(g) \in \Phi$, so we have $Z(f) \in \Phi$ or $Z(g) \in \Phi$. Thus, $f \in \check{\Phi}$ or $g \in \check{\Phi}$, so $\check{\Phi}$ is a prime ideal of $\mathscr{C}^\infty(\mathbb{R}^E)$; moreover, by item (a), $\check{\Phi}$ is $\mathscr{C}^\infty$-radical.

Let I be a $\mathscr{C}^\infty$-radical prime ideal of $\mathscr{C}^\infty(\mathbb{R}^E)$, that is, if $f, g \in \mathscr{C}^\infty(\mathbb{R}^E)$ are such that $f \cdot g \in I$ then $f \in I$ or $g \in I$. We need to show that $\widehat{I} = \{Z(h) | h \in I\}$ is a prime filter of zerosets of $\mathbb{R}^E$.

Let $F = Z(f), G = Z(g), H = Z(h)$ be zerosets of $\mathbb{R}^E$ such that $F \cup G = H \in \widehat{I}$, $h \in I$. Select $E' \subseteq_{\text{fin}} E$ and $f', g', h' \in \mathscr{C}^\infty(\mathbb{R}^{E'})$ such that $f = \mu_{E'E}(f'), g = \mu_{E'E}(g'), h = \mu_{E'E}(h')$. Let $F' = Z(f'), G' = Z(g'), H' = Z(h') \subseteq \mathbb{R}^{E'}$ then $F' \cup G' = H' \in \widehat{I'}$, where $h' \in I' := \mu_{E'E}^{\dashv}[I]$. Since I is a $\mathscr{C}^\infty$-radical prime ideal of $\mathscr{C}^\infty(\mathbb{R}^E)$, then I' is a $\mathscr{C}^\infty$-radical prime ideal of $\mathscr{C}^\infty(\mathbb{R}^{E'})$, see **Proposition 1.15.(d)**. *If* we show that $\widehat{I'}$ is a prime filter of zerosets of $\mathbb{R}^{E'}$ then, we may assume w.l.o.g. that $Z(f') = F' \in \widehat{I'}$ and $f' \in \sqrt[\infty]{I'} = I' = \mu_{E'E}^{\dashv}[I]$; thus $f = \mu_{E'E}(f') \in I$ and $F = Z(f) \in \widehat{I}$, finishing the proof.

We will prove that $\widehat{I'}$ is a prime filter. We have $Z(f'.g') = Z(f') \cup Z(g') = F' \cup G' = H' = Z(h')$, where $h' \in I'$. Then, $f'.g' \in \sqrt[\infty]{I'} = I'$. Since I' is a prime ideal, $f' \in I'$ or $g' \in I'$. Thus $F' = Z(f') \in \widehat{I'}$ or $G' = Z(g') \in \widehat{I'}$.

□

3 The Order Theory of $\mathscr{C}^\infty$-Reduced $\mathscr{C}^\infty$-Rings

The results established in the previous section are fundamental to develop an order theory over a broader class of $\mathscr{C}^\infty$-rings. In fact, in order to get nice results, we need to assume some technical conditions: the $\mathscr{C}^\infty$-rings must be non-trivial (i.e. $0 \neq 1$) and $\mathscr{C}^\infty$- **reduced** (see **Section 1**).

The fundamental notion here is the following (see p. 328 of [16]):

Definition 3.1. *Let A be a $\mathscr{C}^\infty$–ring. The* canonical relation *on A is*

$$\prec_A = \{(a, b) \in A \times A \mid (\exists u \in A^\times)(b - a = u^2)\}$$

Remark 3.2. *Note that the canonical relation is preserved by $\mathscr{C}^\infty$-homomorphism. In more detail: let A, A' be $\mathscr{C}^\infty$-rings and $h : A \to A'$ be a $\mathscr{C}^\infty$-homomorphism. Then for each $a, b \in A$:*

$$a \prec_A b \Rightarrow h(a) \prec_{A'} h(b).$$

Proposition 3.3. *Let A be any $\mathscr{C}^\infty-$ring. The canonical relation on A, $\prec_A$, is compatible with the sum and with the product of A.*

Proof. Let $a, b \in A$ such that $a \prec b$ and let $c \in A^\times$ such that $(b - a = c^2)$.

Given any $x \in A$, we have:

$$(b + x) - (a + x) = b - a = c^2,$$

thus $a + x \prec b + x$.

Given $x \in A$ such that $0 \prec x$, one has $(\exists d \in A^\times)(x = d^2)$. We have, thus:

$$b \cdot x - a \cdot x = (b - a) \cdot x = c^2 \cdot d^2 = (c \cdot d)^2.$$

Since both c amd d are invertible, it follows that $c \cdot d$ is invertible, and $a \cdot x \prec b \cdot x$. □

Proposition 3.4. *If A is a non trivial $\mathscr{C}^\infty$-ring, then $\prec$, defined as above, is irreflexive, that is,*

$$(\forall a \in A)(\neg(a \prec a)).$$

Proof. Suppose there is some $a_0 \in A$ such that $a_0 \prec a_0$. By definition, this happens if, and only if there is some $c \in A^\times$ such that $0 = a_0 - a_0 = c^2$, so 0 would be invertible and $0 = 1$. □

In order to obtain further information about the canonical relation $\prec_A$, we need to pass to its specific "spatial" characterizations, by the aid of the results developed in the previous section. We start this enterprise by the following:

Proposition 3.5. *Let $A = \frac{\mathscr{C}^\infty(\mathbb{R}^n)}{I}$ be a finitely generated $\underline{\mathscr{C}^\infty-\text{reduced}}$ $\mathscr{C}^\infty-$ring. Then $\prec$ equals:*

$$\left\{(f + I, g + I) \in \frac{\mathscr{C}^\infty(\mathbb{R}^n)}{I} \times \frac{\mathscr{C}^\infty(\mathbb{R}^n)}{I} \mid (\exists h \in \left(\frac{\mathscr{C}^\infty(\mathbb{R}^n)}{I}\right)^\times)(g - f + I = h^2 + I)\right\}$$

$$(f + I \prec g + I) \iff ((\exists \varphi \in I)(\forall x \in Z(\varphi))(f(x) < g(x))).$$

Proof. Despite this is a direct application of **Corollary 2.6**, we register here a detailed proof, since it stresses the centrality of the $\mathscr{C}^\infty-$reducedness hypothesis.

Suppose $f + I \prec g + I$, so there is some $h + I \in \left(\frac{\mathscr{C}^\infty(\mathbb{R}^n)}{I}\right)^\times$ such that $g - f + I = h^2 + I$. Since $h + I$ is invertible, by **Proposition 2.5** there is some $\psi \in I$ such that:

$$(\forall x \in Z(\psi))(h(x) \neq 0)$$

Since $g - f + I = h^2 + I$, by **Proposition 2.4**, there is some $\phi \in I$ such that:

$$(\forall x \in Z(\phi))(g(x) - f(x) = h^2(x)),$$

that is, such that:

$$(\forall x \in Z(\phi))(g(x) = f(x) + h^2(x))$$

Taking $\varphi = \phi^2 + \psi^2 \in I$ we have, for every $x \in Z(\psi) \cap Z(\phi) = Z(\varphi)$ both:

$$g(x) - f(x) = h^2(x)$$

and

$$h^2(x) > 0$$

Hence,

$$(\forall x \in Z(\varphi))(f(x) < g(x))$$

Conversely, suppose $f, g \in \mathscr{C}^\infty(\mathbb{R}^n)$ are such that there is some $\varphi \in I$ with satisfying:

$$(\forall x \in Z(\varphi))(f(x) < g(x)).$$

Since f and g are continuous functions, there is an open subset $U \subseteq \mathbb{R}^n$ such that $Z(\varphi) \subseteq U$ and

$$(\forall x \in U)(f(x) < g(x)).$$

The $\mathscr{C}^\infty$-function:

$$\begin{array}{rccc} m: & \mathbb{R}^n & \to & \mathbb{R} \\ & x & \mapsto & g(x) - f(x) \end{array}$$

is such that $(\forall x \in U)(m(x) > 0)$. Thus $m \upharpoonright_{Z(\varphi)}$ is smooth and $(\forall x \in Z(\varphi))(m(x) > 0)$, so by **Fact 2.3** there is a smooth function $\widetilde{m} \in \mathscr{C}^\infty(\mathbb{R}^n)$ such that $\widetilde{m} \upharpoonright_{Z(\varphi)} = m \upharpoonright_{Z(\varphi)}$ and $(\forall x \in \mathbb{R}^n)(\widetilde{m}(x) > 0)$.

Now the function $h := \sqrt{\widetilde{m}} : \mathbb{R}^n \to \mathbb{R}$ is a smooth function and since $h \in (\mathscr{C}^\infty(\mathbb{R}^n))^\times$, by **Proposition 2.5**,

Since $I = \sqrt[\infty]{I}$, by **Proposition 2.4** it follows that $(g - f) + I = h^2 + I$ with $h + I$ invertible. Thus, $f + I < g + I$. □

Proposition 3.6. *Given any $\mathscr{C}^\infty-$reduced $\mathscr{C}^\infty-$ring A. Then there is a directed system of its finitely generated $\mathscr{C}^\infty-$rings $(A_i, \alpha_{ij} : A_i \to A_j)_{\substack{i \in I \\ i \leqslant j}}$ such that:*

1. *$A \cong \varinjlim_{i \in I} A_i$;*
2. *For each i, j, if $i \leqslant j$ then $\alpha_{ij} : A_i \to A_j$ and $\alpha_i : A_i \to A$ are injective;*
3. *For each $i \in I$, A_i is a $\mathscr{C}^\infty$-reduced $\mathscr{C}^\infty-$ring;*
4. *For each $a, b \in A$, $a \prec_A b$ iff $\exists i \in I, \exists a_i, b_i \in A_i(\alpha_i(a_i) = a, \alpha_i(b_i) = b$ and $a_i \prec_{A_i} b_i)$.*

Proof. Note that any $\mathscr{C}^\infty$-reduced $\mathscr{C}^\infty$-ring can be presented as $A \cong \mathscr{C}^\infty(\mathbb{R}^E)/I$, where $I = \sqrt[\infty]{I}$ and the latter can be described as a directed colimit of finitely generated $\mathscr{C}^\infty$-reduced $\mathscr{C}^\infty$-rings. Indeed, we have that

$$\mathscr{C}^\infty(\mathbb{R}^E)/I \cong (\varinjlim_{E' \subseteq_{\text{fin}} E} \mathscr{C}^\infty(\mathbb{R}^{E'}))/I \cong \varinjlim_{E' \subseteq_{\text{fin}} E} (\mathscr{C}^\infty(\mathbb{R}^{E'})/\mu^{\dashv}_{E',E}[I]).$$

It is clear that $\alpha_{E'} : \mathscr{C}^\infty(\mathbb{R}^{E'})/\mu^{\dashv}_{E',E}[I] \to \mathscr{C}^\infty(\mathbb{R}^E)/I$ is injective, for each $E' \subseteq_{\text{fin}} E$. Thus if $E'' \subseteq E' \subseteq_{\text{fin}} E$, then $\alpha_{E''E'} : \mathscr{C}^\infty(\mathbb{R}^{E''})/\mu^{\dashv}_{E'',E}[I] \to \mathscr{C}^\infty(\mathbb{R}^{E'})/\mu^{\dashv}_{E',E}[I]$ is injective too.

We combine the results in **Proposition** 1.14.(a) and **Proposition** 1.15.(d) to conclude that $\mathscr{C}^\infty(\mathbb{R}^{E'})/\mu^{\dashv}_{E',E}[I]$ is a $\mathscr{C}^\infty$-reduced $\mathscr{C}^\infty$-ring.

Now item (4) follows directly from the definition of canonical relation, since for each $f, g \in \mathscr{C}^\infty(\mathbb{R}^E)$:

$$\exists u \in \mathscr{C}^\infty(\mathbb{R}^E)((g - f) + I = u^2 + I; u + I \in (\mathscr{C}^\infty(\mathbb{R}^E)/I)^\times)$$

iff $\exists E' \subseteq_{\text{fin}} E$, $\exists f', g', u' \in \mathscr{C}^\infty(\mathbb{R}^{E'}), \mu_{E',E}(f') = f, \mu_{E',E}(g') = g, \mu_{E',E}(u') = u$ such that:

$$(g' - f') + \mu^{\dashv}_{E',E}[I] = u'^2 + \mu^{\dashv}_{E',E}[I] \text{ and } u' + \mu^{\dashv}_{E',E}[I] \in (\mathscr{C}^\infty(\mathbb{R}^{E'})/\mu^{\dashv}_{E',E}[I])^\times).$$

□

We are ready to state and proof the following (very useful) general characterization result of $\prec$:

Theorem 3.7. *Let $A = \frac{\mathscr{C}^\infty(\mathbb{R}^E)}{I}$ be a "general" $\mathscr{C}^\infty-$reduced $\mathscr{C}^\infty-$ring. We have: $(f + I \prec g + I) \iff ((\exists \varphi \in I)(\forall x \in Z(\varphi))(f(x) < g(x)))$.*

Proof. Suppose $f + I \prec g + I$, so there is some $h + I \in \left(\frac{\mathscr{C}^\infty(\mathbb{R}^n)}{I}\right)^\times$ such that $g - f + I = h^2 + I$. Since $h + I$ is invertible, by **Proposition 2.13** there is some $\psi \in I$ such that:

$$(\forall x \in Z(\psi))(h(x) \neq 0).$$

Since $g - f + I = h^2 + I$, by **Proposition 2.12**, there is some $\phi \in I$ such that:

$$(\forall x \in Z(\phi))(g(x) - f(x) = h^2(x)),$$

Taking $\varphi = \phi^2 + \psi^2 \in I$ we have, for every $x \in Z(\psi) \cap Z(\phi) = Z(\varphi)$ both $g(x) - f(x) = h^2(x)$ and $h^2(x) > 0$. Thus,

$$(\forall x \in Z(\varphi))(f(x) < g(x)).$$

Conversely, suppose $f, g \in \mathscr{C}^\infty(\mathbb{R}^E)$ are such that there is some $\varphi \in I$ with satisfying:

$$(\forall x \in Z(\varphi))(f(x) < g(x)).$$

Pick $E' \subseteq_{\text{fin}} E$ and $f', g', \varphi' \in \mathscr{C}^\infty(\mathbb{R}^{E'})$ such that $f = \mu_{E'E}(f'), g = \mu_{E'E}(g')$, $\varphi = \mu_{E'E}(\varphi')$.

Then $I' := \mu_{E'E}^{\dashv}[I]$ is a $\mathscr{C}^\infty$-radical ideal of $\mathscr{C}^\infty(\mathbb{R}^{E'})$, $\varphi' \in I'$ and

$$\forall x' \in \mathbb{R}^{E'}(x' \in Z(\varphi') \to f'(x') < g'(x')).$$

By the finitely generated case (i.e. **Proposition 3.5**),

$$f' + I' \prec g' + I'.$$

By (the proof of) **Proposition 3.6.(4)** we obtain

$$f + I \prec g + I.$$

□

Now, having available a characterization of the canonical relation $\prec$, we can establish many of its properties.

Proposition 3.8. *Let A be any $\mathscr{C}^\infty$-reduced $\mathscr{C}^\infty$–ring. The canonical relation $\prec_A$ is transitive.*

Proof. The $\mathscr{C}^\infty$-reduced $\mathscr{C}^\infty$-ring A can be presented as $A \cong \mathscr{C}^\infty(\mathbb{R}^E)/I$, for some set E and some $\mathscr{C}^\infty$-radical ideal $I = \sqrt[\infty]{I} \subseteq \mathscr{C}^\infty(\mathbb{R}^E)$. Thus let $f, g, h \in \mathscr{C}^\infty(\mathbb{R}^E)$ such that $f + I \prec g + I \prec h + I$.

Now apply the characterization **Theorem 3.7** and consider $\alpha, \beta \in I$ such that $f(x) < g(x), \forall x \in Z(\alpha)$ and $g(x) < h(x), \forall x \in Z(\beta)$.

Then $\gamma := \alpha^2 + \beta^2 \in I$ and $Z(\gamma) = Z(\alpha) \cap Z(\beta)$.

Thus $f(x) < g(x) < h(x), \forall x \in Z(\gamma)$ and since $\gamma \in I$, applying again the **Theorem 3.7** we obtain $f + I \prec h + I$. □

Proposition 3.9. *Let A be any non-trivial $\mathscr{C}^\infty$-ring and $a, b \in A$. Then the relation $\prec_A$ is asymmetric, i.e. it holds at most one of the following conditions: $a \prec b$, $b \prec a$.*

Proof. Suppose that holds simultaneously both the conditions: $a \prec b$, $b \prec a$. Since $\prec$ is transitive (**Proposition 3.8**) we have $a \prec a$, but this contradicts **Proposition 3.4**, since A is non-trivial.

□

By a combination of **Propositions 3.4, 3.8** (and **3.9**), the canonical relation $\prec$ on every $\mathscr{C}^\infty$-ring A that is non-trivial and $\mathscr{C}^\infty$-reduced is is irreflexive, transitive (and asymmetric) bynary relation on A : thus it defines a strict partial order. This motivates the following:

Definition 3.10. *Let A be a non-trivial, $\mathscr{C}^\infty$-reduced $\mathscr{C}^\infty-$ring. Then the canonical bynary relation on A, $\prec_A$, (***Definition 3.1***) will be called the "canonical strict partial order on A".*

Moreover, by **Proposition 3.3**, it holds:

Theorem 3.11. *Let A be any non-trivial $\mathscr{C}^\infty-$reduced $\mathscr{C}^\infty-$ring. The canonical partial order on A, $\prec_A$, is compatible with the sum and with the product of A.*

Note that, due to the above result, to prove the trichotomy of $\prec$ it suffices to prove that holds the "restricted form of trichotomy": given any $a \in A$ one has either $a = 0$, $a \prec 0$ or $0 \prec a$. But, clearly, this is not true in general:

Example 3.12. *Let $A = \mathscr{C}^\infty(\mathbb{R}^n)$ and consider the $\mathscr{C}^\infty$-function:*

$$f(x_1, \cdots, x_n) := e^{(x_1 + \cdots + x_n)} - 1.$$

If $x_1 + \cdots + x_n > 0$, then $f(x_1, \cdots, x_n) > 0$; if $x_1 + \cdots + x_n = 0$, then $f(x_1, \cdots, x_n) = 0$ and if $x_1 + \cdots + x_n < 0$, then $f(x_1, \cdots, x_n) < 0$. Thus the assertion ($f \prec 0$ or $f = 0$ or $0 \prec f$) is false.

On the other hand, the restricted trichotomy holds for invertible members of some classes of $\mathscr{C}^\infty$-reduced $\mathscr{C}^\infty$-ring:

Proposition 3.13. *Given a $\mathscr{C}^\infty$-reduced $\mathscr{C}^\infty$-ring, A, one has:*

$$A^\times = (A^\times)^2 \cup (-A^\times)^2$$

provided A satisfies some of the conditions below:

1. *A is a free $\mathscr{C}^\infty$-ring;*

2. *A is a $\mathscr{C}^\infty$-reduced $\mathscr{C}^\infty$-domain.*

Proof. This hols trivially if $0 = 1$. We will prove that for a non trivial $\mathscr{C}^\infty$-reduced $\mathscr{C}^\infty$-ring, A, the non obvious inclusion: $A^\times \subseteq (A^\times)^2 \cup (-A^\times)^2$ holds.

Item (1): First recall that any free $\mathscr{C}^\infty$-ring is $\mathscr{C}^\infty$-reduced (see **Example 1.12**). Let $f \in \mathscr{C}^\infty(\mathbb{R}^E)^\times$, then there is $E' \subseteq_{\text{fin}} E$ and $f' \in \mathscr{C}^\infty(\mathbb{R}^{E'})^\times$ such that $f = \mu_{E'E}(f') = f' \circ \pi_{EE'}$. Since $f' : \mathbb{R}^{E'} \to \mathbb{R}$ is continuous and $\mathbb{R}^{E'}$ is connected, then $\text{range}(f) = \text{range}(f')$ is a connected subset of $\mathbb{R}$, so it is an interval. Since $0 \notin \text{range}(f)$, then exactly one of the following alternatives holds: (i) $\text{range}(f) \subseteq]-\infty, 0[$ or (ii) $\text{range}(f) \subseteq]0, \infty[$. If (i) holds then $f \in -(\mathscr{C}^\infty(\mathbb{R}^E)^\times)^2$ and if (ii) holds then $f \in (\mathscr{C}^\infty(\mathbb{R}^E)^\times)^2$.

Item (2): We take a presentation of A as $A \cong \mathscr{C}^\infty(\mathbb{R}^E)/I$, for some set E and some (ring theoretical) ideal $I \in \text{Spec}^\infty(A)$. Let $f \in \mathscr{C}^\infty(\mathbb{R}^E)$ such that $\left(f + I \in \dfrac{\mathscr{C}^\infty(\mathbb{R}^E)}{I}^\times\right)$. By **Proposition 2.13**, there exists $\varphi \in I$ such that:

$$(\forall x \in \mathbb{R}^E)(x \in Z(\varphi) \to f(x) \neq 0).$$

Let $E' \subseteq_{\text{fin}} E$ such that $\varphi = \varphi' \circ \pi_{EE'}$ and $f = f' \circ \pi_{EE'}$, for some $\varphi', f' \in \mathscr{C}^\infty(\mathbb{R}^{E'})$. Then

$$(\forall x' \in \mathbb{R}^{E'})(x' \in Z(\varphi') \to f'(x') \neq 0).$$

Thus,

$$Z(\varphi') \subseteq [f' > 0] \cup [-f' > 0],$$

where: $[\pm f' > 0] := \{x' \in \mathbb{R}^{E'} : \pm f'(x') > 0\}$.

Note that:

$$Z(\varphi') \cap [\pm f' \geqslant 0] = Z(\varphi') \cap [\pm f' > 0]$$

Since f' is a continuous function, $[\pm f' \geqslant 0] = (\pm f')^{\dashv}[[0,\infty[]$ is a closed subset of $\mathbb{R}^{E'}$, and by **Fact 2.1**, there is some $\chi_{\pm} \in \mathscr{C}^{\infty}(\mathbb{R}^{E'})$ such that

$$Z(\chi_{\pm}) = [\pm f' \geqslant 0].$$

Thus,

$$Z(\varphi') = Z(\varphi') \cap \mathbb{R}^{E'} = Z(\varphi') \cap (Z(\chi_-) \cup Z(\chi_+)) =$$

$$= (Z(\varphi') \cap Z(\chi_-)) \cup (Z(\varphi') \cap Z(\chi_+)) = Z(\varphi'^2 + \chi_-^2) \cup Z(\varphi'^2 + \chi_+^2).$$

Since I is a $\mathscr{C}^{\infty}$-radical (proper) prime ideal of $\mathscr{C}^{\infty}(\mathbb{R}^E)$, then $I' := \mu_{E'E}^{\dashv}[I] \subseteq \mathscr{C}^{\infty}(\mathbb{R}^{E'})$ is a $\mathscr{C}^{\infty}$-radical (proper) prime ideal of $\mathscr{C}^{\infty}(\mathbb{R}^{E'})$ (see **Proposition 1.15.(d)**).

By **Proposition 2.20**, I' corresponds to a prime filter (of zero sets) $\widehat{I'}$ and since $Z(\varphi'^2 + \chi_-^2) \cup Z(\varphi'^2 + \chi_+^2) = Z(\varphi')$ and $\varphi' \in I'$, then some of the subsets $Z(\varphi'^2 + \chi_-^2), Z(\varphi'^2 + \chi_+^2)$ belong to the $\mathscr{C}^{\infty}$-radical ideal I'. By **Proposition 2.18**: $\check{\widehat{I'}} = \sqrt[\infty]{I'} = I'$, thus some of the functions $(\varphi'^2 + \chi_-^2), (\varphi'^2 + \chi_+^2)$ belongs to I'.

Now recall that:

$$Z(\varphi'^2 + \chi_{\pm}^2) = Z(\varphi') \cap Z(\chi_{\pm}) = Z(\varphi') \cap [\pm f' \geqslant 0] = Z(\varphi') \cap [\pm f' > 0]$$

and consider $\alpha_{\pm} := (\varphi'^2 + \chi_{\pm}^2) \circ \pi_{EE'} \in \mathscr{C}^{\infty}(\mathbb{R}^E)$.

Then some of the alternatives holds:

(i) $\alpha_- \in I$ and $(\forall x \in Z(\alpha_-))(-f(x) > 0)$;

(ii) $\alpha_+ \in I$ and $(\forall x \in Z(\alpha_+))(f(x) > 0)$.

Applying **Theorem 3.7**, if (i) holds then $f + I \in -((\mathscr{C}^{\infty}(\mathbb{R}^E)/I)^{\times})^2$ and if (ii) holds then $f + I \in ((\mathscr{C}^{\infty}(\mathbb{R}^E)/I)^{\times})^2$.

This establishes the desired inclusion $A^{\times} \subseteq (A^{\times})^2 \dot{\cup} (-A^{\times})^2$.

□

There is another natural way to consider that a partial order $\preceq$ (where $a \preceq b$ iff $a \prec b$ or $a = b$) is compatible with sums: if $0 \preceq x$ and $0 \preceq y$, then $0 \preceq x + y$. This one also holds, as it follows (directly) from the results obtained below.

Proposition 3.14. *Given any $\mathscr{C}^\infty$-reduced $\mathscr{C}^\infty$-ring A, denote by $(A^\times)^2 = A^2 \cap A^\times = (A^2)^\times$. Then the following hold:*

1. $(\sum A^2) \cap A^\times = (A^\times)^2$
2. $(A^\times)^2 + \sum A^2 = (A^\times)^2$
3. $\sum (A^\times)^2 = (A^\times)^2$

Proof. **First equality:**

One easily checks that:

$$\left(\sum A^2\right) \cap A^\times \supseteq A^2 \cap A^\times,$$

so we only need to prove the opposite inclusion.

We know that $A \cong \dfrac{\mathscr{C}^\infty(\mathbb{R}^E)}{I}$ for some set E and some $\mathscr{C}^\infty$-radical ideal $\sqrt[\infty]{I} = I \subseteq \mathscr{C}^\infty(\mathbb{R}^E)$. Let $f \in \mathscr{C}^\infty(\mathbb{R}^E)$ such that $q_I(f) = f + I \in \left(\sum \dfrac{\mathscr{C}^\infty(\mathbb{R}^E)}{I}^2\right) \cap \dfrac{\mathscr{C}^\infty(\mathbb{R}^E)}{I}^\times$.

Since $q_I(f) \in \left(\dfrac{\mathscr{C}^\infty(\mathbb{R}^E)}{I}\right)^\times$, by **Proposition 2.13**, there is some $\varphi \in I$ such that:

$$(\forall x \in Z(\varphi))(f(x) \neq 0)$$

and since $q_I(f) \in \sum \dfrac{\mathscr{C}^\infty(\mathbb{R}^E)}{I}^2$, by **Proposition 2.12**, there are $f_1, \cdots, f_k \in \mathscr{C}^\infty(\mathbb{R}^E)$ and $\psi \in I$ such that:

$$(\forall x \in Z(\psi))(f(x) = f_1(x)^2 + \cdots + f_k^2(x) \geqslant 0),$$

Thus $\varphi^2 + \psi^2 \in I$ and

$$(\forall x \in Z(\varphi^2 + \psi^2) = Z(\varphi) \cap Z(\psi))(f(x) = f_1^2(x) + \cdots + f_k^2(x) > 0).$$

Applying **Theorem 3.7**, we have:

$$0 + I < f + I,$$

so $f + I = u^2 + I$, for some $u \in \left(\dfrac{\mathscr{C}^\infty(\mathbb{R}^E)}{I}\right)^\times$, establishing the equality in item (1).

Second and third equalities:

One easily checks that:

$$(A^\times)^2 + \sum A^2 \supseteq \sum (A^\times)^2 \supseteq (A^\times)^2,$$

so, to establish the items (2) and (3), we only need to prove that

$$(A^\times)^2 + \sum A^2 \subseteq (A^\times)^2.$$

Present A as $A \cong \dfrac{\mathscr{C}^\infty(\mathbb{R}^E)}{I}$ for some set E and some $\mathscr{C}^\infty$-radical ideal $\sqrt[\infty]{I} = I \subseteq \mathscr{C}^\infty(\mathbb{R}^E)$. Let $f \in \mathscr{C}^\infty(\mathbb{R}^E)$ such that $q_I(f) = f + I \in \left(\dfrac{\mathscr{C}^\infty(\mathbb{R}^E)}{I}^\times\right)^2 + \left(\sum \dfrac{\mathscr{C}^\infty(\mathbb{R}^E)}{I}^2\right)$. I.e., there are $g, h_1, \cdots, h_k \in \mathscr{C}^\infty(\mathbb{R}^E)$ such that $f + I = (g^2 + h_1^2 + \cdots h_k^2) + I$ and $g + I \in \left(\dfrac{\mathscr{C}^\infty(\mathbb{R}^E)}{I}^\times\right)$.

Applying **Proposition 2.12** and **Proposition 2.13**, we conclude that there is $\theta \in I$ such that:

$$(\forall x \in Z(\theta))(f(x) = g(x)^2 + h_1(x)^2 + \cdots + h_k^2(x) \text{ and } g(x) \neq 0)$$

Thus

$$(\forall x \in Z(\theta))(f(x) > 0).$$

Since $\theta \in I$, applying **Theorem 3.7**, we have:

$$0 + I < f + I,$$

so $f + I = u^2 + I$, for some $u \in \left(\dfrac{\mathscr{C}^\infty(\mathbb{R}^E)}{I}\right)^\times$, establishing the desired inclusion. □

From the second equality above, it follows directly the:

Corollary 3.15. *Every $\mathscr{C}^\infty$-ring A has the "weak bounded inversion property" (definition 7.1. in [7]), i.e.* $1 + \sum A^2 \subseteq A^\times$.

4 On the Order Theory of $\mathscr{C}^\infty$-Fields and Applications

In this section we present concrete examples of finitely generated $\mathscr{C}^\infty$−fields and some important facts about general $\mathscr{C}^\infty$−fields. Then we use these facts, along with the results presented in the previous section, to deepen the study of the order theory of $\mathscr{C}^\infty$-fields initiated by Moerdijk and Reyes in [16]. We address some results concerning $\mathscr{C}^\infty$-fields that will be useful in the sequel.

Now we consider two explicit situations where we have finitely generated $\mathscr{C}^\infty$−fields.

Example 4.1. *Let M be a smooth manifold. By* ***Theorem 2.3*** *of [18], $\mathscr{C}^\infty(M) \cong \frac{\mathscr{C}^\infty(\mathbb{R}^k)}{J}$ for some $k \in \mathbb{N}$ and some finitely generated ideal J. Thus, for any ideal $I \subset \mathscr{C}^\infty(M)$, $\mathscr{C}^\infty(M)/I$ is a finitely generated $\mathscr{C}^\infty$−ring.*

Note that for each $x \in M$, $\mathfrak{m}_x = \{f \in \mathscr{C}^\infty(M) \mid f(x) = 0\}$ is a maximal ideal in $\mathscr{C}^\infty(M)$.

Now suppose that M is compact. If $I \subset \mathscr{C}^\infty(M)$ is a maximal ideal of $\mathscr{C}^\infty(M)$, then $I = \mathfrak{m}_x$ for some unique $x \in M$. In fact, given any ideal $I \subseteq \mathscr{C}^\infty(M)$, one has either $I \subseteq \mathfrak{m}_x$ for some (unique) $x \in M$ or $I = \mathscr{C}^\infty(M)$.

Suppose it is not the case that there is some $x \in M$ such that $I \subseteq \mathfrak{m}_x$, i.e., $(\forall x \in M)(I \nsubseteq \mathfrak{m}_x)$. For every $x \in M$ we can find a function $f_x \in I$ such that $f_x(x) \neq 0$. Consider the open covering $\{M \backslash f_x^{\dashv}[\{0\}] \mid x \in M\}$ of M, which has a finite sub-covering, say $\{M \backslash f_{x_1}^{\dashv}[\{0\}], \cdots, M \backslash f_{x_r}^{\dashv}[\{0\}]\}$. We obtain, thus, the function $f = f_{x_1}^2 + \cdots + f_{x_r}^2 \in I$ such that $(\forall x \in M)(f(x) > 0)$. Hence, $f \in I \cap \mathscr{C}^\infty(M)^\times$ and $I = \mathscr{C}^\infty(M)$.

As for the uniqueness of $x \in M$, suppose that $I \subset \mathfrak{m}_x$ and let $y \in M$ be such that $x \neq y$. By the ***Smooth Tietze's Theorem***, *there is some $f \in \mathscr{C}^\infty(M)$ such that $f(x) = 0$ (so $f \in I$ and $f \in \mathfrak{m}_x$) and $f(y) = 1$, so $f \notin \mathfrak{m}_y$. Thus $\mathfrak{m}_x \not\subset \mathfrak{m}_y$ and $I \not\subset \mathfrak{m}_y$.*

It follows that whenever I is a maximal ideal of $\mathscr{C}^\infty(M)$ - where M is a compact manifold - there is a unique $x \in M$ such that $I \subseteq \mathfrak{m}_x \subset \mathscr{C}^\infty(M)$. Since I is a maximal ideal, then $I = \mathfrak{m}_x$.

Thus, for every maximal ideal $I \subset \mathscr{C}^\infty(M)$, $\mathscr{C}^\infty(M)/I \cong \mathbb{R}$ using the fact that the $\mathscr{C}^\infty$−homomorphism:

$$\begin{array}{rccc} \mathrm{ev}_x: & \mathscr{C}^\infty(M) & \to & \mathbb{R} \\ & f & \mapsto & f(x) \end{array}$$

is surjective and the ***Fundamental Theorem of the $\mathscr{C}^\infty$−Homomorphism*** *that:*

$$\frac{\mathscr{C}^\infty(M)}{\mathfrak{m}_x} = \frac{\mathscr{C}^\infty(M)}{\ker \mathrm{ev}_x} \cong \mathbb{R}$$

Hence, every $\mathscr{C}^\infty$−field obtained as a quotient $\mathscr{C}^\infty(M)/I$ is isomorphic to $\mathbb{R}$.

However, not every finitely generated $\mathscr{C}^\infty$−field is isomorphic to $\mathbb{R}$, as we see in the following:

Example 4.2. *Consider $\mathscr{C}^\infty(\mathbb{R})$ together with the ideal of all compactly supported functions:*

$$I = \{f \in \mathscr{C}^\infty(\mathbb{R}) \mid \text{supp}(f) \subset \mathbb{R} \text{ is compact}\}$$

Naturally the constant function 1 does not belong to I, so there is a maximal ideal $\widehat{I} \subset \mathscr{C}^\infty(\mathbb{R})$ such that $I \subseteq \widehat{I}$. Also, note that for every $x \in \mathbb{R}$, $I \not\subset \mathfrak{m}_x$. In fact, for every $x \in \mathbb{R}$, the smooth characteristic function $\chi_{]x-1,x+1[} : \mathbb{R} \to \mathbb{R}$ is a compactly supported function which does not belong to $\mathfrak{m}_x$. Since $I \subset \widehat{I}$, $\widehat{I} \neq \mathfrak{m}_x$ for every $x \in \mathbb{R}$.

Now, since $\widehat{I}$ is a maximal ideal different from $\mathfrak{m}_x$ for every $x \in \mathbb{R}$, $\mathscr{C}^\infty(\mathbb{R})/\widehat{I} \cong \mathbb{F}$ is a finitely generated $\mathscr{C}^\infty$-field that is different from $\mathbb{R}$.

An explicit description is given as follows. Let $U \subset \wp(\mathbb{N})$ be a non-principal ultrafilter and let:

$$\widehat{I} = \{f \in \mathscr{C}^\infty(\mathbb{R}) \mid \{n \in \mathbb{N} \mid f(n) = 0\} \in U\} \subset \mathscr{C}^\infty(\mathbb{R})$$

It is straightforward to check that $\widehat{I}$ is an ideal of $\mathscr{C}^\infty(\mathbb{R})$. Since U is a non principal ultrafilter, U contains all cofinite subsets of $\mathbb{N}$. Thus, given any $f \in I$ - that is, any $f \in \mathscr{C}^\infty(\mathbb{R})$ with compact support, $K = \text{supp}(f)$, since K is limited there is some $n_0 \in \mathbb{N}$ such that $K \subseteq [-n_0, n_0]$, so $(\forall n > n_0)(f(n) = 0)$. Hence $\{n \in \mathbb{N} \mid f(n) = 0\} \subset \mathbb{N}$ is cofinite and $f \in \widehat{I}$. Thus $I \subset \widehat{I}$.

Finally, in order to show that $\widehat{I}$ is a maximal ideal, we show that $\mathscr{C}^\infty(\mathbb{R})/\widehat{I}$ is a $\mathscr{C}^\infty$-field.

In fact, if $f + \widehat{I} \neq 0 + \widehat{I}$, then $f \notin \widehat{I}$ and $\{n \in \mathbb{N} \mid f(n) = 0\} \notin U$. Since U is an ultrafilter, we have $\{n \in \mathbb{N} \mid f(n) \neq 0\} \in U$.

Now, since $\mathbb{N} \subset \mathbb{R}$ is discrete, we can take, for every $n \in \mathbb{N}$ such that $f(n) \neq 0$, the open neighbourhood $]n - \frac{1}{2}, n + \frac{1}{2}[$ with the smooth characteristic function:

$$\begin{array}{rccl} \chi_{]n-\frac{1}{2},n+\frac{1}{2}[} : & \mathbb{R} & \to & \mathbb{R} \\ & x & \mapsto & \begin{cases} e^{1-\frac{1}{1-4(x-n)^2}}, & \text{if } x \in]n - \frac{1}{2}, n + \frac{1}{2}[\\ 0, & \text{otherwise} \end{cases} \end{array}$$

and then glue them up to get the smooth function:

$$\begin{array}{rccl} h : & \mathbb{R} & \to & \mathbb{R} \\ & x & \mapsto & \begin{cases} \chi_{]n-\frac{1}{2},n+\frac{1}{2}[}(x), & \text{if } x \in]n - \frac{1}{2}, n + \frac{1}{2}[\text{ and } f(n) \neq 0 \\ 0 & \text{otherwise} \end{cases} \end{array}$$

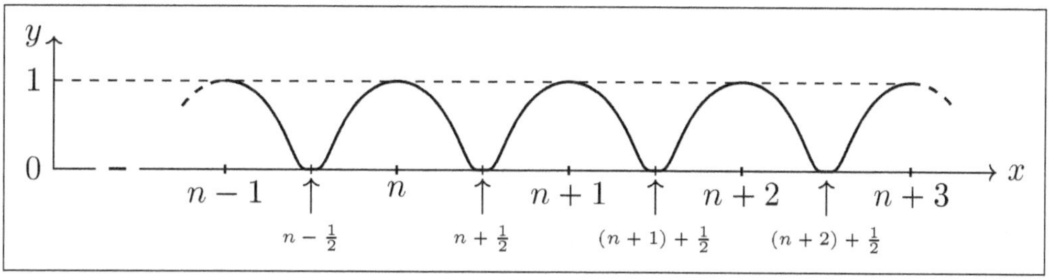

Now consider:

$$\begin{array}{rcl} g: \mathbb{R} & \to & \mathbb{R} \\ x & \mapsto & \begin{cases} \dfrac{h(x)}{f(n)}, & \textit{if } x \in]n-\frac{1}{2}, n+\frac{1}{2}[\textit{ and } f(n) \neq 0 \\ 0 \textit{ otherwise} \end{cases} \end{array}$$

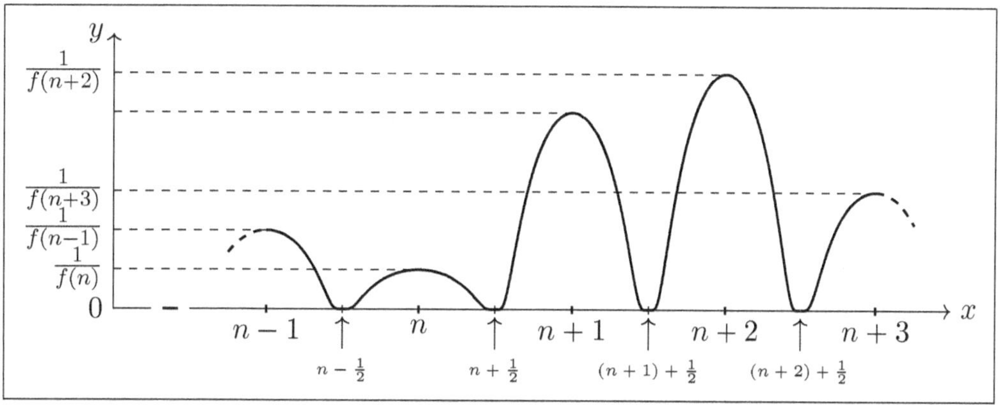

and note that $g \in \mathscr{C}^\infty(\mathbb{R})$.

Also, note that for every n *such that* $f(n) \neq 0$, *we have* $h(n) = 1$, *so* $g(n) = 1/f(n)$. *Thus, since* $\{n \in \mathbb{N} \mid f(n) \cdot g(n) - 1 = 0\} \in U \subset \wp(\mathbb{N})$ *(for* $\{n \in \mathbb{N} \mid f(n) \cdot g(n) - 1 = 0\}$ *is cofinite), it follows that:*

$$f \cdot g - 1 \in \widehat{I},$$

so $f + \widehat{I} \in (\mathscr{C}^\infty(\mathbb{R})/I)^\times$. *It follows that* $\mathscr{C}^\infty(\mathbb{R})/\widehat{I}$ *is a finitely generated* $\mathscr{C}^\infty$*–field which is not isomorphic to* $\mathbb{R}$.

Proposition 4.3. *Let* A *be a* $\mathscr{C}^\infty$*–ring.*

1. *If* A *is a* $\mathscr{C}^\infty$*–field, then* A *is a* $\mathscr{C}^\infty$*–reduced* $\mathscr{C}^\infty$*–domain (see **Proposition 4.6** of [4]).*

2. *If A is a $\mathscr{C}^\infty$−reduced $\mathscr{C}^\infty$−domain, then $A\{(A\backslash\{0\})^{-1}\}$ is a $\mathscr{C}^\infty$-field, $\eta_{A\backslash\{0\}}$: $A \rightarrowtail A\{(A\backslash\{0\})^{-1}\}$ is an injective $\mathscr{C}^\infty$-homomorphism and $\eta_{A\backslash\{0\}}$ is universal among the injective $\mathscr{C}^\infty$-homomorphisms from A into some $\mathscr{C}^\infty$-field (see* ***Proposition 4.51*** *of [4]).*

3. *A is isomorphic to a $\mathscr{C}^\infty$-subring of a $\mathscr{C}^\infty$-field if, and only if A is a $\mathscr{C}^\infty$−reduced $\mathscr{C}^\infty$−domain (see* ***Corollary 4.36*** *and* ***Proposition 4.51*** *of [4]).*

4. *For every proper prime ideal $\mathfrak{p} \subseteq A$ that is $\mathscr{C}^\infty$−radical, we have a canonical $\mathscr{C}^\infty$-field $k_\mathfrak{p}(A) := \dfrac{A}{\mathfrak{p}}\{q_\mathfrak{p}[A\backslash\mathfrak{p}]^{-1}\}$ and a canonical $\mathscr{C}^\infty$-morphism with kernel $\mathfrak{p}$, $A \overset{q_\mathfrak{p}}{\twoheadrightarrow} \dfrac{A}{\mathfrak{p}} \overset{\eta_{q_\mathfrak{p}[A\backslash\mathfrak{p}]}}{\rightarrowtail} k_\mathfrak{p}(A)$ (see p. 102 of [4]).*

Now we are ready to turn our attention to the order theory of $\mathscr{C}^\infty$-fields.

By **Corollary 3.15**, every $\mathscr{C}^\infty$-reduced $\mathscr{C}^\infty$-ring A satisfies the relation $1 + \sum A^2 \subseteq A^\times$. In particular, every $\mathscr{C}^\infty$-field A is formally real, i.e. $-1 \notin \sum A^2$, thus it can be endowed with some linear order relation compatible with its sum and product. In fact, since a $\mathscr{C}^\infty$-field is a non-trivial $\mathscr{C}^\infty$-reduced $\mathscr{C}^\infty$-ring, we have a distinguished linear order relation in A that is compatible with its sum and product:

Theorem 4.4. *Let A be a $\mathscr{C}^\infty$−field and $\prec_A$ be the canonical strict partial order on A, cf.* **Definition 3.10**. *Then $(A, \preceq)$ is a totally/linearly ordered field,* i.e., *$\preceq$ is a reflexive, transitive and anti-symmetric binary relation in A that is compatible with sum and product and, moreover, it holds the* trichotomy law, i.e., *for every $a, b \in A$ we have exactly one of the following $a = b$ or $a \prec b$ or $b \prec a$. Moreover, $0 \preceq a$ iff $a = b^2$ for some $b \in A$.*

Proof. Since $a \preceq b$ iff $a \prec b$ or $a = b$, it follows directly from **Theorem 3.11** that $\preceq$ is a reflexive, transitive and anti-symmetric binary relation in A that is compatible with sum and product.

By the compatibility of $\prec$ with the sum, to obtain the trichotomy law it is enough to show that for every $f \in A\backslash\{0\}$ we have either $(0 \prec f)$ or $(f \prec 0)$.

Since A is a $\mathscr{C}^\infty$-field, $A^\times = A\backslash\{0\}$ and the result follows directly from **Proposition 4.3.(1)** and **Proposition 3.13.(2)**: $A^\times = (A^\times)^2 \cup (-A^\times)^2$.

□

In general, a field could support many linear orders compatible with its sum and product. A field is called **Euclidean** if it has a unique (linear, compatible with $+, \cdot$) ordering: these fields are precisely the ordered fields such that every positive member has a square root in the field. It is clear from the definition of $\prec$ and by **Theorem 4.4** that every $\mathscr{C}^\infty$-field is Euclidean.

Now recall that a totally ordered field $(\mathbb{F}, \leqslant)$ is **real closed** if it satisfies the following two conditions:

(a) $(\forall x \in \mathbb{F})(0 \leqslant x \to (\exists y \in \mathbb{F})(x = y^2))$ (i.e. it is an Euclidean field);

(b) every polynomial of odd degree has, at least, one root;

Equivalently, a totally ordered field $(\mathbb{F}, \leqslant)$ is real closed if, and only if it satisfies the conclusion of **intermediate value theorem** for all *polynomial* functions $h : \mathbb{F} \to \mathbb{F}$.

As pointed out in **Theorem 2.10** of [16], it holds the following:

Fact 4.5. *Every $\mathscr{C}^\infty$–field, A, together with its canonical order $\prec$ is a real closed field.*

In fact, a stronger property holds for every $\mathscr{C}^\infty$-field. We have the $\mathscr{C}^\infty$–analog of the notion of "real closedness":

Fact 4.6 (Theorem 2.10' of [16]). *Let $(\mathbb{F}, \prec)$ be a $\mathscr{C}^\infty$–field. Then $(\mathbb{F}, \prec)$ is $\mathscr{C}^\infty$–**real closed**. I.e., it holds:*

$$(\forall f \in \mathbb{F}\{x\})((f(0) \cdot f(1) < 0)\&(1 \in \langle\{f, f'\}\rangle \subseteq \mathbb{F}\{x\}) \to$$

$$(\exists \alpha \in]0, 1[\subseteq \mathbb{F})(f(\alpha) = 0))$$

Note that the class of $\mathscr{C}^\infty$-fields is an $\mathscr{L}$-elementary (proper) class, where all structures have cardinally at least $2^{\aleph_0}$.

The notion of $\mathscr{C}^\infty$-field is also useful to analyze the order theory of a $\mathscr{C}^\infty$-reduced $\mathscr{C}^\infty$-ring:

Remark 4.7. *Consider $A = \mathscr{C}^\infty(\mathbb{R}^n)$. The inclusion $i : \mathscr{C}^\infty(\mathbb{R}^n) \hookrightarrow \mathrm{Func}(\mathbb{R}^n, \mathbb{R}) = \mathbb{R}^{\mathbb{R}^n}$ obviously preserves and reflects the equality relation $(=)$ and the canonical strict partial order $(\prec)$. Note that, by* **Example 4.1**, *this inclusion can be factored through the $\mathscr{C}^\infty$-homomorphism:*

$$A \overset{(q_\mathfrak{m})_\mathfrak{m}}{\longrightarrow} \prod_{\mathfrak{m} \in \mathrm{Max}(A)} \frac{A}{\mathfrak{m}}$$

Thus, the family of all $\mathscr{C}^\infty$-fields $\{\frac{A}{\mathfrak{m}} : \mathfrak{m} \in \mathrm{Max}(A)\}$ encodes the canonical relation $\prec_A$ on A.

The family of all $\mathscr{C}^\infty$-fields $\{k_\mathfrak{p}(A) : \mathfrak{p} \in \mathrm{Spec}^\infty(A)\}$ also encodes the canonical relation $\prec_A$ on A.

Consider the canonical $\mathscr{C}^\infty$-homomorphism $c_A : A \to \prod_{\mathfrak{p}\in\mathrm{Spec}^\infty(A)} k_{\mathfrak{p}}(A)$, given by:

$$A \stackrel{(q_{\mathfrak{p}})_{\mathfrak{p}}}{\to} \prod_{\mathfrak{p}\in\mathrm{Spec}^\infty(A)} \frac{A}{\mathfrak{p}} \stackrel{(\eta_{q_{\mathfrak{p}}[A\setminus\mathfrak{p}]})_{\mathfrak{p}}}{\rightarrowtail} \prod_{\mathfrak{p}\in\mathrm{Spec}^\infty(A)} k_{\mathfrak{p}}(A)$$

By **Proposition 4.3.(4)**, $\ker(c_A) = \bigcap_{\mathfrak{p}\in\mathrm{Spec}^\infty(A)} \mathfrak{p}$. *Thus, by* **Theorem 1.18.(e)**, $\ker(c_A) = \sqrt[\infty]{(0)} = \{0\}$ *and c_A is an injective $\mathscr{C}^\infty$-homomorphism, i.e., it preserves and* reflects *the equality relation. We will see that c_A also preserves and reflects the canonical relation $\prec$.*

The $\mathscr{C}^\infty$-homomorphism c_A preserves $\prec$ (see **Remark 3.2***). Note that, to establish that $c_A : A \to \prod_{\mathfrak{p}\in\mathrm{Spec}^\infty(A)} k_{\mathfrak{p}}(A)$ reflects $\prec$ it suffices to guarantee that:*

$$A \stackrel{(q_{\mathfrak{m}})_{\mathfrak{m}}}{\to} \prod_{\mathfrak{m}\in\mathrm{Max}(A)} \frac{A}{\mathfrak{m}}$$

reflects $\prec$. In fact, since $(\forall \mathfrak{m} \in \mathrm{Max}(A))(\frac{A}{\mathfrak{m}} \cong k_{\mathfrak{m}}(A))$, *the inclusion* $\mathrm{Max}(A) \subseteq \mathrm{Spec}^\infty(A)$ *(this holds by* **Proposition 4.3.(1)***) induces a canonical "projection":*

$$\pi_A : \prod_{\mathfrak{p}\in\mathrm{Spec}^\infty(A)} k_{\mathfrak{p}}(A) \twoheadrightarrow \prod_{\mathfrak{m}\in\mathrm{Max}(A)} \frac{A}{\mathfrak{m}}$$

and, obviously,

$$(q_{\mathfrak{m}})_{\mathfrak{m}} = \pi_A \circ c_A,$$

Thus, if $a, b \in A$ are such that $a \not\prec b$, implies $(q_{\mathfrak{m}})_{\mathfrak{m}}(a) \not\prec (q_{\mathfrak{m}})_{\mathfrak{m}}(b)$, then also holds $c_A(a) \not\prec c_A(b)$.

Now we will apply the results on $\mathscr{C}^\infty$-fields to describe another approach of the order theory of (general) $\mathscr{C}^\infty$-rings.

Definition 4.8. *Let A be an arbitrary $\mathscr{C}^\infty$-ring. Let $\mathscr{F}$ be the (proper) class of all the $\mathscr{C}^\infty$−homomorphisms of A to some $\mathscr{C}^\infty$−field, that is:*

$$\mathscr{F} = \bigcup_{\mathbb{F}\in\mathrm{Obj}(\mathscr{C}^\infty-\mathbf{Fld})} \mathrm{Hom}_{\mathscr{C}^\infty-\mathbf{Rng}}(A, \mathbb{F})$$

We define the following relation $\mathscr{R}$: given $h_1 : A \to \mathbb{F}_1$ and $h_2 : A \to \mathbb{F}_2$, we say that h_1 is related with h_2 if, and only if, there is some $\mathscr{C}^\infty$−field $\widetilde{\mathbb{F}}$ and some

$\mathscr{C}^\infty$–fields homomorphisms $\mathscr{C}^\infty$ $f_1 : \mathbb{F}_1 \to \widetilde{\mathbb{F}}$ and $f_2 : \mathbb{F}_2 \to \widetilde{\mathbb{F}}$ such that the following diagram commutes:

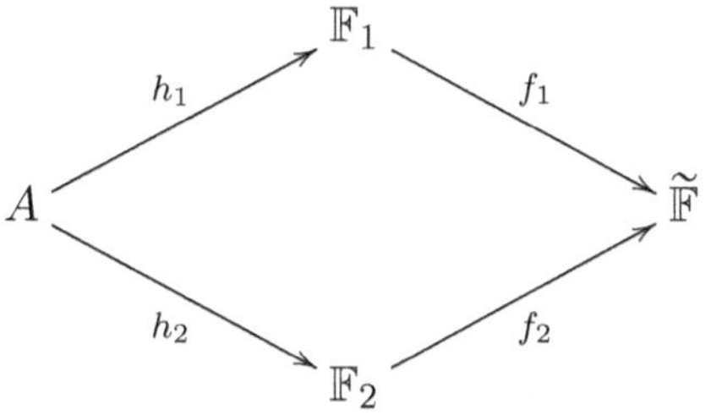

The relation $\mathscr{R}$ defined above is symmetric and reflexive.

The above considerations prove the following:

Proposition 4.9 (see **Proposition 6.11 of** [4]). *If $h_1 : A \to \mathbb{F}_1$ and $h_2 : A \to \mathbb{F}_2$ be two $\mathscr{C}^\infty$–homomorphisms from the $\mathscr{C}^\infty$–ring A to the $\mathscr{C}^\infty$–fields $\mathbb{F}_1, \mathbb{F}_2$ such that $(h_1, h_2) \in \mathscr{R}$, then $\ker(h_1) = \ker(h_2)$.*

Definition 4.10. *Let A be a $\mathscr{C}^\infty$–ring. A $\mathscr{C}^\infty$–**ordering** in A is a subset $P \subseteq A$ such that:*

(O1) $P + P \subseteq P$;

(O2) $P \cdot P \subseteq P$;

(O3) $P \cup (-P) = A$

(O4) $P \cap (-P) = \mathfrak{p} \in \mathrm{Spec}^\infty(A)$

Fact 4.11. *Let $\Sigma(A) := \{(\mathfrak{p}, Q) : \mathfrak{p} \in \mathrm{Sper}^\infty(A), Q \in \mathrm{Spec}^\infty(k_\mathfrak{p}(A))\}$. The mapping $P \in \mathrm{Spec}^\infty(A) \mapsto (\mathfrak{p}_P, Q_P) \in \Sigma(A)$, where $\mathfrak{p}_P := P \cap (-P)$ (or simply $\mathfrak{p}$) and $Q_P := \{\eta_{q_\mathfrak{p}[A\setminus\mathfrak{p}]}(a+\mathfrak{p}).(\eta_{q_\mathfrak{p}[A\setminus\mathfrak{p}]}(b+\mathfrak{p}))^{-1} : b \notin \mathfrak{p}, a.b \in P\}$ is a bijection (this is the analog of* **Proposition 5.1.1** *in [13]).*

Definition 4.12. *Let A be a $\mathscr{C}^\infty$–ring. Given a $\mathscr{C}^\infty$–ordering P in A, the $\mathscr{C}^\infty$–**support** of A is given by:*

$$\mathrm{supp}^\infty(P) := \mathfrak{p}_P = P \cap (-P)$$

Definition 4.13. *Let A be a $\mathscr{C}^\infty$–ring. The $\mathscr{C}^\infty$–**real spectrum of** A is given by:*

$$\mathrm{Sper}^\infty(A) = \{P \subseteq A | P \textit{ is an ordering of the elements of } A\}$$

together with the (spectral) topology generated by the sets:

$$H^{\infty}(a) = \{P \in \mathrm{Sper}^{\infty}(A) | a \in P \backslash \mathrm{supp}^{\infty}(P)\}$$

for every $a \in A$. *The topology generated by these sets will be called "smooth Harrison topology", and will be denoted by* Har^{∞}.

Remark 4.14. *The suitable notion of prime spectrum of a* $\mathscr{C}^{\infty}$*-ring* A, $\mathrm{Spec}^{\infty}(A)$, *appeared for the first time in [17]: this is the main spatial notion to develop "Smooth Algebraic Geometry". On the other hand, in [2] was introduced the notion of smooth real spectrum of a* $\mathscr{C}^{\infty}$*-ring* A, $\mathrm{Sper}^{\infty}(A)$*: this seems to be the suitable spatial notion for the development of "Smooth Real Algebraic Geometry".*

Fact 4.15. *Given a* $\mathscr{C}^{\infty}-$*ring* A*, we have a function given by:*

$$\begin{array}{rccc} \mathrm{supp}^{\infty}: & (\mathrm{Sper}^{\infty}(A), \mathrm{Har}^{\infty}) & \to & (\mathrm{Spec}^{\infty}(A), \mathrm{Zar}^{\infty}) \\ & P & \mapsto & P \cap (-P) \end{array}$$

which is spectral, and thus continuous, since given any $a \in A$*, we have:*

$$\mathrm{supp}^{\infty\,\dashv}[D^{\infty}(a)] = H^{\infty}(a) \cup H^{\infty}(-a).$$

Unlike what happens in ordinary Commutative Algebra, we have the following (and stronger) result in "Smooth Commutative Algebra", as a consequence of the fact that every $\mathscr{C}^{\infty}$-field is ($\mathscr{C}^{\infty}$-)real closed[1] and some separation theorems (see [5] or **Theorem 6.22** of [4]):

Theorem 4.16. *For each* $\mathscr{C}^{\infty}-$*ring* A*, the mapping*

$$\mathrm{supp}^{\infty}: (\mathrm{Sper}^{\infty}(A), \mathrm{Har}^{\infty}) \to (\mathrm{Spec}^{\infty}(A), \mathrm{Zar}^{\infty})$$

is a (spectral) bijection.

5 Concluding Remarks and Future Works

Remark 5.1. *It is natural to ask if the class of* $\mathscr{C}^{\infty}$*-fields is model-complete in the language of* $\mathscr{C}^{\infty}$*-rings or even admits elimination of quantifiers (possibly in the language expanded by a unary predicate for the positive cone of an ordering). If the former holds, then the relation* $\mathscr{R}$ *between pairs of morphisms with the same source and target* $\mathscr{C}^{\infty}$*-fields, that encodes* Sper^{∞}*, is already a transitive relation (as it occurs in the algebraic case).*

[1] In fact, to obtain this result it is enough to know that every $\mathscr{C}^{\infty}$-field is *Euclidean.*

Remark 5.2. *If the class of $\mathscr{C}^\infty$-fields admits quantifier elimination (over a reasonable language), then it is possible to adapt the definition and results provided in [19] on "Model-theoretic Spectra" and describe "logically" the spectral topological spaces* $\mathrm{Spec}^\infty(A)$ *and/or* $\mathrm{Sper}^\infty(A)$ *as certain equivalence classes of homomorphisms from A into models of a "nice" theory T. Moreover, since the techniques in this work provide structural sheaves of "definable functions", we could compare them with other ones previously defined and determine other new natural model-theoretic spectra in $\mathscr{C}^\infty$-structures.*

Remark 5.3. *Another evidence that a systematic model-theoretic analysis of $\mathscr{C}^\infty$-rings, (not only under real algebra perspective but also under differential algebra perspective), should be interesting and deserves a further attention is indicated in [10]. In that work the first steps were taken towards a model-theoretic connection between three kinds of structures: o-minimal structures, Hardy fields and smooth rings. This triple is related to another one – Hardy fields, surreal numbers and transseries – studied in [1]: these are linked by the notion of H-field which provides a common framework for these structures. They present a model-theoretic analysis of the category of H-fields, e.g. the theory of H-closed fields is model complete, and relate these results with the latter triple, that according the authors M. Aschenbrenner, L. v. Dries and J. v. Hoeven ([1]):* "...are three ways to enrich the real continuum by infinitesimal and infinite quantities. Each of these comes with naturally interacting notions of ordering and derivative".

References

[1] M. Aschenbrenner, L. van den Dries, J. van der Hoeven, *On numbers, germs, and transseries*, *Proc. ICM*-2018, Rio de Janeiro, Vol.1, 1–24.

[2] J.C. Berni, *Some Algebraic and Logical Apects of $\mathscr{C}^\infty$–Rings*. PhD thesis, Instituto de Matemática e Estatística (IME), Universidade de São Paulo (USP), Brazil, 2018.

[3] J.C. Berni, H.L. Mariano, *A Universal Algebraic Survey of $\mathscr{C}^\infty$-Rings*, arXiv:1904.02728 [math.RA], 2019.

[4] J.C. Berni, H.L. Mariano, *Topics on Smooth Commutative Algebra*, arXiv:1904.02725 [math.AC], 2nd version, 2019.

[5] J.C. Berni, H.L. Mariano, *Separation Theorems in Smooth Commutative Algebra and Applications*, submitted, 2019.

[6] D. Borisov, K. Kremnizer *Beyond Perturbation 1: de Rham Spaces* https://arxiv.org/pdf/1701.06278.pdf

[7] M. Dickmann, F. Miraglia, *Faithfully Quadratic Rings*, Memoirs AMS **1128**, Providence, R.I., 2015.

[8] L. van den Dries, *Tame Topology and o-minimal Structures.* (English summary) London Mathematical Society Lecture Note Series, 248. Cambridge University Press, Cambridge, 1998. x+180 pp.

[9] L. van den Dries, C. Miller, *Geometric Categories and o-minimal Structures*, Duke Math. J, 84 (1996), no. 2, 497–540.

[10] R. Figueiredo, H. L. Mariano, *Remarks on Expansions of the Real Field: Tameness, Hardy Fields and Smooth Rings*, South American Journal of Logic Vol. 4, n. 2, 373–383, 2018.

[11] D. Joyce, *Algebraic Geometry over $\mathscr{C}^{\infty}-$Rings*, volume 7 of *Memoirs of the American Mathematical Society*, 2019.

[12] A. Kock, *Synthetic Differential Geometry*, Cambridge University Press, 2006.

[13] M. A. Marshall, *Spaces of Orderings and Abstract Real Spectra*, Lecture Notes in Mathematics 1636, Springer-Verlag, Berlin, Germany, 1996.

[14] P. Michor, J. Vanžura. *Characterizing Algebras of Smooth Functions on Manifolds*, Comment. Math. Univ. Carolinae (Prague) 37(3), 519-521, 1996, available at https://www.mat.univie.ac.at/ michor/ci-alg.pdf

[15] C. Miller, *Basics of O-minimality and Hardy Fields*, Lecture Notes on O-minimal Structures and Real Analytic Geometry, Fields Institute Communications, Springer, New York, 2012.

[16] I. Moerdijk, G. Reyes, *Rings of Smooth Functions and Their Localizations, I*, Journal of Algebra 99, 324-336, 1986.

[17] I. Moerdijk, G. Reyes, N.v. Quê. *Rings of Smooth Functions and Their Localizations II.* Mathematical Logic and Theoretical Computer Science, Lecture Notes in Pure and Applied Mathematics 106, 277-300, 1987.

[18] I. Moerdijk, G. Reyes, *Models for Smooth Infinitesimal Analysis*, Springer-Verlag, New York, 1991.

[19] R. O. Robson, *Model Theory and Spectra*, Journal of Pure and Applied Algebra 63 (1990) 301-327, North-Holland 301.

Received 31 January 2020

Strong Normalization for Np-Systems via Mimp-Graphs

Vaston Gonçalves da Costa
Mathematics and Technology Institute, Federal University of Goiás (in Transition), Catalão-GO, Brazil.
vaston@ufg.br

Edward Hermann Haeusler
Informatics Department, Pontifical Catholic University of Rio de Janeiro, Rio de Janeiro-RJ, Brazil.
hermann@puc-rio.br

Abstract

Proof systems such as Frege, Hilbert, Tableaux, Sequent Calculus and Natural Deduction, as well as Resolution, represent (or structure) their proofs either as trees of formulas or as sequents/clauses. Some results show a significant computational reduction of time and space in proofs structured as graphs. Mimp-graph is a particular type of direct graph to represent proofs/derivations whose vertices and edges are labelled. A proof in mimp-graph requires fewer nodes than its Natural Deduction-tree-like representation. Besides its ability to provide compact representation for Natural Deduction proofs/derivations, the usage of mimp-graph has shown that some proof-theoretical investigations may be more straightforward in it than in usual Natural Deduction. For example, strong-normalization for the implicational fragment of Minimal Logic is simpler as a mimp-graph than it is in typical Natural Deduction tree-like. Np-imp is a propositional fragment of the Natural Deduction system that uses the Peirce rule instead of the classical absurd rule. The implicational fragment of the Np-system is decidable has normalization and a kind sub-formula principle, but lacks strong normalization. In this article, we show that Np-system implicational fragment represented as mimp-graph has strong normalization. This new theoretical result attests to the feasibility of building theorem provers that employ only the implication symbol.

The authors gratefully acknowledge the partial financial support by CAPES, CNPq, UFG/UFCat, PUC-Rio, FAPEG and FAPERJ.

1 Introduction

Deductive systems usually structure their proofs as trees. However, studies have shown that a significant reduction of computational time and space in proofs structured as direct acyclic graphs [3,5,8,9].

In [5], we present methods to reduce the weight and the size of proofs when switching from traditional tree-structured to graphs; such reduction occurs by adding unification substitutions. As example, it is presented a logarithmic reduction for the proof of the pigeonhole principle, known to be exponential [1].

In [8] the authors present the *mimp-graphs* structure. Mimp-graph is a special type of direct graph to represent logical proofs with labeled edges and vertices. In mimp-graph, one can distinguish two parts, one representing the proofs derivations and the other the formulas.

Furthermore, proofs representation in mimp-graph requires fewer nodes than a tree representation, and it is possible to represent any natural deduction tree-like proof in mimp-graph-like. Additionally, the minimal formula representation, which separates introducing and eliminating of implication rules, a crucial feature of mimp-graph, allow us to quickly determine the upper bounds on the length of reductions to produce a normal proof.

A normalization process using mimp-graphs structure can be found in [8] for full propositional logic. The normalization procedure consists in identify the number of maximal formulas in the full graph derivation, choose any minimal formula (detailed in the article) and eliminate the minimal formula chosen, repeating this until there is no minimal formula.

In [6] one can find the definition and main results evolve Np-system, it is a natural deduction system that uses the Peirce rule instead of the classical absurd rule. The implicational fragment of Np-system is complete and normalizable. Moreover, it has a kind of sub-formula principle that allow us to implement a theorem prover using only the implication symbol, leaving the classical part, the Peirces rules, exclusively at the end of the derivation. However, the system does not have strong normalization.

Some studies show the feasibility of translating the Np-system into mimp-graphs structure [3].

In this work, we present the gains from this translation; in particular, we show strong normalization for Np-systems structured as mimp-graph, what, as showed in [6], is not possible in tree representation.

One of the motivations of this work is to improve the quality of the theorem prover under development by the TecMF team [4].

The present paper is organized into sections as follows: in section 2 it is presented a review of the mimp-graph structure, in section 3 it is reviewed of Np-system . In section 4, it is presented the Np-system to mimp-graph translation. In section 5, it is presented the main result of the paper, the strong normalization proof of NP-imp-graph. Finally, section 6 presents the conclusions of the paper.

2 Mimp-graphs

In [9], a graphical representation for proofs is presented, called *mimp-graph*, idealized, among other objectives, to extract theoretical proof properties from a proof system.

Mimp-graphs are directed-graphs whose nodes and edges are labelled, in such a way that one can distinguish two parts, one representing the proof inferences, and other the formulas that take part in it. By convention, we use continuous lines to represent formulas in the graph and dotted lines to represent derivations in the proof.

The representation of proof in mimp-graph allows to reuse derivations and formulas by referencing instead of copying. In figure 1 one can see a representation of a propositional variable p and the formula $(p \to q) \to (p \to q)$.

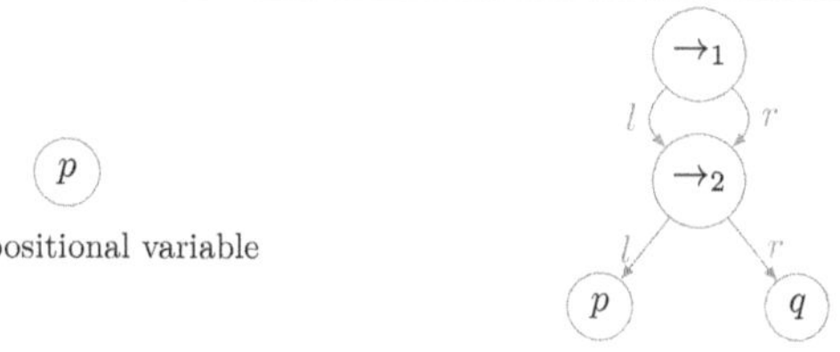

(a) Propositional variable

(b) $(p \to q) \to (p \to q)$

Figure 1: Formulas representation in mimp-graph

Formulas, in mimp-graph, may occur only once and subformulas are indicated by outgoing edges with labels l (left) and r (right).

For each rule in natural deduction there is a associate rule in mimp-graph. In figure 2 is depicted the rules of $\to$.

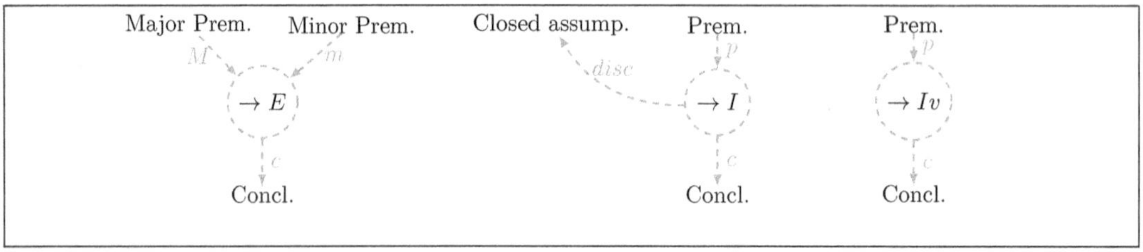

Figure 2: $\to$ rules

In the rules vertices, the premises are indicated by ingoing edges and there are edges from the rule vertices to the conclusion formulas. The vacuo discharging of hypotheses is represented by a disconnected graph, where the discharged formula vertex is not linked to the conclusion of the rule by any directed path.

For the purposes of alphabetical reading, we extracted from [9] the main definitions and properties of mimp-graph, as follows.

Definition 2.1. L *is the union of the four sets of labels types:*

- R-Labels *is the set of inference labels:* $\{\to I_n/n \in \mathbb{Z}\} \cup \{\to E_m/m \in \mathbb{Z}\}$,
- F-Labels *is the set of formula labels:* $\{\to_i /i \in \mathbb{N}\}$ *and the propositional letters* $\{p, q, r, ...\}$,
- E-Labels *is the set of edge labels: {l (left), r (right), conc (final conclusion), hyp (hypothesis)}* $\cup$ *{p_j (premise)/$j \in \mathbb{Z}$}* $\cup$ *{m_j (minor premise)/$j \in \mathbb{Z}$}* $\cup$ *{M_j (major premise)/$j \in \mathbb{Z}$}* $\cup$ *{c_j (conclusion)/$j \in \mathbb{Z}$}* $\cup$ *{$disc_j$ (discharge)/$j \in \mathbb{Z}$}*,
- D-Labels *is the set of delimiter labels:* $\{H_k/k \in \mathbb{Z}\} \cup \{C\}$.

Definition 2.2. *A* mimp-graph G *is a directed graph* $\langle$V, E, L, l_V, $l_E\rangle$ *where:* V *is a set of vertices,* E *is a set of* edges, L *is a set of labels,* $\langle v \in$ V, $t \in$ L, $v' \in$ V$\rangle$, *where* v *is the source and* v' *the target,* l_V *is a labeling function from* V *to* R$\cup$F-Labels, l_E *is a labeling function from* E *to* E-Labels.

Mimp-graphs are defined recursively as follows:

Basis *If* G_1 *is a formula graph with root vertex* α_m [1], *then the graph* G_2 *is defined as* G_1 *with the delimiter vertices* H_n *and* C *and the edges* $(\alpha_m, conc, C)$ *and* (H_n, hyp, α_m) *is a mimp-graph.*

[1]We will use the terms α_m, β_n and γ_r to represent the principal connective of the formula α, β and γ respectively.

$\to$E *If G_1 and G_2 are mimp-graphs, and the graph (intermediate step) obtained by $G_1 \oplus G_2$* [2] *contains the edge $(\to_q, l, \alpha_m)$ and the two vertices $\to_q$ and α_m linked to the delimiter node C, then the graph G_3 is defined as $G_1 \oplus G_2$ with*

1. *the removal of the ingoing edges in the node C which were generated in the intermediate step (see Figure 3, dotted area in $G_1 \oplus G_2$);*
2. *a rule vertex $\to E_i$ at the top level;*
3. *the edges: $(\alpha_m, m_{new}, \to E_i)$, $(\to_q, M_{new}, \to E_i)$, $(\to E_i, c_{new}, \beta_n)$ and $(\beta_n, conc, C)$, where new is a fresh (new) index considering all edges of kind c, m and M ingoing and/or outgoing the formula-vertices α_m, β_n and $\to_q$,*

is a mimp-graph (see Figure 3).

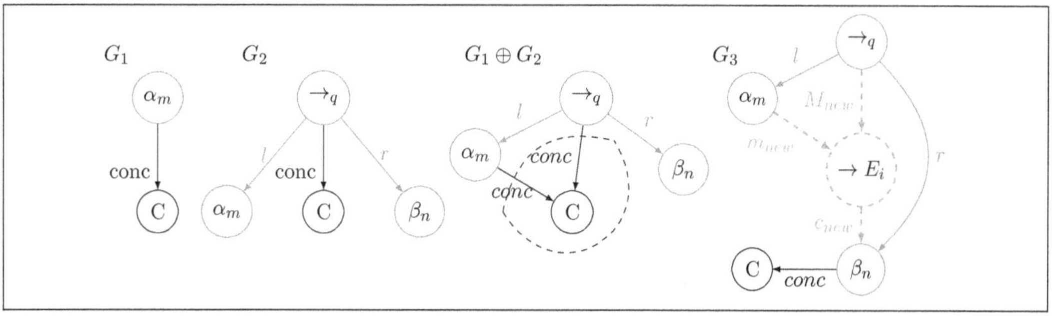

Figure 3: The $\to$E rule of mimp-graph

$\to$I *If G_1 is a mimp-graph and contains a vertex β_n linked to the delimiter node C and the node α_m linked to the delimiter node H_k, then the graph G_2 is defined as G_1 with*

1. *the removal of the edges: $(\beta_n, conc, C)$;*
2. *a rule vertex $\to I_j$ at the top level;*
3. *a formula vertex $\to_t$ linked to the delimiter node C by an edge $(\to_t, conc, C)$;*
4. *the edges: $(\to_t, l, \alpha_m)$, $(\to_t, r, \beta_n)$, $(\beta_n, p_{new}, \to I_j)$, $(\to I_j, c_{new}, \to_t)$, and $(\to I_j, disc_{new}, H_k)$, where new is a fresh index concerning ingoing and outgoing edges of type c and p of the formula-vertices β_n, $\to_t$ and α_m,*

[2]By definition $G_1 \oplus G_2$ equalizes the vertices of G_1 with the vertices of G_2 that have the same label, and equalizes edges with the same source, target and label into one.

is a mimp-graph (see Figure 4; the α_m-vertex is discharged*).*

→I-v [3] *If G_1 is a mimp-graph, and G is a formula graph with root vertex α_m, and G_1 contains a vertex β_n linked to the delimiter vertex C, then the graph G_2 is defined as $G_1 \oplus G$ with*

1. *the removal of the edge: $(\beta_n, conc, C)$;*
2. *a rule vertex $\rightarrow I_j$ at the top level;*
3. *a formula vertex $\rightarrow_t$ linked to the delimiter node C by an edge $(\rightarrow_t, conc, C)$;*
4. *the edges: $(\rightarrow_t, l, \alpha_m)$, $(\rightarrow_t, r, \beta_n)$, $(\beta_n, p_{new}, \rightarrow I_j)$ and $(\rightarrow I_j, c_{new}, \rightarrow_t)$, where new is an index under the same conditions of the previous case,*

is a mimp-graph (see Figure 5).

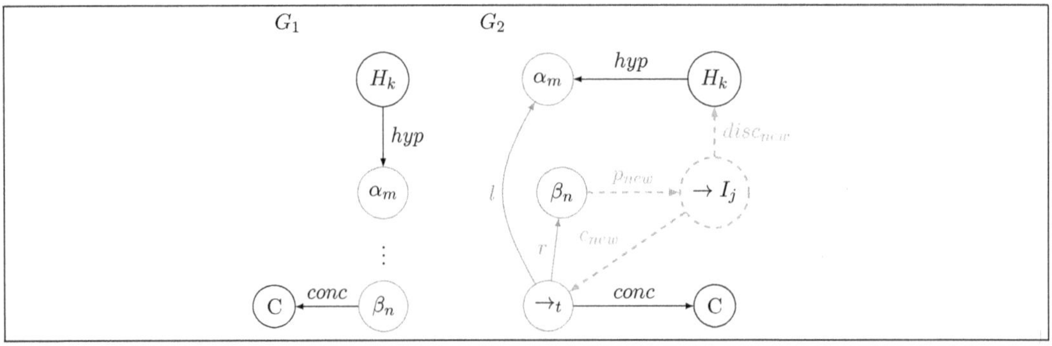

Figure 4: The →I rule of mimp-graph

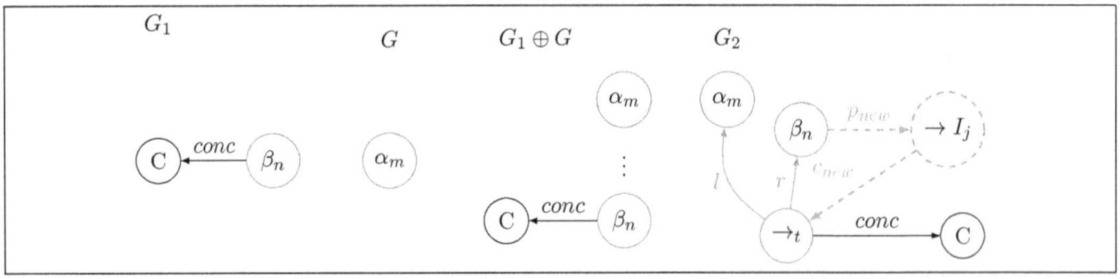

Figure 5: The →I-v rule of mimp-graph

[3] the "v" stands for "vacuous", this case of the rule →I discharges a hypothesis vacuously. This means that α_m has no ingoing Hyp-edge

Example 2.3. *A derivation of $(p \to r) \to (p \to r)$ from $(p \to q)$ in natural deduction can be represented as the following mimp-graph:*

$$\dfrac{\dfrac{\dfrac{\dfrac{[p]^1 \quad p \to q}{q} \qquad [q \to r]^2}{r}\,(\to E_2)}{p \to r}\,(\to I_3, 1)}{(q \to r) \to (p \to r)}\,(\to I_4, 2)$$

$$\nabla$$

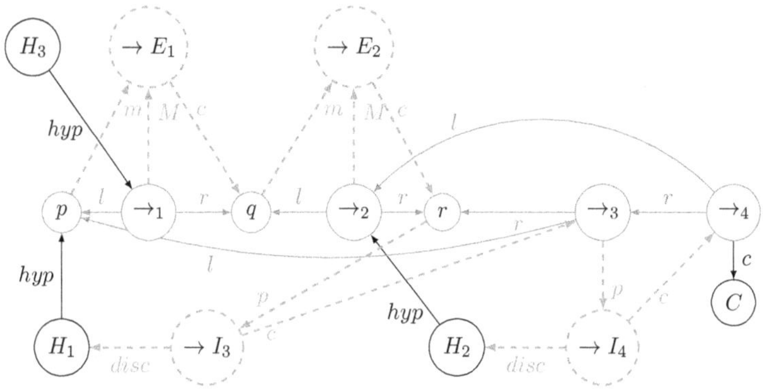

Also in [9] there is a proof that every natural tree-like minimal proof has a mimp-graph representation, as state in theorem 2.5. The definition of inferential order for mimp-graphs (see definition 2.4) is used as a mesure to proof the theorem.

Definition 2.4 (Inferential Ordering). *Let G be a mimp-graph. An inferential order $>$ on vertices of G is a partial ordering of the rule vertices of G, such that, $n < n'$, iff, n and n' are rule vertices, and there is a formula vertex f, such that, $n \xrightarrow{l_1} f \xrightarrow{l_2} n'$ and l_1 is c and l_2 is m, or , l_1 is c and l_2 is M, or, l_1 is c and l_2 is p.*

Theorem 2.5 (F-minimal representation). *Every standard tree-like natural deduction Π has a uniquely determined (up to graph-isomorphism)* F-minimal mimp-like representation G_Π*, i.e. such a one that satisfies the following four conditions.*

1. *G_Π is a mimp-graph whose size does not exceed the size of Π.*
2. *Π and G_Π both have the same (set of) hypotheses and the same conclusion.*
3. *There is graph homomorphism $h : \Pi \to G_\Pi$ that is injective on R-Labels.*
4. *All F-Labels occurring in G_Π denote pairwise distinct formulas.*

There is a nomalization process, as such as in Natural Deduction tree-like representarion, for mimp-graph-like.

A *maximal formula* in mimp-graphs is a $\rightarrow$-I followed by a $\rightarrow$-E of the same formula graph.

Definition 2.6. *A* maximal formula m *in a mimp-graph* G *(see Figure 6) is a sub-graph of* G *consisting of:*

1. *the formula nodes* α_m, β_n, $\rightarrow_q$, *the rule node* $\rightarrow I_i$ *and the delimiter node* H_u;
2. *the rule node* $\rightarrow E_j$ *at the top level;*
3. *the edges:* $(\rightarrow_q, l, \alpha_m)$, $(\rightarrow_q, r, \beta_n)$, $(\beta_n, p, \rightarrow I_i)$, $(\rightarrow I_i, c, \rightarrow_q)$, (H_u, hyp, α_m), $(\rightarrow I_i, disc, H_u)$, $(\alpha_m, m, \rightarrow E_j)$, $(\rightarrow_q, M, \rightarrow E_j)$ *and* $(\rightarrow E_j, c, \beta_n)$;

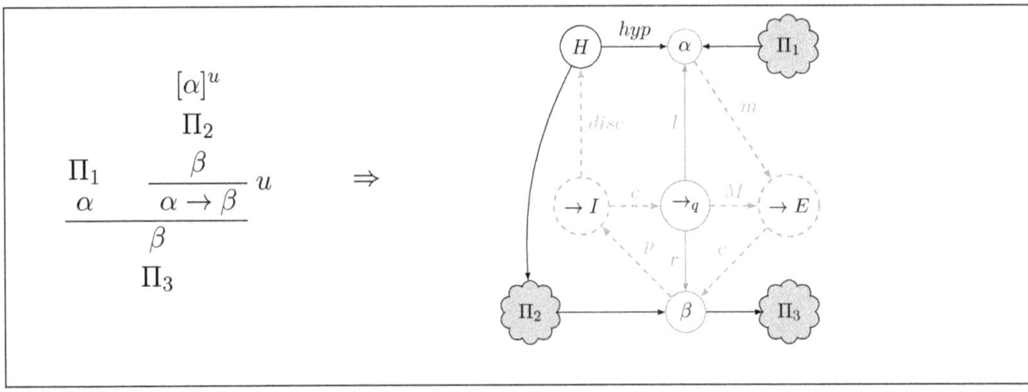

Figure 6: Maximal formula in mimp-graphs

The process of normalization transforms a mimp-graph G into another mimp-graph G' by dropping rule vertices and increasing by one the order of those vertices that have the conclusion as a major premise. The complete proof of such a process is available in [9]. Roughly speaking, the elimination process of a maximal formula that does not has maximal formula between the rule vertices $\rightarrow I$ and $\rightarrow E$ uses the following operations:

1. If the edge $(\rightarrow \mathrm{I}_i, c, \rightarrow_q)$ is the only ingoing edge to $\rightarrow_q$ and the edge $(\rightarrow_q, M, \rightarrow \mathrm{E}_l)$ is the only outgoing edge from $\rightarrow_q$ then remove the edges to and from the formula node $\rightarrow_q$, and the formula node $\rightarrow_q$.
2. Remove the edges to and from the nodes $\rightarrow \mathrm{I}_i$, $\rightarrow \mathrm{E}_l$ and H_u.

3. Remove the nodes $\rightarrow I_i$, $\rightarrow E_l$ and H_u.

4. Apply the re-ordination operation described in [9] resulting a mimp-graph.

Example 2.7. *The mimp-graph in figure 6 is transformed, after applying the reduction process above, into the graph represented in the figure 7.*

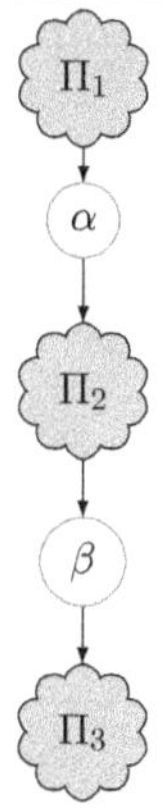

Figure 7: The result of a mimp-graph reduction process.

3 Np-imp system

Np is a Natural Deduction system that uses the Peirce rule instead of classical absurd [2]. Restricting the Natural Deduction system to the three rules depicted in figure 8 generates the minimal implicational Peirce system, denoted by *Np-Imp*.

$$\dfrac{\begin{array}{c}[\alpha \rightarrow \beta]^n \\ \vdots \\ \alpha\end{array}}{\alpha}\, n\; P-rule \qquad \dfrac{\begin{array}{c}\Pi_1 \\ \alpha\end{array} \quad \begin{array}{c}\Pi_2 \\ \alpha \rightarrow \beta\end{array}}{\beta} \rightarrow E \qquad \dfrac{\begin{array}{c}[\alpha]^n \\ \vdots \\ \beta\end{array}}{\alpha \rightarrow \beta}\, n \rightarrow I$$

Figure 8: Np-Imp deductive system.

As shown in [6], the Np-imp fragment is complete, normalizable and has a kind of sub-formula principle. These properties allow us to implement a theorem prover that uses only the implication symbol. Moreover, in Np-imp one can deals with the classical part, composed of Peirce rules, at the end of the derivation focusing only on the atomic formulas that occur in the formula to be proved. This strategy for proving theorems in Np-imp is a consequence of theorem 3.1.

Theorem 3.1. *Let $\{p_1, \ldots, p_n\}$ be the set of atomic formulas that occur in a formula α. So, $\vdash_{Np-Imp} \alpha$ if, and only if, $\vdash_{Np-Imp} (\alpha \rightarrow p_1) \rightarrow ((\alpha \rightarrow p_2) \ldots ((\alpha \rightarrow p_n) \rightarrow \alpha) \ldots)$.*

Theorem 3.1 resembles Glivencko's theorem [10, p. 166], in the sense that, a set of propositional formulas is intuitionistically consistent if and only if it is classically satisfiable.

The normalizations prove of the NP-Imp and, consequently, the theorem 3.1, in [6], uses the reductions RED_1, RED_2, RED_3 and RED_4, respectively, figures 9, 10, 11 and 12.

Remark 3.2. *The expression reduction, used here, has its origin in Prawitz's work on the inversion theorem for natural deduction [7, p. 35].*

Reductions were used by Prawitz, as well as in the present work, to invert rules in the proof's deviation process [7, p. 51].

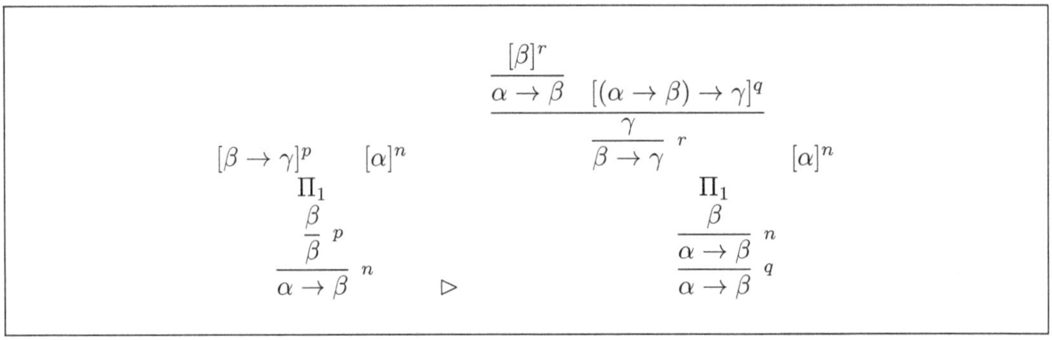

Figure 9: RED_1

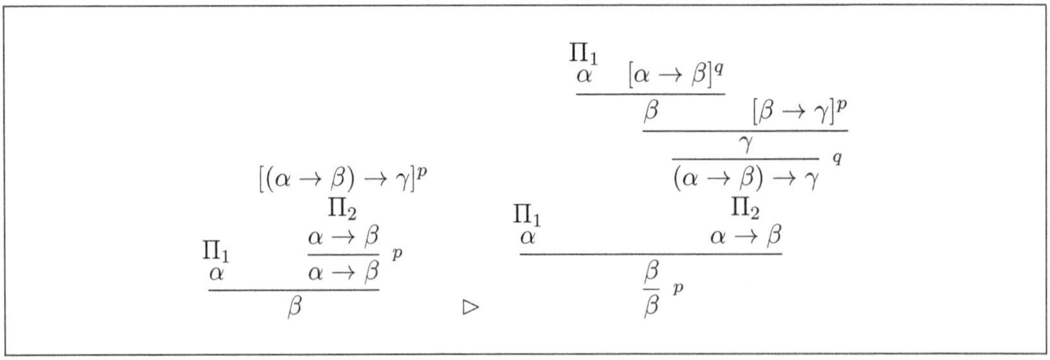

Figure 10: RED_2

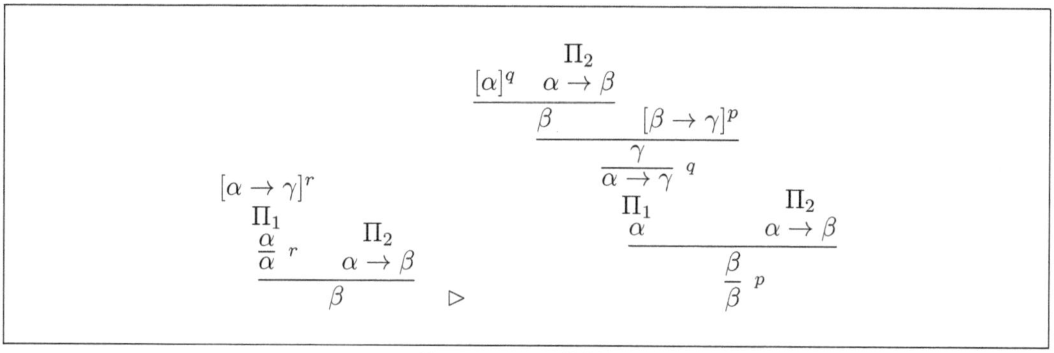

Figure 11: RED_3

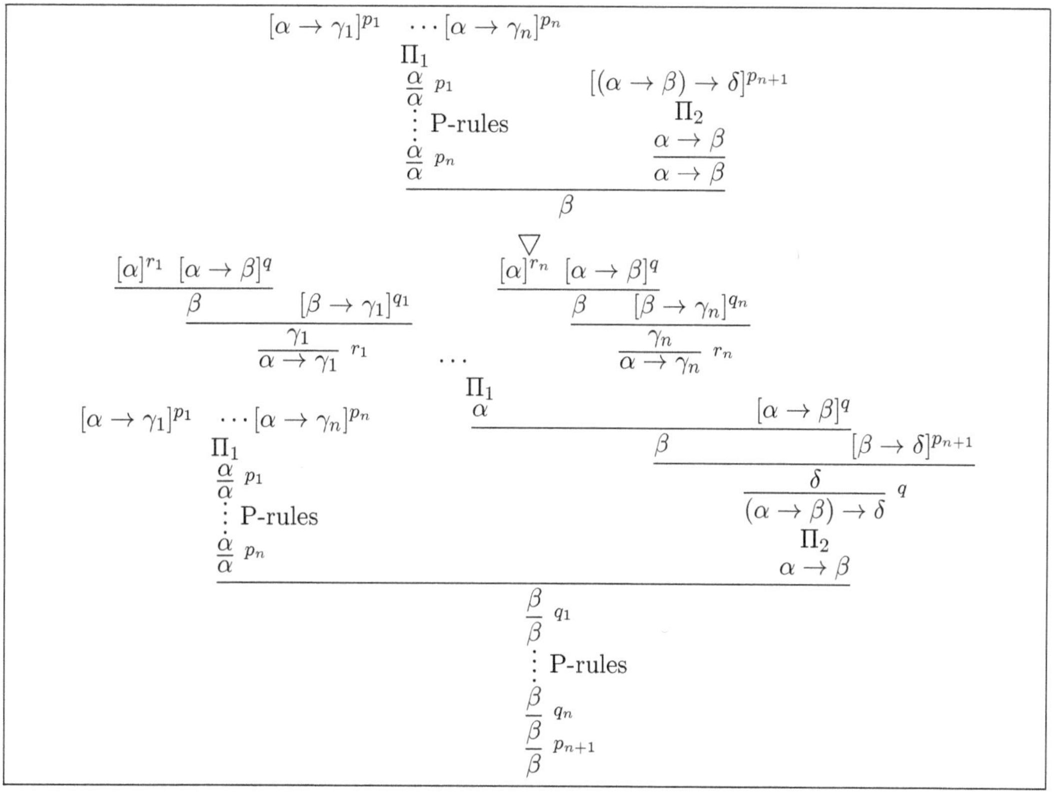

Figure 12: RED_4

As shown in [2], the Np-imp system *has not* strong normalization and, although it seems to be redundant, it is necessary the use of RED_4 to achieve normalization (see example 3.3), since it controls the proper order to descend Pierce's rules when they occur in both derivation's branches.

Example 3.3 (Np strong normalization counter-example, cf. [2]). *Consider the following derivation* Π *of* $\alpha \to \alpha$*:*

$$\dfrac{\dfrac{\dfrac{\dfrac{\dfrac{\dfrac{\dfrac{[\alpha]^1}{\alpha\to\alpha} \quad [(\alpha\to\alpha)\to\alpha]^p}{\alpha} \quad [\alpha\to\alpha]^{p_1}}{\alpha}}{\alpha}\,p_1}{\alpha\to\alpha}\,1}{\alpha\to\alpha}\,p \qquad \dfrac{\dfrac{\dfrac{\dfrac{\dfrac{\dfrac{\dfrac{\dfrac{[\alpha]^{1^*}}{\alpha\to\alpha}}{(\alpha\to\alpha)\to(\alpha\to\alpha)} \quad [((\alpha\to\alpha)\to(\alpha\to\alpha))\to\alpha]^{p_1^*}}{\alpha} \quad [\alpha\to\alpha]^{p^*}}{\alpha}}{\alpha}\,p^*}{\alpha\to\alpha}\,1^*}{(\alpha\to\alpha)\to(\alpha\to\alpha)}}{(\alpha\to\alpha)\to(\alpha\to\alpha)}\,p_1^*}{\alpha\to\alpha}$$

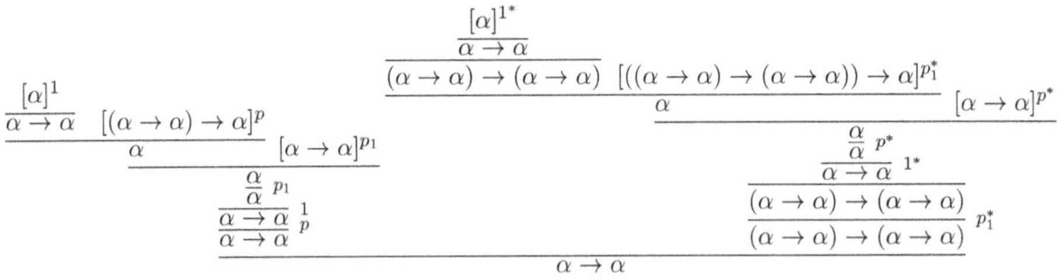

Notice that in derivation of the minor premise there is an $\to$*I rule application between two Peirce's rule applications. The correct strategy for normalizing the derivation is to apply* RED_4 *after* RED_1*. However, if one tries to apply* RED_2 *and* RED_3 *in an alternating way will produce the derivation* Π' *below.*

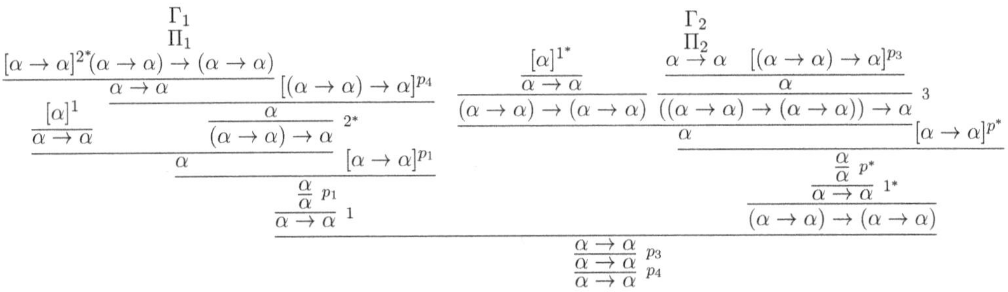

$$\dfrac{\dfrac{\dfrac{\dfrac{\dfrac{\dfrac{\dfrac{[\alpha]^1}{\alpha\to\alpha} \quad \dfrac{\dfrac{\dfrac{[\alpha\to\alpha]^{2^*} \quad \begin{matrix}\Gamma_1\\ \Pi_1\\ (\alpha\to\alpha)\to(\alpha\to\alpha)\end{matrix}}{\alpha\to\alpha} \quad [(\alpha\to\alpha)\to\alpha]^{p_4}}{\alpha}}{(\alpha\to\alpha)\to\alpha}\,2^*}{\alpha} \quad [\alpha\to\alpha]^{p_1}}{\alpha}}{\alpha}\,p_1}{\alpha\to\alpha}\,1 \qquad \dfrac{\dfrac{\dfrac{\dfrac{\dfrac{\dfrac{\dfrac{[\alpha]^{1^*}}{\alpha\to\alpha}}{(\alpha\to\alpha)\to(\alpha\to\alpha)} \quad \dfrac{\dfrac{\begin{matrix}\Gamma_2\\ \Pi_2\\ \alpha\to\alpha\end{matrix} \quad [(\alpha\to\alpha)\to\alpha]^{p_3}}{\alpha}}{((\alpha\to\alpha)\to(\alpha\to\alpha))\to\alpha}\,3}{\alpha} \quad [\alpha\to\alpha]^{p^*}}{\alpha}}{\alpha}\,p^*}{\alpha\to\alpha}\,1^*}{(\alpha\to\alpha)\to(\alpha\to\alpha)}}{\dfrac{\dfrac{\alpha\to\alpha}{\alpha\to\alpha}\,p_3}{\alpha\to\alpha}\,p_4}$$

Where $\begin{matrix}\Gamma_1\\ \Pi_1\\ (\alpha\to\alpha)\to(\alpha\to\alpha)\end{matrix}$*, for* $\Gamma_1 = \{(\alpha\to\alpha), (\alpha\to\alpha)\to\alpha\}$ *is as follow:*

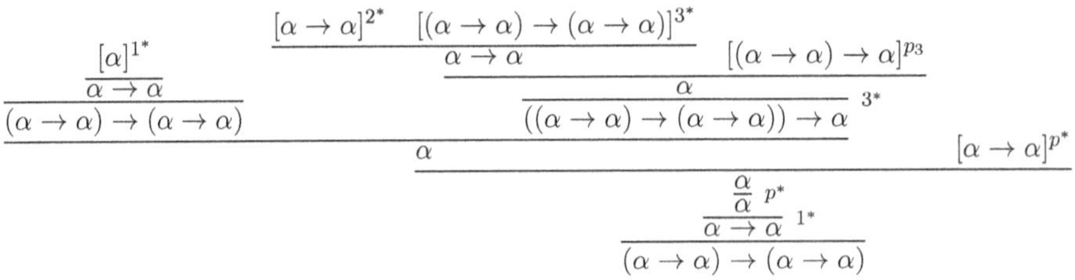

$$\dfrac{\dfrac{\dfrac{\dfrac{\dfrac{\dfrac{[\alpha]^{1^*}}{\alpha\to\alpha}}{(\alpha\to\alpha)\to(\alpha\to\alpha)} \quad \dfrac{\dfrac{\dfrac{[\alpha\to\alpha]^{2^*} \quad [(\alpha\to\alpha)\to(\alpha\to\alpha)]^{3^*}}{\alpha\to\alpha} \quad [(\alpha\to\alpha)\to\alpha]^{p_3}}{\alpha}}{((\alpha\to\alpha)\to(\alpha\to\alpha))\to\alpha}\,3^*}{\alpha} \quad [\alpha\to\alpha]^{p^*}}{\alpha}}{\alpha}\,p^*}{\dfrac{\alpha\to\alpha}{(\alpha\to\alpha)\to(\alpha\to\alpha)}}\,1^*$$

And $\begin{matrix}\Gamma_2\\ \Pi_2\\ (\alpha\to\alpha)\end{matrix}$*, for* $\Gamma_2 = \{(\alpha\to\alpha)\to(\alpha\to\alpha), (\alpha\to\alpha)\to\alpha\}$ *is:*

$$\dfrac{\dfrac{\dfrac{\dfrac{\dfrac{\dfrac{[\alpha]^1}{\alpha\to\alpha}\quad \dfrac{\dfrac{\dfrac{[\alpha\to\alpha]^2\quad[(\alpha\to\alpha)\to(\alpha\to\alpha)]^3}{\alpha\to\alpha}\quad [(\alpha\to\alpha)\to\alpha]^{p_4}}{\alpha}}{(\alpha\to\alpha)\to\alpha}\,2}{\alpha}\quad [\alpha\to\alpha]^{p_1}}{\alpha}}{\alpha}\,p_1}{\alpha\to\alpha}\,1\quad [(\alpha\to\alpha)\to(\alpha\to\alpha)]^3}{(\alpha\to\alpha)}$$

Observe that Π' has more than four P-rule applications above $r(\Pi')$. If we proceed by applying RED_1 to permute down only one P-rule from each branch, we can apply RED_2 and RED_3 again in order to produce a derivation with four P-rules above the lowest $\to$E. Evolving this way, the reduction process does not terminate.

4 The Np-imp-graph System

In section 2 was presented the translation from natural deduction to a proof-graph representation. Therefore, to carry out a translation from Np-imp to mimp-graph, it remains only to present the translation of Peirce's rule, that is showed in figure 13.

$$\begin{array}{c}[\alpha\to\beta]^n\\ \vdots\\ \dfrac{\alpha}{\alpha}\; n\ P-rule\end{array}$$

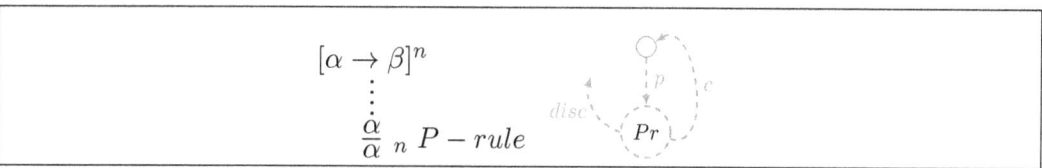

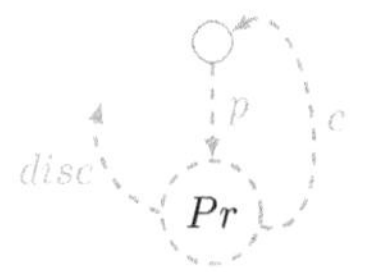

Figure 13: Peirce rule in mimp-graph.

From now on, it will be used *Np-imp-graph* to denote the Np-sytems as a mimp-graph structure.

Example 4.1. *Below is a translation of the derivation presented in example 3.3 into natural deduction for mimp-graph.*

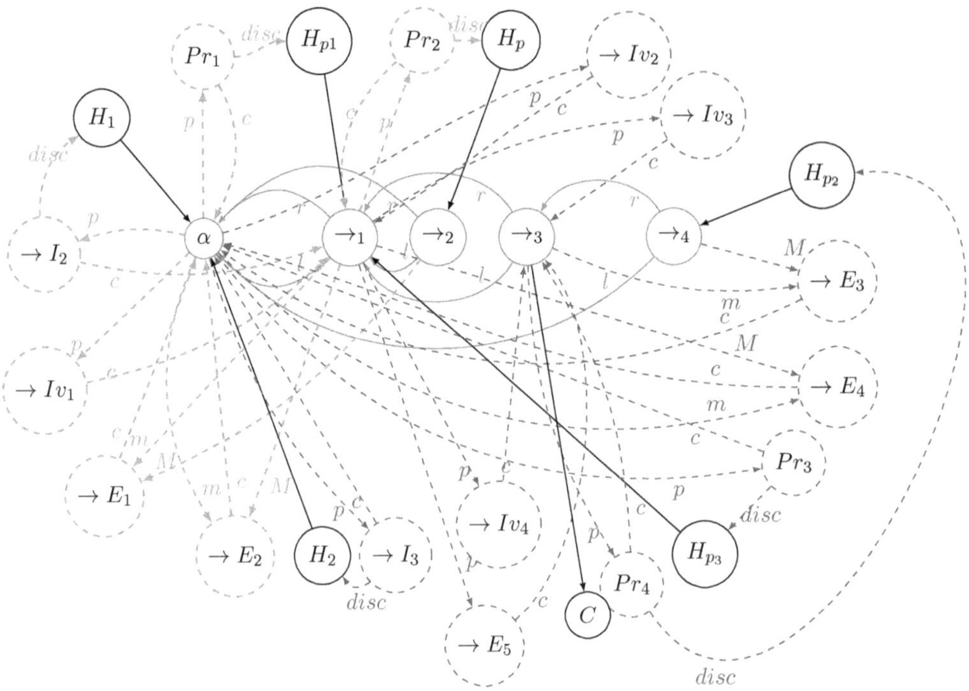

The formal definition of Np-imp-graph follows directly from definitions 2.1 and 2.2. In fact, it is necessary to add to such definitions the proper labels types and the recursive process to build up a Peirce reduction, as follows:

Definition 4.2. L *is the union of the* F-Labels, E-Labels *and* D-Labels *presented in definition 2.1 with the modified* R-labels*:*

- R-Labels *is the set of inference labels:* $\{\to I_n/n \in \mathbb{Z}\} \cup \{\to E_m/m \in \mathbb{Z}\} \cup \{Pr_k/k \in \mathbb{Z}\}$,

Definition 4.3. *A* Np-imp-graph G *is a directed graph* $\langle$V, E, L, l_V, l_E, $\rangle$ *where:* V *is a set of vertices,* E *is a set of* edges, L *is a set of labels,* $\langle v \in$ V, $t \in$ L, $v' \in$ V$\rangle$, *where* v *is the source and* v' *the target,* l_V *is a labeling function from* V *to* R$\cup$F-Labels, l_E *is a labeling function from* E *to* E-Labels.

Np-imp-graphs are defined recursively as follows:

- Basis, $\to$E, $\to$I *and* $\to$ Iv *are the same as in definition 2.2, switching the word mimp-graph to Np-imp-graph.*

- Pr*:*

If G_1 is a Np-imp-graph and contains a vertex α_m linked to the delimiter vertex C and the edge $(\rightarrow_q, l, \alpha_m)$ linked to the delimiter vertex H_k, then the graph G_2 is defined as G_1 with

1. *a rule node Pr_k at the top level;*
2. *the edges: $(H_k, p_{new}, Pr_k), (Pr_k, disc_{new}, H_k)$ and $(Pr_k, c_{new}, \alpha_m)$, where new is a fresh index considering all edges of kind c and p ingoing and/or outgoing the formula-nodes α_m, and $\rightarrow_q$,*

is a mimp-graph (see Figure 14).

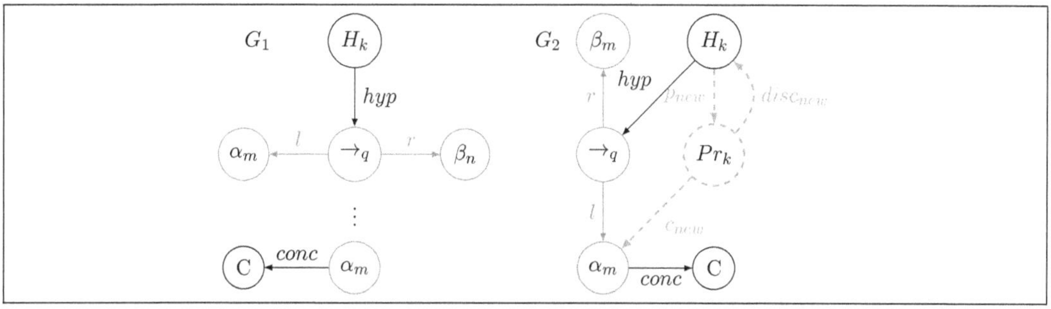

Figure 14: The Pr rule of Np-imp-graph

4.1 Np-imp-graph reductions

The reduction RED_1 in NP-imp, cf. figure 9, moves down [4] the Peirce rule that was introduced below a implication introduction. The purpose of RED1 is to lower the Peirce rules that are neither a minor premise nor a major premise of a maximal formula. Since the goal of the paper is to eliminate maximal formulas, there is no need to present the RED_1 translation for Np-imp-graph. The representation of RED_1 in Np-imp-graph is depicted in figure 15.

RED_2 and RED_3 are associated with maximal formulas introduced by Peirce rule. This is:

1. RED_2, cf. figure 10, moves down the Peirce rule that was introduced in the major premise branch. Contrary to what happens in the Np system's reduction, the translation to RED_2's Np-imp-graph does not duplicate de derivation Π_1 (see figure 16).

[4] More details about the expression "moves down" in the next section when it will be explained the conclusion distance.

2. RED_3, cf. figure 11, moves down the Peirce rule that was introduced in the minor premise branch before been eliminated. Similar to what happens with the RED_2 reduction, the translation of RED_3 to NP-mimp-graph also does not duplicate the derivation Π_2, cf. figure 17.

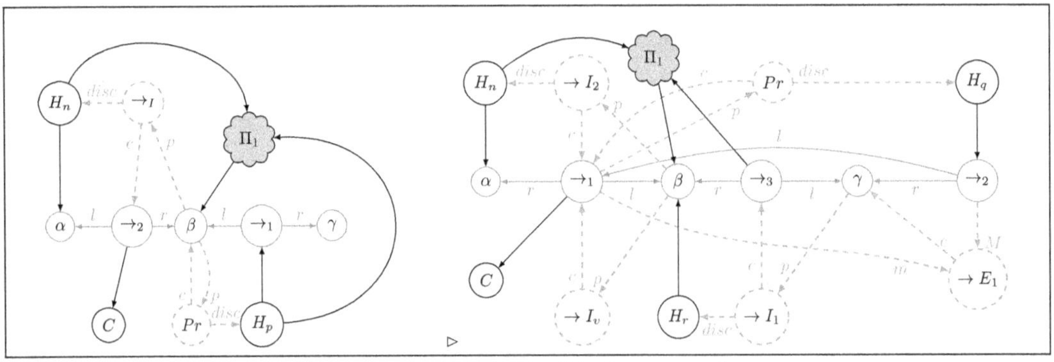

Figure 15: RED_1 in Np-imp-graph.

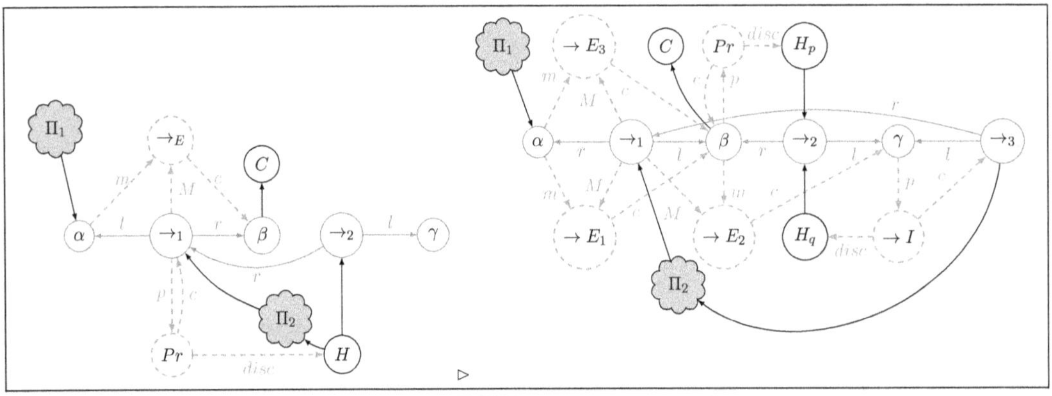

Figure 16: RED_2 in mimp-graph

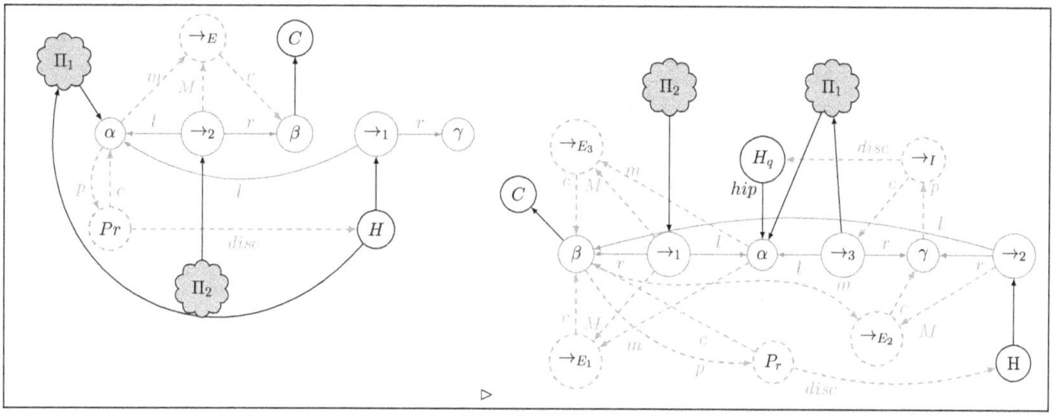

Figure 17: RED_3 in mimp-graph

Remark 4.4. *Since the only purpose of RED_4 was to control the order of the application of other reductions, to prove strong normalization, it is mandatory that such control no longer exists—this is the reason for not presenting a translation for RED_4.*

5 Strong Normalization for Np-imp-graph

A maximal formula in Np-imp-graphs is a $\rightarrow I$ followed by a $\rightarrow E$ of the same formula graph (see Definition 5.8). It is the same notion of maximal formulas that is being used in natural deduction derivations. The normalization process used here is similar to the one presented by Prawitz, which consists of eliminating the maximum formulas using reductions. However, here we are dealing with Np-imp-graphs that have Peirce rules which also contribute to the inclusion of formulas in the proof. As already said, the Np-system is normalizable but not strong normalizable. On the other hand, the mimp-graph system is strongly normalizable.

In this section, we will show the normalization procedure for NP-imp-graph. We want to reduce the original formula by dropping down the vertices and edges that are involved with Pierce rules and then apply the normalization process for maximum formulas for mimp-graphs. As the process of eliminating a maximal formula on mimp-graphs always ends in the elimination of at least one maximal formula, and with the decrease in the number of vertices of the graph, to use the normalization for mimp-graph before or after the reduction of pierce rules in Np-imp-graph will produce a normal formula when the two processes are combined.

Definition 5.1 (Path and branch).

1. *For $n_i \in V$, a* p-path *in a proof-graph is a sequence of vertices and edges of the form:* $n_1 \xrightarrow{l_1} n_2 \xrightarrow{l_2} \ldots \xrightarrow{l_{k-2}} n_{k-1} \xrightarrow{l_{k-1}} n_k$*, such that n_1 is a hypothesis formula vertex, n_k is the conclusion formula node, n_i alternating between a rule node and a formula vertex. The edges l_i alternate between two types of edges: the first is $l_j \in \{m, M, p\}$ and the second $l_j = c$.*

2. *A* branch *is an initial part of a* p-path *which stops at the conclusion formula node or at the first minor premise whose major premise is the conclusion of a rule node.*

The definition 5.2 presents the metric used in the process of descending the Pierce rules in a Np-imp-graph derivation.

Definition 5.2 (Conclusion Distance). *Let Π be a Np-imp-graph derivation from a set o hypotheses H and conclusion C. A conclusion distance from a vertex rule R to C, denoted by $C(R)$, is the number of edges of the p-path's part from R to C.*

Example 5.3. *In the derivation depicted in figure 18.*

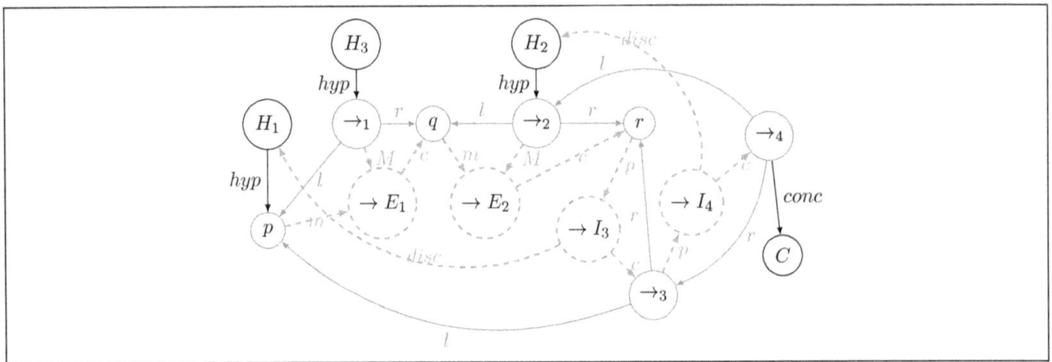

Figure 18: Derivation of $(q \to r) \to (p \to r)$ from p e $p \to q$.

The conclusion distance $C(\to E_1)$ is 8, since the p-path's part from vertex $\to E_1$ to vertex C has 8 edges. As shown in figure 19.

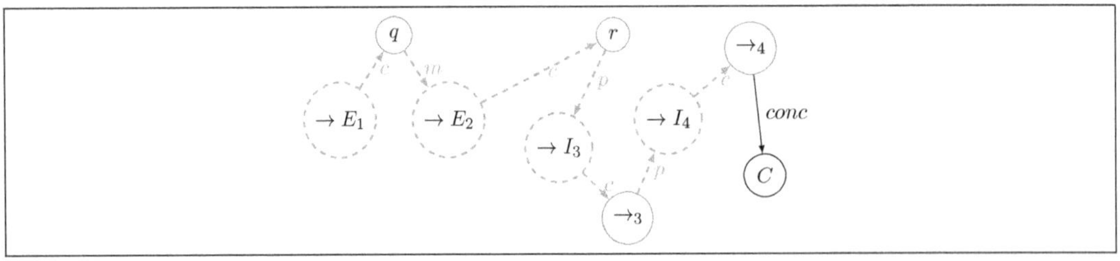

Figure 19: Distance rule in Np-imp-graph.

Definition 5.4 (RED_1 reducible). *A Π derivation in mimp-graph is said to be reducible by RED_1 if there is a formula node that is the conclusion of a Peirce rule and the premise of an implies introduction.*

The figure 20 shows a derivation fragment that is RED_1 reducible.

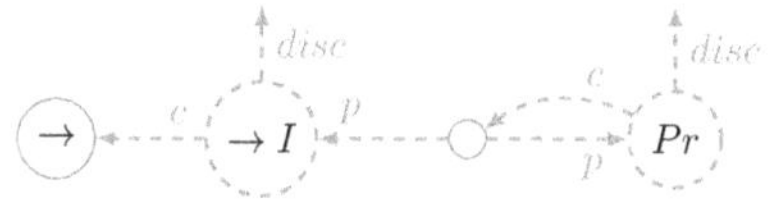

Figure 20: RED_1 reductible in Np-imp-graph.

Definition 5.5 (RED_2 reducible). *A Π derivation in mimp-graph is said to be reducible by RED_2 if there is a formula node that is the conclusion of a Peirce rule and the major premise of an implies elimination.*

The figure 21 shows a derivation fragment that is RED_2 reducible.

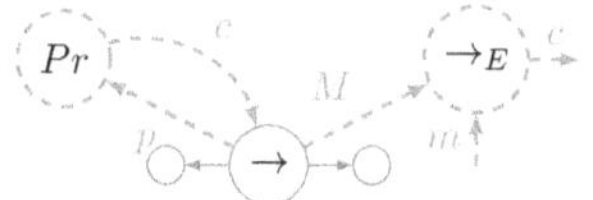

Figure 21: RED_2 reductible in Np-imp-graph.

The figure 22 shows a derivation fragment that is RED_3 reducible.

Definition 5.6 (RED_3 reducible). *A Π derivation in mimp-graph is said to be reducible by RED_3 if there is a formula node that is the conclusion of a Peirce rule and the minor premise of an implies elimination.*

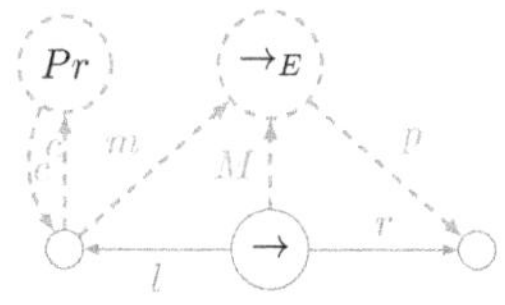

Figure 22: RED_3 reductible in Np-imp-graph.

Lemma 5.7 (Reduction lemma 1). *Let Π be a Np-imp-graph derivation RED_1-reducible and Π' the derivation derived from Π after applying RED_1. If Pr is the Peirce rule in Π reducible by RED_1 then:*

1. *The number of Peirce rules in Π' is less than or equal to the number of Peirce rules in Π;*

2. *If Pr' is the Peirce rule in Π' associated with RED_1 over Π then $C(Pr') \leq C(Pr)$.*

3. *The number of vertices rule's in Π' is equal to that of Π plus 3.*

Proof. In fact, looking at the derivation produced by RED_1, cf. figure 15, Π_1 is maintained in the new derivation and the associated Peirce rule Pr' has conclusion distance equals to 1, which is less than the distance of Pr in Π which is 3. Since Π_1 is kept in both derivations, the amount of rule vertices in Π is 2 while in Π' it becomes 5. □

Definition 5.8. *A maximal formula m in Np-imp-graph G is a sub-graph o G consisting of:*

Traditional *The maximal formula that is $\rightarrow I$ followed by a $\rightarrow E$ of the same formula graph, according to definition 2.6.*

RED_2 **type** *(Graph representation similar to the left side graph of figure 16):*

1. *the formulas vertices α, β, $\rightarrow_m$ and $\rightarrow_n$, the rule Pr and the delimiter vertex H_u;*
2. *the rule vertex $\rightarrow E$ at the top level;*
3. *the edges: $(\rightarrow_m, l, \alpha)$, $(\alpha, m, \rightarrow E)$, $(\rightarrow_m, r, \beta)$, $(\beta, conc, C)$, $(Pr, c, \rightarrow_m)$, $(\rightarrow_m, M, \rightarrow E)$, $(\rightarrow E, c, \beta)$, $(\rightarrow_m, p, Pr)$, $(Pr, disc, H_u)$, $(H_u, hyp, \rightarrow_n)$, $(\rightarrow_n, l, \gamma)$ and $(\rightarrow_n, r, \rightarrow_m)$;*
4. *If a branch will be separated from the inferential order then the $C(Pr)$ must diminish, i.e. the conclusion distance of the pierce rule in the new derivation must be less than the conclusion distance from the old Peirce rule in G.*

RED_3 **type** *(Graph representation similar to the left side graph of figure 17):*

1. *the formulas vertices α, β, $\rightarrow_m$, $t\emptyset_n$, the rule Pr and the rule delimiter H_u;*
2. *the rule vertices $\rightarrow E$ at the top level;*
3. *the edges: $(\rightarrow_m, l, \alpha)$, $(\rightarrow_m, r, \beta)$, $(\alpha, m, \rightarrow E)$, $(\rightarrow E, c, \beta)$, $(\beta, conc, C)$, (α, p, Pr), (Pr, c, α), $(Pr, disc, H_u)$,$(H_u, hyp, \rightarrow_n,)$,$(\rightarrow_n, l, \alpha)$, $(\rightarrow_n, r, \gamma)$, and $(\rightarrow_m, M, \rightarrow E)$;*

4. *If a branch will be separated from the inferential order then the $C(Pr)$ must diminish, i.e. the conclusion distance of the pierce rule in the new derivation must be less than the conclusion distance from the old Peirce rule in G.*

Definition 5.9. *Given a Np-imp-graph G with a maximal formula m, eliminating a maximal formula is the following transformation of a Np-imp-graph, where the maximal formula m satisfies the following requirements:*

Traditional

1. *Between the rule nodes $\rightarrow I_i$ and $\rightarrow E_l$ there are zero or more maximal formulas with inferential orders within the range of these rule nodes.*
2. *There is an edge $(\rightarrow I_i, c, \rightarrow_q)$, and, the formula node $\rightarrow_q$ has zero or more ingoing edges.*
3. *There is an edge $(\rightarrow_q, M, \rightarrow E_l)$, and, the formula node $\rightarrow_q$is the premise of zero or more of another rule nodes.*
4. *If a branch will be separated from the inferential order this branch must be insertable in the following branch, according to the order, i.e. the conclusion of this separated branch is the premise in the following branch.*

RED_2 **type**

1. *Between the rule vertices Pr and $\rightarrow E_i$ there are zero or more maximal formulas of type RED_2 with inferential orders within the range of these rule nodes.*
2. *There is an edge $(Pr, c, \rightarrow_q)$, and, the formula node $\rightarrow_q$ has zero or more ingoing edges.*
3. *There is an edge $(\rightarrow_q, M, \rightarrow E_i)$, and, the formula vertice $\rightarrow_q$ is the premisse of zero or more of another rule vertex.*

RED_3 **type**

1. *Between the rule vertices Pr and $\rightarrow E_i$ there are zero or more maximal formulas of type RED_3 with inferential orders within the range of these rule nodes.*
2. *There is an edge (Pr, c, α), and, the formula node $\rightarrow_q$ has zero or more ingoing edges.*
3. *There is an edge $(\alpha, m, \rightarrow E_i)$, and, the formula vertice $\rightarrow_q$ is the premisse of zero or more of another rule vertex.*

Lemma 5.10 (Reduction lemma 2). *Let Π be a Mp-imp-graph derivation RED_2-reducible and Π' the derivation derived from Π after applying RED_2. If Pr is the Peirce rule in Π reducible by RED_2 then:*

1. *The number of Peirce rules in Π' is less than or equal to the number of Peirce rules in Π;*
2. *If Pr' is the Peirce rule in Π' associated with RED_2 over Π then $C(Pr') \leq C(Pr)$.*
3. *The number of rule vertices in Π' is equal to that of Π plus 3.*

Proof. Analogous to proof of the lemma 5.7. □

Lemma 5.11 (Reduction lemma 3). *Let Π be a Np-imp-graph derivation RED_3-reducible and Π' the derivation derived from Π after applying RED_3. If Pr is the Peirce rule in Π reducible by RED_3 then:*

1. *The number of Peirce rules in Π' is less than or equal to the number of Peirce rules in Π;*
2. *If Pr' is the Peirce rule in Π' associated with RED_3 over Π then $C(Pr') \leq C(Pr)$.*
3. *The number of rule vertices in Π' is equal to that of Π plus 3.*

Proof. Analogous to proof of the lemma 5.7. □

The mimp-graph is normalizable as stated in theorem 5.12.

Theorem 5.12 (mimp-graph Normalization). *Every mimp-graph G can be reduced to a normal mimp-graph G' having the same hypotheses and conclusion as G. Moreover, for any standard tree-like natural deduction Π, if $G := G(\Pi)$ (the mimp-like representation of Π), then the size of G' does not exceed the size of G, and hence also Π.*

The proof of theorem 5.12 can be found in [9]. However, it is worth mentioning, that the maximal formula elimination process on mimp-graphs always ends in the elimination of at least one maximal formula, and with the decrease in the number of vertices on the graph.

Theorem 5.13 (Normalisation Np-imp-graph). *Every Np-imp-graph G can be reduced to a normal Np-imp-graph G' having the same hypotheses and conclusion as G. Moreover, for any standard NP-imp deduction Π, if $G := G_\Pi$ (the Np-imp like representation of Π), then the size of G' exceed the size of G in at most $O(n)$, and hence also Π.*

Proof. The preservation of the premises and conclusions of the derivation is assured by theorem 5.12, when maximal formula does not involve Peirce rule, and by 5.10 or 5.11 involve Peirce rules, that is, when the minor premiss of the maximal formula is the conclusion of pierce rule vertex or when the major premiss of the maximal formula is the conclusion of Peirce rule vertex.

In addition, the demonstration of this theorem has two primary requirements. First, to guarantee that through the application of RED_2 and RED_3 in the Np-imp-graph, it is generate more maximal formulas. The second requirement is to guarantee that during the normalization process, the normalization measure adopted is always reduced.

The first requirement is easily verifiable through an inspection of each one of the lemmas 5.10 and 5.11, in the case involve Peirce rules, and for the theorem 5.12, in the case that not involves.

The second requirement is established through the normalization procedure and demonstrated by analyzing the existing cases in the elimination of maximal formulas in Np-imp-graph. It is adopted a measure of complexity the *Conclusion Distance* of the Peirce rules, C(Pr_i). Besides, working with mimp-graph representations we can use as optional inductive parameter the ordinary size of Np-imp-graph.

Each application of RED_2 or RED_3 increases the Np-imp-graph's size in 3 vertices, i.e., G' increases by $O(n)$. □

5.1 Normalization Process

An Np-imp-graph G can have one or more maximal formulas reducible either by RED_2, RED_3 or by the *tradicional* way, named by $P_1, ..., P_n$. Thus, the normalization procedure is described by the followings steps:

1. Choose a maximal formula P_k.
2. Identify the respective conclusion distance $C(P_k)$.
3. Apply the associate reduction method: Theorema 5.12, lemma 5.10 and lemma 5.11.
    (a) When the maximal formula has the either the minor premiss or the major premiss as conclusion of Peirce rule, $C(P_k)$ decreases and the size of the graph increases in only 3 new vertices.
    (b) When the major premiss of the maximal formula is the conclusion of a introduction rule $C(P_k)$ keeps zero and the size of the decreases.

Example 5.14. *Considering the proof graph presented in the example 4.1, if we choose the Peirce rule* Pr_1 *we find 7 rules that use the conclusion of* Pr_1 *as a premise, namely,* $\rightarrow I_2$, $\rightarrow Iv_1$, $\rightarrow E_1$, $\rightarrow I_3$, $\rightarrow E_4$, $\rightarrow Iv_2$ *and* Pr_3. *Being a minor premise for* $\rightarrow E_1$ *and* $\rightarrow E_4$.

The reduction lemmas make no distinction as to which of the rules can be applied. Let us first consider $\rightarrow Iv_1$. *In this case, we must apply* RED_1. *Below is the graph before applying the rule and the graph generated.*

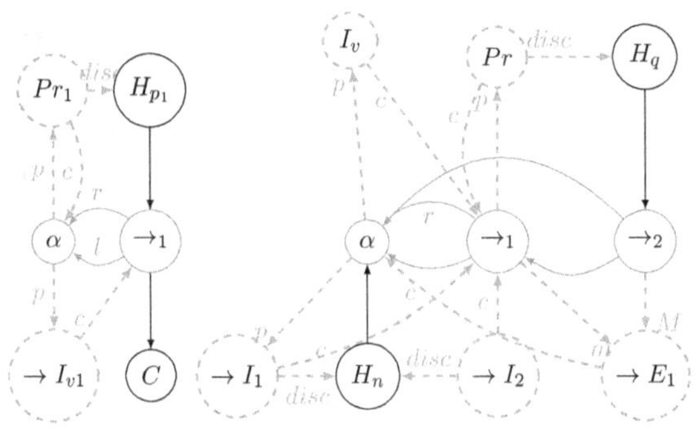

Attaching the generated graph to the original derivation produces the following graph:

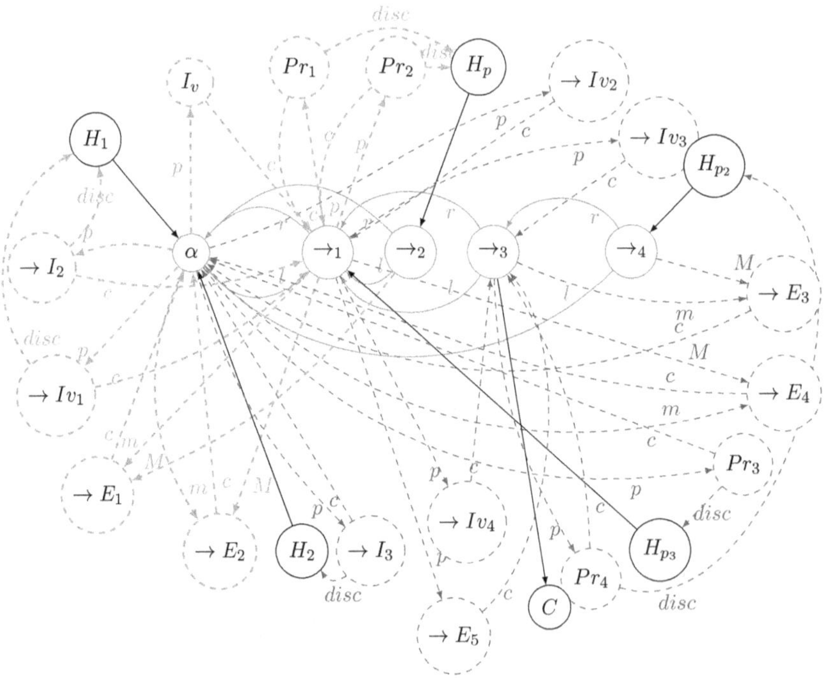

As one can see, the conclusion of Pr_1 *is no longer the premise of* $\rightarrow I_v$.

Likewise, if someone chooses another rule, instead of $\rightarrow I_{v_1}$, *based on the theorem 5.13 conclusion, a similar result is obtained.*

6 Conclusion

Thus, Np-imp-graph was introduced through definitions and examples, preserving the ability to represent Np-system's proofs.

As presented in [6], Np-system is not strong normalizable and for achieving normalization it is mandatory to use four reductions rules, one of those needed for control the rule's applications order (see example 3.3).

In contrast, the mimp-graph structure has a key feature of representing the normal derivations since it is easy to determine maximal formulas and upper bounds in the length of reduction sequences.

A translation from the Np-system to mimp-graph is feasible and, as shown in the present work, also improves the proof process since it adds strong normalization to the system that only had normalization.

To proof the normalization theorem for Np-imp-graph, we measure the distance to the conclusion of the Peirce rules in the original derivation. The strong normalization property result is a direct consequence of such normalization, and the normalization process of mimp-graphs [9] since the application of any reduction decreases the corresponding measures of derivation complexity.

We have gone one step further in the process of ensuring that a graph-based theorem prover can be more efficient, in computational terms of execution time and storage space, than traditional tree-based one's.

References

[1] Stephen A. Cook and Robert A. Reckhow. The relative efficiency of propositional proof systems. *The Journal of Symbolic Logic*, 44(1):36–50, 1979.

[2] Vaston G. Costa, Wagner Sanz, Edward H. Haeusler, and Luiz C. Pereira. Peirce's rule in a full natural deduction system. *Electronic Notes in Theoretical Computer Science*, 256:5 – 18, 2009. Proceedings of the Fourth Workshop on Logical and Semantic Frameworks, with Applications (LSFA 2009).

[3] Vaston Gonçalves da Costa, Edward Hermann Haeusler, Marcela Quispe Cruz, and Jefferson de Barros Santos. Np system and mimp-graph association. In *5th International Workshop on Universal Logic - UNILOG 2015*, pages 346–347, Istambul, Turkey, 2015.

[4] Jefferson de Barros Santos, Bruno Lopes Vieira, and Edward Hermann Haeusler. A unified procedure for provability and counter-model generation in minimal implicational

logic. *Electronic Notes in Theoretical Computer Science*, 324:165 – 179, 2016. WEIT 2015, the Third Workshop-School on Theoretical Computer Science.

[5] Lew Gordeev, Edward Hermann Haeusler, and Vaston G. Costa. Proof compressions with circuit-structured substitutions. *Journal of Mathematical Sciences*, 158(5):645–658, 2009.

[6] Luiz C. Pereira, Edward H. Haeusler, Vaston G. Costa, and Wagner Sanz. A new normalization strategy for the implicational fragment of classical propositional logic. *Studia Logica*, 96(1):95–108, 2010.

[7] Dag Prawitz. *Natural deduction: A proof-theoretical study.* Almqvist & Wiksell, Stockholn, 1965.

[8] Marcela Quispe-Cruz, Edward Haeusler, and Lew Gordeev. On strong normalization in proof-graphs for propositional logic. *Electronic Notes in Theoretical Computer Science*, 323:181 – 196, 2016. Proceedings of the Tenth Workshop on Logical and Semantic Frameworks, with Applications (LSFA 2015).

[9] Marcela Quispe-Cruz, Edward Hermann Haeusler, and Lew Gordeev. Proof-graphs for minimal implicational logic. In Mauricio Ayala-Rincón, Eduardo Bonelli, and Ian Mackie, editors, *Proceedings 9th International Workshop on Developments in Computational Models, DCM 2013, Buenos Aires, Argentina, 26 August 2013*, volume 144 of *EPTCS*, pages 16–29, 2013.

[10] Dirk Van Dalen. *Logic and structure.* Springer, Berlin, 3 edition, 1997.

Received 31 January 2020

The Generalized Continuum Hypothesis and two parametrized families of hit-and-miss games

Samuel G. da Silva
Department of Mathematics, Federal University of Bahia (UFBA)
samuel@ufba.br

Abstract

In this note, we show that for every infinite cardinal κ the statement $2^\kappa = \kappa^+$ is equivalent to say that two of certain hit-and-miss games with parameter κ are equally complex. The games we are dealing with (which are formalized as objects of certain categories) involve elements as much as $\leqslant \kappa$-sized subsets of sets of size 2^κ, where κ ranges over all infinite cardinals. If one assumes, for any κ, that those $\leqslant \kappa$-sized subsets should be regarded as "very small" with respect to the corresponding 2^κ-sized set then it is easily arguable that the intuitive outcome regarding the complexity of those two parametrized types of games should *not* coincide at *any* level. Therefore, **GCH** (the Generalized Continuum Hypothesis) has been shown to be equivalent to a kind of infinitary conjunction of a certain indexed-by-all-infinite-cardinals list of counterintuitive (and possibly surprising) statements.

Keywords: Generalized Continuum Hypothesis; complexity; Dialectica Categories.

1 Introduction

In what follows, we work in **ZFC** (which is Zermelo-Frankel Axiomatic Set Theory **ZF** with the addition of the Axiom of Choice, **AC**), meaning that all results in this paper are theorems of **ZFC**. All set-theoretical terminology and notation are standard, see e.g. [6]. The cardinality of a set A is denoted by $|A|$. The family of

The author acknowledges the referees for their careful reading of the paper and for providing a number of corrections and suggestions which improved the quality of the paper.
Funding: The author was supported by FAPESB, Fundação de Amparo à Pesquisa do Estado da Bahia, Grant APP0072/2016.

all functions from a set A into a set B is denoted by ${}^{A}B$. If $\kappa \leqslant |A|$, we denote by $[A]^{\leqslant \kappa}$ the family of all subsets of A whose cardinality is not greater than κ; it is well-known that $|[A]^{\leqslant \kappa}| = |{}^{\kappa}A| := |A|^{\kappa}$. The cardinality of the real line, $|\mathbb{R}| = 2^{\aleph_0}$, is denoted by $\mathfrak{c}$ and is referred to as *the continuum.* **CH** denotes the *Continuum Hypothesis*, which is the statement "$\mathfrak{c} = \aleph_1$". **GCH** denotes the *Generalized Continuum Hypothesis*, which is the statement "For every infinite cardinal κ one has $2^{\kappa} = \kappa^{+}$" – or, equivalently, "For every ordinal α one has $2^{\aleph_\alpha} = \aleph_{\alpha+1}$". Both **CH** and **GCH** are known to be independent of **ZFC**.

In our previous paper [13], two families of incidence problems were investigated. Those families were denoted by $\mathcal{C}_1$ and $\mathcal{C}_2$ and problems from both families involved real numbers as much as countably infinite subsets of the set $\mathbb{R}$ of all real numbers. Each one of those problems could also be stated (assuming the context of a thought experiment where randomness and arbitrariness are identified) as a sort of *challenge* – or as some kind of one-round hit-and-miss game between two players, on which the second player wins if his/her random move solves the proposed problem. Stated in the language of games, instances of $\mathcal{C}_1$ were as follows: the move of the first player is a real number x, and the move of the second player is a randomly taken countably infinite set of reals A. The second player wins if, and only if, the real number x is an element of A (that is, the solution for the instance x of $\mathcal{C}_1$ is a countably infinite set of reals A such that $x \in A$ – say, A should be such that x *hits* A). Instances of $\mathcal{C}_2$ were as follows: the move of the first player is a countably infinite set of reals A, and the move of the second player is a randomly taken real number x. The second player wins if, and only if, the real number x is *not* an element of A (that is, the solution for the instance A of $\mathcal{C}_2$ is a real number x such that $x \notin A$ – say, x should be such that x *misses* A). As pointed out in [13], our intuition tends to accept that games from $\mathcal{C}_2$ are much more likely to be won by the second player than games from $\mathcal{C}_1$ – or, in terms of *complexity*, problems of $\mathcal{C}_2$ are easier to be solved than problems of $\mathcal{C}_1$. The two main results of [13] were the following (working within the **ZFC** setting): on one hand (and after a suitable formalization), problems of $\mathcal{C}_2$ were shown to be at least as easy to solve as problems of $\mathcal{C}_1$. On the other hand, the statement "Problems of $\mathcal{C}_1$ have the exact same complexity of problems of $\mathcal{C}_2$" was shown to be an equivalent of the Continuum Hypothesis.

In this note, we generalize the results from [13] by exhibiting two classes of parametrized types of problems, which will be denoted by $\mathbf{C}_1$ and $\mathbf{C}_2$, whose parameters will be given by all infinite cardinals, and we show that the Generalized Continuum Hypothesis (**GCH**) is equivalent to a level-by-level equality between the complexities of the problems from each class. As in [13], the comparison of complexities of problems will be done using *GT*-reductions (i.e., morphisms) between objects of the dual Dialectica category $\mathrm{Dial}_2(\mathbf{Sets})^{\mathrm{op}}$.

$\mathrm{Dial}_2(\mathbf{Sets})^{\mathrm{op}}$ is the dual of the simplest case of the Dialectica Categories introduced by Valeria de Paiva in her PhD Thesis ([8],[9]). The morphisms between objects of this category are usually referred to as *Galois-Tukey connections*; this terminology is due to Peter Vojtáš (see [14]). Applications of such morphisms in Set Theory – mostly to prove inequalities between *cardinal invariants of the continuum* – were studied (in a large series of papers) by Andreas Blass in the 90's (see e.g. [1],[2]), and, more recently, were also investigated by the author ([11],[12],[13]).

The objects of $\mathrm{Dial}_2(\mathbf{Sets})^{\mathrm{op}}$ are triples (A, B, E), where A, B and E are sets and E is an usual set-theoretic binary relation E from A to B, i.e. $E \subseteq A \times B$. Next, we introduce certain parametrized versions of the usual category $\mathcal{PV}$ (see [13]). For each infinite cardinal κ we will consider a subcategory $\mathcal{PV}_\kappa$ of the category $\mathrm{Dial}_2(\mathbf{Sets})^{\mathrm{op}}$, whose objects are those which satisfy the following conditions (which will be called the $\mathbf{MHD}_\kappa$ conditions – where **MHD** stands for: Moore, Hrušák and Džamonja, [7]):

(1) $0 < |A|, |B| \leqslant 2^\kappa$.

(2) $(\forall a \in A)(\exists b \in B)[aEb]$

(3) $(\forall b \in B)(\exists a \in A)[\neg (aEb)]$

The morphisms between objects of $\mathcal{PV}_\kappa$ are, precisely, the morphisms of the category $\mathrm{Dial}_2(\mathbf{Sets})^{\mathrm{op}}$ – that is, given objects $o_1 = (A_1, B_1, E_1)$ and $o_2 = (A_2, B_2, E_2)$ of $\mathcal{PV}_\kappa$, a $\mathcal{PV}_\kappa$ morphism from o_2 to o_1 is a pair of functions (φ, ψ), where $\varphi : A_1 \to A_2$ and $\psi : B_2 \to B_1$ should satisfy the following requirement:

$$(\forall a \in A_1)\ (\forall b \in B_2)\ [\varphi(a)\, E_2\, b \longrightarrow a\, E_1\, \psi(b)].$$

Morphisms of $\mathcal{PV}_\kappa$ induce the so-called **(parametrized) Galois-Tukey preorder** $\leqslant_{GT,\kappa}$, which is defined in the following way: if $o_1 = (A_1, B_1, E_1)$ and $o_2 = (A_2, B_2, E_2)$ are objects of $\mathcal{PV}_\kappa$, then we have

$$o_1 \leqslant_{GT,\kappa} o_2 \iff \text{There is a } \mathcal{PV}_\kappa \text{ morphism from } o_2 \text{ to } o_1.$$

The diagram below represents the situation where $o_1 \leqslant_{GT,\kappa} o_2$:

$$\begin{array}{ccc} {}^{a}A_1 & E_1 & B_1{}^{\psi(b)} \\ \varphi\big\downarrow & & \big\uparrow\psi \\ {}_{\varphi(a)}A_2 & E_2 & B_2\ {}_{b} \end{array}$$

Of course, the usual category $\mathcal{PV}$ coincides with $\mathcal{PV}_{\aleph_0}$. Similarly, $o_1 \leqslant_{GT} o_2$ means $o_1 \leqslant_{GT,\aleph_0} o_2$.

Given an object $o = (A, B, E)$ of $\mathcal{PV}_\kappa$, its *dual object* is given by $o^* = (B, A, E^*)$, where bE^*a means that $\neg(aEb)$. A contrapositive argument shows that:

$$\text{If } o_1 \leqslant_{GT,\kappa} o_2, \text{ then } o_2^* \leqslant_{GT,\kappa} o_1^*.$$

The GT-connections are also referred to as *GT-reductions*, accordingly to Blass. In fact, Blass' view on those morphisms corresponds to a notion of *reduction between problems*. More precisely, Blass (see [1]) gives to each object $o = (A, B, E)$ the following interpretation: the object o stands for a certain problem (or a type of problem); A is the set of all instances of the problem represented by o; B is the set of all possible candidates to solve the corresponding problem; and E is the relation "is solved by" – so, aEb says that "b solves a".

As explained in [13], within this interpretation we have that the Galois-Tukey pre-order corresponds to a measure of complexity: if $o_1 \leqslant_{GT,\kappa} o_2$ then the problems of o_1 are not more complicated to solve than problems of o_2. This can be said because instances in o_1 are at least as simple to be solved as instances of o_2 – since every instance in o_1 may have its solution *reduced* to the solution of an instance in o_2 – more precisely, the act of solving an instance of a problem of o_1 may be reduced to the act of solving a (corresponding) instance of a problem of o_2. Indeed: assume $o_1 \leqslant_{GT,\kappa} o_2$. Then, if $b \in B_2$ is a solution for the instance $\varphi(a)$ of o_2 then $\psi(b) \in B_1$ is a solution for the instance a of o_1.

We will present our parametrized families of problems using directly the language of *ideals* (thus, our presentation is slightly different from that initially done at [13]). A family $\mathcal{I}$ of subsets of a non-empty set X is said to be an *ideal of subsets of* X if it is a proper, non-empty subfamily of $\mathcal{P}(X)$ which is closed under taking subsets and under taking finite unions.

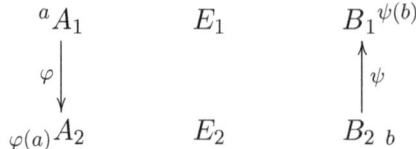

Definition 1.1. *Let κ be an infinite cardinal. We denote by $\mathfrak{I}_\kappa$ the ideal of $\leqslant \kappa$-sized subsets of ${}^\kappa\kappa$, that is:*

$$\mathfrak{I}_\kappa = \{A \subseteq {}^\kappa\kappa : |A| \leqslant \kappa\}.$$

By elementary cardinal arithtmetic, one has $|\mathfrak{I}_\kappa| = |[\kappa]^{\leqslant\kappa}| = \kappa^\kappa = 2^\kappa$ for every infinite cardinal κ.

The link between ideals and the categories $\mathcal{PV}_\kappa$ is given by the notion of *norms* (or *evaluations*) of objects. If $o = (A, B, E)$ be an object of $\mathcal{PV}_\kappa$, its **norm** is the cardinal number $||o|| = ||(A, B, E)||$ given by

$$||o|| = \min\{|Y| : Y \subseteq B \text{ and } (\forall a \in A)(\exists b \in Y)[aEb]\}.$$

Notice that such cardinal is well-defined by **MHD** condition (2) – and, moreover, **MHD** condition (3) ensures that the norm of the dual object o^* is well-defined as well.

The so-called *method of morphisms* to prove inequalities between cardinal invariants of the continuum is, arguably, the more important application of Dialectica categories in Set Theory. As Blass once said ([1]), *it is an empirical fact that proofs of inequalities between cardinal characteristics of the continuum usually proceed by representing the characteristics as norms of objects in $\mathcal{PV}$ and then exhibiting explicit morphisms between those objects.* The existence of morphisms imply inequalities between norms because of the following "folklore" result (mentioned in [2]):

$$\text{If } o_1 \leqslant_{GT} o_2 \text{ then } ||o_1|| \leqslant ||o_2||.$$

So, in fact $o_1 \leqslant_{GT} o_2$ implies both $||o_1|| \leqslant ||o_2||$ and $||o_2^*|| \leqslant ||o_1^*||$. More details (and proofs) on the above results may be found in [4] and [11]. It is easy to see that the analogous results for $\leqslant_{GT,\kappa}$ hold for any infinite cardinal κ.

When it comes to ideals, certain cardinal invariants associated to ideals can also be expressed in terms of norms of objects of $\mathcal{PV}_\kappa$.

Definition 1.2 (Cardinal invariants related to ideals). *Let $\mathfrak{I}$ be an ideal of subsets of an infinite set X.*

(i) add($\mathfrak{I}$) (the **additivity** *of $\mathfrak{I}$) is the smallest size of a subfamily of $\mathfrak{I}$ whose union is not in $\mathfrak{I}$ – that is,*

$$add(\mathfrak{I}) = \min\{|\mathcal{A}| : \mathcal{A} \subseteq \mathfrak{I} \text{ and } \bigcup \mathcal{A} \notin \mathfrak{I}\}.$$

(ii) $cov(\mathfrak{I})$ (the **covering number** *of $\mathfrak{I}$) is the smallest size of a subfamily of $\mathfrak{I}$ which covers X – that is,*

$$cov(\mathfrak{I}) = \min\{|\mathcal{A}| : \mathcal{A} \subseteq \mathfrak{I} \text{ and } \bigcup \mathcal{A} = X\}.$$

(iii) $non(\mathfrak{I})$ (the **uniformity** *of $\mathfrak{I}$) is the smallest size of a subset of X which is not in $\mathfrak{I}$ – that is,*

$$non(\mathfrak{I}) = \min\{|A| : A \subseteq X \text{ and } A \notin \mathfrak{I}\}.$$

(iv) $cof(\mathfrak{I})$ (the **cofinality** *of $\mathfrak{I}$) is the smallest size of a subfamily of $\mathfrak{I}$ which is cofinal in $\mathfrak{I}$ – that is,*

$$cof(\mathfrak{I}) = \min\{|\mathcal{A}| : \mathcal{A} \subseteq \mathfrak{I} \text{ and } (\forall I \in \mathfrak{I})(\exists A \in \mathcal{A})[I \subseteq A]\}.$$

One can easily check that, given an ideal $\mathfrak{I}$ of ideals over an infinite set X, the following equalities hold:

$$\begin{aligned} \mathrm{add}(\mathfrak{I}) &= ||(\mathfrak{I}, \mathfrak{I}, \not\supseteq)||, \\ \mathrm{non}(\mathfrak{I}) &= ||(\mathfrak{I}, X, \not\ni)||, \\ \mathrm{cov}(\mathfrak{I}) &= ||(X, \mathfrak{I}, \in)||, \\ \mathrm{cof}(\mathfrak{I}) &= ||(\mathfrak{I}, \mathfrak{I}, \subseteq)||. \end{aligned}$$

In the following section, we present our parametrized families of problems $\mathbf{C}_1$ and $\mathbf{C}_2$ and it will be clear that one of the implications of our main theorem may be proved using no more than the language of ideals.

2 A level-by-level equivalence with GCH

In this section, we generalize the main results of [13]. Differently from the paper referred to, the language of ideals will be directly used in certain proofs (and not simply presented as a way to state "possible reformulations"). The apparent adequacy of using ideals in this context will be object of further research.

Our parametrized families of problems are given by the following definition. In what follows, let **CARD** denote the class of all infinite cardinals.

Definition 2.1 (The parametrized families $\mathbf{C}_1$ and $\mathbf{C}_2$). *We set the classes of problems*

$$\mathbf{C}_1 = \{\mathcal{C}_{1,\kappa} : \kappa \in \mathbf{CARD}\}$$

and

$$\mathbf{C}_2 = \{\mathcal{C}_{2,\kappa} : \kappa \in \mathbf{CARD}\},$$

where, for each infinite cardinal κ, problems of $\mathcal{C}_{1,\kappa}$ and $\mathcal{C}_{2,\kappa}$ are as follows:

(i) Instances of $\mathcal{C}_{1,\kappa}$ are functions $f : \kappa \to \kappa$ and a solution for a given instance f is an element A of $\mathcal{I}_\kappa$ such that $f \in A$.

(ii) Instances of $\mathcal{C}_{2,\kappa}$ are elements A of $\mathcal{I}_\kappa$ and a solution for a given instance A is a function $f : \kappa \to \kappa$ such that $f \notin A$.

Of course, one may rephrase the preceding problems in the language of hit-and-miss games between two players, in the very same spirit of [13] – that is, the first player gives the initial data and the second player wins if he gives a random response which solves the problem. Wondering on an oriented thought experiment, we could think of our two players standing in front of two jars – one of the jars contains all functions from κ into κ and the other jar contains all $\leqslant \kappa$-sized subsets of ${}^\kappa\kappa$, and the moves of the players correspond to arbitrary (thus, random in the *ad hoc* identification proposed in [13]) pickings of elements of the jars, in accordance with the description of the problems.

Notice that, under Blass' interpretation, we may identify $\mathcal{C}_{1,\kappa}$ and $\mathcal{C}_{2,\kappa}$, respectively, with the objects $({}^\kappa\kappa, \mathcal{I}_\kappa, \in)$ and $(\mathcal{I}_\kappa, {}^\kappa\kappa, \not\ni)$ of $\mathcal{PV}_\kappa$. The same way as in [13], our intuition tends to accept that problems of $\mathcal{C}_{2,\kappa}$ are easier to be solved than problems of $\mathcal{C}_{1,\kappa}$ – since $\leqslant \kappa$-sized subsets of a set of size 2^κ are often understood as being "very small". Indeed, as the cofinality of 2^κ is larger than κ (by the well-known König's Lemma), we can prove within **ZFC** that the comparison of inequalities coincide with our intuition in the mentioned direction.

Proposition 2.2. *Let κ be any infinite cardinal. Then, there is a GT-reduction witnessing*

$$(\mathcal{I}_\kappa, {}^\kappa\kappa, \not\ni) \leqslant_{GT,\kappa} ({}^\kappa\kappa, \mathcal{I}_\kappa, \in).$$

Equivalently: There is a GT-reduction which shows that problems of $\mathcal{C}_{2,\kappa}$ are at least as simple to be solved as problems of $\mathcal{C}_{1,\kappa}$.

Proof: Enumerate ${}^\kappa\kappa = \{f_\alpha : \alpha < 2^\kappa\}$. We associate to each $A \in \mathcal{I}_\kappa$ the ordinal $\gamma(A) < 2^\kappa$ given by

$$\gamma(A) = \sup\{\alpha < 2^\kappa : f_\alpha \in A\} + 1.$$

The preceding function is well-defined, by the previously mentioned König's Lemma on the cofinality of 2^κ. Notice that, for each $A \in \mathcal{I}_\kappa$, $\gamma(A)$ is not the index of an element of A and, moreover, it is the smallest ordinal ζ for which we are sure that the final segment $[\zeta, 2^\kappa[$ is disjoint from $\{\alpha < 2^\kappa : f_\alpha \in A\}$.

Let $\varphi : \mathcal{I}_\kappa \to {}^\kappa\kappa$ be defined in the following way: for every $A \in \mathcal{I}_\kappa$,

$$\varphi(A) = f_{\gamma(A)}.$$

Notice that, as $[\gamma(A), 2^\kappa[$ is surely disjoint from A (identifying, of course, A with its ordinal indexes), every element of A has its ordinal index strictly smaller than the ordinal index of $\varphi(A)$.

Now, we consider the pair of functions given by (φ, φ). In what follows, we check that such pair constitute a morphism of $\mathcal{PV}_\kappa$ showing that the solution of instances of $(\mathcal{I}_\kappa, {}^\kappa\kappa, \not\ni)$ may be reduced to the solution of instances of $({}^\kappa\kappa, \mathcal{I}_\kappa, \in)$ (since it is a morphism of $\mathcal{PV}_\kappa$ from the object $({}^\kappa\kappa, \mathcal{I}_\kappa, \in)$ to the object $(\mathcal{I}_\kappa, {}^\kappa\kappa, \not\ni)$).

$$\begin{array}{ccccc} A \ \mathcal{I}_\kappa & & \not\ni & & {}^\kappa\kappa \quad f_{\gamma(B)} \\ \varphi \big\downarrow & & & & \big\uparrow \varphi \\ f_{\gamma(A)} \ {}^\kappa\kappa & & \in & & \mathcal{I}_\kappa \ B \end{array}$$

To verify the claim, one has only to notice that if A, B are elements of the ideal of $\leqslant \kappa$-sized subsets of ${}^\kappa\kappa$ which satisfy $\varphi(A) = f_{\gamma(A)} \in B$ then necessarily $\gamma(A) < \gamma(B)$, and this clearly implies $f_{\gamma(B)} = \varphi(B) \notin A$. ■

Notice that, as a consequence of the preceding proposition, one can say that, level-by-level, problems of $\mathcal{C}_2$ are, indeed, simpler to solve than problems of $\mathcal{C}_1$.

Now we check that, for a fixed infinite cardinal κ, the existence of a GT, κ-reduction in the opposite direction is equivalent to a consistent statement from cardinal arithmetic[1].

Theorem 2.3. *Let κ be an infinite cardinal. Then, the following statements are*

[1]However, despite of its consistency, item (*iii*) could be argued to be a very counterintuitive mathematical statement under Blass' interpretation of comparison of complexities – since if one accepts that κ should be considered "very small" in comparison with 2^κ, then the corresponding intuitive outcome would be in the direction of problems of $\mathcal{C}_{2,\kappa}$ being regarded as easier to solve (by far) than problems of $\mathcal{C}_{1,\kappa}$.

equivalent:

(*i*) *The Generalized Continuum Hypothesis holds at* κ *– i.e.,* $2^\kappa = \kappa^+$.

(*ii*) $({}^\kappa\kappa, \mathfrak{I}_\kappa, \in) \leqslant_{GT,\kappa} (\mathfrak{I}_\kappa, {}^\kappa\kappa, \not\ni)$.

In words: There is a GT-reduction which shows that problems of $\mathcal{C}_{1,\kappa}$ *are at least as simple to be solved as problems of* $\mathcal{C}_{2,\kappa}$.

(*iii*) $({}^\kappa\kappa, \mathfrak{I}_\kappa, \in) \cong_{GT,\kappa} (\mathfrak{I}_\kappa, {}^\kappa\kappa, \not\ni)$.

In words: There are GT-reductions (in both directions) which show that problems of $\mathcal{C}_{1,\kappa}$ *have the exact same complexity of problems of* $\mathcal{C}_{2,\kappa}$.

Proof. The equivalence between (*ii*) and (*iii*) is a consequence of the preceding absolute **ZFC** proposition – so, it suffices to prove (*i*) $\Rightarrow$ (*ii*) and (*iii*) $\Rightarrow$ (*i*).

Proof of (*i*) $\Rightarrow$ (*ii*)*:* By hypothesis we have $2^\kappa = \kappa^+$ and therefore we may enumerate ${}^\kappa\kappa$ as

$$ {}^\kappa\kappa = \{f_\alpha : \alpha < \kappa^+\}. $$

As κ^+ is precisely the set of all ordinals whose size is not larger than κ, we are able to define $\varphi : {}^\kappa\kappa \to \mathfrak{I}_\kappa$ by putting, for every $\alpha < \kappa^+$,

$$ \varphi(f_\alpha) = \{f_\xi : \xi \leqslant \alpha\}. $$

Now we pick $\psi = \varphi$, i.e., we consider the pair of functions (φ, φ) and we show that such pair is a morphism from $(\mathfrak{I}_\kappa, {}^\kappa\kappa, \not\ni)$ to $({}^\kappa\kappa, \mathfrak{I}_\kappa, \in)$.

To check this, fix arbitrary $f, g \in {}^\kappa\kappa$, say $f = f_\alpha$ and $g = f_\beta$ for $\alpha, \beta < \kappa^+$ and notice that if $f_\beta \notin \varphi(f_\alpha) = \{f_\xi : \xi \leqslant \alpha\}$ then $\beta > \alpha$ – thus, $f_\alpha \in \{f_\zeta : \zeta \leqslant \beta\} = \varphi(f_\beta)$, and so we are done.

Proof of (*iii*) $\Rightarrow$ (*i*)*:* In [13], the proof of the implication for $\mathbb{R}$ and $[\mathbb{R}]^{\aleph_0}$ – which corresponds to the proof for $\kappa = \aleph_0$, in the notation and terminology of the present paper – was given by a contrapositive argument which shows that no pair of functions

(φ, ψ) constitutes a morphism as required. Such argument resembles Freiling's proof of a certain equivalence of $\neg$**CH** in the celebrated paper on *throwing darts at the real line* ([3]). In fact, one could say that the argument is due to Sierpiński, since it is openly recognized in Freiling's paper itself that the mathematical content of this part of [3] comes from the classical Sierpiński's monograph on **CH** ([10]). We refrain for presenting a similar contrapositive argument here, since the language of ideals is enough to prove the desired implication. Indeed, the method of morphisms gives us

$$({}^{\kappa}\kappa, \mathcal{I}_{\kappa}, \in) \cong_{GT,\kappa} (\mathcal{I}_{\kappa}, {}^{\kappa}\kappa, \not\ni) \Rightarrow ||({}^{\kappa}\kappa, \mathcal{I}_{\kappa}, \in)|| = ||(\mathcal{I}_{\kappa}, {}^{\kappa}\kappa, \not\ni)||$$

and, as we have already mentioned that the above norms correspond to specifical cardinal invariants of the ideal $\mathcal{I}_{\kappa}$, it follows that, in **ZFC**, the following implication holds:

$$({}^{\kappa}\kappa, \mathcal{I}_{\kappa}, \in) \cong_{GT,\kappa} (\mathcal{I}_{\kappa}, {}^{\kappa}\kappa, \not\ni) \Rightarrow \mathrm{cov}(\mathcal{I}_{\kappa}) = \mathrm{non}(\mathcal{I}_{\kappa}).$$

However, the absolute **ZFC** values of $\mathrm{cov}(\mathcal{I}_{\kappa})$ and $\mathrm{non}(\mathcal{I}_{\kappa})$ are, precisely, 2^{κ} and κ^{+}. The equality $\mathrm{non}(\mathcal{I}_{\kappa}) = \kappa^{+}$ is obvious, and for the other case notice that if $\lambda < 2^{\kappa}$ then the union of a family of λ subsets of size $\leqslant \kappa$ of ${}^{\kappa}\kappa$ has size not larger than $\max\{\lambda, \kappa\} < 2^{\kappa}$. ■

The following corollary (which comes by simply applying the preceding theorem for all infinite cardinals) is the main result of this paper:

Corollary 2.4. *The Generalized Continuum Hypothesis is equivalent to the existence of a level-by-level equality between the complexities of the problems from the parametrized families* $\mathbf{C}_1$ *and* $\mathbf{C}_2$.

3 Notes and Questions

Our main result shows that **GCH** is equivalent to a kind of infinitary conjunction of certain indexed-by-all-infinite-cardinals list of counterintuitive (and even surprising) statements. In this sense, it corresponds, in the level of intuition, to a very strong failure of **GCH**. Indeed, in one hand, the existence of a single infinite cardinal κ for which the complexities of the problems $\mathcal{C}_{1,\kappa}$ and $\mathcal{C}_{2,\kappa}$ do not coincide was shown to be enough to refute **GCH**. On the other hand, our intuition (nourished by the general belief that κ is always "very small" in size when compared to 2^{κ}) tends to accept that the mentioned complexities *do not coincide at any level.* In a similar way as done in [13], we will not go further on the philosophical discussion of whether the main result of this paper should be understood (or not) as an evidence against **GCH** – but we believe that it is an interesting discussion.

We finish this paper by presenting some "generalized versions" of questions and problems posed in [13]. Some of these problems are related to the precise role of the Axiom of Choice in the presented results. Notice that we have defined several constituent functions of morphisms using enumerations (essentially, well-orderings) of the set ${}^{\kappa}\kappa$ (with order type 2^{κ} or κ^{+}, depending on the specifical hypotheses in each case). In a different way, the morphism which has shown to hold in all models of **ZFC** (which is $(\mathfrak{I}_{\kappa}, {}^{\kappa}\kappa, \not\ni) \leqslant_{GT,\kappa} ({}^{\kappa}\kappa, \mathfrak{I}_{\kappa}, \in)$) also strongly depends on the Axiom of Choice – since the argument is based on the inequality $cf(2^{\kappa}) > \kappa$, which is a consequence of the well-known theorem of König on sums and products of cardinals – and such classical theorem may be rephrased into an equivalent of the Axiom of Choice.

Question 3.1. *Consider the results involving the families of problems* $\mathbf{C_1}$ *and* $\mathbf{C_2}$ *which have been proved in this paper. Is it possible to recover them, if one avoids* **AC** *? I.e., what happens if we restrict ourselves to* **ZF** *?*

The above line of investigation seems very promising, and one of the reviewers has insisted that results of this kind would be very appealing. Further research is needed here.

It is also interesting to investigate what happens if one considers that "having size less than 2^{κ}" corresponds to a reasonable notion of "smallness" for subsets of ${}^{\kappa}\kappa$. Let $\mathfrak{I}_{<2^{\kappa}}$ denote the ideal of subsets of ${}^{\kappa}\kappa$ whose size is less than 2^{κ} (for a given infinite cardinal κ). Then $({}^{\kappa}\kappa, \mathfrak{I}_{<2^{\kappa}}, \in) \leqslant_{GT,\kappa} (\mathfrak{I}_{<2^{\kappa}}, {}^{\kappa}\kappa, \not\ni)$ would follow, in **ZF**, from a well-ordering of ${}^{\kappa}\kappa$, and, in **ZFC**, the morphism in the opposite direction would follow from the statement "2^{κ} is regular", which is easily seen to be consistent.

Problem 3.2. *Determine the precise deductive strengths of*
(*i*) $\forall\kappa \in \mathbf{CARD}[({}^{\kappa}\kappa, \mathfrak{I}_{<2^{\kappa}}, \in) \leqslant_{GT,\kappa} (\mathfrak{I}_{<2^{\kappa}}, {}^{\kappa}\kappa, \not\ni)]$, *relatively to* **ZF**.
(*ii*) $\forall\kappa \in \mathbf{CARD}[(\mathfrak{I}_{<2^{\kappa}}, {}^{\kappa}\kappa, \not\ni) \leqslant_{GT} ({}^{\kappa}\kappa, \mathfrak{I}_{<2^{\kappa}}, \in)]$, *relatively to* **ZFC**.

The set ${}^{\kappa}\kappa$ may be endowed with the so-called *bounded topology* and thus become what is nowadays referred to as the *Generalized Baire Space* – and this new line of research has attracting the attention of several set theorists. One can generalize the notion of *Borel sets* for the Generalized Baire Space as the smallest collection containing the open sets in the bounded topology on ${}^{\kappa}\kappa$ and closed under complements and unions of size $\leqslant \kappa$. Background and information on this subject may be found in the recent survey [5]. It is known that well-orderings of $\mathbb{R}$ are not Borel functions in the usual sense, and this had justified the very last question of [13]. The following question is the natural generalization of it.

Question 3.3. *What happens if one requires the constituent functions of all morphisms (in the context of the results of this paper) to be Borel functions (in the above described generalized sense of the Generalized Baire Space* ${}^{\kappa}\kappa$*) ?*

References

[1] Blass, A., Questions and Answers – A Category Arising in Linear Logic, Complexity Theory, and Set Theory, in: J.-Y. Girard, Y. Lafont, and L. Regnier (Editors), *Advances in Linear Logic*, London Math. Soc. Lecture Notes 222, 61–81, 1995.

[2] Blass, A., Propositional Connectives and the Set Theory of the Continuum, *CWI Quarterly* 9:25–30, 1996.

[3] Freiling, C., Axioms of Symmetry: throwing darts at the real number line. *Journal of Symbolic Logic*, 51(1):190–200, 1986.

[4] Garcia, H.; da Silva, S.G. Identifying small with bounded: unboundedness, domination, ideals and their cardinal invariants. *South American Journal of Logic*, 2(2):425–436, 2016.

[5] Khomskii, Y.; Laguzzi, G.; Löwe, B.; Sharankou, I., Questions on generalised Baire spaces. *MLQ Mathematical Logic Quarterly* 62(4-5): 439–456, 2016.

[6] Kunen, K., *Set theory. An introduction to independence proofs.* Studies in Logic and the Foundations of Mathematics, Vol. 102. Amsterdam, North-Holland Publishing Company, 1980, xvi + 313 pp.

[7] Moore, J. T.; Hrušák, M.; and Džamonja, M., Parametrized $\diamondsuit$ principles, *Transactions of Americal Mathematical Society* 356(6):2281–2306, 2004.

[8] de Paiva, V., A dialectica-like model of linear logic, in: Pitt, D., Rydeheard, D., Dybjer, P., Pitts, A. and Poigne, A. (Editors), *Category Theory and Computer Science*, Springer, 341–356, 1989.

[9] de Paiva, V., *The Dialectica Categories*, Computer Laboratory, University of Cambridge, 1990.

[10] Sierpiński, W., *Hypothèse du Continu.* Monografie Matematyczne, 1 ère ed. PWN, Varsóvia, 1934, v + 192 pp.

[11] da Silva, S. G.; de Paiva, V., Dialectica categories, cardinalities of the continuum and combinatorics of ideals, *Logic Journal of the IGPL* 25(4):585–603, 2017.

[12] da Silva, S. G., The Axiom of Choice and the Partition Principle from Dialectica Categories, 2019. Submitted.

[13] da Silva, S. G., Reductions between certain incidence problems and the Continuum Hypothesis. *Reports on Mathematical Logic* 54:121–143, 2019.

[14] Vojtáš, P., Generalized Galois-Tukey-connections between explicit relations on classical objects of real analysis, in: Judah, H. (Editor), *Set theory of the reals* (Ramat Gan, 1991), Bar-Ilan Univ., Ramat Gan, Israel Math. Conf. Proc. 6, 619–643, 1993.

Received 31 January 2020

$\mathbf{G}'_3$ as the logic of modal 3-valued Heyting algebras

Marcelo Esteban Coniglio
Institute of Philosophy and the Humanities (IFCH) and Centre for Logic, Epistemology and The History of Science (CLE), University of Campinas (UNICAMP), Campinas, SP, Brazil
coniglio@unicamp.br

Aldo Figallo-Orellano
Departamento de Matemática, Universidad Nacional del Sur (UNS), Argentina and Centre for Logic, Epistemology and The History of Science (CLE), University of Campinas (UNICAMP), Campinas, SP, Brazil
aldofigallo@gmail.com

Alejandro Hernández-Tello
Instituto de Física y Matemática, Universidad Tecnológica de la Mixteca (UTM) Huajuapan de Leon, Oaxaca, México
alheran@gmail.com

Miguel Pérez-Gaspar
Facultad de Ingeniería, Universidad Nacional Autónoma de México (UNAM) Ciudad de México, México
miguel.perez@fi-b.unam.mx

Abstract

In 2001, W. Carnielli and Marcos considered a 3-valued logic in order to prove that the schema $\varphi \vee (\varphi \to \psi)$ is not a theorem of da Costa's logic C_ω. In 2006, this logic was studied (and baptized) as $\mathbf{G}'_3$ by Osorio *et al.* as a tool to define semantics of logic programming. It is known that the truth-tables of $\mathbf{G}'_3$ have the same expressive power than the one of Łukasiewicz 3-valued logic as well as the one of Gödel 3-valued logic $\mathbf{G}_3$. From this, the three logics coincide up-to language, taking into acccount that 1 is the only designated truth-value in these logics.

From the algebraic point of view, Canals-Frau and Figallo have studied the 3-valued modal implicative semilattices, where the modal operator is the well-known Moisil-Monteiro-Baaz Δ operator, and the supremum is definable from this. We prove that the subvariety obtained from this by adding a bottom element 0 is term-equivalent to the variety generated by the 3-valued algebra of $\mathbf{G}'_3$. The algebras of that variety are called $\mathbf{G}'_3$-algebras. From this result, we obtain the equations which axiomatize the variety of $\mathbf{G}'_3$-algebras. Moreover, we prove that this variety is semisimple, and the 3-element and the 2-element chains are the unique simple algebras of the variety. Finally an extension of $\mathbf{G}'_3$ to first-order languages is presented, with an algebraic semantics based on complete $\mathbf{G}'_3$-algebras. The corresponding soundness and completeness theorems are obtained.

1 Introduction and preliminaries

In 2001, W. Carnielli *et al.* [4] considered a 3-valued logic in order to prove that the schema $\varphi \vee (\varphi \to \psi)$ is not a theorem of da Costa's logic C_ω. In 2006 this logic was studied (and baptized as $\mathbf{G}'_3$) by Osorio *et al.* [7] as a tool to define semantics of logic programming. They define the connectives $\to$ and $\neg$ of $\mathbf{G}'_3$ logic in terms of some connectives of the three-valued logic of Łukasiewicz Ł$_3$. Conjunction and disjunction, $\wedge$ and $\vee$ respectively, are defined as minimum and maximum. It is known that the truth-tables of $\mathbf{G}'_3$ have the same expressive power than the ones of Łukasiewicz 3-valued logic Ł$_3$ –hence, to the ones of Gödel 3-valued logic **G3**. From this, the three logics coincide up-to language, taking into account that **1** is the only designated truth-value in these logics.

The three-valued Gödel logic **G3**, which is also equivalent to $\mathbf{G}'_3$ and Ł$_3$, is well-suited to express the Stable Model Semantics. $\mathbf{G}'_3$, besides being very close to **G3**, can be used to express non-monotonic reasoning. It is worth mentioning that the negation of **G3** can be reconstructed from connectives of $\mathbf{G}'_3$ by virtue of the formula:

$$\neg_{\mathbf{G3}}\, \varphi = \varphi \to_{\mathbf{G}'_3} (\neg_{\mathbf{G}'_3}\, \varphi \wedge_{\mathbf{G}'_3} \neg_{\mathbf{G}'_3} \neg_{\mathbf{G}'_3}\, \varphi)$$

where the subscripts indicate the underlying logic.

Two different Hilbert-style systems for $\mathbf{G}'_3$ were introduced in [11] and [10], respectively. However, in both approaches it was assumed the validity of the *Deduction Theorem* in the proposed Hilbert calculi for $\mathbf{G}'_3$. As it will discussed in Remark 2.3 below, the Deduction Theorem does not hold in $\mathbf{G}'_3$. This issue in the previous axiomatic approaches to $\mathbf{G}'_3$ justifies proposing a new Hilbert calculus for $\mathbf{G}'_3$, as it will be done in Section 2. Taking into account that $\mathbf{G}'_3$ was introduced as a model of da Costa's logic C_ω, it seems reasonable to define a Hilbert calculus for $\mathbf{G}'_3$ which contains the calculus C_ω.

The paper is organized as follows. In the next section, we present a new Hilbert calculus for $\mathbf{G}'_3$ called $\mathbf{G}'_{3h}$, as an extension of C_ω. In Section 3 we consider the class of $\mathbf{G}'_3$-algebras, proving the soundness and completeness theorem of $\mathbf{G}'_{3h}$ w.r.t the class of $\mathbf{G}'_3$-algebras. After this, in Subsection 3.1 we connect the class of $\mathbf{G}'_3$-algebras with the variety of 3-valued modal implicative semilattices studied by Canals-Frau and Figallo. It will be proved that the subvariety of 3-valued modal implicative semilattices with bottom is term-equivalent to the class of $\mathbf{G}'_3$-algebras. From the latter, we obtain the equations that characterize the class of $\mathbf{G}'_3$-algebras as a variety. From this algebraic analysis, we prove in Subsection 3.2 a second adequacy theorem $\mathbf{G}'_{3h}$ w.r.t. the class of $\mathbf{G}'_3$-algebras. Finally, in Section 4 we present first-order version of $\mathbf{G}'_{3h}$ logic using algebraic tools developed in [5] (see, also, [6]) and our algebraic results of the class of $\mathbf{G}'_3$-algebra presented in the above section.

2 A new Hilbert-style axiomatization of $\mathbf{G}'_3$

Consider from now on the propositional signature $\Sigma = \{\wedge, \vee, \rightarrow, \neg\}$. First of all, let us recall the 3-valued semantics for $\mathbf{G}'_3$ logic. It is obtained from the logical matrix $\mathcal{M} = \langle \mathcal{D}, \mathcal{A}_3 \rangle$, where $\mathcal{D} = \{1\}$ and $\mathcal{A}_3 = \langle \mathcal{V}, \sigma \rangle$ is the 3-valued algebra over Σ with domain $\mathcal{V} = \{0, \frac{1}{2}, 1\}$ such that σ interprets the connectives of Σ as follows:

$\wedge$	**0**	$\mathbf{\frac{1}{2}}$	**1**
0	0	0	0
$\mathbf{\frac{1}{2}}$	0	$\frac{1}{2}$	$\frac{1}{2}$
1	0	$\frac{1}{2}$	1

$\vee$	**0**	$\mathbf{\frac{1}{2}}$	**1**
0	0	$\frac{1}{2}$	1
$\mathbf{\frac{1}{2}}$	$\frac{1}{2}$	$\frac{1}{2}$	1
1	1	1	1

$\rightarrow$	**0**	$\mathbf{\frac{1}{2}}$	**1**
0	1	1	1
$\mathbf{\frac{1}{2}}$	0	1	1
1	0	$\frac{1}{2}$	1

x	$\neg x$
0	1
$\mathbf{\frac{1}{2}}$	1
1	0

The set of well-formed formulas, denoted by $\mathcal{L}_\Sigma$, is constructed as usual from a given denumerable set $Var = \{p_0, p_1, \ldots\}$ of propositional variables. As usual, the bi-implication $\leftrightarrow$ can be defined in $\mathbf{G}'_3$ by $\varphi \leftrightarrow \psi \stackrel{\text{def}}{=} (\varphi \rightarrow \psi) \wedge (\psi \rightarrow \varphi)$. Its truth-table is displayed below.

$\leftrightarrow$	**0**	$\frac{1}{2}$	**1**
0	1	0	0
$\frac{1}{2}$	0	1	$\frac{1}{2}$
1	0	$\frac{1}{2}$	1

The consequence relation of $\mathbf{G_3'}$ induced by the logical matrix $\mathcal{M}$ will be denoted by $\models_{\mathbf{G_3'}}$. Thus: $\Gamma \models_{\mathbf{G_3'}} \varphi$ if and only if, for every valuation h (that is, for every homomorphism $h : \mathcal{L}_\Sigma \to \mathcal{A}_3$ of algebras over Σ), if $h(\psi) = 1$ for every $\psi \in \Gamma$ then $h(\varphi) = 1$.

A formal axiomatic system for $\mathbf{G_3'}$ called $\mathbf{G_{3h}'}$ over the signature Σ will be defined below (see Definition 2.1). Previous to this, some motivations will be given. The implication above is a particular case ($n = 3$) of the family of implicative systems LC_n proposed by Thomas in [12]. This implication, together with Thomas's axiom for 3-valued systems

$$\textbf{(Tho)} \quad ((\varphi \to \psi) \to \gamma) \to (((\gamma \to \varphi) \to \gamma) \to \gamma)$$

was used by L. Monteiro in [8] to introduce the class of 3-valued Heyting algebras. As we shall see, the logic $\mathbf{G_3'}$ is closely related to L. Monteiro's 3-valued Heyting algebras. Because of this, axiom **(Tho)** for 3-valued systems will be considered in $\mathbf{G_{3h}'}$. In addition, axiom

$$\textbf{(CF)} \quad (((\psi \to \neg\neg\psi) \to (\varphi \to \neg\neg\varphi)) \to \neg\neg(\varphi \to \psi)) \leftrightarrow (\neg\neg\varphi \to \neg\neg\psi)$$

which is adapted from an axiom introduced by Canals-Frau and Figallo in [2] to axiomatize the variety of 3-valued implicative semilattices, will be also considered by reasons which will be clear in Section 3.1 below.

Definition 2.1. *The Hilbert calculus* $\mathbf{G_{3h}'}$ *over* Σ *is defined as follows:*

Axiom schemas:

(Ax1)	$\varphi \to (\psi \to \varphi)$
(Ax2)	$(\varphi \to (\psi \to \gamma)) \to ((\varphi \to \psi) \to (\varphi \to \gamma))$
(Ax3)	$(\varphi \wedge \psi) \to \varphi$
(Ax4)	$(\varphi \wedge \psi) \to \psi$
(Ax5)	$\varphi \to (\psi \to (\varphi \wedge \psi))$
(Ax6)	$\varphi \to (\varphi \vee \psi)$
(Ax7)	$\psi \to (\varphi \vee \psi)$
(Ax8)	$(\varphi \to \gamma) \to ((\psi \to \gamma) \to ((\varphi \vee \psi) \to \gamma))$
(Ax9)	$\varphi \vee \neg\varphi$
(Ax10)	$\neg\neg\varphi \to \varphi$
(Ax11)	$\neg\varphi \to (\neg\neg\varphi \to \psi)$
(Ax12)	$\neg\neg(\varphi \vee \psi) \to (\neg\neg\varphi \vee \neg\neg\varphi)$
(Ax13)	$\neg\neg(\varphi \to \psi) \leftrightarrow ((\varphi \to \psi) \wedge (\neg\neg\varphi \to \neg\neg\varphi))$
(Tho)	$((\varphi \to \psi) \to \gamma) \to (((\gamma \to \varphi) \to \gamma) \to \gamma)$
(CF)	$(((\psi \to \neg\neg\psi) \to (\varphi \to \neg\neg\varphi)) \to \neg\neg(\varphi \to \psi)) \leftrightarrow (\neg\neg\varphi \to \neg\neg\psi)$

Inference Rules:

(MP) $\dfrac{\varphi \quad \varphi \to \psi}{\psi}$ ***(imp)*** $\dfrac{\varphi \to \psi}{\neg\psi \to \neg\varphi}$

It is worth noting that axioms **(Ax1)**-**(Ax8)** plus **(MP)** constitute a Hilbert calculus sound and complete for *Positive Intuitionistic Propositional Logic* $\mathbf{IPL}^+$. This means that $\mathbf{IPL}^+$ is contained in $\mathbf{G}_{3h}'$ (this fact will be used later). In addition, the calculus formed by axioms **(Ax1)**-**(Ax10)** plus **(MP)** is exactly da Costa's logic C_ω. Thus, $\mathbf{G}_3'$ is an extension of C_ω, in accordance with the original intuitions mentioned in Section 1.

Definition 2.2. *Let $\Gamma \cup \{\varphi\} \subseteq \mathcal{L}_\Sigma$ be a set of formulas. A derivation of φ from Γ in $\mathbf{G}_{3h}'$ is a finite sequence $\varphi_1 \cdots \varphi_n$ of formulas in $\mathcal{L}_\Sigma$ such that $\varphi_n = \varphi$ and for $1 \leq i \leq n$, it holds:*

1. *φ_i is an instance of some axiom in $\mathbf{G}_{3h}'$, or*
2. *$\varphi_i \in \Gamma$, or*
3. *there exist some $j, k < i$ such that φ_i follows from φ_j and φ_k by applying **MP**, or*
4. *there exist some $j < i$ such that φ_i follows from φ_j by applying **imp**.*

We say that φ is derivable from Γ in $\mathbf{G}_{3h}'$, *denoted as $\Gamma \vdash_{\mathbf{G}_{3h}'} \varphi$ (or simply $\Gamma \vdash \varphi$), if there exists a derivation of φ from Γ in $\mathbf{G}_{3h}'$.*

Remark 2.3. *It is easy to see that the* Deduction Theorem *does not hold* $\mathbf{G}'_3$*: indeed,* $p, \neg p \vDash_{\mathbf{G}'_3} q$ *for every propositional variables* p *and* q *with* $p \neq q$.[1] *However,* $\nvDash_{\mathbf{G}'_3} p \to (\neg p \to q)$*: it is enough to consider a valuation* h *such that* $h(p) = \frac{1}{2}$ *and* $h(q) = 0$*. Alternatively, the failure of the Deduction Theorem in* $\mathbf{G}'_3$ *can be seen by observing that* $p \vDash_{\mathbf{G}'_3} \neg\neg p$ *for every propositional variable* p*, but* $\nvDash_{\mathbf{G}'_3} p \to \neg\neg p$*: it suffices to consider a valuation* h *such that* $h(p) = \frac{1}{2}$*. From this, the Deduction Theorem should not be valid in* $\mathbf{G}'_{3h}$ *(since* $\mathbf{G}'_{3h}$ *is intended to be adequate to* $\mathbf{G}'_3$*). This fact will be proven in Corollary 3.28.*

Despite this, a restricted version of the Deduction Theorem holds in $\mathbf{G}'_{3h}$:

Proposition 2.4 (Restricted Deduction Theorem **(RDT)**). *Let* $\Gamma \cup \{\varphi, \psi\}$ *be a set of formulas in* $\mathcal{L}_\Sigma$*. Assume that* $\Gamma, \varphi \vdash_{\mathbf{G}'_{3h}} \psi$ *such that there is a derivation in* $\mathbf{G}'_{3h}$ *of* ψ *from* $\Gamma \cup \{\varphi\}$ *in which the inference rule* ***imp*** *is not used. Then,* $\Gamma \vdash_{\mathbf{G}'_{3h}} \varphi \to \psi$ *without using* ***imp****.*

Proof. It follows from the fact that, in such derivation of ψ from $\Gamma \cup \{\varphi\}$ in $\mathbf{G}'_{3h}$, axioms **Ax1** and **Ax2** are available, and **MP** is the only inference rule used there. Under these circumstances, the Deduction Theorem holds (see, for instance, [9]). Hence, $\Gamma \vdash \varphi \to \psi$ without using **imp**. □

A different form of the Deduction Theorem will be obtained in Theorem 3.13 below. A direct consequence of Proposition 2.4 is the *Restricted Proof by Cases* property:

Proposition 2.5 (Restricted Proof by Cases **(RPC)**). *Let* $\Gamma \cup \{\varphi, \psi\}$ *be a set of formulas in* $\mathcal{L}_\Sigma$*. Then, the following holds in* $\mathbf{G}'_{3h}$*:*

If $\Gamma, \varphi \vdash_{\mathbf{G}'_{3h}} \gamma$ *and* $\Gamma, \psi \vdash_{\mathbf{G}'_{3h}} \gamma$ *without using* ***imp****, then* $\Gamma, \varphi \vee \psi \vdash_{\mathbf{G}'_{3h}} \gamma$ *without using* ***imp****.*

In particular,

If $\Gamma, \varphi \vdash_{\mathbf{G}'_{3h}} \gamma$ *and* $\Gamma, \neg\varphi \vdash_{\mathbf{G}'_{3h}} \gamma$ *without using* ***imp****, then* $\Gamma \vdash_{\mathbf{G}'_{3h}} \gamma$ *without using* ***imp****.*

Proof. The first part is a direct consequence of Proposition 2.4, **(Ax8)** and **(MP)**. The second part follows from the first one by using **(Ax9)** and **(MP)**. □

[1]By structurality, $\varphi, \neg\varphi \vDash_{\mathbf{G}'_3} \psi$ for every formulas φ and ψ. Hence, the negation $\neg$ is *explosive* in $\mathbf{G}'_3$, so this logic is *not* paraconsistent w.r.t. $\neg$.

Proposition 2.6. *The following schemas are derivable in* $\mathbf{G}'_{\mathbf{3h}}$*:*

(1) $(\alpha \to \beta) \to ((\beta \to \gamma) \to (\alpha \to \gamma))$;

(2) $(\gamma \to \alpha) \to ((\gamma \to \beta) \to (\gamma \to (\alpha \wedge \beta)))$;

(3) $(\alpha \to \alpha') \to ((\beta \to \beta') \to ((\alpha \wedge \beta) \to (\alpha' \wedge \beta')))$;

(4) $(\alpha \to (\beta \to \gamma)) \to ((\alpha \wedge \beta) \to \gamma)$;

(5) $(\alpha' \to \alpha) \to ((\beta \to \beta') \to ((\alpha \to \beta) \to (\alpha' \to \beta')))$.

Proof. It follows from the fact that all these schemas are provable in positive intuitionistic propositional logic $\mathbf{IPL}^+$, which is contained in $\mathbf{G}'_{\mathbf{3h}}$. □

Proposition 2.7. *The following rules*

$$\textit{(Dneg)}\ \frac{\varphi}{\neg\neg\varphi} \qquad\qquad \textit{(exp)}\ \frac{\varphi \quad \neg\varphi}{\psi}$$

are derivable in $\mathbf{G}'_{\mathbf{3h}}$.

Proof. For **(Dneg)**, observe firstly that $(\varphi \to \varphi) \to \neg\neg(\varphi \to \varphi)$ is a theorem in $\mathbf{G}'_{\mathbf{3h}}$. Indeed, since both $(\varphi \to \varphi) \to (\neg\neg\varphi \to \neg\neg\varphi)$ and $(\varphi \to \varphi) \to (\varphi \to \varphi)$ are derivable in $\mathbf{G}'_{\mathbf{3h}}$ (by $\mathbf{IPL}^+$), so is $(\varphi \to \varphi) \to (\varphi \to \varphi) \wedge (\neg\neg\varphi \to \neg\neg\varphi)$. Hence $(\varphi \to \varphi) \to \neg\neg(\varphi \to \varphi)$ follows from this by using **(Ax13)**. Now, consider the following derivation in $\mathbf{G}'_{\mathbf{3h}}$:

1.	φ	Hyp.
2.	$(\varphi \to \varphi) \to \varphi$	$\mathbf{IPL}^+$
3.	$\neg\neg(\varphi \to \varphi) \to \neg\neg\varphi$	**(imp)**, 2 (two times)
4.	$(\varphi \to \varphi) \to \neg\neg(\varphi \to \varphi)$	Observation above
5.	$(\varphi \to \varphi) \to \neg\neg\varphi$	$\mathbf{IPL}^+$, 4,3
6.	$\neg\neg\varphi$	$\mathbf{IPL}^+$, 5

For **(exp)**, consider the following (meta)derivation in $\mathbf{G}'_{\mathbf{3h}}$:

1.	φ	Hyp.
2.	$\neg\varphi$	Hyp.
3.	$\neg\neg\varphi$	**(Dneg)**, 1
4.	$\neg\varphi \to (\neg\neg\varphi \to \psi)$	**(Ax11)**
5.	$\neg\neg\varphi \to \psi$	**(MP)**, 2,4
6.	ψ	**(MP)**, 3,5

□

Proposition 2.8. *The following schemas are derivable in* $\mathbf{G}'_{\mathbf{3h}}$*:*

(1) $\neg\neg(\varphi \wedge \psi) \leftrightarrow (\neg\neg\varphi \wedge \neg\neg\psi)$;

(2) $\neg\neg(\varphi \vee \psi) \leftrightarrow (\neg\neg\varphi \vee \neg\neg\psi)$

(3) $\neg\neg\neg\varphi \leftrightarrow \neg\varphi$.

Proof.

(1) From **(Ax5)** and **(Dneg)** it follows that $\neg\neg(\varphi \to (\psi \to (\varphi \wedge \psi)))$ is a theorem of $\mathbf{G'_{3h}}$. Using **(Ax13)**, **(Ax3)**, **(Ax4)**, Proposition 2.6 items (1), (4) and **(MP)** it follows that $(\neg\neg\varphi \wedge \neg\neg\psi) \to \neg\neg(\varphi \wedge \psi)))$. The converse is proved analogously.

(2) Analogously to the proof of item (1) (but now using **(Ax6)** and **(Ax7)**) it is proved that $\neg\neg(\varphi \vee \psi) \to (\neg\neg\varphi \vee \neg\neg\psi)$ is a theorem of $\mathbf{G'_{3h}}$. The converse is just **(Ax12)**.

(3) By **(Ax10)** it follows that $\neg\neg\neg\varphi \to \neg\varphi$ is a theorem of $\mathbf{G'_{3h}}$. In addition, $\neg\varphi, \neg\neg\varphi \vdash_{\mathbf{G'_{3h}}} \neg\neg\neg\varphi$ without using **imp** (just by using **(Ax10)** and **(MP)**), and also $\neg\varphi, \neg\neg\neg\varphi \vdash_{\mathbf{G'_{3h}}} \neg\neg\neg\varphi$ without using **imp** (by Definition 2.2). Then, $\neg\varphi \vdash_{\mathbf{G'_{3h}}} \neg\neg\neg\varphi$ without **imp**, by Proposition 2.5. Hence, $\vdash_{\mathbf{G'_{3h}}} \neg\varphi \to \neg\neg\neg\varphi$ by Proposition 2.4. □

Instead of proving directly the soundness and completeness of $\mathbf{G'_{3h}}$ w.r.t. $\mathbf{G'_3}$, in the next section an algebraic semantics for $\mathbf{G'_{3h}}$ will be proposed, based on a new class of algebras called $\mathbf{G'_3}$-algebras. After proving the adequacy of $\mathbf{G'_{3h}}$ w.r.t. this algebraic semantics, in Section 3.1 it will be proved that the class of $\mathbf{G'_3}$-algebras is in fact a variety (that is, it can be axiomatized by means of equations) which is term-equivalent to a subvariety of a variety already studied in the literature ([2]). This allows us to show that the algebra underlying the 3-valued matrix of $\mathbf{G'_3}$ generates the variety of $\mathbf{G'_3}$-algebras (see Corollary 3.23). The completeness of $\mathbf{G'_{3h}}$ w.r.t. $\mathbf{G'_3}$ will be obtained easily from this (see Theorem 3.27).

3 The class of $\mathbf{G'_3}$-algebras

Recall that $\Sigma = \{\wedge, \vee, \to, \neg\}$ is the propositional signature for logic $\mathbf{G'_3}$, and that $\mathcal{L}_\Sigma$ is the algebra of formulas of $\mathbf{G'_3}$ generated over Σ by Var. Let $\Sigma_I = \{\wedge, \vee, \to, \mathbf{0}, \mathbf{1}\}$ be the signature of Heyting algebras and let $\Sigma_+ = \{\wedge, \vee, \to, \neg, \mathbf{0}, \mathbf{1}\}$.

Definition 3.1. *A* $\mathbf{G'_3}$*-algebra is an algebra* $\mathcal{A} = \langle A, \wedge, \vee, \to, \neg, 0, 1\rangle$ *of type* $(2, 2, 2, 1, 0, 0)$ *such that*

(i) The reduct $\mathcal{H}_\mathcal{A} = \langle A, \wedge, \vee, \to, 0, 1\rangle$ *is a 3-valued Heyting algebra (see [8]). That is,* $\mathcal{H}_\mathcal{A}$ *is a Heyting algebra such that, for every* $x, y, z \in A$:

$$((x \to y) \to z) \to (((z \to x) \to z) \to z) = 1;$$

(ii) $x \vee \neg x = 1$, *for every* x;

(iii) $\neg x \wedge \neg\neg x = 0$, *for every* x;

(iv) $\neg\neg x \to x = 1$, *for every* x;

(v) $(\neg\neg(x \vee y) \to (\neg\neg x \vee \neg\neg y)) = 1$, *for every* x, y;

(vi) $\neg\neg(x \to y) = (x \to y) \wedge (\neg\neg x \to \neg\neg y)$, *for every* x, y;

(vii) $(((y \to \neg\neg y) \to (x \to \neg\neg x)) \to \neg\neg(x \to y)) = (\neg\neg x \to \neg\neg y)$, *for every* x, y;

(viii) *for every* x, y: *if* $x \to y = 1$ *then* $\neg y \to \neg x = 1$.

The class of $\mathbf{G}_3'$*-algebras will be denoted by* $\mathbb{A}\mathbf{G}_3'$.

Proposition 3.2. *Let* $\mathcal{A}$ *be a* $\mathbf{G}_3'$*-algebra. Then, for any* $x, y \in A$:
(1) $\neg\neg(x \wedge y) = (\neg\neg x \wedge \neg\neg y)$;
(2) $\neg\neg(x \vee y) = (\neg\neg x \vee \neg\neg y)$;
(3) $\neg\neg\neg x = \neg x$;
(4) $\neg\neg x \to 0 = \neg x$;
(5) $\neg 0 = 1$, $\neg\neg 0 = 0$, $\neg 1 = 0$ *and* $\neg\neg 1 = 1$;
(6) $(\neg\neg x \to \neg\neg y) \to \neg\neg x = \neg\neg x$;
(7) *If* $\neg\neg x \leq y \to z$ *then* $\neg\neg x \leq \neg\neg y \to \neg\neg z$;
(8) $\neg\neg x \to \neg\neg y = \neg y \to \neg x$.

Proof. Straightforward, taking into account that $\mathcal{A}$ is an implicative lattice, hence: $x \leq y$ iff $x \to y = 1$. □

Definition 3.3. *Let* $\mathcal{A}$ *be a* $\mathbf{G}_3'$*-algebra. The* logical matrix induced by $\mathcal{A}$ *is* $\mathcal{M}_\mathcal{A} \stackrel{def}{=} \langle \mathcal{A}, \{1\} \rangle$.

A *valuation over* $\mathcal{M}_\mathcal{A}$ is any homomorphism $h : \mathcal{L}_\Sigma \to \mathcal{A}$.[2] If $\Gamma \cup \{\varphi\}$ is a set of formulas in $\mathcal{L}_\Sigma$ we say that φ *is a consequence of* Γ *w.r.t. the logical matrix* $\mathcal{M}_\mathcal{A}$, written as $\Gamma \models_{\mathcal{M}_\mathcal{A}} \varphi$, if the following holds: for every valuation h over $\mathcal{M}_\mathcal{A}$, $h(\varphi) = 1$ whenever $h(\gamma) = 1$ for every $\gamma \in \Gamma$.

Definition 3.4. *Let* $\Gamma \cup \{\varphi\}$ *be a set of formulas in* $\mathcal{L}_\Sigma$. *Then* φ *is said to be a* consequence of Γ w.r.t. $\mathbf{G}_3'$-algebras, *denoted by* $\Gamma \models_{\mathbb{A}\mathbf{G}_3'} \varphi$, *if* $\Gamma \models_{\mathcal{M}_\mathcal{A}} \varphi$ *for every matrix* $\mathcal{M}_\mathcal{A}$ *and every* $\mathbf{G}_3'$*-algebra* $\mathcal{A}$.

[2]To be rigorous, h is a homomorphism from $\mathcal{L}_\Sigma$ to the Σ-reduct of $\mathcal{A}$.

Now, the adequacy of $\mathbf{G'_{3h}}$ with respect to the $\mathbf{G'_3}$-algebras semantics $\models_{\mathbb{A}\mathbf{G'_3}}$ will be proved.

Theorem 3.5 (Soundness of $\mathbf{G'_{3h}}$ w.r.t. $\mathbf{G'_3}$-algebras). *Let $\Gamma \cup \{\varphi\}$ be a set of formulas in $\mathcal{L}_\Sigma$. Then, $\Gamma \vdash_{\mathbf{G'_{3h}}} \varphi$ implies that $\Gamma \models_{\mathbb{A}\mathbf{G'_3}} \varphi$.*

Proof. It is easy to see that every axiom in $\mathbf{G'_{3h}}$ is valid in any $\mathbf{G'_3}$-algebra, that is: for every $\mathcal{A} \in \mathbb{A}\mathbf{G'_3}$ and for every valuation h over $\mathcal{A}$, $h(\varphi) = 1$ for every instance φ of every axiom of $\mathbf{G'_{3h}}$. In addition, satisfaction is preserved by the inference rules. Indeed, suppose that h is a valuation over $\mathcal{A}$ such that $h(\varphi) = h(\varphi \to \psi) = 1$. Then $1 = h(\varphi) \to h(\psi) = 1 \to h(\psi) = h(\psi)$ (recall that, in any Heyting algebra, $1 \to x = x$ for every x). In addition, if h is a valuation such that $h(\varphi \to \psi) = h(\varphi) \to h(\psi) = 1$ then $h(\neg\psi \to \neg\varphi) = \neg h(\psi) \to \neg h(\varphi) = 1$, by Definition 3.1(viii). Using this, the result follows by induction on the length of derivations. □

In order to prove the completeness of $\mathbf{G'_{3h}}$ with respect to $\mathbf{G'_3}$-algebras, some previous definitions and results are needed.

Definition 3.6 (Tarskian Logic). *A logic $\mathcal{L}$ is* Tarskian *if it satisfies the following properties, for every set of formulas $\Gamma \cup \Upsilon \cup \{\alpha\}$:*

(i) if $\alpha \in \Gamma$ then $\Gamma \vdash \alpha$;

(ii) if $\Gamma \vdash \alpha$ and $\Gamma \subseteq \Upsilon$ then $\Upsilon \vdash \alpha$;

(iii) if $\Upsilon \vdash \alpha$ and $\Gamma \vdash \beta$ for every $\beta \in \Upsilon$ then $\Gamma \vdash \alpha$.

The logic $\mathcal{L}$ is finitary *if it satisfies the following property:*

(iv) if $\Gamma \vdash \alpha$ then there exists a finite subset Γ_0 of Γ such that $\Gamma_0 \vdash \alpha$.

Definition 3.7. *Let $\mathcal{L}$ be a Tarskian logic. A set of formulas Γ is* closed in $\mathcal{L}$ *if, for every formula ψ: $\Gamma \vdash \psi$ iff $\psi \in \Gamma$.*

Definition 3.8. *Let $\mathcal{L}$ be a Tarskian logic, and let $\Gamma \cup \{\varphi\}$ be a set of formulas. The set Γ is* maximal non-trivial w.r.t. φ *in $\mathcal{L}$, or* φ-saturated *in $\mathcal{L}$, if $\Gamma \nvdash \varphi$ but $\Gamma, \psi \vdash \varphi$ for any $\psi \notin \Gamma$.*

It is easy to prove that any φ-saturated set of formulas in a Tarskian logic is closed. Recall now the following classical result (see [13, Theorem 22.2]):

Theorem 3.9 (Lindenbaum-Łoś). *Let $\mathcal{L}$ be a Tarskian and finitary logic, and let $\Gamma \cup \{\varphi\}$ be a set of formulas such that $\Gamma \nvdash \varphi$. Then, there exists a set of formulas Υ such that Υ is φ-saturated in $\mathcal{L}$ and $\Gamma \subseteq \Upsilon$.*

Remark 3.10. *Clearly $\mathbf{G}'_{\mathbf{3h}}$ is Tarskian and finitary, then Theorem 3.9 applies to it. Observe that, if Υ is a φ-saturated set in $\mathbf{G}'_{\mathbf{3h}}$ then, for every formula β: $\beta \in \Upsilon$ iff $\neg\neg\beta \in \Upsilon$.*

Theorem 3.11 (Completeness of $\mathbf{G}'_{\mathbf{3h}}$ w.r.t. $\mathbf{G}'_3$-algebras). *Let $\Gamma \cup \{\varphi\}$ be a set of formulas in $\mathcal{L}_\Sigma$. Then, $\Gamma \models_{\mathbb{A}\mathbf{G}'_3} \varphi$ implies that $\Gamma \vdash_{\mathbf{G}'_{\mathbf{3h}}} \varphi$.*

Proof. Suppose that $\Gamma \nvdash_{\mathbf{G}'_{\mathbf{3h}}} \varphi$. By Theorem 3.9 and Remark 3.10 there exists a set Υ which is φ-saturated in $\mathbf{G}'_{\mathbf{3h}}$ such that $\Gamma \subseteq \Upsilon$. Define the following relation in $\mathcal{L}_\Sigma$: $\beta \equiv_\Upsilon \gamma$ iff $\Upsilon \vdash_{\mathbf{G}'_{\mathbf{3h}}} \beta \leftrightarrow \gamma$. By the properties of $\mathbf{IPL}^+$, including the ones listed in Proposition 2.6, it is easy to prove that $\equiv_\Upsilon$ is a congruence over $\mathcal{L}_\Sigma$ with respect to the connectives $\wedge$, $\vee$ and $\rightarrow$. Moreover, $\mathcal{A}_\Upsilon \stackrel{\text{def}}{=} \mathcal{L}_\Sigma/_{\equiv_\Upsilon}$ is an implicative lattice with such operations. In addition, it has a bottom element given by $0 = [\neg\beta \wedge \neg\neg\beta]_\Upsilon$ for any formula β, where $[\psi]_\Upsilon$ denotes the equivalence class of the formula ψ w.r.t. $\equiv_\Upsilon$. This means that $\mathcal{A}_\Upsilon$ is a 3-valued Heyting algebra, by virtue of **(Tho)**. Note also that $\neg[\beta]_\Upsilon \stackrel{\text{def}}{=} [\neg\beta]_\Upsilon$ is a well-defined operation in $\mathcal{A}_\Upsilon$, because of **(imp)**. It is immediate to see that $\mathcal{A}_\Upsilon$ satisfies properties (ii)-(viii) of Definition 3.1. Hence, $\mathcal{A}$ is a $\mathbf{G}'_3$-algebra such that $[\gamma]_\Upsilon = 1$ iff $\Upsilon \vdash_{\mathbf{G}'_{\mathbf{3h}}} \gamma$ iff $\gamma \in \Upsilon$. Consider now the function $h_\Upsilon : \mathcal{L}_\Sigma \rightarrow \mathcal{A}_\Upsilon$ given by $h_\Upsilon(\gamma) = [\gamma]_\Upsilon$. It is easy to see that h_Υ is a valuation over $\mathcal{M}_{\mathcal{A}_\Upsilon}$ such that $h_\Upsilon(\gamma) = 1$ iff $\gamma \in \Upsilon$. Therefore, h_Υ is a valuation over $\mathcal{M}_{\mathcal{A}_\Upsilon}$ such that $h_\Upsilon(\gamma) = 1$ for every $\gamma \in \Gamma$ but $h_\Upsilon(\varphi) \neq 1$, since $\varphi \notin \Upsilon$. This shows that $\Gamma \not\models_{\mathbb{A}\mathbf{G}'_3} \varphi$. □

As a corollary of the completeness theorem above, a special and useful form of the Deduction Theorem can be obtained (see Theorem 3.13 below). Previously, some results must be stated.

Lemma 3.12. *Let $\Gamma \cup \{\varphi, \psi, \gamma\}$ be a set of formulas in $\mathcal{L}_\Sigma$.*
(1) If $\Gamma \vdash_{\mathbf{G}'_{\mathbf{3h}}} \neg\neg\varphi \rightarrow \psi$ and $\Gamma \vdash_{\mathbf{G}'_{\mathbf{3h}}} \neg\neg\varphi \rightarrow (\psi \rightarrow \gamma)$ then $\Gamma \vdash_{\mathbf{G}'_{\mathbf{3h}}} \neg\neg\varphi \rightarrow \gamma$.
(2) If $\Gamma \vdash_{\mathbf{G}'_{\mathbf{3h}}} \neg\neg\varphi \rightarrow (\psi \rightarrow \gamma)$ then $\Gamma \vdash_{\mathbf{G}'_{\mathbf{3h}}} \neg\neg\varphi \rightarrow (\neg\gamma \rightarrow \neg\psi)$.

Proof. (1) It follows by using the hypothesis together with the theorems $(\neg\neg\varphi \rightarrow \psi) \rightarrow ((\neg\neg\varphi \rightarrow (\psi \rightarrow \gamma)) \rightarrow (\neg\neg\varphi \rightarrow (\psi \wedge (\psi \rightarrow \gamma))))$ and $(\psi \wedge (\psi \rightarrow \gamma)) \rightarrow \gamma$ of $\mathbf{G}'_{\mathbf{3h}}$, taking also into account Proposition 2.6(1).
(2) Suppposse that $\Gamma \vdash_{\mathbf{G}'_{\mathbf{3h}}} \neg\neg\varphi \rightarrow (\psi \rightarrow \gamma)$. Then $\Gamma \models_{\mathbb{A}\mathbf{G}'_3} \neg\neg\varphi \rightarrow (\psi \rightarrow \gamma)$, by

Theorem 3.5. Let h be a valuation over a matrix $\mathcal{M}_\mathcal{A}$, for a given $\mathbf{G'_3}$-algebra $\mathcal{A}$, such that $h(\delta) = 1$ for every $\delta \in \Gamma$. Then, $h(\neg\neg\varphi \to (\psi \to \gamma)) = 1$. Let $x = h(\varphi)$, $y = h(\psi)$ and $z = h(\gamma)$. Then $\neg\neg x \leq y \to z$ and so $\neg\neg x \leq \neg\neg y \to \neg\neg z = \neg z \to \neg y$, by Proposition 3.2 items (7) and (8). That is, $h(\neg\neg\varphi \to (\neg\gamma \to \neg\psi)) = 1$. This shows that $\Gamma \models_{\mathbb{A}\mathbf{G'_3}} \neg\neg\varphi \to (\neg\gamma \to \neg\psi)$. By Theorem 3.11, $\Gamma \vdash_{\mathbf{G'_{3h}}} \neg\neg\varphi \to (\neg\gamma \to \neg\psi)$. □

Theorem 3.13 (Special Deduction Theorem (SDT)). *Let $\Gamma \cup \{\varphi, \psi\}$ be a set of formulas in $\mathcal{L}_\Sigma$. Then, $\Gamma, \varphi \vdash_{\mathbf{G'_{3h}}} \psi$ if and only if $\Gamma \vdash_{\mathbf{G'_{3h}}} \neg\neg\varphi \to \psi$.*

Proof.
(*Only if* part). Suppose that $\Gamma, \varphi \vdash_{\mathbf{G'_{3h}}} \psi$. By induction on the length n of a derivation $\varphi_1 \cdots \varphi_n$ of ψ from $\Gamma \cup \{\varphi\}$ in $\mathbf{G'_{3h}}$, it can be proven that $\Gamma \vdash_{\mathbf{G'_{3h}}} \neg\neg\varphi \to \varphi_i$ for every $1 \leq i \leq n$, and so $\Gamma \vdash_{\mathbf{G'_{3h}}} \neg\neg\varphi \to \psi$ (for $i = n$). To do this, it must taken into account the fact that $\Gamma \vdash_{\mathbf{G'_{3h}}} \neg\neg\varphi \to \psi$ if either $\psi \in \Gamma \cup \{\varphi\}$ or ψ is an instance of an axiom (for the base step), and Lemma 3.12 (to deal with the inference rules from the induction hypothesis). The details of the proof are left to the reader.

(*If* part). Suppose that $\Gamma \vdash_{\mathbf{G'_{3h}}} \neg\neg\varphi \to \psi$, and let $\varphi_1 \cdots \varphi_n = \neg\neg\varphi \to \psi$ be a derivation of $\neg\neg\varphi \to \psi$ from Γ in $\mathbf{G'_{3h}}$. Consider now the following (meta)derivation of ψ from $\Gamma \cup \{\varphi\}$ in $\mathbf{G'_{3h}}$:

1.	φ_1	
$\vdots$	$\vdots$	
n.	$\varphi_n = \neg\neg\varphi \to \psi$	
$n+1$.	φ	Hyp.
$n+2$.	$\neg\neg\varphi$	**(Dneg)**, n+1
$n+3$.	ψ	**(MP)**, n, n+2

This shows that $\Gamma, \varphi \vdash_{\mathbf{G'_{3h}}} \psi$. □

3.1 $\mathbf{G'_3}$-algebras as a variety

The aim of this section is proving that $\mathbf{G'_3}$-algebras are three-valued modal implicative semilattices (see [2]) with a bottom element. From this, and from the results obtained in [2], together with Theorem 3.11, the completeness of $\mathbf{G'_{3h}}$ w.r.t. the matrix $\mathbf{G'_3}$ will follow easily (see Theorem 3.27 below).

As mentioned in the Introduction, Canals-Frau and Figallo have studied in [2] the reduct $\{\wedge, \to, \triangle, 1\}$ of the three-valued MV-algebras, where $\to$ is a three-valued Heyting implication, and $\triangle$ is a Moisil operator from the three-valued Łukasiewicz-Moisil algebras (or, equivalently, $\triangle$ is a Monteiro-Baaz Delta-operator). They also consider the operator $\nabla x = (x \to \Delta x) \to \Delta x$. This reduct can be defined as follows:

Definition 3.14 (See [2]). *An algebra $\mathcal{A} = \langle A, \wedge, \to, \Delta, 1\rangle$ of type $(2, 2, 1, 0)$ is a* three-valued modal implicative semilattice *(a MIS_3-*algebra*, for short) if it satisfies the following identities, for every $x, y, z \in A$:*

(IS1) $(x \to x) = 1$,

(IS2) $((x \to y) \wedge y) = y$,

(IS3) $(x \to (y \wedge z)) = ((x \to z) \wedge (x \to y))$,

(IS4) $(x \wedge (x \to y)) = (x \wedge y)$,

(T) $((x \to y) \to z) \to (((z \to x) \to z) \to z) = 1$,

(M1) $(\Delta x \to x) = 1$,

(M2) $(((y \to \Delta y) \to (x \to \Delta\Delta x)) \to \Delta(x \to y)) = (\Delta x \to \Delta\Delta y)$,

(M3) $((\Delta x \to \Delta y) \to \Delta x) = \Delta x$.

It is worth mentioning that any MIS_3-algebra is an ordered structure if we consider $x \leq y$ if and only if $x \to y = 1$ (if and only if $x \wedge y = x$). Moreover, in [5] it was proved that any MIS_3-algebra $\mathcal{A}$ is a distributive lattice where the supremum is given by $x \vee y \stackrel{\text{def}}{=} ((x \to y) \to y) \wedge ((y \to x) \to x)$ for $x, y \in A$.

Recall (see the beginning of Section 2) that $\mathcal{A}_3 = \langle \mathcal{V}, \sigma\rangle$ is the 3-valued algebra of $\mathbf{G}_3'$ with domain $\mathcal{V} = \{0, \frac{1}{2}, 1\}$. Let $\mathcal{B}_3 = \langle \mathcal{V}, \sigma'\rangle$ be the 3-valued algebra over $\{\wedge, \to, \Delta, \mathbf{1}\}$ with domain $\mathcal{V}$ such that σ' interprets the connectives as follows: $\sigma'(\mathbf{1}) = 1$; $\sigma'(\wedge)$ and $\sigma'(\to)$ coincide with the corresponding operators of $\mathcal{A}_3$; and $\sigma'(\Delta)$ is defined by the truth-table below.[3]

x	Δx
$\mathbf{0}$	0
$\frac{\mathbf{1}}{\mathbf{2}}$	0
$\mathbf{1}$	1

It is easy to see that the induced operator $x \vee y \stackrel{\text{def}}{=} ((x \to y) \to y) \wedge ((y \to x) \to x)$ coincides with the $\vee$-operator of $\mathcal{A}_3$. In addition, $0 \leq \frac{1}{2} \leq 1$. On the other hand, $\nabla x = 0$ if $x = 0$, and 1 otherwise.

The following fundamental results can be found in [2] (see also [5]).

[3] As usual, we identify $\sigma'(c)$ with c, for any connective c.

Definition 3.15. *Let $\mathcal{A}$ be a MIS_3-algebra and let $D \subseteq A$. D is said to be deductive system if $1 \in D$, and if $x, x \to y \in D$ imply $y \in D$. Also, we say that D is modal, if $x \in D$ implies $\triangle x \in D$. Besides, we denote by $D_m(\mathcal{A})$ the set of modal deductive systems and by $Con(\mathcal{A})$ the set of congruence relations.*

Lemma 3.16. *([2]) For a given MIS_3-algebra $\mathcal{A}$, the poset $D_m(\mathcal{A})$ is lattice-isomorphic to $Con(\mathcal{A})$.*

Now, for a given MIS_3-algebra $\mathcal{A}$, a deductive systems D of $\mathcal{A}$ is said to be a maximal if for every deductive system M such that $D \subseteq M$ implies $M = A$ or $M = D$.

Theorem 3.17. *([2]) Let M be a non-trivial maximal modal deductive system of MIS_3-algebras $\mathcal{A}$. Let us consider the sets $M_0 = \{x \in A : \nabla x \notin M\}$ and $M_{1/2} = \{x \in A : x \notin M, \nabla x \in M\}$, and the map $h : A \longrightarrow \mathcal{V}$ defined by*

$$h(x) = \begin{cases} 0 & \textit{if } x \in M_0 \\ 1/2 & \textit{if } x \in M_{1/2} \\ 1 & \textit{if } x \in M. \end{cases}$$

Then, h is a MIS_3-homomorphism $h : \mathcal{A} \to \mathcal{B}_3$ such that $h^{-1}(\{1\}) = M$.

Theorem 3.18. *([2]) The variety of MIS_3-algebras is semisimple and it is generated by the 3-valued algebra $\mathcal{B}_3$.*

Remark 3.19. *The above theorem states that an equation $s = t$ holds in every MIS_3-algebra iff it holds in $\mathcal{B}_3$. For instance, since $\triangle\triangle x = \triangle x$ holds in $\mathcal{B}_3$ for every $x \in \mathcal{V}$, it follows that, for every MIS_3-algebra $\mathcal{A}$, $\triangle\triangle x = \triangle x$ for every $x \in A$.*

The next step is to connect the variety of MIS_3-algebras with the class of $\mathbf{G}'_3$-algebras introduced in Definition 3.1. Firstly, observe that any $\mathbf{G}'_3$-algebra has a bottom element 0. This suggest the following definition:

Definition 3.20. *An algebra $\mathcal{A} = \langle A, \wedge, \to, \triangle, 0, 1\rangle$ of type $(2, 2, 1, 0, 0)$ is a* three-valued modal Heyting algebra *(a MIS_3^0-*algebra*, for short) if its reduct $\langle A, \wedge, \to, \triangle, 1\rangle$ is a MIS_3-algebra and, for every $x \in A$:*

(IS5) $(0 \to x) = 1$.

Observe that the expansion $\mathcal{B}_3^0 = \langle \mathcal{V}, \sigma' \rangle$ of $\mathcal{B}_3$ to the signature $\{\wedge, \to, \triangle, \mathbf{0}, \mathbf{1}\}$ such that $\sigma'(\mathbf{0}) = 0$ is a MIS_3^0-algebra. Moreover, the following result holds:

Theorem 3.21. *The variety of MIS_3^0-algebras is generated by the 3-valued algebra $\mathcal{B}_3^0$.*

Proof. According to Theorem 3.18, there is a non-empty set X and a homomorphism $h : \mathcal{A} \to \mathcal{B}_3^X$ for every MIS_3-algebra $\mathcal{A}$. Besides, it is clear that h verify $h(x \vee y) = h(x) \vee h(y)$. Thus, in particular, this representation holds for every MIS_3^0-algebra $\mathcal{A}$. Thus, taking into account axiom (IS5), it is clear that $h(0) \leq h(x)$ for every $x \in A$. But $\mathcal{A}_3$ is a subdirectly irreducible algebra of the variety of MIS_3-algebras. Therefore, every canonical projection $q_i : h(\mathcal{A}) \to \mathcal{B}_3$ is onto and so, $q_i(h(0)) \leq q_i(h(x))$ for every $x \in A$, in particular $q_i(h(0)) \leq 0$. Therefore, $q_i(h(0)) = 0$ for every $i \in X$ and thus, $h(0) = 0$, which completes the proof. □

Theorem 3.22. *The class $\mathbb{A}\mathbf{G}_3'$ of $\mathbf{G}_3'$-algebras and the variety of MIS_3^0-algebras are term-equivalent via $\triangle x \overset{def}{=} \neg\neg x$, on the one hand; and $\neg x \overset{def}{=} \triangle x \to 0$ and $x \vee y \overset{def}{=} ((x \to y) \to y) \wedge ((y \to x) \to x)$, on the other.*

Proof. Let $\mathcal{A} = \langle A, \wedge, \to, \triangle, 0, 1\rangle$ be a MIS_3^0-algebra, and define the following operators:

$\neg x \overset{\text{def}}{=} \triangle x \to 0$ and $x \vee y \overset{\text{def}}{=} ((x \to y) \to y) \wedge ((y \to x) \to x)$, for every $x, y \in A$.

Let $\mathcal{A}^{\neg} \overset{\text{def}}{=} \langle A, \wedge, \vee, \to, \neg, 0, 1\rangle$. It is immediate to see that $\langle A, \wedge, \vee, \to, 0, 1\rangle$ is a 3-valued Heyting algebra. This follows from the fact that $\mathcal{B}_3^0$ satisfies the equations characterizing such class of algebras, and then by using Theorem 3.21. Moreover, by using the same argument it can be proven that $\mathcal{A}^{\neg}$ satisfies properties (ii)-(vii) of Definition 3.1. Finally, suppose that $x \leq y$. Then, $\triangle x \leq \triangle y$ (see [2] and [5]). From this, $\neg y = \triangle y \to 0 \leq \triangle x \to 0 = \neg x$. Hence, $\mathcal{A}^{\neg}$ satisfies property (viii) of Definition 3.1. This means that any MIS_3^0-algebra can be transformed into a $\mathbf{G}_3'$-algebra by using appropriate terms.

Conversely, let $\mathcal{A} = \langle A, \wedge, \vee, \to, \neg, 0, 1\rangle$ be a $\mathbf{G}_3'$-algebra and define the following operation: $\triangle x \overset{\text{def}}{=} \neg\neg x$, for every $x \in A$. Consider the algebra $\mathcal{A}^{\triangle} \overset{\text{def}}{=} \langle A, \wedge, \to, \triangle, 0, 1\rangle$. We shall prove that $\mathcal{A}^{\triangle}$ is a MIS_3^0-algebra. Observe that $\mathcal{A}^{\triangle}$ satisfies properties (IS1)-(IS4) and (T) of Definition 3.14, as well as property (IS5) of Definition 3.20. The algebra $\mathcal{A}^{\triangle}$ satisfies properties (M1) and (M2) since $\mathcal{A}$ satisfies properties (iv) and (vii) of Definition 3.1, and since $\neg\neg\neg\neg x = \neg\neg x$, by Proposition 3.2(3). Finally, $\mathcal{A}^{\triangle}$ satisfies property (M3), by Proposition 3.2(6). This means that any $\mathbf{G}_3'$-algebra can be transformed into a MIS_3^0-algebra by means of a suitable term. This shows that the class $\mathbb{A}\mathbf{G}_3'$ of $\mathbf{G}_3'$-algebras and the variety of MIS_3^0-algebras are term-equivalent via the proposed terms. □

Let $\mathcal{A}_3^0 = \langle \mathcal{V}, \sigma \rangle$ be the expansion of $\mathcal{A}_3$ (recall the beginning of Section 2) to the signature Σ_+ such that $\sigma(\mathbf{0}) = 0$

Corollary 3.23. *The class* $\mathbb{A}\mathbf{G_3'}$ *of* $\mathbf{G_3'}$*-algebras is generated by the 3-valued algebra* $\mathcal{A}_3^0$.

Proof. If follows immediately form theorems 3.22 and 3.21.

□

Corollary 3.24. *Let* φ *be a formula. Then* $\models_{\mathbb{A}\mathbf{G_3'}} \varphi$ *if and only if* $\models_{\mathbf{G_3'}} \varphi$.

Corollary 3.25. *The class* $\mathbb{A}\mathbf{G_3'}$ *of* $\mathbf{G_3'}$*-algebras is a variety defined by the following equations:*

$(\mathbf{G_3'}1)$ $(x \to x) = 1$,

$(\mathbf{G_3'}2)$ $((x \to y) \wedge y) = y$,

$(\mathbf{G_3'}3)$ $(x \to (y \wedge z)) = ((x \to z) \wedge (x \to y))$,

$(\mathbf{G_3'}4)$ $(x \wedge (x \to y)) = (x \wedge y)$,

$(\mathbf{G_3'}5)$ $((x \to y) \to z) \to (((z \to x) \to z) \to z) = 1$,

$(\mathbf{G_3'}6)$ $(\neg\neg x \to x) = 1$,

$(\mathbf{G_3'}7)$ $(\neg x \to \neg\neg\neg x) = 1$,

$(\mathbf{G_3'}8)$ $(((y \to \neg\neg y) \to (x \to \neg\neg x)) \to \neg\neg(x \to y)) = (\neg\neg x \to \neg\neg y)$,

$(\mathbf{G_3'}9)$ $((\neg\neg x \to \neg\neg y) \to \neg\neg x) = \neg\neg x$,

$(\mathbf{G_3'}10)$ $(x \vee y) = (((x \to y) \to y) \wedge ((y \to x) \to x))$,

$(\mathbf{G_3'}11)$ $(x \vee \neg x) = 1$,

$(\mathbf{G_3'}12)$ $(\neg x \wedge \neg\neg x) = 0$,

$(\mathbf{G_3'}13)$ $(0 \to x) = 1$.

Proof. By Definition 3.1 and by Corollary 3.23, it follows that any $\mathbf{G_3'}$-algebra satisfies the equations $(\mathbf{G_3'}1)$-$(\mathbf{G_3'}13)$. Conversely, let $\mathcal{A}$ be an algebra satisfying $(\mathbf{G_3'}1)$-$(\mathbf{G_3'}13)$, and define $\triangle x \stackrel{\text{def}}{=} \neg\neg x$. Then $\triangle\triangle x = \triangle x$ for every $x \in A$, and so $\mathcal{A}^\triangle \stackrel{\text{def}}{=} \langle A, \wedge, \to, \triangle, 0, 1 \rangle$ is a MIS_3^0-algebra. By the proof of Theorem 3.22, the algebra $(\mathcal{A}^\triangle)^\neg$ is a $\mathbf{G_3'}$-algebra. It will be shown that $(\mathcal{A}^\triangle)^\neg = \mathcal{A}$. Indeed, the negation in $(\mathcal{A}^\triangle)^\neg$ is given by $\neg' x \stackrel{\text{def}}{=} \triangle x \to 0 = \neg\neg x \to 0$. But $\neg\neg x \to 0 = \neg x$. The proof of this fact is analogous to the one given for Proposition 3.2(4) above. Using this, it follows that $\neg' x = \neg x$. In addition, the disjunction

$x \vee' y \stackrel{\text{def}}{=} ((x \to y) \to y) \wedge ((y \to x) \to x)$ in $(\mathcal{A}^{\triangle})^{\neg}$ coincides with $x \vee y$, by ($\mathbf{G}_3'$10). This shows that $(\mathcal{A}^{\triangle})^{\neg} = \mathcal{A}$, hence $\mathcal{A}$ is a $\mathbf{G}_3'$-algebra, by Theorem 3.22. $\square$

3.2 Adequacy of $\mathbf{G}_{3h}'$ w.r.t. $\mathbf{G}_3'$

Finally, we can prove the adequacy of the Hilbert calculus $\mathbf{G}_{3h}'$ with respect to the intended 3-valued semantics $\mathbf{G}_3'$. Firstly, a technical result will be stated:

Proposition 3.26. *If $\varphi \models_{\mathbf{G}_3'} \psi$ then $\models_{\mathbf{G}_3'} \neg\neg\varphi \to \psi$.*

Proof. Supose that $\varphi \models_{\mathbf{G}_3'} \psi$, and let h be a valuation over $\mathbf{G}_3'$. If $h(\varphi) = 1$ then $h(\psi) = 1$, by hypothesis, hence $h(\neg\neg\varphi \to \psi) = 1 \to 1 = 1$. Otherwise, if $h(\varphi) \neq 1$ then $h(\neg\neg\varphi) = 0$ and so $h(\neg\neg\varphi \to \psi) = 0 \to h(\psi) = 1$. In any case, $h(\neg\neg\varphi \to \psi) = 1$. This shows that $\models_{\mathbf{G}_3'} \neg\neg\varphi \to \psi$. $\square$

Theorem 3.27 (Soundness and completeness of $\mathbf{G}_{3h}'$ w.r.t. $\mathbf{G}_3'$). *For every finite set $\Gamma \cup \{\varphi\} \subseteq \mathcal{L}_\Sigma$:*

$$\Gamma \vdash_{\mathbf{G}_{3h}'} \varphi \quad \textit{if and only if} \quad \Gamma \models_{\mathbf{G}_3'} \varphi.$$

Proof.

Only if part (Soundness): It follows from Theorem 3.5 and the fact that $\mathcal{A}_3^0$ is a $\mathbf{G}_3'$-algebra.

If part (Completeness): Suppose that $\Gamma \models_{\mathbf{G}_3'} \varphi$. If $\Gamma = \emptyset$ then $\models_{\mathbb{A}\mathbf{G}_3'} \varphi$, by Corollary 3.24. From this, $\vdash_{\mathbf{G}_{3h}'} \varphi$, by Theorem 3.11. Otherwise, if $\Gamma = \{\psi_1, \ldots, \psi_n\}$ for $n \geq 1$ let $\psi = (\ldots((\psi_1 \wedge \psi_2) \wedge \psi_3) \wedge \ldots) \wedge \psi_n$ if $n > 1$, and $\psi = \psi_1$ if $n = 1$. Since $\psi \models_{\mathbf{G}_3'} \varphi$ then $\models_{\mathbf{G}_3'} \neg\neg\psi \to \varphi$, by Proposition 3.26. From this $\models_{\mathbb{A}\mathbf{G}_3'} \neg\neg\psi \to \varphi$, by Corollary 3.24. Then $\vdash_{\mathbf{G}_{3h}'} \neg\neg\psi \to \varphi$, by Theorem 3.11. By Theorem 3.13, $\psi \vdash_{\mathbf{G}_{3h}'} \varphi$. By using the properties of the conjunction in $\mathbf{G}_{3h}'$ it follows from here that $\Gamma \vdash_{\mathbf{G}_{3h}'} \varphi$. $\square$

Corollary 3.28. *The Deduction Theorem does not hold in $\mathbf{G}_{3h}'$.*

Proof. It is an immediate consequence of Remark 2.3 and Theorem 3.27. $\square$

4 The first-order $\mathbf{G}_3'$-logic

In this section, a first-order version of $\mathbf{G}_3'$, called $\mathcal{Q}\mathbf{G3}'$, will be proposed. The semantics will be given by structures defined over $\mathbf{G}_3'$-algebras which are complete (as lattices), in order to interpret the quantifiers.

Recall that Σ denotes the propositional signature $\{\wedge, \vee, \to, \neg\}$ for $\mathbf{G}_3'$.

Definition 4.1. *Consider the symbols* $\forall$ *(universal quantifier) and* $\exists$ *(existential quantifier), together with commas and parentesis as the punctuation marks. Let* $IVar = \{v_1, v_2, \ldots\}$ *be a denumerable set of* individual *variables. A first-order signature is a triple* $\Theta = \langle \mathcal{C}, \{\mathcal{F}_n\}_{n\in\mathbb{N}}, \{\mathcal{P}_n\}_{n\in\mathbb{N}}\rangle$ *such that:*

- $\mathcal{C}$ *is a set of individual constants;*
- *for each* $n \geq 1$, $\mathcal{F}_n$ *is a set of function symbols of arity* n,
- *for each* $n \geq 1$, $\mathcal{P}_n$ *is a set of predicate symbols of arity* n.[4]

The notions of bound and free variables inside a formula, closed terms, closed formulas (or sentences), and of term free for a variable in a formula are defined as usual (see, for instance, [9]). We denote by Ter_Θ and $\mathfrak{Fm}_\Theta$ the set of terms and the set of first-order formulas over Θ (by using the connectives in Σ), respectively. Given a formula φ, the formula obtained from φ by substituting every free occurrence of a variable x by a term t will be denoted by $\varphi(x/t)$.

Definition 4.2. *Let* Θ *be a first-order signature. The logic* $\mathcal{Q}\mathbf{G3}'$ *over* Θ *is defined by the Hilbert calculus obtained by extending* $\mathbf{G}'_3$ *expressed in the language* $\mathfrak{Fm}_\Theta$ *by adding the following:*

Axiom schemas:

(Ax14) $\varphi(x/t) \to \exists x\varphi$ *if* t *is a term free for* x *in* φ

(Ax15) $\forall x\varphi \to \varphi(x/t)$ *if* t *is a term free for* x *in* φ

Inference Rules:

$$(\exists - In)\quad \frac{\varphi \to \psi}{\exists x\varphi \to \psi}\quad \text{where } x \text{ does not occur free in } \psi$$

$$(\forall - In)\quad \frac{\varphi \to \psi}{\varphi \to \forall x\psi}\quad \text{where } x \text{ does not occur free in } \varphi$$

Definition 4.3. *A* Θ*-structure for* $\mathcal{Q}\mathbf{G3}'$ *is a triple* $\mathfrak{A} = \langle U, \mathcal{A}, \cdot^{\mathfrak{A}}\rangle$ *such that* U *is a non-empty set,* $\mathcal{A}$ *is a complete* $\mathbf{G}'_3$*-algebra and* $\cdot^{\mathfrak{A}}$ *is an interpretation map which assigns:*

- *to each individual constant* $c \in \mathcal{C}$, *an element* $c^{\mathfrak{A}}$ *of* U;

[4] It will be assumed, as usual, that Θ has at least one predicate symbol, in order to have a non-empty set of formulas.

- *to each function symbol f of arity n, a function $f^{\mathfrak{A}} : U^n \to U$;*
- *to each predicate symbol P of arity n, a function $P^{\mathfrak{A}} : U^n \to A$.*

Given a Θ-structure $\mathfrak{A}$ for $\mathcal{Q}\mathbf{G3}'$, an *assignment* over $\mathfrak{A}$ is a function $s : IVar \to U$. Given s and $a \in U$ let $s[x \to a]$ be the assignment such that $s[x \to a](x) = a$ and $s[x \to a](y) = s(y)$ for every $x \neq y$. A Θ-structure $\mathfrak{A}$ and an assignment s induce an interpretation map $[\![\cdot]\!]^{\mathfrak{A}}_s$ for terms and formulas defined as follows:

$$[\![x]\!]^{\mathfrak{A}}_s = s(x) \text{ if } x \in IVar,$$
$$[\![c]\!]^{\mathfrak{A}}_s = c^{\mathfrak{A}} \text{ if } c \in \mathcal{C},$$
$$[\![f(t_1,\ldots,t_n)]\!]^{\mathfrak{A}}_s = f^{\mathfrak{A}}([\![t_1]\!]^{\mathfrak{A}}_s,\ldots,[\![t_n]\!]^{\mathfrak{A}}_s), \text{ if } f \in \mathcal{F}_n,$$
$$[\![P(t_1,\ldots,t_n)]\!]^{\mathfrak{A}}_s = P^{\mathfrak{A}}([\![t_1]\!]^{\mathfrak{A}}_s,\ldots,[\![t_n]\!]^{\mathfrak{A}}_s), \text{ if } P \in \mathcal{P}_n,$$
$$[\![\phi\#\varphi]\!]^{\mathfrak{A}}_s = [\![\phi]\!]^{\mathfrak{A}}_s \# [\![\varphi]\!]^{\mathfrak{A}}_s \text{ for } \# \in \{\wedge,\vee,\to\},$$
$$[\![\neg\varphi]\!]^{\mathfrak{A}}_s = \neg[\![\varphi]\!]^{\mathfrak{A}}_s,$$
$$[\![\forall x\varphi]\!]^{\mathfrak{A}}_s = \bigwedge_{a\in U} [\![\varphi]\!]^{\mathfrak{A}}_{s[x\to a]},$$
$$[\![\exists x\varphi]\!]^{\mathfrak{A}}_s = \bigvee_{a\in U} [\![\varphi]\!]^{\mathfrak{A}}_{s[x\to a]}.$$

We say that $\mathfrak{A}$ and s *satisfy* a formula φ, denoted by $\mathfrak{A} \vDash \varphi[s]$, if $[\![\varphi]\!]^{\mathfrak{A}}_s = 1$. On the other hand, φ is *true in* $\mathfrak{A}$ if $\mathfrak{A} \vDash \varphi[s]$ for every s. We say that φ is a *semantical consequence of* Γ *in* $\mathcal{Q}\mathbf{G3}'$, denoted by $\Gamma \vDash_{\mathcal{Q}\mathbf{G3}'} \alpha$, if, for any structure $\mathfrak{A}$: if every $\psi \in \Gamma$ is true in $\mathfrak{A}$ then α is true in $\mathfrak{A}$. Observe that, if $\mathfrak{A}$ is a structure and φ is a closed formula, then $[\![\varphi]\!]^{\mathfrak{A}}_s = [\![\varphi]\!]^{\mathfrak{A}}_{s'}$, for every assignments s and s'. This being so, either $\mathfrak{A} \vDash \varphi[s]$ for every s or $\mathfrak{A} \nvDash \varphi[s]$ for every s. Thus, if $\Gamma \cup \{\varphi\}$ is a set of sentences then: $\Gamma \nvDash_{\mathcal{Q}\mathbf{G3}'} \varphi$ iff there is a structure $\mathfrak{A}$ such that every $\psi \in \Gamma$ is true in $\mathfrak{A}$ but $\mathfrak{A} \nvDash \varphi[s]$ for any assignment s.

In order to prove the soundness of $\mathcal{Q}\mathbf{G3}'$ w.r.t. the given semantics, an important technical result, the *substitution lemma*, must be established:

Proposition 4.4 (Substitution lemma). *Let φ be a formula, t a term free for x in φ, $\mathfrak{A}$ an structure and s and assignment. Then:* $[\![\alpha]\!]^{\mathfrak{A}}_{s[x\to[\![t]\!]^{\mathfrak{A}}_s]} = [\![\alpha(x/t)]\!]^{\mathfrak{A}}_s$.

Proof. It is easily proved by induction on the complexity of the formula α. □

Theorem 4.5 (Soundness of $\mathcal{Q}\mathbf{G3}'$). *Let $\Gamma \cup \{\varphi\} \subseteq \mathfrak{Fm}_\Theta$. If $\Gamma \vdash_{\mathcal{Q}\mathbf{G3}'} \varphi$ then $\Gamma \vDash_{\mathcal{Q}\mathbf{G3}'} \varphi$.*

Proof: Consider a given structure $\mathfrak{A} = \langle U, \mathcal{A}, \cdot^{\mathfrak{A}}\rangle$. It is enough to prove the following facts: the new axioms **(Ax14)** and **(Ax15)** are true in $\mathfrak{A}$, and the new inference rules $(\exists - In)$ and $(\forall - In)$ preserve trueness in $\mathfrak{A}$.

(Ax14) and **(Ax15)**: Suppose that φ is $\alpha(x/t) \to \exists x\alpha$, and let s be an assignment. Then, by Proposition 4.4, $[\![\varphi]\!]^{\mathfrak{A}}_s = [\![\alpha]\!]^{\mathfrak{A}}_{s[x\to[\![t]\!]^{\mathfrak{A}}_s]} \to [\![\exists x\alpha]\!]^{\mathfrak{A}}_s$. It is clear that $[\![\alpha]\!]^{\mathfrak{A}}_{s[x\to[\![t]\!]^{\mathfrak{A}}_s]} \leq \bigvee_{a\in U} [\![\alpha]\!]^{\mathfrak{A}}_{s[x\to a]}$, hence $[\![\alpha(x/t)]\!]^{\mathfrak{A}}_s \leq [\![\exists x\alpha]\!]^{\mathfrak{A}}_s$. Therefore $[\![\alpha(x/t) \to \exists x\alpha]\!]^{\mathfrak{A}}_s = 1$. The validity of **(Ax15)** is proved analogously.

$(\exists - In)$ and $(\forall - In)$: Let $\alpha \to \beta$ such that x is not free in β, and let $\varphi = \exists x\alpha \to \beta$. Suppose that that $[\![\alpha \to \beta]\!]^{\mathfrak{A}}_s = 1$ for every s, and fix an assignment s. By definition, $[\![\varphi]\!]^{\mathfrak{A}}_s = [\![\exists x\alpha]\!]^{\mathfrak{A}}_s \to [\![\beta]\!]^{\mathfrak{A}}_s = \bigvee_{a\in U} [\![\alpha]\!]^{\mathfrak{A}}_{s[x\to a]} \to [\![\beta]\!]^{\mathfrak{A}}_s$. By hypothesis, $[\![\alpha]\!]^{\mathfrak{A}}_{s'} \leq [\![\beta]\!]^{\mathfrak{A}}_{s'}$ for every s'. In particular, $[\![\alpha]\!]^{\mathfrak{A}}_{s[x\to a]} \leq [\![\beta]\!]^{\mathfrak{A}}_{s[x\to a]} = [\![\beta]\!]^{\mathfrak{A}}_s$ for every $a \in U$, since x is not free in β. So, $\bigvee_{a\in U} [\![\alpha]\!]^{\mathfrak{A}}_{s[x\to a]} \to [\![\beta]\!]^{\mathfrak{A}}_s = [\![\exists x\alpha \to \beta]\!]^{\mathfrak{A}}_s = [\![\varphi]\!]^{\mathfrak{A}}_s = 1$. The preservation of trueness by the rule $(\forall - In)$ is proved analogously. □

Now, let us consider the relation $\equiv$ defined by $\alpha \equiv \beta$ iff $\vdash_{\mathcal{Q}\mathbf{G3}'} \alpha \to \beta$ and $\vdash_{\mathcal{Q}\mathbf{G3}'} \alpha \to \beta$. Then, we have that the algebra $\mathfrak{Fm}_\Theta/_\equiv$ is a $\mathbf{G}'_3$-algebra (the proof is exactly the same as in the propositional case). It is clear that the algebra of formulas is an absolutely free algebra generated by the atomic formulas. The equivalence class of a formula α w.r.t. $\equiv$ will be denoted by $\overline{\alpha}$.

It is clear that $\mathcal{Q}\mathbf{G3}'$ is a Tarskian logic, see Definition 3.6. Besides, it is possible to consider the notion of set of formulas maximal non-trivial w.r.t to some formula φ (see Definition 3.8) and the notion of closed theories is defined in the same way as the propositional case, see Definition 3.7. Therefore, we have that the Lindenbaum-Łoś's Theorem holds for $\mathcal{Q}\mathbf{G3}'$. Then, we have the following

Lemma 4.6. *Let $\Gamma \cup \{\varphi\}$ be a set of formulas with Γ maximal non-trivial w.r.t. φ in* $\mathbf{QG3}'$. *Let $\Gamma/_\equiv = \{\overline{\alpha} : \alpha \in \Gamma\}$ be a subset of $\mathbf{G}'_3$-algebra $\mathfrak{Fm}_\Theta/_\equiv$, then:*

1. *If $\alpha \in \Gamma$ and $\overline{\alpha} = \overline{\beta}$, then $\beta \in \Gamma$. If $\overline{\alpha} \in \Gamma/_\equiv$, then $\overline{\forall x\alpha}, \overline{\exists x\alpha} \in \Gamma/_\equiv$.*
2. *$\Gamma/_\equiv$ is a modal deductive system of $\mathfrak{Fm}_\Theta/_\equiv$. Also, if $\overline{\varphi} \notin \Gamma/_\equiv$ then, for any closed modal deductive system $\overline{D}$ containing properly to $\Gamma/_\equiv$, it is the case that $\overline{\varphi} \in \overline{D}$.*

Proof. Suppose that $\alpha \in \Gamma$ and $\alpha \equiv \beta$. Then, $\vdash_{\mathcal{Q}\mathbf{G3}'} \alpha \to \beta$ and $\vdash_{\mathcal{Q}\mathbf{G3}'} \beta \to \alpha$. Therefore, $\beta \in \Gamma$. It is not hard to see that the conditions of Definition 3.15 are verified by $\Gamma/_\equiv$.

On the other hand, let $\overline{D} \subseteq \mathfrak{Fm}_\Sigma/_\equiv$ be a closed modal deductive system that properly contains $\Gamma/_\equiv$ and so, there is $\overline{\gamma} \in \overline{D}$ such that $\overline{\gamma} \notin \Gamma/_\equiv$. Now, we have that

$\gamma \notin \Gamma$ and therefore, $\Gamma \cup \{\gamma\} \vdash_{\mathcal{Q}\mathbf{G3}'} \varphi$. From the latter and taking $D = \{\alpha : \overline{\alpha} \in \overline{D}\}$, we can infer that $D \vdash_{\mathcal{Q}\mathbf{G3}'} \varphi$. Now, since D is closed we obtain that $\overline{\varphi} \in \overline{D}$. □

It is worth mentioning that item 2. of last lemma states that $\Gamma/_{\equiv}$ is a maximal modal deductive system. Besides, we know that $\mathfrak{Fm}_\Theta/_{\equiv}$ is a $\mathbf{G}'_3$-algebra, and for every Γ maximal non-trivial w.r.t. φ we have that $\Gamma/_{\equiv}$ is a maximal modal deductive system of $\mathfrak{Fm}_\Sigma/_{\equiv}$. Then, from Theorems 3.17 and 3.21, there is a homomorphism $h : \mathfrak{Fm}_\Theta/_{\equiv} \to \mathcal{B}_3^0$ such that $h^{-1}(\{1\}) = \Gamma/_{\equiv}$. Thus, if we consider the canonical projection $\pi : \mathfrak{Fm}_\Theta \to \mathfrak{Fm}_\Theta/_{\equiv}$, there is a homomorphism $f : \mathfrak{Fm}_\Theta \to \mathcal{B}_3^0$ defined by $f = h \circ \pi$ such that $f^{-1}(\{1\}) = \Gamma$. Observe that $f(\alpha) = h(\overline{\alpha})$.

Proposition 4.7. *Let Γ be a set of formulas which is maximal non-trivial w.r.t. φ in $\mathbf{G}'_3$. Let $\mathfrak{Fm}_\Theta/_{\equiv_\Gamma}$ be the quotient algebra obtained by the following congruence: $\alpha \equiv_\Gamma \beta$ iff $(\alpha \leftrightarrow \beta) \in \Gamma$. Then $\mathfrak{Fm}_\Theta/_{\equiv_\Gamma}$ is isomorphic to a subalgebra of $\mathcal{B}_3^0$ and so is a simple $\mathbf{G}'_3$-algebra.*

Proof. Let $\bar{\pi} : \mathfrak{Fm}_\Theta \to \mathfrak{Fm}_\Theta/_{\equiv_\Gamma}$ be the canonical projection. By considering the homorphism $f : \mathfrak{Fm}_\Theta \to \mathcal{B}_3^0$ defined above and by adapting the first isomorphism Theorem from Universal Algebra (see [1]), there is a monomorphism $\bar{f} : \mathfrak{Fm}_\Theta/_{\equiv_\Gamma} \to \mathcal{B}_3^0$ such that $\bar{f} \circ \bar{\pi} = f$. This means that $\mathfrak{Fm}_\Theta/_{\equiv_\Gamma}$ is isomorphic to a subalgebra of $\mathcal{B}_3^0$; that is to say, $\mathfrak{Fm}_\Theta/_{\equiv_\Gamma}$ is a simple algebra. □

Observe that $\bar{f}([\alpha]_\Gamma) = f(\alpha) = h(\overline{\alpha})$, where $[\alpha]_\Gamma$ denotes the equivalence class of α in $\mathfrak{Fm}_\Theta/_{\equiv_\Gamma}$. Since $\mathcal{B}_3^0$ is finite, we have the following:

Corollary 4.8. *Let Γ be a set of formulas which is maximal non-trivial w.r.t. φ in $\mathbf{G}'_3$. Then, $\mathfrak{Fm}_\Theta/_{\equiv_\Gamma}$ is finite, hence it is a complete lattice.*

In fact, $\mathfrak{Fm}_\Theta/_{\equiv_\Gamma}$ is isomorphic to either the 2-element chain $\{0, 1\}$ or the 3-element chain $\mathcal{B}_3^0$.

Theorem 4.9 (Completeness (for sentences) of $\mathcal{Q}\mathbf{G3}'$). *Let $\Gamma \cup \{\varphi\}$ be a set of closed formulas over Θ. Then: $\Gamma \vDash_{\mathcal{Q}\mathbf{G3}'} \varphi$ implies that $\Gamma \vdash_{\mathcal{Q}\mathbf{G3}'} \varphi$.*

Proof. Let us suppose that $\Gamma \nvdash_{\mathcal{Q}\mathbf{G3}'} \varphi$. Then, there is M maximal non-trivial w.r.t. φ such that $\Gamma \subseteq M$. Hence, $\alpha \in M$ for every $\alpha \in \Gamma$ and $\varphi \notin M$. Now, let us consider the algebra $\mathcal{A} := \mathfrak{Fm}_\Theta/_{\equiv_M}$ defined by the congruence $\alpha \equiv_M \beta$ iff $(\alpha \leftrightarrow \beta) \in M$. By Corollary 4.8, $\mathcal{A}$ is a complete $\mathbf{G}'_3$-algebra. It is easy to see that $[\alpha]_M \leq [\beta]_M$ iff $\alpha \to \beta \in M$.

Now, let us consider the canonical structure $\mathfrak{A} = \langle U, \mathcal{A}, \cdot^{\mathfrak{A}}\rangle$ such that U is the set Ter_Θ of terms over Θ and $\mathcal{A}$ is as above, for every term t we consider its name $\hat{t}$ as a constant of Θ. Assume that, if $\hat{t}$ is a constant then $\hat{t}^{\mathfrak{A}} := t$, and if $f \in \mathcal{F}_n$ then $f^{\mathfrak{A}}(t_1, \ldots, t_n) := f(t_1, \ldots, t_n)$. From this, it follows that, for any $t \in U$ and any assignment s, $[\![t]\!]^{\mathfrak{A}}_s = t$. On the other hand, if $P \in \mathcal{P}_n$, assume that the mapping $P^{\mathfrak{A}}$ is defined as follows: $P^{\mathfrak{A}}(t_1, \ldots, t_n) = [P(t_1, \ldots, t_n)]_M$. By induction on the complexity of the formula, it can be proven that, for every closed formula α and every s, $[\![\alpha]\!]^{\mathfrak{A}}_s = [\alpha]_M$. Indeed, the case for α atomic holds by definition of $\mathfrak{A}$. The cases $\alpha = \beta \# \psi$ and $\alpha = \neg\beta$ hold by induction hypothesis, the definition of $[\![\cdot]\!]^{\mathfrak{A}}_s$ and the definition of the operations in the $\mathbf{G}'_3$-algebra $\mathcal{A}$; moreover, it is not hard to see that for every formula $\psi(x)$ and every term t we have that $[\![\psi(x/\hat{t})]\!]^{\mathfrak{A}}_s = [\![\psi(x/t)]\!]^{\mathfrak{A}}_s$.

Suppose now that α is $\exists x\beta$. By axiom **(Ax14)**, for every $t \in U$, $\beta(x/\hat{t}) \to \alpha \in M$ and so $[\beta(x/\hat{t})]_M \leq [\alpha]_M$. By induction hypothesis, we have that $[\beta(x/\hat{t})]_M = [\![\beta(x/\hat{t})]\!]^{\mathfrak{A}}_s = [\![\beta(x/t)]\!]^{\mathfrak{A}}_s$ (by Proposition 4.4). Thus, $[\beta(x/t)]_M \leq [\alpha]_M$, for every $t \in U$. Now, let ψ be a sentence such that $[\beta(x/t)]_M \leq [\psi]_M$ for every term $t \in U$ and so $[\beta(x/\hat{t})]_M \leq [\psi]_M$ for every term $t \in U$. In particular, $[\beta(x/\hat{x})]_M \leq [\psi]_M$ and then, $[\beta(x)]_M \leq [\psi]_M$. This means that $\beta(x) \to \psi \in M$. Since x does not occur free in ψ then, by $(\exists - In)$, $\alpha \to \psi \in M$. This means that $[\alpha]_M \leq [\psi]_M$ and so $[\alpha]_M = \bigvee_{t \in U} [\beta(x/t)]_M$. Analogously, but now by using **(Ax15)** and $(\forall - In)$, it is proved that $[\![\alpha]\!]^{\mathfrak{A}}_s = [\alpha]_M$ for $\alpha = \forall x\beta$. This shows that $[\![\alpha]\!]^{\mathfrak{A}}_s = [\alpha]_M$ for every closed formula α and every s.

Thus, $\mathfrak{A}$ is a Θ-structure for $\mathcal{Q}\mathbf{G3}'$ such that, for every closed formula α, α is true in $\mathfrak{A}$ iff $\alpha \in M$. From this we have that $\Gamma \not\models_{\mathcal{Q}\mathbf{G3}'} \varphi$.

□

Given a formula α such that the set of variables occurring free in α is $\{x_1, \ldots, x_n\}$. The *universal closure* of α is the closed formula $(\forall\alpha)$ given by α (if $n = 0$) or $\forall x_1 \ldots \forall x_n \alpha$ otherwise. Then, the completeness theorem of $\mathcal{Q}\mathbf{G3}'$ for arbitrary formulas can now be easily obtained from the last result:

Theorem 4.10 (Completeness of $\mathcal{Q}\mathbf{G3}'$). *Let $\Gamma \cup \{\varphi\}$ be a set of formulas over Θ. Then: $\Gamma \models_{\mathcal{Q}\mathbf{G3}'} \varphi$ implies that $\Gamma \vdash_{\mathcal{Q}\mathbf{G3}'} \varphi$.*

Proof. By **(Ax15)** and $(\forall - In)$ it is easy to prove that $\alpha \vdash_{\mathcal{Q}\mathbf{G3}'} (\forall)\alpha$ and $(\forall)\alpha \vdash_{\mathcal{Q}\mathbf{G3}'} \alpha$, for every formula α. On the other hand, by definition of $\models_{\mathcal{Q}\mathbf{G3}'}$, it is immediate to see that $\alpha \models_{\mathcal{Q}\mathbf{G3}'} (\forall)\alpha$ and $(\forall)\alpha \models_{\mathcal{Q}\mathbf{G3}'} \alpha$, for every formula α. Then, for every $\Gamma \cup \{\varphi\}$: $\Gamma \vdash_{\mathcal{Q}\mathbf{G3}'} \varphi$ iff $(\forall)\Gamma \vdash_{\mathcal{Q}\mathbf{G3}'} (\forall)\varphi$, and $\Gamma \models_{\mathcal{Q}\mathbf{G3}'} \varphi$ iff $(\forall)\Gamma \models_{\mathcal{Q}\mathbf{G3}'} (\forall)\varphi$, where $(\forall)\Gamma = \{(\forall)\beta \ : \ \beta \in \Gamma\}$. From this, the desired result follows immediately from Theorem 4.9. □

Acknowledgments

Coniglio was financially supported by an individual research grant from CNPq, Brazil (308524/2014-4 and 306530/2019-8). Figallo-Orellano acknowledges the support of a post-doctoral grant 2016/21928-0 from São Paulo Research Foundation (FAPESP), Brazil. This paper was elaborated while Hernández-Tello and Pérez-Gaspar were visiting PhD students at CLE/UNICAMP with financial support from Benemérita Universidad Autónoma de Puebla (BUAP), México.

References

[1] Burris, S. and Sankappanavar, H. P., *A Course in Universal Algebra*, Graduate Texts in Mathematics volume 78. Springer, 1981.

[2] Canals Frau, M. and Figallo, A. V., *Modal 3-valued implicative semilattices*, Preprints del Instituto de Ciencias Básicas. Universidad Nacional de San Juan, Argentina, pp. 1–24, 1992.

[3] Canals Frau, M., Figallo, A. V. and Saad, S., *Modal three valued Hilbert algebras*, Preprints del Instituto de Ciencias Básicas. Universidad Nacional de San Juan, Argentina, p.1 - 21, 1990.

[4] Carnielli, W. A. and Marcos, J., *A taxonomy of C-systems*, Paraconsistency: The Logical Way to the Inconsistent. Proceedings of the 2nd World Congress on Paraconsistency (WCP 2000), (eds. W. Carnielli, M. Coniglio and I. D'Ottaviano), Lecture Notes in Pure and Applied Mathematics volume 228, pp. 1–94, Marcel Dekker, 2002.

[5] Figallo-Orellano, A. and Slagter, J., *An algebraic study of the first order some implicational fragments of 3-valued Łukasiewicz logic*, submitted, 2019.

[6] Figallo-Orellano, A. and Slagter J., *Algebraic Monteiro's notion of maximal consistent theory for tarskian logics*, submitted, 2019.

[7] Osorio, M. and Navarro, J., Arrazola, J. and Borja, V., *Logics with common weak completions*, Journal of Logic and Computation, 16:6, 867–890, 2006

[8] Monteiro, L., *Algèbre du calcul propositionnel trivalent de Heyting*, Fundamenta Mathematicae, 74:2, 99–109, 1972.

[9] Mendelson, E., *Introduction to Mathematical Logic*, 2009, CRC Press.

[10] Osorio, M. and Carballido, J., *Brief study of* $\mathbf{G}_3'$ *logic*, Journal of Applied Non-Classical Logics, 18:4, 475–499, 2008.

[11] Osorio, M., Ramírez, J., Arrazola, R., Carballido, J. and Estrada, O., *An Axiomatization of* $\mathbf{G}_3'$., LoLaCOM, 2006.

[12] Thomas, I., *Finite limitations on Dummett's LC*, Notre Dame Journal of Formal Logic, 3:3, 170–174, 1962.

[13] Wójcicki, R., *Lectures on propositional calculi*, Ossolineum, Warsaw, 1984.

Received 01 February 2020

A LOGICAL FRAMEWORK TO REASON ABOUT REO CIRCUITS

ERICK GRILO
Instituto de Computação, Universidade Federal Fluminense
simas_grilo@id.uff.br

DANIEL TOLEDO
Instituto de Computação, Universidade Federal Fluminense
danieltoledo@id.uff.br

BRUNO LOPES
Instituto de Computação, Universidade Federal Fluminense
bruno@ic.uff.br

Abstract

Reo is a graphic-based coordination modelling language which aims to capture and model the interaction between pieces of software, using structures known as channels. The fact that Reo has been used to model many real-world situations, from software components to Smart Cities entities, has attracted attention from researchers, resulting in a great effort directed in its formalization in order to verify and certify properties of Reo circuits. This work presents a constructive formalization in Coq of Reo's formal semantics (based on Constraint Automata) and a formalization in the nuXmv model checker, both with a composition operation and with tools to automate the process. We describe the formalizations and present some usage examples with experimental results.

1 Introduction

In recent years, many software developers have researched new techniques on how to develop software. Technologies and methods have emerged since the nineties, such as service-oriented computing [ ] and model-driven development [ ], where the first

The authors thank CAPES, CNPq and FAPERJ for supporting this work.

approach advocates the idea of composing software out of other software and the latter, developing software based on previous models.

Reo [,] is a graphical-based coordination modelling language which aims to capture and model the interaction between pieces of software, using structures known as channels. A Reo connector may be built from one or more channels and its goal is to integrate these communicating entities. Each channel has its predefined behavior and may be combined with other channels to build more complex Reo models. Reo's main objective is to provide a model of the code regarding how the integrated entities within a Reo connector interact between themselves, generating the model of the code which coordinates this interaction, known as "glue-code". Reo has proven to be successful in modeling the orchestration of concurrent systems' interaction, being employed in a wide range of applications, from process modeling to Web-Services integration [,].

The usage of channel-based models is considered an alternative to advocating model-driven development of such "glue code". A Reo channel is an entity that connects two distinct ends with its unique behavior. Channels can be seen as primitives for modelling concurrent systems. The usage of channel-based models brings several advantages by efficiently modelling primitives of concurrent systems (remote function calls, message passing, and shared memory, to name a few), enabling the capture of properties like the efficiency of how messages are exchanged.

The advent of formal techniques in software development such as proof assistants and model checkers led to a new way of dealing with formal verification of software. Proof assistants' advantages include reducing the human effort involved in the proof-construction process and reducing human errors that could be introduced in such proof, such as invalid proof steps, thus making the task of building up a proof easier to perform. Model checkers enable the verification of properties by exploring all possible states of a model in order to assure that the model indeed satisfies some property, returning a counterexample if the model is fully verified and fails to satisfy such property.

Modelling software components as Reo connectors may bring some advantages, namely a modelling approach that may bring better insights on how the involved components interact with each other and the usage of one of its formal semantics [ ] to formally certify Reo models. Constraint Automata [ ] is the first operational formal semantics for Reo [ ], proposed by the language's creators and it is the formal semantics employed in the present work.

Constraint Automata enable users to extend Reo connectors to suit their needs, given that their corresponding automaton is supplied. It also allows the model and verification of properties regarding connectors' scenarios and how they change their configuration when data flows through them, by interpreting their states as the connector's configuration, and their transitions as "how the connector itself changes its configuration". While Timed Data Streams (TDS) provide an intuitive way to understand data flow between Reo connectors, Constraint Automata provides more suitable means to analyze these connectors [ ], giving an overview on "how data flowing through the connector changes its configuration".

To verify whether a system complies with its requirements may be a hard task. Certifying liveness, absence of deadlocks and that any important constraint is fulfilled may demand a huge effort that may not solvable by standard test or simulation techniques. Employing Constraint Automata as a formal semantics for Reo leads to the possibility of reducing the verification of Reo circuits to the satisfiability problem.

Given a Reo circuit $\mathcal{R}$ we provide a translation $\mathcal{R} \overset{Coq}{\Longrightarrow} \mathcal{R}'$ and also propose a translation $\mathcal{R} \overset{SMV}{\Longrightarrow} \mathcal{R}''$ to logic-based models. Using $\mathcal{R}'$ it is possible to use an Interactive Theorem Prover (namely Coq,[1] detailed in Section 5.1) to prove properties about the circuit, compare the behaviour of two circuits through bisimulation, and extract certified code. Verifying if some constraint C is complied by $\mathcal{R}''$ is equivalent to the problem $\mathcal{R}'' \models C$, where $\models$ is the usual logical entailment relation. This model checking process is performed by nuXmv,[2] a symbolic model checker capable of providing counterexamples (detailed in Section 5.2).

This paper proposes an automatic approach to model, reason, and certify Reo circuits by means of proof theory employing Coq, and perform model checking evaluations using nuXmv, providing logic-based models. We provide an automatic framework which compiles a textual representation of Reo circuits (denoted by a subset of Treo [ ] language) which takes as input Reo connectors and port names denoting the interfaces they connect and generate both Coq and nuXmv models. Then, one may use all tooling available to automatically generate models in both systems for reasoning and compile certified code corresponding to the given Reo circuit supported by its formal semantics. We also present a Graphical User Interface to model and apply all of these translations.

The organization of this paper is as follows. Section 2 presents the related works available in the literature pointing out their main features and Section 3 gives an overview of the formal background used in this work. Reo's characteristics and formal semantics are presented in Section 4 followed by the presentation of the translations and their implementations in Section 5, with usage examples. Section 6 discuss the final remarks and ongoing work.

[1] `https://coq.inria.fr`

[2] `https://nuxmv.fbk.eu/`

2 Related work

The fact that Reo can be used to model many real-world situations has attracted attention from researchers all around the world, resulting in a great effort directed in formalizing means in order to verify properties of Reo models [, , , , ,]. Reo's formal studies also resulted in the proposal of many formal semantics for this modelling language [ ].

The approach presented by Klein et al. [ ] provides a platform to reason about Reo models using Vereofy,[3] a model checker for component-based systems, while Pourvatan et al. [ ] employ Constraint Automata in reasoning about Reo models by means of symbolic execution. Kokash & Arbab [ ] formally verify Long-Running Transactions (LRTs) modelled as Reo connectors using Vereofy, enabling expressing properties of these connectors in logics such as Linear Temporal Logic (LTL) or a variant of Computation Tree Logic (CTL) named Alternating-time Stream Logic (ASL). Kokash et al. [ ] encode Reo in mCRL2 model checker using Constraint Automata and its main variants, encoding their behaviour as mCRL2 processes and enabling the expression of properties regarding deadlocks and data constraints which depend upon time. Mouzavi et al. [ ] propose an approach based on Maude in order to model checking Reo models, encoding Reo's operational semantics of the connectors, and Li et al. [ ] propose a real-time extension to Reo, implementing new channels and relying on Stochastic Timed Automata for Reo (STA) as its formal semantics, also providing a translation of STA to PRISM[4] for model checking. UPPAAL [5] model checker has also been employed in the verification of Reo connectors employing the usage of Timed Constraint Automata [ ] to build the corresponding UPPAAL model, and in the simulation of Hybrid Reo Connectors [ ].

Among the tools used by researchers to formally reason about Reo connectors, proof assistants are employed towards this objective [, , , , ,]. When restricting these implementations to the formalization of Constraint Automata as Reo's formal semantics, only a few approaches aim to implement them, although no implementation of such formalism in a proof assistant as Coq was found in the literature.

The approaches adopted by Li et al. [,] are among the ones that employ Coq to verify Reo models formally. The first work formalizes four of the Reo canonical connectors along with an LTL-based language defined as an inductive type in Coq. While the latter proposes the formalization of some of the channels depicted by Figure 1 as logical propositions in Coq, where their behavior are defined as conjunctions

[3] http://www.vereofy.de

[4] https://www.prismmodelchecker.org

[5] http://www.uppaal.org/

regarding data and time constraints on streams denoting input and output of the automaton. Both formalizations implement the notion of Timed Data Streams as it is the first formalization of semantics of Reo connectors [ ].

The implementation proposed by Li et al. [ ] enables the verification of timed properties of connectors: such properties may be proven considering the data flow a connector takes as input. The formalized LTL-based language enables bounded model checking on these connectors. Although, it lacks any automatic composition operation for formalized connectors. Therefore, complex channels may be manually written. The approach employed by Li et al. [ ] implements composition of Reo connectors employing logical conjunction of connectors' behaviour, denoted by their respective TDS.

The usage of Constraint Automata as formal semantics for Reo is advantageous when considering a systematic notion of a Reo connector. One can use the fact that states of an automaton are its possible configurations to validate certain properties. Informally, Constraint Automata enables one to see "How the system behaves internally". When not considering only Reo, Constraint Automata provides a formal basis to reason about general coordination modeling languages, not only Reo [ ]. Although Constraint Automata does not suit the verification of timed properties of connectors as well as Timed Data Streams, the fact that Constraint Automata can be interpreted as TDS-language acceptors and the notion of acceptance regarding Timed Data Streams can be used to prove timed properties regarding the connector (considering the input Timed Data Stream) and possible configurations that the connector can achieve (considering the Constraint Automata itself). More specifically, implementing such formalism in a system such as Coq enables the automatic generation of certified code from the model to languages such as Haskell, Scheme or OCaml.

The adopted approach in here proposed is twofold. This paper presents (i) a framework to model and reason about Reo circuits by means of the construction of a proof-theory in Coq which leads to certifying and compiling Reo code and (ii) a detailed and compacted translations to a nuXmv model, which has many optimizations to achieve a competitive performance. The Coq and nuXmv formalizations deviates from existing ones by encoding the Automata model, enabling deductive and also explicit reasoning on states, taking advantage of operations like the product construction to build more complex models, and also making notions of bisimulation more natural by implementing them.

3 Background

This section is bound to give an overview of core aspects regarding the employed systems and their logical background. We discuss the main aspects regarding CTL, nuXmv and Coq and show usage example of these systems.

3.1 Computation Tree Logic

Computation Tree Logic (CTL) [ ] is a branching-time modal logic where the modalities refer to time. In CTL, time is modeled as a tree-like structure in which the future is not determined, there are different paths in the future, which any of those might become the actual path that is realized. CTL's syntax is presented in Definition 3.1.

Definition 3.1 (CTL's formulae). *A formula in CTL is* $\varphi ::= p \mid \top \mid \varphi \wedge \varphi \mid \neg\varphi \mid EG\ \varphi \mid EX\ \varphi \mid E\ \varphi\ U\ \varphi \mid AG\ \varphi \mid AX\ \varphi \mid A\ \varphi\ U\ \varphi$, *where* $p \in \Phi$ *is an atomic proposition, E denotes for the* existence *of a path were some property* φ *holds Globally, in the neXt state or Until some other property holds. The analogous for all paths is achieved replacing E for A.*

We use the standard abbreviations $\bot \equiv \neg\top$, $\varphi \vee \phi \equiv \neg(\neg\varphi \wedge \neg\phi)$, $\varphi \rightarrow \phi \equiv \neg(\varphi \wedge \neg\phi)$, $EF\ \varphi \equiv E\ \top\ U\ \varphi$.

A CTL model is a transition system $\mathcal{M} = \langle S, R, V\rangle$ where S is an enumerable set of states, $R \subseteq S \times S$ is a binary relation and $V\colon S \rightarrow 2^{\Phi}$ is a valuation function. A formula φ is satisfied in a state $s \in S$ of a model $\mathcal{M}$ (says $\mathcal{M},s \models \varphi$) iff

- $\mathcal{M},s \models p$ iff $p \in V(s)$
- $\mathcal{M},s \models \top$ always
- $\mathcal{M},s \models \neg\varphi$ iff $\mathcal{M},s \not\models \varphi$
- $\mathcal{M},s \models \varphi \wedge \psi$ iff $\mathcal{M},s \models \varphi$ and $\mathcal{M},s \models \psi$
- $\mathcal{M},s \models EG\ \varphi$ iff there exists $sRs_1, \cdots, s_{n-1}Rs_n$ such that in all $\mathcal{M},s_i \models \varphi$
- $\mathcal{M},s \models EX\ \varphi$ iff there exists sRs_1 such that $\mathcal{M},s_1 \models \varphi$
- $\mathcal{M},s \models E\ \varphi\ U\ \psi$ iff there exists $sRs_1, \cdots, s_{n-1}Rs_n$ such that in all $\mathcal{M},s_i \models \varphi$ until that $\mathcal{M},s_n \models \psi$
- $\mathcal{M},s \models AG\ \varphi$ iff for every $sRs_1, \cdots, s_{n-1}Rs_n$ in all $\mathcal{M},s_i \models \varphi$
- $\mathcal{M},s \models AX\ \varphi$ iff for every sRs_1 $\mathcal{M},s_1 \models \varphi$

- $\mathcal{M}, s \models A\ \varphi\ U\ \psi$ iff for every $sRs_1, \cdots, s_{n-1}Rs_n$ in all $\mathcal{M}, s_i \models \varphi$ until that $\mathcal{M}, s_n \models \psi$.

3.2 nuXmv

Model checkers are tools used to verify the properties of systems formally modeled. That verification is done exploring all of the model states [ ], in other words, the checker examines each possible system scenario to systematically verify that the property holds in the model.

Those properties to be verified can represent safety characteristics, that is, properties that should always hold so that something undesirable never happens, for example, the absence of deadlocks. They can also represent liveness characteristics that guarantee the program evolution, that is, they define what the system must do to execute.

nuXmv is a symbolic model checker [ ] that extends NuSMV [ ], having all of its functionalities. NuSMV combines Binary Decision Diagram (BDD) verification with SAT-based verification. nuXmv extends NuSMV in two major directions: it improves finite models verification with state of the art SAT-based algorithms; with infinite models, it presents verification techniques based on Satisfiability Modulo Theory (SMT).

As a model checker, nuXmv was used in various projects such as AUTOGEF [ ], validating aerospace critical systems, EuRailCheck [ ] was a project that developed a tool to formalize and validate specifications for The European Train Control System (ETCS).

One of the ways in which nuXmv verifies a model is based on BDD. Binary Decision Diagrams are directed, acyclic and rooted graphs [ ] that represent a boolean function. The other way of verification is based on the Propositional Satisfiability Problem (SAT). That problem consists of finding any interpretation that satisfies a given boolean formula. The notion of using SATsolvers in model checking [ ] to solve the state explosion problem exists since the years 2000. As the SAT problem is NP-complete, various implementations minimizes its costs in different ways, nuXmv uses both zChaff[6] and MiniSat[7] implementations.

A nuXmv program is structured as a list of **MODULEs** in which one of those is the `main`, which is from where the model will begin. A **MODULE** declaration is where the variables and transitions are declared. Those declarations follow a block structure, as follows:

[6] `http://www.princeton.edu/~chaff/index.html`

[7] `http://minisat.se/Main.html`

MODULE encapsulates others declarations, it is declared as **MODULE identifier** (module_parameters), where **identifier** is the **MODULE** identifier and **module_parameters** are the optional **MODULE** parameters.

VAR encapsulates the list of variables in the **MODULE**. A variable is declared as identifier : type_specifier;, where **identifier** is its identifier, **type_specifier** is its type. A **MODULE** can be used as a variable type, that way, it is possible to instantiate a **MODULE** as a variable.

FROZENVAR encapsulates a variable list declared as above, the difference is that these variable will not have its first value changed.

ASSIGN encapsulates the attribution list, an attribution can be as **init**(identifier) := simple_expr, where the variable with the **identifier** receives the simple_expr as its initial value; or as **next**(identifier) := simple_expr, where **identifier** will receive **simple_expr** in the next state.

TRANS encapsulates a transitions list, those transitions are declared as boolean expressions that must contain the **next()** attribution.

INVAR encapsulates a list of boolean expressions that restrict the model states.

CTLSPEC encapsulates a CTL expression to be verified in the model.

LTLSPEC encapsulates a LTL expression to be verified in the model.

Listing 1 presents an example of a nuXmv model, **MODULE** has a **VAR** block where variable **a** is declared as a **boolean** and **b** as a instance of **foo** that receives **a** as a parameter. **MODULE foo** receives a variable as a parameter and initializes it with the **TRUE** value in its **ASSIGN** block.

Listing 1: nuXmv model example

```
MODULE main
...
VAR
  a : boolean;
  b : foo(a);
...
MODULE foo(x)
ASSIGN
  init(x) := TRUE;
```

3.3 Coq

Coq [ ] is a proof assistant based on an implementation of Calculus of Inductive Constructions, a type theory based on Calculus of Constructions [ ]. It is a powerful system which aims at representing both functional programs and proofs in higher-order logic using only one programming language named Gallina [ ]. Terms of Gallina can represent programs as well as properties of these programs and proofs of these properties.

Coq offers a centralized environment to write programs, algorithms, and prove properties regarding these objects. All expressions formalized in Coq are named terms, and all terms have a type. Hence, every object handled in Coq is typed. There are types for propositions, programs (or functions), data types (natural numbers, booleans, lists, pairs, and many others). The types of types are called `sort`. All sorts have a type, and there is an infinite well-founded typing hierarchy of sorts, whose base sorts are Prop, Set and Type, to avoid inconsistencies [ ].

Coq provides a built-in language to work both with program definitions and with the proof process. In what follows some of its keywords are briefly introduced. For more insights regarding these keywords, we suggest the reader refer to the system's reference manual.[8]

`Lemma` *id* : *Prop* denotes the binding of the type of a proposition to the variable *id*, enabling the interactive proving of *id* by employing Coq tactics.

`Qed/Defined`. `Qed` defines the proof term in Coq as an opaque term (a term which can be unfolded in tactic applications), whereas the usage of `Defined` closes the proof term as a transparent term, enabling it to be unfolded in posterior programs.

`Inductive` ***ident* : *type* := {| ident : type}** defines an inductive type whose constructors are defined by each {| ident : type} clause. The type of *ident* is *type* (which can be omitted as Coq's type checker is capable of deducting the term's type from its constructors). Inductive definitions are closed by types (i.e., their constructors have the same type of the definition) and they can be parametrized.

`Definition` ***id*** lets their users bind functions, theorems, (co-)inductive definitions and the evaluation of an expression (basically any well-typed term) to a variable named *id*.

[8]`https://coq.inria.fr/distrib/current/refman/`

`Record` ***ident* : *sort?* := *ident?* { *ident binders* : type }** defines a macro which constructs records as in many programming languages, similar to C's "struct" keyword. The first identifier *ident* is the name of the defined record. The keyword *sort?* is the record's type (which may be omitted), *ident?* is an optional identifier which defines the record's constructor. A record may have one or more fields (denoted by the *ident* within the curly brackets) separated by ";".

`Fixpoint` ***param* {*struct id*}** is the command that allows the definition of functions by pattern-matching over an inductive structure which is one of the `param` provided, defining recursive functions in Coq. These definitions need to meet syntactical criteria on an argument called decreasing argument. Thus, the idea of the criteria is to have a structure that tells Coq such definitions always terminate. The decreasing argument can be specified by using *struct id* or automatically guessed by Coq.

`Instance` ***id class_id binders* : *type* := { *id* := *term*}** declares a class instance identified by *id*, with non-obligatory parameters *binders* and the fields declared within the scope of {*id* := *term*}.

`Extraction` ***id*** enables the extraction of definition *id* to either Haskell, OCaml or Scheme. Variants like *Extraction "file.v" destFile* extracts all definitions within the file *file.v* (where *.v* is Coq's source code file extension) to the specified target language in a file named *destFile*. This language can be set with the command `Extraction Language lang`, where `lang` is either `Scheme`, `Haskell` or `OCaml` (the default extraction language).

A simple yet representative usage example is introduced as follows. One can use Coq to obtain certified code in other languages, such as Haskell or Scheme. Suppose one would like to formalize weekdays and then reason about the next day. This can be achieved by formalizing *weekdays* as an inductive type with its constructors denoting days of the week.

```
Inductive weekdays :=
  | monday | tuesday | wednessday | thursday | friday | saturday | sunday.
```

Then *nextDay* is a function that takes a weekday and returns the next day according to the current calendar.

```
Definition nextDay (day : weekdays) : weekdays :=
  match day with
  | monday ⇒ tuesday| tuesday ⇒ wednessday| wednessday ⇒ thursday
  | thursday ⇒ friday| friday ⇒ saturday| saturday ⇒ sunday
```

```
| sunday => monday
end.
```

Properties about *nextDay* can then be formalized, proved and the specified algorithm can be extracted to the aforementioned target languages.

```
Lemma nextDayMonday : ∀ day: weekdays, nextDay day = monday ↔ day = sunday.
Proof.
split.
- intros. destruct day. all: inversion H. reflexivity.
- intros. rewrite H. reflexivity.
Defined.
```

The extraction of these definitions may be done with the command `Extraction Language Scheme`, given that the desired target extraction language is Scheme. By formalizing these definitions within a module named *example*, the command `Extraction` *example usageEx* generates a .scm file named *usageEx*, containing all the aforemetioned definitions as Scheme code.

4 Reo

In this section, a succinct overview of Reo [,] is presented, considering its main characteristics with two usage examples. We also briefly introduce the main aspects of one popular formal semantics for Reo, Constraint Automata as proposed by Baier et al. [ ] We employ Constraint Automata as the formal semantics employed in the presented framework as means to reason about Reo connectors. We also introduce two usage examples of Reo, which will later be discussed in Section 5, bound to show how one can formalize and certify Reo circuits employing Coq and nuXmv.

4.1 The modelling language

Reo plays a central role in integrating software components, especially considering Component Based Software Engineering, where it is expected that software components are independent of each other, being more adapted to the environment they were conceived for. In recent times, software development has shifted from building large, single instances of a system to building systems by reuse of already existing pieces of software, where the full application (system) itself is generated by means of the orchestrated interaction of these software components, where Reo may be adequate in orchestrating such interaction.

As a coordination model, Reo focuses on connectors, their composition and how they behave, not focusing on particular details regarding the entities that are connected, communicate and interact through those connectors. Connected entities may be modules of sequential code, objects, agents, processes, web services and any other software component where its integration with other software can be used to build a system [ ]. Such entities are defined as component instances in Reo.

Component instances are defined as a non-empty set P that denotes a set of entities involved in an instance (process, services, actors, usually denoted by capital letters) and a predefined set of I/O operations associated with each of those entities, where they only interact with each other by the channel that connect these instances. A software component is a software implementation which may execute in physical or logical devices. Therefore, software components are abstract entities that describes the behavior of its instances.

Channels in Reo are defined as a point-to-point link between two distinct nodes, where each channel has its unique predefined behavior. Each channel in Reo has exactly two ends, which can be of the following types: the source end, which accepts data into the channel, and the sink end, which disperses data out of the channel. Channels are used to compose more complex connectors, being possible to combine user-defined channels amongst themselves and with the canonical connectors provided by Baier et al. [ ]. Figure 1 shows the basic set of connectors as presented by Kokash et al. [ ].

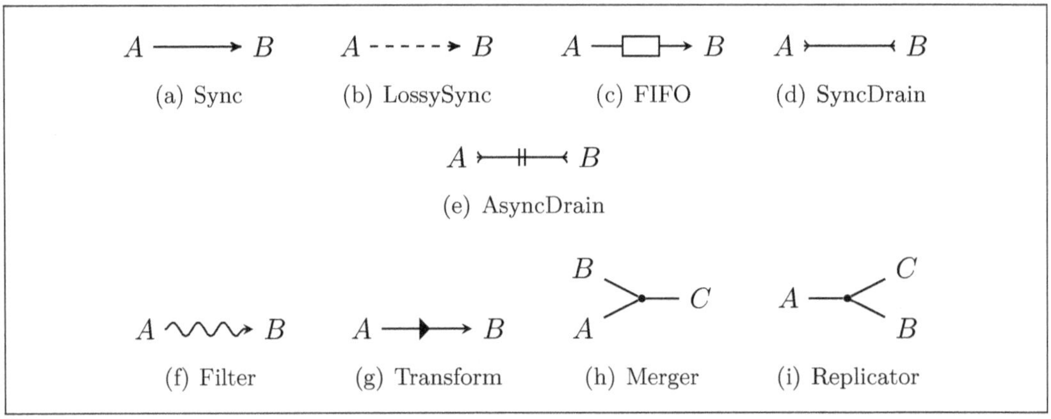

Figure 1: Canonical Reo connectors

A node in Reo is defined as a logical organization denoting the structure of how channel ends are linked to each other in Reo connectors. Nodes composing channel ends in Reo can be either source nodes, sink nodes or mixed nodes. Source nodes are nodes that accept data into the channel, i.e., nodes that serve as a gateway to data flow into the channel, while sink nodes are nodes where data flows out of the channel and mixed nodes are nodes that act both as source nodes and sink nodes.

Channel ends can be used by any entity to disperse/receive data, given that the entity belongs to an instance that knows these ends. In other words, entities may use channels only if the instance they belong to is connected to one of the channel ends, enabling either sending or receiving data (depending on the kind of the channel end the entity has access to).

The bound between a software instance and a channel end is a logical connection which does not rely on properties such as the location of the involved entities. Channels in Reo have the sole objective to enable the data exchange by means of I/O operations predefined for each entity in an instance. A channel can be known by zero or more instances at a time, but its ends can be used by at most one entity at the same time.

Figure 2 presents a variant of a Reo circuit that models a Sequencer.[9] Such circuit models the sequencing of processes which are interconnected by means of this connector. Therefore, properties such as how data flow between the connected entities can be stated and proven using its corresponding constraint automaton.

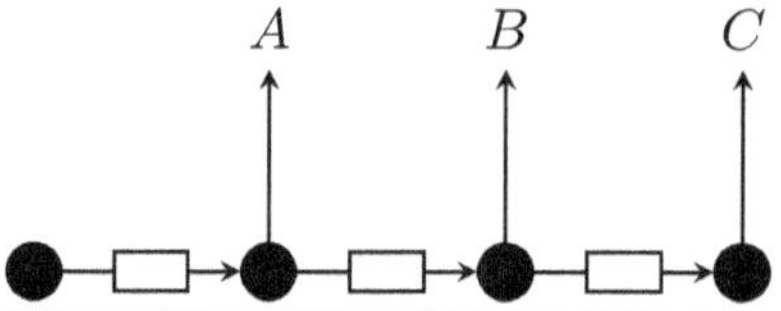

Figure 2: Modelling of a variant of the Sequencer in Reo

Figure 3 models a simplification of a scenario containing two Smart traffic lights A and B in a crossroad [ ]. Their default functioning follows a timed schedule: while one of them is green, the other is red. In addition to this timed behaviour, a controlling station have a sensor (i.e., a camera, denoted by the upper dot) which monitors the crossroad and identifies whether there is a heavy traffic waiting for the green light on one of the traffic lights.

Intuitively, The circuit controls the effective time a traffic light may be green or red depending on the amount of cars waiting to pass. This may be done by verifying which data item is coming from both the timer and the sensor, and when is these data incoming. The circuit filters this data, in order to mutually exclude one of the traffic lights. The destination node (denoted by the leftmost dot in Figure 3) will receive the data item (0 or 1) and, based on this item decide which traffic light gains the priority to go green.

The data incoming from the uppermost dot denotes a property which the sensor has detected (i.e., many cars waiting for the traffic light to be green), while d is a data

[9] http://reo.project.cwi.nl/v2/#examples-of-complex-connectors

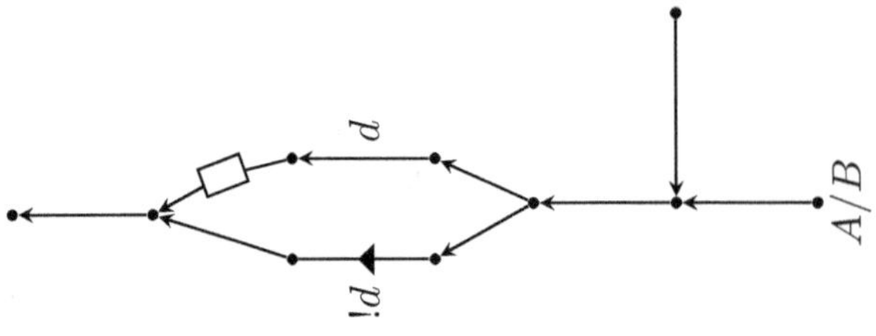

Figure 3: A Reo model for Smart traffic lights and a controller of a crossroad

item denoting that the semaphores will alternate between open and closed, enabling the interchange of which traffic light will be either open or closed (the interchange between d and $!d$ forced by the circuit renders unable the scenario where one of the traffic lights is always open).

The analysis of Reo circuits by means of its semantics enable the verification of properties employing already existing mathematical theories and tools. By modelling them as constraint automata, one can state that a circuit contains properties regarding specific data flows (i.e., given a TDS, we want to prove that it correctly describes the circuit or not) or even to guarantee that specific scenarios never happen (by assuring that some desired TDS will not be accepted by the automaton). Besides these properties, one may also guarantee that the circuit itself satisfies properties regarding its structure (i.e., in order to some port "B" to have data flow, it will always require that the same data have to flow another port "A"). Such properties can be reflected into real-world requirements, such as liveness and synchronization between software.

4.2 Formal semantics

Constraint Automata [ ] are defined as the most basic operational models for Reo, although there are many other formal semantics for Reo [ ]. The present work focuses on Constraint Automata as proposed by Reo creators as it is the first operational semantics to reason about Reo connectors [ ]. Proposed as a formalism to denote and reason about coordination models (mainly described as Reo models), Constraint Automata can be seen as a variation of Finite Automata where the transitions are influenced by ports containing data and data constraints over those ports.

Constraint Automata essentially capture the semantics of communication between interconnected entities on a Reo connector, mainly enabling reasoning over specific communication patterns and which piece(s) of data flow to/from each connected entity. Specific properties such as timed behaviour, actions taken by some entity or even probability of communication between entities to happen are not captured by standard CA and require more expressive formalisms, such as Timed Constraint Automata [ ], Action Constraint Automata [ ] or Probabilistic Constraint Automata [ ], respectively.

Constraint automata compose a basis on modeling and verifying the specification of such coordination mechanisms by the usage of formal methods (e.g., by means of model checking against temporal-logic specifications [ ]). By using Constraint Automata as formal semantics for Reo, automata states depict the possible configurations of a channel (e.g., the data within a connector at a given time), while transitions of the automaton denote how data in the connector flow and how it changes the configuration of the automaton.

Definition 4.1 (Constraint Automata). *A Constraint Automaton (CA) is a tuple* $\mathcal{A} = (Q, \mathcal{N}ames, \rightarrow, Q_0)$ *where*

Q *is a finite set of states, denoting possible configurations of* $\mathcal{A}$

$\mathcal{N}ames$ *is a finite set of port names,*

$\rightarrow\ \subseteq Q \times 2^{\mathcal{N}ames} \times DC \times Q$ *is the transition relation with* DC *a set of (propositional) Data Constraints, and*

$Q_0 \subseteq Q$ *is the set of initial states.*

Constraint automata are seen as Timed Data Stream (TDS) acceptors. To understand how constraint automata relate to Timed Data Streams, we recover the main definitions from Baier et al. [ ] regarding Timed Data Streams and Constraint Automata.

Let A be any set. Streams are defined as a set A^ω containing all infinite sequences over A. Therefore, $A^\omega = \{\alpha \mid \alpha \colon \{0,1,2,\dots\} \rightarrow A\}$. Individual streams are described as $\alpha = \alpha(0), \alpha(1), \alpha(2), \dots$ and the derivative of a stream α is denoted as the stream initiating in the next value, namely $\alpha' = \alpha(1), \alpha(2), \alpha(3), \dots$. $\alpha^{(i)}$ denotes the i-th derivative, where $\alpha'(k) = \alpha(k+1)$ and $\alpha^{(i)}(k) = \alpha(i+k), \forall i, \forall k > 0$. Hence, TDS are composed by a stream $\alpha \in Data^\omega$, $Data$ a non-empty finite set and a time stream $a \in \mathbb{R}_+^\omega$ as a stream of increasing positive real numbers.

The behavior of Reo channels modeled as Timed Data Streams (TDS) [ ] introduces the notion of a channel node's behavior being a relation $R \subseteq TDS \times TDS$. TDS are composed of two streams, one denoting the data items that will flow through a given port and the other one denoting the time instant that the port observes this data flow. TDS are used to model how data flows through a Reo connector by discriminating the flow through the relation R. TDS are formally presented by Definition 4.2.

Definition 4.2 (Timed Data Streams).
A Timed Data Stream is defined as a pair of functions (α, a) *as follows.*
$TDS = \{(\alpha, a) \in Data^\omega \times \mathbb{R}_+^\omega \colon \forall k \geq 0 \colon a(k) \leq a(k+1) \text{ and } \lim_{k\to\infty} a(k) = \infty\}$

Hence, TDS are composed by a stream $\alpha \in Data^\omega$ with $Data$ as a non-empty finite set and a time stream $a \in \mathbb{R}_+^\omega$, a stream of increasing positive real numbers. A TDS can be intuitively seen as a "controller" which denotes, for each data item $\alpha(k)$, the moment $a(k)$ it is flowing.

In order to formalize the concept of input/output behavior of Constraint Automata by means of TDS, a set of names $\mathcal{N}ames$ is used, where $\mathcal{N}ames$ consists of a finite set of names $\mathrm{A}_1, \mathrm{A}_2, \ldots, \mathrm{A}_n$ used to identify the input/output ports that connect different components or a whole system within the environment it is inserted. For each port $A_i \in \mathcal{N}ames$, a TDS is defined. Intuitively, each TDS depicts the behavior of how data flow in a port denoted by a port name $A \in \mathcal{N}ames$. Definition 4.3 formalizes the notion of $TDS^{\mathcal{N}ames}$ as the set of all TDS-tuples containing one TDS for each port.

Definition 4.3 ($\mathrm{TDS}^{\mathcal{N}ames}$).
$TDS^{\mathcal{N}ames} = \{((\alpha_1, a_1), (\alpha_2, a_2), \ldots, (\alpha_n, a_n)) \colon (\alpha_i, a_i) \in TDS, i = 1, 2, \ldots, n,$
$\quad \text{with } \mathcal{N}ames = A_1, A_2, \ldots, A_n. \}$

A Data Assignment denotes which data element is in each port that belongs to a non-empty subset of ports $N \subseteq \mathcal{N}ames$. Hence, Data Assignments for ports are defined as functions $\delta \colon N \to Data$, and Definition 4.4 presents the notation's definition.

Definition 4.4 (Data Assignment).
A Data Assignment δ *describes the assignment of some data item* $\delta_A \in Data$ *to a port name* $A \in \mathcal{N}ames$. *Shortly,* $\delta = [A \mapsto \delta_A \colon A \in N]$

By defining $\theta = ((\alpha_1, a_1), (\alpha_2, a_2), \ldots, (\alpha_n, a_n)) \colon (\alpha_i, a_i)) \in TDS^{\mathcal{N}ames}$, $\theta.time$ is defined in Definition 4.5 as the time stream obtained by merging all timed streams $a_1, a_2, \ldots, a_n$ increasingly. At each iteration, $\theta.time$'s value is recalculated as the minimum time value obtained by such merging, considering θ' as the derivative of θ.

Definition 4.5 ($\theta.time(k)$). *The merging of time streams in increasing order denotes* $\theta.time(k)$ *as*

$$\theta.time(0) = \min\{a_i(0) \colon i = 1, 2, \ldots, n\},$$

$$\theta.time(m+1) = \min\{a_i(k) \colon a_i(k) > \theta.time(m), i \in 1, 2, \ldots, n, k \in 1, 2, \ldots\}.$$

With $\theta.time(k)$ as the k-th minimum time in where data starts to flow in a port, the next definition captures the idea of selecting all ports that are in $\theta.time(k)$, $\theta.N = \theta.N(0), \theta.N(1), \theta.N(2), \ldots$, as a stream over $2^{\mathcal{N}ames}$ as follows.

Definition 4.6 ($\theta.N(k)$). *$\theta.N(k)$ denotes all ports that contains data in time instant $\theta.time(k)$:*
$\theta.N(k) = \{A_i \in \mathcal{N}ames\colon a_i(l) = \theta.time(k)$ for some $l \in \{0, 1, 2, \ldots\}, i = 1, 2, \ldots, n\}$.

The first derivative of θ is written θ' as the TDS-tuple that is obtained by calculating the derivatives of all TDS (α_i, a_i) with its associated port $A_i \in \theta.N(0)$. Then, it can be inductively defined as streams' derivatives [ ]. As an example, let $\theta = ((\alpha_1, a_1), (\alpha_2, a_2), (\alpha_3, a_3))$. If $\theta.N(0) = \{A_1\}$, $\theta' = (\alpha_1', a_1'), (\alpha_2, a_2), (\alpha_3, a_3)$.

Following the same idea presented in Definition 4.6, the concept of a stream over the data flow in ports in $\theta.time$ is defined as $\theta.\delta = \theta.\delta(0), \theta.\delta(1), \theta.\delta(2), \ldots$ as a stream over the set of data assignments for each port $A_i \in \theta.N$. Intuitively, $\theta.\delta(k)$ holds all observed data flow at time instant $\theta.time(k)$ and is defined as follows.

Definition 4.7 ($\theta.\delta(k)$). *The stream $\theta.\delta(k)$ over the set of Data Assignments is defined as*
$\theta.\delta(k) = [A_i \mapsto \alpha_i(l_i)\colon A_i \in \theta.N(k)]$
where $l_i \in [0, 1, 2, \ldots]$ is the unique index with $a_i(l_i) = \theta.time(k)$.

A TDS language (for $\mathcal{N}ames$) denotes any subset of $TDS^{\mathcal{N}ames}$ where TDS languages are used as a formalism to describe the possible data flow of a coordination model (namely, data flow of a Reo circuit).

Constraint Automata uses a finite set $\mathcal{N}ames$, where $\mathcal{N}ames$ can be a set as $\{A_1, A_2, \ldots, A_n\}$, and the i-th name stands for a I/O port of a Reo connector or component. As depicted by Definition 4.1, transitions of Constraint Automata are labeled with pairs containing a non-empty subset $N \subseteq \mathcal{N}ames$ and a data constraint g. Data constraints are seen as a symbolic representation of data assignments in the sense of denoting which data item may be observed at a given port, being propositional formulae built from atomic propositions such as $d_A = d$, meaning "at port A the data item observed must be d", with $A \in \mathcal{N}ames$ and $d \in Data$. Hence, Definition 4.8 formally introduces the grammar to describe data constraints.

Definition 4.8 (Data constraints). *A data constraint (DC) g is formally defined by the following grammar:*

$$g ::= true \mid d_A = d \mid g_1 \vee g_2 \mid \neg g.$$

We follow the same notation presented by Baier et al. [ ], where a transition is denoted by $q \xrightarrow{N,g} p$ rather than $(q, N, g, p) \in \rightarrow$, $q, p \in Q$ are states of the automaton,

g a data constraint which must be satisfied to enable the transition, and $N \subseteq \mathcal{N}ames$. Given θ, a transition $q \xrightarrow{N,g} p$ can be fired at the k-th iteration iff $N \neq \emptyset$ and g is satisfiable by $\theta.\delta(k)$ (i.e., $\theta.\delta(k) \models g$, where $\models$ stands for the satisfaction relation for classical propositional logic).

The intuitive meaning of Constraint Automata as formal semantics for Reo models can be understood by interpreting the states as the configuration of the connector and the transitions as how the connector's behavior can change in a single step. Hence, $q_0 \xrightarrow{N,g} q_1$ means that, in order for the automaton to change its configuration from q_0 to q_1, it must have data flow only in ports $A \in N$, where such data flow must satisfy the data constraint denoted by g, while in the other ports $\mathcal{N}ames \setminus N$ there must not have any data flow (i.e., $\mathcal{N}ames \setminus N \not\subseteq \theta.N(k)$).

Hence, the idea of Constraint Automata being TDS acceptors can be interpreted as follows. Given a TDS-tuple $\theta \in TDS^{\mathcal{N}ames}$ as input to a Constraint Automaton $\mathcal{A}$, the automaton tries to figure out whether θ denotes a possible data flow of $\mathcal{A}$ the same way a finite automaton would get as input a finite word and decides whether it describes an accepting run. Nevertheless, since Constraint Automata does not have final states as a criterion for acceptance, all accepting runs are infinite runs since θ is infinite.

Formally, an accepting run in a Constraint Automaton is defined as follows.

Definition 4.9 (Accepting runs on Constraint Automata).
Given a TDS-tuple $\theta \in TDS^{\mathcal{N}ames}$ as input, an accepting infinite run on a constraint automaton $\mathcal{A}$ is denoted by the stream $q = q_0, q_1, q_2, \ldots$ over Q where:

(i) There exists a transition $q_0 \xrightarrow{N,g} q_1, q_0 \in Q_0$;

(ii) $N = \theta.N(0)$;

(iii) $\theta.\delta(0) \models g$;

(iv) q' (an infinite stream initiating from the resulting state obtained from (i)) is an accepting q_1-run on θ' in $\mathcal{A}$;

Therefore, Definition 4.9 respectively states the following: it is necessary to have at least one transition that can be fired from the actual state in the run, the other ports other than the ones involved in a firing transition contains data, the data on those ports must satisfy g, and that these conditions may hold for the remainder of θ, denoted by its derivative θ'. Such conditions establish the notion of accepting runs on Constraint Automata. Alternatively, Definition 4.10 formally introduces the notion of rejecting runs on Constraint Automata.

Definition 4.10 (Rejecting runs on Constraint Automata).
Given a TDS-tuple $\theta \in TDS^{\mathcal{N}ames}$ as input, a finite rejecting run of a constraint automaton $\mathcal{A}$ is denoted by a finite sequence $q = q_0, q_1, q_2, \ldots, q_n$ over Q where either one of the following conditions hold:

(i) from q_0 there is no transition $q_0 \xrightarrow{N,g} q_1$ with $N = \theta.N(0)$ and $\theta.\delta(0) \models g$;

(ii) from q_n there is a transition $q_0 \xrightarrow{N,g} q_1$ with $N = \theta.N(0)$ and $\theta.\delta(0) \models g$, and $q_1, q_2, \ldots, q_n$ denotes a rejecting run in $\mathcal{A}$ with initial state q_1;

(iii) n=0 and there is no transition from q_0 to any state.

Subsuming Definitions 4.9 and 4.10, an accepting run for θ in an automaton $\mathcal{A}$ is an infinite run which satisfies Definition 4.9 starting at some initial state $q_0 \in Q_0$, while a rejecting run is a run for θ in $\mathcal{A}$ where at some point k there is no transition to be fired.

Arbab [ ] provides the canonical set of Reo connectors that may be used to compose more complex channels. Because Constraint Automata are a theory that provides a formal semantics for Reo connectors, a constraint automaton for each canonical connector (depicted in Figure 1) is also provided, each following its respective channel's behavior. Table 1 summarizes basic channels provided by Arbab and their corresponding Constraint Automata. The label depicted above the edges between {} are the ports that can "observe" data for the transition to be fired, while the label below it stands for (possible) data constraints upon observed data. The absence of this second label means that there are no data constraints for this transition.

The idea of compositionally building out more complex Reo connectors out of canonical ones is to join source nodes in Reo with other nodes (sink, source or mixed) by the usage of a product construction between automata. Thus, the natural join of two languages L_1 and L_2, respectively the languages of Constraint Automata $\mathcal{A}_1$ and $\mathcal{A}_2$ is done by composing the product automata of $\mathcal{A}_1$ and $\mathcal{A}_2$ as a product operation. This natural join is analogue to the operation defined for relational databases [ ]. Definition 4.11 summarizes such operation.

Definition 4.11 (Product Automata). *Given two Constraint Automata: $A_1 = (Q_1, \mathcal{N}ames_1, \rightarrow_1, Q_{0,1})$ and $A_2 = (Q_2, \mathcal{N}ames_2, \rightarrow_2, Q_{0,2})$, the Product Automaton $A_1 \bowtie A_2$ is formally defined as $A_1 \bowtie A_2 = (Q_1 \times Q_2, \mathcal{N}ames_1 \cup \mathcal{N}ames_2, \rightarrow, Q_{0,1} \times Q_{0,2})$, where $\rightarrow$ is the resulting transition relation, defined by the following rules.*

$$(i)\quad \frac{q_1 \xrightarrow{N_1, g_1} p_1,\, q_2 \xrightarrow{N_2, g_2} p_2,\, N_1 \cap \mathcal{N}ames_2 = N_2 \cap \mathcal{N}ames_1}{(q_1, q_2) \xrightarrow{N_1 \cup N_2, g_1 \wedge g_2} (p_1, p_2)}$$

$$(ii)\quad \frac{q_1 \xrightarrow{N, g} p_1,\, N \cap \mathcal{N}ames_2 = \emptyset}{(q_1, q_2) \xrightarrow{N, g} (p_1, q_2)}$$

$$(iii)\quad \frac{q_2 \xrightarrow{N, g} p_2,\, N \cap \mathcal{N}ames_1 = \emptyset}{(q_1, q_2) \xrightarrow{N, g} (q_1, p_2)}$$

Intuitively, the rules for constructing the resulting product automaton's transitions as the natural join of languages of both automata is expressed as follows. Let $\mathcal{A}_1$ and $\mathcal{A}_2$ constraint automata. The product of $\mathcal{A}_1$ with $\mathcal{A}_2$ generates a product automaton which joins transitions from both automata where its behavior affect equally both automata (rule (*i*))), or are disjoint transitions (rules (*ii*) and its symmetric, (iii)).

Baier et al. [ ] also define bisimulation for constraint automata. In this context, we are interested in the TDS languages that are induced by Reo circuits (and consequently by their constraint automata). As Baier et al. [ ] state, bisimulation defines an alternative to verify whether two automata are language equivalent, or if the language of an automaton is included in another automaton's language. Therefore, Definition 4.12 introduces bisimulation for constraint automata.

Definition 4.12 (Bisimulation). *Let $\mathcal{A} = (Q, \mathcal{N}ames, \rightarrow, Q_0)$ be a constraint automaton. An equivalence relation $\mathcal{R}$ on Q is a bisimulation if, for all pairs of states $(q_1, q_2) \in \mathcal{R}$, for all $N \subseteq \mathcal{N}ames$, and for every $\mathcal{R}$-equivalence classes $P \in Q/R$:*
$dc(q_1, N, P) \equiv dc(q_2, N, P)$.

where $dc(q, N, P)$ is a notation to denote the disjunction of all data constraints of a $\mathcal{N}$-transition from q to some state $p \in P$, defined as follows:
$dc(q, N, P) = \bigvee\{g \colon q \xrightarrow{N, g} p, p \in P\}$.

Following Definition 4.12, the notion of bisimulation is defined for constraint automata as follows. Let two constraint automata with the same set of port names $\mathcal{A}_1 = (Q_1, N, \rightarrow_1, Q_{0,1})$ and $\mathcal{A}_2 = (Q_2, N, \rightarrow_2, Q_{0,2})$. They are bisimulation-equivalent (notation : $\mathcal{A}_1 \sim \mathcal{A}_2$) iff, for all initial states $q_{0,1} \in Q_{0,1}$, there exists an initial state $q_{0,2} \in Q_{0,2}$ where $q_{0,1}$ and $q_{0,2}$ are bisimilar, and vice versa.

Channel	**Reo**	**Constraint automaton**
Sync		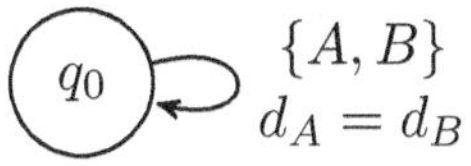
LossySync		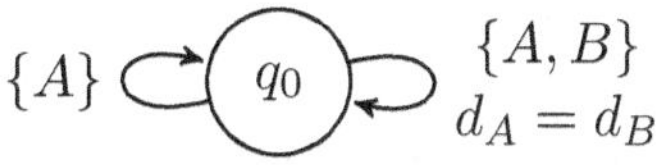
FIFO	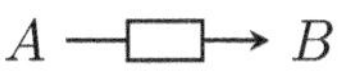	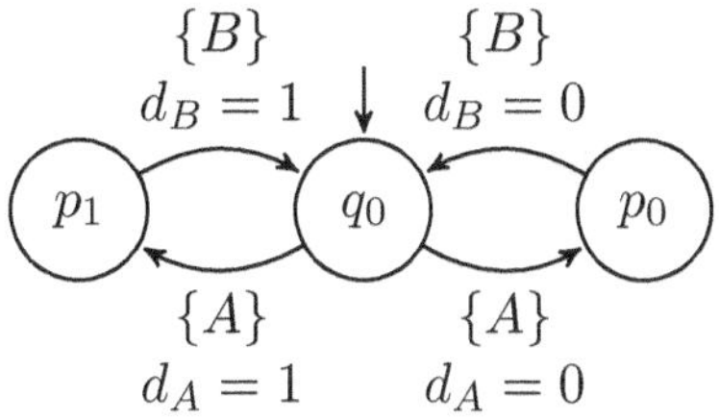
SyncDrain		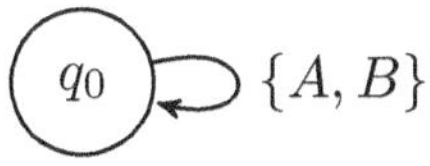
AsyncDrain	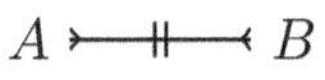	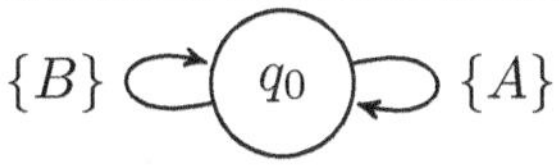
Filter		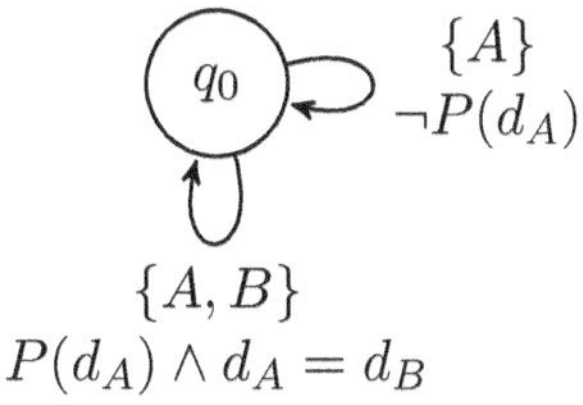
Transform		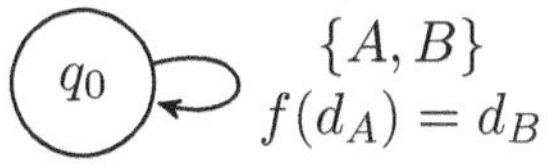
Merger	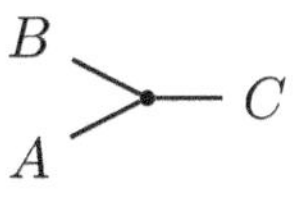	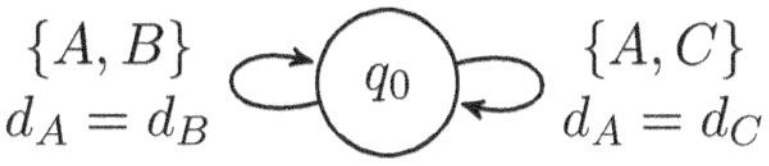
Replicator	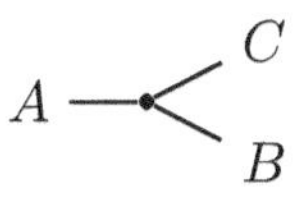	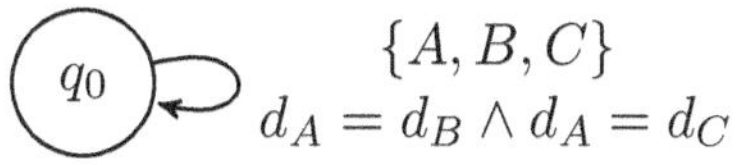

Table 1: Basic Reo channels and their respective constraint automata

5 Formalizing and reasoning about Reo circuits

This Section presents ReoXplore: a modelling and reasoning framework for Reo circuits. We present the formalization of concepts presented in Section 4 in Coq, an interactive proof-assistant, and nuXmv, a symbolic model checker for the analysis of synchronous finite-state and infinite-state systems. The development forks Reo Graphical Editor[10] in order to provide a Graphical User Interface (GUI) to graphically model Reo circuits, compile them to the aforementioned logic-based models, and compile it to Haskell code.

We present two usage examples as the formalizations of Figure 2 and Figure 3. We show examples of properties that may be verified and show an example of the generation of certified code of a model by means of Coq, along with experimental results concerning execution time and memory.

5.1 A constructive formalization of Constraint Automata in Coq

The formalization of this theory in an environment such as Coq leads to the possibility of certifying instances of models which formal semantics relies on Constraint Automata (namely Reo connectors), considering its desired properties and consequently retrieve certified code using Coq's extraction mechanism. Besides formalizing the aspects presented in Section 4.2, we also have deployed a C program which takes as input a textual representation of a Reo connector and returns the corresponding Coq model, with all of its necessary definitions.

Constraint Automata in Coq are introduced by Definition 5.1 as a `Record` which maps the formalism presented by Definition 4.1. Q, N, T and $Q0$ are, respectively, the set of states, set of port names, transition relation and the set of initial states of the automaton. Variables *state* and *name* stand for the type of states and the type of port names of the automaton.

Definition 5.1. *Constraint Automata in Coq*

```
Record constraintAutomata : Type := CA {
        Q : set state;
        N : set name;
        T : state → set (set (name) × DC × state);
        Q0 : set state;
    }.
```

The definition of an automaton's transition relation slightly deviates from the one provided by Baier et al. [ ] (depicted by Definition 4.1): instead of formalizing

[10] https://github.com/ReoLanguage/reo-graphical-editor

$T \subseteq (Q \times 2^{\mathcal{N}ames} \times DC \times Q)$, it is defined as a function *state* $\rightarrow$ `set` (`set` (*name*) $\times$ *DC* $\times$ *state*) only to ease automata specifications, and to enable modelling of non-determinism straightforward.

5.1.1 Timed Data Streams

The formalization of Reo connectors in Coq enables reasoning on these connectors using their corresponding constraint automata. Properties regarding the connector's structure or how data may be exchanged between entities in different scenarios may be carried on employing Coq's proof mechanism. Though, it may also be useful to reason about specific data flow on a Reo connector, which may denote possible real scenarios this connector may face.

We define inputs for constraint automata the same way Definition 4.2 and Definition 4.3 formalize. While Definition 5.2 formalizes a single TDS in Coq, input θ are seen as a set containing tuples of TDS formalized as *tds*. It is a Coq record which respectively contains the automaton's port name to which it refers to, data and time streams, a proof that its time stream is always crescent, and a field used to calculate its derivative. As far as the authors know, there is no constructive representation of real numbers in Coq (by default). Therefore, we denote time streams as a function $\mathbb{N} \rightarrow \mathbb{Q}$ rather than depicted by Definition 4.2.

Definition 5.2. *Timed Data Streams in Coq*

```
Record tds := mktds {
      id : name;
      dataAssignment : nat → data;
      timeStamp : nat → QArith_base.Q ;
      tdsCond : ∀ n:nat, Qle (timeStamp n) (timeStamp (S n));
      index : nat
   }.
```

5.1.2 Data Constraints

Data constraints (DCs) are formalized in Coq mapping Definition 4.8 to an inductive type which encapsulates logical constraints' definitions. They will be later evaluated to Coq's boolean datatypes in runtime. By defining Data Constraints as *DC*, a way to evaluate DCs in Coq is needed in order to verify whether a given port A has a data item d (the propositional formula $d_A = d$) and the composite data constraints. This is provided by the following definitions.

The definition of *evalCompositeDc* provides boolean semantics for the inductive type *DC*. Its objective is to evaluate data constraint of transitions either to "true" or "false" by decomposing data constraints which are conjunctions or disjunctions, to recover data constraints $d_A = d, f(d_a) = d$ and $P(d_A) = d$.

Definition *evalDC* provides the evaluation of whether a given port *po* has the data item *d*. The reason for *po* to be of type *option port* is that Coq needs a way to ensure that every function always terminate, where for *retrievePortFromInput*, Coq also considers the case where no port matching an existing port in the automaton is found. From Baier et al. [ ], input $\theta \in \text{TDS}^{\mathcal{N}ames}$ for constraint automata must not be empty; therefore, *retrievePortFromInput*'s usage always have an non-empty set provided as input.

The evaluation of *eqDc* by means of *evalCompositeDc* relies on the idea of traversing the input TDS θ in order to check whether the ports in *eqDc* contains the same data item at the k-th step in a run on constraint automata. Therefore *eqDataPorts* is a function that, given two port names *n1*, *n2* and a set of TDS tuples *tds*, it returns *true* if both ports *n1* and *n2* have the same data item at the same moment. This definition evaluates data constraints $d_{n_1} = d_{n_2}$.

DCs formalized by means of *trDc* are evaluated by *transformDC*, a function that given a transformation function *transform*, two port names *n1* and *n2* and a set of ports *s*, it relies on *retrievePortFromInput* to retrieve TDSs defined for port names *n1* and *n2*. Then *transformDC* returns *true* if the data item flowing in *n2* is equal to the data item flowing in *n1* with *transform* applied to it, and false otherwise. In short, it evaluates data constrains $f(d_{n_1}) = d_{n_2}$.

5.1.3 θ.*time*

The next step in the formalization process is to formalize θ.*time* as depicted by Definition 4.5. We formalize means to retrieve the possible values from each port's time stream in order to return θ.*time*. We start with the function *getThetaTimeCandidate* which returns the current time stamp of a TDS *p* calculated on its index.

Then *getAllThetaTimes* is defined as a function that, with a set of ports *s*, it returns all θ.*time* candidates for the current step, retrieving all time stamps of each port $p \in s$ by means of *getThetaTimeCandidate*.

The last function used to formalize θ.*time* is *minimum*, a function that returns the smallest timestamp in the set of θ.*time* candidates *l* by means of *returnSmallerNumber*. It compares the first element of the set of θ.*time* candidates with the rest of its elements employing *hd*, a function which takes as parameters a default value and a set, returning the first element of the list if it is not empty, and the default value otherwise.

Therefore, θ.*time* denoted by Definition 4.5 is *nextTime* as the minimum value of all time stamps of TDS in θ.

```
Definition nextTime (theta:set tds) := minimum(getAllThetaTimes s).
```

5.1.4 $\theta.N$

The formalization of $\theta.N$ as represented by Definition 4.6 begins with the definition of *timeStampEqThetaTime*, a function that with a set of TDS *theta* and a single TDS *a* returns *true* if *a*'s current time equals nextTime(*theta*).

$\theta.N$ is then formalized by means of *thetaN* as follows. This function takes as input *theta* and *theta2* as the automaton's input θ. The first *theta* is used by *TimeStampEqThetaTime* in order to traverse the whole input with each TDS $A_i \in \theta$, retrieving all port names $N \in \theta.time$ with their corresponding TDS's time stamp equals $\theta.time$, following Definition 4.6.

```
Fixpoint nextNames (theta: set tds) (theta2:set tds) : set name :=
  match theta2 with
  | a::t ⇒ if (timeStampEqThetaTime theta a) then id a :: thetaN theta t
           else thetaN theta t
  | [] ⇒ []
  end.
```

The function *thetaN* is a function that takes a set of TDS tuples *theta* as the input θ and calculates $\theta.N(k)$ by means of *nextNames*. This function is necessary to assure that the ports which can fire at the k-th iteration are indeed the ones with their timestamp equal to *theta.time*(k).

```
Definition thetaN (theta: set tds) := nextNames (theta) (theta).
```

We define $\theta.\delta$ similarly to the definitions presented above. *nextData* is a function that implements the same idea as *nextNames*, but returning pairs (portName, portData) instead of only the port name.

5.1.5 Manipulating Timed Data Streams

With the basic definitions required to relate constraint automata with TDS, some functionalities for Constraint Automata, such as runs were formalized. Therefore, the notion of derivative of a TDS now formalized follows the one proposed by Baier et al. [ ]: we define the calculus of derivative of a single TDS, only calculating the entire derivative θ' of θ in runtime (considering the port names $p \in \theta.N$). Intuitively, a derivative of a TDS p is p evaluated (i.e., p's streams) in its next index. The function *derivative* returns a TDS with its updated index (i.e., the same way Definition 4.2 introduces derivative for TDS), incrementing its value by one. Function *mktds* is a constructor of the record *tds*.

```
Definition derivative (p: tds) := mktds (id p) (dataAssignment p)
```

(*timeStamp p*)(*tdsCond p*) (*S* (*index p*)).

This definition will later be used in order to calculate the derivative of TDS used in each transition during a run. This is done with *derivativePortInvolved*, a function that takes a set of names *s* and a TDS *a*, verifying whether *a*'s port identifier *id* matches one of the names in *s*. If it does, then it calculates *a*'s derivative.

Then, *allDerivativesFromPortsInvolved* is a function that extends the behavior denoted by *derivativePortInvolved* to a set of ports *port*. Hence, *allDerivativesFromPortsInvolved* with a set of names *names* and a set of ports *ports* applies *derivativePortInvolved* for each port $a \in port$.

```
Definition allDerivativesFromPortsInvolved (names: set name) (theta:set
    tds) : set tds := flat_map (derivativePortInvolved names) theta .
```

5.1.6 Runs

By formalizing all concepts so far, definitions regarding runs can now be formalized. We incrementally define runs through auxiliary functions. The first auxiliary definition is *step'*, a function that with *theta* as input θ, a set of port names *portNames*, and *transitions* as the transitions of an automaton, *step'* verifies whether the transitions $a \in$ *transitions* are eligible to be fired. For each fired transition, the reached state of the transition is returned.

The idea of *step'* is to store all possible paths from a single state at a k-th iteration of a run. It defines how a single step in a run works: exploring all transitions departing from a state, *step'* will verify whether the following requirements are fulfilled:

- the transition has its name set equal to the name set given as parameter (which will later be supplied with the name set returned by *thetaN*, denoting the name of ports with data in $\theta.N(k)$);
- the data constraint depicted in the transition is currently satisfied by the available data at the ports.

The topics mentioned above are necessary to a transition to fire according to Definition 4.9. The requirement that no other ports outside a transition's set of port names N must have data at $\theta.time(k)$ is guaranteed by the first item. Hence, the function *step'* is defined as below.

```
Fixpoint step' (theta : set tds) (portNames: set name)
(transitions: set(set name × DC × state)) : set state :=
  match transitions with
  | [] ⇒ []
```

| $a::t \Rightarrow$ if (set_eq ($portNames$)((fst(fst(a))))) && ($evalCompositeDc$ ($theta$) (snd(fst(a))))
then $snd(a)$::$step'$ $theta$ $portNames$ t
else $step'$ $theta$ $portNames$ t
end.

A single step comprising the automaton's current configuration is defined by *step*, which takes a Constraint Automaton *ca*, a set of states *s* denoting the automaton's current configuration, and the automaton's input θ, and applies *stepa* with its name set parameter as the name set of ports $portNames \subseteq \theta.N(k)$, which are the ports that currently have their k-th time stamp equal to $\theta.time$. *stepa* is a function that applies *step'* to a set of states, considering *ca*'s transition relation, returning a pair (*portNames*,*states*), where *states* are the states reached by *step'*. Intuitively, *step* extends *step'*'s behavior considering a set of states and its corresponding transitions by means of *stepa*, a definition which applies *step'* to a set of states and with port names as *thetaN*.

Definition *step* (*ca*:*constraintAutomata*) (*s*: **set** *state*) (*theta*: **set** *tds*) : **set** *name* × *data* := *stepa ca s theta* (*thetaN theta*).

The idea behind *run'*'s implementation is to implement (finite) runs on constraint automata, where a run may last k iterations. In each iteration, it applies *step* with the same parameters as provided to *run* (which in turn calls *run'* with its parameters), storing the resulting states obtained with *step* in *trace*, calculating the derivatives of all ports involved in the transition fired by means of *derivativePortsInvolved* before recursively calling *run*. Parameter *theta* denotes the input θ an automaton takes, while k denotes the number of iterations a bounded run will last, *currentStates* are the current states considered to discover and fire transitions at the current step, and *trace* denotes the execution trace (all states a run has reached).

Definition *run'* (*ca*:*constraintAutomata*) :
set *tds* → *nat* → **set** *state* → **set** (**set** *state*) → **set** (**set** *state*) :=
fix rec *theta k currentStates trace* :=
match k with
| 0 ⇒ *trace*
| S m ⇒ *trace* ++ [*snd* (*step ca currentStates theta*)]
| > rec(*flat_map*(*derivativePortInvolved*(*fst*
((*step ca currentStates theta*)))) *theta*) m (*snd* (*step ca currentStates theta*))
end.

Therefore, bounded runs on constraint automata are achieved by *run*, a function that, with a constraint automaton *ca*, *theta* as the TDS θ and a natural number k denoting the number of steps of a (finite) run, calls *run'* with its respective parameters: both the set of states denoted by *currentStates* and *trace* with *ca*'s initial states, which respectively denote the ones where a run may start and the automaton's starting states as the first reached states of a run.

```
Definition run (ca:constraintAutomata) (theta: set tds) (k : nat) :=
    run' ca theta k (Q0 ca) [Q0 ca].
```

It is important to point out that the notion of runs in constraint automata formalized above is slightly different from the one discussed in Definition 4.9. Baier et al. [ ] define runs in constraint automata as infinite. Roughly speaking, a (bounded) run is an accepting run if the input TDS $\theta \in TDS^{\mathcal{N}ames}$ always fire at least one transition in the automaton during the run, and it is a rejecting run otherwise (if it reaches a state where no transitions can be fired). By algorithmic aspects, *run* formalizes the notion of a finite run bounded to k steps. Infinite runs can be specified upon a TDS and how it always satisfies conditions presented by Definition 4.9 as Coq propositions.

This framework allows the reasoning over infinite runs by following the idea depicted by Definition 4.9: given a TDS-tuple $\theta \in TDS^{\mathcal{N}ames}$ as input, it denotes an accepting infinite run in a CA if the CA can always fire a transition, for some sequence of states reached during its execution, satisfying the required conditions in Definition 4.9. Intuitively, for all indexes k, for some state of the automaton, $\theta^{(k)}$ as the k-th derivative of the input θ must trigger at least one transition in the current state of the automaton.

Reasoning over infinite runs may require a bit of extra effort for the user, as it will require the user to provide the property to reason about a specific connector and its input θ. An example of how this can be achieved is denoted in Section 5.1.12. The idea is to define a property which verifies that, for all index k and considering all states of the automaton, a transition can be fired (i.e., at any step of an execution, the k-th derivative $\theta^(k)$ fires a transition. To help this definition, we provide an additional Coq definition which generalizes the index k of a TDS as follows, basically copying the TDS with a new index k.

```
Definition calcIndex (k: nat) (p : tds) := mktds (id p) (dataAssignment p)
(timeStamp p)(tdsCond p) (k).
```

A more generic definition of this property might rely on an external definition which generates its corresponding Coq code, considering the TDS-tuples in θ and all possible combinations of different indexes k for each TDS-tuple in θ, thus verifying all possibilities of indexes of ports at the k-th derivative of θ in a fashion similar to the one below.

```
Definition accepting (ca: constraintAutomata) (theta1: tds) (theta2): tds)
...(thetan: tds) :=
    ∀ q,∀ k1,k2, ...kn, stepAux ca [(calcIndex k1 theta1); (calcIndex k2 theta2);...;
```

(*calcIndex* k_n *theta*$_n$)] (*thetaN* [(*calcIndex* k_1 *theta*$_1$); (*calcIndex* k_2 *theta*$_2$);...; (*calcIndex* k_n *theta*$_n$)) $q \neq$ [].

The idea behind *accepting* is to provide a means to reason over infinite runs considering each possible k_i index of each TDS-tuple, hence considering all possible scenarios by induction on these indexes. The reasoning may proceed over each index, showing that it holds for each possible combination of each TDS-tuple's derivative. Such reasoning may take advantage of the behavior of the connector, limiting its analysis to only relevant cases of the derivatives and consequently easing the process. By employing *stepAux*, we employ *step'* with the automaton's transitions for a state of the automaton.

Definition *stepAux* (*ca*:*constraintAutomata*) (*theta*:`set` *tds*) (*portNames*:`set` *name*) (*s*: *state*) : `set` *state* := *step' theta portNames* (*T ca s*).

Conversely, following Definition 4.10 an infinite run is a rejecting one if, at some point of the execution (i.e., a state q_n and at θ's k-th derivative), there are no transitions that can be fired, yielding a finite execution trace. A possible means to denote this property is to state that there is a number k denoting the actual execution step in which the absence of valid transitions is detected. This idea is captured by the following definitions, in which *lastReachedStates* is a function that returns the last set of states reached by *run*. If its return is an empty set of states, it is as a consequence that no transitions were fired in this execution step.

Definition *rejecting* (*ca*: *constraintAutomata*) (*theta*: `set` *tds*) : Prop :=
$\exists$ *k*, (*lastReachedStates ca theta k*) = [].

5.1.7 Bisimulation

We also formalize the definition to build the equivalence relation $\mathcal{R}$ which denotes the bisimulation of two constraint automata $\mathcal{A}_1$ and $\mathcal{A}_2$ as introduced by Definition 4.12. Our formalization provides the bisimilar states relation if it exists. We compute the relation on both *a1* and *a2*'s set of states as the candidate relation $\mathcal{R}$ which is a bisimulation. The procedure follows the definitions of the required operations following Definition 4.12. We first compute $\mathcal{R}$ as the candidate relation. Then, we compute Q/R in order to compute if the required conditions to $\mathcal{R}$ to be an bisimulation on Q_1 and Q_2 as the set of states of *a1* and *a2* are met.

We define a function that effectively returns the set containing pairs of equivalent states through *bisimulationAux* as a definition that takes two constraint automata *ca1* and *ca2*, and returns the corresponding relation $\mathcal{R}$. We recall that the current implementation of bisimulation requires that two automata to be compared to have the same type for their states. The first usage of *iterateOverA1States* is bound to compute the subset of the bisimulation relation from *a1* to *a2*, while the latter does the same from *a2* to *a1*.

`Definition` *bisimulationAux* (*ca1*: *constraintAutomata*)
(*ca2*: *constraintAutomata*) :=
(*iterateOverA1States ca1 ca2* (*ConstraintAutomata.Q0 ca1*)
(*ConstraintAutomata.Q0 ca2*) []) ++ (*iterateOverA1States ca2 ca1* (*ConstraintAutomata.Q0 ca2*) (*ConstraintAutomata.Q0 ca1*) []).

By means of *bisimulationAux*, we define *bisimulation* as the definition that returns the bisimulation if $\mathcal{R}$ is an equivalence relation on Q. Definition *isBisim* is a function that computes whether $\mathcal{R}$ is a bisimulation, returning an empty relation if it fails to be. This empty relation means there is no relation $\mathcal{R}$ that is a bisimilar relation on (Q_1, Q_2) as respectively the states of *a1* and *a2*.

`Definition` *bisimulation* (*ca1*: *constraintAutomata*) (*ca2*: *constraintAutomata*) := *isBisim* (*bisimulationAux ca1 ca2*).

Based on $\mathcal{R}$, it is possible to compute the set $Q \setminus \mathcal{R}$ as the partition that contains states $q \in Q$ that are related through $\mathcal{R}$. This is needed to calculate the equivalence of data constraints of transitions that leave a state q with a set of port names $\mathcal{N}$ arriving on states $p \in P, P \subseteq Q \setminus \mathcal{R}$, following the notation introduced by Definition 4.12.

We formalize the same notation ($dc(q, N, P)$) presented by Definition 4.12 as follows. Function *dcsOfState* takes a constraint automaton *ca*, a state *q*, a set of port names *portNames*, and a set of states P returns the disjunction of all transitions departing from q to some state $p \in P$.

We also define functions that provide a means to iterate over $\mathcal{R}$ considering all possible set of port names $\mathcal{N} \subseteq 2^{\mathcal{N}ames}$(*compareDcBisimPortName*), all $\mathcal{R}$-equivalence classes P (*compareDcBisimPartition*), and to iterate over all pairs of states $(q_1, q_2) \in \mathcal{R}$ where the last one is *compareBisimStates*, the top-level definition which employs *compareDcBisimPartition* and *compareDcBisimPortName* to calculate $dc(q_1, N, P) \equiv dc(q_2, N, P)$ by means of *dcsOfState* as Definition 4.12 requires.

The last auxiliary function is *checkBisimilarity* which takes two constraint automata *ca1* and *ca2*, a set of pairs of states *bisimRelation* (to which will be supplied the relation $\mathcal{R}$ obtained by *bisimulation*), returning true if both automata have the same set of port names $\mathcal{N}$ and are bisimilar, and false otherwise. The names-space check is done in order to avoid unnecessary computation of whether *a1* and *a2* are bisimilar (as they cannot be bisimilar if their set of port names are different).

`Definition` *checkBisimilarity* (*ca1*: *constraintAutomata*)
(*ca2*: *constraintAutomata*) (*bisimRelation* : **set** (*state* × *state*)) :=
`if` (*set_eq* (*ConstraintAutomata.N ca1*) (*ConstraintAutomata.N ca2*))
`then` *compareDcBisimStates ca1 ca2* (*powerset* (*ConstraintAutomata.N ca1*)) (*getQrelR* (*bisimRelation*)) (*bisimRelation*)

`else` *false*.

Using *checkBisimilarity*, we deploy a definition that may be used to check if *a1* and *a2* are bisimilar. This definition may be useful to validate whether two Reo connectors are bisimilar (i.e., if their corresponding constraint automaton recognize the same language), which may bring advantages in the sense that it is possible to exchange more complex Reo circuits by simpler ones [ ]. Function *areBisimilar* takes two constraint automaton *ca1* and *ca2*, returning true if they are bisimilar (and consequently, language-equivalent) and false otherwise.

`Definition` *areBisimilar* (*ca1*: *constraintAutomata*) (*ca2*: *constraintAutomata*) := *checkBisimilarity ca1 ca2* (*bisimulation ca1 ca2*).

5.1.8 Basic Connectors Formalization in CACoq

We also provide the formalization of the default behavior of canonical constraint automata as depicted by Baier et al. [ ]. Then we provide means to parametrize automata definitions. Function *ReoCABinaryChannel* introduces the parametric definition of binary canonical Reo connectors, as a function that takes two-port names, a set of states to be the set of states of the automaton, a set of initial states which maps to the automaton's initial state, and its transition relation, while *ReoCATernaryChannel* similarly builds ternary Reo connectors, but expecting three port names instead of two.

An example of the formalization of constraint automata for the canonical Reo connectors through CACoq can be obtained as Table 2 shows. Regarding the basic channels, we provide *transitionRelation* as their constraint automaton's transition relation as Table 1 shows.

The transitions of an automaton rely on the data type of the data to be observed by the ports, the automaton's state, and port names types. A brief explanation on the definition of the canonical connectors' constraint automata as presented in Tables 2 is given as follows, respectively regarding their states, port names, transition relation, and formalization. For the sake of simplicity, we explain only the formalization of the LossySync connector. The same idea is replicated to each Reo connector. Their formalization can be found at the project's repository in file "ReoCa.v".

LossySync The LossySync channel enables the modelling of data synchronization between two entities (denoted by two Reo ports having the same data item simultaneously), and the modelling of scenarios where data was lost on its way from the source to the sink node (in this scenario, only the source node has data). In short, it models synchronization as well as scenarios where data may not reach the destination Reo node. This is the idea implemented by *lossySyncCaBehavior* which is

Reo Channel	Constraint automaton in CACoq
Sync,LossySync,FIFO, SyncDrain,AsyncDrain, Filter,Transform	`Variable` *name state data*: `Set.` `Definition` *ReoCABinaryChannel* (*a b*: *name*) (*states*: **set** *state*) (*initialStates* : **set** *state*) (*transitionRelation* : *state* → **set** (**set** *name* × *ConstraintAutomata.DC name data* × *state*)):= {| *ConstraintAutomata.Q* := *states*; *ConstraintAutomata.N* := [*a*;*b*]; *ConstraintAutomata.T* := *transitionRelation*; *ConstraintAutomata.Q0* := *initialStates* |}.
Merger,Replicator	`Definition` *ReoCATernaryChannel* (*a b c*: *name*) (*states*: **set** *state*) (*initialStates* : **set** *state*) (*transitionRelation*: *state* → **set** (**set** *name* × *ConstraintAutomata.DC name data* × *state*)) := {| *ConstraintAutomata.Q* := *states*; *ConstraintAutomata.N* := [*a*;*b*;*c*]; *ConstraintAutomata.T* := *transitionRelation*; *ConstraintAutomata.Q0* := *initialStates* |}.

Table 2: CA formalized in Coq for the canonical Reo connectors

```
Inductive lossySyncStates : Type := q0.
Inductive lossySyncPorts : Type := A | B.
Definition lossySyncCaBehavior (s: lossySyncStates) : set
(set lossySyncPorts × ConstraintAutomata.DC lossySyncPorts nat ×
    lossySyncStates) :=
    match s with
    | q0 ⇒ [([A;B] , ConstraintAutomata.eqDc nat A B, q0);
              ([A], (ConstraintAutomata.tDc lossySyncPorts nat), q0)]
    end.
Definition paramLossySync := ReoCa.ReoCABinaryChannel A B [q0] [q0]
lossySyncCaBehavior.
```

5.1.9 Product Automata

We also formalize the product operation proposed by Baier et al. [ ] and presented in Definition 4.11. This implementation is done within a section/module named *ProductAutomata*, following a similar approach used in the formalization of Constraint Automata.

Product automata are also constraint automata. Therefore, the same *record* defined for constraint automata will be the product automaton's definition. From Definition 4.11, a product automaton is a tuple $\mathcal{A}_1 \bowtie \mathcal{A}_2 = (Q_1 \times Q_2, \mathcal{N}ames_1 \cup \mathcal{N}ames_2, \rightarrow, Q_{0,1} \times Q_{0,2})$. In what follows, let $\mathcal{A}_1$ be *a1* and $\mathcal{A}_2$ as *a2*, and *a1* and *a2* as constraint automata introduced by Definition 5.1. Variables *state name*, *data*, and *state2* have type `Set`. We parametrize both *state* and *state2* respectively as the type of states of $\mathcal{A}_1$ and $\mathcal{A}_2$, enabling the formalization of different automata with different state types that can be used in the composition process.

The set of states $Q_1 \times Q_2$ and $Q_{0,1} \times Q_{0,2}$ are formalized respectively by means of *statesSet* and *initialStates*. Both functions take two constraint automata $\mathcal{A}_1$ and $\mathcal{A}_2$, where *statesSet* returns the product set of $\mathcal{A}_1$'s set of states with $\mathcal{A}_2$'s set of states, and *initialStates* returns the product set of $\mathcal{A}_1$'s set of initial states with $\mathcal{A}_2$'s set of initial states by means of *list_prod*, a function from Coq's standard library that computes the product of two lists.

The implementation of $\mathcal{N}ames_1 \times \mathcal{N}ames_2$ is obtained with *nameSet*, a function that with two constraint automata *a1* and *a2* returns the union of the set of names of *a1* with the set of names of $\mathcal{A}_2$ by means of *set_union*, a function that denotes the union of two sets defined in library *ListSet*.

The resulting transition relation is more complex to obtain than other components of a product automaton. Its implementation is split into several functions that comprehend the process of deriving the transition rules as depicted in Definition 4.11, ending with the function that enables the usage of the produced set of rules as a function with the same support as Definition 4.1's transition relation. We develop a few auxiliary functions to traverse the set of states of the automaton, applying the product rules introduced in Definition 4.11. Section 5.1.10 describes the implementation of the first product rule, and Section 5.1.11 discuss the implementation of the second and third rules.

5.1.10 First rule

The condition test required in order to calculate whether two transitions (one from *a1* and the other from *a2*) satisfy the necessary conditions is formalized as *condR1*, which takes as arguments two transitions *t1* and *t2* (respectively denoting a single transition of *a1* and *a2*), and two set of port names *names1* and *names2* (respectively depicting the set of port names of *a1* and *a2*), and returns true if the intersection of the port names of *a1* with the port names of *t2* equals the intersection of the port names of *t1* equals the port names of *a2*.

The creation of a single transition regarding the first product rule employs *condR1* in order to verify whether the required conditions are met: given two states *Q1* and *Q2*, *transition1* and *transition2* denoting a single transition of A_1 and

A_2 respectively *names1* and *names2* denoting respectively the name set of A_1 and A_2, *singleTransitionR1* returns a single transition as denoted by the first rule in Definition 4.11.

```
Definition singleTransitionR1 (Q1: state) (Q2: state2)
(transition1: (set (name) × DC × (state))) (transition2: (set (name) × DC
  × (state2))) (names1 : set name) (names2: set name) :
(set (state × state2 × ((set name × DC) × (state × state2)))) :=
        if (condR1 (transition1) (transition2) (names1) (names2)) then
           [((Q1,Q2),(((set_union equiv_dec (fst(fst(transition1)))
           (fst(fst(transition2)))),ConstraintAutomata.andDc
           (snd(fst(transition1))) (snd(fst(transition2)))),(snd(transition1),
           (snd(transition2)))))]
        else [].
```

Now we need to extend the aforementioned idea to traverse the set of states of both *a1* and *a2*. The next definition is *moreTransitionsR1*, which follows the idea of applying the first product rule on a transition set employing *singleTransitionR1* with its corresponding parameters, where *transition2* is now a set of transitions instead of a single transition. Intuitively, *moreTransitionsR1* pins a single transition of *a1*, while iterating over another set of transitions, namely all transitions departing from a single state of *a2*.

We then extend *moreTransitionsR1* to traverse all transitions of *a1* departing from *q1* as *transitionsForOneStateR1*, *iterateOverA2R1* as a function that employs *transitionsForOneStateR1* considering a single state *q1* of *a1* and all states of *a2*, and *allTransitionsR1* as the auxiliary function that applies *transitionsForOneStateR1* to *a1*'s set of states.

These definitions culminate in *transitionsRule1* as the function used to build the transition relation (regarding the first rule) with data incoming from *a1* and *a2*. Therefore *transitionsRule1* takes two constraint automata *a1* and *a2* and returns a set containing all created product transitions regarding the first rule of the product automaton's transition relation.

5.1.11 Second and third rules

Because the second and third rules are symmetrical, we provide details regarding only the second rule's implementation. All definitions now presented have also been similarly formalized for the third rule. We opted to formalize separate functions for each rule to support automata with states of different types.

The following steps formalize a procedure that builds transitions according to the second rule. We formalize the condition test of this rule directly from Definition 4.11

as *condR2*, which considers a single transition from *a1* (denoted by *tr*) and a set of port names *names2* denoting *a2*'s set of port names.

The next step towards this formalization is achieved by a definition that comprises the notion of constructing a resulting transition's origin state as denoted by Definition 4.11. The resulting state (as depicted in the second product rule) for a transition of *a1* that leaves a state q_1 to p_1 will be (q_1, q_2) and (p_1, q_2), $\forall q_2 \in Q_2$ where Q_2 is *a2*'s states set. In order to build the resulting transition rule as depicted by Definition 4.11, we need to bind the states $q_2 \in Q_2$ in the resulting states of the newly built transition.

The function *singleTransitionR2* is conceived as the definition that has the aforementioned idea driving it. Hence *singleTransitionR2*, with a state *q1* as a single state of *a1*, a set of transitions *transitions*, a set of states of states Q_2 of *a2* denoted by *a2States* and a set of names *a2Names* that stands for *a2*'s names set, it returns a set containing resulting a single transition as the second rule in Definition 4.11 by applying this rule with q_1.

```
Fixpoint singleTransitionR2 (q1:state) (transition : (set (name) × DC
× (state))) (a2States : set state2) (a2Names: set name)
: set (state × state2 × ((set name × DC) × (state × state2))) :=
 match a2States with
 | [] ⇒ []
 | q2::t ⇒ if (condR2 (transition) (a2Names)) then
          ((q1,q2),((fst(transition)), ( (snd(transition)), (q2)))):: 
          singleTransitionR2 q1 transition t a2Names
          else singleTransitionR2 q1 transition t a2Names
 end.
```

Similar to the formalization discussed in Section 5.1.10, we employ *transitionR2* to traverse all transitions of *a1* considering a single state *q1*, which will be used by *allTransitionsR2* in order to iterate over all states of *a2*. Finally, *transitionsRule2* is the top-level function that creates all transitions regarding the second rule.

The set of transitions of the resulting product automata $\mathcal{A}_1 \bowtie \mathcal{A}_\in$ is then produced by *buildTransitionRuleProductAutomaton*, which with *a1* and *a2* builds the resulting transition relation by means of *transitionsRule1*, *transitionsRule2*, and *transitionsRule3*.

```
Definition buildTransitionRuleProductAutomaton
   (a1: constraintAutomata) (a2: constraintAutomata) :=
   (transitionRule1 a1 a2)++(transitionsRule2 a1 a2)++(transitionsRule3 a1
a2).
```

Recall that states in the resulting product automaton have type (*state* × *state2*). Then, the resulting transition relation for product automata is $T_{\mathcal{A}_1 \bowtie \mathcal{A}_2}$: (*state* ×

state2) $\rightarrow$ **set**(**set**(*name*) $\times$ *DC* $\times$ (*state* $\times$ *state2*). In this context, *recoverResultingStatesPA* is a function that takes a state *st* ($st \in Q_{\mathcal{A}_1 \bowtie \mathcal{A}_2}$) and a set of transitions *t*, returning the set of transitions that origins in *st*.

```
Fixpoint recoverResultingStatesPA (st: (state × state2))
    (t:set (state × state2 × ((set name × DC) × set (state × state2)))):=
    match t with
        | [] ⇒ []
        | a::tx ⇒ if st == fst((a)) then (snd((a))::recoverResultingStatesPA st
tx)
                    else recoverResultingStatesPA st tx
    end.
```

The actual transition relation of the resulting product automaton is denoted by *transitionPA*, a function that, with a state of type (*state* $\times$ *state2*) as its input *s*, returns all transitions that origins in *s*, where the set of transitions is computed by *buildTransitionRuleProductAutomaton*.

```
Definition transitionPA (a1: constraintAutomata) (a2: constraintAutomata2)
(s: (state × state2)) := Eval compute in recoverResultingStatesPA s (buildTransitionRuleProductAutomaton a1 a2).
```

Definition 5.3 introduces the product automaton. It is created as a constraint automaton, using *CA* as the alias of the constructor of the Record *constraintAutomata* (as depicted in Definition 4.1). Therefore, *buildPA* takes two constraint automaton *a1* and *a2* creating a product automaton based on *resultingStatesSet*, *resultingNameSet*, *transitionPA* and *resultingInitialStatesSet*, which respectively builds the set of states, set of names, the transition relation and the set of initial states of the resulting automaton as Definition 4.11 presents.

Definition 5.3. *Product Automata in Coq*

```
Definition buildPA (a1: constraintAutomata)(a2:constraintAutomata2) :=
        ConstraintAutomata.CA (statesSet a1 a2) (nameSet a1 a2)
        (transitionPA a1 a2) (initialStates a1 a2).
```

We also have developed a C program which compiles a textual representation of Reo to ease the modelling process. Therefore, a textual file containing primitives regarding the connectors' structure will generate the corresponding Reo circuit code in Coq as its constraint automaton's model. Listings 2 and 4 denote sample files regarding their respective usage examples. Some instances formalized in Coq v. 8.9.1 and nuXmv v. 1.1.1 are executed on a machine equipped with an Intel i7-5930k with 3.50GHz, and 32GB of RAM memory under Ubuntu GNU/Linux kernel 4.15.0-72.

5.1.12 Introducing a basic Reo connector

This first example introduces the definition of a Reo connector in the proposed framework in some detail. It is in the project's repository in file "SyncCA.v". We start by defining a TDS-tuple and an input θ for its constraint automaton. From Table 1, the Sync connector sends data over a channel from an entity to another entity. Its constraint automaton holds a transition where both port names must have the same data flow at the same time to fire. In what follows we define the required items to formalize a TDS in this framework.

We proceed by formalizing the states of the automaton as *automatonStates* and its corresponding port names as *automatonPorts*.

```
Inductive automatonStates := q0.
Inductive automatonPorts := A | B.
```

The definition of a data stream and a time stream can be done respectively as follows.

```
Definition dataAssignmentA (n:nat) :=
 match n with
 | 0 ⇒ 1
 | 1 ⇒ 1
 | 2 ⇒ 0
 | S n ⇒ 0
 end.
Definition timeStampAutomatonA (n:nat) : QArith_base.Q :=
 match n with
 | 0 ⇒ 1#1
 | 1 ⇒ 5#1
 | 2 ⇒ 8#1
 | 3 ⇒ 11#1
 | 4 ⇒ 14#1
 | 5 ⇒ 17#1
 | S n ⇒ Z.of_nat (S n) + 19#1
 end.
```

The notation "a # b" is a shorthand to construct rational numbers in Coq by using the standard library *QArith_base*. The following definition states that *timeStampAutomatonA* is a crescent (and consequently, divergent) time stream. This is required by Definition 4.2 and captured by Definition 5.2 to denote that time streams are always crescent. Its proof may employ a lemma named *orderZofnat* which is available in file "CaMain.v" and available after compiling this file. *Qlt* is the definition of "lesser than" relation for rational numbers in *QArith_base*.

Lemma *timeStampAutomatonAHolds* : ∀ *n*,
Qlt (*timeStampAutomatonA n*) (*timeStampAutomatonA* (*S n*)).

A TDS denoting the port name *A* is defined as follows. A TDS for *B* may be defined by following the same steps. The transition relation of this constraint automaton can be formalized as *automatonTransition*.

```
Definition portA := {|
    ConstraintAutomata.id := A;
    ConstraintAutomata.dataAssignment := dataAssignmentA;
    ConstraintAutomata.timeStamp := timeStampAutomatonA;
    ConstraintAutomata.tdsCond := timeStampAutomatonAHolds;
    ConstraintAutomata.index := 0 |}.
Definition automatonTransition (s:automatonStates):=
 match s with
 | q0 ⇒ [([A;B], (ConstraintAutomata.eqDc nat A B), q0)]
 end.
```

The automaton is defined as *reoSync*, employing the definitions to ease connectors' construction introduced in Section 5.1.8. Its input θ as a set of TDS-tuples can be formalized as *theta*.

```
Definition reoSync := ReoCa.ReoCABinaryChannel A B [q0] [q0] automatonTransition.
Definition theta := [portA;portB].
```

We may then carry out a finite execution over *theta*, stating that this is an accepting execution for 11 iterations or even reason over *theta* to verify that it denotes an infinite run in this constraint automaton as the following two definitions state.

```
Eval compute in ConstraintAutomata.run reoSync theta 11.
Theorem acceptingRun : accepting reoSync theta.
```

The idea behind *accepting* is a property which employs the idea denoted in the final lines of Section 5.1.6 to reason about infinite runs, in which we formalize the idea that a transition can always be fired, given that the required conditions for a transition to fire are indeed satisfied by any of θ's derivative. In this example, we take advantage of the Sync semantics to write a shorthand for the presented idea, considering only the cases where the derivatives of each TDS-tuple in θ are the same, which is the only possible scenario in Sync.

```
Definition accepting (ca: constraintAutomata) (theta: set tds) :=
 ∀ q,∀ k, stepAux ca (map(calcIndex k) (theta)) (thetaN (map(calcIndex k) (theta))) q ≠ [].
```

5.1.13 Sequencing entities' communication with Reo

As another example, we recover and model the Reo connector depicted in Figure 2. We only provide details to the leftmost FIFO-sync connector, namely the Sync containing A as its sink node. Its corresponding source code can be found in the project's repository in file "Sequencer.v". We focus on the first FIFO-Sync pair of connectors of Figure 2 to simplify the modelling explanation. This idea is extended to the whole circuit to obtain its corresponding constraint automaton. In this example, we also show how one can generate Haskell code from the certified Coq model using Coq's extraction mechanism.

Listing 2: Treo representation of Figure 2

```
fifo1(d,e)
sync(e,a)
fifo1(e,f)
sync(f,b)
fifo1(f,g)
sync(g,c)
```

```
Definition dToEFIFOrel (s:sequencerStates) :=
  match s with
  | q0a ⇒ [([D], (ConstraintAutomata.dc D 0), p0a);
            ([D], (ConstraintAutomata.dc D 1), p1a)]
  | p0a ⇒ [([E], (ConstraintAutomata.dc E 0), q0a)]
  | p1a ⇒ [([E], (ConstraintAutomata.dc E 1), q0a)]
  | s0 ⇒ []
  end.

Definition dToEFIFOCA := ReoCa.ReoCABinaryChannel D E
    ([q0a;p0a;p1a]) ([q0a]) (dToEFIFOrel).

Definition syncEACaBehavior (s: sequencerStates) :=
  match s with
  | s0 ⇒ [([E;A] , ConstraintAutomata.eqDc nat E A, s0)]
  | _ ⇒ []
  end.

Definition EAsyncCA := ReoCa.ReoCABinaryChannel E A ([s0])
  ([s0]) syncEACaBehavior.
```

As an example of a desired property, the sequencer may hold the data item first before distributing to its interconnected nodes. Concerning the initial state, there is no way that one of the sink nodes (A, B, and C) will receive data before it passes through the first FIFO (which source node is denoted by D). In short, data need to flow into the circuit to the connected entities to receive data.

Lemma *firstPortToHavaDataIsD* : $\forall$ *state*, *In* (*state*) (*ConstraintAutomata.Q0 resultingSequencerProduct*) $\rightarrow$ *In* (*state*) (*ConstraintAutomata.Q resultingSequencerProduct*) $\wedge$*In* (*D*) (*ConstraintAutomata.portsOfTransition resultingSequencerProduct state*) $\wedge$ $\neg$ *In* (*A*) (*ConstraintAutomata.portsOfTransition resultingSequencerProduct state*) $\wedge$ $\neg$ *In* (*B*) (*ConstraintAutomata.portsOfTransition resultingSequencerProduct state*) $\wedge$ $\neg$ *In* (*C*) (*ConstraintAutomata.portsOfTransition resultingSequencerProduct state*).

We may also reason on specific data flow for a Reo circuit by means of its constraint automaton. Therefore, let us use *resultingSequencerProduct* and *singleExecInput* as an input $\theta \in TDS^{\mathcal{N}ames}$ for *resultingSequencerProduct* and, as an example, suppose one would like to verify if a given scenario (depicted by *singleExecInput*) is modelled by the connector. In what follows, *run1* denotes a bounded run on *resultingSequencerProduct* which at least one transition could be fired. The input θ in *run1* denotes a flow which sequentially distributes the data item to the connected entities. It also depicts that certain configurations were reached (namely, the ones that denote that data would go to each one of the connected entities).

Lemma *acceptingRunAllPortsWData* : $\neg$ *In* [] (*run1*) $\wedge$ *In* [(*p1a*, *s0*, *q0a*, *s0*, *q0a*, *s0*)] (*run1*) $\wedge$ *In* [(*q0a*, *s0*, *p1a*, *s0*, *In* [(*q0a*, *s0*, *q0a*, *s0*, *p1a*, *s0*)] (*run1*).

Conversely, we can reason over data flows which must not be accepted by the connector. This modelling may be useful to state that the connector is free of undesired properties or unwanted scenarios. We define *run2* as an run on an input θ denoting a scenario where both ports A and B are the ones which initially have data (i.e., $A, B \in \theta.N(0)$), a situation which cannot be described by this connector. The following lemma states that this run is not accepted by the circuit's corresponding constraint automaton. The existence of an empty set of states in the resulting trace denotes that at a specific set of states (i.e., a configuration of the connector), no transitions could be fired.

Lemma *nonAcceptingRun* : *In* [] (*run2*).

We can also state that a run described by input θ as *secondExinput* is a rejecting run following the idea explained in section 5.1.6. Indeed, there is a index when executing this automaton with *secondExInput* in which the data flow do not trigger a transition to fire, due to its TDS failing to satisfy the required conditions.

Lemma *rejectingRun* : *ConstraintAutomata.rejecting resultingSequencerProduct secondExInput*.

When the required properties are met by the model, it is possible to use Coq's code extraction apparatus to retrieve certified code regarding it, in languages such as Haskell or Scheme. The following sequence of commands returns a file containing all definitions which *resultingSequencerProduct* relies on, to a Haskell source code file named "SequencerCertified". Listing 3 presents the definition of the corresponding product automaton in Haskell.

```
Require Extraction.
Extraction Language Haskell.
Extraction "SequencerCertified" resultingSequencerProduct.
```

Listing 3: Haskell counterpart for *resultingSequencerProduct*

```
resultingSequencerProduct :: ConstraintAutomata
(Prod (Prod (Prod (Prod (Prod SequencerStates SequencerStates)
SequencerStates) SequencerStates)
SequencerStates) SequencerStates) SequencerPorts
Nat resultingSequencerProduct = buildPA sequencerPortsEq
(pair_eqdec (pair_eqdec (pair_eqdec (pair_eqdec sequencerStatesEq
    sequencerStatesEq) sequencerStatesEq) sequencerStatesEq) sequencerStatesEq
    ) sequencerStatesEq fifo4Product hCsyncCA
```

5.1.14 Smart crossroads' modelling in Coq

We also recover the Reo circuit depicted by Figure 3 and model it by means of the process described by Section 5.3. Its generated Treo code is depicted by Listing 4. This generates the whole Coq code of the example, which can be found in the repository in a file named "trafficLights.v". The formalization of its rightmost Reo channel (a Merger) can be done as follows.

Listing 4: Treo representation of Figure 3

```
merger(a,b,y)
sync(y,x)
sync(x,i)
sync(x,j)
sync(i,k)
transformer(j,l)
sync(l,m)
fifo1(k,n)
merger(m,n,c)
```

```
Definition merger1Automaton := {|
        ConstraintAutomata.Q := [q0];
        ConstraintAutomata.N := [a;y;b];
        ConstraintAutomata.T := merger1rel;
        ConstraintAutomata.Q0 := [q0]
|}.
```

The only state of this automaton is denoted by *q0*, while *a*, *y*, and *b* are port names of the automaton. Our C program that generates the model's code creates an inductive type containing all port names of all automata that compose the circuit. The formalization of both the type of states and port names are shown as follows.

```
Inductive merger1StatesType : Type := q0
Inductive modelPortsType := Inductive modelPortsType :=
    a | y | b | x | i | j | l | k | m | n | c.
```

Definition *merger1rel* is the automaton's transition relation, namely the Merger channel's behavior reflected as a constraint automaton's transition relation. This is achieved as follows.

```
Definition merger1rel (s: merger1StatesType) :=
    match s with
        | q0 ⇒ [([a;y], ConstraintAutomata.eqDc nat a y , q0);([b;y],
ConstraintAutomata.eqDc nat b y , q0)]
    end.
```

We formalize the constraint automaton of the Reo circuit depicted by Figure 3 by employing the aforementioned idea to obtain the remainder of corresponding constraint automata for each Reo channel that compose the circuit. Then, the composition operation is done between every two automata. It will comprehend all automata, which result in the product automata of the whole Reo Circuit. Definition *buildPA* is the top-level function that takes two automata and returns their product.

We present the formalization of one of the connectors below (the others are analogous). The full model of this example contains 8 automata, and we briefly discuss their formalization as follows. *merger1Automaton* models the rightmost data transmission, controlling the data input of the circuit. The rest of the circuit is denoted by *sync2Automaton* and *sync3Automaton* as the channel that replicates data incoming from *merger1Automaton*, *sync4Automaton* is the Sync channel parallel to the transformer channel (*transformer5Automaton*). Then, *sync6Automaton* is the Sync that connects the Transformer to the last node, and *fifo8Automaton* the channel that connects the Sync parallel to the Transform channel. Lastly, *sync7Automaton* joins the sink node of the FIFO with *merger12Automatom*, linking the circuit with the data receiver as the controller that decides which traffic light may be green, based on data transmitted.

```
Definition merger1sync2Product := ProductAutomata.buildPA
  merger1Automaton sync2Automaton.
```

In this scenario, it is interesting to validate properties regarding how the interaction of the traffic lights may affect both cars and pedestrians passing by this crossroad. We may certify that this model is free from situations that one of the traffic lights is always open, by means of the sensor (denoted by port name *a* in the rightmost node). As an example, the following Coq lemmas state respectively that, from any initial state, the automaton always change its configuration to another state if data has departed from the sensor and arrived at the controller, and that from any state of the automaton, this always holds.

Lemma *possibleTrafficLightToBeOpenedNotTheSame* : ∀ *state*,∀ *transition*,
In (*state*) (*ConstraintAutomata.Q0 fifo8merger9Product*) ∧
In (*transition*) (*ConstraintAutomata.T fifo8merger9Product state*) ∧
In (*a*) (*fst*(*fst*(*transition*))) ∧ *In* (*c*) (*fst*(*fst*(*transition*))) →
snd(*transition*) ≠ (*q0*,*q1*,*q2*,*q3*,*q4*,*q5*,*q6*,*q7*,*q8*).
Lemma *possibleTrafficLightToBeOpenedNotTheSame2* : ∀ *state*,∀ *transition*,
In (*state*) (*ConstraintAutomata.Q fifo8merger9Product*) ∧
In (*transition*) (*ConstraintAutomata.T fifo8merger9Product state*) ∧
In (*a*) (*fst*(*fst*(*transition*))) ∧ *In* (*c*) (*fst*(*fst*(*transition*))) →
snd(*transition*) ≠ (*state*).

By executing the following commands, Coq generates a Haskell file named "trafficLightsCertified", containing all definitions which the resulting automaton (*fifo8merger9Product*) relies on. Listing 5 presents the resulting automaton of the whole circuit as Haskell source code. Line 1 builds the resulting automata and the following lines denotes the construction of the transitions.

Require Extraction.
Extraction *Language Haskell*.
Extraction "trafficLightsCertified" *replicator1merger12Product*.

Listing 5: Haskell counterpart for *replicator1merger12Product*

```
1 fifo8merger9Product =
2 buildPA modelPortsEqDec
3 (pair_eqdec (unsafeCoerce pair_eqdec
4 (pair_eqdec (pair_eqdec (pair_eqdec (pair_eqdec (pair_eqdec (\_ _ ->
      merger1EqDec) (\_ _ -> sync2EqDec))
5 (\_ _ -> sync3EqDec)) (\_ _ -> sync4EqDec)) (\_ _ ->
6 transformer5EqDec)) (\_ _ -> sync6EqDec)) (\_ _ -> sync7EqDec))
7 fifo8EqDec) (\_ _ -> merger9EqDec) sync7fifo8Product merger9Automaton
```

5.1.15 Connectors' equivalence by bisimilarity

Let us define a 2-bounded FIFO connector as a single Reo connector itself, manually defining its corresponding constraint automaton as follows. This example may be found in the project's repository in file "2-boundedFIFO.v".

Definition *twoBoundedFIFOrel* (*s*:(*fifoStates* × *fifoStates*)) :**set** (**set** *fifoPorts* × *ConstraintAutomata.DC fifoPorts* (*option nat*) ×(*fifoStates* × *fifoStates*)) :=
match *s* with
| (*q0a*, *q0a*) ⇒ [([*A*], (*ConstraintAutomata.dc A* (*Some* 0)), (*p0a*,*p0a*));
([*A*], (*ConstraintAutomata.dc A* (*Some* 1)), (*p1a*,*p1a*))]
| (*p0a*,*p0a*) ⇒ [([*B*], (*ConstraintAutomata.dc B* (*Some* 0)), (*q0b*,*q0b*))]
| (*p1a*,*p1a*) ⇒ [([*B*], (*ConstraintAutomata.dc B* (*Some* 1)), (*q0b*,*q0b*))]

```
| (q0b,q0b) ⇒ [([B], (ConstraintAutomata.dc B (Some 0)), (p0b,p0b));
                ([B], (ConstraintAutomata.dc B (Some 1)), (p1b,p1b))]
| (p0b,p0b) ⇒ [([C], (ConstraintAutomata.dc C (Some 0)), (q0b,q0b))]
| (p1b,p1b) ⇒ [([C], (ConstraintAutomata.dc C (Some 1)), (q0b,q0b))]
| _ ⇒ []
end.

Definition twoBoundedFIFOCA:= {|
  ConstraintAutomata.Q := [(q0a,q0a);(p0a,p0a);(p1a,p1a);(q0b,q0b);(p0b,p0b);
                            (p1b,p1b)];
  ConstraintAutomata.N := [A;B;C];
  ConstraintAutomata.T := twoBoundedFIFOrel;
  ConstraintAutomata.Q0 := [(q0a,q0a)]
|}.
```

The definition of the states of *twoBoundedFIFOCA* as a pair of states (q_a, q_b) is required to use the bisimilarity notions formalized in this framework (as it requires both automata to have the same state type) as the product construction will yield states as $(q1, q2)$. This does not affect the whole behavior at all, merely serving as a label for the states of the automaton above.

The same connector's automaton can be obtained by "attaching" two 1-bounded FIFO connectors (as Table 1 presents), where the corresponding constraint automaton can be obtained by means of the product operation.

```
Definition oneBoundedFIFOrel (s:fifoStates) :
  set (set fifoPorts × ConstraintAutomata.DC fifoPorts (option nat) × fifoStates) :=
  match s with
  | q0a ⇒ [([A], (ConstraintAutomata.dc A (Some 0)), p0a) ;
           ([A], (ConstraintAutomata.dc A (Some 1)), p1a)]
  | p0a ⇒ [([B], (ConstraintAutomata.dc B (Some 0)), q0a)]
  | p1a ⇒ [([B], (ConstraintAutomata.dc B (Some 1)), q0a)]
  | q0b | p0b | p1b ⇒ []
  end.
Definition oneBoundedFIFOCA:= {|
  ConstraintAutomata.Q := [q0a;p0a;p1a];
  ConstraintAutomata.N := [A;B];
  ConstraintAutomata.T := oneBoundedFIFOrel;
  ConstraintAutomata.Q0 := [q0a]
|}.
Definition oneBoundedFIFOrel2 (s:fifoStates) :=
```

```
match s with
| q0b ⇒ [([B], (ConstraintAutomata.dc B (Some 0)), p0b) ;
          ([B], (ConstraintAutomata.dc B (Some 1)), p1b)]
| p0b ⇒ [([C], (ConstraintAutomata.dc C (Some 0)), q0b)]
| p1b ⇒ [([C], (ConstraintAutomata.dc C (Some 1)), q0b)]
| q0a | p0a | p1a ⇒ []
end.
Definition oneBoundedFIFOCA2 := {|
  ConstraintAutomata.Q := [q0b;p0b;p1b];
  ConstraintAutomata.N := [B;C];
  ConstraintAutomata.T := oneBoundedFIFOrel2;
  ConstraintAutomata.Q0 := [q0b]
|}.
Definition twoBoundedFifo := ProductAutomata.buildPA oneBoundedFIFOCA
oneBoundedFIFOCA2.
```

We proceed by employing the bisimilarity apparatus to check that they indeed yield the same behavior as expected as follows. The execution of the below code returns true, denoting that both automata are indeed bisimilar.

```
Eval compute in ConstraintAutomata.areBisimilar twoBoundedFifo twoBoundedFIFOCA.
```

The performance of the usage examples are measured in Table 3. We calculate the average time and the amount of RAM memory employed of 10 executions of each of the code regarding the usage examples, which contains the definitions and properties evaluated.

Usage example	Execution time (s)	Max used memory (KB)
Sequencer	152.373	450,306.4
Smart Crossroads	13.242	564,254.8

Table 3: Time and memory used in processing the usage examples

5.2 Model checking Reo circuits with nuXmv

Model checkers are powerful tools used to verify properties of logically modeled systems. That verification is made exploring every possible state of the model. nuXmv is a symbolic model checker used for the analysis of synchronous systems [ ] which, among others, suports CTL logic [ ].

The model translation in here proposed is automatized by a compiler available at https://github.com/frame-lab/Reo2nuXmv. This compiler translates Reo circuits to Constraint Automata in nuXmv following a representation where a **MODULE** will represent a CA condensing its states and transitions representations.

The states are denoted by a variable and the transitions determine how that variable can change. nuXmv has a **VAR** block, where the variables are declared; those variables are `cs`, which represents the current state, and `ports`, which is a instance of the **MODULE** that models the ports. There is also a **TRANS** block, where transitions for those variables are declared and an **ASSIGN** block, where the variables can be initialized with a value.

The canonical channels are modeled and Listing 6 presents the Sync channel, the circuit in which the data on both ports must be the same. The constraint automaton that models it consists of one state and one transition, so line 4 declares `cs` with only one possible value. Line 6 models the transition, stating that the data on both ports must be equal and can't be `NULL`. Notice that the model states are equal to all states from the CA, and only them, similar to the transitions.

Listing 6: nuXmv MODULE that represents the Sync channel

```
1 MODULE sync1(time)
2 VAR
3   ports: portsModule;
4   cs: {q0};
5 TRANS
6   ((cs = q0 &  ports.a[time] != NULL & ports.a[time] = ports.b[time]) <-> next(
       cs) = q0);
```

Listing 7 represents the lossySync channel, it has the same rule as the sync channel plus the behavior where the data can be lost. Its CA has one state and two transitions, in line 4 the `cs` is declared with only one possible value. The transition in line 7 is the same as the one in the sync **MODULE**, the first transition in line 6 represents the behavior where the data in the first port is lost, it says that the second port has no data while the first one has. Hence, all of the automaton states are represented, and all of its transitions are also modeled, and nothing more is added.

Listing 7: nuXmv MODULE that represents the LossySync channel

```
1 MODULE lossySync1(time)
2 VAR
3   ports: portsModule;
4   cs: {q0};
5 TRANS
6   ((cs = q0 & ports.b[time] = NULL &  ports.a[time] != NULL)
7   | (cs = q0 &  ports.a[time] != NULL & ports.a[time] = ports.b[time]) <-> next
       (cs) = q0);
```

The other connectors modelling are analogous. In addition to the canonical Reo channels' models, the compiler can also create models of more complex circuits using the composition of those channels (product automata). The result of the product operation is also modeled in a nuXmv **MODULE** but, as another option, the compiler also provides a fully-described model where each application of the product gives rise to a **MODULE**. Despite this second option leads to a more complex and with more states model, it allows traceability: during reasoning, it is possible to recover the Reo channel that has some observed behavior by checking on the verification of the corresponding CA of the basic Reo connector.

As an example, Listing 8 presents the compact model and the components model from the Sequencer in Figure 2. In the components model, it is possible to trace the state of each channel that was composed in the product. In the compact representation, it is only represented the circuit state `q0q0q0q0q0q0` and in the components representation we have the circuit state in `finalAutomata.cs = q0q0q0q0q0q0` but we can also check that this state is composed from the state of it's components seen in `finalAutomata.prod1.cs` and `finalAutomata.prod2.cs` and `finalAutomata.prod1.cs` is composed from it's components and so on.

Another advantage of the components model is that it can reason about the canonical channels. If it is required to infer that the last FIFO of the Sequencer never reaches it's state `p1`, in the compact model it would be:

```
AG(automato.cs!= q0q0q0q0p1q0 & automato.cs != q0q0p1q0p1q0 &
automato.cs != p0q0q0q0p1q0 & automato.cs != p0q0p0q0p1q0 &
automato.cs != p0q0p1q0p1q0 & automato.cs != p1q0q0q0p1q0 &
automato.cs != p1q0p0q0p1q0 & automato.cs != p1q0p1q0p1q0)
```

covering every state of the final circuit where the corresponding state of the final FIFO is `p1`; whilst in the components model it could simply be:

`AG(finalAutomata.prod1.prod2.cs != p1)`,

directly referencing the last FIFO model.

Listing 8: Snapshot of a state from the components product and it's similar im the compact product

```
—Components
   finalAutomata.prod1.prod1.prod1.prod1.prod1.cs = q0
   finalAutomata.prod1.prod1.prod1.prod1.cs = q0q0
   finalAutomata.prod1.prod1.prod1.prod2.cs = q0
   finalAutomata.prod1.prod1.prod1.cs = q0q0q0
   finalAutomata.prod1.prod1.cs = q0q0q0q0
   finalAutomata.prod1.prod2.cs = q0
   finalAutomata.prod1.cs = q0q0q0q0q0
   finalAutomata.prod2.cs = q0
   finalAutomata.cs = q0q0q0q0q0q0
—Compact
```

```
automato.cs = q0q0q0q0q0q0
```

The later model is generated following Algorithm 1, which takes two automata as input which are already modeled in nuXmv, and provides the necessary restrictions to build the nuXmv model of their product following the Constraint Automata product rules.

Lines 2 to 4 initialize with empty sets the resulting ones. The loop in line 5 iterates over the states of the first automaton and the one in line 6 over the second automaton. Those iterations are used to generate the product states in line 7 and line 8 checks if both states are initial states; if so then the product state will also be an initial state.

The loop in line 11 iterates over the generated product states, line 12 initializes the set of unreachable states of that product state with the states of the product minus the current state. Line 13 is a loop that iterates over every transition of the first automaton that starts with the first part of the product state. Similarly, line 14 iterates over every transition of the second automaton that starts with the second part of the product state.

Line 15 has the verification of the first rule that gives the product transitions, and if that holds, then the endpoint of both transitions will be removed from the unreachable states, line 16. Line 17, following the first rule of transition, creates a new transition and adds it to the set of new transitions.

Line 20 verifies the second rule of the transition, and if it holds then the state that represents the endpoint of that transition is removed from the unreachable states. Observe that the transition is not added to the set of new transitions, that happens because the product will use the transition already modeled in the component. But if the condition does not hold, then line 21 blocks the transition in the component.

Line 23 iterates over the transitions of the second component that starts with the second part of the product state, line 24 verifies the third rule of the transition, and similarly to the previous case, if it holds, the endpoint state is removed from the set of unreachable states, and if it is not, line 25 blocks that transition.

The algorithm generates a set that denotes the states of the product, a set that defines the initial states of the product, a set of new transition rules that need to be modeled in the product `MODULE`, and a set of sets, that represents the unreachable states for each state of the product. Those sets will be used to create the `MODULE` of the result of the product.

Listing 9 is an example of a product result of a Sync channel with a FIFO channel. It shows the composition of the Sync channel and the FIFO channel, the **MODULE** of the composition is `finalAutomata` and it instantiates its components through the variables `prod1` and `prod2`. The set of states of `cs` inside the **MODULE** is equal

Algorithm 1: Generation of the product model from two automata

Data: $A_1 = (Q_1, \mathcal{N}ames_1, \rightarrow_1, Q_{0,1})$ and $A_2 = (Q_2, \mathcal{N}ames_2, \rightarrow_2, Q_{0,2})$

Result: Q_p: Product automaton states; $Q_{0,p}$: Product automaton initial states; $Trans_p$: New transitions created following the product transitions rules; $Q_{inalc}(q_1q_2)$: Unreacheable states from each state q_1q_2 of product automaton;

begin
    $Q_p \leftarrow \emptyset$;
    $Q_{0,p} \leftarrow \emptyset$;
    $Trasn_p \leftarrow \emptyset$;
    foreach $q_1 \in Q_1$ **do**
        foreach $q_2 \in Q_2$ **do**
            $Q_p \leftarrow Q_p \cup \{q_1q_2\}$;
            if $(q_1 \in Q_{0,1}) \wedge (q_2 \in Q_{0,2})$ **then** $Q_{0,p} \leftarrow Q_{0,p} \cup \{q_1q_2\}$;
        end
    end
    foreach $q_1q_2 \in Q_p$ **do**
        $Q_{inalc}(q_1q_2) \leftarrow Q_p \setminus \{q_1q_2\}$;
        foreach $q_1 \xrightarrow{N_1,g_1} p_1 \in \rightarrow_1$ **do**
            foreach $q_2 \xrightarrow{N_2,g_2} p_2 \in \rightarrow_2$ **do**
                if $(N_1 \cap \mathcal{N}ames_2) = (N_2 \cap \mathcal{N}ames_1)$ **then**
                    $Q_{inalc}(q_1q_2) \leftarrow Q_{inalc}(q_1q_2) \setminus \{p_1p_2\}$;
                    $Trans_p \leftarrow Trans_p \cup q_1q_2 \xrightarrow{N_1 \cup N_2, g_1 \wedge g_2} p_1p_2$
                end
            end
            if $(N_1 \cap \mathcal{N}ames_2 = \emptyset)$ **then** $Q_{inalc}(q_1q_2) \leftarrow Q_{inalc}(q_1q_2) \setminus \{p_1q_2\}$;
            else $g_1 \leftarrow$ FALSE;
        end
        foreach $q_2 \xrightarrow{N_2,g_2} p_2 \in \rightarrow_2$ **do**
            if $(N_2 \cap \mathcal{N}ames_1) = \emptyset)$ **then**
             $Q_{inalc}(q_1q_2) \leftarrow Q_{inalc}(q_1q_2) \setminus \{q_1p_2\}$;
            else $g_2 \leftarrow$ FALSE;
        end
    end
end

to the set of states created from the composition rule. The transitions generated by the first rule of the product transition are in the **TRANS** block. Rule 2 and its symmetric of the transition will be carried over from the product components, such that to fulfill the model, all of the components transitions that do not follow these rules were removed from the model. The **INVAR** block guarantees consistency between the product and components.

Listing 9: nuXmv model of the composition of Sync with FIFO

```
MODULE sync1(time)
VAR
  ports: portsModule;
  cs: {q0};

MODULE fifo2(time)
VAR
  ports: portsModule;
  cs: {q0,p0,p1};
 ASSIGN
  init(cs) := {q0};
 TRANS
  ((cs = p1 & ports.a[time] = NULL & ports.b[time] = NULL &  ports.d[time] = 1)
  | (cs = p0 & ports.a[time] = NULL & ports.b[time] = NULL &  ports.d[time] =
      0) <-> next(cs) = q0) &
  ((cs = p0) -> ((next(cs) != p1))) &
  ((cs = p1) -> ((next(cs) != p0)));

MODULE finalAutomata(time)
VAR
  prod1: sync1(time);
  prod2: fifo2(time);
  ports: portsModule;
  cs: {q0q0,q0p0,q0p1};
 ASSIGN
  init(cs) := {q0q0};
 TRANS
  ((cs = q0q0 & ports.d[time] = NULL &  ports.a[time] != NULL & ports.a[time] =
        ports.b[time] & ports.b[time] = 0) <-> next(cs) = q0p0) &
  ((cs = q0q0 & ports.d[time] = NULL &  ports.a[time] != NULL & ports.a[time] =
        ports.b[time] & ports.b[time] = 1) <-> next(cs) = q0p1) &
  ((cs = q0p0) -> ((next(cs) != q0p1))) &
  ((cs = q0p1) -> ((next(cs) != q0p0)));
 INVAR
  (((prod1.cs = q0) & (prod2.cs = q0)) <-> (cs = q0q0)) &
  (((prod1.cs = q0) & (prod2.cs = p0)) <-> (cs = q0p0)) &
  (((prod1.cs = q0) & (prod2.cs = p1)) <-> (cs = q0p1));
```

The **MODULE** `portsModule` is a representation of the circuit ports, in which each port of the circuit is represented as an array. The value of each port in a given `time` instance can also be represented there, that way, we can have control over the values of the ports to represent an specific TDS. Those values can be omitted, allowing the model to check for any possible TDS over the possible values.

Besides that, both Filter and Transformer representations are parameterized, but the current version does not support passing the predicate in the Treo model, the user must manually add the predicate to the nuXmv model by replacing the `FALSE` statement.

The example of Figure 3 was modeled using the compiler, passing each of the Reo channels that compose the circuit (described in Treo representation). Hence those channels were composed in pairs that resulted in the `finalAutomata` that models the whole circuit. Two models were generated, one where the components are represented as in Listing 9 (this way all the transitions will be modeled in its **MODULE**, leading to traceability) and the other where the result of the composition is modeled (with fewer states, requiring less computational resources to verify).

The model allows verification of possible TDS combinations and other properties. As an example, it is possible to verify that the automaton will only leave its initial state if one of its input ports has data or the output port has data through the following CTL specification "`A [automato.cs = q0q0q0q0q0q0q0q0q0q0q0q0q0q0 U (automato.ports.a[time]!=NULL | automato.ports.b[time]!=NULL)]` " where `A` denotes a global verification that the current state is the defined `Until` the input ports have some not null value. The model in figure 2 was also modeled. Table 4 presents a comparison of the verification of the specifications run time and memory usage between compact and components models considering the average result of ten executions. Notice that the compact model takes way less time and memory to reason.

Usage example	**Execution time (s)**	**Max used memory (KB)**
Smart crossroads (compact)	0.05	40,229.2
Smart crossroads (components)	1.937	182,794
Sequencer (compact)	0.09	44,652
Sequencer (components)	0.21	69,785.6

Table 4: Experimental results of compact and components based models

5.3 A Graphical User Interface to integrate ReoXplore

All of the framework tools are unified by a graphical interface (extending the Reo Graphical Editor), composing ReoXplore framework, developed using Node JS[11]. The source code and installation instructions are available at https://github.com/frame-lab/ReoXplore.

[11] https://nodejs.org/

ReoXplore allows the construction of Reo circuits graphically and their conversion to a nuXmv model, a Coq model, or a Haskell program. Figure 5 presents ReoXplore's interface. In the center of the GUI there is a canvas where the circuit can be graphically modeled. On the left side, there is a text area, denoting a textual version of the Reo circuit (using a subset of Treo language [ ]) generated from the graphical model. The right menu consists of a list of buttons with each Reo channel and several options to export the model, either as Coq, nuXmv code, or even images.

The framework is conceived to be used in a flow summarized in Figure 4 and detailed below:

(i) The user can graphically model a Reo connector by drawing and linking the channels using the GUI Figure 5 displays. One can also directly write the corresponding code in the text area in the leftmost part of the GUI, where the Treo code is automatically generated from the graphical model. It can be exported as a text or image file.

(ii) The inputted model in the above step can be converted to its counterpart in Coq or nuXmv code employing the framework described in Sections 5.1 and 5.2.

(iii) The converted model to CA in the desired framework is generated: The Coq model can be downloaded, and the same model can be generated as code in Haskell using the Coq code extraction mechanism. The nuXmv code generation enables the user to either:

- use the export option "Compact nuXmv", which executes the ZZZ compiler and returns a SMV file of the compact product for download.
- use the export option "Components nuXmv", which does the same as the "Compact nuXmv" option but returns the components' product.

Each button triggers a HTTP request to the server, sending the Treo model generated in the GUI. Depending on the action triggered, a specific method is executed, returning the target file based on the button's action. All export buttons can be used right away, except for the Haskell code generation, which requires Coq to be installed (version 8.9+) as it relies on the Coq code extraction mechanism.

6 Conclusions and future work

This work presented ReoXplore: a framework to model and reason about Reo circuits. ReoXplore's background uses (i) Coq for proof-theoretical reasoning and

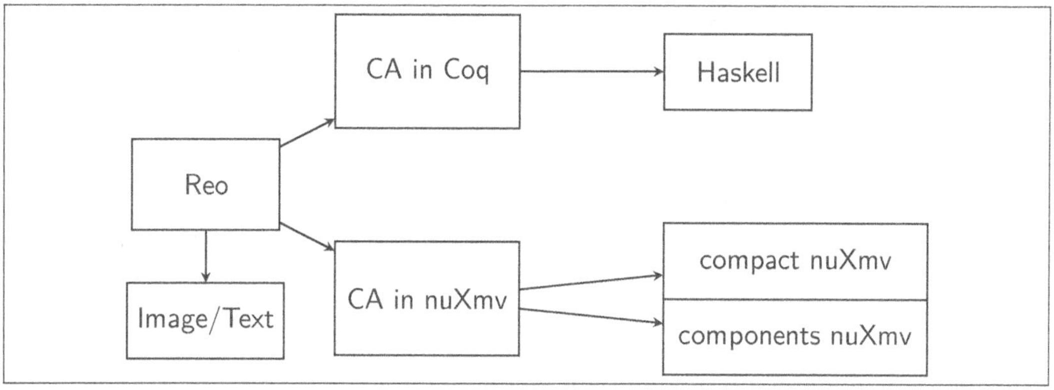

Figure 4: Framework execution schema

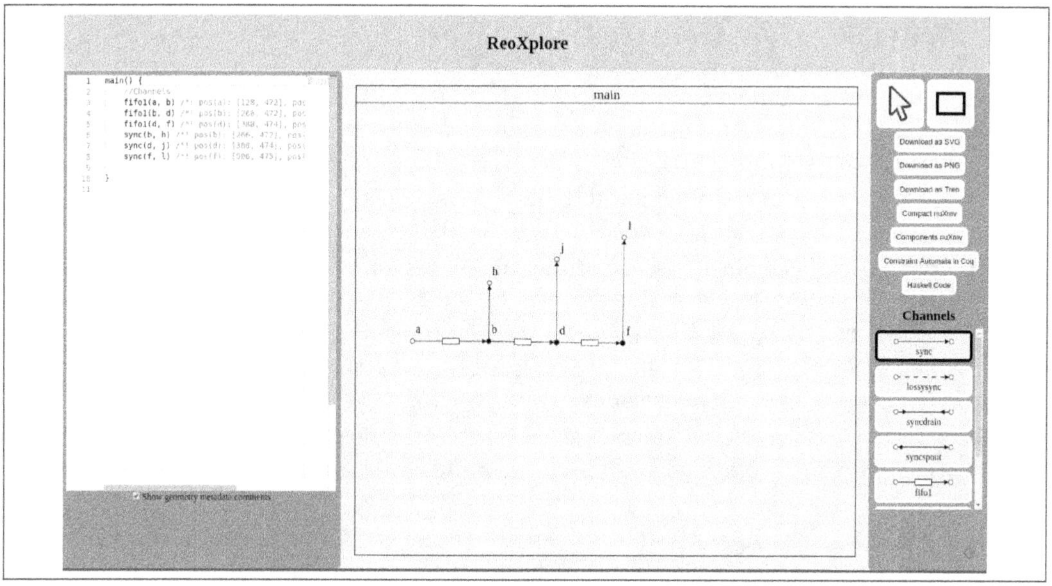

Figure 5: Sequencer modelled in ReoXplore GUI

generation of certified Haskell code, (ii) nuXmv for symbolic model checking Reo circuits, and extends (iii) Reo Graphical Editor to use a canvas for modelling Reo circuits (also converted to a variant of Treo language) and integrate all of the presented subsystems. We present some experimental results concerning time and memory used in the reasoning process.

The proof-theoretical subsystem is depicted in a Coq model, employing a consolidated system to verify and prove properties of Reo connectors using Constraint Automata. Coq specifically lets their users extract certified code to programming languages and also check for bisimilarity. This formalization is followed by the modelling in nuXmv with the product pre or post-processed. The pre-processed version provides a smaller nuXmv model; the post-processed creates a bigger model, but allows better traceability. It leads to a simpler way for identifying some channel in question. Some usage examples are presented and performance issues are discussed concerning time and state-space.

Further work includes the adoption of coalgebric models aiming a more expressive formalization of systems. Concerning the implementations, the integration with the representation of finite and infinite (by means of functions) Timed Data Streams (TDS) to compose the input data of the system to be incorporated in nuXmv and Coq generated models (ongoing work).

References

[1] E. Alaña, H. Naranjo, Y. Yushtein, M. Bozzano, A. Cimatti, M. Gario, R. De Ferluc, and G. Gérald. Automated generation of fdir for the compass integrated toolset (autogef). *European Space Agency, (Special Publication) ESA SP*, 701, 01 2012.

[2] H. R. Andersen. An introduction to binary decision diagrams. *Lecture notes, available online, IT University of Copenhagen*, page 5, 1997.

[3] F. Arbab. Reo: a channel-based coordination model for component composition. *Mathematical Structures in Computer Science*, 14(3):329–366, 2004.

[4] F. Arbab. Coordination for component composition. *Electronic Notes in Theoretical Computer Science*, 160:15 – 40, 2006. Proceedings of the International Workshop on Formal Aspects of Component Software (FACS 2005).

[5] F. Arbab, C. Baier, F. de Boer, and J. Rutten. Models and temporal logical specifications for timed component connectors. *Software & Systems Modeling*, 6(1):59–82, 2007.

[6] F. Arbab, N. Kokash, and S. Meng. Towards using reo for compliance-aware business process modeling. In *International Symposium On Leveraging Applications of Formal Methods, Verification and Validation*, pages 108–123. Springer, 2008.

[7] F. Arbab and J. J. Rutten. A coinductive calculus of component connectors. In *International Workshop on Algebraic Development Techniques*, pages 34–55. Springer, 2002.

[8] E. Ardeshir-Larijani, A. Farhadi, and F. Arbab. Simulation of hybrid reo connectors. In *2020 CSI/CPSSI International Symposium on Real-Time and Embedded Systems and Technologies (RTEST)*, pages 1–10. IEEE, 2020.

[9] C. Atkinson and T. Kuhne. Model-driven development: a metamodeling foundation. *IEEE software*, 20(5):36–41, 2003.

[10] C. Baier. Probabilistic models for reo connector circuits. *J. UCS*, 11(10):1718–1748, 2005.

[11] C. Baier and J.-P. Katoen. *Principles of Model Checking.* The MIT Press, 2008.

[12] C. Baier, M. Sirjani, F. Arbab, and J. Rutten. Modeling component connectors in reo by constraint automata. *Science of computer programming*, 61(2):75–113, 2006.

[13] R. Cavada, A. Cimatti, M. Dorigatti, A. Griggio, A. Mariotti, A. Micheli, S. Mover, M. Roveri, and S. Tonetta. The nuxmv symbolic model checker. In A. Biere and R. Bloem, editors, *CAV*, volume 8559 of *Lecture Notes in Computer Science*, pages 334–342. Springer, 2014.

[14] R. Cavada, A. Cimatti, A. Mariotti, C. Mattarei, A. Micheli, S. Mover, M. Pensallorto, M. Roveri, A. Susi, and S. Tonetta. Supporting requirements validation: The eurailcheck tool. In *2009 IEEE/ACM International Conference on Automated Software Engineering*, pages 665–667, Nov 2009.

[15] A. Cimatti, E. Clarke, E. Giunchiglia, F. Giunchiglia, M. Pistore, M. Roveri, R. Sebastiani, and A. Tacchella. Nusmv 2: An opensource tool for symbolic model checking. In E. Brinksma and K. G. Larsen, editors, *Computer Aided Verification*, pages 359–364, Berlin, Heidelberg, 2002. Springer Berlin Heidelberg.

[16] E. Clarke, A. Biere, R. Raimi, and Y. Zhu. Bounded model checking using satisfiability solving. *Formal Methods in System Design*, 19(1):7–34, Jul 2001.

[17] E. M. Clarke and E. A. Emerson. Design and synthesis of synchronization skeletons using branching time temporal logic. In *Workshop on Logic of Programs*, pages 52–71. Springer, 1981.

[18] T. Coquand. *Une théorie des constructions*. PhD thesis, Paris 7, 1985.

[19] T. Coquand. *An analysis of Girard's paradox*. PhD thesis, INRIA, 1986.

[20] K. Dokter and F. Arbab. Treo: Textual syntax for reo connectors. In S. Bliudze and S. Bensalem, editors, *Proceedings of the 1st International Workshop on Methods and Tools for Rigorous System Design*, volume 272 of *Electronic Proceedings in Theoretical Computer Science*, pages 121–135. Open Publishing Association, 2018.

[21] G. Dowek, A. Felty, H. Herbelin, G. Huet, C. Murthy, C. Parent, C. Paulin-Mohring, and B. Werner. *The Coq Proof Assistant: User's Guide: Version 5.6*. INRIA, 1992.

[22] E. A. Emerson and J. Y. Halpern. "sometimes" and "not never" revisited: On branching versus linear time temporal logic. *J. ACM*, 33(1):151–178, Jan. 1986.

[23] E. Grilo and B. Lopes. Modelling and certifying smart cities in reo circuits. In *2020 International Conference on Systems, Signals and Image Processing (IWSSIP)*, pages 453–458. IEEE, 2020.

[24] S.-S. T. Jongmans and F. Arbab. Overview of thirty semantic formalisms for reo. *Scientific Annals of Computer Science*, 22(1), 2012.

[25] J. Klein, S. Klüppelholz, A. Stam, and C. Baier. Hierarchical modeling and formal verification. an industrial case study using reo and vereofy. In *International Workshop on Formal Methods for Industrial Critical Systems*, pages 228–243. Springer, 2011.

[26] N. Kokash and F. Arbab. Formal design and verification of long-running transactions with extensible coordination tools. *IEEE Transactions on Services Computing*, 6(2):186–200, 2011.

[27] N. Kokash, B. Changizi, and F. Arbab. A semantic model for service composition with

coordination time delays. In *International Conference on Formal Engineering Methods*, pages 106–121. Springer, 2010.

[28] N. Kokash, C. Krause, and E. De Vink. Reo+ mcrl2: A framework for model-checking dataflow in service compositions. *Formal Aspects of Computing*, 24(2):187–216, 2012.

[29] Y. Li and M. Sun. Modeling and verification of component connectors in coq. *Science of Computer Programming*, 113:285–301, 2015.

[30] Y. Li, X. Zhang, Y. Ji, and M. Sun. Capturing stochastic and real-time behavior in reo connectors. In *Formal Methods: Foundations and Applications - 20th Brazilian Symposium, SBMF 2017, Recife, Brazil, November 29 - December 1, 2017, Proceedings*, pages 287–304, 2017.

[31] Y. Li, X. Zhang, Y. Ji, and M. Sun. A formal framework capturing real-time and stochastic behavior in connectors. *Science of Computer Programming*, 2019.

[32] M. R. Mousavi, M. Sirjani, and F. Arbab. Formal semantics and analysis of component connectors in reo. *Electronic Notes in Theoretical Computer Science*, 154(1):83–99, 2006.

[33] S. Navidpour and M. Izadi. Linear temporal logic of constraint automata. In *Advances in Computer Science and Engineering*, pages 972–975. Springer, 2008.

[34] M. S. Nawaz and M. Sun. Reo2pvs: Formal specification and verification of component connectors. In *The 30th International Conference on Software Engineering and Knowledge Engineering, Hotel Pullman, Redwood City, California, USA, July 1-3, 2018.*, pages 391–390, 2018.

[35] M. P. Papazoglou. Service-oriented computing: Concepts, characteristics and directions. In *Web Information Systems Engineering, 2003. WISE 2003. Proceedings of the Fourth International Conference on*, pages 3–12. IEEE, 2003.

[36] C. Paulin-Mohring. Introduction to the calculus of inductive constructions, 2015.

[37] B. Pourvatan, M. Sirjani, H. Hojjat, and F. Arbab. Automated analysis of reo circuits using symbolic execution. *Electronic Notes in Theoretical Computer Science*, 255:137–158, 2009.

[38] M. Sun and Y. Li. Formal modeling and verification of complex interactions in e-government applications. In *Proceedings of the 8th International Conference on Theory and Practice of Electronic Governance*, pages 506–507. ACM, 2014.

[39] S. Tasharofi and M. Sirjani. Formal modeling and conformance validation for ws-cdl using reo and casm. *Electronic Notes in Theoretical Computer Science*, 229(2):155–174, 2009.

[40] X. Zhang, W. Hong, Y. Li, and M. Sun. Reasoning about connectors in coq. In *International Workshop on Formal Aspects of Component Software*, pages 172–190. Springer, 2016.

[41] X. Zhang, W. Hong, Y. Li, and M. Sun. Reasoning about connectors using coq and z3. *Science of Computer Programming*, 170:27–44, 2019.

 Received 20 December 2019

Defeasible Reasoning in Navya-Nyāya

Guhe
Fudan University/School of Philosophy, 220 Handan Road, 200433 Shanghai, P.R. China.
guhe@fudan.edu.cn

Abstract

The present paper is about affinities between John L. Pollock's theory of defeasible reasoning and the doctrine of the so-called *upādhi* in Navya-Nyāya, a school of classical Indian philosophy. We will show that the defeasible character of the five-membered inferences (*anumāna*), which Navya-Naiyāyikas regard as a knowledge source, can be explained in terms of enumerative induction and nomic generalization. Moreover, our analysis of an *upādhi*'s function as a "vitiator" (*dūṣaṇa*) and of Navya-Nyāya definitions of the concept of *upādhi* will give a clue to relations between *upādhi*s and "defeaters" in the sense of Pollock's theory.

Although the Navya-Nyāya doctrine of *upādhi* can be assimilated to Pollock's theory of defeasible reasoning, *upādhi*s should not be confused with defeaters. *upādhi*s are objects of the domain. The equivalents of defeaters in the Navya-Nyāya doctrine of *upādhi* are rather certain propositions which can be gleaned from definitions of the *upādhi* and from specifications of an *upādhi*'s vitiating function in Navya-Nyāya sources. Some of these propositions are only rebutters and some are rebutters and undercutters with respect to different prima facie reasons which are involved in a five-membered inference. It is important to note that the defeater-related vitiating function of an *upādhi*, which aims at overruling an inference, applies only to the so-called "ascertained *upādhi*s" (*niścitopādhi*). The so-called "dubious *upādhi*s" (*saṃdigdhopādhi*) are relevant to situations which Pollock describes as "collective defeat". We will see that the distinction between a skeptical and a credulous reasoner can help to understand in what sense this type of *upādhi* is also regarded as a vitiator in Navya-Nyāya.

Keywords: Indian Logic, Navya-Nyāya, *upādhi*, defeasible reasoning, Pollock

Thanks go to the organizers of the 19th Brazilian Logic Conference for having me as a keynote speaker. I am particularly indebted to Prof. Ricardo Silvestre for his great hospitality. Moreover, I would like to express my gratitude to Prof. Parimal Patil for mentoring my work on Navya-Nyāya during my 2018-19 sabbatical at Harvard as a Harvard-Yenching Research Scholar. The present article takes its cue from many inspiring conversations with Parimal.

1 Introduction

The idea that the Navya-Nyāya concept of *upādhi* is somehow akin to John L. Pollock's concept of a defeater hails from Stephen Phillips' groundbreaking study of Gaṅgeśa's Upādhivāda (cf. [13]). The defeasible character of the five-membered inferences (*anumāna*) which Navya-Naiyāyikas regard as a knowledge source (*pramāṇa*) is owing to an inductive leap in the third member. Although Navya-Naiyāyikas clearly refer to induction as a method of identifying the so-called "pervasion" (*vyāpti*), a relation between two properties functioning as probans (*hetu* or *sādhana*) and probandum (*sādhya*) in an inference, it should not be overlooked that induction is not accepted as a knowledge source sui generis in Navya-Nyāya. From a Navya-Nyāya point of view it is perception (*pratyakṣa*) in combination with multiple observation (*bhūyodarśana*) and the resulting mental impressions (*saṃskāra*) which functions as a knowledge source for the cognition of a pervasion. Nevertheless, the procedure involves an extrapolation from known cases of a co-presence (*anvaya*) and a co-absence (*vyatireka*) of probans and probandum. As noted by Matilal (cf. [10]: 17f), the reference to agreeing and disagreeing instances (*sapakṣa* and *vipakṣa*) in the third member of a five-membered inference is reminiscent of Mill's "Joint Method of Agreement and Difference" (cf. [11]: 402f). Since the Navya-Naiyāyikas were well aware that the cognition of a pervasion is fallible, they became keenly interested in the *upādhi* as a vitiator which targets precisely this weak point of a five-membered inference.[1]

2 A Recap of Pollock's Theorie of Defeasible Reasoning

2.1 Nondefeasible (Conclusive) Reasons, Defeasible (Prima Facie) Reasons and Defeaters

"There are two kinds of reasons – defeasible and nondefeasible. Nondefeasible reasons are those reasons that logically entail their conclusions. For instance, $(P\&Q)$ is a nondefeasible reason for P. Such reasons are conclusive reasons. Everyone has

[1] Although Phillips is well aware that an *upādhi* is not the same as a defeater in the sense of Pollock's theory, his translation of *upādhi* as "undercutting condition" can easily give rise to the misunderstanding that the vitiating function of an *upādhi* consists merely in undercutting, not in rebutting. Literal translations like "associate condition" ([10]: 166) or "adjunct" ([3]: 5) might work with reference to theories according to which the *upādhi*'s function of vitiating is intertwined with the function of a corrector. However, as noted by Phillips, Gaṅgeśa's concept of *upādhi* cannot be adequately captured by means of these translations, since Gaṅgeśa conceives of an *upādhi* merely as a vitiator. In order to avoid the impression of an undue narrowness in our presentation of the Navya-Nyāya doctrine of *upādhi*, we will leave the term *upādhi* untranslated.

always recognized the existence of nondefeasible reasons, but defeasible reasons are a relatively new discovery in philosophy, as well as in allied disciplines like AI. Focusing first upon reasons that are beliefs, P is a defeasible reason for Q just in case P is a reason for Q, but adding additional information may destroy the reason connection. Such reasons are called 'prima facie reasons'." ([14]: 484) The additional information which destroys the reason connection is called a "defeater".

"For example, somethingâĂŹs looking red to me may justify me in believing that it is red, but if I subsequently learn that the object is illuminated by red lights and I know that that can make things look red when they are not, then I cease to be justified in believing that the object is red." ([14]: 481) In this example "X (= the object referred to) looks red" is a prima facie reason, "X is red" is the conclusion. "X is illuminated by red light" is a defeater.

Definition (Prima facie reason): "P is a *prima facie reason* for S to believe Q if and only if P is a reason for S to believe Q and there is an R such that R is logically consistent with P but $(P \& R)$ is not a reason for S to believe Q." ([14]: 484)

The purpose of the consistency condition is to ensure that the existence of such an R would not be trivially fulfilled. If R would not have to be consistent with P, one might, e.g., choose $R :\leftrightarrow \neg P$. In this case $P \& R$ is false and then it is trivially not a reason for S to believe Q. Hence, without the consistency condition every reason would be a prima facie reason.

Definition (Defeater): "R is a *defeater* for P as a prima facie reason for Q if and only if P is a reason for S to believe Q and R is logically consistent with P but $(P \& R)$ is not a reason for S to believe Q." ([14]: 484)

2.2 Rebutter and Undercutter

Definition (Rebutting defeater): "R is a *rebutting defeater* for P as a prima facie reason for Q if and only if R is a defeater and R is a reason for believing $\sim Q$." ([14]: 485)

Definition (Undercutting defeater): "R is an *undercutting defeater* for P as a prima facie reason for S to believe Q if and only if R is a defeater and R is a reason for denying that P wouldnâĂŹt be true unless Q were true." ([14]: 485)

Denying that P wouldnâĂŹt be true unless Q were true means to claim that P can be true, even though Q is false. In Pollock's "red light"-example the defeater is an undercutter: "This is a defeater, but it is not a reason for denying that X is red (red things look red in red light too). Instead, this is a reason for denying that X wouldn't look red to me unless it were red." ([14]: 485)

3 Defeasible Reasoning in Navya-Nyāya

3.1 What Is Navya-Nyāya?

Navya-Nyāya (âĂIJNew LogicâĂİ?) is a school of classical Indian philosophy, which succeeded the earlier Nyāya School (*prācīnanyāya*). The beginnings of Navya-Nyāya date back to the 12th or 13th century with authors such as Śaśadhara and Maṇikaṇṭha Miśra. There is, however, good reason to believe that the advent of Navya-Nyāya is already foreboded in the works of Udayana (11th century), who is mostly still considered to be a representative of the old Nyāya School (however, cf. [25]: 9f). GaṅgeśaâĂŹs magnum opus Tattvacintāmaṇi (14th century) was seminal for the development of the typical style of the Navya-Naiyāyikas' approach to logical and epistemological issues. In order to define their concepts with utmost precision they designed an ideal language, a kind of Leibnizian characteristica universalis based on a canonical form of Sanskrit, which serves to explicate the objective content of verbalized and unverbalized cognitions and to disambiguate sentences formulated in ordinary Sanskrit. The school reached its peak in the works of authors such as Raghunātha Śiromaṇi (16th century), Jagadīśa and Gadādhara (17th century) and has remained active through to the present day, although the scholarly work of contemporary Navya-Naiyāyikas is mostly confined to exegetical endeavours.

3.2 The Navya-Nyāya Theory of Inference

3.2.1 The Five-Membered Inference

According to the theory of inference in late Nyāya and in Navya-Nyāya the verbalized form of an inference (*anumāna*) consists of five members, a thesis, three members which are supposed to corroborate the claim formulated in the thesis, and a conclusion which restates the thesis as a result of the inference. We will study this type of inference with special regard to the third member, which seems to contain an inductive leap. Deductive components are involved in the five-membered inference as well, although from a Navya-Nyāya perspective the logical framework is essentially inductive.

In the case of a valid inference the content of each of the five members must be a veridical awareness (*pramā*), where truth is usually understood in the sense of correspondence to facts. However, we will see that accepting the third member as true might involve a pragmatic concept of truth as an evidentially constrained property.

Although Indian logicians did not symbolize the components of a correct five-membered inference, it is clear that they conceived of it as an instance of formally valid reasoning, since all the components are identified by certain technical terms and the way they are related to each other and the order in which they are supposed to appear is strictly determined. In Keśava Miśra's Tarkabhāṣā, a work which according to Dineśchandra Bhattacharya dates from the 12th century A.D. (cf. [2]: 64), one can find the following stock example of a valid five-membered inference:

[1] *parvato 'yam agnimān*
[2] *dhūmavattvāt.*
[3] *yo dhūmavān so 'gnimān yathā mahānasaḥ.*[2]
[4] *tathā cāyam.*
[5] *tasmāt tathā.* ([22]: 40, 2f)

[1. Thesis (*pratijñā*)]	This mountain possesses fire.
[2. Reason (*hetu*)]	For, it possesses smoke.
[3. Example (*dṛṣṭānta*)]	Whatever possesses smoke, possesses fire, like the kitchen.
[4. Application (*upanaya*)]	This [mountain] is so (i.e., it possesses smoke as a token of fire).
[5. Conclusion (*nigamana*)]	Therefore, [it is] so (i.e., the mountain possesses fire).

In this example the mountain functions as the subject of the inference (*pakṣa*). Fire is the probandum (*sādhya*), a property which according to the first member can be proved to be present on the subject of the inference. Smoke is the probans or prover (*hetu* or *sādhana*), a property whose presence on the subject of the inference is an established fact according to the second member of the inference (which happens to be called *hetu* as well). The probans functions as an indicator of the probandum. According to the third member this is possible, because there is a universal relation

[2]A few lines later Keśava Miśra adds the following to the third member: *evaṃ yatrāgnir nāsti tatra dhūmo 'pi nāsti yathā mahāhrada iti vyatirekeṇa vyāptiḥ.* ([22]: 40, 8f) – "Similarly, 'Wherever there is no fire, there is also no smoke, such as in a big lake.' [This] is pervasion in terms of the negative correlation."

called "pervasion" (*vyāpti*) between the probans and the probandum in the sense that every locus of the probans is a locus of the probandum. The reference to examples in the third member indicates that the cognition of a pervasion involves an inductive generalization. The cognition is supported by wide experience of positive correlations such as smoke and fire on a kitchen hearth and negative correlations such as absence of smoke on a lake where there is absence of fire. Thus, the cognition of a pervasion is the result of an extrapolation from certain known cases, i.e., it involves an inductive leap to all unobserved cases, past, present and future, including the case at issue in an inferential situation. Due to an epistemic process called "reflective grasping" (*parāmarśa*), which is indicated in the fourth member, the probans's presence on the subject of an inference is associated with the pervasion relation and it is this that finally gives rise to the "inferential awareness" (*anumiti*), the conclusion expressed in the fifth member, i.e., the cognition that the probandum is present on the subject of the inference.

3.2.2 The Third Member as an Instance of Enumerative (as Opposed to Statistical) Induction

An adequate interpretation of the third member needs to warrant the universality of the pervasion at issue without compromising the requirement that in the case of a valid inference a reasoner needs to have cognitive access to the truth of a pervasion in some sense. A possible solution to this problem is to interpret the pervasion at issue as the conclusion of a rule, viz. the rule of enumerative induction. (The precise definition will be given below.)

The conclusion of this type of induction is a so-called "nomic generalization". Pollock's concept of a nomic generalization (cf. section 3.2.3) seems to be a close approximation to what Navya-Naiyāyikas understand by "pervasion". In contrast to generics like "Birds fly" nomic generalizations are not compatible with the existence of exceptions. At this point a formal explication is still expendable. We will rather make do with informal renderings like "All loci of the probans are loci of the probandum" and tacitly assume that even future instances of the probans and the probandum fall within the scope of such formulations. Hence, the following first-order formalization is only a prima facie approximation to what the Navya-Naiyāyikas actually understand by pervasion:

($\mathrm{V}_1^{\mathrm{pf}}$) $\forall x(Hx \rightarrow Sx)$ (where Hx translates into "x is a locus of the probans" and Sx translates into "x is a locus of the probandum")[3]

[3]Since Naiyāyikas conceive of pervasion as a relation (*saṃbandha*), one might object that the denotatum of ($\mathrm{V}_1^{\mathrm{pf}}$) is not a relation, but a proposition, which expresses a condition on the in-

Our interpretation of the third member of a five-membered inference in the sense of an application of the rule of enumerative induction settles the question how a cognitive agent can reasonably regard a pervasion as an object of a veridical awareness, even without being able to check whether every locus of the probans is a locus of the probandum. According to the rule of enumerative induction the evidential support provided by the examples is a reason for a cognitive agent $\mathcal{A}$ to believe that all H's are S, in fact that even future H's are S. This would be impossible if $\mathcal{A}$ knew about counterexamples, or if $\mathcal{A}$ had any strong reasons to believe that there will be counterexamples in the future. (The defeasible character of the rule admits only the logical possibility of there being counterexamples.) Hence, the pervasion at issue as the conclusion of an enumerative induction can be said to be true in a pragmatic sense, inasmuch as sustained inquiry leads to a dependable final verdict.[4] In this case cognitive access to the truth of the pervasion at issue

stantiating pairs. However, it is also customary in Nyāya to talk about pervasion in the sense of a proposition. Thus, Jinavardhana Sūri, a commentator on ŚivādityamiśraâĂŹs Saptapadārthī, defines pervasion as follows: *vyāptir yatra yatra sādhanaṃ tatra tatra sādhyam, yatra sādhyaṃ nāsti tatra sādhanam api nāstītilakṣaṇā.* ([20]: 69, 24) – "Pervasion is defined as: 'Wherever there is the probans, there is the probandum. Wherever there is not the probandum, there is not the probans either'." Hereafter, we will use the term "pervasion" with reference to a relation and to propositions alike. If necessary, the context will help to disambiguate the intended meaning.

Although expressions like "the pervasion of smoke by fire" seem to suggest that pervasion should be construed as a second-order relation, it is also common in Nyāya to regard the relata as objects of the domain, which includes properties. Thus, the property "being a locus of smoke" (*dhūmavattva*) is said to be pervaded by the property "being a locus of fire" (*vahnimavattva*), so that pervasion can be construed as a first-order relation. In George Bealer's property theories properties are expressed by means of special terms (cf. [1]: 43f). Instead of the predicates S ("...is a locus of smoke") and F ("...is a locus of fire") the terms $[Sx]_x$ ("being an x such that x is a locus of smoke") and $[Fx]_x$ ("being an x such that x is a locus of fire") can be regarded as denotations of relata of the pervasion relation. However, instead of construing pervasion as a first-order relation, Navya-Naiyāyikas rather conceive of it as a "relation-in-intension" (cf. [9]: 170f). Hence, a semi-formal representation by means of Bealer's formalization methods, such as $[x \text{ is pervaded by } y]_{xy}$ ("being an x and a y such that x is pervaded by y"), comes closer to the Navya-Nyāya understanding of pervasion as a relation.

[4]According to one version of pragmatism, which attempts to preserve truth's objectivity, truth is what will be accepted in the end of inquiry. This approach identifies "being true with being warrantedly assertible under epistemically ideal conditions" ([18]: 220). Such a pragmatic concept of truth, which Putnam had advocated at least for some time, seems to be close to what Pollock understands by "warrant": "In contrast to justification, *warrant* is what the system of reasoning is ultimately striving for. A proposition is warranted in a particular epistemic situation iff (if and only if), starting from that epistemic situation, an ideal reasoner unconstrained by time or resource limitations would ultimately be led to believe the proposition. Warranted propositions are those that would be justified 'in the long run' if the system were able to do all possible relevant reasoning. A proposition can be justified without being warranted, because although the system has done everything correctly up to the present time and that has led to the adoption of the belief, there may

does not depend on verifying empirically whether every locus of the probans is, has always been and forever will be a locus of the probandum.[5]

Enumerative induction is an inference relation between a prima facie reason and a conclusion. In our formal reconstruction of the third member of a five-membered inference the claim of evidential support in the form of positive or negative examples functions as a prima facie reason and the pervasion at issue functions as the conclusion.

Enumerative induction has to be distinguished from statistical induction: "The simplest kind of induction is enumerative induction, which proceeds from the observation that all members of a sample X of A's are B's, and makes a defeasible inference to the conclusion that all A's are B's. This is what is known as the Nicod Principle. Goodman [1955] observed that the principle requires a projectibility constraint. With this constraint, it can be formulated as follows:

(7.1) If B is projectible with respect to A, then ⌜X is a sample of A's all of which are B's⌝ is a prima facie reason for ⌜All A's are B's⌝." ([17]: 77)

"Enumerative induction has been a favorite topic of philosophers, but statistical induction is much more important for the construction of a rational agent. It is rare that we are in a position to confirm exceptionless universal generalizations. Induction usually leads us to statistical generalizations that either estimate the probability of an A being a B, or the proportion of actual A's that are B's, or more simply, may

be further reasoning waiting to be done that will mandate the retraction of the belief. Similarly, a proposition can be warranted without being justified, because although reasoning up to the present time may have failed to reveal adequate reasons for adopting the proposition, further reasoning may provide such reasons. Analogously, reasoning up to the present may mandate the adoption of defeaters which, upon further reasoning, will be retracted. So justification and warrant are two importantly different notions, although they are closely related." ([16]: 5)

[5]Pragmatic theories focus on what people are doing when describing statements as true. They link truth to verifiability, assertibility, usefulness, or long-term durability. The author of the Upādhidarpaṇa, an unpublished anonymous pre-Gaṅgeśa Navya-Nyāya manuscript (cf. [2]: 167), seems to refer to such a pragmatic theory of truth when he explains in what sense an inference can be accepted as valid, even though an acquaintance with all loci of the probans which are involved in the pervasion relation is beyond the scope of a reasoner's limited cognitive abilities: *jīvanavilopa-prasaṅgāt. pratidinam āhārārthina āhārasya hitasādhanatāṃ vijñāya samanantaraṃ bhāvino 'pi tad anumāyaiva pravartate. anyathā pravṛttivirodhaprasaṅgāt. yas tv anumānaṃ nāṅgīkaroti sa kathaṃ prativādībhavet.* ([24]: 4b, 14f) – "[Proponent of the right view:] [Inferences which are not based on the knowledge of all substrates of a positive or negative correlation can also be valid], because [otherwise] there would be the undesirable consequence of a disturbance of everyday life. Observing day by day that for someone who seeks food eating brings about benefit, one thereupon acts [in accordance with that] after having inferred that also for the future. Because, otherwise, there would be an obstacle to active life. Who, however, does not agree to [this] inference, how should he respond?"

lead us to the conclusion that most A's are B's. Such statistical generalizations are still very useful, because the statistical syllogism enables a cognizer to make defeasible inferences from them to nonprobabilistic conclusions." ([17]: 78)

Since according to Pollock the conclusion of an enumerative induction is a nomic generalization, this type of induction is preferable to statistical induction for the purpose of a formal explication of the third member. Nevertheless, some refinement of the enumerative inductive inference pattern seems appropriate in order to assimilate it to the theory of inference in Navya-Nyāya. Since the Navya-Naiyāyikas also take into account negative examples of a co-absence of probans and probandum, one might want to include a second sample of negative examples in the prima facie reason. Thus, the prima facie reason would be the claim of evidential support in the form of a sample X exemplifying the co-presence of the properties "being an H" and "being an S" and of a sample Y exemplifying the co-absence of "being an H" and "being an S" (cf. the kitchen hearth as a locus of smoke and fire and the lake as a locus of the co-absence of smoke and fire in the above-mentioned smoke/fire-example). However, such a set Y does not exist in cases in which the probandum is a property whose presence in all loci other than the inferential subject is warranted. Thus, the above-mentioned smoke/fire-example should not be taken to be representative of all kinds of five-membered inferences. Moreover, there are cases in which the occurrence of the probandum is confined to the inferential subject, so that only the previously observed instances of a co-absence of "being an H" and "being an S" can serve as an inductive support for the inferential subject's being H and S. Thus, from the perspective of Navya-Nyāya it seems desirable to have three enumerative induction principles, one for *anvayavyatirekyanumānas*, i.e., inferences with positive and negative inductive support ($\text{EI}^{\pm}$), one for *kevalānvayyanumānas*, i.e., inferences with only positive inductive support (EI^{+}), and one for *kevalavyatirekyanumānas*, i.e., inferences with only negative inductive support (EI^{-}).[6]

($\text{EI}^{\pm}$) If S is projectible with respect to H, then ⌜There is a set X of HâĂŹs, and all the members of X are S, and there is a set Y of things which are not S and they are all not H⌝ is a prima facie reason for ⌜All HâĂŹs are S⌝.

[6]Here are two examples which Navya-Naiyāyikas regard as shortcuts of valid inferences with only positive inductive support (1) and only negative inductive support (2):

(1) *ghaṭo 'bhidheyaḥ prameyatvāt.* ([23]: 40) – "The pot is nameable, because it is knowable."

(2) *pṛthivītarebhyo bhidyate gandhavattvāt.* (ibid.) – "Earth is different from other things (than earth), because it possesses smell."

Interpreting (1) and (2) as examples of inductive inferences seems rather odd at first sight. However, in section 3.2.3 we will see that such an interpretation stands to reason if we take into account the strong realist footing of Navya-Nyāya epistemology.

(EI$^+$) If S is projectible with respect to H, then ⌜There is a set X of HâĂŹs, and all the members of X are S⌝ is a prima facie reason for ⌜All HâĂŹs are S⌝.

(EI$^-$) If S is projectible with respect to H, then ⌜There is a set Y of things which are not S and they are all not H⌝ is a prima facie reason for ⌜All HâĂŹs are S⌝.

The projectibility constraint is actually well noted by Navya-Naiyāyikas. In Gaṅgeśa's Upādhivāda and in the UD there are long discussions about the possibility of admitting "being different from the inferential subject" (*pakṣetaratva*) as an inductively inferable property of all things exhibiting the probandum. If this property were projectible with respect to the probandum, it would undermine all sound inferences based on (EI$^\pm$). Since the probandum of such inferences is supposed to occur in other loci than the inferential subject as well, there will always be a set X of examples which exhibit the probandum and the property "being different from the inferential subject". Hence, by (EI$^+$) every locus of the probandum would be different from the inferential subject, i.e., the inferential subject would not be an instance of the probandum.

3.2.3 Pervasion as a Nomic Generalization

In Navya-Nyāya pervasion is an ontological relation. It can be an ordinary causal relation as in the case of the smoke/fire-example. Since the Navya-Naiyāyikas are realists, they also regard a pervasion relation as ontological in cases in which it is based on a genus-species relationship. Waterness, e.g., was supposed to be pervaded by substanceness: Whatever is a locus of waterness, i.e., whatever is water, is a locus of substanceness, i.e., it is a substance. In this case probans and probandum were conceived of as universals, the ontological correlates of the terms "water" and "substance".

"Realism is important along every dimension of Nyāya epistemology. A pervasion is held to obtain in nature, and for this reason, strictly speaking, one could be grasped from a single exhibited connection. For example, from a single instance of sight of smoke rising from fire, the pervasion, wherever smoke there fire, could be grasped, or from viewing a horse and a cow, one could know that a horse is not a cow and (to express the pervasion, here a negative one) whatever is a cow is not a horse and conversely. In other words, inductive generalization may proceed from a single instance, what we might call extrapolation (Fs as generally G from a single observation of an F as a G). However, repeated experience of the relation (along with no experience of an F as not a G) is said normally to be necessary for a firm impression to take hold in one's memory, an impression capable of informing or guiding an action including speech. In any case, recurrent experience normally means an increase in epistemic confidence." ([13]: 7f)

Since the relata of a genuine pervasion were always supposed to be grounded in the empirical world, all statements of pervasion relations could be regarded as positive statements that can be tested, amended or rejected by referring to the available evidence. Thus, all five-membered inferences could be regarded as inductive, not only those based on causal pervasion relations. According to some interpreters Gaṅgeśa even seems to have construed pervasion relations in general as causal relations, although in cases like, e.g., the pervasion of waterness by substanceness this seems possible only if we considerably extend our ordinary understanding of causality beyond the limits (cf. [13]: 11 and 99f and [3]: 217).

A pervasion relation is a law-like relation in the sense that due to ontological regularities every locus of the probans is, has always been and forever will be a locus of the probandum. The presence of the probans necessarily entails the presence of the probandum. In order to formalize this kind of invariable concomitance, we can rely on the standard way of expressing metaphysical necessity, viz. by means of an S5 $\Box$-operator. By prefixing it to (V_1^{pf}) we obtain the following more accurate prima facie formalization:

(V_2^{pf}) $\Box\forall x(Hx \to Sx)$

Pollock's formalization of a nomic generalization is a bit more complex, because it is designed to be applicable to certain cases which are probably not relevant to an adequate formal explication of the concept of pervasion, viz. so-called "counterlegal nomic generalizations", i.e., nomic generalizations whose antecedents are counterlegal (cf. [15]: 43 and [5]: 432f). Newton's First Law is an example:

(N) All objects unaffected by an outside force travel in a straight line at a constant velocity.

According to Newton's Law of Gravitation all objects exert a gravitational force on all other objects (albeit a small one for objects far from each other). Hence, there are no objects which are unaffected by an outside force. However, (N) is not vacuously true, because it is not equally true that an object which is unaffected by an outside force would not travel in a straight line at a constant velocity. Moreover, an adequate formalization of counterlegal nomic generalizations should be compatible with our intuition that "laws" like (N*) are false:

(N*) All objects unaffected by an outside force turn into a hippopotamus.

According to Pollock (N) and (N*) can be formalized as ...

$(\mathrm{N_f})\ \Diamond_p \exists x Fx > \Box_p \forall x(Fx \rightarrow Gx)$ and

$(\mathrm{N_f^*})\ \Diamond_p \exists x Fx > \Box_p \forall x(Fx \rightarrow Hx)$, where ...

... Fx, Gx and Hx translate into "x is unaffected by an outside force", "x travels in a straight line at a constant velocity" and "x turns into a hippopotamus", respectively, ">" is a counterfactual conditional, and "$\Diamond_p$" and "$\Box_p$" are S5 modalities translating into "It is physically possible that" and "It is physically necessary that", respectively (cf. [15]: 43).

Pollock regards nomic generalizations (including counterlegal ones) as counterfactuals, i.e., he reads (N) and (N*) as: "If objects unaffected by an outside force did possibly exist, it would necessarily be the case that they travel in a straight line at a constant velocity (or: that they turn into a hippopotamus)." According to this explication (N) can reasonably be regarded as true and (N*) as false.

The semantics of Pollock's counterfactual conditional ">" is similar to that of David Lewis' "$\Box\!\!\rightarrow$"-conditional: Pollock adopts Lewis' sphere conception in order to model the intuition that the accessible worlds are the most similar worlds at which the antecedent is true. However, according to Pollock similarity is cashed out in terms of consistency with the laws of nature, whereas for Lewis it is a comparative overall similarity which accounts for the fact that a world w' is regarded as more (or less) similar to a world w than a world w'' (cf. [15]: 41 and [8]: 91).

The idea of a counterlegal nomic generalization was by no means alien to the Naiyāyikas. Chakrabarti notes the following example: "An eternal entity that is independently productive is productive for ever." ([3]: 40) However, the instances of pervasion which Navya-Naiyāyikas take into account are normally non-counterlegal nomic generalizations. Gaṅgeśa, e.g., explicitly excludes counterlegal pervasions by including the constraint of existential import in his definition of pervasion, i.e., the requirement that the probans should be instantiated at the actual world (cf. [6]: 109). Hence, according to Gaṅgeśa pervasion should rather be formalized as ...

(V) $\Box \forall x(Hx \rightarrow Sx) \wedge \exists x Hx$

..., where the quantifiers should be understood in an actualist sense (cf. [5]: 249 and 252f). A possibilist reading of the existential quantifier in the second conjunct would strip the constraint of existential import of its ontological commitment, which is surely associated with it. Thus, an interpretation of (V) by means of a varying domain model (cf. [5]: 254f and [7]: 274f) seems to be preferable to a constant domain approach, which presupposes a single domain (consisting of actual as well as possible objects) as the range of quantification.

A formalization of pervasion as a nomic generalization which is equivalent to (V) can be obtained from (V) by replacing the first conjunct by $\Diamond\exists x Hx > \Box\forall x(Hx \to Sx)$, the corresponding formula for counterlegal and non-counterlegal nomic generalizations alike:

$$(\mathrm{V}')\ \underbrace{(\Diamond\exists x Hx > \Box\forall x(Hx \to Sx))}_{\leftrightarrow:\, H\Rightarrow S} \wedge \exists x Hx$$

Since the present modal framework is S5, $\exists x Hx \to \Diamond\exists x Hx$. Moreover, according to Pollock's Theorem (3.15) $\Diamond\exists x Hx \to ((H \Rightarrow S) \leftrightarrow \Box\forall x(Hx \to Sx))$ (cf. [15]: 44), where $H \Rightarrow S$ is used as an abbreviation of $\Diamond\exists x Hx > \Box\forall x(Hx \to Sx)$. The "$\Rightarrow$" between predicate symbols is Pollock's standard way of abbreviating nomic generalizations.

The utility of a (V)- or (V′)-like pervasion in an inferential situation becomes obvious if we take into account that in S5 $\Box\forall x(Hx \to Sx) \to \forall x(Hx \to Sx)$, or that nomic generalizations entail material generalizations, i.e., $(H \Rightarrow S) \to \forall x(Hx \to Sx)$ (cf. [15]: 36). Hence, a (V)- or (V′)-like pervasion and the second member of a five-membered inference entail that the inferential subject is a locus of the probandum. In a particular inferential situation the material implication $\forall x(Hx \to Sx)$ gives a clue to the probandum's occurrence on the inferential subject, since the latter is an element of the domain of existing objects. Thus, the following wff ($\mathrm{V}_{\mathrm{act}}$) can be regarded as the adaptation of the pervasion at issue to a contextually relevant application:

$$(\mathrm{V}_{\mathrm{act}})\ \forall x(Hx \to Sx) \wedge \exists x Hx$$

The quantifiers in ($\mathrm{V}_{\mathrm{act}}$) are to be understood in the same way as in (V) and (V′), i.e., in an actualist sense. Since ($\mathrm{V}_{\mathrm{act}}$) is a relativization of (V) and (V′) to the actual state of affairs, we will refer to ($\mathrm{V}_{\mathrm{act}}$) as "the actualized concept of pervasion". Fig. A is its representation as a Venn diagram:

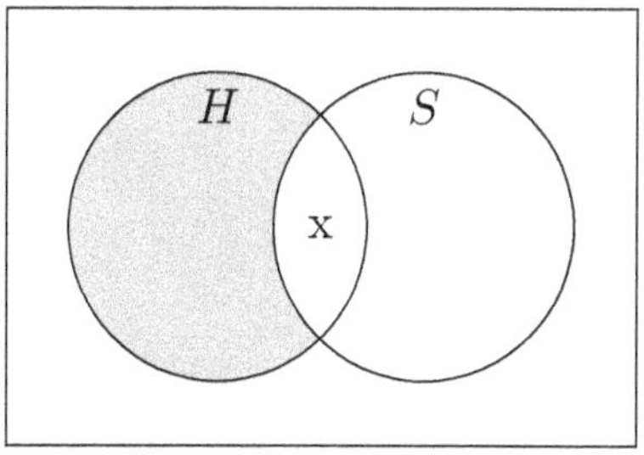

Fig. A

(V), (V′) and (V_{act}) are all conjunctions with the same conjunct $\exists x Hx$. Hence, the "non-pervasion" (*avyāpti*) of "being an H" by "being an S" might be said to apply to a case in which only this conjunct is false. From the perspective of the actual world, the region with a cross in Fig. A would have to be shaded and H would be an empty term then:

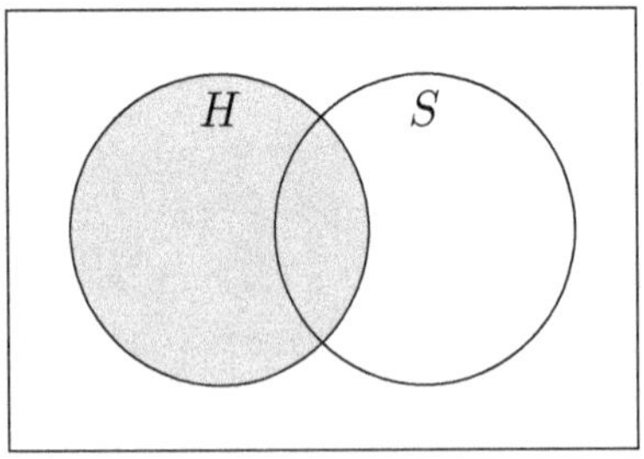

Fig. B

However, Navya-Naiyāyikas normally understand "non-pervasion" in the sense of a "deviation" (*vyabhicāra*) of the probans from the probandum, i.e., in the sense of the existence of a counter-example to the pervasion relation at the actual world. Thus, the non-pervasion of "being an H" by "being an S" can be expressed as the negation of (V_1^{pf}), i.e., as ...

(V_{neg}) $\exists x(Hx \wedge \neg Sx)$

..., where the quantifier should be read in an actualist sense, i.e., the domain $\mathcal{D}_{w_0}$ of objects at the actual world w_0 has to be regarded as the range of quantification, represented as a box in the corresponding Venn diagram Fig. C:

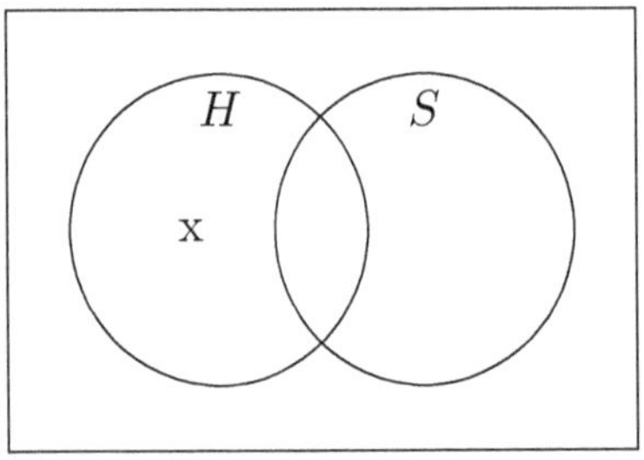

Fig. C

3.3 The Doctrine of *upādhi* in Navya-Nyāya

3.3.1 An Example

An *upādhi* is said to be a "vitiator" (*dūṣaṇa*) of assailable pervasions and, hence, of assailable inferences. However, the vitiating function is rather attributed to the *upādhi*'s absence in certain loci. Here is a classical example: Although smoke is supposed to be pervaded by fire, the putative pervasion of fire by smoke is unwarranted. The sample from which this pseudo-pervasion might be extrapolated turns out to be not a fair one if we find out that in the case of all hitherto observed instances of fire the co-occurrence with smoke was owing to the presence of an *upādhi*, viz. wet fuel. Wherever there is fire without wet fuel, as in the case of molten metal, there is no smoke. Of course, the identification of a counter-example would suffice to refute the universality of a putative pervasion. The *upādhi* is supposed to account for the existence of counter-examples: The reason why there are loci of fire without smoke is that fire produces smoke only in the presence of wet fuel.

3.3.2 The *upādhi*'s Function of Vitiating

In the above-mentioned stock example of an *upādhi* the absence of the *upādhi* in a locus of the probans entails the absence of the probandum. However, an *upādhi* may also vitiate in the sense that its absence in a locus of the probans is a reason to doubt the presence of the probandum. Two basic senses of "vitiating" have to be distinguished:

1.) Vitiating in the sense of "overruling": We will see that in the case of the so-called ascertained *upādhi* (*niścitopādhi*) the claim of its absence in certain loci can be regarded as a defeater in the sense of Pollock's theory. A putative pervasion or inference are overruled by means of an ascertained *upādhi*.

2.) Vitiating in the more general sense of "blocking an inference": A dubious *upādhi* (*saṃdigdhopādhi*) or rather its absence in certain loci vitiates in the sense that it prevents us from drawing a conclusion if there are equally good reasons for and against denying the thesis of the inference or the universality of a putative pervasion relation. Thus, the Navya-Naiyāyika's way of resolving epistemic ties dovetails with the behaviour of a skeptical reasoner who acknowledges epistemic ignorance and withholds belief rather than choosing randomly like a credulous reasoner.

On the surface, an *upādhi* seems to be similar to a so-called "undercutting defeater" in the sense of Pollock's theory of defeasible reasoning. Stephen Phillips translates *upādhi* as "undercutting condition", although he is well aware that there are differences between an *upādhi* and an undercutting defeater in the sense of Pollock's theory (cf. [13]: 14f): An *upādhi* is an object of the domain such as, e.g., wet fuel, whereas Pollock's defeater is a mental state. In most cases it is a belief. If we identify a belief with its propositional content, the equivalents of a defeater in the Navya-Nyāya doctrine of *upādhi* can be said to be certain propositions, which are true of an *upādhi*, and not all of them are undercutters. Some of them are rebutters and some of them are rebutters and undercutters with respect to different prima facie reasons. These defeaters can be gleaned from specifications of an *upādhi*'s vitiating function and from the UD's so-called "specific defining characteristic" (*viśeṣalakṣaṇa*) of an *upādhi*, which serves to determine whether a certain property is an *upādhi* relative to a specific pair of probans and probandum. In the strict sense of Pollock's terminology it is inappropriate to say that in the case of the putative pervasion of fire by smoke smoke is a prima facie reason, which is defeated by wet fuel, or that the property "looking red to me" is a prima facie reason for the property "being red", which is defeated by the property "absence of red light".[7] If we want to assimilate the theory of the *upādhi* to his theory of defeasible reasoning, it cannot be done in quite the same way as Stephen Phillips suggests.

A prima facie reason in Pollock's sense cannot be the probans of an assumed pervasion. There are actually two other candidates for a prima facie reason which Navya-Naiyāyikas seem to take into account and which can be said to be veritable prima facie reasons in Pollock's sense:

(R_1) There is evidential support for the pervasion at issue in the form of a sample $\{e_1, \ldots, e_n\}$ of positive or negative examples. (This claim of evidential support is part of the third member of a five-membered inference.)

(R_2) $\underbrace{\text{The inferential subject is a locus of the probans.}}_{\text{=the second member}} \wedge$

$$\underbrace{(\mathrm{R}_1) \wedge ((\mathrm{R}_1) \text{ is a prima facie reason for "All } H\text{'s are } S\text{".})}_{\text{=the third member}}$$

The pervasion at issue ("All H's are S") can be regarded as the conclusion drawn from (R_1). The thesis of an inference (Ss: "The inferential subject is a locus of the probandum.") can be regarded as the conclusion drawn from (R_2).

[7]The counterpart of wet fuel in the smoke/fire-example is the absence of red light in Pollock's example, since the absence of the absence of red light, i.e., the presence of red light, is a vitiator in his example, in the same way as the absence of wet fuel functions as a vitiator in the smoke/fire-example.

In the Navya-Nyāya doctrine of *upādhi* we find also genuine equivalents of a defeater in the sense of Pollock's theory. Let us look again at Pollock's "red light"-example: Apart from the defeater-proposition (marked as $\sqrt[D]{\ldots}$ below) it contains a piece of background information (marked as $\sqrt[B]{\ldots}$ below), which warrants the latter's functionality as a defeater:

(E0) "For example, somethingâĂŹs looking red to me may justify me in believing that it is red, but if I subsequently learn that $\sqrt[D]{\text{the object is illuminated by red lights}}$ and I know that $\sqrt[B]{\text{that can make things look red when they are not}}$, then I cease to be justified in believing that the object is red."

Similar propositions can be identified in Navya-Nyāya specifications of the vitiating function of an *upādhi*. Two propositions are involved in such specifications, viz. the statement of the *upādhi*'s absence in a locus of the probans, which in an inferential situation may be identical to the inferential subject (also marked as $\sqrt[D]{\ldots}$ below), and the statement that the absence of the *upādhi* in a locus of the probans entails the absence of the probandum (also marked as $\sqrt[B]{\ldots}$ below).

Gaṅgeśa defines an *upādhi* according to its "essence" (*svarūpa*) as follows:

(E1) *yad vā yaḥ sādhanavyabhicārī sādhyavyabhicāronnāyakaḥ, sa upādhiḥ.* ([4]: 13, 19f) – "$\sqrt[B]{\text{What indicates the deviation (of the probans) from the probandum}}$[8] if $\sqrt[D]{\text{it has the probans as something deviating from it}}$, that is an *upādhi*."

Gaṅgeśa cites also Maṇikaṇṭha, whose definition expresses basically the same: *anye tu yadvyāvṛttyā yasya sādhanasya sādhyaṃ nivartate, sa dharmas tatra hetāv upādhiḥ.* ([4]: 14, 30f) – "Others, however, [say]: 'That property on account of whose absence the probans is lacking the probandum is an *upādhi* for that probans.' " We can paraphrase this as follows:

(E2) "If $\sqrt[D]{\text{the } upādhi \text{ is absent from a locus of the probans}}$, $\sqrt[B]{\text{its absence there entails the absence of the probandum}}$."

The UD specifies the vitiating function of an *upādhi* with respect to an inferential context:

[8] According to the "If"-clause, i.e., the $\sqrt[D]{}$-part, this can be explicated as: "What indicates the absence of the probandum due to its own absence in a locus of the probans, ..."

(E3) *asmin pakṣa upādhāv avidyamāne tanniṣṭhavyāpyatvahetor viruddhavyabhicārisādhāraṇabhūtasādhyābhāvasādhakatvaṃ tadvyatirekasya paryavasyatīti* ([24]: 7b, 20f) – "If $\sqrt[\mathrm{D}]{\text{an } \textit{upādhi} \text{ is not present on this inferential subject}}$, then it (i.e., the property of being a vitiator) amounts to the fact that the absence of that (= the absence of the *upādhi* on the inferential subject) is like an incoherent deviating [pseudo-probans] and proves the absence of the probandum (from the inferential subject),[9] because of the pervadedness (by the absence of the probandum) resident in that (absence of the *upādhi*, i.e., because $\sqrt[\mathrm{B}]{\text{the absence of the } \textit{upādhi} \text{ is pervaded by the absence of the probandum in the range of the probans}}$)."

By means of first-order formalization techniques and the symbolization key ...

Hx: "x is a locus of the probans"
Sx: "x is a locus of the probandum"
Ux: "x is a locus of the *upādhi*"
s: "the inferential subject"

...we can render the $\sqrt[\mathrm{B}]{\ldots}$-proposition in (E1) – (E3) as ...

(B) $\forall x(Hx \wedge \neg Ux \rightarrow \neg Sx)$[10]

...and the $\sqrt[\mathrm{D}]{\ldots}$-proposition as ...

(D_1) $\exists x(Hx \wedge \neg Ux)$

...in the case of (E1) and (E2) and as ...

(D_2) $Hs \wedge \neg Us$

[9]The "incoherent" (*viruddha*) probans is a special type of "pseudo-probans" (*hetvābhāsa*). In Nyāya a probans is called "incoherent" if "it is in contradiction to something which the proponent has already accepted or is known to hold" ([12]: 98), such as the probans in the inference "Sound is eternal, because it is produced". This is "a *hetu* which is constantly accomapanied (sic!) by the absence of the *sādhya*" (ibid.: 115). The absence of the *upādhi* is related to the probandum in the same way as an incoherent probans is related to the probandum in an invalid inference. If the *upādhi* pervades the entire probandum, every locus of the absence of the *upādhi* is a locus of the absence of the probandum. According to Gaṅgeśa's defining characteristic of the *upādhi* (cf. section 3.3.3), the latter needs to pervade the probandum only in the range of the probans. In this case every locus of the absence of the *upādhi* in the range of the probans is a locus of the absence of the probandum.

[10](D_1) and (D_2) (cf. below) ensure that the quantification in (B) is not vacuous. Hence, (B) is an adequate rendering of the actualized concept of pervasion expressed in the $\sqrt[\mathrm{B}]{\ldots}$-proposition in (E3).

…in the case of (E3).

If (D_1) and (D_2) are conjoined with (B), they actually function as defeaters of the prima facie reasons (R_1) and (R_2). (D_1) and (D_2) rebut (R_1), because the conjunction of each of them with (B) entails $\exists x(Hx \wedge \neg Sx)$, which means that the conclusion drawn from (R_1) is false, i.e., that the purported pervasion relation does not hold:

$$\vdash \exists x(Hx \wedge \neg Ux) \wedge \forall x(Hx \wedge \neg Ux \rightarrow \neg Sx) \rightarrow \exists x(Hx \wedge \neg Sx)$$

$$\vdash Hs \wedge \neg Us \wedge \forall x(Hx \wedge \neg Ux \rightarrow \neg Sx) \rightarrow \exists x(Hx \wedge \neg Sx)$$

The rebutting function of (D_1) and (D_2) with respect to (R_1) can be represented as follows:

$$\underbrace{\text{claim of evidential support}}_{\text{prima facie reason } (R_1)} \rightsquigarrow \underbrace{\text{pervasion}}_{\text{conclusion}}$$

$\uparrow$ rebuts in combination with $\forall x(Hx \wedge \neg Ux \rightarrow \neg Sx)$

$$\exists x(Hx \wedge \neg Ux)/Hs \wedge \neg Us$$

(R_2) is rebutted by (D_2), because the conjunction with (B) entails $\neg Ss$, which means that the conclusion drawn from (R_2), viz. the thesis of the inference, is false:

$$\vdash Hs \wedge \neg Us \wedge \forall x(Hx \wedge \neg Ux \rightarrow \neg Sx) \rightarrow \neg Ss$$

(R_2) is undercut by (D_1), because the conjunction of (D_1) and (B) only implies that $\exists x(Hx \wedge \neg Sx)$. So, Ss may be true or false. (R_2) is just not a proper reason for Ss. The rebutting function of (D_1) and the undercutting function of (D_2) with respect to (R_2) can be represented as follows:

$$\underbrace{\text{second member} \wedge \text{third member}}_{\text{prima facie reason } (R_2)} \rightsquigarrow \underbrace{\text{thesis}}_{\text{conclusion}}$$

$\uparrow$ rebuts in combination with $\forall x(Hx \wedge \neg Ux \rightarrow \neg Sx)$

$$Hs \wedge \neg Us$$

$$\underbrace{\text{second member} \wedge \text{third member}}_{\text{prima facie reason } (R_2)} \rightsquigarrow \underbrace{\text{thesis}}_{\text{conclusion}}$$

$\uparrow$ undercuts in combination with $\forall x(Hx \wedge \neg Ux \rightarrow \neg Sx)$

$$\exists x(Hx \wedge \neg Ux)$$

The following chart outlines the ways in which (D_1) and (D_2) are related to (R_1) and (R_2):

	rebuts	undercuts
(D_1)	(R_1)	(R_2)
(D_2)	(R_1), (R_2)	–

(D_2) can only be a rebutter, whereas (D_1) can be a rebutter and an undercutter, but with respect to different prima facie reasons. Since (D_2) implies (D_1), (D_2) rebuts everything that (D_1) rebuts. If we understand an undercutter as a proposition which can be true, while the conclusion may be true or false, only (D_1) can be an undercutter, because (D_1) does not give a clue to the locus of the deviation of the probans from the probandum. Hence, (D_1) is not a reason to deny the conclusion drawn from (R_2).

Let us look at an example: We assume that an agent $\mathcal{A}$ observes fire on a distant mountain in the dark and concludes that there is also smoke hidden in the dark. $\mathcal{A}$'s pseudo-inference involves the pseudo-pervasion of fire by smoke. In this situation $\mathcal{A}$ might get to know from a reliable source that there is no wet fuel on the mountain, and $\mathcal{A}$ knows that the absence of wet fuel entails the absence of smoke. The fire on the mountain actually emanates from a red-hot iron ball. Thus, the proposition that the mountain is a locus of fire without wet fuel is a (D_2)-like rebutter, because it is for $\mathcal{A}$ a reason to deny (i) the thesis of the inference as the conclusion drawn from an (R_2)-like prima facie reason and (ii) the universality of the putative pervasion of fire by smoke as the conclusion drawn from an (R_1)-like prima facie reason.

Now, let us assume that $\mathcal{A}$ gets to know from a reliable source that there are loci of fire without wet fuel, and $\mathcal{A}$ knows that the absence of wet fuel entails the absence of smoke. However, $\mathcal{A}$ does not know whether the mountain is a locus of the absence of wet fuel. In this situation the proposition that there is a locus of fire without wet fuel is a (D_1)-like rebutter, because it is for $\mathcal{A}$ a reason to deny (ii). Moreover, it is a (D_1)-like undercutter, because it is for $\mathcal{A}$ a reason to deny that (R_2) would not be true unless the thesis of the inference were true. The mountain may well be a locus of the probans and there may well be a sample of positive or negative examples providing evidential support for the pervasion at issue. Nevertheless, the thesis of the inference may be false, because the sample providing evidential support might not be a fair one.

It is important to note that a (D_1)-like undercutter attacks only the connection between (R_2) and the conclusion, viz. the thesis of the inference. The latter may be true or false. Thus, (D_1) can also function as an undercutter with respect to a pseudo-inference whose thesis is undeniably true. If, e.g., an agent $\mathcal{A}$ argues that a pot is produced, because it is nameable like a piece of cloth etc., one might refute the argument by defeating $\mathcal{A}$'s choice of a sample for drawing the conclusion that whatever is nameable is also produced. Although the thesis of the inference is uncontroversial, $\mathcal{A}$'s sample is apparently not a fair one, since only things which are nameable *and non-eternal* are produced. The proposition expressing the absence of the property "being non-eternal" in a locus of the property "nameability" functions here as a (D_1)-like undercutter.

The types of rebutting and undercutting defeaters which we are distinguishing here correspond to different characterizations of an *upādhi*'s vitiating functions which are discussed by Gaṅgeśa in the Tattvacintāmaṇi. Gaṅgeśa prefers to characterize this vitiating function as "leading to the fallacy of the conditional probans (i.e., a type of pseudo-probans which only in combination with an *upādhi* allows us to infer the probandum)" (*vyāpyatvāsiddhyāpādaka*). This might be understood as a portrayal of (D_1)'s and (D_2)'s function of rebutting (R_1) and of (D_1)'s function of undercutting (R_2). Alternatively, these rebutting and undercutting functions might be characterized as "indicating a deviation" (*vyabhicāronnāyaka*), i.e., indicating that the purported universality of the pervasion at issue is unwarranted. Moreover, (D_2)'s function of rebutting (R_2) might be characterized as "amounting to [the affirmation of] the counter-thesis" (*pratipakṣaparyavasāna*).

It should be noted that the rebutting or undercutting function of (D_1) and (D_2) depends essentially on an appropriate background information. If (B) is true, (D_1) and (D_2) fulfill the criteria of a rebutting or undercutting defeater according to Pollock's definition. First of all, they are defeaters of (R_1) and (R_2), because ...

a) they are consistent with (R_1) and (R_2): Due to (B), the absence of the *upādhi* in a locus of the probans entails the absence of the probandum. But despite the absence of the probandum (from the inferential subject or from any other locus of the probans) there might be a sample of instances of a co-presence or co-absence of probans and probandum (in accordance with (R_1)). Moreover, despite the absence of the probandum (from the inferential subject or from any other locus of the probans) the inferential subject might be a locus of the probans and one might reasonably claim that a sample of instances of a co-presence or co-absence of probans and probandum provides inductive support for the pervasion at issue (in accordance with (R_2)).

b) (R_1) & (D_1) and (R_1) & (D_2) are no reasons to believe that all H's are S, and (R_2) & (D_1) and (R_2) & (D_2) are no reasons to believe that Ss, since due to (B) the absence of the *upādhi* in a locus of the probans entails the absence of the probandum.

Finally, (D_1) and (D_2) are rebutters for (R_1), because due to (B) they are a reason for believing that not all H's are S. (D_2) is a rebutter for (R_2), because due to (B) it is a reason for believing $\neg Ss$. (D_1) is an undercutter for (R_2), because due to (B) it is a reason to deny that (R_2) would not be true unless the conclusion Ss were true. (s might be a locus of the probans and there might be a sample X which provides positive or negative inductive support for the pervasion at issue. Nevertheless, the sample might not be a fair one, since the H's in X are S, only because they are all U. However, s might be an H which is not U and in that case $\neg Ss$ due to (B).)

It is important to note that (B) is sufficient, but not necessary for (D_1)'s and (D_2)'s functionality as defeaters. (D_2), e.g., could also function as an undercutter for (R_2) if the absence of the *upādhi* in a locus of the probans co-occurs with both, the absence and the presence of the probandum, i.e., in the case of ...

(B') $\exists x(Hx \wedge \neg Ux \wedge \neg Sx) \wedge \exists x(Hx \wedge \neg Ux \wedge Sx)$.

Due to (B') (D_1) is also a reason to deny that (R_2) would not be true unless the conclusion Ss were true. However, if the *upādhi* is related to probans and probandum in the sense of (D_1) and (B'), it is only a so-called "apparent *upādhi*" (*upādhyābhāsa*) according to Gaṅgeśa, since he insists that the absence of the *upādhi* in a locus of the probans entails the absence of the probandum. Although the Navya-Nyāya doctrine of an *upādhi* can be assimilated to Pollock's theory of defeasible reasoning in the manner outlined here, the latter is applicable to a far wider range of defeasible inferences. Thus, Pollock's example, the inference of "X is red" from "X looks red", is rather based on a defeasible rule of thumb than on an inductive generalization. Even if we regard "looking red" in this example as a probans and "being red" as a probandum, the object which comes closest to an *upādhi* here, viz. "the absence of red light", is only an apparent *upādhi*, since "looking red", "being red" and "the absence of red light" are related to each other in terms of (D_1) and (B'). (B) is not warranted in this case, because red light (= the absence of the apparent *upādhi*, i.e., the absence of the absence of red light) co-occurs with both, the absence of the probandum and the presence of the probandum: "Red objects look red in red light too." ([14]: 485)

3.3.3 The Definition of the Concept of an *upādhi*

The definition of the *upādhi* is a controversial topic in Navya-Nyāya. Let us compare (D_1) to the following adaptation of Gaṅgeśa's definition of the *upādhi* to the canonical form of a specific defining characteristic which defines an *upādhi*'s relation to probans and probandum in terms of pervasion and non-pervasion:

(G_1) The *upādhi* does not pervade the probans and

(G_2) it pervades the probandum in the range of the probans.[11]

If we understand pervasion in terms of the actualized concept of pervasion (cf. (V_{act})), (G_1) and (G_2) can be formalized as follows:

(G_1') $\exists x(Hx \wedge \neg Ux)$

(G_2') $\forall x(Hx \wedge Sx \rightarrow Ux) \wedge \exists x(Hx \wedge Sx \wedge Ux)$

(G_1') is identical to (D_1). In (G_2') the first conjunct is equivalent to (B), i.e., $\forall x(Hx \wedge \neg Ux \rightarrow \neg Sx)$. From (B) and ($G_1'$) we can conclude that $\exists x(Hx \wedge \neg Sx)$. Hence, the first conjunct of (G_2') warrants the rebutting or undercutting function of (G_1'). The second conjunct, which serves to warrant existential import for the universally quantified formula, is merely collateral to the formulation of the first conjunct of (G_2') in terms of the pervasion relation. Thus, Gaṅgeśa's specific defining characteristic of an *upādhi* implies that (G_1') is a (D_1)-like rebutting or undercutting defeater in the sense of Pollock's theory.

Fig. D below is a diagrammatic representation of this definition. The next diagram, Fig. E, represents the classical fire/smoke/wet fuel-example. It shows that in this case the *upādhi* fulfills the criteria (G_1) and (G_2).

[11]Gaṅgeśa actually formulates his defining characteristic in a different way: *yadvyabhicāritvena sādhanasya sādhyavyabhicāritvaṃ sa upādhiḥ.* ([4]: 12 [§6], 36 = [13]: 80) – "That is an *upādhi* due to whose deviation (from the probans) the probans deviates from the probandum." The difference between (G_1) and (G_2) and Gaṅgeśa's original formulation is, however, rather a matter of form than of content. Gaṅgeśa expresses (G_1) in terms of a "deviation" (*vyabhicāra*). As indicated by the instrumental *yadvyabhicāritvena*, this deviation entails the deviation of the probans from the probandum. Phillips understands Gaṅgeśa's formulation in the sense of $\forall x(Hx \wedge \neg Ux \rightarrow Hx \wedge \neg Sx)$, which is equivalent to ($*$) $\forall x(Hx \wedge Sx \rightarrow Ux)$ (cf. ibid. 80f). One might object that there is still a difference in meaning between ($*$) and (G_2), since ($*$) does not express the requirement of existential import which is associated with a pervasion. However, in the Upādhivāda-section where Gaṅgeśa distinguishes between ascertained and dubious *upādhis* he explicitly attaches the credibility criterion involved in this distinction to the non-pervasion condition and the pervasion condition in a defining characteristic of the *upādhi* (cf. [4]: 15 [§8], 7f). Hence, it seems plausible that he understands the entailment between the two deviations in his own defining characteristic in the sense of (G_2).

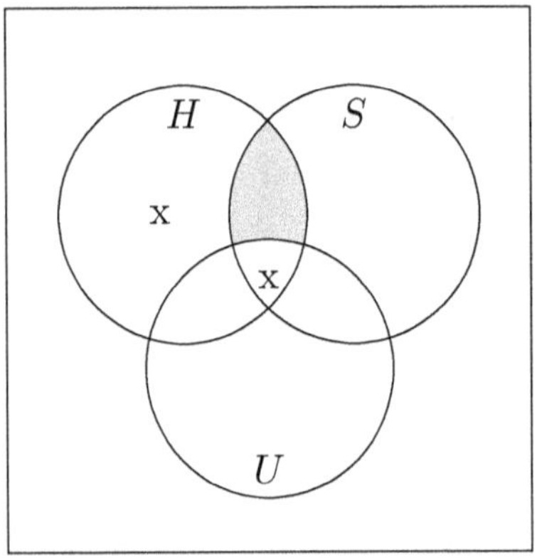

Fig. D

H = the class of loci of the probans
S = the class of loci of the probandum
U = the class of loci of the *upādhi*

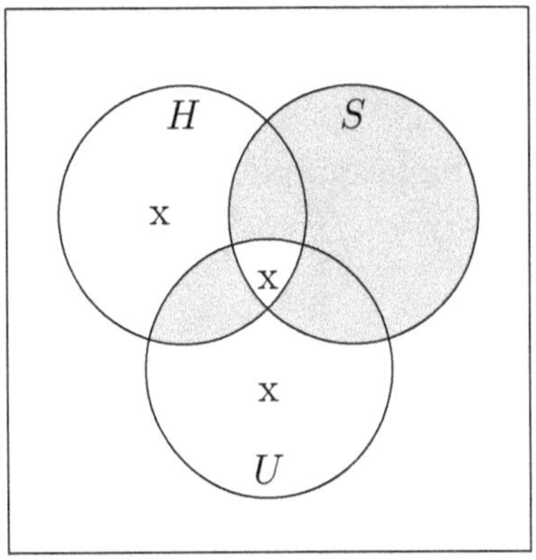

Fig. E

H = the class of loci of the probans "fire"
S = the class of loci of the probandum "smoke"
U = the class of loci of the *upādhi* "wet fuel"

molten metal $\in H \setminus (S \cup U)$; a locus of wet fuel which is not ignited $\in U \setminus (H \cup S)$; a locus of fire + smoke + wet fuel $\in H \cap S \cap U$

The fire/smoke/wet fuel-example matches also Udayana's specific defining characteristic of an *upādhi*:

(U_1) The *upādhi* does not pervade the probans and

(U_2) it pervades the probandum.

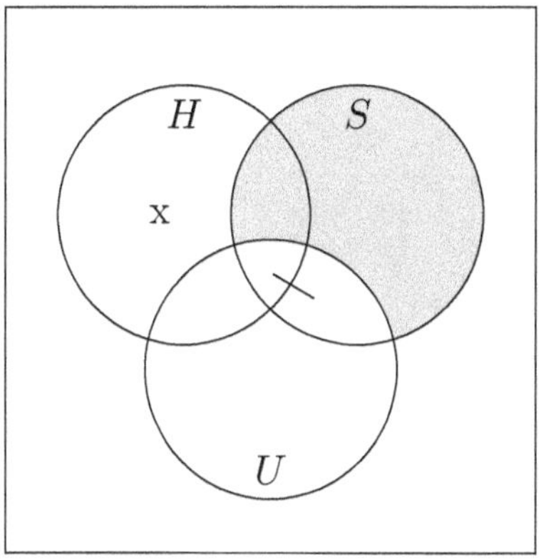

Fig. F[12]

In symbols:

(U_1') $\exists x(Hx \wedge \neg Ux)$ and

(U_2') $\forall x(Sx \rightarrow Ux) \wedge \exists x(Sx \wedge Ux)$

As indicated by the "x" $\in H \setminus S \cup U$ in Fig. F, (U_1) and (U_2) entail that $\exists x(Hx \wedge \neg Sx)$, and thus that an original apparent inference fails. In other words, the *upādhi* vitiates the apparent inference by showing that the required pervasion does not hold.

It is not difficult to see how Udayana might have derived his definition. In order to satisfy the condition of existential import, let us assume that the classes of the loci of the probans and the probandum, which we will symbolize as H and S, respectively, are non-empty. Then the material generalization involved in a pervasion can be expressed set-theoretically as:

$H \subseteq S$

Now, the equivalence ...

$H \subseteq S \leftrightarrow \forall U(S \subseteq U \rightarrow H \subseteq U)$

... is easily provable: The direction from left to right follows from the transitivity of the pervasion relation. The direction from right to left can be obtained by applying

[12]The bar indicates non-emptiness of a compound region (cf. [19]: 79f).

the rule of universal instantiation: If $S \subseteq U \rightarrow H \subseteq U$ is true of all U, it is especially true of S. Now, let us negate both sides of the equivalence:

$$\frac{H \subseteq S \leftrightarrow \forall U(S \subseteq U \rightarrow H \subseteq U)}{\therefore H \nsubseteq S \leftrightarrow \exists U(S \subseteq U \wedge H \nsubseteq U)}$$

Hence, there is no pervasion *iff* there is an *upādhi* in the sense of Udayana's definition.

For Gaṅgeśa this definition is too narrow, because it does not include cases where the *upādhi* does not pervade a probandum except in conjunction with a putative probans. Consider the following example of an assumed inference: The child to be born will be dark-complexioned, since it is MitrāâĂŹs child. ("Mitrā" is the name of a woman.) An appropriate *upādhi* would be "being prenatally nourished on vegetables". This is actually a dubious *upādhi*, since the non-pervasion-condition and the pervasion-condition are supposed to be uncertain. The example will be analyzed in greater detail in section 3.3.4. In order to simplify matters, we will treat both conditions here as if they were ascertained. Thus, we are to imagine that Mitrā has five sons who are all dark-complexioned, because Mitrā's diet during her previous pregnancies consisted of vegetables, which is a necessary and sufficient cause for a human baby's dark complexion. Only in the case of her sixth son Mitrā did not stick to her vegetable diet, so that he is not going to be dark-complexioned. It is then easy to see that the example matches Gaṅgeśa's definition, whereas it is beyond the scope of Udayana's definition, since in Fig. G an unbaked pot is $\in S \setminus (H \cup U)$, whereas in the Venn diagram for Udayana's definition the corresponding segment is shaded.

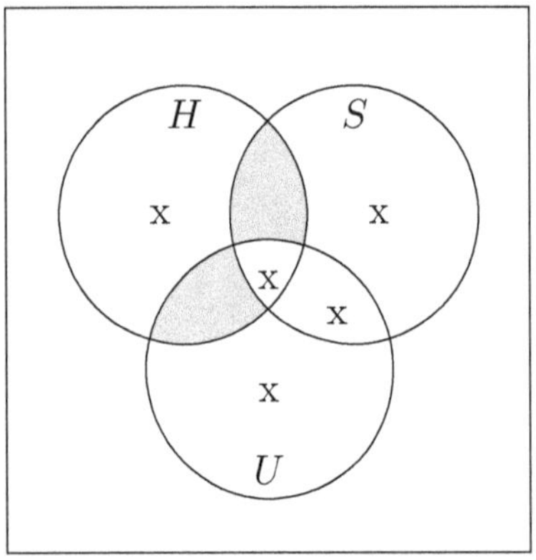

Fig. G

H = the class of loci of the probans *mitrātanayatva* ("being a son of Mitrā")

S = the class of loci of the probandum *śyāmatva* ("being dark-complexioned")
U = the class of loci of the *upādhi śākādyāhārapariṇatipūrvakatva* ("being prenatally nourished on vegetables etc.")

Mitrā's 6th son $\in H \setminus (S \cup U)$; Mitrā's first 5 sons $\in H \cap S \cap U$; an unbaked pot $\in S \setminus (H \cup U)$; someone else's dark-complexioned son prenatally nourished on vegetables $\in (S \cap U) \setminus H$; a white rabbit (*śvetaśaśa*) $\in U \setminus (H \cup S)$

In Fig. D and Fig. F the segment corresponding to $(H \cap U) \setminus S$ is not shaded. So, according to Gaṅgeśa's and Udayana's definitions the putative probans and the *upādhi* can co-occur in a locus where the probandum is absent. But this should not happen if the *upādhi* is supposed to be a corrector. In order to make sure that the *upādhi* functions as a corrector which secures the probandum in an assumed inference, the author of the UD adds a requirement to (G_2), the pervasion-condition of an *upādhi*: The *upādhi* should not only pervade the conjunction of probandum and putative probans (i.e., $\forall x(Hx \wedge Sx \rightarrow Ux)$, as in ($G_2$)). The conjunction of the *upādhi* and the putative probans should also be pervaded by the probandum (i.e., $\forall x(Hx \wedge Ux \rightarrow Sx)$). Thus, the pervasion relation between *upādhi* and probandum has to be symmetrical in the range of the putative probans. There should be a "coextensive pervasion" (*samavyāpti*) between both relata.

3.3.4 Ascertained vs. Dubious *upādhi*s

According to the UD an attack on (R_1) or (R_2) by means of the corresponding UD-like defeater is successful if it is based on an ascertained *upādhi*, i.e., if the *upādhi* at issue is known to fulfill the criteria of the specific defining characteristic. The proposition "There are loci of fire without wet fuel", e.g., actually rebuts the claim of evidential support consisting in a sample of co-occurring instances of fire and smoke as a prima facie reason for the putative pervasion of fire by smoke as the conclusion, because a locus like molten metal provides evidence for the existence of a locus of fire without wet fuel. So, the probans "fire" is actually not pervaded by the *upādhi* "wet fuel". Moreover, the probandum "smoke" is actually pervaded by the *upādhi* in the range of the probans, i.e., wherever there is both, fire and smoke, there is wet fuel, which is equivalent to: Wherever there is fire devoid of wet fuel, there is not smoke. This is the background information which in combination with the defeater "There are loci of fire without wet fuel" rebuts the prima facie reason for the pseudo-pervasion of fire by smoke.

A doubt may arise if the required *upādhi*'s absence in a locus of the probans is an open issue. Another occasion for doubt is the *upādhi*'s pervasion of the probandum. The Mitrā-example, of which we presented a simplified version above, is understood by Gaṅgeśa in such a way that both, the non-pervasion of the probans by the probandum and the pervasion of the conjunction of probans and probandum by the *upādhi*, are dubious (cf. [4]: 59 and [13]: 19f and 111f). It is assumed that eating vegetables during pregnancy is a sufficient cause for the baby's dark complexion. Hence, $(H \cap U) \setminus S = \emptyset$ in both diagrams in Fig. H, i.e., the pervasion of the *upādhi* by the probandum in the range of the probans is warranted. However, the vegetable diet is not a necessary condition for the baby's dark complexion. The dark complexion may as well be genetically determined by the father's dark complexion. Moreover, Mitrā may or may not have eaten vegetables while she was pregnant with her first five sons. Therefore, her first five sons, who are dark-complexioned, are $\in (H \cap S) \setminus U$ or $\in H \cap S \cap U$ (cf. the bar across the boundary between the corresponding areas in Fig. H, which indicates an alternative). We are also not sure about Mitrā's diet during her current pregnancy. Mitrā might not have eaten vegetables during her current pregnancy, or she has so far eaten vegetables, but will not stick to her diet, until she gives birth to her sixth son. Thus, the unborn sixth son may turn out to be fair-complexioned (cf. the leftmost diagram in Fig. H). If he turns out to be dark-complexioned (cf. the rightmost diagram in Fig. H), he is like the other sons $\in H \cap S$. Following a convention introduced by Shin, we connect the two diagrams by means of a straight line, which encodes disjunctive information. It is reminiscent of the bar across the boundary of two adjacent regions as a means to express that at least one of them is non-empty (cf. [21]).

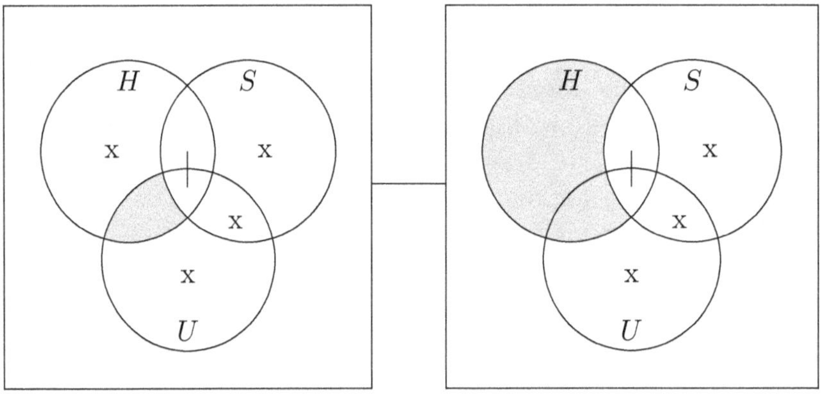

Fig. H

H = the class of loci of the probans *mitrātanayatva* ("being Mitrā's son")
S = the class of loci of the probandum *śyāmatva* ("being dark-complexioned")
U = the class of loci of the associate condition *śākādyāhārapariṇatipūrvakatva* ("being prenatally nourished on vegetables etc.")

Mitrā's 6th son is fair-complexioned and $\in H \setminus (S \cup U)$, or he is dark-complexioned and $\in H \cap S$; Mitrā's first 5 sons $\in H \cap S$; an unbaked pot $\in S \setminus (H \cup U)$; someone else's dark-complexioned son nourished on vegetables $\in (S \cap U) \setminus H$; a white rabbit $\in U \setminus (H \cup S)$

3.3.5 Credulous vs. Skeptical Reasoning

The UD agrees with Gaṅgeśa that if there is a tie between the alternatives involved in the doubt concerning the criteria of the defining characteristic of an *upādhi*, the latter can still vitiate. Thus, being faced with the alternatives between drawing the conclusion that Mitrā's unborn sixth son is dark-complexioned or not doing so, Gaṅgeśa and the UD hold that a rational agent would not choose randomly, i.e., they would distance themselves from what a credulous reasoner would do. Their position clearly coincides with that of a skeptical reasoner, i.e., they argue that the *upādhi* prevents us from drawing any conclusion. This is a situation which Pollock describes as "collective defeat" ([17]: 62). A typical example is the Nixon diamond, with two arguments "Nixon is a pacifist, because he is a Quaker" and "Nixon is not a pacifist, because he is a Republican". If there are no grounds for preferring one argument over the other, they intuitively defeat each other:

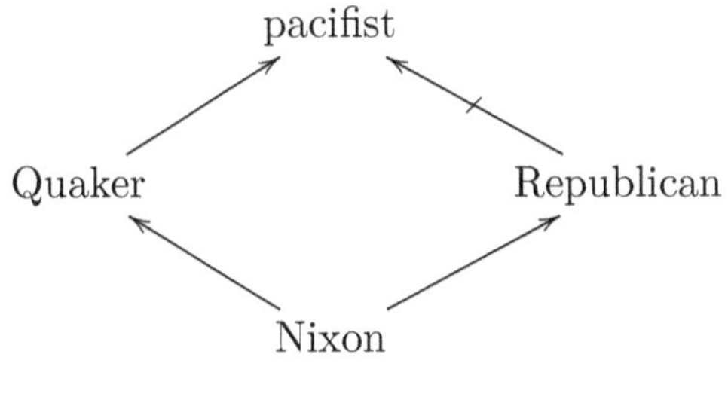

The Mitrā-example can also be understood along the lines of the situation of collective defeat. We will assume that Mitrā may well refrain from eating the vegetables during her current pregnancy and that not eating them may well have the effect that the baby turns out to be fair-complexioned. In this case the following diamond can be said to represent a collective defeat:

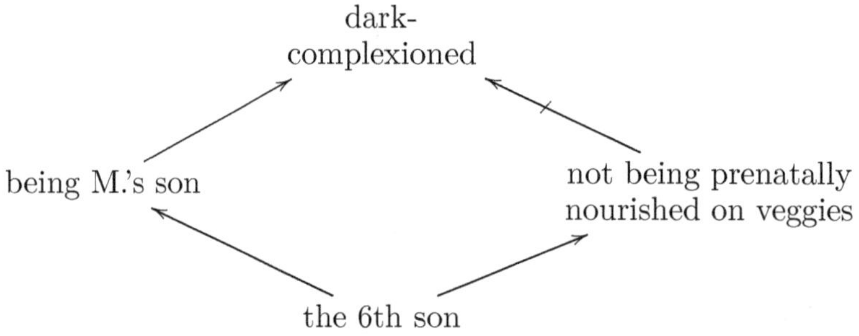

According to Gaṅgeśa and the UD, choosing randomly does not seem to be an option for responding to epistemic ignorance. Similarly, Pollock argues that a rational agent would rather resolve epistemic ties skeptically and withhold belief when there are equally good reasons for and against a conclusion: "The preceding considerations suggest that the controversy over skeptical and credulous reasoning stems from a confusion of epistemic reasoning with practical reasoning. In practical reasoning, if one has no basis for choosing between two alternative plans, one should choose at random. The classical illustration is the medieval tale of Buridan's ass who starved to death standing midway between two equally succulent bales of hay because he could not decide from which to eat. This marks an important difference between practical reasoning and epistemic reasoning. An agent making practical decisions must first decide what to believe and then use those beliefs in deciding what to do, but these are two different matters. If the evidence favoring two alternative hypotheses is equally good, the agent should record that fact and withhold belief. Subsequent practical reasoning can then decide what to do given that epistemic conclusion. In some cases it may be reasonable to choose one of the hypotheses at random and act as if it is known to be true, and in other cases more caution will be prescribed. But what must be recognized is that the design of the system of practical reasoning is a separate matter from the design of the system of epistemic reasoning that feeds information to the practical reasoner. Epistemic reasoning should acknowledge ignorance when it is encountered rather than drawing conclusions at random. This is what the principle of collective defeat mandates." ([17]: 64)

From the point of view of a skeptical reasoner an *upādhi* which is dubious in the sense that the *upādhi*'s non-pervasion of the probans or its pervasion of the probandum is dubious can still vitiate, insofar as it blocks an inference. However, $\exists x(Hx \wedge \neg Ux)$ and $Hs \wedge \neg Ss$ might not rebut if they are dubious. The rebutter should be a reason to believe the negation of the conclusion drawn from the rebutted prima facie reason. For a skeptical reasoner a doubt attached to these propositions incapacitates their rebutting force if there are equally good reasons for the conclusion that the probans is pervaded by the probandum or that the inferential subject is a locus of the probandum. Moreover, if $\exists x(Hx \wedge \neg Ux)$ is dubious and there are equally good reasons for the conclusion that the inferential subject is a locus of the probandum, a skeptical reasoner would not rely on $\exists x(Hx \wedge \neg Ux)$ as a reason to deny that (R_2) would not be true unless the thesis of the inference were true. So, for a skeptical reasoner $\exists x(Hx \wedge \neg Ux)$ is not an undercutter in this case.

Even if they are ascertained, the propositions $\exists x(Hx \wedge \neg Ux)$ and $Hs \wedge \neg Ss$ might not be able to function as defeaters, because their role as a rebutter or undercutter depends on the background information $\forall x(Hx \wedge \neg Ux \rightarrow \neg Sx)$, which follows from the second criterion of the defining characteristic. If the latter is dubious, there might again be equally good reasons for the conclusion that the probans is pervaded by the probandum or that the inferential subject is a locus of the probandum.

Thus, due to the dubiety of either of the two criteria of an *upādhi*'s defining characteristic, viz. its non-pervasion of the probans or its pervasion of the probandum, $\exists x(Hx \wedge \neg Ux)$ and $Hs \wedge \neg Ss$ may lose their rebutting or undercutting force. Nevertheless, the *upādhi* can still vitiate in a situation of collective defeat. Instead of overruling an inference, it blocks an inference, i.e., it prevents a skeptical reasoner from drawing any conclusion.

References

[1] Bealer, George 1982. *Quality and Concept.* Oxford: Clarendon Press.

[2] Bhattacharya, Dineshchandra 1958. *History of Navya-Nyāya in Mithilā.* Mithila Institute Series Studies 2. Darbhanga: Mithilā Institute of Post-Graduate Studies and Research in Sanskrit Learning.

[3] Chakrabarti, Kisor Kumar 2010. *Classical Indian Philosophy of Induction.* Lanham/Boulder/New York/Toronto/Plymouth: Lexington Books.

[4] Frauwallner, Erich 1970. *Die Lehre von der zusätzlichen Bestimmung (upādhi) in Gaṅgeśas Tattvacintāmaṇi.* Sitzungsberichte der Österreichischen Akademie der Wissenschaften, philosophisch-historische Klasse 266, 2 (= Veröffentlichungen der Kommission für Sprachen und Kulturen Südasiens 9). Wien/Köln/Graz: Böhlau.

[5] Garson, James W. 2013. *Modal Logic for Philosophers.* New York: Cambridge University Press.

[6] Goekoop, Cornelius 1967. *The Logic of Invariable Concomitance in the Tattvacintāmaṇi.* Dordrecht: Reidel.

[7] Hughes, George Edward, and Maxwell John Cresswell 2005. *A New Introduction to Modal Logic.* London/New York: Routledge.

[8] Lewis, David 1973. *Counterfactuals.* Malden, Massachusetts/Oxford: Blackwell.

[9] Matilal, Bimal Krishna 1990. *Logic, Language and Reality.* Indian Philosophy and Contemporary Issues. Delhi: Motilal Banarsidass.

[10] Matilal, Bimal Krishna 1998. *The Character of Logic in India*, ed. J. Ganeri and H. Tiwari. Albany, NY: State University of New York Press.

[11] Mill, John Stuart 1851. *A System of Logic, Ratiocinative and Inductive, Being a Connected View of the Principles of Evidence, and the Methods of Scientific Investigation. In Two Volumes.* Vol. 1. London: John W. Parker, West Strand.

[12] Pandeya, Raghavendra 1984. *Major Hetvābhāsas. A Formal Analysis (With Reference to Nyāya and Buddhism).* Delhi: Eastern Book Linkers.

[13] Phillips, Stephen H. 2002. *Gaṅgeśa on the Upādhi, the "Inferential Undercutting Condition". Introduction, Translation and Explanation (with N. S. Ramanuja Tatacharya).* Delhi: Indian Council of Philosophical Research.

[14] Pollock, John L. 1987. Defeasible Reasoning. Cognitive Science 11: 481 – 518.

[15] Pollock, John L. 1990. *Nomic Probability and the Foundations of Induction.* New York/Oxford: Oxford University Press.

[16] Pollock, John L. 1992. How to reason defeasibly. Artificial Intelligence 57: 1 – 47.

[17] Pollock, John L. 1995. *Cognitive Carpentry: A Blueprint for How to Build a Person.* Cambridge (Massachusetts)/London: MIT Press.

[18] Putnam, Hilary 2013. Comments on Russell Goodman's "Some Sources of Putnam's Pluralism", in: *Reading Putnam*, ed. M. Baghramian. 219 – 224. London/New York: Routledge.

[19] Quine, Willard van Orman 1966. *Methods of Logic.* New York/Chicago/San Francisco/Toronto: Holt, Rinehart and Winston.

[20] Saptapadārthī: *Śivāditya's Saptapadārthī with a Commentary by Jinavardhana Sūri.* Ed. J. S. Jetley. Lalbhai Dalpatbhai Series 1. Ahmedabad 1963: L. D. Institute of Indology.

[21] Shin, Sun-Joo, Oliver Lemon, and John Mumma 2018. Diagrams, in: *The Stanford Encyclopedia of Philosophy*, ed. E. Zalta. URL: `<https://plato.stanford.edu/archives/win2018/entries/diagrams/>`.

[22] Tarkabhāṣā: *Tarka-Bhāṣā of Keśava Miśra with the Commentary Tarkabhāṣāprakāśikā of Cinnaṃbhaṭṭa.* Ed. Devadatta Ramkrishna Bhandarkar and Kedarnath Sāhityabhūṣaṇa. Bombay Sanskrit and Prakrit Series 84. Bombay 1937: Aryabhushan Press.

[23] Tarkasaṃgraha: Tarkasaṃgraha of Annambhaṭṭa with the Autor's Own Dīpikā and Govardhana's Nyāyabodhini. Edited with Critical and Explanatory Notes by the Late Yashwant Vasudev Athalye, Together with Introduction and English Translation of the Text by Mahadev Rajaram Bodas. Revised and Enlarged Second Edition. Bombay Sanskrit Series 55. Poona 1930: Bhandarkar Institute Press.

[24] Upādhidarpaṇa: Bhandarkar Oriental Research Institute Manuscript No. 6. 1898âĂŞ99.

[25] Wada, Toshihiro 2007. *The Analytical Method of Navya-Nyāya.* Gonda Indological Studies XIV. Groningen: Egbert Forsten.

Received 03 February 2020

Exponentially Huge Natural Deduction proofs are Redundant: Preliminary results on $M_{\supset}$

Edward Hermann Haeusler*
Departamento de Informática, Pontifical Catholic University of Rio de Janeiro, PUC-Rio, Brazil
hermann@inf.puc-rio.br

Abstract

We estimate the size of a labelled tree by comparing the amount of (labelled) nodes with the size of the set of labels. Roughly speaking, an exponentially big labelled tree, is any labelled tree that has an exponential gap between its size, number of nodes, and the size of its labelling set. The amount of sub-formulas from a formula is linear on its size. Thus, exponentially big proofs have a size a^n, where $a > 1$ and n is the size of its conclusion. In this article, we show that any linearly height labelled tree whose size have an exponential gap with the size of their labelling set possess at least one sub-tree that occurs exponentially many times in them. Natural Deduction proofs and derivations in minimal implicational logic ($\mathbf{M}_{\supset}$) are essentially labelled trees. By the sub-formula principle any normal derivation of a formula α from a set of formulas $\Gamma = \{\gamma_1, \ldots, \gamma_n\}$ in $\mathbf{M}_{\supset}$, establishing $\Gamma \vdash_{\mathbf{M}_{\supset}} \alpha$, has only sub-formulas of the formulas $\alpha, \gamma_1, \ldots, \gamma_n$ occurring in it. By this relationship between labelled trees and derivations in $\mathbf{M}_{\supset}$, we show that any normal proof of a tautology in $\mathbf{M}_{\supset}$ that is exponential on the size of its conclusion has a sub-proof that occurs exponentially many times in it. Thus, any normal and linearly height bounded proof in $\mathbf{M}_{\supset}$ is inherently redundant. Finally, we briefly point out how this redundancy leads us towards a highly efficient compression method for propositional proofs. We also provide some examples that serve to convince us that exponentially big proofs are more frequent than one can imagine.

The author thanks L. Gordeev and L.C. Pereira for the fruitful discussion during the elaboration of this article. Special thanks to the referees. Their comments improved the paper, indeed.
*Sponsored by CNPq, CAPES and FAPERJ

1 Introduction

The estimation of the size of a labelled tree can proceed by comparing the amount of (labelled) nodes with the size of the set of labels. Roughly speaking, an exponentially big labelled tree, is any labelled tree that has an exponential gap between its size, number of nodes, and the size of its labelling set. Labelled trees can be the underlying structure of logical proofs, Natural Deduction (ND) proofs, for example. Thus, exponentially big labelled trees are the underlying structure of exponentially big proofs or derivations[1]. The exponentially big proof are natural candidates of what we can call "huge proofs", or "huge validity certificates" in the context of computational complexity. The minimal propositional purely implicational logic, denoted by $\mathbf{M}_{\supset}$ in this article, is $PSPACE$-complete for tautology checking. Moreover, $\mathbf{M}_{\supset}$ can polynomially simulate Classical and Intuitionistic provability. Two polynomial maps yield $\mathbf{M}_{\supset}$ formulas from formulas in the full language $\{\supset, \bot\vee, \wedge\}$ preserving Classic and Intuitionistic provability. See [17], or [12] for a general approach to propositional logics complexity and their relation to $\mathbf{M}_{\supset}$. Due to this, we can state that $\mathbf{M}_{\supset}$ is the hardest propositional logic among the three mentioned[2].

In conclusion, a good computational implementation of $\mathbf{M}_{\supset}$ provide us with good implementations for the other propositional logics. An efficient implementation should take into account the size of the proofs that it is dealing with. This investigation is important from the theoretical point of view, $NP = PSPACE$ is related to the question whether every $\mathbf{M}_{\supset}$ tautology has a short (polynomial) proof or not, while, any investigation on "huge proofs" in $\mathbf{M}_{\supset}$ can shed some light on how to deal with them, including problems of storing exponential objects in a computational environment. In this article, we show that any "huge proof", under some conditions that are not so restrictive, is highly redundant, i.e., has at least one subproof that repeats at least exponentially many times in it. As a consequence, we obtain an important result from Proof-Theory. Almost all article is devoted to proving this result. However, at section 6, we give an idea how, in [11], we can use the main result in this paper to obtain a compressing method that overtakes the superexponential feature of $\mathbf{M}_{\supset}$ by compressing every $\mathbf{M}_{\supset}$ proof into a subexponential DAG proof. This introductory section shows the main reasons for having the redundancy for huge proofs and a more technical justification on why to carry on an investigation on Natural Deduction in the realm of binary labelled trees.

[1] Following the usual terminology of proof-theory for Natural Deduction, a proof is a derivation that has no open assumption occurrence. Every assumption occurrence in the proof is discharged by some rule application in it.

[2] In [12] we show that it is the hardest propositional logic among the logics that satisfy a general form of the subformula principle.

The amount of sub-formulas from a formula α is linear on the size of α. A (super)exponentially big proof, for example, has a size at least a^n, where $a > 1$ and n is the size of its conclusion. In this article, we show that the linearly height labelled trees whose sizes have an exponential gap with the size of their labelling sets posses at least one sub-tree that occurs at least exponentially many times in them. The main reason for this is that Natural Deduction proofs and derivations in minimal implicational logic ($\mathbf{M}_{\supset}$) are essentially labelled trees. By the sub-formula principle any normal derivation of a formula α from a set of formulas $\Gamma = \{\gamma_1, \ldots, \gamma_n\}$ in $\mathbf{M}_{\supset}$, establishing $\Gamma \vdash_{\mathbf{M}_{\supset}} \alpha$, has only sub-formulas of the formulas $\alpha, \gamma_1, \ldots, \gamma_n$ occurring in it. By this relationship between labelled trees and derivations in $\mathbf{M}_{\supset}$, we show that any normal proof of a tautology in $\mathbf{M}_{\supset}$ that is exponential on the size of its conclusion has a sub-proof that occurs exponentially many times in it. Thus, any normal and linearly height bounded proof in $\mathbf{M}_{\supset}$ is inherently redundant.

Natural Deduction system, as conceived by Gentzen ([5]), is given by a set of rules that settle the concept of a deduction for some (logic) language. The system of Natural Deduction, as used and considered here, is determined by the logic language and this set of rules also called inference rules. Language and inference rules can be viewed as a logical calculus, as defined by Church ([1]). In contrast with the main formulations of logical calculus for some logics by Hilbert ([14]), Natural Deduction does not have axioms. Moreover, Natural Deduction implements in the level of the logical calculus the (meta)theorem of deduction, namely from $\Gamma, A \vdash A \supset B$, employing the discharging mechanism. The introduction rule for the $\supset$ application, as it follows, shows how this discharging mechanism implements in the logic calculus the deduction theorem.

$$\dfrac{\begin{array}{c}[A]\\ \Pi \\ B\end{array}}{A \supset B}\ \supset\text{-Intro}$$

We embrace, with []s, some of the top-formula occurrences A in a derivation Π means that we are discharging these occurrences from Π. To embrace formula occurrences in a proof means that from the application of the $\supset$-Intro rule applied down to the conclusion of the derivation, the inferred formulas do not depend anymore on these embraced occurrences of A. The choice of which formula occurrences an application of a $\supset$-Intro embraces, as a consequence of an application of a $\supset$-intro rule, is arbitrary. The range of this choice goes from every occurrence of A until none of them. The derivations in figure 1 show two different ways of deriving $A \supset (A \supset A)$. Observe that in both deductions or derivations, we use numbers to indicate which is the application of the $\supset$-Intro that discharged the marked formula occurrence. For example, in the right derivation, the upper application discharged the marked occurrences of A, while in the left derivation, it is the lowest application that discharges the formula occurrences A. There is a third derivation that both applications do not discharge any A, and the conclusion $A \supset (A \supset A)$ keep depending on A. This third alternative appears in figure 2. Natural Deduction systems can provide logical calculi without any need to use axioms. In this article, we focus on the system formed only by the $\supset$-Intro rule and the $\supset$-Elim rule, as shown below, also known by modus ponens. The logic behind this logical calculus is the purely minimal implicational logic, $M_{\supset}$.

$$\frac{A \qquad A \supset B}{B}\ \supset\text{-Elim}$$

One thing to observe is that we can substitute liberal discharging mechanism by a greedy discipline of discharging that discharges every possible formula occurrence whenever the $\supset$-Intro is applied. Observe that, in this case, the derivation in figure 2 would not be possible anymore. Completeness regarding derivability would be lost. However, when considering proofs, i.e., derivations with no assumption undischarged, the greedy version of the $\supset$-Intro is enough to ensure the demonstrability of valid formulas.

$$\dfrac{\dfrac{[A]^1}{A \supset A}}{A \supset (A \supset A)}\,{\scriptstyle 1} \qquad\qquad \dfrac{\dfrac{[A]^1}{A \supset A}\,{\scriptstyle 1}}{A \supset (A \supset A)}$$

Figure 1: Two diferent derivations that discharge assumptions in different ways

$$\dfrac{\dfrac{A}{A \supset A}}{A \supset (A \supset A)}$$

Figure 2: Two vacuous $\supset$-Intro applications

With the sake of providing simpler proofs of our results, we take Natural Deduction as trees. From any derivation in ND, there is a binary tree having nodes labelled by the formulas and edges linking premises to conclusion, such that the root of the tree would be the conclusion of the derivation, and the leaves are its assumptions. For example, the derivation in figure 3 has the tree in figure 4 representing it. The set of labels (formulas) that label formula of u depends on the label formula of v labels the edge from a node v to a node u. This set of formulas is called the dependency set of the label of u from the label of v. In this way, the $\supset$-intro, in fact, its greedy version, removes the discharged formula from the dependency set, as shown in figure 4. Note that due to this labelling of edges by dependency sets, we need one more extra edge and the root node. The dependency set of the conclusion labels this new edge. That is the reason for the edge linking the conclusion to the dot in figure 4.

$$1\;\frac{\dfrac{\dfrac{[A]^1 \qquad A \supset B}{B} \qquad B \supset C}{C}}{A \supset C}$$

Figure 3: A derivation in $M_{\supset}$

As a matter of computational representation of ND proofs as trees, we use bitstrings induced by an arbitrary linear ordering of formulas in order to have a more compact representation of the dependency sets. Taking into account that only subformulas of the conclusion can be in any dependency set, we only need bitstrings of the size, i.e. length, of the formula that it is the conclusion of the proof. In figure 4b we show this final form of the tree representing the derivation in figure 3 and 4a when the linear order $\prec$ is $A \prec B \prec C \prec A \supset B \prec B \supset C \prec A \supset C$.

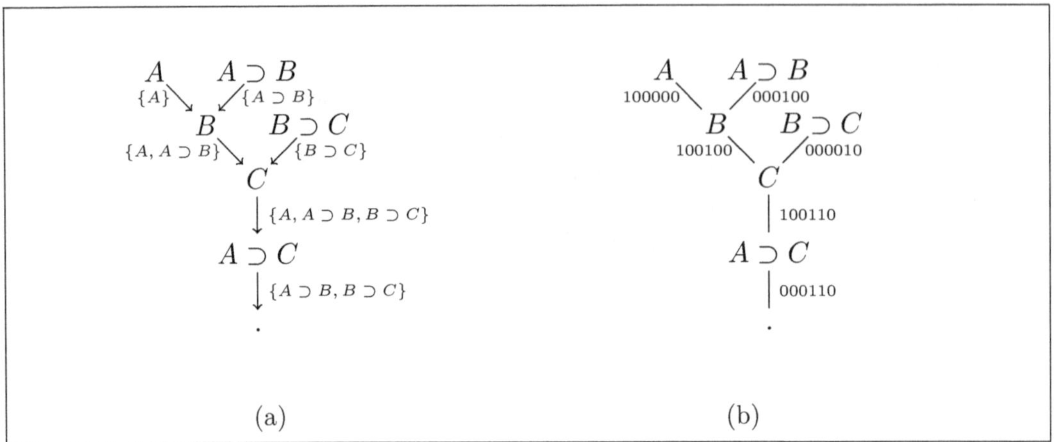

Figure 4: (a) The tree representing derivation in figure 3 and (b) the tree with bitstrings representing the same derivation.

The previous paragraphs describe how to get rid of the need of explicitly representing the discharge function relative to $\supset$-Intro applications in Natural Deduction proofs for $\mathbf{M}_{\supset}$. We consider the representation of Natural Deduction proofs in $\mathbf{M}_{\supset}$ using labelled trees with bitstrings labelling the edges. In this article, we consider this representation, such that, the results obtained for labelled trees can be extended to Natural Deduction proofs accordingly. Without any generality loss, in what follows, most of the results will be stated and proved for labelled trees. The mention of Natural Deduction proofs and derivation will be explicit only when the result is worth for logic or proof-theory.

Proof-theory is the branch of logic, the foundation of Mathematics and Computer Science that studies proofs. It has a well-established set of results[3] and tools devoted to formal proofs and consistency proofs of formalized mathematical theories and metatheories. However, because of the scope of this paper, we only briefly list the minimal results and definitions in the next section, such that the reader can make the connections between trees and Natural Deduction proofs needed to understand the main result of this article. We briefly explain, in the next paragraph, the intuition that motivates our result.

The Natural Deduction system for $\mathbf{M}_{\supset}$ satisfies the subformula principle. It states that if $\Gamma \vdash_{\mathbf{M}_{\supset}} \alpha$ then there is a derivation in Natural Deduction, for $\mathbf{M}_{\supset}$, of

[3]The proof of the consistency of Arithmetic by Gentzen in the '30s is one of the champion results.

α from $\Gamma' \subseteq \Gamma$ and every formula in this derivation is either a subformula of α or some $\gamma \in \Gamma'$. In the particular case of α being a tautology, there is a proof, i.e. a derivation without open assumptions, of α having only occurrences of sub-formulas of α in it. The subformula for Natural Deduction is a corollary of the Normalization theorem for Natural Deduction (see [16], page 42, corollary I) a central result, and tool of proof-theory. Well, $\mathbf{M}_{\supset}$satisfies the normalization and hence the sub-formula principle. The normalization for $\mathbf{M}_{\supset}$is a particular case of the normalization for Classical, and Intuitionistic logic as well. Only $\supset$-reductions are applied in a non-normal derivation to obtain a normal one. The inductive measure and strategy of critical derivation used in the Classical Logic case are enough, see Theorem 2, page 4, in [16].

We note that the amount of sub-formulas of any formula is linear on its size. Instead of the denomination 'exponentially big proofs' we use the denomination 'huge proofs' merely. We consider a proof huge when its size is larger than or equal to any exponential on the size of its conclusion. However, this is not the correct denomination[4]. Thus, if a proof is exponentially big, i.e., **huge**, its corresponding labelled tree is also exponential. That is, it is at least of size a^n, $a > 1$ and n is the size/length of the formula[5] that labels its root. We remind that each sub-formula is a possible label node in the tree. We have then that an exponentially big normal proof of size a^n, $a > 1$ is labelled with n labels only. We remember that n is the size/length of the string that labels its root or conclusion. This configuration allows us to say that at least one label repeats exponentially many times in the tree under the additional consideration that the tree is linearly-height bounded. We show that this repetition happens in a way that a sub-tree repeats exponentially many times.

The additional hypothesis on the linear bound on height of the proof of $\mathbf{M}_{\supset}$tautologies can be taken into account without loss of generality if we consider the complexity class $CoNP$ (see appendix A). Moreover, in [7], we show that any tautology in $\mathbf{M}_{\supset}$has a Natural Deduction normal proof of height bound by the square of the size of this tautology. However, it is not easy to extend the reasoning in this article to the case of polynomially height-bound trees.

In the next section, Section 2, we provide the main terminology and definitions used in the article. Section 3 presents some useful properties regarding normal proofs and syntax trees of formulas and their size. Section 4 examines three examples of classes of huge proofs and shows concrete cases of the redundancy in huge proofs.

[4]If we follow Cook-Karp conjecture that says that computationally easy to verify and compute objects are of polynomial-size, huge proofs include the hard proofs for verification, namely, the super-polynomial ones.

[5]The size/length of a formula is the length of it viewed as a string, the number of occurrences of characters in it.

These examples serve to convince us that exponentially big proofs are more frequent than we can imagine. Section 5 states and proves the main lemma. Section 6 provides the sketch on the compression method and its main properties. Finally, section 7 discusses the consequences of what we show here and the method for compressing propositional proofs.

2 Terminology and definitions

In this article the notation $i = 1, k$ denotes $i \in \{1, 2, \ldots, k\}$. Following the usual terminology in Natural Deduction and proof-theory, we briefly describe what we use in this article.

The left premise of a $\supset$-Elim rule is called a minor premise, and the right premise is called the major premise. We should note that the conclusion of this rule, as well as its minor premise, are sub-formulas of its major premise. We also observe that the premise of the $\supset$-Intro is the sub-formula of its conclusion. A derivation is a tree-like structure built using $\supset$-Intro and $\supset$-Elim rules. We have some examples depicted in the last section. The conclusion of the derivation is the root of this tree-like structure, and the leaves are what we call top-formulas. A proof is a derivation that has every top-formula discharged by a $\supset$-Intro application in it. The top-formulas are also called assumptions. An assumption that it is not discharged by any rule $\supset$-Intro in a derivation is called an open assumption. If Π is a derivation with conclusion α and $\delta_1, \ldots, \delta_n$ as all of its open assumptions then we say that Π is a derivation of α from $\delta_1, \ldots, \delta_n$.

Definition 1. *A branch of a derivation or proof* Π *is any sequence* $\beta_1, \ldots, \beta_k$ *of formula occurrences in* Π*, such that:*

- β_1 *is a top-formula, and;*
- *For every* $i = 1, k-1$*, either* β_i *is a* $\supset$*-Elim major premise of* β_{i+1} *or* β_i *is a* $\supset$*-Intro premise of* β_{i+1}*, and;*
- β_k *either is the conclusion of the derivation or the minor premise of a* $\supset$*-Elim.*

A normal derivation/proof in $\mathbf{M}_{\supset}$is any derivation that does not have any formula occurrence that is simultaneously a major premise of a $\supset$-Elim and the conclusion of a $\supset$-Intro. A formula occurrence that is at the same time a conclusion of a $\supset$-Intro and a major premise of $\supset$-Elim is called a maximal formula. In [16] there is the proof of the following theorem for the Natural Deduction for the full[6]

[6]The full propositional fragment is $\{\vee, \wedge, \supset, \neg, \bot\}$.

propositional fragment of minimal logic. The proof of this theorem uses a strategy of application of a set of reduction rules that eliminates maximal formulas. It is out of the scope of this article to provide more details on the proof of the normalization theorem. Except for the observation that the normalization proof for the full language of minimal logic, as we can found in [16], obtains a normalization for the language with $\supset$ only by solely restricting the reductions to $\supset$. The inductive measure keeps unchanged.

Theorem 1 (Normalization). *Let* Π *be a derivation of* α *from* $\Delta = \{\delta_1, \dots, \delta_n\}$. *There is a normal proof* Π' *of* α *from* $\Delta' \subseteq \Delta$.

In any normal derivation/proof, the format of a branch is essential and provides worth information on why huge proofs are redundant, as we will see in the next sections. Since no formula occurrence can be a major premise of $\supset$-Elim and conclusion of a $\supset$-Intro rule in a branch we have that the conclusion of a $\supset$-Intro can only be the minor premise of a $\supset$-Elim, premise of an $\supset$-Intro or it is not a premise at all. In this last case, it is the conclusion of the derivation. In any case, it is the last formula in the branch. Any conclusion of a $\supset$-Intro, if it is a premise of an $\supset$-Elim rule, it is the minor premise of this rule, and hence the last formula in the branch, otherwise it is the premise of an $\supset$-Intro. Hence, any branch in a normal derivation is divided into two parts (possibly empty). The Elim-part begins the branch with the top-formula and, every formula occurrence in it is the major premise of a $\supset$-Elim. There is a formula occurrence that is the conclusion of a $\supset$-Elim and can be premiss of a $\supset$-Intro rule that is called minimal formula of the branch. The minimal formula begins the I-part of the branch. In the I-part, every formula is the premise of a $\supset$-Intro, except for the last formula of the branch. From the format of the branches, we can conclude that the sub-formula principle holds for normal proofs in Natural Deduction for $\mathbf{M}_{\supset}$, in fact, for many extensions of it.

Corollary 2 (Sub-formula principle). *Let* Π *be a normal derivation of* α *from the set* $\Delta = \{\delta_1, \dots, \delta_m\}$. *It is the case that for every formula occurrence* β *in* Π, β *is a sub-formula of either* α *or of some of* δ_i.

This corollary ensures that without loss of generality, any Natural Deduction proof of a $\mathbf{M}_{\supset}$tautology has only sub-formulas of it occurring in it. Normal proofs/derivations offer the *EOL*-tree abstraction in forms of the trees associated with derivations in Natural Deduction for $\mathbf{M}_{\supset}$as we show in the sequence. The definition of *EOL*-tree facilitates the proof of the main result of this article. With labelled trees, we can focus on the combinatorial aspects rather than the proof-theoretical ones.

We assume the standard definition of a tree and a (possibly) incomplete binary tree. The size of a, possibly incomplete, tree $\langle V, E\rangle$ is $|V|$, the number of vertexes of the tree. As a tree is a simple graph, the number of edges is upper-bounded by $|V|^2$. The root of a tree is the unique node $r \in V$, such that, there is no $v \in V$, such that $\langle v, r\rangle \in E$. Given a tree $\mathcal{T} = \langle V, E\rangle$, the level of the node v, $lev(v)$, is the number of nodes in the path from v to the root of the tree. This can be defined in a recursive/inductive way as: (basis) $lev(r) = 0$; (rec) if $lev(u) = n$ and $\langle u, v\rangle \in E$ then $lev(v) = n + 1$.

Natural Deduction derivation trees for $\mathbf{M}_{\supset}$inspire the following definition of trees. When applying lemma 11 below, we can think of Natural Deduction derivation trees in $M_{\supset}$. We reinforce that a vertex with two children plays the role of an instance of a $\supset$-Elimination role application having it as the conclusion, and a vertex with one child plays the role of an instance of $\supset$-Introduction with it as the conclusion. The leaves are either hypothesis, also called open assumptions, of the derivation or discharged assumptions. In this section scope, we remember that there is no representation for the discharging function attached to each instance of $\supset$-Introduction application. In the concrete case, the labels of the nodes, set B below, are formulas and the order may be the sub-formula ordering between propositional implicational formulas. We advise the reader to not confuse this partial ordering abstracted from the sub-formula ordering, with the linear and arbitrary ordering that the bitstrings mentioned at the introduction uses. For the sake of simplicity, we consider an arbitrary partial order in the main results below. Considering an arbitrary partial order facilitates the reading and lastly can be applied to the concrete.

If A is a set, then we use $card(A)$ to denote the number of elements in A.

Definition 2 (EOL-Binary tree). *An edge-ordered-labelled binary tree $\mathcal{T}$ is a structure $\langle V, E_L \cup E_R \cup E_U, \ell, B\rangle$, where:*

1. *$\langle V, E_L \cup E_R \cup E_U\rangle$ is a (possibly incomplete) binary tree and;*
2. *$\ell : V \to B$ is the labeling function, with B a finite and partially ordered set of labels, with a partial operation $\odot$ and;*
3. *E_L, E_R and E_U are mutually disjoint, and;*
4. *Whenever $\langle v, v_1\rangle \in E_R$ and $\langle v, v_2\rangle \in E_R$ then $v_1 = v_2$;*
5. *Whenever $\langle v, v_1\rangle \in E_L$ and $\langle v, v_2\rangle \in E_L$ then $v_1 = v_2$;*
6. *$\langle v, v_1\rangle \in E_L$, if and only if, $\langle v, v_2\rangle \in E_R$, $v_2 \neq v_1$, and;*

7. *If* $\langle v, v' \rangle \in E_U$ *and* $\langle v, v'' \rangle \in E_U$ *then* $v' = v''$, *and;*

8. *If* $\langle v, v_1 \rangle \in E_L$ *and* $\langle v, v_2 \rangle \in E_R$ *then* $\ell(v) \prec_B \ell(v_2)$ *and* $\ell(v_1) \prec_B \ell(v_2)$ *and for each* $\ell(v_2)$, *such that,* $\ell(v) \prec_B \ell(v_2)$, *there is only one* $b \in B$, *such that* $b \odot \ell(v)$ *is equal to* $\ell(v_2)$. *This* b *should be equal to* $\ell(v_1)$*;*

9. *If* $\langle v, v' \rangle \in E_U$ *then there is* $q \in B$, *such that,* $q \odot \ell(v') = \ell(v)$ *and, of course,* $\ell(v') \prec_B \ell(v)$

Given a tree T, the height of T, denoted as $h(T)$, is the length of the longest path linking a leaf of T to the root of T. In this article, we are interested in the kind of EOL-tree that is linearly-bounded on the height regarding the size of its labelling set. That is, T is linearly bounded on the height when $h(T) \leq k \times card(B(T))$, for some $0 < k \in \mathbb{R}$ where $h(T)$ is the height of the tree T[7]. We call these trees linearly height B-labelled trees or linear-height EOL-trees. In section 5, we prove that for any linearly-height B-labelled tree of exponential or bigger size, there is a tree that occurs exponentially many times as a subtree in it. The content of lemma 11 states this. In the sequel, we provide more definitions that we use. In section 5, we also comment on the non-triviality of the extension of this main result to polynomially height bounded EOL-trees.

A skeletal-tree is defined as any non-empty B-labelled edge-labelled tree with edge labels U, L and R. In the sequel, we ask the reader to remind herself (himself) the concept of injective tree-mapping from labelled trees into labelled trees (see any book on graph theory or theory of computation, for example, [4]).

Definition 3 (Skeletal-tree occurrence). *A skeletal-tree instance* $\mathcal{S}$ *occurs in a* B*-labelled tree* $\mathcal{T} = \langle V, E, l, B \rangle$, *iff, there is an injective* B*-labelled tree mapping* f *from* $\mathcal{S}$ *into* $\mathcal{T}$, *such that, for each* $v, u \in V_{\mathcal{S}}$, *if* $\langle v, u \rangle \in E_\lambda$ *then* $\langle f(v), f(u) \rangle \in E_\lambda$, $\lambda = U, L, R$*;* $\ell(v) = \ell(f(v))$ *and* $\ell(u) = \ell(f(u))$.

Whenever a skeletal tree instance $\mathcal{S}$ occurs in a tree $\mathcal{T}$, we say that there is a sub-tree of the skeletal form $\mathcal{S}$ in the tree. We also say simply that $\mathcal{S}$ is a sub-tree occurring in $\mathcal{T}$. We can conclude that any sub-tree of an instance of a skeletal occurring in a tree is also a skeletal sub-tree instance of this first tree. A skeletal-tree occurrence/instance is full whenever if it is not possible to extend it to other sub-tree by adding any contiguous vertex from the tree to it. Sometimes we use the term "Skeletal-tree instance', instead of "skeletal-tree occurrence". When a Skeletal-tree instance $\mathcal{Y}'$ of $\mathcal{Y}$ occurs in a tree $\mathcal{T}$ and $\mathcal{Y}$'s root is at level k then we say that $\mathcal{Y}$ occurs at level k in $\mathcal{T}$.

[7]We can say also that $h(T) \in \mathcal{O}(B(T))$.

Concerning the computational complexity of propositional proofs, we count the size of proof as to the number of symbol occurrences used to write it. If we put all the symbol occurrences used to write a Natural Deduction derivation Π side by side in a long string then the size of the derivation, denoted by $|\Pi|$, is the length of this string. The function $|\,|$: *Strings* $\longrightarrow \mathbb{N}$, the size-of-string function, denotes the mapping of strings to their corresponding sizes[8]. For derivations α from $\Delta = \{\delta_1, \ldots, \delta_n\}$ we estimate the complexity of the derivation by means of a function of $|\alpha| + \sum_{i=1,n} |\delta_i|$ into the size of the derivation itself. Thus, we should not take the complexity of a derivation individually. It is taken together with the set of all derivations.

A set $\mathcal{S}$ of EOL-trees is unlimited, if and only if, for every $n > 0$ there is $T \in \mathcal{S}$, such that, $|T| > n$.

Definition 4. *An unlimited set $\mathcal{S}$ of EOL-trees is huge (exponentially big or $\mathcal{EB}$ for short) iff there are $a \in \mathbb{R}$, $a > 1$, $n_0, p \in \mathbb{N}$, $p > 1$, $c \in \mathbb{R}$, $c > 0$, such that, for every $n > n_0$ and for every $T \in \mathcal{S}$, if $card(B(T)) = n$ then $|T| \geq c \times a^{n^p}$.*

In this article, we use an alternative, equivalent, and more applicable definition for use in the demonstration of our result than the above one. We use the following auxiliary definitions to define it.

Definition 5. *Let $\mathcal{S}$ be an unlimited set of EOL-trees and $|\,|$ the size-of-string function. The function $len_{\mathcal{S}} : \mathcal{S} \rightarrow \mathbb{N}$ is the defined as $len_{\mathcal{S}}(T) = |T|$.*

In the definition above, we advise the reader that the size of the alphabet used to write the strings is at least 2. Unary strings cannot be consistently used in computational complexity estimations, since its use trivializes[9] the conjecture $NP = P$. We use to call an alphabet reasonable whenever it has at least two symbols.

Definition 6. *A function $f : \mathbb{N} \longrightarrow \mathbb{N}$ is exponential or bigger if and only if there are $a \in \mathbb{R}$, $a > 1$, $n_0, p \in \mathbb{N}$, $c \in \mathbb{Q}$, $p > 1$, $c > 0$, such that, $\forall n > n_0$, $f(n) \geq c \times a^{n^p}$.*

Technically, the above definition says that a function is exponential or bigger whenever it has a tight exponential lower bound.

Consider a property $\Phi(x)$ on EOL-trees. This property is used to select, from a set $\mathcal{S}$, all the EOL-trees satisfying it. This defines a subset $\{T \in \mathcal{S} : \Phi(T)\}$ of $\mathcal{S}$. As an example we can set a particular $\Phi_{\Gamma,\alpha}(x)$, where Γ is a set of labels and α is a label, to be true only on EOL-trees T, such that $leaves(T) = \Gamma$ and $r(T) = \alpha$. Thus, given a set $\mathcal{S}$ of EOL-trees, the set $\{T \in \mathcal{S} : \Phi_{\Gamma,\alpha}(T)\}$ is the subset of all

[8]Some authors use the term length instead of size.

[9]If there is a NP-complete Formal Language $L \subseteq \Sigma^\star$, where Σ is a singleton, then $NP = P$, see for example [3] (theorem 5.7, page 87).

trees from $\mathcal{S}$ that have Γ as labelling the leaves and α labelling the respective root of each of them. We use properties as $\Phi_{\Gamma,\alpha}(x)$ to specify the set of all trees that correspond to Natural Deduction derivations of a formula α from a set of hypothesis Γ. We further refine this to get the set of all minimal trees (derivations) of α from Γ. For example

$$Min_{\mathcal{S}}(\Gamma,\alpha) = \{T \in \mathcal{S} : \Phi_{\Gamma,\alpha}(T) \ \wedge \ \forall T'(\Phi_{\Gamma,\alpha}(T') \ \rightarrow \ |T| \leq |T'|)\}$$

is the set of all smallest EOL-trees satisfying $\Phi_{\Gamma,\alpha}(x)$. We can see these EOL-trees as Natural Deduction derivations in $\mathbf{M}_{\supset}$, having then the set of all smallest derivations of α from Γ in $\mathbf{M}_{\supset}$. In the general case, where the predicate $\Phi(x)$ is arbitrary, we denote the set above by $Min_{\mathcal{S}}(\Phi)$, that is:

$$Min_{\mathcal{S}}(\Phi) = \{T \in \mathcal{S} : \Phi(T) \ \wedge \ \forall T'(\Phi(T') \ \rightarrow \ |T| \leq |T'|)\}$$

Definition 7. *Let $\mathcal{S}$ be an unlimited set of EOL-trees. Let $\Phi(x)$ represent a property on EOL-trees of $\mathcal{S}$ and let $\Phi_{\mathcal{S},m}(x)$ be defined as $(x \in \mathcal{S} \wedge \Phi(x) \wedge card(B(x)) \leq m)$ with $0 < m \in \mathbb{N}$. We define the function $F_{\mathcal{S},\Phi} : \mathbb{N} \longrightarrow \mathbb{N}$ that associates do each natural number m the size of a minimal tree satisfying $\Phi_{\mathcal{S},m}(x)$.*

$$F_{\mathcal{S},\Phi}(m) = \begin{cases} 0 & \textit{if } m = 0 \\ |T| & T \in Min_{\mathcal{S}}(\Phi_{\mathcal{S},m}) \ \textit{if } m > 0 \end{cases}$$

We point out that depending on Φ, the above function $F_{\mathcal{S},\Phi}$ can be quite uninteresting. For example, if Φ is satisfiable by every tree in $\mathcal{S}$ then $F_{\mathcal{S},\Phi}(m) = 1$, for every $m > 0$. Any tree T with only one node, such that, it is labelled by any element of $B(T)$, is a smallest[10] tree that satisfies Φ. On the other hand, we can have $\Phi_{\mathcal{S},\Phi}(m)$ true only when T is a tree that represents a proof of a $\mathbf{M}_{\supset}$tautology α, $m = |\alpha|$ and $\Phi(T)$ is true whenever T is a tree representing a proof of α.

The following proposition points out an alternative and more adequate definition for a family of exponential or bigger sized trees as already previously mentioned. Observe that if $\mathcal{A}$ is the set of all EOL-trees and $\Phi(x)$ is a property defining a subset $\mathcal{S}$ of $\mathcal{A}$ and $\Phi_{\mathcal{S},m}$ is defined as in definition 7 then $\mathcal{S} = \Phi(\mathcal{A}) = \bigcup_{m \in \mathbb{N}} \Phi_{\mathcal{A},m}(\mathcal{A})$. The reader should note that we use $\Phi(\mathcal{A})$ as an abbreviation of $\{T : \Phi(T) \wedge T \in \mathcal{A}\}$. Observing what is discussed in the last paragraphs, we have the following proposition.

Proposition 3. *Let $\mathcal{S} \subset \mathcal{A}$ be an unlimited set of EOL-trees. Let $\Phi(x)$ be the defining property of $\mathcal{S}$. We have then that $\mathcal{S}$ is $\mathcal{EB}$ if and only if $F_{\mathcal{A},\Phi}$ is an exponential or bigger function from $\mathbb{N}$ in $\mathbb{N}$.*

[10]We do not take the null tree in this work since it represents no meaningful representation of data in our case.

We note that the size of an EOL-tree considers the representation of it when coded in a string under a reasonable alphabet[11].

The following definition is quite useful in the statement of the main lemma of this article.

Definition 8 (Set of nodes at level i labelled with q). *Given an EOL-tree $T = \langle V, E_L \cup E_R \cup E_U, l, B\rangle$, a natural number $i \in \mathbb{N}$ and a label $q \in B(T)$. We use the notation $V_T^{i,q}$ to denote:*

$$\{v \in V(T) : lev_T(v) = i \text{ and } l_T(v) = q\}$$

Definition 9. *Let $\mathcal{S}$ be an unlimited set of EOL-trees and $|\,|$ the size-of-string function. The functions $occ_{\mathcal{S}}^{i,q} : \mathcal{S} \to \mathbb{N}$, $i \in \mathbb{N}$, and $q \in SYMB$, are defined below, where $SYMB = \bigcup_{T \in \mathcal{S}} B(T)$ is the symbol set*[12] *that labels the trees in $\mathcal{S}$:*

$$occ_{\mathcal{S}}^{i,q}(T) = \left|V_T^{i,q}\right|$$

Note that for any $T \in \mathcal{S}$, if $i > h(T)$ then $occ_{\mathcal{S}}^{i,q}(T) = 0$ and that for any $q \notin B(T)$, $occ_{\mathcal{S}}^{i,q}(T) = 0$ too.

The following lemma is essential in the proof of the results shown in this article.

Lemma 4 (Super-Exponential Pigeonhole Principle). *Let $k : \mathbb{N} \longrightarrow \mathbb{N}$ be a polynomial and let $f_i : \mathbb{N} \longrightarrow \mathbb{N}$ be a function, for each $i \in \mathbb{N}$.Consider $m \in \mathbb{N}$ and the function $g_m : \mathbb{N} \longrightarrow \mathbb{N}$ defined as:*

$$g_m(n) = \sum_{i \leq k(m)} f_i(n)$$

For each m, if g_m is $\mathcal{EB}$ on m then there is $0 \leq j \leq k(m)$, such that, f_j is $\mathcal{EB}$, on m, too.

Proof. Consider some $m \in \mathbb{N}$ and suppose that there is no j, $0 \leq j \leq k(m)$, such that f_j is exponential or bigger than it on m. Thus, for every j, f_j is polynomially bounded. Since any polynomial sum of polynomials is a polynomial too, we conclude that g_m, for this m, is polynomially bounded, that is a contradiction.

□

[11]An alphabet is reasonable if it has more than one symbol.

[12]We allow the existence of an infinite set of symbols, but this is not essencial at this point of our formalization.

3 Some useful properties in normal proofs in $\mathbf{M}_\supset$

In this section, we show some definitions and properties used in section 4 to prove that every exponential tree has a subtree that is repeated exponentially many times in the original tree. For example, the parsing tree helps to relate the size of formula in $\mathbf{M}_\supset$, the number of atomic propositions in a formula and the height of it. Here we also relate the lenght of elimination parts in proofs with the lenght of branchs in the parsing tree. We show in the sequel these facts.

Definition 10. *Let α be a formula in $\mathbf{M}_\supset$, the abstract syntax tree of α is the tree T_α defined by induction as:*

- *If α is a propositional letter A then $T_A = A$, and;*
- *If α is $\alpha_1 \supset \alpha_2$ then T_α is* 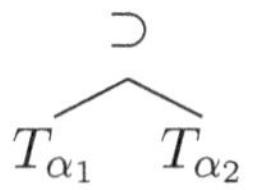

It is easy to see that, for every α in $\mathbf{M}_\supset$, T_α is a full binary tree[13]. For example, below we can find the abstract syntax trees of $((A \supset B) \supset A) \supset A)$ and $(A \supset B) \supset ((B \supset C) \supset (A \supset C))$ respectively.

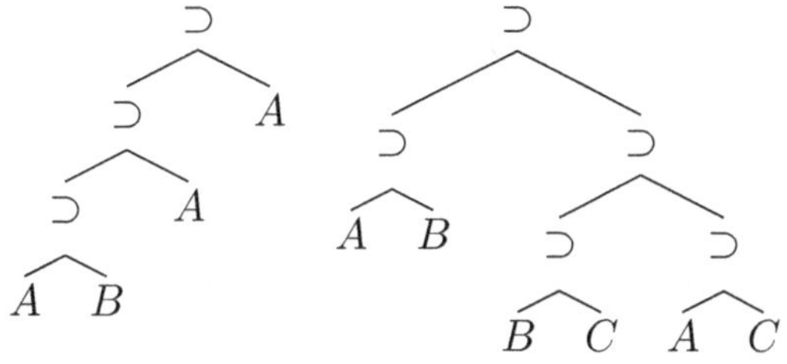

We remind the reader that $A \supset (B \supset C)$ can be written $A \supset B \supset C$ under the convention that the parenthesis is associative to the right. We also observe that the tree representation seems to be fully adequate to be used here in this work on computational complexity analysis of the size of proofs. We only have to observe that for any α, the length of α is, in general, larger than the number the subformulas of α. When α has parenthesis, they hold positions that do not determine any subformula from α. On the other hand, each node in T_α determines a subformula from α. Instead of estimating the size of a proof as a function of the length of the formula α, we take into account the size of the abstract parsing tree $|T_\alpha|$, i.e., how many nodes it has.

[13]A binary tree, such that, every node that is not a leaf has two childrens.

Given a tree T, the height of T, denoted by $h(T)$, is the length of the longest path, in T, from the root of T to its leaves. It is well-known that the height of a balanced tree with a size of n is $\log(n)$. On the other hand, the longest path from the root to any leaf in a tree T is bounded by $\left\lfloor \frac{|T|}{2} \right\rfloor$. For example, when T is the abstract syntax tree of $a_1 \supset (a_2 \supset (a_3 \ldots (a_n \supset b)) \ldots)$, as shown in the tree in figure 5a. Another thing to observe is that, because of the way the implications are nested by the right, we can consider the right-hand side of the rightmost implication as being a propositional variable. This is summarized in observation 1

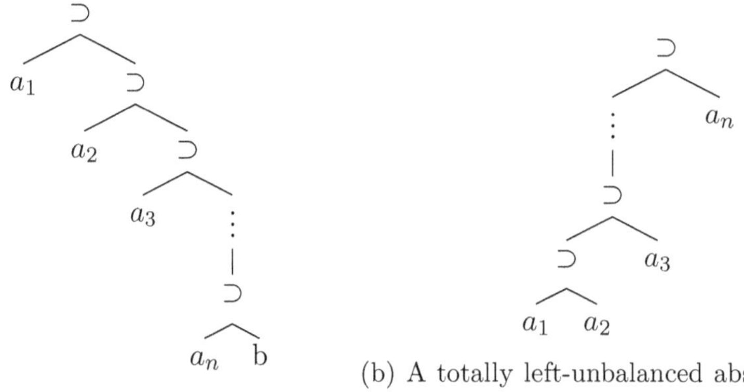

(a) A totally right-unbalanced abstract syntax tree

(b) A totally left-unbalanced abstract syntax tree

Figure 5

Observation 1. *For every tree T, $\lfloor \log(|T|) \rfloor \leq h(T) \leq \left\lfloor \frac{|T|}{2} \right\rfloor$*

In any normal proof Π of a formula α, the subformulas of it that can be top-formulas of some branch in Π are antecedents of α. We observe that any formula α in $\mathbf{M}_{\supset}$ is of the form $\alpha_1 \supset (\alpha_2 \supset \ldots (\alpha_n \supset q) \ldots)$, with q being a propositional variable. Moreover, the antecedents of α are, hence, α_i, $i = 1, n$. These are the formulas that can be the top-formula of the main branch, due to the fact that in order to prove α, some $\supset$-introductions can be needed, and one of them has to discharge an assumption α_j, for some j, that it is top-formula of the main branch. For the secondary branches, each of them is determined by an antecedent of the top-formula α_j of the main branch, and, that is the minor premise of the corresponding $\supset$-Elim application that has a subformula derived from α_j, or itself in the topmost $\supset$-Elim application, as the major premise.

In virtue of theorem 1 and definition 1, any branch in a normal proof has an Elim-part, where there are only $\supset$-Elim rule application rules, and an Intro-part, containing only $\supset$-Intro rule applications. The minimal formula in the branch is the conclusion of the last $\supset$-Elim rule and premise of the first $\supset$-Intro rule if any of them. This fact shows us that the Elim-part of any branch in proof of T_α, i.e. proof of α, is a sequence of subformulas of the respective top-formula of this branch. Any branch in T_α can be identified by its level and its top-formula. Moreover, the sequence of formulas of the Elim-part of this branch is the sequence of subformulas of the top-formula that appears in the abstract syntax tree T_α. Thus, we have the following proposition.

Proposition 5. *In a normal proof of α, the Elim-part of any branch is uniquely determined by the formula β that is the major premiss of the topmost $\supset$-Elim rule application of this branch. Any subtree T_β of T_α determines each formula occurrence in this Elim-part of the branch. These formulas are the sequence of right descendants of β in T_β.*

A corollary of the proposition 5 is that there is a one-to-one correspondence between sub-formulas of α and possible Elim-parts of the branches that occur in any normal proof of α. This is stated by the following lemma.

Lemma 6. *Let Π be a normal proof of α. Then each Elim-part of a branch in Π is the sequence of formulas occurring in the right-branch of T_β, where β is the major premise of the topmost Elim-rule application in the branch.*

One last thing to observe is that the number of formulas in a normal proof of α that is a major premise in a topmost Elim-rule application in a branch is upper-bounded by $\frac{|T_\alpha|}{3}$, since the major premise of a $\supset$-Elim rule must be of degree at least 3.

4 Examining some huge normal proofs and their inherent redundancy

The following sets can be seen as examples of $\mathcal{EB}$ sets of EOL-trees, whenever we take Natural Deduction derivations in the format of EOL-trees.

4.1 A family of huge proofs in $M_\supset$: Fibonacci numbers

We use the notation $\begin{pmatrix} \phi_1 \\ \vdots \\ \phi_k \end{pmatrix} \supset \alpha$ to denote the formula $\phi_1 \supset (\phi_2 \supset (\ldots(\phi_k \supset \alpha)\ldots)))$

We show here a family of formulas that have only huge proofs as least normal proofs. Consider the following formulas.

- $\eta = A_1 \supset A_2$
- $\sigma_k = A_{k-2} \supset (A_{k-1} \supset A_k)$, $k > 2$.

In this section, if Π is a $\mathbf{M}_{\supset}$ ND proof then $len(\Pi)$ denotes its length or size. Any normal proof Π of $\chi_n = \begin{pmatrix} \eta \\ \sigma_3 \\ \vdots \\ \sigma_n \end{pmatrix} A_1 \supset A_n$ is such that $|\Pi| \geq Fibonnaci(n)$. In what follows, due to have a more economical presentation, we omit all, but the first, $\supset$-Intro applications in the I-part of the main branch of the proofs. Thus, we have the following derivation of $A_1 \supset A_n$ from $\eta, \sigma_1, \ldots, \sigma_n$. The formulas $\eta, \sigma_1, \ldots, \sigma_n$ that are discharged by $n+1$ $\supset$-rules to complete the proof of χ_n from this derivation are omitted. With the sake of a more elegant presentation on the size of the proofs of χ_n we call the derivation of $A_1 \supset A_n$ from $\eta, \sigma_1, \ldots, \sigma_n$ as Π_n. Note that Π_1 is A_1 Π_2 is η, and for Π_k, $k > 2$, we have the derivation shown below. We have that:

$$\dfrac{\dfrac{\dfrac{\begin{array}{c}[A_1] \\ A_1 \supset A_2 \\ A_1 \supset (A_2 \supset A_3) \\ \Pi_3 \\ A_3\end{array} \quad \dfrac{\dfrac{[A_1] \quad A_1 \supset A_2}{A_2} \quad A_2 \supset (A_3 \supset A_4)}{A_3 \supset A_4}}{A_4} \quad \dfrac{\begin{array}{c}[A_1] \\ A_1 \supset A_2 \\ A_1 \supset (A_2 \supset A_3) \\ \Pi_3 \\ A_3\end{array} \quad A_3 \supset (A_4 \supset A_5)}{A_4 \supset A_5}}{A_5}}{A_1 \supset A_5}$$

In general, for each $3 \leq k$

$$\dfrac{\dfrac{\begin{array}{c}[A_1] \\ \eta \\ \sigma_3, \ldots, \sigma_{k-1} \\ \Pi_{k-1} \\ A_{k-1}\end{array} \quad \dfrac{\begin{array}{c}[A_1] \\ \eta \\ \sigma_3, \ldots, \sigma_{k-2} \\ \Pi_{k-2} \\ A_{k-2}\end{array} \quad A_{k-2} \supset (A_{k-1} \supset A_k)}{A_{k-1} \supset A_k}}{A_k}}{A_1 \supset A_k}$$

So we have that:

$$\begin{array}{rcl} len(\Pi_2) & = & 1 \\ len(\Pi_3) & = & len(\Pi_2) + 1 \\ len(\Pi_k) & = & len(\Pi_{k-2}) + len(\Pi_{k-1}) + 2 \end{array}$$

Thus, by a well-known fact about Fibonacci[14] numbers, we see that:

$$\frac{\phi^k}{\sqrt{5}} \approx Fibonacci(k) \leq len(\Pi_k)$$

where $\phi = 1.618$

Thus, the proof of χ_n has size $len(\Pi_n) + n$, that is lower-bounded by an exponential function on n, as above. From definitions in Section 2, the set Fib of all trees that correspond with the proofs of $A_1 \supset A_n$ is a $\mathcal{EB}$ set of trees. A fascinating phenomenon that happens with this set of huge proofs concerns the main result of this article. Almost all of these proofs have levels, above the level of the minimal formula of the main branch, where there are exponentially (or more) many repetitions of a formula, labelling a node of the underlying tree, and these formulas are conclusions of sub-trees that occurs exponentially (or more) many times in these underlying trees.

We have the following proposition on these derivations.

Proposition 7. *Let Π_n be the derivation of A_n, we drop-off the $\supset$-Intro last rule, from η and σ_k, $k = 3, n$. We have thus that $occ_{Fib}^{l,A_{n-l}}(\Pi_n) = Fibonacci(l+1)$, for $l = 0, n-1$*

Proof. By induction on n:

(Basis) $n = 1$, $\Pi_1 = A_1$ and $occ_{Fib}^{0,A_1}(\Pi_1) = 1 = Fibonacci(1)$ and for $n = 2$ we have that

$$\frac{A_1 \qquad A_1 \supset A_2}{A_2}$$

and hence $occ_{Fib}^{0,A_2}(\Pi_2) = 1 = Fibonacci(1)$ and $occ_{Fib}^{1,A_1}(\Pi_2) = 1 = Fibonacci(2)$
(Inductive step) Let $n > 2$ and $0 \leq l < n-1$. Π_n is the following derivation:

$$\dfrac{\begin{matrix}\Pi_{n-1}\\ A_{n-1}\end{matrix} \qquad \dfrac{\begin{matrix}\Pi_{n-2}\\ A_{n-2}\end{matrix} \qquad A_{n-2} \supset (A_{n-1} \supset A_n)}{A_{n-1} \supset A_n}}{A_n}$$

For $l = 0$, $l = 1$ and $l = 2$ it is straightforward, consider $l > 2$ then

$$occ_{Fib}^{l,A_{n-l}}(\Pi_n) = occ_{Fib}^{l-1,A_{n-1-(l-1)}}(\Pi_{n-1}) + occ_{Fib}^{l-2,A_{n-2-(l-2)}}(\Pi_{n-2})$$

[14]The Fibonacci numbers, Fib(n), or Fibonacci(n), are defined according the following recursive definition: $Fib(0) = Fib(1) = 1$ and $Fib(n+2) = Fib(n+1) + Fib(n)$.

what is, by inductive hypothesis:

$$\begin{array}{l} occ^{l,A_{n-l}}_{Fib}(\Pi_n) = \\ Fibonacci((l-1)+1) + Fibonacci((l-2)+1) = \\ Fibonacci(l) + Fibonacci(l-1) = Fibonacci(l+1) \end{array}$$

□

We also can prove by induction that $h(\Pi_n) = n$ and that $B(\Pi_n) = 4 \times (n-1)$, for $n > 2$. Thus this set Fib of proofs is a set of huge proofs, linear height bounded. By the proposition 7 for almost all derivations in Fib, there is at least one sub-derivation that occurs exponentially (or more) many times in it. To see that, observe that almost all Π_n, which are of exponential size on n, in fact, are of exponential-size on $|B(\Pi_n)| = 4 \times (n-1)$ too. Of course, all of Π_n are normal derivations in $\mathbf{M}_{\supset}$.

4.2 Proving in $\mathbf{M}_{\supset}$ that a graph is not Hamiltonian

In Appendix A we remember that a well-known propositional coding of the hamiltonianicity of graphs can be used, in purely implicational propositional logic, to have a Natural Deduction proof for the non-hamiltonianicity of graphs, by negating the previous statement. For any non-hamiltonian graph G the sentence $\neg\alpha_G$ is a certificate for its non-hamiltonianicity. The set of all derivations α_G for G non-hamiltonian is a set of huge proofs. These derivations are linearly height bounded in $\mathbf{M}_{\supset}$ as we demonstrate in the appendix. By counting formulas instead of symbols, if a simple non-hamiltonian graph G has n nodes, then the normal proof of α_G showing that it has no Hamiltonian cycle has n^n formulas, in the worst case. Of course, the size of the normal proof depends on the topology of the graph. For the Petersen graph, for example, the size is approximately 10^{10}. Petersen graph has ten nodes. We call this set of huge normal proofs $NHam$, for Non-hamiltonian.

$NHam$ has proofs Pi that has sub-derivations that repeats exponentially many times in Π. The reason for that relies on the fact that $NHam$ is the set of naive proofs of non-hamiltonianicity. These proofs consider all possible paths. In a graph of n nodes, a (possible) path is any sequence of n positions. The proof is an upside-down decision tree that checks the consistency of each possible path for being a correct. Well, we have many repetitions of sub-paths, so the proof reflects this repetition too. We invite the reader to see this in detail in the Appendix. We included this material as a matter of completeness and illustration. It firstly appears in [10].

4.3 Normal Proofs that need exponentially many assumptions

In [13] we show a family of purely implicational formulas ξ_n, such that, each of them needs 2^n assumptions of the same formula, we have to discharge at the end of the proof, in their proofs. Almost all of the normal proofs of ξ_n are hence of exponential. All of them are linearly height bounded. This set of huge proofs, denoted by *Exp* also has the property that almost every proof in it has sub-proofs that are polynomially many times repeated, in a level of the normal proof of ξ_n, let us say, in it. This is one more concrete example of what we prove in the next section and as the main result of the article. Moreover, moving from $\mathbf{M}_{\supset}$to classical logic does not always work to get subexponential classical versions of any $\mathbf{M}_{\supset}$theorems. There are counter-examples, many of them based on the exponential lower-bound for Normal Proofs of the implicational version of the Pigeonhole Principle that comes from Haken's lower-bound for the resolution system. In conclusion, section 7, we briefly comment on general reasons that turn the moving from $\mathbf{M}_{\supset}$to Classical logic ineffective as a basis to compressing proofs in $\mathbf{M}_{\supset}$.

5 Huge proofs are redundant

In this section, we show the main result of this article, roughly stated as "Almost every linearly height-bounded huge proof is redundant". We show that for any unlimited set $\mathcal{S}$ of EOL-trees if $\mathcal{S}$ is $\mathcal{EB}$ then for almost all trees $T \in \mathcal{S}$, there is a subtree T' of T that occurs exponentially (or more) many times at the same level in T. The formal statement of this assertion is lemma 11. The following lemma is an initial step in the proof of lemma 11. We need one more auxiliary definition before. The function $OCC^{l,q}_{\mathcal{A}}(m)$ provides the least number of occurrences of the label q in the level l among the trees in $\mathcal{A}$ that are labelled with at most m labels. The function $OCC^{l}(m)$ is the maximum number of occurrences among all the $OCC^{l,q}_{\mathcal{A}}(m)$ for all q's. They are used to count the number of repetitions of occurrences of subtrees. This counting is used to characterize when the amount of subtrees is comparable to the size of the tree, in lemma 8.

Definition 11. *Let $\mathcal{A}$ be a set of EOL-trees and $occ^{l,q}_{\mathcal{A}}$ as in definition 9. We define the function $OCC^{l,q}_{\mathcal{A}} : \mathbb{N} \longrightarrow \mathbb{N}$ as:*

$$OCC^{l,q}_{\mathcal{A}}(m) = Min(\{occ^{l,q}_{\mathcal{A}}(T) : card((B(T)) = m \wedge (occ^{l,q}_{\mathcal{A}}(T) > 0)\})$$

A straightforward consequence of the definition above is that for every tree $T \in \mathcal{A}$, for every $m \in \mathbb{N}$, for every $l \leq h(T)$, if $card(B(T)) = m$ then for every $q \in B(T)$, $OCC^{l,q}_{\mathcal{A}}(m) \leq occ^{l,q}_{\mathcal{A}}(T)$.

Observation 2. *We first observe that for all trees, the first levels have few nodes, starting with level 1 that has at most two nodes, 2 with at most 4, and so on. That is, for every tree T, there is at least one level $i \leq h(T)$ and $0 < p \in \mathbb{N}$, such that, for all sets $V^{i,q}(T) \leq card(B(T))^p$, for each $q \in B(T)$, obviously with $V^{i,q}(T) \neq \emptyset$.*

Definition 12. *Considering the conditions in definition 11, we define:*

$$OCC^l_{\mathcal{A}}(m) = Max(\{OCC^{l,q}_{\mathcal{A}}(m) : card(B(T)) = m \wedge q \in B(T)\}$$

In contrast with the observation above, huge proofs have levels i, such that $V^{i,q}(T)$, for some $q \in B(T)$, is lower-bounded by an exponential or faster-growing function. The proof of this resembles the proof of lemma 4.

Lemma 8. *Let $\mathcal{A}$ be the set of all EOL-trees and $\Phi(T)$ the predicate that holds only when T is linearly height-bounded, on $card(B(T))$, tree. If $\Phi(\mathcal{A})$ is $\mathcal{EB}$ then $OCC^l_{\Phi(\mathcal{A})}$ is $\mathcal{EB}$, on the argument (m in the definition 11) that bounds $card(B(T))$ for almost all $l \in \mathbb{N}$.*

Proof. Since $\Phi(\mathcal{A})$ is $\mathcal{EB}$ then by proposition 3 $F_{\mathcal{A},\Phi}$ is an exponential or faster growing function from $\mathbb{N}$ into $\mathbb{N}$. We observe that, for any tree T,

$$|V(T)| = \left| \bigcup_{i \leq h(T)} \bigcup_{q \in B(T)} V_T^{i,q} \right|$$

, hence, we have that

$$|V(T)| = \sum_{i \leq h(T)} \left| \bigcup_{q \in B(T)} V_T^{i,q} \right| = \sum_{i \leq h(T)} \sum_{q \in B(T)} \left| V_T^{i,q} \right|$$

The above equality holds for $T \in \Phi(\mathcal{A})$ and for all m, such that $card(B(T)) = m$. Thus, we have $|V(T)| \geq |Mins_{\mathcal{S}}(\Phi_{\mathcal{S},m})|$. By the fact that $h(T)$ is linearly bounded by $card(B(T))$ we have that $h(T)$ is linearly bounded by m. Let $k \times m$ be such bound. Hence, we have that:

$$F_{\mathcal{A},\Phi}(m) \leq \sum_{i \leq h(T)} \sum_{q \in B(T)} \left| V_T^{i,q} \right| = \sum_{i \leq k(m)} \sum_{q \in B(T)} \left| V_T^{i,q} \right|$$

for almost all m, and almost all T. In the right-hand side (RHS) of the last equation we have to remember $card(B(T)) = m$, so the inner summation is linearly bounded by m. Moreover, by hypothesis, $\Phi(\mathcal{A})$ is $\mathcal{EB}$. Thus, $F_{\mathcal{A},\Phi}$ is $\mathcal{EB}$ and there are $a \in \mathbb{Q}$, $a > 1$, and, $c \in \mathbb{Q}$, $c > 0$, such that:

$$a^{cm} \leq \sum_{i \leq k(m)} \sum_{q \in B(T)} \left| V_T^{i,q} \right|$$

for almost all m and all T, such that $card(B(T)) = m$. The above RHS can be seen as a function from $\Phi(\mathcal{A}) \times \mathbb{N}$ into $\mathbb{N}$. If for almost all m, for all $l \leq k \times m$ and $q \in B(T)$, with $card(B(T)) \leq m$ we have that $V_T^{i,q}$ is polynomially bounded then we have that the mentioned RHS is polynomially bounded also. Hence, there must be $i \leq k \times m$ and $q \in B(T)$, with $card(B(T)) = m$, such that $\left| V_T^{i,q} \right| > a^{cm}$. Thus, $occ_{\mathcal{A}}^{l,q}(T) > a^{cm}$, for almost all m and all T with $card(B(T)) = m$. Then, by definition 11, there is m_0, such that for all $m > m_0$, there are k, l, $l \leq k \times m$, such that $a^{cm} \leq OCC_{\Phi(\mathcal{A})}^{l}(m)$. So $OCC_{\Phi(\mathcal{A})}^{l}$ is $\mathcal{EB}$ too. □

The above lemma is, indeed, stronger than what we need to show that almost all normal proof is redundant. However, from it, we can start our reasoning on the intrinsic redundancy of huge proofs. It states that any huge proof T has a level $i \leq h(T)$ and a label $q \in B(T)$ such $a^{cm} \leq \left| V_T^{i,q} \right|$, for $m = card(B(T))$, and some $a > 1$ and $c > 0$, $a, c \in \mathbb{Q}$. In fact, there is an m_0, such that for all $m > m_0$, the previous statement holds.

We know that $a^{cm} = 2^{cm \log(a)}$ Before we prove the main result of this article, we need one more lemma.

Lemma 9. *Let q be a label node in an EOL-tree $T = \langle V, E_L \cup E_R \cup E_U, l, B \rangle$ derived from a normal proof of some formula α, such that $B(T)$ is the set of formulas associated to each subtree (non-leaf) T_α. Let*

$$Label_{suc}^2(q) = \{(c, d) : \langle u, v_1 \rangle \in E_L,\ \langle u, v_2 \rangle\, E_R,\ \ell(u) = q,\ \ell(v_1) = c \text{ or } \ell(v_2) = d\}$$

be as above. Then $\left| Label_{suc}^2 \right| < \frac{1}{3} \times card(B(T))$

Proof. By the definition of EOL-trees, whenever we have three nodes u, v_1, v_2, such that, $\langle u, v_1 \rangle \in E_L$ and $\langle u, v_2 \rangle \in E_R$, we have that $\ell(u) \prec \ell(v_2)$ and $\ell(v_1)$ is the unique label such $\ell(v_1) \odot \ell(u) = \ell(v_2)$. Since T is derived from a normal proof, $\ell(v_2)$ is $\ell(v_1) \supset \ell(u)$ and by lemma 6 there are at most $\frac{1}{3} \times card(B(T))$ different possible branches from any u to the topmost associated to the major premise of the topmost binary rule in T. □

Using the above lemma, we can draw the following result.

Lemma 10. *Let $\mathcal{S}$ be a set of EOL-trees. Suppose that there are $a \in \mathbb{R}$, $a > 1$, $n_0, p \in \mathbb{N}$, $c \in \mathbb{Q}$, $p > 1$, $c > 0$, such that, $\forall n > n_0$, for all $T \in \mathcal{S}$, If $card(B(T)) = n$ then there is a level $i > 0$ in T and a label $q \in B(T)$, such that, $\left|V_T^{i,q}\right| \geq c \times a^{n^p}$, i.e., $V_T^{i,q}$ has exponentially many elements on $card(B(T))$. Thus, there is at least one pair of labels $l_1, l_2 \in B(T)$, such that $l_2 = l_1 \odot q$, and $\left|V_T^{i+1,l_1}\right| = \left|V_T^{i+1,l_2}\right| \geq (a^{\frac{a-1}{a}})^{n^p}$. Moreover, for every $u_1 \in V_T^{i+1,l_1}$ there is only one $u_2 \in V_T^{i+1,l_2}$ and only one $u \in V_T^{i,q}$, such that, $\langle u, u_1\rangle \in E_L(T)$ and $\langle u, u_2\rangle \in E_R(T)$.*

Proof. By the hypothesis of the lemma, we have that for all $n > n_0$, and, for every $T \in \mathcal{S}$, such that $card(B(T)) = n$, there are i and q, such that, $\left|V_T^{i,q}\right| \geq c \times a^{n^p}$, with $a > 1$, $c > 0$, $p > 1$. By lemma 9, there are at most $\frac{1}{2} \times card(B(T)) = \frac{1}{2} \times n$ possible different pairs of labels, i.e., the sets $Left = \{v : \langle u, v\rangle \in E_L(T) \text{ and } u \in V_T^{i,q}\}$ and $Right = \{v : \langle u, v\rangle \in E_R(T) \text{ and } u \in V_T^{i,q}\}$ have the same cardinality :

$$\frac{\left|V_T^{i,q}\right|}{\frac{n}{2}}$$

This is lower-bounded by $\frac{c \times a^{n^p}}{\frac{n}{2}}$, that is the same of $(2c)\frac{a^{n^p}}{n}$, that is equal to $(2c)\frac{a^{n^p}}{a^{\log_a n}} = (2c)a^{(n^p - \log_a n)}$. Since, for all $n > a$, we have that $n^a > a^n$, thus $\log_a n^a < \log_a a^n = n < n^p$, and hence, $\log_a n < \frac{n^p}{a}$. Thus $n^p - \log_a n > n^p - \frac{n^p}{a}$, and finally

$$\frac{\left|V_T^{i,q}\right|}{\frac{n}{2}} > (2c)a^{(n^p - \log_a n)} > (2c)a^{(n^p - \frac{n^p}{a})} = (2c)a^{\frac{a-1}{a}n^p} = (2c)(\sqrt[a]{a}^{a-1})^{n^p}$$

We observe that as $a > 1$ then $\sqrt[a]{a}^{a-1} > 1$. □

Finally, by the last observation above, we can conclude that if there is an unlimited set of trees that each tree has a level and a label that repeats exponentially often, then in the levels above this level there are labels that repeat exponentially often too. We have, hence, the main lemma below.

Lemma 11 (Every linear height bounded huge tree is inherently redundant). *Let $\mathcal{A}$ be the set of all EOL-trees and $\Phi(T)$ the predicate that only holds when T is of linear height, on $card(B(T))$, bounded tree. If $\Phi(\mathcal{A})$ is $\mathcal{EB}$ then, for almost all T, such that $\Phi(T)$, there is a sub-tree T' of T, and a level $i \leq h(T)$, such that $occ^{i,\ell(r(T'))}$ is $\mathcal{EB}$, and there is no level $j < i$ such that $occ^{j,q}$ is $\mathcal{EB}$; for any $q \in B(T)$.*

Proof. The essential reasoning in this proof is to use the observation 2 together with lemma 8 above. The observation and lemma contribute to proving that the set of levels that has no label repeated exponentially or more many times is not empty. The lemma 8 encapsulate this reasoning. A first observation is that by lemma 8 for almost all trees $T \in \mathcal{A}$ there is a level $i \leq h(T)$, such that $\sum_{q \in B(T)} V^{i,q}$ is exponential. The last phrase is a kind of abuse of language, but it facilitates the understanding of the argumentation. Formally, we had to state that given an exponential lower-bound a^{cm}, we prove that there is i, such that, $a^{cm} \leq \sum_{q \in B(T)} V^{i,q}$, from the hypothesis, that $a^{cm} \leq V(T)$, but we can use the lemma to infer this existence directly.

Turning back to the proof, since $\sum_{q \in B(T)} V^{i,q}$ is exponential (or longer than), using lemma 4 we can conclude that there is $q \in B(T)$, such that, $V_T^{i,q}$ is exponential or longer than exponential. We choose, i and q among the least possible values. Thus, any level j, $j < i$, does have sub-exponentially many nodes labelled by each label $q' \in B(T)$. Any node v in level i, with $\ell(v) = q$, can belong to one and only one of three disjoint sets Bin, Un and $Zero$, according if it has two, only one, or none children in T. Since $V_T^{i,q} = Bin + Un + Zero$ is exponentially lower-bounded, then by lemma 4 at least one of them is exponentially lower-bounded too. Of course, more than one can be exponentially lower-bounded, but we proceed the proof, without generality loss, for each case separately. We prove by induction of the minimal distance of the elements of $V_T^{i,q}$.

- The basis case is when one of the subsets of $V_T^{i,q}$ is the set $Zero$. In this case, the distance is 0, for the elements of $Zero$ are leaves themselves. Then, the sub-tree T' is formed by each $v \in Zero$ itself, labelled with q. They occur exponentially or more than exponentially many times in T.

- The subset, provided by lemma 4 is Un, hence the set Un^- defined as $\{u : \langle v, u \rangle \in E(T) \text{ and } v \in Un\}$ is exponentially or longer too, by the item 9 of definition 2. Thus, by the inductive hypothesis, there is a sub-tree T'' that has the elements of Un^- as roots. Thus, we have exponentially or longer many sub-trees of the form T'' occurring in T, with roots in level $i+1$. By adjoining the nodes in Un to these trees, we obtain exponentially or longer many trees T' occurring in T. We remind the induction value of Un^- is smaller than the induction value of Un.

- The case when Bin is an exponentially bigger subset of $V_T^{i,q}$ is analogous to the above. There are at most $\frac{1}{3}\left|V_T^{i,q}\right|$ right-handed branches T_{right} of formulas labelled by q in level i. By using iterated inductive hypothesis, in each of the levels of T_{right}, we obtain sub-trees T_{left} that occurs exponentially, or more, many times in T in the level $i+1$. We join them in trees with root in the level of i. We obtain the desired trees occurring with roots in i that repeat at least exponentially. We have only to observe the use of lemma 6 to be sure about the fact that there are only linearly many possible labels for the children of the nodes in Bin and lemma 10 does all the job.

□

Using Lemma 11 we obtain the following Theorem, the main result of this article.

Theorem 12. *Every huge proof in $\mathbf{M}_{\supset}$ is redundant. There is a sub-derivation of it that repeats at least exponentially many times in some level i of the proof.*

Proof. The proof is an application of Lemma 11 on EOL-trees derived from Natural Deduction proofs. All conditions for being a valid EOL-tree hold on the derived trees from ND of $\mathbf{M}_{\supset}$. Including the fact that the $\odot$ concrete operation is such that $l_1 \odot l_2 = l_1 \supset l_2$. We only have to observe that a sub-derivation of a Natural Deduction proof/derivation is any full sub-tree of the underlying tree of the Natural Deduction derivation. □

6 Using Theorem 12 to obtain subexponential proofs

This section contains a brief explanation of the application of Theorem 12 to show that every tautology in $\mathbf{M}_{\supset}$ that has a superexponentially sized proof with height linearly bounded has DAG-based proof that it is subexponential on the size of its conclusion. The proof of the following proposition and some of the companion results listed here are in [11]. In this section, we argue in favour of their validity. For a complete, formal and detailed definitions and proofs see [11]. We notice that the DAG-proofs are rooted, and, the conclusion.is labels the root node

Proposition 13. *Let α be a $\mathbf{M}_{\supset}$ tautology that has a superexponential normal proof of height linearly bounded by $|\alpha|$. There is a DAG-like proof of α of size subexponential on $|\alpha|$.*

Proof. **Sketch of the Proposition 13**. The production of a DAG-like proof obtains through the method described by algorithm 1. DAG-like proofs, rooted by their respective conclusions, are proofs that are the result of the Collapse of two or more identical full subtrees of them. Its root completely determines a full subtree of a DAG-like proof. Given a root r, the set of vertexes reachable from r is the set of nodes of the subtree. In line 3 is where Lemma 11 is called to return the list of subtrees of $\mathcal{T}$ that occurs exponentially many times in the lowest level possibles.

That is, if l is the lowest level of occurrence of exponentially many instances of the same subtree, then below l in $\mathcal{T}$ there is no occurrence of a root of a subtree that occurs exponentially many times in $\mathcal{T}$. We have the operation $Collapse(\mathcal{D}, \mathcal{Y}, D_{\mathcal{Y}})$, where $\mathcal{Y}$ is a list of nodes, each of them is the root of an instance of the subtree $\mathcal{C}_{\mathcal{Y}}$ that occurs in the same level of $\mathcal{D}$, the ambient DAG-like proof. The rooted DAG-like proof $D_{\mathcal{Y}}$ is the result of the application of $Compress()$ to $\mathcal{C}_{\mathcal{Y}}$, a statment executed in line 8. The procedure $Collapse()$ links the root of $\mathcal{D}_{\mathcal{Y}}$ to the respective targets of each $y \in \mathcal{Y}$ in $(\mathcal{D} - (\bigcup_{y \in \mathcal{Y}} \mathcal{C}_{\mathcal{Y}})) \cup \bigcup_{y \in \mathcal{Y}} \{y\}$. We obtain the linking in a way that the path from the premisses to conclusion remains unchanged regarding the previous instance of $\mathcal{C}_{\mathcal{Y}}$. The result of the Collapse is such that, all of the instances of $\mathcal{C}_{\mathcal{Y}}$ are compressed and replaced by their respective instances in $\mathcal{D}$. Obviously, the size of the new construction is smaller. This operation is depicted in figures 6 and 7, original from [11]. In the algorithm 1, $c(\mathcal{T})$ is the conclusion of the proof/derivation $\mathcal{T}$, $h(\mathcal{T})$ is the height of $\mathcal{T}$, $|s|$ is the length of the string representation of s. An r-DagProof is a rooted DAG-proof, formally defined in [11]. We have then the following results proved and state in [11]. We can see, by the definition of Algorithm 1, that this proposition is proved by induction on the number of recursive calls to $Compression$. The basis is when the size of the proof-tree is smaller than any exponential on $a > 1$, $a \in \mathbb{Q}$. In each recursive call, the size of the new recursive argument is smaller than the current one, regarding the fixed m, determined at the starting call and fixed as a global value. Thus, for any starting argument, $\mathcal{T}$, $Compress(\mathcal{T})$ is defined, i.e., it halts as a running sequence of statements. Last but not least, we can prove that the tree that remains when we remove all the subtrees returned by the statement in line 3 of algorithm 1 is subexponential. Denote by $\mathcal{T}_0$ the tree resulted when we cut out all the subtrees in LOCALLOWESTLEVELS from $\mathcal{T}$. We remember that the removal of a subtree includes the removal of the root of the removed tree. Thus, if $\mathcal{T}_0$ is exponential, then we can apply Lemma 11 to it, and we have a list of subtrees that occurs exponentially many times in $\mathcal{T}_0$. However, all of them are below the respective least occurrences in $\mathcal{T}$ that we removed to obtain $\mathcal{T}_0$. By the hypothesis, i.e., previous application of Lemma 11 to $\mathcal{T}$, we have a contradiction. So $\mathcal{T}_0$ is subexponential. □

We have only to remember that the sub-trees may not be sub-proofs, for the discharging of assumptions is below them. One last, but not least, thing to remind the reader is the fact that since the sub-trees are the same, the edges have the same dependency set. The bitstring that is used to represent, formally, of Natural Deduction proofs and derivation in as EOL-trees do not even need to be altered or manipulated in the demonstrations we made in this article.

We have the following proposition that it is proved by induction on the number of recursive calls to *Compress*(), analogous to the above proof. In the proof below we use a lemma that states that each application of $Collapse(\mathcal{D}, \mathcal{Y}, D_{\mathcal{Y}})$ produces a valid DAG proof/derivation of the root of $\mathcal{D}$ from valid DAG proofs/derivations of $\mathcal{D}_Y$. The validity notion is the usual semantics for $\mathbf{M}_{\supset}$.

Proposition 14. *Let α be a $\mathbf{M}_{\supset}$ tautology that has a superexponential normal proof $\mathcal{T}$ of height linearly bounded by $|\alpha|$. The DAG-like proof of $Compress(\mathcal{T})$ is a DAG-like valid proof of its conclusion (root)*

Finally, there is an algorithm, that runs in linear time on the size of a rooted DAG-proof $\mathcal{D}$ that verifies whether $\mathcal{D}$ is valid or not. Due to space limits, we cannot detail this anymore, and, we refer to [11].

Algorithm 1 Compress a **normal** proof $\mathcal{T}$ using repeatly Lemma 11.

Precondition: Uses the global variable m with value $|c(\mathcal{T})|$ and $a > 1$, $a \in \mathbb{Q}$
Precondition: $\mathcal{T}$ is a normal proof, $h(\mathcal{T}) \leq |c(\mathcal{T})|$ and $|\mathcal{T}| > a^m$
Ensure: a r-DagProof $\mathcal{D}$ proving $c(\mathcal{T})$ of size subexponential in $|c(\mathcal{T})|$

Function Compress($\mathcal{T}$)
if $(|\mathcal{T}|) > a^m$ **then**
  $LocalLowestLevels \leftarrow MinLevel(Lemma\ 11(\mathcal{T}))$
  $\mathcal{D} \leftarrow \mathcal{T}$
  for $lev \in LocalLowestLevels$ upwards $h(\mathcal{T})$ **do**
    $L \leftarrow ExpSubProofs(\mathcal{T}, lev)$
    for $\langle \mathcal{Y}, \mathcal{C}_{\mathcal{Y}} \rangle \in L$ **do**
      $\mathcal{D}_{\mathcal{Y}} \leftarrow Compress(\mathcal{C}_{\mathcal{Y}})$
      $\mathcal{D} \leftarrow Collapse(\mathcal{D}, \mathcal{Y}, D_{\mathcal{Y}})$
    end for
  end for
  Return$\mathcal{D}$
else
  Return$\mathcal{T}$
end if

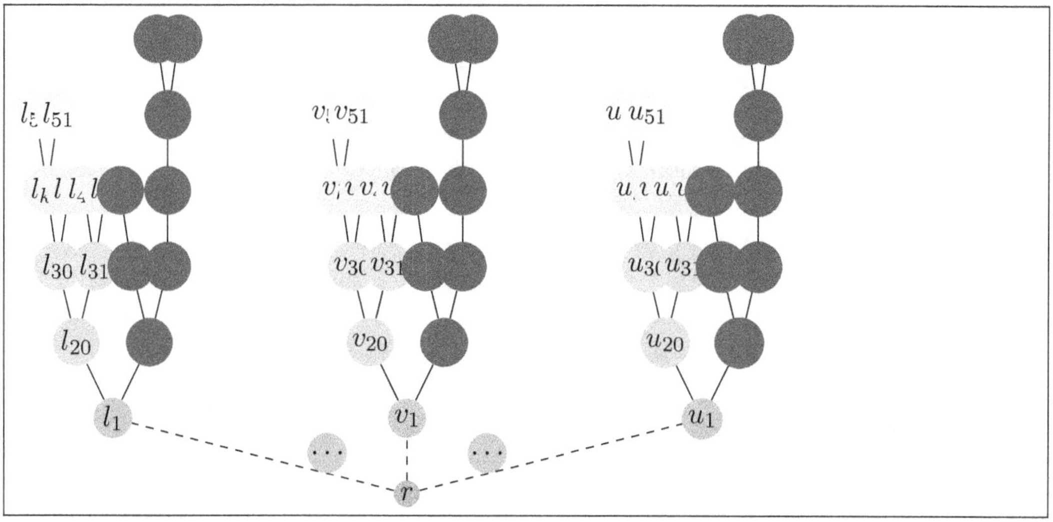

Figure 6: Some instances of the matrix $\mathcal{C}$ in the ambient r-DagProof $\mathcal{D}$

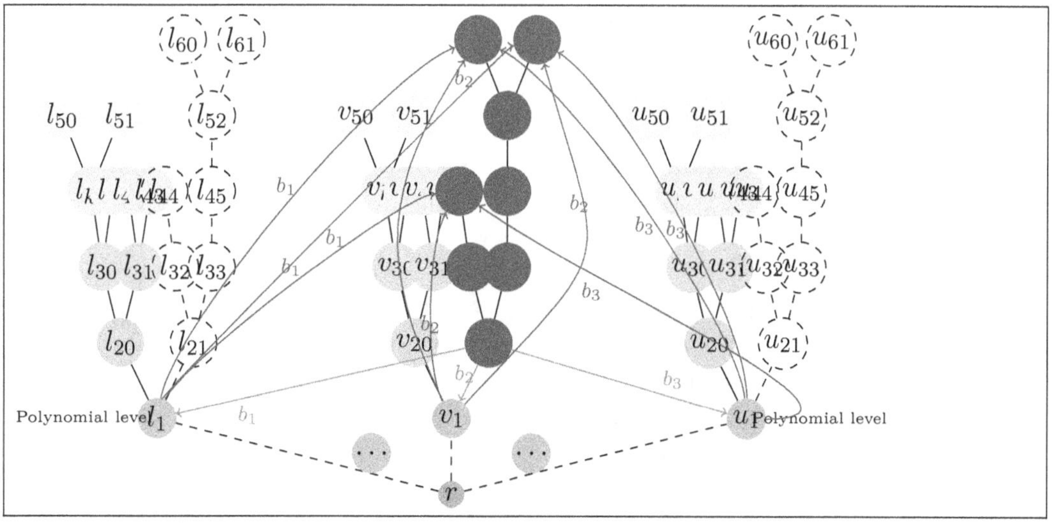

Figure 7: Three DetachLink were applied in the ambient r-DagProof $\mathcal{D}$ of fig. 6

7 Conclusion

In this article, we proved that redundancy is an inherent property of huge normal proofs in Natural Deduction for $\mathbf{M}_{\supset}$. The extension of this result to the full language of propositional logic seems to hold. The proof of this result, however, can be more complicated and may not be so relevant as what we have proved. [17, 12] provides a translation of any formula α from the full language of propositional logic, namely $\{\wedge, \vee, \neg, \supset, \bot\}$, to a purely implicational formula α^*, such that $|\alpha^*| \leq |\alpha|^3$ and $\vdash_{\mathbf{M}_\supset} \alpha^*$ if and only if $\vdash_{Int} \alpha$. Moreover, given a proof Π of α of height $h(\Pi)$ in the intuitionistic Natural Deduction, there is a proof Π^* of α^*, in $\mathbf{M}_\supset$, such that $h(\Pi^*) \leq 3 \times h(\Pi)$. Thus, any intuitionistic tautology α that has a huge normal proof and it is linearly height-bounded has an implicational translation α^*, equiprobable, that has a highly redundant proof too. We conclude that some huge certificates (proofs) for the validity of Intuitionistic tautologies, based on Natural Deduction proofs, can be chosen to be (highly redundant). It seems easier to obtain shorter proofs/certificates of validity from redundant proofs than from non-redundant ones. In [13], we have shown that in some particular cases, the use of Classical reasoning, i.e., to move to a Classical Logic Natural Deduction system, can shorten an exponentially sized proof to the linear size. The cases reported in [13] are of linear hight, and hence the proofs are highly redundant. Moving to the Classical Logic system does have nothing to do with the redundancy of the proofs. It has to do with the fact that what makes the proofs shown in [13] exponentially big, in this specific case, is a weaker form of the iterated Peirce formula, that needs at least exponentially many assumptions in the smallest normal proofs.

The compression of $\mathbf{M}_\supset$ normal ND proofs, in the general case, is not only a matter of moving to Classical Logic as it is in [13]. We notice that we have examples of exponentially sized Natural Deduction proofs, with linear height, in Classical logic itself. In the appendix A we show a class of Classical tautologies that lies in this case. However, we can use the inherent redundancy property from this article. The compression of implicational minimal proofs uses a device that we firstly have shown in ref [7] and improved in [8]. In [7], we collapse all nodes, in each level, of a linearly height-bounded proof-tree that have the same label (formula), from bottom up. The collapse of nodes in the way we just mentioned is what we call Horizontal Collapse ($\mathcal{HC}$). Using $\mathcal{HC}$ in exponentially sized and linear height proofs we provide a sequence of arbitrary NP-approximations for a PSPACE-complete problem, and in [8], we prove that NP=SPACE by using a certificate to verify that the DAG obtained by $\mathcal{HC}$. $\mathcal{HC}$ produces DAG-proofs of polynomial-size, and the certificates to verify that these DAG-proofs are correct are of polynomial-size too. That is, for each implicational proof that is exponentially sized and of linear-bounded height, we provide a pair of certificates, i.e., a double certificate, for the $\mathbf{M}_\supset$-tautology that it proves. By a result proved in [7], that shows that any $\mathbf{M}_\supset$-tautology has a proof in Natural Deduction that has the height linearly bounded, and, since $\mathbf{M}_\supset$ is PSPACE-complete, we have a proof that $NP = PSPACE$. We advise that the technique used in [7] and [8] does not rely on the normality of the huge linearly height-bounded proof, the huge proof can be non-normal. Thus, the range of a compression method based on the inherent redundancy theorem is narrower than what we have in [7] and [8]. However, we can observe that the use of the technique of collapsing sub-trees in any huge proof Π in the set $NHam$ (see section 4 and appendix A) produces a sub-exponential certificate by compressing Π into a Dag-proof, as shown in section 6. For all proofs in $NHam$ are normal and linearly height-bounded. This application proves the existence of sub-exponential certificates for any $CoNP$-complete problem since $NHam$ is a $CoNP$-complete problem. The algorithm 1 describes the main steps in a recursively defined compression. In [9] we have a preliminary version of the redundancy theorem for super-polynomially sized proofs that are linearly height-bounded. With [9] and an analogous compression method, we can provide an alternative proof of $NP = CoNP$ by providing polynomial certificates for every member of $NHam$.

There are other reasons to consider an alternative compression method based on the inherent redundancy theorems. Firstly, the use of a short double certificate, as in [8], to prove the validity of $\mathbf{M}_{\supset}$ tautology indeed introduces additional complexity in finding or building short DAG-based proofs of them. The definition of an automatic prover that always provides short DAG proofs (certificates) is a natural continuation of the investigations reported in [7, 8]. We do not want to consider a prover that applies the compression after obtaining a, possibly huge, proof as an efficient propositional prover. The provision of a simpler compression algorithm that does not produce a double certificate can be a step towards automatic generation of short proofs. A more straightforward compression method, i.e., one that produces a single certificate, can be based on the collapse of sub-derivations, instead of nodes. This method would provide us with a mathematically simpler proof of the soundness of the compressed proof or certificate than its double certificate counterpart proof of soundness. The definition of an automatic prover to generate shorter proofs of $\mathbf{M}_{\supset}$ tautologies based on the existence of the single short certificates seems to be easier to define than its double certificate counterpart.

A Proofs of non-hamiltonicity in ND

In this appendix, we show how to use $\mathbf{M}_{\supset}$ to generate certificates of the non-hamiltonicity, non-existence of a Hamiltonian cycle, employing Natural Deduction proofs that are linear height-bound.

A simple directed graph is a directed graph having no multiple edges, that is, for every pair of nodes (v_1, v_2) from the graph there is at most one edge from v_1 to v_2. Given a simple directed graph $G = \langle V, E \rangle$, with $card(V) = n$, a hamiltonian path in G is a sequence of nodes $v_1 v_2 \ldots v_n$, such that $v_i \in V$, $i = 1, n$, and for each $i, j \in V$, if $v_i = v_j$ then $i = j$, i.e, a path has no repetition of nodes. Moreover, if $v_i v_{i+1}$ is in the path then exists the edge $(v_i, v_{i+1} \in E$. The (decision) problem of knowing whether there is or not a hamiltonian path in a graph is known to be NP-complete. Thus, given a graph $\langle V, E\rangle$, with $card(V) = n$, to verify that a sequence of n nodes is a hamiltonian path it is enough to verify that: (1) There is no repeated node in the sequence; (2) No element $v \in V$ is out of the sequence, and; (3) For each pair $v_i v_j$ in the sequence there is an edge $(v_i, v_j) \in E$. We can see that these verifications require polynomial time on the size of the graph and that any path is linearly bounded by n. Thus any sequence of nodes of a graph can be viewed as a polynomially verified certificate for this graph hamiltonicity. Consider the following reduction of a Hamiltonian path to SAT, quite well-known from the literature in computational complexity (see [4]).

Definition 13. *Given $G = \langle V, E\rangle$, $card(V) = n > 0$. Let $X_{i,v}$, $i = 1, n$ e $v \in V$ be the proposicional language. Intuitively, $X_{i,v}$ express that the vertex v is visited in the step i in a path on G. Consider the formulas in the following definition:*

1. $A = \bigwedge_{v \in V}(X_{1,v} \vee \ldots \vee X_{n,v})$ *indicating that every vertex can be visited in any step in a hamiltonian path/cycle;*
2. $B = \bigwedge_{v \in V} \bigwedge_{i \neq j}(\neg(X_{i,v} \wedge X_{j,v}))$ *indicanting that there are no repetitions in any hamiltonian path/cycle;*
3. $C = \bigwedge_{i=1,n} \bigvee_{v \in V} X_{i,v}$ *that says that in each step one vertex should be visited;*
4. $D = \bigwedge_{v \neq w} \bigwedge_{i=1,n} \neg(X_{i,v} \wedge X_{i,w})$ *that indicates that each step can visit at most one vertex, and;*
5. $E = \bigwedge_{(v,w) \notin E} \bigwedge_{i=1,n-1}(X_{i,v} \supset \neg X_{i+1,w})$ *that indicates that if there is no edge from v to w then w cannot be visited immediately after v;*

We can see that G has a hamiltonian path if and only if $\alpha_G = A \wedge B \wedge C \wedge D \wedge E$ is satisfiable. Any hamiltonian path $v_1 \ldots v_n$ induces a truth-assignment T, such that $T(X_{i,w}) = true$ if and only if $w = v_i$, that satisfies α_G. Conversely, any truth-assignment that satisfies α_G induces a hamiltonian path in G. If we denote SAT_{Cla} the set of satisfiable formulas for the classical propositional logic and as $TAUT_{Int}$ the set of tautologies for the intuitionistic propositional logic, we can observe that the following statements are equivalent:

$$(1) G \text{ is not hamiltonian if and only if } \alpha_G \notin SAT_{Cla}$$

(2)G is not hamiltonian if and only if α_G is unsatisfiable

(3)G is not hamiltoniano if and only if $\neg\alpha_G \in TAUT_{Cla}$

(4)G is not hamiltonian if and only if $\neg\alpha_G \in TAUT_{Int}$

Hence, G is non-hamiltonian graph if and only if there is an intuitionistic proof (positive certificate) for $\neg\alpha_G$. Such proof is a certificate for non-hamiltonicity of graph G. To go from statement (3) to (4), we use Glyvenko theorem. In [17, 12] it is described a translation from formulas in the full language $\{\bot, \neg, \wedge, \vee, \supset\}$ to the purely implicational formulas, .i.e, formulas containing only the constant symbol $\supset$ and propositional variables. From any formula γ, the formula $\gamma^\star$ from purely implicational minimal logic is provable in the minimal logic if and only if γ is provable in intuitionistic logic. Moreover, concerning the sizes of the formulas, we have that $size(\alpha^\star) \leq size^3(\alpha)$ ([12]). The main idea described in [17, 12] is the use of implicational schemata that simulate the introduction and elimination of Natural Deduction rules. This simulation employs the use of new/fresh propositional variables. For example, for each pair of formulas A and B, we add the propositional variable $q_{A\vee B}$ and the formulas $A \supset q_{A\vee B}$, $B \supset q_{A\vee B}$ are used to simulate the $\vee$-introduction rules. Hence, any application of the rule:

$$\frac{A}{A \vee B}$$

is replaced by the following derivation in $ND_\supset$

$$\frac{A \qquad A \supset q_{A\vee B}}{q_{A\vee B}}$$

In this way, the new derivation is normal too. Remember that the changing of language, i.e., replacing the formula $A \vee B$ by $q_{A\vee B}$ occurs in all formulas of the original derivation. The formulas $(A \supset \beta) \supset ((B \supset \beta) \supset (q_{A\vee B} \supset \beta))$, for each β sub-formula from $\neg\alpha$, simulate the $\vee$-elimination. The fact that the original derivation is normal ensures that any application of an $\vee$-elimination has minor premisses as sub-formulas of the hypotheses or the conclusion, for the sub-formula principle holds for normal derivation. The ND proofs that we show here, for hamiltonicity, use a more economic translation, as we see below.

A (normal) proof of $\neg\alpha_G$, where G is a non-hamiltonian graph, is a proof of $\bot$ from α_G. Since α_G is a conjunction we can consider the certificate of non-hamiltonicity as any proof of $\bot$ from the set of formulas that form the components of the conjunctions. Thus, we consider (A) the disjunctions of the form $(X_{1,v}\vee\ldots\vee X_{n,v})$, with $v \in V$, from the item 1 of the definition of α_G; (B) the formulas $\neg(X_{i,v}\wedge X_{j,v})$, with $i = 1,n$ and $v \in V$, from item 2; (C) the formulas $\bigvee_{v\in V} X_{i,v}$, with $i = 1,n$, from item 3; (D) the formulas $\neg(X_{i,v} \wedge X_{i,w})$, $i = 1,n$ and $w \in V$, from item 4, and; (E) finally, the formulas $X_{i,v} \supset \neg X_{i+1,w}$, with $i = 1, n-1$ and $(v,w) \notin E$, from item 5. Let us examine a bit more the proof of $\neg\alpha^\star$ as a proof of $\bot$ from this set of formulas S_G just detailed above. A naive proof of non-hamiltonicity considers every possible path starting with every possible node of the graph, and then, considering every possible node of the graph to be visited at the second step and so on. Of course, as G has no hamiltonian path, every possible choice ends up in a contradiction. Either there is no unvisited node possible to be visited, and we use formulas from the conjunctions in B, and D to obtain the absurdity $\bot$ or there is no node to visit at all, and we use conjunctions from E to prove the $\bot$. Each step's choice accomplishes a $\vee$ elimination having a disjunction from the formulas A or C as major premises. The proof is a kind of decision tree upwards-down. It is upper-bounded by n^n, while its height is linear on n. We observe that each $\vee$-elimination is replaced by a combination of $\supset$-Introductions and $\supset$-eliminations. This combination increases the height of the tree by two for each application of $\vee$-elimination. See schematic proof below. There is no need for $\vee$-introduction. There is no need of the $\bot$ intuitionistic rule. The translation in [17] that replaces the $\bot$ by a new/fresh propositional variable q is used formally with the sake of uniformity.

Summing up, for each non-hamiltonian simple and directed graph G, with n nodes, there is a set of formulas S_G with complexity (length) at most n^3 and proof of size n^n rules and height at most $3n$ rules that have q as the conclusion. This proof is the certificate of non-hamiltonicity of G. This proof is also a normal proof, as defined by Gentzen and Prawitz. Hence, it satisfies the sub-formula principle.

A.1 Normal $ND_\supset$ proofs of non-hamiltonicity of simple directed graphs

The material in this subsection is taken from [10] and a short version of it also appears in an addendum to the article [8], to appear, eventually. In this addendum, and here, it has the purpose of pointing out a way to prove that $NP = CoNP$ without using a translation from linearly height-bounded Sequent Calculus proofs to linearly height-bounded Natural Deduction proofs, but not necessarily normal proofs. The redundancy theorem does not apply to non-normal proofs, but it applies to normal proofs of non-hamiltonicity of graphs as described in the sequel. The goal is to obtain compressed and polynomial certificates of validity (proofs, or DAG-proofs) for non-hamiltonicity. Here, in this paper, we reach only subexponential certificates.

Consider a non-hamiltonian graph $G = \langle V, E \rangle$ and the formula $\neg\alpha_G$ as stated in the previous section. Thus, $\alpha_G = A \wedge B \wedge C \wedge D \wedge E$, with A, B, C, D and E as defined in items 1 until 5 from definition 13. Let p be any sequence of nodes from V of length $card(V)$. From p we have the set $\{X_{1,p[1]}, \ldots, X_{n,p[n]}\}$ of propositional variables from the language of α_G. This set, as the sequence p, represents a potential path in the graph G, namely, the path that starts by visiting vertex $p[1]$, this is $X_{1,p[1]}$ holds, visits vertex $p[2]$, i.e. $X_{2,p[2]}$ holds, until ending with the visit of $p[n]$. However, this sequence is not checked as a valid path. The set $\{X_{1,p[1]}, \ldots, X_{n,p[n]}\}$ is a valid path if and only if it does not inconsistent with the formula α_G. As G does not have any hamiltonian path then we known that $\{X_{1,p[1]}, \ldots, X_{n,p[n]}\}$ is inconsistent with α. We consider a mapping that given a sequence p the set $X_p = \{X_{1,p[1]}, \ldots, X_{n,p[n]}\}$ is inconsistent with $alpha_G$, $p \mapsto X_p$. Thus, for any p, there is a derivation of $\perp$ from α in Natural Deduction by completeness. Using the translations described in [17, 12] we have a normal derivation of q from $alpha^\star_G$ and the set X_p. This is the content of lemma 15. Considering the set P of all sequences of length n and the lemma 15 we have a set of normal proofs $\{\Pi_p : p \in P\}$, where each Π_p is of the form:

$$\begin{array}{c} X_p \\ \Pi_p \\ q \end{array}$$

where sometimes we use Π_p to denote the above derivation. Moreover, in order to have an easier understanding, the derivation Π_p is taken as depending from the whole set X_p even when not every formula $X_{i,p[i]}$, $i = 1, n$, occurs in Π_p. We consider the set of nodes (vertexes) ordered as in $\{v_1, \ldots, v_k\}$, where $k = card(V)$. We introduce the following notations:

Definition 14. *Given a sequence of vertexes $p : \{1, \ldots, n\} \mapsto V$, we denote the sub-sequence $p[1] \ldots p[j]$ of p as $p[1..j]$, where $j \in \{1, \ldots, n\}$. Of course, $p[1] = p[1..1]$ and $p[1..n] = p$. Moreover, we denote by $p[-j]$ the sub-sequence $p[1..n-j]$, $j \leq n-1$. Obviously, $p[-(n-1)] = p[1]$. The concatenation of sequences p and q is denoted by $p; q$.*

In the sequel, given a set $X_p = \{X_{1,p[1]}, \ldots, X_{n,p[n]}\}$, we denote by $X_{p[-1]}$ the set $\{X_{1,p[1]}, \ldots, X_{n-1,p[n-1]}\}$. When dealing with sets of formulas A, the union $A \cup \{\beta\}$ can be denoted by A, β.

Consider now the following derivations $\Pi_{p[-1]}$, for each sequence $p \in P$, that use the set of normal proofs $\{\Pi_p : p \in P\}$ defined as above.

$$\Pi_{p[-1]} =$$

$$\dfrac{X_{n,v_1} \vee \ldots \vee X_{n,v_n} \qquad \begin{array}{c} X_{p[-1]}, X_{n,v_1} \\ \Pi_{p[-1];X_{n,v_1}} \\ q \end{array} \qquad \ldots \qquad \begin{array}{c} X_{p[-1]}, X_{n,v_n} \\ \Pi_{p[-1];X_{n,v_n}} \\ q \end{array}}{q}$$

If we consider that each Π_p is a derivation in $ND_{\supset}$, cf. lemma 15, then the following derivation is the derivation $\Pi_{p[-1]}$ translated to purely implicational minimal logic Natural Deduction, i.e. $ND_{\supset}$. We used, here, an n-ary version of the $\vee$-elimination rule, with the sake of a shorter presentation.

$$\Pi_{p[-1]} =$$

$$\dfrac{ORX_n \qquad \dfrac{\begin{array}{c} \dfrac{\dfrac{\begin{array}{c} X_{p[-1]},[X_{n,v_2}] \\ \Pi_{p[-1];X_{n,v_2}} \\ q \end{array}}{X_{n,v_2} \supset q} \qquad \dfrac{\dfrac{\begin{array}{c} X_{p[-1]},[X_{n,v_1}] \\ \Pi_{p[-1];X_{n,v_1}} \\ q \end{array}}{X_{n,v_1} \supset q} \quad (X_{n,v_1} \supset q) \supset ((((X_{n,v_2} \supset q) \supset \ldots ((X_{n,v_n} \supset q) \supset (ORX_n \supset q)))}{(\ldots (X_{n,v_2} \supset q) \ldots ((X_{n,v_n} \supset q) \supset (ORX_n \supset q)))}}{(\ldots (X_{n,v_3} \supset q) \ldots ((X_{n,v_n} \supset q) \supset (ORX_n \supset q)))} \\ \vdots \end{array}}{ORX_n \supset q}}{q}$$

The propositional variable ORX_n is the translation os the disjunction $X_{n,v_1} \vee \ldots \vee X_{n,v_n}$, as indicated by the translations schemata described in [17] and [12]. Moreover, we can see that the derivations $\Pi_{p[-1]}$, for each $p \in P$ are normal derivations in $ND_{\supset}$. We build proof of non-hamiltonicity from the last step to the first. The j-th element in the sequence p is related to the choice of visiting vertexes of the graph during step $n - j + 1$. The following derivation is, in analogy with the previous derivation, regarded to this j-th choice in the sequence, and is denoted by $\Pi_{p[-j]}$. Use use $k = n - j + 1$ to have a cleaner derivation. Observe that the propositional variables ORX_n to ORX_k were introduced as the translations of the corresponding disjunctions $X_{n,v_1} \vee \ldots \vee X_{n,v_n}$ to $X_{k,v_1} \vee \ldots \vee X_{k,v_n}$.

$$\Pi_{p[-j]} =$$

$$\dfrac{\dfrac{\begin{matrix} ORX_n, \\ \dots, \\ ORX_{k+1}, \\ X_{p[-j],[X_{k,v_2}]} \\ \Pi_{p[-j];X_{k,v_2}} \\ q \end{matrix}}{X_{k,v_2} \supset q} \qquad \dfrac{\dfrac{\begin{matrix} X_{p[-j],[X_{k,v_1}]} \\ \Pi_{p[-j];X_{k,v_1}} \\ q \end{matrix}}{X_{k,v_1} \supset q} \quad (X_{k,v_1} \supset q) \supset ((((X_{k,v_2} \supset q) \supset \dots ((X_{k,v_n} \supset q) \supset (ORX_k \supset q)))}{(\dots (X_{k,v_2} \supset q) \dots ((X_{k,v_n} \supset q) \supset (ORX_k \supset q)))}}{(\dots (X_{k,v_3} \supset q) \dots ((X_{k,v_n} \supset q) \supset (ORX_k \supset q)))}$$

$$\dfrac{ORX_k \qquad \dfrac{\vdots}{ORX_k \supset q}}{q}$$

The derivation that is our goal is the application of the schemata above to the first step. We can see that there are only n $p[-(n-1)]$ sequences, and by the recursive definitions we used above, there are only n derivations $\Pi_{p[-(n-1)]}$ normal derivations of q. Finally, by the last application of the implicational schema for the $\vee$-elimination, we obtain a proof of q. q is the (new) propositional variable used to translate the $\perp$, from ORX_1 to ORX_n that are in fact parts of A and from the formulas D, E and B used by lemma 15 in producing the derivations Π_p, for each $p \in P$. Now, by an iterated application of a series of $\supset$-introduction rules, we obtain a proof the translation of the formula $\neg\alpha$ in the purely implicational minimal logic Natural Deduction.

Lemma 15. *Let $G = \langle V, E \rangle$ be a simple and directed non-hamiltonian graph and $p = p[1] \dots p[n]$ be a sequence of vertexes from V. Then there is a (not necessarily unique) derivation*

$$\begin{matrix} X_p \\ \Pi_p \\ q \end{matrix}$$

where $X_p = \{X_{1,p[1]}, \dots, X_{n,p[n]}\}$ in $ND_{\supset}$.

The above derivation mentioned in the above lemma can use, and must use at least one of, the formulas B, C, D and E from items 2,3 and 4 from definition 13

Proof of the lemma: Since the graph G is not hamiltonian then any sequence of n vertexes cannot be a valid path. So using one of the formulas B to D , in their purely implicational form, we derive q (the translation of the absurdity logical constant). Since p is not a valid path on G, then at least one of the items must hold abut p:

Visiting a vertex more than once There are i, j, $1 \leq i < j \leq n$, such that $p[i] = p[j] = v \in V$. In this case consider i_1 and i_2 the least pair, $i_1 < i_2$, such that $p[i_1] = p[i_2]$. Π_p is the following derivation:

$$\dfrac{X_{i_2,v} \qquad \dfrac{X_{i_1,v} \qquad X_{i_1,v} \supset (X_{i_2,v} \supset q)}{X_{i_2,v} \supset q}}{q}$$

where $X_{i_1,v} \supset (X_{i_2,v} \supset q)$ is the translation of $(\neg(X_{i_1,v} \wedge X_{i_2,v})$ to purely implicational minimal logic. Observe that $\{X_{i_1,v}, X_{i_2,v}\} \subset X_p$ and $(\neg(X_{i_1,v} \wedge X_{i_2,v})$ is a component of the conjunction B from α.

Visiting a vertex without any linking edge There is i, $1 \leq i < n$, such that $p[i] = y \in V$, $p[i+1] = z \in V$ and there is no $(y, z) \in E$. In this case consider j the least number $1 \leq j < n$ such that $p[j] = y$, $p[j+1] = z$ and $(y, z) \notin E$. Π_p is the following derivation:

$$\dfrac{X_{j+1,z} \qquad \dfrac{X_{j,y} \qquad X_{j,y} \supset (X_{j+1,z} \supset q)}{X_{j+1,z} \supset q}}{q}$$

where $X_{j,y} \supset (X_{j+1,z} \supset q)$ is the translation of $X_{j,y} \supset \neg X_{j+1,z}$ to purely implicational minimal logic. Observe that $\{X_{j,y}, X_{j+1,z}\} \subset X_p$ and $X_{j,y} \supset \neg X_{j+1,z}$ is a component of the conjunction E from α.

Since the two items above are the only two possible reasons for a sequence p, with lengh $n = card(V)$, not being a valid path, we are done.

References

[1] A.Church. A set of postulates for the foundations of logic. Annals of Mathematics, 33(2):346–366, 1932. Corrections in [2].

[2] A.Church. A set of postulates for the foundations of logic (second paper). Annals of Mathematics, 34(4):839–864, 1933.

[3] Bovet, D. and Crescenzi, P. Introduction to the Theory of Complexity, Prentice-Hall, 1994.

[4] S. Arora and B. Barak. Computational Complexity: A Modern Approach. Cambridge University Press, 2009.

[5] G.Gentzen. Untersuchungen über das logische schließsen I. Mathematische Zeitschrift, 39:176–210, 1935. English translation in [6].

[6] G.Gentzen. Investigations into logical deduction I. American Philosophical Quarterly, 1(4):288–306, 1964. English translation by M. E. Szabo.

[7] L.Gordeev and E. H. Haeusler. Proof Compression and NP versus PSPACE. Studia Logica 107, 53-83 (2019).

[8] Lew Gordeev and Edward Hermann Haeusler. Proof Compression and NP Versus PSPACE II. Bulletin of the Section of Logic, Volume 49/3, pp. 213-230. https://doi.org/10.18778/0138-0680.2020.16

[9] E.H.Haeusler, On the Intrinsic Redundancy in Huge Natural Deduction proofs II: Analysing $\mathbf{M}_{\supset}$Super-Polynomial Proofs CoRR, abs/2009.09802v2, 2020, link: https://arxiv.org/abs/2009.09802v2/

[10] E.H.Haeusler, On the Intrinsic Redundancy in Huge Natural Deduction proofs: Analysing Purely Implicational Minimal Logic Proofs, CoRR, abs/2004.10659v2, link: https://arxiv.org/abs/2004.10659v2

[11] E.H.Haeusler, Going from the huge to the small: Efficient succint representation of proofs and derivations in Minimal implicational logic CoRR, submmitted, 2020.

[12] E. H. Haeusler, Propositional Logics Complexity and the Sub-Formula Property, in Proceedings Tenth International Workshop on Developments in Computational Models, DCM 2014, Vienna, Austria, 13th July 2014.

[13] E. H. Haeusler, How Many Times do We Need an Assumption to Prove a Tautology in Minimal Logic? Examples on the Compression Power of Classical Reasoning, Electronic Notes in Theoretical Computer Science, v.315, 2015, pp 31-46.

[14] D. Hilbert und W. Ackermann, *Grundzüge der theoretischen Logik.* Julius Springer, Berlin 1928. English version *Principles of Mathematical Logic*, Chelsea, New York 1950, ed. by R. E. Luce is based on 2nd German edition, 1938.

[15] J. Hudelmaier, An $O(n \log n)$-space decision procedure for intuitionistic propositional logic, J. Logic Computat. (3): 1–13 (1993)

[16] D. Prawitz, **Natural deduction: a proof-theoretical study**. Almqvist & Wiksell, 1965

[17] R. Statman, Intuitionistic propositional logic is polynomial-space complete, Theor. Comp. Sci. (9): 67–72 (1979)

 Received 03 February 2020

Extensive Measurement with Unrestricted Concatenation and No Maximal Elements

Gregory A. Kyriazis
Instituto Nacional de Metrologia, Qualidade e Tecnologia (Inmetro), Brazil
gakyriazis@inmetro.gov.br

Abstract

First-order predicate logic is consistently used here to prove the representation theorem for extensive measurement with unrestricted concatenation and no maximal elements stated by P. Suppes in his 1951 article "A set of independent axioms for extensive quantities." Suppes regarded as unnecessary to present a detailed proof of the theorem. He limited himself to showing that a function for any structure satisfying his axioms exists that satisfies the desired properties. He added that its proof follows along standard lines, as given, for instance, by O. Hölder in 1901. Our proof here follows where feasible Hölder's arguments and requires an Archimedean axiom which is slightly different from that provided by Suppes. Dedekind's theory of irrational numbers, simplified and described in great detail by E. Landau, is adopted because it is the most convenient for our purposes.

1 Introduction

Three parallel research mainstreams before 1950 were the abstract work on the part of mathematicians concerned with the axiomatic foundations of geometry, in special [6] (English version of that article available in [10]), the logical analysis of empirical procedures of measurement by philosophers of science, in particular [2], and the work on scale types and transformations by psychologists, in special [12]. Suppes showed in [13] that the ratio scale listed in [12] can be derived from purported qualitative laws about observed relations among certain entities, some of those laws already advanced in [2], thus unifying the three research mainstreams.

The general framework requires that a representation theorem, which provides an exact analysis of how we may infer quantitative assertions from fundamentally qualitative observations, and a uniqueness theorem, which specifies the conditions for regarding such inferences as unique, be both stated and proved [4] [7] [11] [15].

Suppes' approach in [13] differs in some ways from Hölder's in [6], including the axioms which were weakened to apply to both denumerable and nondenumerable relation structures, the fact that it covers the case where equivalence is not identity, and the use of "modern" abstract algebra to express the results. He regarded as unnecessary to present a detailed proof of the representation theorem. He limited himself to showing that a function for any structure satisfying his axioms exists that satisfies the desired properties. He added that its proof follows along standard lines, as given, for instance, in [6].

First-order predicate logic [3] [5] [14] is consistently used here to prove the representation theorem for the measurement of extensive quantities with unrestricted concatenation and no maximal elements. Dedekind's theory of irrational numbers, simplified and described in great detail in [8], is adopted because it is the most convenient for our purposes. Our proof follows where feasible Hölder's arguments and requires an Archimedean axiom which is slightly different from [13]. The Archimedean condition adopted here is essentially the one used by Behrend in [1]. Incidentally, Behrend's work also does not cover the case where equivalence is not identity.

The article is organized as follows. The axioms are presented and discussed in section 2. Elementary lemmas are listed and proved in an abbreviated way in section 3. The representation theorem is stated and proved in section 4. The uniqueness theorem is stated and proved in section 5. An outline is given in section 6. The conclusions are drawn in section 7.

2 Axioms

Definition 1. *A structure $\langle K, Q, \circ\rangle$, where K is a nonempty set of arbitrary elements x, y, z, ..., Q is a binary relation defined over K, and $\circ$ is a binary operation from $K \times K$ to K, is an* empirical extensive structure *if it satisfies the following six axioms for all x, y, and z:*

A.I. $xQy \wedge yQz \rightarrow xQz$

A.II. $(x \circ y) \circ zQx \circ (y \circ z)$

A.III. $xQy \rightarrow x \circ zQz \circ y$

A.IV. $\neg xQy \rightarrow \exists z(xQy \circ z \wedge y \circ zQx)$

A.V. $\neg x \circ yQx$

A.VI. $\exists n(\neg nxQy)$, *where the notation nx is defined recursively below.*

Notation 1. *Let $1x = x$ and $nx = (n-1)x \circ x$, where n is a positive integer.*

Consider for instance the comparison of masses of material objects x, y, and z. When x and y are placed respectively on each pan of a two-pan, equal-arm balance, the expression xQy symbolizes the observation that the pan containing x stabilizes itself at least as low as that containing y. The concatenation operation $x \circ y$ symbolizes the observation that both x and y are placed on the same pan. The axioms express qualitative laws about observable relations among the objects. The first axiom describes the relation as transitive. The second axiom asserts that the relation does not depend on the arrangement we choose to place the objects on the pans. The third axiom says the relation between x and y is preserved when z is simultaneously placed on both pans together with the existing objects. The fourth axiom asserts that if x stabilizes at a lower level than y, then an object z can be found which placed together with y exactly balances x. The fifth axiom says every object has a positive mass; that is, the pan containing both x and y stabilizes at a lower level than that containing only x. The sixth axiom is the Archimedean property which differs slightly from

$$xQy \rightarrow \exists n(yQnx),$$

which was provided in [13]. The version of the Archimedean property utilized here is essentially the one used by Behrend in [1]. We shall see that its use is required to obtain a proof of the representation theorem that follows where feasible Hölder's arguments in [6]. Some theorems that will be proved in the next sections refer to corresponding sections of that Hölder's article. We recommend the reader to follow the referred sections of [6] (or [10]) when reference to his article is made in the next sections.

3 Elementary lemmas

All lemmas and theorems in this article assume that all elements are in K. Proofs in this section are abbreviated due to lack of space. Lemmas 1 to 5 are respectively equivalent to Ths. 1 to 5 listed in [13] and proofs of those theorems are given therein. In particular, the proof of Th. 3 in [13] was corrected in a republished version of that article in [16].

Lemma 1. xQx

Lemma 2. $x \circ yQy \circ x$

Lemma 3. $xQy \wedge uQv \rightarrow x \circ uQy \circ v$

Lemma 4. $x \circ (y \circ z)Q(x \circ y) \circ z$

Lemma 5. $xQy \vee yQx$

Lemma 6a. $x \circ uQy \circ u \to xQy$ (see Th. 6 in [13] for a proof of this lemma).

Full proofs are given for the following two lemmas, where T denotes one of those tautologies listed in [14] and C.P. denotes conditional proof.

Lemma 6b. $xQy \to x \circ uQy \circ u$

Proof.

1.	xQy	Hypothesis
2.	uQu	Lemma 1
3.	$xQy \wedge uQu$	1, 2 T
4.	$x \circ uQy \circ u$	3 Lemma 3
5.	Lemma 6b	1, 4 C.P.
	Q.E.D.	

□

Lemma 6c. $xQy \to u \circ xQu \circ y$

Proof.

1.	uQu	Lemma 1
2.	xQy	Hypothesis
3.	$uQu \wedge xQy$	2, 1 T
4.	$u \circ xQu \circ y$	3 Lemma 3
5.	Lemma 6c	2, 4 C.P.
	Q.E.D.	

□

Lemmas 7 and 8 are not used here; they are listed just to follow where feasible the numbering of theorems adopted in [13].

Lemma 7. $y \circ zQu \wedge xQy \to x \circ zQu$

Lemma 8. $uQx \circ z \wedge xQy \to uQy \circ z$

A full proof of Lemma 9 is given below since it requires Lemma 4 instead of Lemma 6 (see proof of Th. 9 in [13] where reference is made to Th. 6).

Lemma 9. $mx \circ nxQ(m+n)x$

Proof. We use mathematical induction on n.

1.	$mx \circ xQmx \circ x$	Lemma 1
2.	$mx \circ 1xQmx \circ x$	1 Notation 1
3.	$mx \circ 1xQ(m+1)x$	2 ($n = 1$)
4.	$mx \circ kxQ(m+k)x$	Hyp. ($n = k$)
5.	xQx	Lemma 1
6.	$mx \circ kxQ(m+k)x \wedge xQx$	4, 5 T
7.	$(mx \circ kx) \circ xQ(m+k)x \circ x$	6 Lemma 3
8.	$mx \circ (kx \circ x)Q(mx \circ kx) \circ x$	Lemma 4
9.	$mx \circ (kx \circ x)Q(m+k)x \circ x$	8, 7 A. I.
10.	$mx \circ (k+1)xQ(m+k+1)x$	9 Notation 1
11.	$mx \circ kxQ(m+k)x \rightarrow mx \circ (k+1)xQ(m+k+1)x$	4, 10 C.P.
12.	Lemma 9	3, 11
	Q.E.D. (may repeat mathematical induction for m)	

□

Lemmas 10 to 17 are respectively equivalent to Ths. 10 to 17 listed in [13] and summarized proofs of those theorems are given therein.

Lemma 10. $(m+n)xQmx \circ nx$

Lemma 11. $n(mx)Q(nm)x$

Lemma 12. $(nm)xQn(mx)$

Lemma 13. $n(x \circ y)Qnx \circ ny$

Lemma 14. $nx \circ nyQn(x \circ y)$

Lemma 15. $xQy \rightarrow nxQny$

Lemma 16. $nxQny \rightarrow xQy$

Lemma 17. $m \leq n \rightarrow mxQnx$

Lemma 18a. $\neg xQz \wedge yQz \rightarrow \neg xQy$

Proof. Assume xQy. Using hypothesis, namely $\neg xQz \wedge yQz$, and A. I. we arrive at a contradiction. Q.E.D. □

Lemma 18b. $\neg xQz \wedge \neg zQy \rightarrow \neg xQy$

Proof. Assume xQy. Using hypothesis, namely $\neg xQz \wedge \neg zQy$, and Lemma 18a, we get: $\neg xQz \wedge \neg zQx$, which contradicts Lemma 5. Q.E.D. □

Lemma 18c. $\neg zQx \wedge zQy \rightarrow \neg yQx$

Proof. Similar to Lemma 18a. Q.E.D. □

Lemma 19. $\neg xQy \wedge \neg uQv \rightarrow \neg x \circ uQy \circ v$

Proof. Using hypothesis, namely $\neg xQy \wedge \neg uQv$, and the contraposition of Lemma 6a, we get: $\neg x \circ uQy \circ u$ and $\neg u \circ yQv \circ y$. Then using Lemma 2, Lemma 18c and Lemma 18a on these, we get: $\neg y \circ uQy \circ v$ and, with Lemma 5, $y \circ vQy \circ u$. Finally, using Lemma 18a on $\neg x \circ uQy \circ u$ and $y \circ vQy \circ u$, we get lemma. Q.E.D. □

4 Representation theorem

The exact formulation and proof of the representation theorem requires a remark on the notion of equivalence classes or cosets. For x in K, we define the C-equivalence class, $\mathfrak{x}$, in the standard manner, (see [14], §10.4, or [15]), as the class of all elements which stand in the relation C to x. Intuitively, $\mathfrak{x}$ is the class of all elements whose magnitude is equivalent to that of x. Every element of K must belong to one and only one such coset; and $\mathfrak{K}$ is the set of all such cosets, or the partition of K into C-equivalence classes. We define relations among equivalence classes of $\mathfrak{K}$ (e.g., Q^*, $\circ^*$) corresponding to all the primitive and defined relations among elements of K. Thus, we say that $\mathfrak{x}Q^*\mathfrak{y}$ if and only if every element of $\mathfrak{x}$ stands in the relation Q to every element of $\mathfrak{y}$. Similarly we say that $\mathfrak{x} \circ^* \mathfrak{y}$ if and only if for every x_1, y_1, if $x_1 \in \mathfrak{x}$ and $y_1 \in \mathfrak{y}$, then $x_1 \circ y_1$. It is an immediate consequence of these definitions that every such relation between equivalence classes of $\mathfrak{K}$ holds if and only if the corresponding elementary relation holds between the elements of K which generate the equivalence classes so related. For example, we can show that $\mathfrak{x} \circ^* \mathfrak{y}$ if and only if $x \circ y$. Hence, to every of the original definitions, axioms and lemmas in sections 2 and 3, there corresponds a coset definition, axiom or lemma. To avoid repetition in the proof, we have adopted the convention that when the number of any definition, axiom or lemma is prefixed by the letters "CS" the designation is to be understood as referring to the coset definition, axiom or lemma corresponding respectively to the numbered definition, axiom or lemma. Thus "CS Lemma 6c" names the coset lemma corresponding to Lemma 6c, namely CS Lemma 6c asserts: $\mathfrak{x}Q^*\mathfrak{y} \to \mathfrak{u} \circ^* \mathfrak{x}Q^*\mathfrak{u} \circ^* \mathfrak{y}$.

Definition 2. *Let* $\mathfrak{x}C\mathfrak{y}$ *if* $\mathfrak{x}Q^*\mathfrak{y} \wedge \mathfrak{y}Q^*\mathfrak{x}$

Lemma 20a. $\mathfrak{x}C\mathfrak{y} \wedge \mathfrak{x}Q^*\mathfrak{z} \to \mathfrak{y}Q^*\mathfrak{z}$

Proof. Use hypothesis, namely $\mathfrak{x}C\mathfrak{y} \wedge \mathfrak{x}Q^*\mathfrak{z}$, Def. 2 and CS A. I. Q.E.D. □

Lemma 20b. $\mathfrak{x}C\mathfrak{y} \wedge \neg\mathfrak{y}Q^*\mathfrak{z} \to \neg\mathfrak{x}Q^*\mathfrak{z}$

Proof. Use hypothesis, namely $\mathfrak{x}C\mathfrak{y} \wedge \neg\mathfrak{y}Q^*\mathfrak{z}$, Def. 2 and CS Lemma 18c. Q.E.D. □

Lemma 20c. $\mathfrak{x}C\mathfrak{y} \wedge \neg\mathfrak{x}Q^*\mathfrak{z} \to \neg\mathfrak{y}Q^*\mathfrak{z}$

Proof. Similar to Lemma 20b. Q.E.D. □

Lemma 20d. $\mathfrak{x}C\mathfrak{z} \wedge \neg\mathfrak{y}Q^*\mathfrak{z} \to \neg\mathfrak{y}Q^*\mathfrak{x}$

Proof. Similar to Lemma 20b. Q.E.D. □

Lemma 20e. $\mathfrak{x}C\mathfrak{y} \wedge \mathfrak{y}C\mathfrak{z} \to \mathfrak{x}C\mathfrak{z}$

Proof. Use hypothesis, namely $\mathfrak{x}C\mathfrak{y} \wedge \mathfrak{y}C\mathfrak{z}$, Def. 2, CS A. I. and Def. 2. Q.E.D. □

Lemma 21. $(\mathfrak{x} \circ^* \mathfrak{y}) \circ^* \mathfrak{z}C\mathfrak{x} \circ^* (\mathfrak{y} \circ^* \mathfrak{z})$

Proof. Use CS A. II., CS Lemma 4 and Def. 2. Q.E.D. □

Lemma 22. $\mathfrak{x}C\mathfrak{y} \to \mathfrak{z} \circ^* \mathfrak{x}C\mathfrak{z} \circ^* \mathfrak{y}$

Proof. Use hypothesis, namely $\mathfrak{x}C\mathfrak{y}$, Def. 2, CS Lemma 6c and Def. 2. Q.E.D. □

Lemma 23. $\neg\mathfrak{x}Q^*\mathfrak{y} \rightarrow \exists\mathfrak{z}(\mathfrak{x}C\mathfrak{y} \circ^* \mathfrak{z})$

Proof. Use hypothesis, namely $\neg\mathfrak{x}Q^*\mathfrak{y}$, CS A. IV. and Def. 2. Q.E.D. □

Definition 3. *Let N be a nonempty set of cuts, i.e., of positive real numbers. Let $\leq$ be the usual numerical binary relation and $+$ the usual numerical binary operation of addition, both relations restricted to the set N. Then we call $\langle N, \leq, +\rangle$ a* numerical extensive structure *if N is closed under addition and subtraction of smaller numbers from larger numbers, that is, if $\xi, \psi \in N$ and $\xi > \psi$, then $(\xi + \psi) \in N$ and $(\xi - \psi) \in N$.*

We now give the formal statement of the representation theorem. The formulation of the representation theorem in terms of cosets and their relations has the advantage that elements of $\mathfrak{K}$ are mapped into N in a one-one correspondence.

Definition 4. *An ordered triple $\langle\mathfrak{K}, Q^*, \circ^*\rangle$ is* isomorphic *to a numerical extensive structure $\langle N, \leq, +\rangle$ if, and only if, for each $\mathfrak{z}$ in $\mathfrak{K}$ there is a function $f_\mathfrak{z}$ such that*

(i) *$f_\mathfrak{z}$ maps $\mathfrak{K}$ into N* (Ths. 1a to 1d)*;*
(ii) $f_\mathfrak{z}(\mathfrak{x} \circ^* \mathfrak{y}) = f_\mathfrak{z}(\mathfrak{x}) + f_\mathfrak{z}(\mathfrak{y})$ (Th. 1e)*;*
(iii) *$f_\mathfrak{z}$ is a one-one function* (Th. 1f)*;*
(iv) $\mathfrak{x}Q^*\mathfrak{y} \leftrightarrow f_\mathfrak{z}(\mathfrak{x}) \leq f_\mathfrak{z}(\mathfrak{y})$ (Th. 1g).

Theorem 1. (Representation theorem) *If $\langle K, Q, \circ\rangle$ is an empirical extensive structure, then there is a numerical extensive structure $\langle N, \leq, +\rangle$ which is isomorphic to $\langle\mathfrak{K}, Q^*, \circ^*\rangle$.*

Proof. We prove first isomorphism condition (i) (see [6], or [10], §10). Let the two magnitudes $\mathfrak{x}$ and $\mathfrak{z}$ be given. Consider the comparison between $\mathfrak{x}$ and $\mathfrak{z}$. We need to prove that the set of all rational numbers m/n such that $\neg n\mathfrak{x}Q^*m\mathfrak{z}$ is a cut. If we wish to show that a set of rational numbers is a cut, we need only to show the following (see comments to Th. 120 in [8]):

(a) The set is not empty (Th.1a);
(b) There is a rational number not belonging to it (Th. 1b);
(c) With every number it contains, the set also contains all numbers smaller than that number (Th. 1c);
(d) With every number it contains, the set also contains a greater one (Th. 1d).

Full proofs are given for the following four theorems, where T denotes one of those tautologies listed in [14], C.P. denotes conditional proof, C. contraposition, EI existential instantiation, EG existential generalization, and UI universal instantiation. The proofs of Ths. 1a, 1b and 1d are the only ones where the Archimedean axiom is employed in this article.

Theorem 1a. $\exists m \exists n(\neg n\mathfrak{x}Q^*m\mathfrak{z})$

Proof.

1.	$\exists n(\neg n\mathfrak{x}Q^*\mathfrak{z})$	CS A. VI.
2.	$\exists n(\neg n\mathfrak{x}Q^*1\mathfrak{z})$	1 CS Notation 1
3.	Theorem 1a	2 EG
	Q.E.D.	

□

Theorem 1b. $\exists n \exists m(n\mathfrak{x}Q^*m\mathfrak{z})$

Proof.

1.	$\exists m(\neg m\mathfrak{z}Q^*\mathfrak{x})$	CS A. VI.
2.	$\neg r\mathfrak{z}Q^*\mathfrak{x}$	1 EI
3.	$r\mathfrak{z}Q^*\mathfrak{x} \vee \mathfrak{x}Q^*r\mathfrak{z}$	CS Lemma 5
4.	$\mathfrak{x}Q^*r\mathfrak{z}$	2, 3 T
5.	$\exists m(\mathfrak{x}Q^*m\mathfrak{z})$	4 EG
6.	$\exists m(1\mathfrak{x}Q^*m\mathfrak{z})$	5 CS Notation 1
7.	Theorem 1b	6 EG
	Q.E.D.	

□

Theorem 1c. $\neg n\mathfrak{x}Q^*m\mathfrak{z} \wedge (m'/n' < m/n) \rightarrow \neg n'\mathfrak{x}Q^*m'\mathfrak{z}$

Proof.

1.	$\neg n\mathfrak{x}Q^*m\mathfrak{z} \wedge (m'/n' < m/n)$	Hypothesis
2.	$\neg n\mathfrak{x}Q^*m\mathfrak{z}$	1 T
3.	$m'/n' < m/n$	1 T
4.	$m'n < mn'$	3
5.	$(m'n)\mathfrak{x}Q^*(mn')\mathfrak{x}$	4 CS Lemma 17
6.	$\neg m'(n\mathfrak{x})Q^*m'(m\mathfrak{z})$	2 CS Lemma 16 (C.)
7.	$m'(n\mathfrak{x})Q^*(m'n)\mathfrak{x}$	6 CS Lemma 11
8.	$\neg(m'n)\mathfrak{x}Q^*m'(m\mathfrak{z})$	6, 7 CS Lemma 18c
9.	$(m'm)\mathfrak{z}Q^*m'(m\mathfrak{z})$	CS Lemma 12
10.	$\neg(m'n)\mathfrak{x}Q^*(m'm)\mathfrak{z}$	8, 9 CS Lemma 18a
11.	$\neg(mn')\mathfrak{x}Q^*(m'm)\mathfrak{z}$	10, 5 CS Lemma 18c
12.	$m(m'\mathfrak{z})Q^*(mm')\mathfrak{z}$	CS Lemma 11
13.	$m(m'\mathfrak{z})Q^*(m'm)\mathfrak{z}$	12

14.	$\neg(mn')\mathfrak{x}Q^*m(m'\mathfrak{z})$	11, 13 CS Lemma 18a
15.	$(mn')\mathfrak{x}Q^*m(n'\mathfrak{x})$	CS Lemma 12
16.	$\neg m(n'\mathfrak{x})Q^*m(m'\mathfrak{z})$	14, 15 CS Lemma 18c
17.	$\neg n'\mathfrak{x}Q^*m'\mathfrak{z}$	16 CS Lemma 15 (C.)
18.	Theorem 1c	1, 17 C.P.
	Q.E.D.	

□

Hölder proved instead the following version of Theorem 1c (see [6], or [10], §10): $n'\mathfrak{x}Q^*m'\mathfrak{z} \wedge \neg n\mathfrak{x}Q^*m\mathfrak{z} \to (m/n < m'/n')$. One may easily prove that Hölder´s version entails Th. 1c.

Theorem 1d. $\neg n\mathfrak{x}Q^*m\mathfrak{z} \to \exists n'\exists m'(\neg n'\mathfrak{x}Q^*m'\mathfrak{z} \wedge (m'/n' > m/n))$

Proof.

1.	$\neg n\mathfrak{x}Q^*m\mathfrak{z}$	Hypothesis
2.	$\exists\mathfrak{w}(n\mathfrak{x}Q^*m\mathfrak{z} \circ^* \mathfrak{w} \wedge m\mathfrak{z} \circ^* \mathfrak{w}Q^*n\mathfrak{x})$	1 CS A. IV
3.	$n\mathfrak{x}Q^*m\mathfrak{z} \circ^* \alpha \wedge m\mathfrak{z} \circ^* \alpha Q^*n\mathfrak{x}$	2 EI
4.	$n\mathfrak{x}Q^*m\mathfrak{z} \circ^* \alpha$	3 T
5.	$m\mathfrak{z} \circ^* \alpha Q^*n\mathfrak{x}$	3 T
6.	$l(n\mathfrak{x})Q^*l(m\mathfrak{z} \circ^* \alpha)$	4 CS Lemma 15
7.	$l(m\mathfrak{z} \circ^* \alpha)Q^*l(n\mathfrak{x})$	5 CS Lemma 15
8.	$l(m\mathfrak{z} \circ^* \alpha)Q^*l(m\mathfrak{z}) \circ^* l\alpha$	CS Lemma 13
9.	$l(m\mathfrak{z}) \circ^* l\alpha Q^*l(m\mathfrak{z} \circ^* \alpha)$	CS Lemma 14
10.	$l(n\mathfrak{x})Q^*l(m\mathfrak{z}) \circ^* l\alpha$	6, 8 CS A. I.
11.	$l(m\mathfrak{z}) \circ^* l\alpha Q^*l(n\mathfrak{x})$	9, 7 CS A. I.
12.	$l(m\mathfrak{z})Q^*(lm)\mathfrak{z}$	CS Lemma 11
13.	$(lm)\mathfrak{z}Q^*l(m\mathfrak{z})$	CS Lemma 12
14.	$l\alpha Q^*l\alpha$	CS Lemma 1
15.	$l(m\mathfrak{z}) \circ^* l\alpha Q^*(lm)\mathfrak{z} \circ^* l\alpha$	12, 14 CS Lemma 3
16.	$(lm)\mathfrak{z} \circ^* l\alpha Q^*l(m\mathfrak{z}) \circ^* l\alpha$	13, 14 CS Lemma 3
17.	$l(n\mathfrak{x})Q^*(lm)\mathfrak{z} \circ^* l\alpha$	10, 15 CS A. I.
18.	$(lm)\mathfrak{z} \circ^* l\alpha Q^*l(n\mathfrak{x})$	16, 11 CS A. I.
19.	$(ln)\mathfrak{x}Q^*l(n\mathfrak{x})$	CS Lemma 12
20.	$l(n\mathfrak{x})Q^*(ln)\mathfrak{x}$	CS Lemma 11
21.	$(ln)\mathfrak{x}Q^*(lm)\mathfrak{z} \circ^* l\alpha$	19, 17 CS A. I.
22.	$(lm)\mathfrak{z} \circ^* l\alpha Q^*(ln)\mathfrak{x}$	18, 20 CS A. I.
23.	$(lm+1)\mathfrak{z}Q^*(lm)\mathfrak{z} \circ^* \mathfrak{z}$	CS Lemma 10
24.	$(lm)\mathfrak{z} \circ^* \mathfrak{z}Q^*(lm+1)\mathfrak{z}$	CS Lemma 9

25.	$(lm)\mathfrak{z} \circ^* l\alpha C(ln)\mathfrak{x}$	22, 21 Def. 2
26.	$(lm+1)\mathfrak{z}C(lm)\mathfrak{z} \circ^* \mathfrak{z}$	23, 24 Def. 2
27.	$\exists n(\neg n\alpha Q^*\mathfrak{z})$	CS A. VI
28.	$\neg r\alpha Q^*\mathfrak{z}$	27 EI
29.	$\exists\mathfrak{u}(r\alpha C\mathfrak{z} \circ^* \mathfrak{u})$	28 Lemma 23
30.	$r\alpha C\mathfrak{z} \circ^* \beta$	29 EI
31.	$(rm)\mathfrak{z} \circ^* r\alpha C(rm)\mathfrak{z} \circ^* (\mathfrak{z} \circ^* \beta)$	30 Lemma 22
32.	$(rm)\mathfrak{z} \circ^* (\mathfrak{z} \circ^* \beta)C((rm)\mathfrak{z} \circ^* \mathfrak{z}) \circ^* \beta$	Lemma 21
33.	$(rm)\mathfrak{z} \circ^* r\alpha C((rm)\mathfrak{z} \circ^* \mathfrak{z}) \circ^* \beta$	31, 32 Lemma 20e
34.	$\neg((rm)\mathfrak{z} \circ^* \mathfrak{z}) \circ^* \beta Q^*(rm)\mathfrak{z} \circ^* \mathfrak{z}$	CS A. V.
35.	$\neg(rm)\mathfrak{z} \circ^* r\alpha Q^*(rm)\mathfrak{z} \circ^* \mathfrak{z}$	33, 34 Lemma 20b
36.	$(rm)\mathfrak{z} \circ^* r\alpha C(rn)\mathfrak{x}$	25 UI
37.	$\neg(rn)\mathfrak{x}Q^*(rm)\mathfrak{z} \circ^* \mathfrak{z}$	36, 35 Lemma 20c
38.	$(rm+1)\mathfrak{z}C(rm)\mathfrak{z} \circ^* \mathfrak{z}$	26 UI
39.	$\neg(rn)\mathfrak{x}Q^*(rm+1)\mathfrak{z}$	38, 37 Lemma 20d
40.	$r' = rn$	Def.
41.	$s' = rm+1$	Def.
42.	$\neg r'\mathfrak{x}Q^*s'\mathfrak{z}$	39-41 T
43.	$s'/r' > m/n$	40, 41
44.	$\neg r'\mathfrak{x}Q^*s'\mathfrak{z} \wedge (s'/r' > m/n)$	42, 43 T
45.	$\exists n'\exists m'(\neg n'\mathfrak{x}Q^*m'\mathfrak{z} \wedge (m'/n' > m/n))$	44 EG
46.	Theorem 1d	1, 45 C.P.
	Q.E.D.	

□

For each given magnitude $\mathfrak{z}$ in $\mathfrak{K}$ we may then define a function $f_\mathfrak{z}$ which maps $\mathfrak{K}$ into N. The element of N will be denoted $f_\mathfrak{z}(\mathfrak{x})$ when belonging to the comparison between $\mathfrak{x}$ and $\mathfrak{z}$. One can also call the cut, $f_\mathfrak{z}(\mathfrak{x})$, the measured (numerical) value obtained when magnitude $\mathfrak{x}$ is compared with magnitude $\mathfrak{z}$, in which case $\mathfrak{z}$ is called the unit. The cut belonging to the comparison between $\mathfrak{x}$ and $\mathfrak{x}$ represents the number 1.

Now we prove isomorphism condition (ii) (see [6], or [10], §12). Let the three magnitudes $\mathfrak{x}$, $\mathfrak{y}$ and $\mathfrak{z}$ be given. The comparison between magnitudes $\mathfrak{x}$ and $\mathfrak{z}$ defines the first cut, $f_\mathfrak{z}(\mathfrak{x})$, namely, the set of all rational numbers m/n such that $\neg n\mathfrak{x}Q^*m\mathfrak{z}$. The comparison between magnitudes $\mathfrak{y}$ and $\mathfrak{z}$ defines the second cut, $f_\mathfrak{z}(\mathfrak{y})$, namely, the set of all rational numbers m'/n' such that $\neg n'\mathfrak{y}Q^*m'\mathfrak{z}$. From [8], Th. 129 and Def. 34, these two cuts determine a third cut which is their sum, $f_\mathfrak{z}(\mathfrak{x})+f_\mathfrak{z}(\mathfrak{y})$, namely, the set of all rational numbers $m/n+m'/n'$ such that $\neg n\mathfrak{x}Q^*m\mathfrak{z} \wedge \neg n'\mathfrak{y}Q^*m'\mathfrak{z}$.

Every rational number of the third cut can be represented by $m/n + m'/n' = (n'm+nm')/nn'$, where m/n and m'/n' are rational numbers of the first and second cuts, respectively.

Lemma 24. $\neg n\mathfrak{x}Q^*m\mathfrak{z} \wedge \neg n'\mathfrak{y}Q^*m'\mathfrak{z} \rightarrow \neg(nn')(\mathfrak{x} \circ^* \mathfrak{y})Q^*(n'm + nm')\mathfrak{z}$

Proof.

1.	$\neg n\mathfrak{x}Q^*m\mathfrak{z} \wedge \neg n'\mathfrak{y}Q^*m'\mathfrak{z}$	Hypothesis
2.	$\neg n\mathfrak{x}Q^*m\mathfrak{z}$	1 T
3.	$\neg n'\mathfrak{y}Q^*m'\mathfrak{z}$	1 T
4.	$\neg n'(n\mathfrak{x})Q^*n'(m\mathfrak{z})$	2 CS Lemma 16 (C.)
5.	$(n'm)\mathfrak{z}Q^*n'(m\mathfrak{z})$	3 CS Lemma 16 (C.)
6.	$(n'm)\mathfrak{z}Q^*n'(m\mathfrak{z})$	CS Lemma 12
7.	$(nm')\mathfrak{z}Q^*n(m'\mathfrak{z})$	CS Lemma 12
8.	$\neg n'(n\mathfrak{x})Q^*(n'm)\mathfrak{z}$	4, 6 CS Lemma 18a
9.	$\neg n(n'\mathfrak{y})Q^*(nm')\mathfrak{z}$	5, 7 CS Lemma 18a
10.	$n'(n\mathfrak{x})Q^*(n'n)\mathfrak{x}$	CS Lemma 11
11.	$n(n'\mathfrak{y})Q^*(nn')\mathfrak{y}$	CS Lemma 11
12.	$\neg(n'n)\mathfrak{x}Q^*(n'm)\mathfrak{z}$	8, 10 CS Lemma 18c
13.	$\neg(nn')\mathfrak{y}Q^*(nm')\mathfrak{z}$	9, 11 CS Lemma 18c
14.	$\neg(n'n)\mathfrak{x} \circ^* (nn')\mathfrak{y}Q^*(n'm)\mathfrak{z} \circ^* (nm')\mathfrak{z}$	12, 13 CS Lemma 19
15.	$(n'n)\mathfrak{x} \circ^* (nn')\mathfrak{y}Q^*(nn')(\mathfrak{x} \circ^* \mathfrak{y})$	CS Lemma 14
16.	$\neg(nn')(\mathfrak{x} \circ^* \mathfrak{y})Q^*(n'm)\mathfrak{z} \circ^* (nm')\mathfrak{z}$	14, 15 CS Lemma 18c
17.	$(n'm + nm')\mathfrak{z}Q^*(n'm)\mathfrak{z} \circ^* (nm')\mathfrak{z}$	CS Lemma 10
18.	$\neg(nn')(\mathfrak{x} \circ^* \mathfrak{y})Q^*(n'm + nm')\mathfrak{z}$	16, 17 CS Lemma 18a
19.	Lemma 24	1, 18 C.P.
	Q.E.D.	

□

Lemma 24 asserts nothing more that every rational number for the third cut is also a rational number for the cut defined by the comparison between magnitudes $\mathfrak{x} \circ^* \mathfrak{y}$ and $\mathfrak{z}$. The latter cut is the set of all rational numbers $(n'm + nm')/nn'$ such that $\neg(nn')(\mathfrak{x} \circ^* \mathfrak{y})Q^*(n'm + nm')\mathfrak{z}$.

Every rational number which is not contained in the third cut can be expressed in the form $r/s+r'/s' = ((s'r+sr'))/(ss')$, where r/s and r'/s' are rational numbers not contained in the first and second cuts, respectively. From [8], Th. 129, no number of the third cut can be written in the form $r/s + r'/s'$ where $s\mathfrak{x}Q^*r\mathfrak{z} \wedge s'\mathfrak{y}Q^*r'\mathfrak{z}$.

Lemma 25. $s\mathfrak{x}Q^*r\mathfrak{z} \wedge s'\mathfrak{y}Q^*r'\mathfrak{z} \to (ss')(\mathfrak{x} \circ^* \mathfrak{y})Q^*(s'r + sr')\mathfrak{z}$

Proof. Similar to Lemma 24, except that CS Lemma 15, CS Lemma 11, CS A. I., CS Lemma 12, CS A. I., CS Lemma 3, CS Lemma 13, CS A. I., CS Lemma 9, and CS A. I. are substituted for CS Lemma 16 (contraposition), CS Lemma 12, CS Lemma 18a, CS Lemma 11, CS Lemma 18c, CS Lemma 19, CS Lemma 14, CS Lemma 18c, CS Lemma 10, and CS Lemma 18a, respectively. Q.E.D. □

Lemma 25 implies that every rational number which is not contained in the third cut is also not contained in the cut defined by the comparison between magnitudes $\mathfrak{x} \circ^* \mathfrak{y}$ and $\mathfrak{z}$.

Taking the cut corresponding to the comparison between magnitudes $\mathfrak{x} \circ^* \mathfrak{y}$ and $\mathfrak{z}$ as the fourth cut, $f_\mathfrak{z}(\mathfrak{x} \circ^* \mathfrak{y})$, then what has been shown so far is that any rational number for the third cut is also a rational number for the fourth, and every rational number which is not contained in the third cut is also not contained in the fourth.

Theorem 1e. $f_\mathfrak{z}(\mathfrak{x} \circ^* \mathfrak{y}) = f_\mathfrak{z}(\mathfrak{x}) + f_\mathfrak{z}(\mathfrak{y})$

Proof. From [8], Th. 123, exactly one is the case:
(a) $f_\mathfrak{z}(\mathfrak{x}) + f_\mathfrak{z}(\mathfrak{y}) = f_\mathfrak{z}(\mathfrak{x} \circ^* \mathfrak{y})$
(b) $f_\mathfrak{z}(\mathfrak{x}) + f_\mathfrak{z}(\mathfrak{y}) > f_\mathfrak{z}(\mathfrak{x} \circ^* \mathfrak{y})$
(c) $f_\mathfrak{z}(\mathfrak{x}) + f_\mathfrak{z}(\mathfrak{y}) < f_\mathfrak{z}(\mathfrak{x} \circ^* \mathfrak{y})$

From [8], Def. 30, case (b) means that there exists a rational number for the third cut which is not contained in the fourth. But this contradicts the conclusion of Lemma 24. From [8], Def. 31, case (c) means that there exists a rational number not contained in the third cut which is a rational number of the fourth. But this contradicts the conclusion of Lemma 25. Therefore, the third cut is identical with the fourth and consequently, from [8], Th. 117, the fourth cut is identical with the third. Q.E.D. □

The cut corresponding to the comparison between magnitudes $\mathfrak{x} \circ^* \mathfrak{y}$ and $\mathfrak{z}$ is the sum of the cuts belonging to the comparisons between $\mathfrak{x}$ and $\mathfrak{z}$ and between $\mathfrak{y}$ and $\mathfrak{z}$. Put in other words this means: The measured (numerical) value of the concatenation of magnitudes $\mathfrak{x}$ and $\mathfrak{y}$, $\mathfrak{x} \circ^* \mathfrak{y}$, is the arithmetical sum of the measured (numerical) values of $\mathfrak{x}$ and $\mathfrak{y}$, provided that $\mathfrak{x}$, $\mathfrak{y}$ and $\mathfrak{x} \circ^* \mathfrak{y}$ are all measured relative to the same arbitrary unit $\mathfrak{z}$.

Then we prove isomorphism condition (iii) (see [6], or [10], §12 and §15).

Lemma 26. $\neg n\mathfrak{x}Q^*m\mathfrak{z} \wedge \mathfrak{x}C\mathfrak{y} \rightarrow \neg n\mathfrak{y}Q^*m\mathfrak{z}$

Proof. Use hypothesis, namely $\neg n\mathfrak{x}Q^*m\mathfrak{z} \wedge \mathfrak{x}C\mathfrak{y}$, Def. 2, CS Lemma 15, Def. 2, and Lemma 20c. Q.E.D. □

Lemma 26 asserts nothing more that if $\mathfrak{x}C\mathfrak{y}$, then every rational number for the first cut is also a rational number for the second.

Lemma 27. $s\mathfrak{x}Q^*r\mathfrak{z} \wedge \mathfrak{x}C\mathfrak{y} \rightarrow s\mathfrak{y}Q^*r\mathfrak{z}$

Proof. Similar to Lemma 26, except that Lemma 20a is substituted for Lemma 20c. Q.E.D. □

Lemma 27 implies that if $\mathfrak{x}C\mathfrak{y}$, then every rational number which is not contained in the first cut is also not contained in the second.

Lemma 28. $\mathfrak{x}C\mathfrak{y} \rightarrow f_{\mathfrak{z}}(\mathfrak{x}) = f_{\mathfrak{z}}(\mathfrak{y})$

Proof. Use Lemma 26, Lemma 27 and a rationale similar to that adopted in Th. 1e to prove that the first cut is identical with the second. Q.E.D. □

Lemma 29. $\neg\mathfrak{x}Q^*\mathfrak{y} \rightarrow f_{\mathfrak{z}}(\mathfrak{x}) > f_{\mathfrak{z}}(\mathfrak{y})$

Proof. Using hypothesis, namely $\neg\mathfrak{x}Q^*\mathfrak{y}$, Lemma 23 (and EI) and Lemma 28, we get: $f_{\mathfrak{z}}(\mathfrak{x}) = f_{\mathfrak{z}}(\mathfrak{y} \circ^* \alpha)$. Then using Th. 1e, we get lemma. Q.E.D. □

Lemma 30. $f_{\mathfrak{z}}(\mathfrak{x}) \leq f_{\mathfrak{z}}(\mathfrak{y}) \rightarrow \mathfrak{x}Q^*\mathfrak{y}$

Proof. Contraposition of Lemma 29. Q.E.D. □

Theorem 1f. $\neg\mathfrak{x}C\mathfrak{y} \rightarrow f_{\mathfrak{z}}(\mathfrak{x}) \neq f_{\mathfrak{z}}(\mathfrak{y})$

Proof. Use hypothesis, namely $\neg\mathfrak{x}C\mathfrak{y}$, Def. 2 and Lemma 29. Q.E.D. □

Theorem 1f implies that, for each given magnitude $\mathfrak{z}$ in $\mathfrak{K}$, the function $f_{\mathfrak{z}}$ which maps $\mathfrak{K}$ into N is one-one. Put in other words this means: if the unit is fixed, then there is exactly one cut for each given magnitude and there is exactly one magnitude for each given cut.

Finally, we prove isomorphism condition (iv) (see [6], or [10], §12).

Lemma 31. $\mathfrak{y}Q^*\mathfrak{x} \wedge \neg\mathfrak{x}Q^*\mathfrak{y} \rightarrow f_{\mathfrak{z}}(\mathfrak{x}) > f_{\mathfrak{z}}(\mathfrak{y})$

Proof. Obvious from Lemma 29. Q.E.D. □

Lemma 32. $\mathfrak{x}Q^*\mathfrak{y} \rightarrow f_{\mathfrak{z}}(\mathfrak{x}) \leq f_{\mathfrak{z}}(\mathfrak{y})$

Proof. Using hypothesis, namely $\mathfrak{x}Q^*\mathfrak{y}$, and the excluded middle assumption, we may write $\mathfrak{x}Q^*\mathfrak{y} \wedge (\neg\mathfrak{y}Q^*\mathfrak{x} \vee \mathfrak{y}Q^*\mathfrak{x})$, so that $(\mathfrak{x}Q^*\mathfrak{y} \wedge \neg\mathfrak{y}Q^*\mathfrak{x}) \vee (\mathfrak{x}Q^*\mathfrak{y} \wedge \mathfrak{y}Q^*\mathfrak{x})$. Then use Def. 2, Lemma 31 and Lemma 28. Q.E.D. □

Theorem 1g. $\mathfrak{x}Q^*\mathfrak{y} \leftrightarrow f_{\mathfrak{z}}(\mathfrak{x}) \leq f_{\mathfrak{z}}(\mathfrak{y})$

Proof. Obvious from Lemma 30 and Lemma 32. Q.E.D. □

Theorem 1g implies that, for each given magnitude $\mathfrak{z}$ in $\mathfrak{K}$, the function $f_{\mathfrak{z}}$ which maps $\mathfrak{K}$ into N is order preserving.

Theorem 1 then follows from Theorems 1a-1g. Q.E.D. □

5 Uniqueness theorem

This theorem is equivalent to saying that in the measurement of extensive quantities only the choice of a unit is arbitrary (see [6], or [10], §13; see also [13] for an alternative proof).

Let the three magnitudes $\mathfrak{x}$, $\mathfrak{y}$ and $\mathfrak{z}$ be given. Consider two numerical extensive structures, $\langle N_1, \leq, + \rangle$ and $\langle N_2, \leq, + \rangle$, isomorphic to $\langle \mathfrak{K}, Q^*, \circ^* \rangle$. From Theorem 1, we may define a function $f_{\mathfrak{z}}$ which maps $\mathfrak{K}$ into N_1. The measured (numerical) value of $\mathfrak{x}$ relative to $\mathfrak{z}$ as a unit defines the first cut, $f_{\mathfrak{z}}(\mathfrak{x})$, namely, the set of all rational numbers m/n such that $\neg n\mathfrak{x}Q^*m\mathfrak{z}$. We may also define a function $g_{\mathfrak{y}}$ which maps $\mathfrak{K}$ into N_2. The measured (numerical) value of $\mathfrak{z}$ relative to $\mathfrak{y}$ as a unit defines the second cut, $g_{\mathfrak{y}}(\mathfrak{z})$, namely, the set of all rational numbers k/l such that $\neg l\mathfrak{z}Q^*k\mathfrak{y}$. From [8], Th. 141 and Def. 36, these two cuts determine a third cut which is their product, $f_{\mathfrak{z}}(\mathfrak{x})g_{\mathfrak{y}}(\mathfrak{z})$, namely, the set of all rational numbers $(m/n)(k/l)$ such that $\neg n\mathfrak{x}Q^*m\mathfrak{z} \wedge \neg l\mathfrak{z}Q^*k\mathfrak{y}$.

Every rational number of the third cut can be represented by $(m/n)(k/l) = mk/nl$, where m/n and k/l are rational numbers of the first and second cuts, respectively.

Lemma 33. $\neg n\mathfrak{x}Q^*m\mathfrak{z} \wedge \neg l\mathfrak{z}Q^*k\mathfrak{y} \rightarrow \neg(ln)\mathfrak{x}Q^*(mk)\mathfrak{y}$

Proof.

1.	$\neg n\mathfrak{x}Q^*m\mathfrak{z} \wedge \neg l\mathfrak{z}Q^*k\mathfrak{y}$	Hypothesis
2.	$\neg n\mathfrak{x}Q^*m\mathfrak{z}$	1 T
3.	$\neg l\mathfrak{z}Q^*k\mathfrak{y}$	1 T
4.	$\neg l(n\mathfrak{x})Q^*l(m\mathfrak{z})$	2 CS Lemma 16 (C.)
5.	$\neg m(l\mathfrak{z})Q^*m(k\mathfrak{y})$	3 CS Lemma 16 (C.)
6.	$(lm)\mathfrak{z}Q^*l(m\mathfrak{z})$	CS Lemma 12
7.	$(mk)\mathfrak{y}Q^*m(k\mathfrak{y})$	CS Lemma 12
8.	$\neg l(n\mathfrak{x})Q^*(lm)\mathfrak{z}$	4, 6 CS Lemma 18a
9.	$\neg m(l\mathfrak{z})Q^*(mk)\mathfrak{y}$	5, 7 CS Lemma 18a
10.	$l(n\mathfrak{x})Q^*(ln)\mathfrak{x}$	CS Lemma 11
11.	$m(l\mathfrak{z})Q^*(ml)\mathfrak{z}$	CS Lemma 11
12.	$\neg(ln)\mathfrak{x}Q^*(lm)\mathfrak{z}$	8, 10 CS Lemma 18c
13.	$\neg(ml)\mathfrak{z}Q^*(mk)\mathfrak{y}$	9, 11 CS Lemma 18c
14.	$\neg(lm)\mathfrak{z}Q^*(mk)\mathfrak{y}$	13
15.	$\neg(ln)\mathfrak{x}Q^*(mk)\mathfrak{y}$	12, 14 CS Lemma 18b
16.	Lemma 33 Q.E.D.	1, 15 C.P.

□

Lemma 33 asserts nothing more that every rational number for the third cut is also a rational number for the cut defined by the comparison between magnitudes $\mathfrak{x}$ and $\mathfrak{y}$. The latter cut is the set of all rational numbers mk/nl such that $\neg(ln)\mathfrak{x}Q^*(mk)\mathfrak{y}$.

Every rational number which is not contained in the third cut can be expressed in the form $(m'/n')(k'/l') = (m'k')/(n'l')$, where m'/n' and k'/l' are rational numbers not contained in the first and second cuts, respectively. From [8], Th. 141, no number of the third cut can be written in the form $(m'/n')(k'/l')$ where $n'\mathfrak{x}Q^*m'\mathfrak{z} \wedge l'\mathfrak{z}Q^*k'\mathfrak{y}$.

Lemma 34. $n'\mathfrak{x}Q^*m'\mathfrak{z} \wedge l'\mathfrak{z}Q^*k'\mathfrak{y} \rightarrow (l'n')\mathfrak{x}Q^*(m'k')\mathfrak{y}$

Proof. Similar to Lemma 33, except that CS Lemma 15, CS Lemma 11, CS A. I., CS Lemma 12, CS A. I. and CS A. I. are substituted for CS Lemma 16 (contraposition), CS Lemma 12, CS Lemma 18a, CS Lemma 11, CS Lemma 18c and CS Lemma 18b, respectively. Q.E.D. □

Lemma 34 implies that every rational number which is not contained in the third cut is also not contained in the cut defined by the comparison between magnitudes $\mathfrak{x}$ and $\mathfrak{y}$.

Taking the cut corresponding to the comparison between magnitudes $\mathfrak{x}$ and $\mathfrak{y}$ as the fourth cut, $g_{\mathfrak{y}}(\mathfrak{x})$, then what has been shown so far is that any rational number for the third cut is also a rational number for the fourth, and every rational number which is not contained in the third cut is also not contained in the fourth.

Lemma 35. $g_{\mathfrak{y}}(\mathfrak{x}) = f_{\mathfrak{z}}(\mathfrak{x})g_{\mathfrak{y}}(\mathfrak{z})$

Proof. Use Lemma 33, Lemma 34 and a rationale similar to that adopted in Th. 1e to prove that the fourth cut is identical with the third. Q.E.D. □

The cut corresponding to the comparison between magnitudes $\mathfrak{x}$ and $\mathfrak{y}$ is the product of the cuts belonging to the comparisons between $\mathfrak{x}$ and $\mathfrak{z}$ and between $\mathfrak{z}$ and $\mathfrak{y}$. Put in other words this means: the measured (numerical) value of $\mathfrak{x}$ relative to $\mathfrak{y}$ as a unit is obtained when the measured (numerical) value of $\mathfrak{x}$ relative to unit $\mathfrak{z}$ is multiplied by the measured (numerical) value of $\mathfrak{z}$ relative to unit $\mathfrak{y}$.

Putting Lemma 35 in this form,

$$f_{\mathfrak{z}}(\mathfrak{x}) = g_{\mathfrak{y}}(\mathfrak{x})/g_{\mathfrak{y}}(\mathfrak{z})$$

allows the calculation of the measured (numerical) value of any magnitude $\mathfrak{x}$ relative to any magnitude $\mathfrak{z}$ chosen as unit, provided that the measured (numerical) values of both magnitudes relative to any other unit $\mathfrak{y}$ are known.

Theorem 2. (Uniqueness theorem) *If an ordered triple $\langle K, Q, \circ\rangle$ is an empirical extensive structure, then any two numerical extensive structures, $\langle N_1, \leq, +\rangle$ and $\langle N_2, \leq, +\rangle$, isomorphic to $\langle \mathfrak{K}, Q^*, \circ^*\rangle$ are related by a similarity transformation.*

Proof. Since $g_{\mathfrak{y}}(\mathfrak{z})$ is a constant for fixed $\mathfrak{y}$ and $\mathfrak{z}$, Theorem 2 follows from Lemma 35. Q.E.D. □

6 Outline

Some comments about the axioms listed in section 2 are in order. First, note that axiom A. V. together with axiom A. I. and the definition of $\circ$ as an operation from $K \times K$ to K imply that the set K is infinite. In measuring mass with a pan balance, one obviously cannot concatenate freely without either damaging the balance or running out of space. A generalization of [13] which eliminates this defect and also considers the existence of a maximal element is discussed in [9].

Second, note that axioms A. II., A. III., A. V., and many lemmas in section 3 have no empirical meaning, since the same object cannot be placed at the same time on the two pans of the balance, except if x, y, and z are to be regarded as equivalence classes as discussed in section 4.

Note also that axiom A. IV. essentially assumes that K contains arbitrary small objects so that an object z can be found which placed together with y exactly balances x. It is assumed here that the existence of such objects in the empirical set is a necessary requirement for the construction of a "system of standards" sufficiently refined to permit the exact measurement of extensive quantities. An alternative set of axioms which weakens this condition is given in [4].

As a last comment, note that the relation Q must be a perfectly transitive relation, which entails that the measuring instrument used to determine whether or not two objects stand in the relation Q must possess infinite resolution and perfect stability. It is beyond the scope of the present article to discuss the ways proposed in the literature to circumvent this limitation.

7 Conclusion

There are two fundamental problems for an analysis of any procedure of measurement: the representation theorem and the uniqueness theorem. The first problem requires the characterization of the formal properties of empirical operations and relations used in the procedure and the proof that they are isomorphic to appropriately chosen numerical operations and relations. We solved it here for extensive quantities with unrestricted concatenation and no maximal elements. In particular, the following was proved:

- For each given magnitude in the set of all magnitudes, we may define a function which maps that set into the set of cuts or positive real numbers. One can call the cuts the measured (numerical) values obtained when magnitudes are compared with the given magnitude, in which case the given magnitude is called the unit.
- The measured (numerical) value of the concatenation of two magnitudes is the arithmetical sum of the measured (numerical) values of each magnitude, provided that each magnitude and their concatenation are all measured relative to the same arbitrary unit.
- For each given magnitude in the set of all magnitudes, the function which maps that set into the set of cuts is one-one. Put in other words this means: if the unit is fixed, then there is exactly one measured (numerical) value for each given magnitude and there is exactly one magnitude for each given measured (numerical) value.

• For each given magnitude in the set of all magnitudes, the function which maps that set into the set of cuts is order preserving.

The second problem is to determine the scale type of the measurements resulting from the procedure. We also solved it here for extensive quantities with unrestricted concatenation and no maximal elements. In particular, the following was proved:

• Consider three magnitudes: the first, the second and the third. The measured (numerical) value of the first magnitude relative to the third as a unit is obtained when the measured (numerical) value of the first magnitude relative to the second as a unit is multiplied by the measured (numerical) value of the second magnitude relative to the third as a unit. This allows the calculation of the measured (numerical) value of any magnitude relative to any magnitude chosen as unit, provided that the measured (numerical) values of both magnitudes relative to any other unit are known. This is the familiar rule for converting units of extensive quantities.

• In the measurement of extensive quantities only the choice of a unit is arbitrary. There are other ways of stating this result: (a) the measurement of extensive quantities is unique up to multiplication by a positive number (the number corresponding to an arbitrary choice of unit), (b) the measurement of extensive quantities is unique up to a similarity transformation, or (c) the measurement of extensive quantities is on a ratio scale.

First-order predicate logic was consistently used here to prove the representation theorem for the deterministic measurement of extensive quantities assuming unrestricted concatenation and no maximal elements. Though being a highly idealized problem, its solution yields the familiar results routinely used by those directly involved with the measurement of extensive quantities.

References

[1] Behrend, F. A. 1956. A contribution to the theory of magnitudes and the foundations of analysis. Mathematische Zeitschrift 63: 345-362.

[2] Campbell, N. R. 1920. Physics - The Elements. London: Cambridge.

[3] Copi, I. M. 1973. Symbolic Logic. New York: Macmillan, 4th Ed.

[4] Falmagne, J. –C. 1975. A set of independent axioms for positive Hölder systems. Philosophy of Science 42: 137-151.

[5] Hegenberg, L. 2012. Lógica. Rio de Janeiro: GEN, 3rd Ed.

[6] Hölder, O. 1901. Die axiome der quantität und die lehre vom mass. Berichte über die Verhandlungen der Königlich Sächsischen Gesellschaft der Wissenschaften zu Leipzig, Mathematisch-Physikalische Klasse 53: 1-64.

[7] Krantz, D. H, R. Duncan Luce, P. Suppes, and A. Tversky. 1971. Foundations of Measurement. Vol. 1 - Additive and Polynomial Representations. New York: Academic Press.

[8] Landau, E. 1966. Foundations of Analysis – The Arithmetic of Whole, Rational, Irrational and Complex Numbers, translated by F. Steinhardt. New York: Chelsea, 3rd Ed. Chapter 3.

[9] Luce, R. Duncan and A. A. J. Marley. 1969. Extensive measurement when concatenation is restricted and maximal elements may exist. In Philosophy, Science, and Method: Essays in Honor of Ernst Nagel, ed. S. Morgenbesser, P. Suppes, and M. G. White, 235-249. New York: St Martin´s Press.

[10] Michell, J. and C. Ernst. 1996. The axioms of quantity and the theory of measurement, Part I. Journal of Mathematical Psychology 40: 235-252.

[11] Roberts, F. S. 1985. Measurement Theory. In Encyclopedia of Mathematics and its Applications, vol. 7, ed. Gian-Carlo Rota. New York: Cambridge University Press.

[12] Stevens, S. S. 1946. On the theory of scales and measurement. Science 103: 677-680.

[13] Suppes, P. 1951. A set of independent axioms for extensive quantities. Portugaliae Mathematica 10: 163-172.

[14] Suppes, P. 1957. Introduction to Logic. New York: Van Nostrand Reinhold.

[15] Suppes, P. and J. L. Zinnes. 1963. Basic measurement theory. In Handbook of Mathematical Psychology, vol: 1, ed. R. Duncan Luce, R. R. Bush and E. Galanter, 1-76. New York: Wiley.

[16] Suppes, P. 1969. Studies in the Methodology and Foundations of Science - Selected Papers from 1951 to 1969. New York: Springer.

Received 14 November 2019

An algebraic (set) theory of surreal numbers, I

Dimi Rocha Rangel
Institute of Mathematics and Statistics, University of So Paulo, Brazil
dimi@ime.usp.br

Hugo Luiz Mariano
Institute of Mathematics and Statistics, University of So Paulo, Brazil
hugomar@ime.usp.br

Abstract

The notion of surreal number was introduced by J.H. Conway in the mid 1970's: the surreal numbers constitute a linearly ordered (proper) class No containing the class of all ordinal numbers (On) that, working within the background set theory NBG, can be defined by a recursion on the class On. Since then, have appeared many constructions of this class and was isolated a full axiomatization of this notion that has been subject of interest due to large number of interesting properties they have, including model-theoretic ones. Such constructions suggests strong connections between the class No of surreal numbers and the classes of all sets and all ordinal numbers.

In an attempt to codify the universe of sets directly within the surreal number class, we have founded some clues that suggest that this class is not suitable for this purpose. The present work, that expounds parts of the PhD thesis of the first author ([28]), establishes a basis to obtain an "algebraic (set) theory for surreal numbers" along the lines of the Algebraic Set Theory - a categorial set theory introduced in the 1990's based on the concept of ZF-algebra: to establish abstract and general links between the class of all surreal numbers and a universe of "surreal sets" similar to the relations between the class of all ordinals (On) and the class of all sets (V), that also respects and expands the links between the linearly ordered class of all ordinals and of all surreal numbers.

In the present work we introduce the notion of (partial) surreal algebra (SUR-algebra) and we explore some of its category theoretic properties, including (relatively) free SUR-algebras (SA, ST).

We want to express our sincere gratitude to the referee for his/her careful reading and valuable suggestions that have improved significantly this revised version.

In a continuation of this work ([29]) we will establish links, in both directions, between SUR-algebras and ZF-algebras (the keystone of Algebraic Set Theory) and develop the first steps of a certain kind of set theory based (or ranked) on surreal numbers, that expands the relation between V and On.

Keywords: Surreal numbers; Algebraic Set Theory; SUR-algebras.

Introduction

The notion of surreal number was introduced by J.H. Conway in the mid 1970's: the surreal numbers constitute a linearly ordered (proper) class No containing the class of all ordinal numbers (On) that, working within the background set theory NBG, can be defined by a recursion on the class On. Since then, have appeared many constructions of this class and was isolated a full axiomatization of this notion.

Surreal numbers have been subject of interest in many areas of Mathematics due to large number of interesting properties that they have:
- In Algebra, through the concept of field of Hahn series and variants (see for instance [27], [11], [30], [14], [24]);
- In Analysis (see the book [2]);
- In Foundations of Mathematics, particularly in Model Theory, since the surreal number line is, for proper class linear orders, what the rational number line $\mathbb{Q}$ is for the countable linear orders: surreal numbers are the (unique up to isomorphism) proper class Fraïssé limit of the finite linearly ordered sets, they are set-homogeneous and universal for all proper class linear orders.

The plethora of aspects and applications of the surreals maintain the subject as an active research field. To emphasize this point, the 2016 edition of the "Joint Mathematics Meetings AMS" –the largest Math. meeting in the world– have counted 14 talks in its "AMS-ASL Special Session on Surreal Numbers".
http://jointmathematicsmeetings.org/meetings/national/jmm2016/
`2181_program_ss16.html`

Here we try to develop, from scratch, a new (we hope!) and complementary foundational aspect of the Surreal Number Theory: to establish, in some sense, a set theory based on the class of surreal numbers.

Set/class theories are one of the few fundamental mathematical theories that holds the power to base other notions of mathematics (such as points, lines, and real numbers). This is basically due to two aspects of these theories: the first is that the basic entities and relations are very simple in nature, relying only on the primitive notions of set/class and a (binary) membership relation ("$X \in Y$"), the second aspect is the possibility that this theory can perform constructions of sets by several methods. This combination of factors allows to achieve a high degree of flexibility, in such a way that virtually all mathematical objects can be realized as being of some kind of set/class, and it has the potential to define arrows (category) as entities of the theory. In particular, the set/class theories traditionally puts as a principle (the Axiom of Infinity) the existence of an "infinite set" - the smallest of these would be the set of all natural numbers - thus, such numbers are a derived (or a posteriori) notion, which encodes the essence of the notion of "to be finite", that is apparently more intuitive.

The usual set/class theories (as ZFC or NBG) have the power of "encode" (syntactically) its Model Theory: constructions of models of set theory by the Cohen forcing method or through boolean valued models method are done by a convenient encoding of the fundamental binary relations $\in$ and $=$.

Let us list below some other fundamental theories:

• Set theories with additional predicates for non-Standard Analysis, as the E. Nelson's set theory named IST.

• P. Aczel's "Non-well-founded sets" ([1]), where sets and proper classes are replaced by directed graphs (i.e., a class of vertices endowed with a binary relation).

• K. Lopez-Escobar "Second Order Propositional Calculus" ([25]), a theory with three primitive terms, that encodes the full Second Order Intuitionistic Propositional Calculus also includes Impredicative Set Theory.

• Toposes, a notion isolated in the 1970's by W. Lawvere and M. Tierney, provide generalized set theories, from the category-theoretic point of view.

• Algebraic Set Theory (AST), another categorial approach to set/class theory, introduced in the 1990's by A. Joyal and I. Moerdijk ([8]).

Algebraic Set Theory replaces the traditional use of axioms on membership by categorial relations, proposing the general study of "abstract class categories" endowed with a notion of "small fibers maps". In the same way that the notion of " being finite " is given a posteriori in ZFC, after guaranteeing an achievement of the Peano axioms - which axiomatizes the algebraic notion of free monoid in 1 generator - the notions of "to be a set" and "be an ordinal" are given a posteriori in AST. The class of all sets is determined by a universal property, that of ZF-free algebra, whereas the class of all ordinals is characterized globally by the property of constituting ZF-free algebra with inflationary/monotonous successor function. In the same direction, the (small fibers) rank map, $\rho : V \to On$, is determined solely by the universal property of V, and the inclusion map, $i : On \to V$, is given by an adjunction condition.

In the present work we introduce the notion of (partial) surreal algebra (SUR-algebra) and we explore some of its category theoretic properties, including (relatively) free SUR-algebras (SA, ST).

In a continuation of this work ([29]) we will establish links, in both directions, between SUR-algebras and ZF-algebras (the keystone of Algebraic Set Theory) and develop the first steps of a certain kind of set theory based (or ranked) on surreal numbers, that expands the relation between V and On.

The main aim of these works is to obtain an "algebraic theory for surreals" along the lines of the Algebraic Set Theory: to establish abstract and general links between the class of all surreal numbers and a universe of "surreal sets" similar to (but expanding it) the (ZF-algebra) relations between the classes On and V, giving the first steps towards a certain kind of (alternative) "relative set theory" (see [21] for another presentation of this general notion).

In more details:

We want to perform a construction (within the background class theory NBG) of a "class of all surreal sets", V^*, that satisfies, as far as possible, the following requirements:

• V^* is an expansion of the class of all sets V, via a map $j^* : V \to V^*$.

• V^* is ranked in the class of surreal numbers No, in an analogous fashion that V is ranked in the class of ordinal numbers On. I.e., expand, through the injective map $j : On \to No$, the traditional set theoretic relationship $V \underset{i}{\overset{\rho}{\rightleftarrows}} On$ to a new setting $V^* \underset{i^*}{\overset{\rho^*}{\rightleftarrows}} No$.

Noting that:

(i) the (injective) map $j : On \to No$, is a kind of "homomorphism", partially encoding the ordinal arithmetic;

(ii) the traditional set-theoretic constructions (in V) keep some relation with its (ordinal) complexity (e.g., $x \in y \;\to\; \rho(x) < \rho(y)$, $\rho(\{x\}) = \rho(P(x)) = \rho(x) + 1$, $\rho(\bigcup_{i \in I} x_i) = \bigcup_{i \in I} \rho(x_i)$);

then we wonder about the possibility of this new structured domain V^* determines a category, by the encoding of arrows (and composition) as objects of V^*, in an analogous fashion to the way that the class V of all sets determines a category, i.e. by the encoding of some notion of "function" as certain surreal set (i.e. an objects of V^*); testing, in particular, the degree of compatibility of such constructions with the map $j^* : V \to V^*$ and examining if this new expanded domain could encode homomorphically the cardinal arithmetic.

We list below three instances of communications that we have founded in our bibliographic research on possible themes relating surreal numbers and set theory: we believe that they indicate that we are pursuing a right track.

(I) The Hypnagogic digraph and applications

J. Hamkins have defined in [22] the notion of "hypnagogic digraph", $(Hg, \rightharpoonup)$, an acyclic digraph graded on $(No, <)$, i.e., it is given a "rank" function $v : Hg \to No$

such that for each $x, y \in Hg$, if $x \rightharpoonup y$, then $v(x) < v(y)$. The hypnagogic digraph is a proper-class analogue the countable random $\mathbb{Q}$-graded digraph: it is the Fraïssé limit of the class of all finite No-graded digraphs. It is simply the On-saturated No-graded class digraph, making it set-homogeneous and universal for all class acyclic digraphs. Hamkins have applied this structure, and some relativized versions, to prove interesting results concerning models of ZF set theory.

(II) Surreal Numbers and Set Theory
https://mathoverflow.net/questions/70934/surreal-numbers-and-set-theory
Asked July 21, 2011, by Alex Lupsasca:
I looked through MathOverflow's existing entries but couldn't find a satisfactory answer to the following question:
What is the relationship between No, Conway's class of surreal numbers, and V, the Von Neumann set-theoretical universe?
In particular, does V contain all the surreal numbers? If so, then is there a characterization of the surreal numbers as sets in V? And does No contain large cardinals?
I came across surreal numbers recently, but was surprised by the seeming lack of discussion of their relationship to traditional set theory. If they are a subclass of V, then I suppose that could explain why so few people are studying them.

(III) Surreal Numbers as Inductive Type?
https://mathoverflow.net/questions/63375/surreal-numbers-as-inductive-type?rq=1
Asked in April 29, 2011, by Todd Trimble:
Prompted by James Propp's recent question about surreal numbers, I was wondering whether anyone had investigated the idea of describing surreal numbers (as ordered class) in terms of a universal property, roughly along the following lines.
In categorical interpretations of type theories, it is common to describe inductive or recursive types as so-called initial algebras of endofunctors. The most famous example is the type of natural numbers, which is universal or initial among all sets X which come equipped with an element x and an function $f : X \to X$. In other words, initial among sets X which come equipped with functions $1 + X \to X$ (the plus is coproduct); we say such sets are algebras of the endofunctor F defined by $F(X) = 1 + X$. Another example is the type of binary trees, which can be described as initial with respect to sets that come equipped with an element and a binary operation, or in other words the initial algebra for the endofunctor $F(X) = 1 + X^2$. In their book Algebraic Set Theory, Joyal and Moerdijk gave a kind of algebraic description of the cumulative hierarchy V. Under some reasonable assumptions on a background category (whose objects may be informally regarded as classes, and equipped with a structure which allows a notion of "smallness"), they define a ZF-structure as an ordered object which has small sups (in particular, an empty sup with which to get started) and with a "successor" function. Then, against such a background, they define the cumulative hierarchy V as the initial ZF-structure, and show that it satisfies the axioms of ZF set theory (the possible back-

grounds allow intuitionistic logic). By tweaking the assumptions on the successor function, they are able to describe other set-theoretic structures; for example, the initial ZF-structure with a monotone successor gives On, the class of ordinals, relative to the background.
Now it is well-known that surreal numbers generalize ordinals, or rather that ordinals are special numbers where player R has no options, or in different terms, where there are no numbers past the "Dedekind cut" which divides L options from R options. In any case, on account of the highly recursive nature of surreal numbers, it is extremely tempting to believe that they too can be described as a recursive type, or as an initial algebraic structure of some sort (in a background category along the lines given by Joyal-Moerdijk, presumably). But what would it be exactly?
I suppose that if I knocked my head against a wall for a while, I might be able to figure it out or at least make a strong guess, but maybe someone else has already worked through the details?

Overview of the paper:

Section 1:

This initial section establishes the notations and contains the preliminary results needed for the sequel of this work. It begins establishing our set theoretic backgrounds – that we will use freely in the text without further reference – which is founded in NBG class theory, and contains mainly the definitions and basic results on some kinds of binary relations, in particular on well-founded relations, and "cuts" as certain pairs of subsets of a class endowed with a binary irreflexive relation. The second subsection is dedicated to introduce the linearly ordered class of surreal numbers under many equivalent constructions and to present a characterization and some of its main structure, including its algebraic structure and its relations with the class (or ZF-algebra) of all ordinal numbers.

Section 2:

Motivated by properties of the linearly ordered class $(No, <)$, we introduce in this section the notion of *Surreal Algebra (SUR-algebra)*: an structure $\mathcal{S} = (S, <, *, -, t)$, where $<$ is an acyclic relation on S, $*$ is a distinguished element of S, $-$ is an involution of S and t is a function that chooses an intermediary member between each small (Conway) cut in $(S, <)$, satisfying some additional compatibility properties between them[1]. Every SUR-algebra turns out to be a proper class. We verify that No provides naturally a SUR-algebra and present new relevant examples: the free surreal algebra (SA) and the free transitive SUR-algebra (ST).

Section 3:

[1]Recently, we came across with a study of surreal (sub)structures [5], that explores the theme under a different perspective.

This section is dedicated to a generalization of this concept of SUR-algebra: we introduce the notion of partial SUR-algebra and consider two kinds on morphisms between them. This relaxation is needed to perform constructions as products, sub partial-SUR-algebra and certain kinds of directed colimits. Some more examples are provided.

Section 4:

As an application of the partial SUR-algebras theory previously worked out, we are able to prove in this section some universal properties satisfied by SA and ST (and generalizations), that justifies its names of (relatively) free SUR-algebras.

Section 5:

In this final section, we briefly comment on the sequel of this paper ([29]) and present a list of questions that have occurred to us during the elaboration of the work that we intend to address in the future.

1 Preliminaries

This section establishes the notations and contains the preliminary results needed for the sequel of this work. It begins establishing our set theoretic backgrounds – that we will use freely in the text without further reference – which is founded in NBG class theory, and contains mainly the definitions and basic results on some kinds of binary relations. After, we present the class of surreal numbers, and some of its main structure, under many equivalent constructions.

1.1 Set theoretic backgrounds

This preliminary subsection is devoted to establishing our set theoretic backgrounds which is founded in NBG class theory[2], and contains mainly the definitions and basic results on the binary relations that will appear in the sequel of this work as: (strict) partial order relations, acyclic relations, extensional relations, well founded relations, and "cuts" as certain pairs of subsets of a class endowed with a binary irreflexive relation.

1.1.1 NBG

In this work, we will adopt the (first-order, with equality) theory NBG as our background set theory, where the unique symbol in the language is the binary relation $\in$. We will use freely the results of NBG, in the sequel, we just recall below some

[2]In some parts of this article, we will need some category-theoretic tools and reasoning, thus we will use an expansion of NBG by axioms asserting the existence of Grothendieck universes.

notions and notations. We recall also the basic notions and results on some kinds of binary relations needed for the development of this work. Standard references of set/class theory are [9] and [10].

1. On NBG:

Recall that the primitive concept of NBG is the notion of *class*. A class is *improper* when it is a member of some class, otherwise the class is *proper*. The notion of *set* in NBG is defined: a set is a improper class.

We will use V to denote the universal class – whose members are all sets –; On will stand for the class of all ordinal numbers and Tr denote the class of all transitive sets. $On \subseteq Tr \subseteq V$ and all the three are proper classes.

Given classes $\mathcal{C}$ and $\mathcal{D}$, then $\mathcal{C}$ is a subclass of $\mathcal{D}$ (notation: $\mathcal{C} \subseteq \mathcal{D}$), when all members of $\mathcal{C}$ are also members of $\mathcal{D}$. Classes that have the same members are equal. Every subclass of a set is a set.

$\emptyset$ stands for the unique class without members. $\emptyset$ is a set.

Given a class $\mathcal{C}$, denote $P_s(\mathcal{C})$ the class whose members are all the *subsets* of $\mathcal{C}$. If $\mathcal{C}$ is a set, then $P_s(\mathcal{C})$ is a set too. There is no class that has as members all the *subclasses* of a proper class[3].

Given classes $\mathcal{C}$ and $\mathcal{D}$, and a function $f : \mathcal{C} \to \mathcal{D}$, then the (direct) image $f[\mathcal{C}] = \{d \in \mathcal{D} : \exists c \in \mathcal{C}, d = f(c)\}$ is a subset of D, whenever $\mathcal{C}$ is a set.

Since NBG satisfies the axiom of global choice (i.e., there is a choice function on $V \setminus \{\emptyset\}$) and then every class (proper or improper) can be well-ordered, which implies nice cardinality results: as in ZFC, any set X is equipotent to a unique cardinal number (= initial ordinal), called the its cardinality of X (notation: $card(X)$); moreover, all the proper classes are equipotent – we will denote $card(C) = \infty$ the cardinality of the proper class C – ∞ can be seen as a representation of the well-ordered the proper class On. □

2. Binary relations:

A relation R is a class whose members are ordered pairs[4]. The domain (respect., range) of R is the class of all first (respect., second) components of the ordered pair in the relation. The support of the relation R (notation: $supp(R)$) is the class obtained by the reunion of its domain and range. We will say that a binary relation is defined *on/over* its support class.

A relation R is reflexive when $(x, x) \in R$ for each x in the support of R; on the other hand, R is irreflexive, when $(x, x) \notin R$ for each x in the support of R. We

[3]This is a "metaclass" in NBG, i.e., an equivalence class of formulae in one free variable, modulo the NBG-theory: any such formula is not collectivizing.

[4]In the sequel, we will use both notations for an ordered pair: (x, y) and $\langle x, y\rangle$, but we have some preference on the second notation to denote a "cut", see 6 below.

will use $<, \prec, \lhd$ to denote general irreflexive relations; $\leq, \preceq, \sqsubseteq$ will stand for reflexive relations.

A pre-order is a reflexive and transitive relation. A partial order is a antisymmetric pre-order. A *strict* partial order is a irreflexive and transitive relation. There are well known processes of: obtain a strict partial order from a partial order and conversely.

Let R be a binary relation and let $s, s' \in supp(R)$. Then s and s' are R-comparable when: $s = s'$ or $(s, s') \in R$ or $(s', s) \in R$. A relation R is total or linear when every pair of members of its support are comparable.

Every pre-order relation $\preceq$ on a class $\mathcal{C}$ gives rise to an equivalence relation $\sim$ on the same support: for each $c, c' \in \mathcal{C}$, $c \sim c'$ iff $c \preceq c'$ and $c' \preceq c$.

Let $n \in \mathbb{N}$, a n-cycle of the relation R is a $n+1$-tuple $(x_0, x_1, \cdots, x_n)$ such that $x_n = x_0$ and, for each $i < n$, $(x_i, x_{i+i}) \in R$. A relation is *acyclic* when it does not has cycles. Every acyclic relation is irreflexive. A binary relation is a strict partial order iff it is a transitive and acyclic relation. Note that a binary relation is acyclic and total iff it is a strict linear order. □

3. Induced binary relations:

Given a binary relation R on a class $\mathcal{C}$. For each $c \in \mathcal{C}$, denote $c^R := \{d \in C : (d, c) \in R\}$.

Define a new binary relation on $\mathcal{C}$: for each $c, c' \in \mathcal{C}$, $c \sqsubseteq_R c'$ iff holds $\forall x((x, c) \in R \to (x, c') \in R)$ or, equivalently, $c^R \subseteq c'^R$. Clearly, $\sqsubseteq_R$ is pre-order relation on $\mathcal{C}$.

Denote $\equiv_R$, the equivalence relation associated to the pre-order $\sqsubseteq_R$. We will say that the binary R is extensional when $\equiv_R$ is the identity relation on $\mathcal{C}$ or, equivalently, when $\sqsubseteq_R$ is a partial order. The axiom of extensionality in NBG ensures that $(V, \in_{\restriction V \times V})$ is a class endowed with an extensional relation and, since members of ordinal numbers are ordinal numbers[5] □

1.1.2 Well founded relations

In this subsection we recall basic properties and constructions concerning general well-founded relations. Also, we introduce a special kind of well-founded relation suitable for our purposes.

4. On well-founded relations:

Recall that a binary relation $\prec$ on a class $\mathcal{C}$ is well-founded when:

(i) The subclass $x^{\prec} = \{y \in \mathcal{C} : y \prec x\}$ is a set.

[5]If $\alpha \in On$, then $\alpha^{\in} = \{\beta \in On : \beta \in \alpha\} = \{x \in V : x \in \alpha\} = \alpha$.), then $(On, \in_{\restriction On \times On})$ is class endowed with an extensional relation.

(ii) For each $X \subseteq \mathcal{C}$ that is a non-empty subset, there is $u \in X$ that is a $\prec$-minimal member of X (i.e., $\forall v \in \mathcal{C},\ v \prec u \Rightarrow v \notin X$).[6]

Let $\prec$ be an well-founded relation on a class $\mathcal{C}$. Since for each $n \in \mathbb{N}$, the (non-empty) subset $\{x_0, \cdots, x_n\} \subseteq \mathcal{C}$ has a $\prec$-minimal member, then $\prec$ is an acyclic relation and, in particular, $\prec$ is irreflexive.

If $\mathcal{D} \subseteq \mathcal{C}$, then $(\mathcal{D}, \prec_{\restriction \mathcal{D} \times \mathcal{D}})$ is an well-founded class.

An well-founded relation that is a strict linear/total order is a well-order relation.

The axiom of regularity in NBG, guarantees that the binary relation $\in$ over the universal class V is an well-founded relation. $(On, \in)$ is an well-ordered proper class.

Let $\prec$ be an well-founded relation on a class $\mathcal{C}$. Then it holds:

The induction principle: Let $X \subseteq \mathcal{C}$ be a subclass. If, for each $c \in \mathcal{C}$, the inclusion $c^{\prec} \subseteq X$ entails $c \in \mathcal{C}$, then $X = \mathcal{C}$.

The recursion theorem: Let H be a (class) function such that $H(c, g)$ is defined for each $c \in \mathcal{C}$ and g a (set) function with domain $c^{\prec}$. Then there is a unique (class) function F with domain $\mathcal{C}$ such that $F(c) = H(c, F_{\restriction c^{\prec}})$, for each $c \in \mathcal{C}$. □

5. Rooted well-founded relations:

Remark: Let $(\mathcal{C}, \prec)$ be a well-founded class; the subclass $root(\mathcal{C})$ of its *roots* has as members its $\prec$-minimal members. Note that:

• If $\mathcal{C}$ is a non-empty class, then $root(\mathcal{C})$ is a non-empty class.

• If $\sqsubseteq$ denotes the pre-order on $\mathcal{C}$ associated to $\prec$ (i.e., $c \sqsubseteq d$ iff $\forall x \in \mathcal{C}(x \prec c \Rightarrow x \prec d)$, then: $root(\mathcal{C}) = \{a \in C : a \sqsubseteq c$, for all $c \in \mathcal{C}\}$.

Definition: A well-founded class $(\mathcal{C}, \prec)$ will be called *rooted*, when it has a unique root Φ. If it is the case, then the structure $(\mathcal{C}, \prec, \Phi)$ will be called a rooted well-founded class.

If $\prec$ is an extensional well-founded relation on a non-empty class $\mathcal{C}$, then $(\mathcal{C}, \prec)$ is rooted: indeed, if $r, r' \in root(\mathcal{C})$, then $r \sqsubseteq r'$ and $r' \sqsubseteq r$, thus $r = r'$. However, to emphasize the distinguished element in a structure of rooted well-founded class, we will employ the redundant expression "rooted extensional well-founded class".

Examples and counter-examples:

$(V, \in, \emptyset)$ is a rooted extensional well-founded class

$(On, \in, \emptyset)$ is a rooted extensional well-ordered class.

Every well-ordered set $(X, \leq)$ gives rise to a rooted extensional well-ordered set $(X, <, \Phi)$, where $\Phi = \bot$ is the least element of X and the strict relation, $<$, associated to $\leq$, is an well-founded relation, since for each $x, y \in X$, $x^{<} \subseteq y^{<}$ iff $x \leq y$.

[6] By the global axiom of choice (for classes), this condition is equivalent of a apparently stronger one:
(ii)' For each $X \subseteq \mathcal{C}$ that is a non-empty *subclas*, there is $u \in X$ that is a $\prec$-minimal member of X.

$(\mathbb{N}, \leq, 0)$ is a well-ordered set, thus it gives rise to a rooted extensional well-ordered set. $(\mathbb{N} \setminus \{0\}, |, 1)$ is determines a rooted well-founded set that is not extensional. Note that $(\mathbb{N} \setminus \{0, 1\}, |)$ is an well-founded set that is not rooted since its subset of minimal elements is the infinite set of all prime numbers. □

1.1.3 Cuts and densities

Many useful variants of the concept of Dedekind cut were already been defined on the setting strict linear order on a given set (see for instance [2]). In this preliminary subsection we present expansions of these notions in two different direction: we consider binary relations that are only irreflexive (instead of being a strict linear order) and defined on general classes instead of improper classes (= sets). We also generalize the notions of density a la Hausdorff to this new setting.

Through this subsection, $\mathcal{C}$ denote a class and $<$ stands for a irreflexive binary relation whose support is $\mathcal{C}$.

6. Cuts

A **Conway cut** in $(\mathcal{C}, <)$ is a pair $\langle A, B\rangle$ of *arbitrary subclasses*[7] of $\mathcal{C}$ such that $\forall a \in A, \forall b \in B, a < b$ (notation $A < B$). Since $<$ is a irreflexive relation on $\mathcal{C}$, then $A \cap B = \emptyset$. A Conway cut $\langle A, B\rangle$ will be called *small*, when A and B are *subsets* of $\mathcal{C}$. We can define in theory NBG the class $C_s(\mathcal{C}, <) := \{\langle A, B\rangle \in P_s(\mathcal{C}) \times P_s(\mathcal{C}) : A < B\}$, formed by all the small Conway cuts in $(\mathcal{C}, <)$.

A **Cuesta-Dutari cut**[8] in $(\mathcal{C}, <)$ is a Conway cut $\langle A, B\rangle$ such that $A \cup B = \mathcal{C}$. Note that $\langle \emptyset, \mathcal{C}\rangle$ and $\langle \mathcal{C}, \emptyset\rangle$ are always Cuesta-Dutari cuts in $(\mathcal{C}, <)$. On the other hand, if $\mathcal{C}$ is a proper class, then the class $CD_s(\mathcal{C}, <)$ of all small Cuesta-Dutari cuts in $(\mathcal{C}, <)$ is the empty class.

A **Dedekind cut** in $(\mathcal{C}, <)$ is a Cuesta-Dutari cut $\langle A, B\rangle$ such that A and B are non-empty subclasses. If $\mathcal{C}$ is a set, then $\langle A, B\rangle$ is a Dedekind cut in $(\mathcal{C}, <)$ iff $\langle A, B\rangle$ is a Conway Cut such that the set $\{A, B\}$ is a partition of $\mathcal{C}$. □

7. Densities

Let α be an ordinal number. Then $(\mathcal{C}, <)$ will be called an η_α-class, when for each *small Conway cut* (A, B) in $(\mathcal{C}, <)$, such that $card(A), card(B) < \aleph_\alpha$, there is some $t \in \mathcal{C}$ such that $\forall a \in A, \forall b \in B$, $(a < t,\, t < b)$ (notation: $A < t < B$).

Let $(\mathcal{C}, <)$ be an η_α-class. Taking cuts $(\emptyset, \{c\})$ (respec. $(\{c\}, \emptyset)$), for all $c \in \mathcal{C}$, we can conclude that an η_α-class $(\mathcal{C}, <)$ does not have $<$-minimal (respec. $<$-maximal)

[7] A and/or B could be the empty set.

[8] The spanish mathematician Norberto Cuesta Dutari (1907-1989) constructed his career on studies of generalized real numbers, continuum and order theory.

elements. Taking cuts $(\emptyset, X)$ (or $(X, \emptyset)$), for all $X \subseteq \mathcal{C}$ such that $card(X) < \aleph_\alpha$, we see that an η_α-class $(\mathcal{C}, <)$ has $card(\mathcal{C}) \geq \aleph_\alpha$.

An η_0-class $(\mathcal{C}, <)$ is just a "dense and without extremes" class.

If $(\mathcal{C}, <)$ is an η_α-class and $\beta \in On$ is such that $\beta \leq \alpha$, then clearly $(\mathcal{C}, <)$ is an η_β-class.

$(\mathcal{C}, <)$ will be called an η_∞-class, when it is an η_α-class for all ordinal number α: this means that for each small Conway cut (A, B) in $(\mathcal{C}, <)$ there is some $t \in \mathcal{C}$ such that $A < t < B$. Every η_∞-class is a proper class. We will see that the strictly linearly ordered proper class of all surreal numbers $(No, <)$ is η_∞. We will introduce in Section 2 the notion of SUR-algebra: every such structure is a η_∞-class. □

From now on, we will use the notion of Conway cut (respectively, Cuesta-Dutari cut) only in the *small* sense.

1.2 On Surreal Numbers

This subsection is dedicated to present the class of surreal numbers – a concept introduced by J.H. Conway in the mid 1970's – under many (equivalent) constructions within the background set class theory NBG, its order and algebraic structure and its relations with the class (or the ZF-algebra) of all ordinal numbers.

1.2.1 The Conway's construction formalized in NBG

We begin with the Conway's construction following the appendix his book [6], in which he gave a more formal construction.

We start defining, recursively, the sets G_α in order to define class of games.

(i) $G_0 = \{\langle \emptyset, \emptyset \rangle\}$

(ii) $G_\alpha = \{\langle A, B \rangle : A, B \subseteq \bigcup_{\beta<\alpha} G_\beta\}$

The class G of Conway games is given by $G = \bigcup_{\alpha<On} G_\alpha$.

If $x = \langle A, B \rangle \in G$, it is usual to write x^L (respectively x^R) to denote a member of the left side (respectively, right side) of x.

We can define a preorder $\leqslant$ in G:

$$x \leqslant y \text{ iff no } x^L \text{ satisfies } x^L \geqslant y \text{ and no } y^R \text{ satisfies } x \geqslant y^R.$$

The second step of the construction is the definition of the class of pre-numbers. We will again define the ordinal steps P_α recursively:

- $P_0 = \{\langle \emptyset, \emptyset \rangle\}$
- $P_\alpha = \{\langle A, B \rangle : A, B \in \bigcup_{\beta<\alpha} P_\beta \text{ and } B \not> A\}$

The class P of the pre-numbers is given by $P = \bigcup_{\alpha<On} P_\alpha$.

Finally, the class No is defined as the quotient of the class of pre-numbers by the equivalence relation induced by $\leqslant$. To avoid problems with equivalence classes that are proper classes, we can make a Scott's Trick.

Following Conway's notation, we will denote a class $\langle X, Y \rangle / \sim$ by $\{X|Y\}$ and given a surreal number x, we will denote $x = \{L_x|R_x\}$, where $\langle L_x, R_x \rangle$ is a pre-number that represents x. We will also use the notation x^L for an element of L_x and x^R for an element of R_x.

The birth function b is defined as $b(x) = min\{\alpha : \exists(L, R) \in P_\alpha \; x = \{L|R\}\}$

We can also define, for any ordinal α, the sets O_α, N_α and M_α ("old", "made" and "new"):
- $O_\alpha = \{x \in No : b(x) < \alpha\}$
- $N_\alpha = \{x \in No : b(x) = \alpha\}$
- $M_\alpha = \{x \in No : b(x) \leqslant \alpha\}$

To end this subsection we will now define, recursively, the operations $+, -, \cdot$ in P:
- $x + y = \{x^L + y, x + y^L | x^R + y, x + y^R\}$;
- $xy = \{x^L y + xy^L - x^L y^L, x^R y + xy^R - x^R y^R | x^L y + xy^R - x^L y^R, x^R y + xy^L - x^R y^L\}$;
- $-x = \{-x^R | -x^L\}$;
- $0 = \{\emptyset|\emptyset\}$;
- $1 = \{0|\emptyset\}$.

Proposition 8. *With this operations, No is a real-closed field. In addition, every (set) field has an isomorphic copy inside No. If Global Choice is assumed, this is valid also for class fields.*

1.2.2 The Cuesta-Dutari cuts construction

Given an order $(T, <)$, we can make a Cuesta-Dutari "completion" of T, denoted by $\chi(T)$, which is defined by

$$\chi(T) = (T \cup CD_s(T), <'),$$

with $<'$ defined as follows:

(i) If $x, y \in T$, then the order is the same as in T;

(ii) If $x = (L, R), y = (L', R') \in CD_s(T)$, then $x <' y$ if $L \subsetneq L'$;

(iii) If $x \in T$ and $y = (L, R) \in CD_s(T)$, then $x <' y$ if $x \in L$ or $y <' x$ if $x \in R$.

The idea of that construction is basically the iteration of the Cuesta-Dutari completion starting from the empty set until the we obtain a η_{On} class.

By recursion we define the sets T_α:

- $T_0 = \emptyset$;
- $T_{\alpha+1} = \chi(T_\alpha)$;
- $T_\gamma = \bigcup_{\beta<\gamma} T_\beta$.

And finally we have

$$No = \bigcup_{\alpha \in On} T_\alpha$$

In that construction, the birth function b is given by the map that assigns to each surreal number x, the ordinal $b(x)$ which corresponds to the set $T_{b(x)}$ that x belongs.

Note that in this construction the sets "old", "made" and "new" can be presented in a simpler way:

- $O_\alpha = \bigcup_{\beta<\alpha} T_\alpha$
- $M_\alpha = \bigcup_{\beta\leqslant\alpha} T_\alpha$
- $N_\alpha = T_\alpha$

1.2.3 The binary tree construction or the space of signs construction

Consider the class $\Sigma = \{f : \alpha \to \{-, +\} : \alpha \in On\}$. We can define in this class an relation $<$ as follows:

- $f < g \iff f(\alpha) < g(\alpha)$, where α is the least ordinal such that f and g differs, with the convention $- < 0 < +$ ($f(\alpha) = 0$ iff f is not defined in α).

With this relation, Σ is a linearly ordered class isomorphic to $(No, \leqslant)$.

In this construction, the birth function is given by the map $b : \Sigma \to On$, $f \mapsto \operatorname{dom} f$.

1.2.4 The axiomatic approach

It is an well-known fact that the notion of real numbers ordered field can be completely described (or axiomatized) as a certain structure –of complete ordered field – and every pair of such kind of structure are isomorphic under a unique ordered field isomorphism (in fact, there is a unique ordered field morphism between each pair of complete ordered fields and it is, automatically, an isomorphism). In this subsection, strongly based on section 3 of the chapter 4 in [2], we present a completely analogous description for the ordered class (or ordered field) of surreal numbers.

Definition 9. *A full class of surreal numbers is a structure $\mathcal{S} = (S, <, b)$ such that:*
(i) $(S, <)$ is a strictly linearly ordered class;
(ii) $b : S \to On$ is a surjective function;
(iii) For each Conway cut[9] *(L, R) in $(S, <)$, the class $I_S(L, R) = \{x \in S : L < \{x\} < R\}$ is non-empty and its subclass $m_S(L, R) = \{x \in I_S(L, R) : \forall y \in S, b(y) < b(x) \to y \notin I_S(L, R)\}$ is a singleton;*
(iv) For each Conway cut (L, R) in $(S, <)$ and each ordinal number α such that $b(z) < \alpha$, $\forall z \in L \cup R$, $b(\{L|R\}) \leqslant \alpha$, where $\{L|R\}$ its unique member of $m_S(L, R)$.
□

Remark 10. Let $\mathcal{S} = (S, <, b)$ be a full class of surreal numbers.
• Condition (ii) above entails that S is a *proper* class.
• Condition (iii) above guarantees that $(S, <)$ is a η_∞-class.
• Since the order relation in On is linear (is an well-order), according the notation in condition (iii), $m_S(L, R) = \{x \in I_S(L, R) : \forall y \in S, y \in I_S(L, R) \to b(x) \leq b(y)\}$.
• By condition (iv), $b(\{\emptyset|\emptyset\}) = 0$. □

As mentioned in section 3 of chapter 4 in [2], by results proven in Conway's book [6], the constructions of surreal numbers classes presented above (by Conway cuts, by Cuesta-Dutari cuts and by the space of sign-expansions), endowed with natural "birthday" functions, are all full classes of surreal numbers. It is a natural question ask if these constructions are equivalent in some sense.

Definition 11. *Let $\mathcal{S} = (S, <, b)$ and $\mathcal{S}' = (S', <', b')$ be full classes of surreal numbers. A surreal (mono)morphism $f : \mathcal{S} \to \mathcal{S}'$ is a function $f : S \to S'$ such that:*
(i) $\forall x, y \in S,\ x < y \leftrightarrow f(x) <' f(y)$;
(ii) $\forall x \in S,\ b'(f(x)) = b(x)$. □

Remark 12. Let $\mathcal{S} = (S, <, b)$ and $\mathcal{S}' = (S', <', b')$ be full classes of surreal numbers.

[9] Recall our previously established convention that all cuts are assumed to be *small*.

• Since $<$ and $<'$ are linear order, a surreal morphism is always injective and condition (i) is equivalent to:
(i)' $\forall x, y \in S,\ x < y \to f(x) <' f(y)$.
• Naturally, we can define a ("very-large") category whose objects are the full classes of surreal numbers and the arrows are surreal morphisms, with obvious composition and identities. Clearly, an isomorphism in such category is just a surjective morphism. □

Proposition 13. *Let $\mathcal{S} = (S, <, b)$ and $\mathcal{S}' = (S', <', b')$ be full classes of surreal numbers. Then:*
(i) There is a unique surreal (mono)morphism $f : \mathcal{S} \to \mathcal{S}'$ and it is an isomorphism.
(ii) For each ordinal number α, $b^{-1}([0, \alpha))$ is a set. Or, equivalently, b is a locally small function.
(iii) The function $(L, R) \in C_s(S, <) \stackrel{t}{\mapsto} \{L|R\} \in S$ is surjective. □

In particular, all the constructions of surreal numbers classes presented in our Subsection 3.1, endowed with natural birthday functions, are canonically isomorphic, through a unique isomorphism.

In the section 4 of chapter 4 in [2], named "Subtraction in No", we can find the following result:

Proposition 14. *Let $\mathcal{S} = (S, <, b)$ be a full class of surreal numbers. Then there is a unique function $- : S \to S$ such that:*
(i) $b(-x) = b(x), \forall x \in S$;
(ii) $-(-x) = x, , \forall x \in S$;
(iii) $x < y \leftrightarrow -y < -x, \forall x, y \in S$;
(iv) $-\{L|R\} = \{-R| - L\}, \forall (L, R) \in C_s(S, <)$. □

Remark 15. Let $\mathcal{S} = (S, <, b)$ be a full class of surreal numbers.
• In the presence of condition (ii), condition (iii) is equivalent to:
(iii)' $x < y \to -y < -x, \forall x, y \in S$.
• By condition (iii), condition (iv) makes sense, since $L < R \Rightarrow -R < -L$.
• By condition (iv), $-\{\emptyset|\emptyset\} = \{\emptyset|\emptyset\}$. □

We finish this Subsection registering the following useful results whose proofs can be found in [2], pages 125, 126.

Fact 16. *Let $\mathcal{S} = (S, <, b)$ be a full class of surreal numbers. Let $A, A', B, B', \{x\}, \{x'\} \subseteq S$ be subsets such that $A < B$ and $A' < B'$ and $x = \{A|B\}, x' = \{A'|B'\}$. Then:*

(a) If A and A' are mutually cofinal and B and B' are mutually coinitial, then $\{A|B\} = \{A'|B'\}$.

(b) Suppose that (A, B) and (A', B') are timely representations of x and x' respectively, i.e $b(z) < b(x), \forall z \in A \cup B$ and $b(z') < b(x'), \forall z' \in A' \cup B'$. If $x = x'$ then A and A' are mutually cofinal and B and B' are mutually coinitial. □

1.2.5 Ordinals in No

The results presented in this Subsection can be found in the chapter 4 of [2].

The ordinals can be embedded in a very natural way in the field No. The function that makes this work is recursively defined as follows:

Definition 17. $j(\alpha) = \{j[\alpha]|\emptyset\}$, $\alpha \in On$.

The following result establishes a relation between the function j and the birthday function:

Proposition 18. $b \circ j = id_{On}$

That map j encodes completely the ordinal order into the surreal order:

Proposition 19. $\alpha < \beta$ *iff* $j(\alpha) < j(\beta)$, $\forall \alpha, \beta \in On$.

We have also that $j(0) = 0$, $j(1) = 1$. In fact, that embedding preserves also some algebraic structure. Although the sum and product of ordinals are not commutative, we have alternative definitions sum and product in On closely related to the usual operations that are commutative:

Remark 20. Recall that there is another (natural) ordinal arithmetic given by the Hessenberg sum and product of ordinals α and β. These operations have the advantage that they are associative and commutative, and product distributes over sum.

A simple way to define the Hessenberg sum and product of two ordinals α and β is to use their Cantor normal forms: consider sequence of ordinals $\gamma_0 > \cdots > \gamma_{n-1}$ and two sequences $(a_0, \cdots, a_{n-1})$ and $(b_0, \cdots, b_{n-1})$ of natural numbers (including zero) such that

$$\alpha = \omega^{\gamma_0} \cdot a_0 + \cdots + \omega^{\gamma_{n-1}} \cdot a_{n-1}$$

$$\beta = \omega^{\gamma_0} \cdot b_0 + \cdots + \omega^{\gamma_{n-1}} \cdot b_{n-1}.$$

The Hessenberg sum of α and β is the ordinal:

$$\alpha \oplus \beta := \omega^{\gamma_0} \cdot (a_0 + b_0) + \cdots + \omega^{\gamma_{n-1}} \cdot (a_{n-1} + b_{n-1}).$$

The Hessenberg product, denoted by $\alpha \odot \beta$, is defined to be the ordinal arising by multiplication (using distributive and associative laws) from the Cantor normal forms of α and β and by using the rule:

$$\omega^\gamma \odot \omega^\delta := \omega^{\gamma \oplus \delta}$$

to multiply powers of ω.

Fact 21. *The Hessenberg sum and products of ordinals are mapped by j to the surreal sum and product.*

In other words, the semi-ring $(On, \dot{+}, \dot{\times}, 0, 1)$ has an isomorphic copy in No given by the image of j.

2 Introducing Surreal Algebras

Motivated by structure definable in the class No of all surreal numbers, we introduce in this section the notion of surreal algebra (SUR-algebra) as a (higher-order) structure $\mathcal{S} = (S, <, *, -, t)$, satisfying some properties where, in particular, $<$ is an acyclic relation on S where $t : C_s(S) \to S$ is a function that gives a coherent choice of witness of η_∞ density of $(S, <)$. Every SUR-algebra turns out to be a proper class. Besides the verification of No, endowed with previously mentioned structure, is an instance of a SUR-algebra, we have defined two distinguished SUR-algebras SA and ST, respectively the "free surreal algebra" and the "free transitive surreal algebra" that will be very useful in the sequel of this work ([29]).

2.1 Axiomatic definition

Definition 22. *A surreal algebra (or SUR-algebra) is an structure $\mathcal{S} = (S, <, *, -, t)$ where:*
- *$<$ is a binary relation in S;*
- *$* \in S$ is a distinguished element;*
- *$- : S \to S$ is a unary operation;*
- *$t : C_s(S) \to S$ is a function, where $C_s(S) = \{(A, B) \in P_s(S) \times P_s(S) : A < B\}$.*

Satisfying the following properties:

(S1) *$<$ is an acyclic relation.*

(S2) *$\forall x \in S,\ -(-x) = x$.*

(S3) *$-* = *$.*

(S4) $\forall a, b \in S$, $a < b$ *iff* $-b < -a$.

(S5) $\forall (A, B) \in C_s(S)$, $A < t(A, B) < B$.

(S6) $\forall (A, B) \in C_s(S)$, $-t(A, B) = t(-B, -A)$.

(S7) $* = t(\emptyset, \emptyset)$.

□

Remark 23.
• The choice of the ingredients in the structure and the constrains imposed by the axioms in the definition above are largely inspired by the properties of the structure on the class No of all surreal numbers, where we retain only the ingredients that does not strongly depends on the linearity of $<$. The "correct" balance seems to be the choice of $<$ as an acyclic relation, that is a requirement between assuming a irreflexive relation and a strict partial order relation. This acyclicness requirement is also an weak form of well-foundness and is inspired on the idea that the complex term $t(A, B)$ is determined by simpler terms, A, B, with $A < B$; moreover $A < t(A, B) < B$ also suggests that we are dealing with a some kind of "ternary membership relation", where $t(A, B)$ is determined by information from below (by A) and from above (by B).
• Let $(S, <)$ be the underlying relational structure of a surreal algebra $\mathcal{S}$. Then $<$ is an irreflexive relation, by condition (S1), and by (S5), $(S, <)$ is a η_∞-relational structure. As a consequence S is a *proper* class: see 7 in the Subsection 1.1.3. The other axioms establish the possibility of choice of witness for the η_∞ property satisfying additional coherence conditions.
• Note that (S3) follows from (S7) and (S6) : $-* = -t(\emptyset, \emptyset) = t(-\emptyset, -\emptyset) = t(\emptyset, \emptyset) = *$.
• Axiom (S7) establish that the SUR-algebra structure is "an extension by definitions" of a simpler (second-order) language: without a symbol for constant $*$.
• In the presence of (S2), condition (S4) is equivalent to:
(S4)' $\forall a, b \in S$, $a < b \Rightarrow -b < -a$.
• By condition (S4), condition (S6) makes sense, since $A < B \Rightarrow -B < -A$ (and if A, B are sets, then $-A$, $-B$ are sets). □

A morphism of surreal algebras is a function that preserves all the structure on the nose. More precisely:

Definition 24. *Let $\mathcal{S} = (S, <, -, *, t)$ and $\mathcal{S}' = (S', <', -', *', t')$ be SUR-algebras. A morphism of SUR-algebras $h : \mathcal{S} \to \mathcal{S}'$ is a function $h : S \to S'$ that satisfies the conditions below:*

(Sm1) $h(*) = *'$.

(Sm2) $h(-a) = -'h(a), \forall a \in S$.

(Sm3) $a < b \implies h(a) <' h(b), \forall a, b \in S$.

(Sm4) $h(t(A, B)) = t'(h[A], h[B]), \forall (A, B) \in C_s(S)$.[10]

□

Definition 25. The category of SUR-algebras:
We will denote by $SUR - alg$ the ("very-large") category such that $Obj(SUR - alg)$ is the class of all SUR-algebras and $Mor(SUR - alg)$ is the class of all partial SUR-algebras morphisms, endowed with obvious composition and identities. □

Remark 26.
Of course, we have the same "size issue" in the categories $ZF - alg$, of all ZF-algebras, and in $SUR - alg$: each object can be (respect., is a) proper class, thus it cannot be represented in NBG background theory this "very large" category. The mathematical (pragmatical) treatment of this question , that we will adopt in the present work, is to assume a stronger background theory: NBG (or ZFC) and also three Grothendieck universes $U_0 \in U_1 \in U_2$. The members of U_0 represents "the sets"; the members of U_1 represents "the classes"; the members of U_2 represents "the meta-classes". Thus a category $\mathcal{C}$ is: (i) "small", whenever $\mathcal{C} \in U_0$; (ii) "large", whenever $\mathcal{C} \in U_1 \setminus U_0$; (iii) "very large", whenever $\mathcal{C} \notin U_2 \setminus U_1$. □

2.2 Examples and constructions

2.2.1 The surreal numbers as SUR-algebras

The structure $(No, <, b)$ of full surreal numbers class, according the Definition 9 in the Subsection 1.2.4, induces a unique structure of SUR-algebra $(No, <, -, *, t)$, where:
- The function $t : C_s(No, <) \to No$ is such that $(A, B) \mapsto t(A, B) := \{A|B\}$;
- The distinguished element $* \in No$ is given by $* := \{\emptyset|\emptyset\}$;
- The function $- : No \to No$ is the unique function satisfying the conditions in Proposition 14 and Remark 15.

This SUR-algebra has two distinctive additional properties:
- t is a surjective function;
- $<$ is a strict linear order (equivalently, since $<$ is acyclic, $<$ is a total relation).

[10]Note that, by property (Sm3), $(A, B) \in C_s(S) \implies (h[A], h[B]) \in C_s(S')$, thus (Sm4) makes sense.

2.2.2 The free surreal algebra

We will give now a new example of surreal algebra, denoted SA[11], which is not a linear order and satisfies a nice universal property on the category of all surreal algebras (see Section 4). The construction is based on a cumulative Conway's cuts hierarchy over a family of binary relations. [12]

We can define recursively the family of **sets** SA_α as follows:

Suppose that, for all $\beta < \alpha$, we have constructed the sets SA_β and $<_\beta$, binary relations on SA_β, and denote $SA^{(\alpha)} = \bigcup_{\beta<\alpha} SA_\beta$ and $<^{(\alpha)} = \bigcup_{\beta<\alpha} <_\beta$. Then, for α we define:

- $SA_\alpha = SA^{(\alpha)} \cup \{\langle A, B\rangle : A, B \subseteq SA^{(\alpha)} \text{ and } A <^{(\alpha)} B\}$.[13]
- $<_\alpha = <^{(\alpha)} \cup \{(a, \langle A, B\rangle), (\langle A, B\rangle, b) : \langle A, B\rangle \in SA_\alpha \setminus SA^{(\alpha)} \text{ and } a \in A, b \in B\}$.
- The class SA[14] is the union $SA := \bigcup_{\alpha\in On} SA_\alpha$.
- $< := \bigcup_{\alpha\in On} <_\alpha$ is a binary relation on SA.

Fact: Note that that:
(a) $SA^{(0)} = \emptyset$, $SA^{(1)} = SA_0 = \{\langle\emptyset, \emptyset\rangle\}$ and $SA_1 = \{\langle\emptyset, \emptyset\rangle, \langle\emptyset, \{\langle\emptyset, \emptyset\rangle\}\rangle, \langle\{\langle\emptyset, \emptyset\rangle\}, \emptyset\rangle\}$. By simplicity, we will denote $* := \langle\emptyset, \emptyset\rangle = 0$, $-1 := \langle\emptyset, \{*\}\rangle$ and $1 := \langle\{*\}, \emptyset\rangle$ thus $SA_1 = \{0, -1, 1\}$.
(b) $<_0 = \emptyset$ and $<_1 = \{(-1, 0), (0, 1)\}$.
(c) -1 and 1 are $<$-incomparable.
(d) $SA^{(\alpha)} \subseteq SA_\alpha$, $\alpha \in On$.
(e) $SA_\beta \subseteq SA_\alpha$, $\beta \leq \alpha \in On$.
(f) $SA^{(\beta)} \subseteq SA^{(\alpha)}$, $\beta \leq \alpha \in On$.
(g) $<^{(\alpha)} = <_\alpha \cap SA^{(\alpha)} \times SA^{(\alpha)}$, $\alpha \in On$ (by the definition of $<_\alpha$).
(h) $<_\beta = <^{(\alpha)} \cap SA_\beta \times SA_\beta$, $\beta < \alpha \in On$.
(i) $<_\beta = <_\alpha \cap SA_\beta \times SA_\beta$, $\beta \leq \alpha \in On$ (by items (g) and (h) above).
(j) $<_\alpha = < \cap SA_\alpha \times SA_\alpha$, $\alpha \in On$.
(k) $C_s(SA_\alpha, <_\alpha) = C_s(SA, <) \cap P_s(SA_\alpha) \times P_s(SA_\alpha)$, $\alpha \in On$ (by item (j)).
(l) $C_s(SA^{(\alpha)}, <^{(\alpha)}) = C_s(SA, <) \cap P_s(SA^{(\alpha)}) \times P_s(SA^{(\alpha)})$, $\alpha \in On$.

[11] The "A" in SA is to put emphasis on **a**cyclic.

[12] Starting from the emptyset, and performing a cumulative construction based on Cuesta-Dutari completion of a linearly ordered set, we obtain No: see for instance [2].

[13] The expression $\langle A, B\rangle$ is just an alternative notation for the ordered pair (A, B), used for the reader's convenience.

[14] Soon, we will see that SA is a *proper* class.

We already have defined $<$ and $*(=\langle\emptyset,\emptyset\rangle)$ in SA, thus we must define $t: C_s(SA)\to SA$ and $-: SA\to SA$ to complete the definition of the structure SA: this will be carry out by recursion on well-founded relations on SA and $C_s(SA)$[15] that will be defined below.

For each $x\in SA$, we define its rank as $r(x):=min\{\alpha\in On : x\in SA_\alpha\}$. Since for each $\beta<\alpha$, $SA_\beta\subseteq SA^{(\alpha)}\subseteq SA_\alpha$, it is clear that $r(x)=\alpha$ iff $x\in SA_\alpha\setminus SA^{(\alpha)}$.

Following Conway ([6], p.291), we can define for the SA setting the notions of: "old members", "made members" and "new members". More precisely, for each ordinal α:
• The set of **old** members w.r.t. α is the subset of SA of all members "born **before** day α". $O(SA,\alpha):=SA^{(\alpha)}$;
• The set of **made** members w.r.t. α is the subset of SA of all members "born **on or before** day α". $M(SA,\alpha):=SA_\alpha$;
• The set of **new** members w.r.t. α is the subset of SA of all members "born **on** day α". $N(SA,\alpha):=SA_\alpha\setminus SA^{(\alpha)}$.

We will denote $x\prec y$ in SA iff $r(x)<r(y)$ in On.

Claim 1: $\prec$ is an well-founded relation in SA.
Proof. Let $y\in SA$ and let $\alpha:=r(y)$. Given $x\in SA$, $r(x)<\alpha$ iff $x\in O(SA,\alpha)=SA^{(\alpha)}$. Therefore, the subclass $\{x\in SA : x\prec y\}$ is a subset of SA.
Now let X be a non-empty subset of SA. Then $r[X]$ is a non-empty subset of On and let $\alpha:=minr[X]$. Consider any $a\in r^{-1}[\{\alpha\}]\cap X$, then clearly a is a $\prec$-minimal member of X. □

We have an induced "rank" on the class (of small $<$-Conway cuts) $C_s(SA)=\{(A,B)\in P_s(A)\times P_s(B) : A<B\}$ $R(A,B):=min\{\alpha\in On : A\cup B\subseteq SA^{(\alpha)}\}$. We can also define a binary relation on the class $C_s(SA)$:
$(A,B)\lhd(C,D)$ in $C_s(SA)$ iff $R(A,B)<R(C,D)$ in On.

Claim 2: $\lhd$ is an well-founded relation in $C_s(SA)$.
Proof. Let $(C,D)\in C_s(SA)$ and let $\alpha:=R(C,D)$. Given $(A,B)\in C_s(SA)$, $R(A,B)<\alpha$ iff $\exists\beta<\alpha, A\cup B\subseteq O(SA,\beta)=SA^{(\beta)}$. Therefore, the subclass $\{(A,B)\in C_s(SA) : (A,B)\lhd(C,D)\}$ is a subset of $C_s(SA)$.
Now let Y be a non-empty subset of $C_s(SA)$. Then $R[Y]$ is a non-empty subset

[15] From now on, we will omit the binary relation on a class when it is clear from the setting.

of On and let $\alpha := min(R[Y])$. Consider any $(A, B) \in R^{-1}[\{\alpha\}] \cap Y$, then clearly (A, B) is a $\lhd$-minimal member of Y. □

Let H be the (class) function $H(p, g)$ where, for each $p = (C, D) \in C_s(SA)$ and g a (set) function with domain $p^{\lhd} := \{(A, B) \in C_s(SA) : (A, B) \lhd p\}$, given by $H(p, g) := \langle C, D\rangle$ (i.e., H is just first coordinate projection). Then H is a class function and we can define by $\lhd$-recursion a unique (class) function $t : C_s(SA) \to SA$ by $t(p) = H(p, t_{\restriction p^{\lhd}})$, i.e. $t(C, D) = \langle C, D\rangle$. The range of t is included in SA: since A and B are *subsets* of SA such that $A < B$, there exists $\alpha \in On$ such that $A, B \subseteq SA^{(\alpha)}$; since $<$ is the reunion of the increasing compatible family of binary relations $\{<_\beta : \beta \in On\}$, then we have that $A <^{(\alpha)} B$, thus $\langle A, B\rangle \in SA_\alpha \subseteq SA$.

Claim 3: $\forall \alpha \in On, M(SA, \alpha) = C_s(O(SA, \alpha))$. Thus $N(SA, \alpha) = C_s(SA^{(\alpha)}) \setminus SA^{(\alpha)}$.
Proof. Since $M(SA, \alpha) = O(SA, \alpha) \cup C_s(O(SA, \alpha))$, we just have to prove that, $SA^{(\alpha)} \subseteq C_s(SA^{(\alpha)})$, for each $\alpha \in On$. Suppose that $SA^{(\beta)} \subseteq C_s(SA^{(\beta)})$ for each $\beta \in On$ such that $\beta < \alpha$. By the assumption, we have $SA^{(\alpha)} = \bigcup_{\beta<\alpha} SA_\beta = \bigcup_{\beta<\alpha} C_s(SA^{(\beta)})$. Since $SA^{(\beta)} \subseteq SA^{(\alpha)}$ and $<^{(\beta)} = <^{(\alpha)} \cap (SA^{(\beta)} \times SA^{(\beta)})$[16], we have $C_s(SA^{(\beta)}) \subseteq C_s(SA^{(\alpha)})$, thus $\bigcup_{\beta<\alpha} C_s(SA^{(\beta)}) \subseteq C_s(SA^{(\alpha)})$. Summing up, we conclude that $SA^{(\alpha)} \subseteq C_s(SA^{(\alpha)})$ and the result follows by induction. □

Claim 4: $C_s(SA) = SA$ and $t : C_s(SA) \to SA$ is the identity map, thus, in particular, t is a bijection.
Proof. By items (k) and (l) in the Fact above, $C_s(SA, <) = \bigcup_{\alpha\in On} C_s(SA_\alpha, <_\alpha) = \bigcup_{\alpha\in On} C_s(SA^{(\alpha)}, <^{(\alpha)})$. By Claim 3 above, $SA_\alpha = C_s(SA^{(\alpha)}, <^{(\alpha)}), \forall \alpha \in On$, thus $\bigcup_{\alpha\in On} C_s(SA^{(\alpha)}, <^{(\alpha)}) = \bigcup_{\alpha\in On} SA_\alpha = SA$. Summing up, we obtain $C_s(SA) = SA$. Then $t : C_s(SA) \to SA$, given by $(A, B) \mapsto \langle A, B\rangle$ is the identity map. □

For each $x \in SA$, denote $(L_x, R_x) \in C_s(SA)$ the unique representation of x: in fact, $x = (L_x, R_x)$.

Claim 5: $r \circ t = R$.
Proof. The functional equation is equivalent to:
$\forall (A, B) \in C_s(SA), \forall \gamma \in On, \ \langle A, B\rangle \in SA_\gamma$ iff $A \cup B \subseteq SA^{(\gamma)}$.
If $(A, B) \in C_s(SA)$ and $A \cup B \subseteq SA^{(\gamma)}$ then, since $A < B$ we have $A <^{(\gamma)} B$,

[16] The non trivial inclusion $<^{(\beta)} \supseteq <^{(\alpha)} \cap (SA^{(\beta)} \times SA^{(\beta)})$ holds since for every pair (x, y) in the right side there are $\delta < \beta$ and $\gamma < \alpha$ (that we can assume $\gamma \geq \delta$) such that $(x, y) \in <_\gamma \cap SA_\delta \times SA_\delta = <_\delta \subseteq <_\beta$.

thus $\langle A, B\rangle \in SA_\gamma$ by the recursive definition of SA_γ. On the other hand, if $(A, B) \in C_s(SA)$ and $\langle A, B\rangle \in SA_\gamma$, then by Claim 3 above, $(A, B) \in C_s(SA, <) \cap C_s(SA^{(\gamma)}), <^{(\gamma)}) = C_s(SA^{(\gamma)}), <^{(\gamma)})$, thus $A \cup B \subseteq SA^{(\gamma)}$. □

Claim 6: Let $(A, B) \in C_s(SA)$ and $\alpha \in On$, then: $\forall a \in A, \forall b \in B$, $r(a), r(b) < \alpha$ iff $r(t(A, B)) \leq \alpha$. In particular: $\forall a \in A, \forall b \in B$, $r(a), r(b) < r(t(A, B))$.
Proof. The equivalence is just a rewriting of the equivalence proved above:
$A \cup B \subseteq SA^{(\alpha)}$ iff $\langle A, B\rangle \in SA_\alpha$. □

Claim 7: $\forall x, y \in SA$, $x < y \Rightarrow r(x) \neq r(y)$. In particular, the relation $<$ in SA is irreflexive.
Proof. Suppose that there are $x, y \in SA$ such that $x < y$ and $r(x) = r(y) = \alpha \in On$. Thus $x, y \in SA_\alpha \setminus SA^{(\alpha)}$ and, since $x, y \in SA_\alpha$ and $(x, y) \in <$, we get $(x, y) \in <_\alpha \setminus <^{(\alpha)}$. Thus $(x, y) = (a, (L_y, R_y))$ for some $a \in L_y \subseteq SA^{(\alpha)}$ or $(x, y) = ((L_x, R_x), d)$ for some $d \in R_x \subseteq SA^{(\alpha)}$. In both cases we obtain $x = a \in SA^{(\alpha)}$ or $y = d \in SA^{(\alpha)}$, contradicting our hypothesis. □

Claim 8: Let $A, B \subseteq SA$ be subclasses such that $A < B$, then $r[A] \cap r[B] = \emptyset$.
Proof. Suppose that $A < B$ and that there are $a \in A$ and $b \in B$ such that $r(a) = r(b) \in r[A] \cap r[B]$. Then $a < b$ and $r(a) = r(b)$, contradicting the Claim 7 above. □

Claim 9: Let $(A, B), (C, D) \in C_s(SA)$. Then $\langle A, B\rangle < \langle C, D\rangle$ iff $\langle A, B\rangle \in C$ (then $r(\langle A, B\rangle) < r(\langle C, D\rangle)$) or $\langle C, D\rangle \in B$ (then $r(\langle C, D\rangle) < r(\langle A, B\rangle)$).
Proof. ($\Leftarrow$) If $\langle A, B\rangle = c \in C$ and $r(\langle C, D\rangle) = \alpha$, then $(c, \langle C, D\rangle) \in <_\alpha \subseteq <$, thus $\langle A, B\rangle < \langle C, D\rangle$. The other case is analogous.
($\Rightarrow$) Suppose that $\langle A, B\rangle < \langle C, D\rangle$. By Claim 7 above we have $\alpha = r(\langle A, B\rangle) \neq r(\langle C, D\rangle) = \gamma$. If $\alpha < \gamma$ we have $SA_\alpha \subseteq SA^{(\gamma)}$ and $\langle C, D\rangle \in SA_\gamma \setminus SA_{(\gamma)}$, thus $(\langle A, B\rangle, \langle C, D\rangle) \in <_\gamma \setminus <^{(\gamma)}$ and we have $\langle A, B\rangle \in C$. If $\gamma < \alpha$ we conclude, by an analogous reasoning, that $\langle C, D\rangle \in B$. □

Claim 10: Let $(A, B), (C, D) \in C_s(SA)$. Then $A < \langle C, D\rangle < B$ and $R(A, B) \leq R(C, D)$ iff $A \subseteq C$ and $B \subseteq D$.
Proof. ($\Leftarrow$) Let $R(C, D) = \alpha$, then $\forall c \in C, \forall d \in D$, $(c, \langle C, D\rangle), (\langle C, D\rangle, d) \in <_\alpha \subseteq <$. If $A \subseteq C$ and $B \subseteq D$, then $A < \langle C, D\rangle < B$ and $A \cup B \subseteq C \cup D \subseteq SA^{(\alpha)}$, i.e. $R(A, B) \leq \alpha = R(C, D)$.
($\Rightarrow$) Let $A < \langle C, D\rangle < B$ and suppose that there is $a \in A \setminus C$, then $\langle L_a, R_a\rangle = a < \langle C, D\rangle$. Since $\langle L_a, R_a\rangle \notin C$, then by Claim 9 above, we have $\langle C, D\rangle \in R_a$, thus:

$R(C, D) =_{Claim5} r(\langle C, D\rangle) <_{Claim6}= r(\langle L_a, R_a\rangle) = r(a) < r(\langle A, B\rangle) = R(A, B)$. Analogously, if $A < \langle C, D\rangle < B$ and $B \setminus D \neq \emptyset$, we obtain $R(C, D) < R(A, B)$. □

Claim 11: For each $(A, B) \in C_s(SA, <)$, $A < \langle A, B\rangle < B$. In particular, $(SA, <)$ is a $\eta_i nfty$ *proper* class.
Proof. By Claim 7 above, $<$ is an irreflexive relation. By Claim 10 above, for each $(A, B) \in C_s(SA, <)$, $A < \langle A, B\rangle < B$, thus $(SA, <)$ is a $\eta_i nfty$ class. It follows from 7 in the Subsection 1.1.3, that SA is proper class. □

Claim 12: For each $(A, B) \in C_s(SA, <)$ and each $z \in SA$ such that $A < z < B$, then $r(t(A, B)) \leq r(z)$.
Proof. Suppose that the result is false and let α the least ordinal such that there are $(A, B) \in C_s(SA)$ and $z \in SA$ such that $A < z < B$, but $r(z) < r(t(A, B)) = R(A, B) = \alpha$: thus $\alpha > 0$. By a simple analysis of the cases α ordinal limit and α successor, we can see that there are $A' \subseteq A, B' \subseteq B$ such that $R(A', B') = \alpha' < \alpha$ and $A' < z < B'$, contradicting the minimality of α[17]. □

Define, by recursion on the well-ordered proper class $(On, <)$, a function $s : On \to SA$ by $s(\alpha) := \langle s[\alpha], \emptyset\rangle$, $\alpha \in On$.

Claim 13: $r \circ s = id_{On}$. In particular, the function $r : SA \to On$ is surjective and SA is a proper class.
Proof. We will prove the result by induction on the well-ordered proper class $(On, <)$. Let $\alpha \in On$ and suppose that $r(s(\beta) = \beta$, for all ordinal $\beta < \alpha$. Then: $r(s(\alpha)) = r(\langle s[\alpha], \emptyset\rangle) = r(t(s[\alpha], \emptyset)) =_{Claim\ 5} R(s[\alpha], \emptyset) = min\{\gamma \in On : s[\alpha] \cup \emptyset \subseteq SA^{(\gamma)}\}$.
By the induction hypothesis, we have:
(IH) $s(\beta) \in SA_\beta \setminus SA^{(\beta)}$, for all ordinal $\beta < \alpha$.
Since $s(\beta) \in SA_\beta$, we have $s(\beta) \in SA^{(\alpha)}$, $\forall \beta < \alpha$. If $s[\alpha] \cup \emptyset \subseteq SA^{(\gamma)}$ for some $\gamma < \alpha$, then $s(\gamma) \in SA^{(\gamma)}$, in contradiction with (IH). Summing up, we conclude that $r(s(\alpha) = \alpha$, and the result follows by induction. □

Claim 14: There is a unique function $- : SA \to SA$, such that:
(i) $\forall x \in SA, r(-x) = r(x)$;
(ii) $\forall x \in SA, -(-x) = x$;
(iii) $\forall x, y \in SA$, $x < y$ iff $-y < -x$;
(iv) $\forall (A, B) \in C_s(SA), -t(A, B) = t(-B, -A)$.
Proof. Let $z \in SA$ and suppose that a function $-$ is defined for all $x, y \in SA$,

[17]Hint: in the case $\alpha = \gamma + 1$, use Claim 7.

such that $x, y \prec z$, satisfying the conditions (i)–(iv) adequately restricted to the subset $SA^{(\alpha)}$, for $\alpha := r(z) \in On$. Then $z = \langle L_z, R_z \rangle = t(L_z, R_z) \in SA_\alpha \setminus SA^{(\alpha)}$ for a unique $(L_z, R_z) \in C_s(SA)$ (Claim 4) and α is the least $\gamma \in On$ such that $L_z \cup R_z \subseteq SA^{(\gamma)}$ (Claim 5), thus $\forall x \in R_z \cup L_z, r(x) < r(z)$. Then $-x$ is defined $\forall x \in L_z \cup R_z$, satisfying the conditions (i)–(iv) restricted to the subset $SA^{(\alpha)}$. Since $x < y$, $\forall x \in L_z \forall y \in R_z$, it holds, by condition (iii), $-y < -x$ then $-R_z < -L_z$ and since $-R_z, -L_z$ are the images of a function on sets, $(-R_z, -L_z) \in C_s(SA)$. Moreover, by condition (i), α is the least $\gamma \in On$ such that $-R_z \cup -L_z \subseteq SA^{(\gamma)}$, i.e. $t(-R_z, -L_z) = \langle -R_z, -L_z \rangle \in SA_\alpha \setminus SA^{(\alpha)}$. Define $-z := t(-R_z, -L_z)$.
Now we will prove that the conditions (i)–(iv) still holds for all members in $SA_\alpha \supsetneq SA^{(\alpha)}$.
(i) Let $x \in SA_\alpha$. If $x \in SA_{(\alpha)}$, this condition holds by hypothesis. If $x \in SA_\alpha \setminus SA^{(\alpha)}$, then by the recursive definition above, $-x = \langle -R_x, -L_x \rangle \in SA_\alpha \setminus SA^{(\alpha)}$, thus $r(-x) = \alpha = r(x)$. Thus (i) holds in SA_α.
(ii) Let $x \in SA_\alpha \setminus SA^{(\alpha)}$, then $-x, --x \in SA_\alpha \setminus SA^{(\alpha)}$ (by the validity of condition (i) on SA_α established above). $-(-x) = -(-t(L_x, R_x)) = -t(-R_x, -L_x) = t(-(-L_x), -(-R_x)) = t(L_x, R_x) = x$, since by hypothesis the conditions (iii) and (ii) holds for members of $SA^{(\alpha)}$. Thus (ii) holds in SA_α.
(iii) We suppose that $\forall x, y \in SA^{(\alpha)}$, $x < y$ iff $-y < -x$. Let $x, y \in SA_\alpha$ such that $x < y$. If both $x, y \in SA^{(\alpha)}$ then, by hypothesis $-y < -x$. Otherwise, by Claim 7, there is exactly one between x, y that is a member of $SA_\alpha \setminus SA^{(\alpha)}$. By Claim 9: if $r(y) < r(x) = \alpha$ then $y \in R_x$; if $r(x) < r(y) = \alpha$ then $x \in L_y$. Thus: if $r(y) < r(x) = \alpha$ then $-y \in -R_x = L_{-x}$, thus $-y < -x$; if $r(x) < r(y) = \alpha$ then $-x \in -L_y = R_{-y}$, thus $-y < -x$. Then we have proved that $\forall x, y \in SA_\alpha$, $x < y \Rightarrow -y < -x$. Since the conditions (i) and (ii) have already be established on SA_α, we also have $\forall x, y \in SA_\alpha$, $-y < -x \Rightarrow x = -(-x) < -(-y) = y$.
(iv) Suppose that $t(A, B) = \langle A, B \rangle = \langle -B, -A \rangle = t(-B, -A)$ holds for all $(A, B) \in \bigcup_{\beta<\alpha} C_s(SA) \cap P_s(SA^{(\beta)}) \times P_s(SA^{(\beta)}) = \bigcup_{\beta<\alpha} C_s(SA^{(\beta)}) = \bigcup_{\beta<\alpha} SA_\beta = SA^{(\alpha)}$. We must prove that the condition still holds for all $(C, D) \in C_s(SA) \cap P_s(SA^{(\alpha)}) \times P_s(SA^{(\alpha)}) = C_s(SA^{(\alpha)}) = SA_\alpha$. Let $z = (C, D) = \langle C, D \rangle = t(C, D) \in SA_\alpha \setminus SA^{(\alpha)}$. Then just by the recursive definition of $-z$, we have $-z = -t(A, B) = t(-B, -A)$, as we wish. □

Finally, we will prove that SA satisfies all the 7 axioms of SUR-algebra:

(S7) $* = t(\emptyset, \emptyset)$.
This holds by our definition of $*$.

(S5) $\forall (A, B) \in C_s(SA)$, $A < t(A, B) < B$.

This holds by the Claim 10 above.

(S6) $\forall (A, B) \in C_s(SA)$, $-t(A, B) = t(-B, -A)$.
This holds by Claim 14.(iv).

(S3) $-* = *$.
Since $-* = -t(\emptyset, \emptyset) = t(-\emptyset, -\emptyset) = t(\emptyset, \emptyset) = *$.

(S2) $\forall x \in SA$, $-(-x) = x$.
This holds by Claim 14.(ii).

(S4) $\forall a, b \in SA$, $a < b$ iff $-b < -a$.
This holds by Claim 14.(iii).

(S1) $<$ is an acyclic relation.
Suppose that $<$ is not acyclic and take $x_0 < ... < x_n < x_0$ a cycle in $(SA, <)$ of minimum length $n \in \mathbb{N}$. Since $<$ is an irreflexive relation (see the Claim 7), $n > 0$. Let $\alpha = max\{r(x_i) : i \leq n\}$ and let j be the least $i \leq n$ such that $r(x_j) = \alpha$.
If $j = 0$: Since $x_0 < x_1$ and $x_n < x_0$, then by Claim 7, $r(x_1), r(x_n) < r(x_0)$. Writing $x_0 = \langle L_{x0}, R_{x0} \rangle$ (since, by Claim 4, $SA = C_s(SA)$), we obtain from Claim 9 that $x_n \in L_{x0}$ and $x_1 \in R_{x0}$. As $L_{x0} < R_{x0}$, we have $x_n < x_1$ and then $x_1 < ... < x_n < x_1$ is a cycle of length $n - 1 < n$, a contradiction.
If $j > 0$: Then define $j^- := j - 1$ and $j^+ := j + 1$ (respec. $j^+ = 0$), if $j < n$ (respec. $j = n$).
Then by Claim 7, $r(x_{j^-}), r(x_{j^+}) < r(x_j)$ and by Claim 9: $x_{j^-} \in L(x_j)$ and $x_{j^+} \in R_{x_j}$. As $L_{x_j} < R_{x_j}$, we have $x_{j^-} < x_{j^+}$ and then we can take a sub-cycle of the original one omitting x_j: this new cycle has of length $n - 1 < n$, a contradiction.

2.2.3 The free transitive surreal algebra

We will give now a new example of surreal algebra, denoted ST[18], which is a strict partial order[19] that is not linear and satisfies a nice universal property on the category of all **transitive** surreal algebras (see Section 4). The construction is similar to the construction of SA in the previous subsection: it is based on a cumulative Conway's cuts hierarchy over a family of binary (transitive) relations.

We can define recursively the family of **sets** ST_α as follows:

[18] The "T" in ST is to put emphasis on transitive.

[19] Recall that a binary relation is that is a strict partial order iff it is a transitive and acyclic relation.

Suppose that, for all $\beta < \alpha$, we have constructed the sets ST_β and $<_\beta$, binary relations on ST_β, and denote $ST^{(\alpha)} = \bigcup_{\beta<\alpha} ST_\beta$ and $<^{(\alpha)} = \bigcup_{\beta<\alpha} <_\beta$. Then, for α we define:

- $ST_\alpha = ST^{(\alpha)} \cup \{\langle A, B\rangle : A, B \subseteq ST^{(\alpha)} \text{ and } A <^{(\alpha)} B\}$.
- $<_\alpha=$ the transitive closure of the relation $<'_\alpha$, where $<'_\alpha :=$ $(<^{(\alpha)} \cup \{(a, \langle A, B\rangle), (\langle A, B\rangle, b) : \langle A, B\rangle \in ST_\alpha \setminus ST^{(\alpha)} \text{ and } a \in A, b \in B\})$.
- The (proper) class ST is the union $ST := \bigcup_{\alpha\in On} ST_\alpha$.
- $< := \bigcup_{\alpha\in On} <_\alpha$ is a binary (transitive) relation on ST.

The following result is straightforward an completely analogous to the corresponding items in the Fact in the previous subsubsection on SA:

Fact 1: Note that that:
(a) $ST^{(0)} = \emptyset$, $ST^{(1)} = ST_0 = \{\langle\emptyset, \emptyset\rangle\}$. By simplicity, we will denote $0 := \langle\emptyset, \emptyset\rangle$, $1 := \langle\emptyset, \{0\}\rangle$, $-1 := \langle\{0\}, \emptyset\rangle$. Thus: $ST_0 = \{0\}$, $SA_1 = \{0, 1, -1\}$.
(b) $<_0 = \emptyset$, $<_1 = \{(-1, 0), (0, 1), (-1, 1)\}$.
(c) $-1 < 0 < 1$, $-1 < \langle\{-1\}, \{1\}\rangle < 1$, but $0, \langle\{-1\}, \{1\}\rangle$ are $<$-incomparable.
(d) $ST^{(\alpha)} \subseteq ST_\alpha$, $\alpha \in On$.
(e) $ST_\beta \subseteq ST_\alpha$, $\beta \leq \alpha \in On$.
(f) $ST^{(\beta)} \subseteq ST^{(\alpha)}$, $\beta \leq \alpha \in On$. □

Analogously to in the SA case, we can define rank functions $r : ST \to On$[20] and $R : C_s(ST) \to On$ that induces well-founded relations on ST and on $C_s(ST)$.

The results below are almost all (the exception are the items (m), (n), (o)) analogous to corresponding items in the Fact in the previous subsection on SA. However, the techniques needed in the proofs are different than in SA case and deserve a careful presentation.

Fact 2:
(g) $<^{(\alpha)} = <_\alpha \cap SA^{(\alpha)} \times SA^{(\alpha)}$, $\alpha \in On$.
(h) $<_\beta = <^{(\alpha)} \cap SA_\beta \times SA_\beta$, $\beta < \alpha \in On$.
(i) $<_\beta = <_\alpha \cap SA_\beta \times SA_\beta$, $\beta \leq \alpha \in On$.
(j) $<_\alpha = < \cap SA_\alpha \times SA_\alpha$, $\alpha \in On$.
(k) $C_s(ST_\alpha, <_\alpha) = C_s(ST, <) \cap P_s(ST_\alpha) \times P_s(ST_\alpha)$, $\alpha \in On$.

[20] For each $x \in ST$, $r(x) = \alpha \in On$ iff $x \in ST_\alpha \setminus ST^{(\alpha)}$.

(l) $C_s(ST^{(\alpha)}, <^{(\alpha)}) = C_s(ST, <) \cap P_s(ST^{(\alpha)}) \times P_s(ST^{(\alpha)})$, $\alpha \in On$.
(m) $\forall \alpha \in On$, $<_\alpha$ is a transitive and a acyclic relation on ST_α.
(n) $<$ is a transitive and acyclic relation (or, equivalently, it is a strict partial order) on ST.
(o) Let $x, y \in ST$ and denote $\alpha := max\{r(x), r(y)\}$. Then are equivalent:
• $x < y$.
• Exists $n \in \mathbb{N}$, exists $\{z_0, \cdots, z_{n+1}\} \subseteq ST_\alpha$ such that: $x = z_0, y = z_{n+1}$; $z_j \in L_{z_{j+1}}$ or $z_{j+1} \in R_{z_j}$, for all $j \leq n$; $\{z_1, \cdots, z_n\} \subseteq ST^{(\alpha)}$.

Proof. Item (i) follows from items (g) and (h). Items (k) and (l) follows from item (j). Items (n) and (o) are direct consequences of item (m), since $<= \bigcup_{\alpha \in On} <_\alpha$.

(g) Clearly $<^{(\alpha)} \subseteq <_\alpha \cap SA^{(\alpha)} \times SA^{(\alpha)}$. To show the converse inclusion let $x, y \in SA^{(\alpha)}$ be such that $x <_\alpha y$ and let $x = x_0 <'_\alpha \ldots <'_\alpha x_n = y$ be a sequence in $(ST_\alpha, <'_\alpha)$ with the number $k = card(\{i \leqslant n : r(x_i) = \alpha\}$ being minimum. We will show that $k = 0$, thus the sequence is just $x = x_0 <^{(\alpha)} \ldots <^{(\alpha)} x_n = y$ and then $x <^{(\alpha)} y$ because $<^{(\alpha)}$ is a transitive relation (since $<_\beta$, $\beta \in On$ is a transitive relation, by construction). Suppose, by absurd, that $k > 0$ and let j be the least $i \leq n$ such that $r(x_j) = \alpha$. By our hypothesis on x, y we have $0 < j < n$. Since $x_{j-1} <'_\alpha x_j <'_\alpha x_{j+1}$, we have $r(x_{j-1}), r(x_{j+1}) < r(x_j) = \alpha$ and $x_{j-1} \in L(x_j)$, $x_{j+1} \in R_{x_j}$. As $L_{x_j} <^{(\alpha)} R_{x_j}$, we have $x_{j-1} <^{(\alpha)} x_{j+1}$ and then we can take a sub-cycle of the original one omitting x_j: this new cycle has $k - 1 < k$ members with rank α, a contradiction.

(h) We only prove the non-trivial inclusion. Let $x, y \in SA_\beta$ be such that $x <^{(\alpha)} y$. Since $<^{(\alpha)} = \bigcup_{\gamma < \alpha} <_\gamma$, let β' be the least $\gamma < \alpha$ such that $x <_{\beta'} y$. We will prove that $\beta' \leq \beta$, thus we obtain $x <_\beta y$, as we wish. Suppose, by absurd, that $\beta' > \beta$. Then $(x, y) \in <_{\beta'} \cap SA^{(\beta')} \times SA^{(\beta')}$, and by the item (g) proved above $(x, y) \in <^{(\beta')}$. Thus there is some $\gamma < \beta'$ such that $x <_\gamma y$, contradicting the minimality of β'.

(j) Let $x, y \in SA_\alpha$ be such that $x < y$. Since $<= \bigcup_{\gamma \in On} <_\gamma$, let α' be the least $\gamma \in On$ such that $x <_{\alpha'} y$. We will prove that $\alpha' \leq \alpha$, thus we obtain $x <_\alpha y$, as we wish. Suppose, by absurd, that $\alpha' > \alpha$. Then $(x, y) \in <_{\alpha'} \cap SA^{(\alpha')} \times SA^{(\alpha')}$, and by the item (g) proved above $(x, y) \in <^{(\alpha')}$. Thus there is some $\gamma < \alpha'$ such that $x <_\gamma y$, contradicting the minimality of α'.

(m) By definition of $<_\gamma$, $<_\gamma$ is a transitive relation, $\forall \gamma \in On$.
Suppose that the statement is false and let $\alpha \in On$ be the least ordinal such that $(ST_\alpha, <_\alpha)$ has some cycle. Then $\forall \beta < \alpha$, $<_\beta$ is an acyclic relation but $<_\alpha$ has some cycle (or, equivalently, $<'_\alpha$ has some cycle). Let $x_0 <'_\alpha \ldots <'_\alpha x_n <'_\alpha x_0$ be a cycle in $(ST_\alpha, <'_\alpha)$ with the number $k = card(\{i \leqslant n : r(x_i) = \alpha\}$ being minimum. Note that $k > 0$, otherwise $x_0, \ldots, x_n \in SA^{(\alpha)}$ and the cycle is $x_0 <^{(\alpha)} \ldots <^{(\alpha)} x_n <^{(\alpha)} x_0$, thus there is a $\beta < \alpha$ and a cycle $x_0 <_\beta \ldots <_\beta x_n <_\beta x_0$ in $(ST_\beta, <_\beta)$, contradicting

our hypothesis.
Let j be the least $i \leq n$ such that $r(x_j) = \alpha$.
If $j = 0$: Since $x_0 <'_\alpha x_1$ and $x_n <'_\alpha x_0$, then $r(x_1), r(x_n) < r(x_0) = \alpha$. Writing $x_0 = \langle L_{x_0}, R_{x_0}\rangle$, we have that $x_n \in L_{x_0}$ and $x_1 \in R_{x_0}$. As $L_{x_0} <^{(\alpha)} R_{x_0}$, we have $x_n <^{(\alpha)} x_1$, and then $x_1 <'_\alpha ... <'_\alpha x_n <'_\alpha x_1$ is a cycle in $(ST_\alpha, <'_\alpha)$ with $k - 2 < k$ members with rank α, a contradiction.
If $j > 0$: Then define $j^- := j - 1$ and $j^+ := j + 1$ (respect. $j^+ = 0$), if $j < n$ (respect. $j = n$).
Then $r(x_{j^-}), r(x_{j^+}) < r(x_j) = \alpha$ and: $x_{j^-} \in L(x_j)$, $x_{j^+} \in R_{x_j}$. As $L_{x_j} <^{(\alpha)} R_{x_j}$, we have $x_{j^-} <^{(\alpha)} x_{j^+}$ and then we can take a sub-cycle of the original one omitting x_j: this new cycle has $k - 1 < k$ members with rank α, a contradiction. □

Since the harder part was already done, we just sketch the construction of the SUR-algebra structure $(ST, <, -, *, t)$:
• As in the SA case, from the well founded relation on $C_s(ST)$ we can define recursively a function with range ST, $t : C_s(ST) \to ST$ by $t(A, B) = \langle A, B\rangle$. We can prove, by induction, that $ST_\alpha = C_s(ST^{(\alpha)})$, $\alpha \in On$. Thus t is a bijection (is the identity function). Moreover, if $(A, B) \in C_s(ST)$, then $A < t(A, B) < B$.
• We define $* := 0 = t(\emptyset, \emptyset)$.
• As in the SA case, we can define (recursively) the function $- : ST \to ST$ by $-\langle A, B\rangle := \langle -B, -A\rangle$.

The verification of the satisfaction of the SUR-algebra axioms (S2)–(S7) are analogous as in the SA case. The satisfaction of (S1) was proved in item (m) of Fact 2 above.

2.2.4 The cut surreal algebra

In this subsection we present a generalization of the SA, ST constructions. Given a surreal algebra S, we can define a new surreal algebra whose domain is $C_s(S)$ with the following relations and operations:

Definition 27. *Let $(S, <, -, *, t)$ be a surreal algebra. Consider the following structure in $C_s(S)$*
• $*' = (\emptyset, \emptyset)$
• $-'(A, B) = (-B, -A)$
• $(A, B) <' (C, D) \iff t(A, B) < t(C, D)$
• $t'(A, B) = (t[A], t[B])$

Proposition 28. *With this operations $(C_s(S), <', -', *', t')$ is a surreal algebra.*

Proof.

(S1) $<'$ is acyclic because any cycle $(A_0, B_0) <' \ldots <' (A_n, B_n)$ induces a cycle $t(A_0, B_0) < \ldots < t(A_n, B_n)$ in S, which is acyclic.

(S2) $-' -' (A, B)) = -'(-B, -A) = (- - A, - - B) = (A, B)$.

(S3) $-'*' = -'(\emptyset, \emptyset) = (-\emptyset, -\emptyset) = (\emptyset, \emptyset)$.

(S4) $(A, B) <' (C, D)$ iff $t(A, B) < t(C, D)$ iff $-t(C, D) < -t(A, B)$ iff $t(-D, -C) < t(-B, -A)$ iff $(-D, -C) <' (-B, -A)$ iff $-'(C, D) <' -'(A, B)$.

(S5) Let $(A, B) \in C_s(C_s(S))$. Then $A <' B$ and thus $t[A] < t[B]$. Since S satisfies **(S5)**, $t[A] < t(t[A], t[B]) < t[B]$. By the definition of $<'$, $A <' (t[A], t[B]) <' B$ and then $A <' t'(A, B) <' B$.

(S6) $-'t'(A, B) = -'(t[A], t[B]) = (-t[B], -t[A]) = (t[-'B], t[-'A]) = t'(-'B, -'A)$

(S7) $t'(\emptyset, \emptyset) = (t[\emptyset], t[\emptyset]) = (\emptyset, \emptyset) = *'$

□

Some properties of X are transferred to $C_s(X)$ as we can see in the above proposition:

Proposition 29.

(a) If X is transitive then $C_s(X)$ is transitive.

(b) If X is linear then $C_s(X)$ is pre-linear, i.e., denote $\sim_t$ the equivalence relation on $C_s(S)$ given by $(A, B) \sim_t (C, D)$ iff $t(A, B) = t(C, D)$. Then it holds exactly one between of the alternatives: $(A, B) <' (C, D)$; $(A, B) \sim_t (C, D)$; $(C, D) <' (A, B)$.

Proof.

(a) Suppose that we have $(A_1, B_1), (A_2, B_2), (A_3, B_3) \in C_s(X)$ satisfying $(A_1, B_1) <' (A_2, B_2) <' (A_3, B_3)$. Then, by definition, $t(A_1, B_1) < t(A_2, B_2) < t(A_3, B_3)$. Since $<$ is transitive, we have that $t(A_1, B_1) < t(A_3, B_3)$ and then $(A_1, B_1) <' (A_3, B_3)$.

(b) Is straightforward.

If follows almost directly by the definition of the structure in $C_s(S)$ that:

Proposition 30. *$t : C_s(S) \to S$ is a morphism of surreal algebras.*

Remark 31. In the case of the three principal examples of SUR-algebras we have that $t : C_s(SA) \to SA$ and $t : C_s(ST) \to ST$ are bijections and $t : C_s(No) \to No$ is a surjection. □

Proposition 32. *If $f : S \to S'$ is a morphism then $C_s(f) : C_s(S) \to C_s(S') : (A, B) \mapsto (f[A], f[B])$ is a morphism.*

Proposition 33. *C_s determines $C_s(f)(A, B) = (f[A], f[B])$ a functor from SUR to SUR, and t determines a natural transformation $t : Id_{SUR-alg} \to C_s$*

From a direct application of the Proposition 29, we obtain the following:

Proposition 34. *Let $\mathcal{S} = (S, <, -, *, t)$ a SUR-algebra.*

1. *If $\mathcal{S}$ is an initial object in the category $SUR - alg$ then the following diagram commutes:*
$$(S \xrightarrow{!} C_s(S) \xrightarrow{t} S) = (S \xrightarrow{id_S} S)$$

2. *If $\mathcal{S}$ is an object of full subcategory $SUR_T - alg \hookrightarrow SUR - alg$, of all* **transitive** *SUR-algebra, and is an initial object in the this (sub)category $SUR_T - alg$, then the following diagram commutes:*
$$(S \xrightarrow{!} C_s(S) \xrightarrow{t} S) = (S \xrightarrow{id_S} S)$$

Remark 35. Note that: $C_s(SA) = SA$ and $C_s(ST) = ST$.

3 Partial Surreal Algebras and morphisms

In several recursive constructions, the intermediate stages play an important role in the comprehension of the object constructed itself. As we have seen in the Subsection 2.1, all surreal algebra is a *proper class* but, on the other hand, the intermediate stages of the constructions of No, SA, ST are sets. To gain some flexibility and avoid technical difficulties, we introduce in this Section the (more general and flexible) notion of *partial* surreal algebra: every SUR-algebra is a partial SUR-algebra and this new notion can be supported by a set. Besides simple examples, that contains in particular the intermediate stages of No, SA, ST, and a relativized notion of Cut (partial) SUR-algebra, we are interest on general constructions of partial SUR-algebras: for that we will consider two kinds on morphisms between them. We will perform general constructions as products, sub partial-SUR-algebra and certain kinds of directed colimits. As an application of the latter construction, we are able to prove some universal properties satisfied by SA and ST (and natural generalizations), that justifies its names of (relatively) free SUR-algebras.

Definition 36. *A* **partial surreal algebra (pSUR-algebra)** *is a structure* $\mathcal{S} = (S, *, -, <, t)$ *where* S *is a class (proper or improper),* $* \in S$, $-$ *is an unary function in* S, $<$ *is a binary relation in* S *and* $t : C_s^t(S) \to S$ *is a* partial function, *i.e.,* $C_s^t(S) \subseteq C_s(S)$, *satisfying:*

(pS1) $<$ *is an acyclic relation.*

(pS2) $\forall x \in S$, $-(-x) = x$

(pS3) $-* = *$.

(pS4) $\forall a, b \in S$, $a < b$ *iff* $-b < -a$

(pS5) *If* $(A, B) \in C_s^t(S)$, *then* $A < t(A, B) < B$.

(pS6) *If* $(A, B) \in C_s^t(S)$, *then* $(-B, -A) \in C^t(S)$ *and* $-t(A, B) = t(-B, -A)$.

(pS7) $(\emptyset, \emptyset) \in C_s^t(S)$ *and* $* = t(\emptyset, \emptyset)$.

□

Note that (pS1), (pS2), (pS3) and (pS4) coincide, respectively, with the SUR-algebra axioms (S1), (S2), (S3) and (S4). The statements (pS5), (pS6) and (pS7) are relative versions of, respectively, the SUR-algebra axioms (S5), (S6) and (S7). SUR-algebras are precisely the pSUR-algebras $\mathcal{S}$ such that $C_s^t(S) = C_s(S)$.

Definition 37. *Let* $\mathcal{S} = (S, <, -, *, t)$ *and* $\mathcal{S}' = (S', <', -', *', t')$ *be partial SUR-algebras. Let* $h : S \to S'$ *be (total) function and consider the conditions below:*

(Sm1) $h(*) = *'$.

(Sm2) $h(-a) = -'h(a)$, $\forall a \in S$.

(Sm3) $a < b \implies h(a) <' h(b)$, $\forall a, b \in S$.

(pSm4) $(A, B) \in C_s^t(S) \implies (h[A], h[B]) \in C_s^{t'}(S')$ *and* $h(t(A, B)) = t'(h[A], h[B])$, $\forall (A, B) \in C_s^t(S)$.

(fpSm4) $(A, B) \in C_s(S) \implies (h[A], h[B]) \in C_s^{t'}(S')$ *and* $h(t(A, B)) = t'(h[A], h[B])$, $\forall (A, B) \in C_s^t(S)$.

We will say that $h : \mathcal{S} \to \mathcal{S}'$ is:
• *a* **partial SUR-algebra morphism (pSUR-morphism)** *when it satisfies: (Sm1), (Sm2), (Sm3) and (pSm4);*
• *a* **<u>full</u> partial SUR-algebra morphism (fpSUR-morphism)** *when it satisfies: (Sm1), (Sm2), (Sm3) and (fpSm4).* □

Remark 38.

- Note that the property (Sm3) entails: $(A, B) \in C_s(S) \implies (h[A], h[B]) \in C_s(S')$.
- The conditions (Sm1), (Sm2) and (Sm3) are already present in the definition of SUR-algebra morphism. The property:

(Sm4) $h(t(A, B)) = t'(h[A], h[B])$, $\forall (A, B) \in C_s(S)$;
completes the definition of SUR-algebra morphism.

- Every full partial SUR-algebra morphism is partial SUR-algebra morphism.
- Let S, S' be partial SUR-algebras and $h : S \to S'$ is a map. Suppose that S or S' is a SUR-algebra, then h is a pSUR morphism iff h is a fpSUR-morphism.
- If S is a partial SUR-algebra, then: $id_S : S \to S$ is a pSUR-morphism and $id_S : S \to S$ is a fpSUR-morphism iff S is a SUR-algebra.
- Let $h : S \to S'$, $h' : S' \to S''$ be pSUR morphisms:
  • Then $f' \circ f$ is a pSUR-morphism.
  • If f is fpSUR-morphism, then $f' \circ f$ is a fpSUR-morphism. In particular, the composition of fpSUR-morphisms is a fpSUR-morphism.

□

Definition 39. The category of partial SUR-algebras:
We will denote by $pSUR-alg$ the ("very-large") category such that $Obj(pSUR-alg)$ is the class of all partial SUR-algebras and $Mor(pSUR-alg)$ is the class of all partial SUR-algebras morphisms, endowed with obvious composition and identities. □

Remark 40.
(a) Of course, we have in the category $pSUR-alg$ the same "size issue" presented in the categories of $ZF - alg$ and $SUR - alg$: we will adopt the same "solution"

explained in Remark 26. An alternative is to consider only "small" partial SUR-algebras (and obtain "large" category –instead of very large– $pSUR_s{-}alg$, of all small partial SUR-algebras) since we will see that there are set-size partial SUR-algebras: we will not pursue this track because our main concern in considering partial SUR-algebras is get flexibility to make (large indexed) categorial constructions with small partial SUR-algebras to obtain a total SUR-algebra as a (co)limit process, i.e., we want $pSUR \supseteq SUR$.

(b) We saw above that, even if the class of full morphism of partial SUR-algebras is closed under composition, it does not determines a category under composition, since it lacks the identities for the small partial SUR-algebras. However this notion will be useful to perform constructions of total SUR-algebra as colimit of a large diagram small partial SUR-algebras and fpSUR-morphisms between them (see Subsection 3.4). □

41. Denote Σ-str the (very large) category such that:

(a) The objects of Σ-str are the structures $\mathcal{S} = (S, *, -, <, t)$ where S is a class, $* \in S$, $-$ is an unary function in S, $<$ is a binary relation in S and $t : D^t \to S$ is a function such that $D^t \subseteq P_s(S) \times P_s(S)$.

(b) Let $\mathcal{S} = (S, <, -, *, t)$ and $\mathcal{S}' = (S', <', -', *', t')$ be partial SUR-algebras. A Σ-morphism, $h : \mathcal{S} \to \mathcal{S}'$, is a (total) function $h : S \to S'$ satisfying the conditions below:

(Σm1) $h(*) = *'$.

(Σm2) $h(-a) = -'h(a)$, $\forall a \in S$.

(Σm3) $a < b \implies h(a) <' h(b)$, $\forall a, b \in S$.

(Σm4) $(h \times h)[D^t] \subseteq D^{t'}$ and $h(t(A, B)) = t'(h[A], h[B])$, $\forall (A, B) \in D^t$.

(c) Endowed with obvious composition and identities, Σ-str is a very large category and

$$SUR - alg \hookrightarrow pSUR - alg \hookrightarrow \Sigma - str$$

are inclusions of full subcategories. □

3.1 Simple examples

In this short subsection we just present first examples of partial SUR-algebras and its morphisms.

Example 42. Let $(G, +, -, 0, <)$ be a linearly ordered group. For each and select $a \in G$ such that $a \geq 0$ (respect. $a \in G \cup \{\infty\}$ such that $a > 0$) then $X_a := [-a, a] \subseteq G$ (respect. $X_a :=]-a, a[\subseteq G$), is a partial SUR-algebra, endowed with obvious definitions of $*, -, <$ and such that:
(1) $C_s^t(X_a) := \{(x^<, x^>) : x \in X_a\}$, $t(x^<, x^>) := x \in X_a$ (t is bijective);
or, alternatively,
(2) $C_s^t(X_a) := \{(L, R) \in C_s(X_a) : \exists(!)x \in X_a\ L^\leq = x^<,\ R^\geq = x^>\}$, $t(L, R) := x \in X_a$ (t is surjective).

Note that if $b \geq a$, then the inclusion $X_a \hookrightarrow X_b$ is a $pSUR$-morphism, if X_a, X_b are endowed with the second kind of t-map. □

Another simple (and useful) class of examples are given by the ordinal steps of the recursive constructions of the SUR-algebras SA, ST and No.

Example 43. For any ordinal α we have that the Σ-structure $(SA_\alpha, <_\alpha, -_\alpha, *_\alpha, t_\alpha)$ is a partial SUR-algebra with the above definitions:

- $*_\alpha = *$
- $-_\alpha = - \restriction_{SA_\alpha}$
- $<_\alpha = < \restriction_{SA_\alpha \times SA_\alpha}$
- $C_s^t(SA_\alpha) = C_s(SA^{(\alpha)})$ and $t_\alpha = t \restriction_{C_s(SA^{(\alpha)})}$

□

Just like in the previous example, we have:

Example 44.

- $*_\alpha = *$
- $-_\alpha = - \restriction_{ST_\alpha}$
- $<_\alpha = < \restriction_{ST_\alpha \times ST_\alpha}$
- $C_s^t(ST_\alpha) = C_s(ST^{(\alpha)})$ and $t_\alpha = t \restriction_{C_s(ST^{(\alpha)})}$

□

Example 45. For any given $\alpha \in On$, the Σ-structure $(No_\alpha, *_\alpha, -_\alpha, <_\alpha, t_\alpha)$ is a partial SUR-algebra with the operations defined above:

- $*_\alpha = 0$
- $-_\alpha = - \restriction_{No_\alpha}$
- $<_\alpha = < \restriction_{No_\alpha \times No_\alpha}$
- $C_s^t(No_\alpha) = C_s(No^{(\alpha)})$ and $t_\alpha = t \restriction_{C_s(No^{(\alpha)})}$

□

Remark 46.

• Note that in the three examples above $S = SA, ST, No$, the inclusion $S_\alpha \hookrightarrow S_\beta$ is a pSUR-morphism, where $\alpha \leq \beta \leq \infty$ are "extended" ordinals, with the convention $S_\infty := S$.

• We can also define partial SUR-algebras on the sets $SA^{(\alpha)}, ST^{(\alpha)}, No^{(\alpha)}$, for each $\alpha \in On \setminus \{0\}$ (this is useful!).

• Note that $i_\alpha : SA^{(\alpha)} \hookrightarrow SA_\alpha$ is a fpSUR-algebra morphism, for each $\alpha \in On \setminus \{0\}$. It can be established, by induction on $\alpha \in On \setminus \{0\}$ that for each $\gamma < \alpha$ $i_{\gamma\alpha} : SA_\gamma \hookrightarrow SA_\alpha$ is a fpSUR-morphism. An analogous situation occurs to the partial SUR-algebras $ST_\gamma \hookrightarrow ST^{(\alpha)} \hookrightarrow ST_\alpha$. □

3.2 Cut partial Surreal Algebras

In this short subsection we present an adaption/generalization of the notion of "Cut Surreal Algebra", introduced in the Subsection 2.4, to the realm of *partial* SUR-algebra.

Definition 47. *Let $\mathcal{S} = (S, <, -, *, t)$ be a partial SUR-algebra. The* Cut structure *of $\mathcal{S}$ is the Σ-structure $\mathcal{S}^{(t)} = (S', <', -', <', t')$, where:*

1. $S' := C_s^t(S)$
2. $*' := (\emptyset, \emptyset)$
3. $-'(A, B) := (-B, -A)$
4. $(A, B) <' (C, D) \iff t(A, B) < t(C, D)$
5. $\forall \alpha, \beta \subseteq C_s^t(S)$, $(\alpha, \beta) \in dom(t')$ *iff* $\alpha <' \beta$ *and* $(t[\alpha], t[\beta]) \in dom(t)$
6. $t' : C_s^{t'}(C_s^t(S)) \to C_s^t(S)$, $(\alpha, \beta) \mapsto t'(\alpha, \beta) := (t[\alpha], t[\beta])$

□

The list below a sequence of results on Cut Partial SUR-algebras that extend the results presented in the Subsection 2.4 on Cut SUR-algebras: its proofs will be omitted.

Proposition 48. *Let $\mathcal{S} = (S, <, -, *, t)$ be a partial SUR-algebra. Then:*

(a) $\mathcal{S}^{(t)} = (S', <', -', *', t')$ as defined above is a partial SUR-algebra. Moreover, if $\mathcal{S}$ is a SUR-algebra, i.e. $C_s^t(S) = C_s(S)$, then $\mathcal{S}^{(t)}$ is a SUR-algebra, i.e. $C_s^{t'}(C_s^t(S)) = C_s(C_s(S))$.

(b) $t : C_s^t(S) \to S$ is a morphism of partial SUR-algebras. Moreover, if $\mathcal{S}$ is a SUR-algebra, then t is a fpSUR-algebra morphism. □

Proposition 49. *Let $\mathcal{S} = (S, <, -, *, t)$ be a partial SUR-algebra. Then:*

(a) If S is transitive, then $C_s^t(S)$ is transitive.

(b) If S is linear, then $C_s^t(S)$ is pre-linear[21]. □

Proposition 50.

(a) If $f : \mathcal{S} \to \mathcal{S}'$ is a morphism of partial SUR-algebras then $C_s^t(f) : C_s^t(S) \to C_s^t(S')$, given by: $(A, B) \mapsto (f[A], f[B])$ is a morphism of partial SUR-algebras.

(b) The cut partial SUR-algebra construction determines a (covariant) functor $C_s^t : pSUR \to pSUR$:

$$(S \xrightarrow{f} S') \mapsto (C_s^t(S) \overset{C_s^t(f)}{\to} C_s^t(S'))$$

(c) The t-map determines a natural transformation between functors on $pSUR-alg$, $t : Id_{pSUR-alg} \to C_s^t$. □

3.3 Simple constructions on pSUR

In this subsection, we will verify the full subcategory $pSUR-alg \hookrightarrow \Sigma-str$ is closed under some simple categorial constructions: as (Σ-)substructure and non-empty products. We also present some results on initial objects and (weakly) terminal objects.

We can also define a notion of substructure in the category $pSUR$:

Definition 51. *Let $\mathcal{S} = (S, <, -, *, t)$ and $\mathcal{S}' = (S', <', -', *', t')$ be Σ-structures. $\mathcal{S}$ will be called a Σ-substructure of $\mathcal{S}$ whenever:*
(s1) $S \subseteq S'$;
(s2) $< = <'_{\restriction S \times S}$;

[21]I.e., denote $\sim_t$ the equivalence relation on $C_s^t(S)$ given by $(A, B) \sim_t (C, D)$ iff $t(A, B) = t(C, D)$. Then it holds exactly one between of the alternatives: $(A, B) <' (C, D)$; $(A, B) \sim_t (C, D)$; $(C, D) <' (A, B)$.

(s3) $- = -'_{\restriction S\times S}$;
(s4) $* = *'$;
(s5) $dom(t) = t'^{-1}[S] \cap (P_s(S) \times P_s(S)) := \{(A,B) \in dom(t') \cap (P_s(S) \times P_s(S)) : t'(A,B) \in S\} \subseteq dom(t')$ *and* $t = t'_{\restriction} : \mathrm{dom}(t) \to S$. □

Remark 52.

(a) The inclusion $i : S \hookrightarrow S'$ determines a Σ-morphism.

(b) By conditions (s1) and (s2) above note that $C_s(S,<) = C_s(S',<') \cap (P_s(S) \times P_s(S))$.

(c) By item (b): if $dom(t') \subseteq C_s(S',<')$, then $dom(t) \subseteq C_s(S,<)$.

(d) By the results presented in the Subsections 2.2 and 2.3, for any two extends ordinals $\alpha \leq \beta \leq \infty$ we have:

- SA_α is a Σ-substructure of SA_β.
- ST_α is a Σ-substructure of ST_β.

(e) An useful generalization of the notion of Σ-substructure is the notion of Σ-embedding: a Σ-morphism $j : \mathcal{S} \to \mathcal{S}'$ is a Σ-embedding when:
(e1) it is injective;
(e2) $\forall a, b \in S, (a < b \Leftrightarrow j(a) <' j(b))$;
(e3) $\forall (A,B) \in P_s(S) \times P_s(S), ((A,B) \in dom(t) \Leftrightarrow t'(j[A], j[B]) \in range(j))$.

(f) An inclusion $i : S \hookrightarrow S'$ determines a Σ-embedding precisely when $\mathcal{S}$ is a Σ-substructure of $\mathcal{S}'$. Note that the Σ-embeddings $j : \mathcal{S} \to \mathcal{S}'$ are precisely the Σ-morphisms described (uniquely) as $j = i \circ h$, where $i : \mathcal{S}^j \hookrightarrow \mathcal{S}'$ is a Σ-substructure inclusion and $h : \mathcal{S} \to \mathcal{S}^j$ is a Σ-isomorphism.

(g) For technical reasons, we consider an even more general notion: a Σ-morphism $j : \mathcal{S} \to \mathcal{S}'$ is a $\Sigma-$*quasi*-embedding whenever it satisfies the conditions (e1) and (e3) above. □

By a straightforward verification we obtain the:

Proposition 53. *Let* $j : \mathcal{S} \to \mathcal{S}'$ *be a* Σ*-embedding of* Σ*-structures. If* $\mathcal{S}'$ *is a partial SUR-algebra, then* $\mathcal{S}$ *is a partial SUR-algebra.* □

Definition 54. *Given a* non-empty *indexed* set *of partial* Σ*-structure* $\mathcal{S}_i = (S_i, <_i, -_i, *_i, t_i)$, $i \in I$, *we define the* Σ*-structure product* $\mathcal{S} = (S, <, -, *, t)$ *as follows: Let* $\mathcal{S} = (S, <, -, *, t)$ *and* $\mathcal{S}' = (S', <', -', *', t')$ *be* Σ*-structures.* $\mathcal{S}$ *will be called a* Σ*-substructure of* $\mathcal{S}$ *whenever:*
(a) $S = \prod_{i \in I} S_i$;
(b) $<= \{((a_i)_{i\in I}, (b_i)_{i\in I}) : a_i <_i b_i, \forall i \in I\}$;
(c) $-(a_i)_{i\in I} = (-_i a_i)_{i\in I}$;
(d) $* = (*_i)_{i\in I}$;

(e) $dom(t) = \bigcap_{i \in I}(\pi_i \times \pi_i)^{-1}[dom(t_i)] = \{((A_i)i \in I, (B_i)_{i \in I}) \in P_s(S) \times P_s(S)) : (A_i, B_i) \in dom(t_i), \forall i \in I\}$ *and* $t((A_i)i \in I, (B_i)_{i \in I})) = (t_i(A_i, B_i))_{i \in I}$. □

Note that: For each $i \in I$, the projection $\pi_i : S \to S_i$ is a Σ-structure morphism. By a straightforward verification we obtain:

Proposition 55. *Keeping the notation above.*

(a) The pair $(\mathcal{S}, (\pi)_{i \in I})$ *above defined constitutes a(the) categorial product in* Σ-*str. I.e., for each diagram* $(\mathcal{S}', (f_i)_{i \in I})$ *in* Σ-*str such that* $f_i : \mathcal{S}' \to \mathcal{S}_i$, $\forall i \in I$, *there is a unique* Σ-*morphism* $f : \mathcal{S}' \to \mathcal{S}$ *such that* $\pi_i \circ f = f_i, \forall i \in I$.

(b) Suppose that $\{\mathcal{S}_i : i \in I\} \subseteq$ *pSUR-alg. Then* $\mathcal{S} \in$ *pSUR-alg and* $(\mathcal{S}, (\pi)_{i \in I})$ *is the product in the category pSUR-alg.* □

Proposition 56. *Let* $f : \mathcal{S} \to \mathcal{S}'$ *be a pSUR-alg morphism. If* $(S, <)$ *is strictly linearly ordered, then:*
(a) $\forall a, b \in S$, $a < b \iff f(a) <' f(b)$;
(b) f *is an injective function.*

Proof. If $a < b$, then $f(a) <' f(b)$, since f is a Σ-structure morphism. Suppose that $f(a) <' f(b)$ but $a \not< b$, then $a = b$ or $b < a$, thus $f(a) = f(b)$ or $f(b) <' f(a)$. In the case, we get a contradiction with $f(a) <' f(b)$, since $<'$ is an acyclic relation. This establishes item (a). Item (b) is similar, since $<$ satisfies trichotomy and $<'$ is acyclic. □

The result above yields some information concerning the empty product (= terminal object) in pSUR-algebras.

Proposition 57. *If there exists a* weakly *terminal object*[22] $\mathcal{S}_1$ *in the category pSUR-alg then* $\mathcal{S}_1$ *must be a proper class.*

Proof. Suppose that $\mathcal{S}_1$ is an weakly terminal object in pSUR-alg. Since the (proper class) SUR-algebra No is strictly linearly ordered, then by Proposition 56 above anyone of the existing morphisms $f : No \to \mathcal{S}_1$ is injective. Then $\mathcal{S}_1$ (and $C_s^t(\mathcal{S}_1)$) must be a proper class. □

If we consider the small size version of pSUR, we can guarantee by an another application of Proposition 56, that this (large but not very-large) category does not have (weakly) terminal objects: there are small abelian linearly ordered abelian groups (or even the additive part of a ordered/real closed field) of arbitrary large cardinality, and we have seen in Example 42 how to produce small pSUR-algebras from that structures.

[22]Recall that an object in a category is weakly terminal when it is the target of *some* arrow departing from each object of the category.

Concerning initial objects we have the following:

Proposition 58.

(a) Consider the Σ-structure $\mathcal{S}_0 = (S_0, *, -, <, t)$ over a singleton set $S_0 := \{*\}$, with $< := \emptyset$, $D^t = dom(t) := \emptyset$ (thus $\mathcal{S}_0 \notin pSUR - alg$) and with $- : S_0 \to S_0$ and $t : D^t \to S_0$ the unique functions available. Then $\mathcal{S}_0$ is the (unique up to unique isomorphism) initial object in Σ-str.

(b) Consider the Σ-structure $\mathcal{S}_0^p = (S_0, *, -, <, t^p)$ over a singleton set $S_0 := \{*\}$, with $< := \emptyset$, $D^{t^p} = dom(t^p) := \{(\emptyset, \emptyset)\} \subseteq C_s(S_0, <)$ and with $- : S_0 \to S_0$ and $t^p : D^{t^p} \to S_0$ the unique functions available. Then $\mathcal{S}_0^p$ is the (unique up to unique isomorphism) initial object in pSUR-alg.

Proof.

(a) Let $\mathcal{S}'$ be a Σ-structure and let $h : \{*\} \to S'$ be the unique function such that $h(*) = *' \in S'$, then clearly h is the unique Σ-structure morphism from $\mathcal{S}_0$ into $\mathcal{S}'$: note that $(h \times h)_\restriction : dom(t) = \emptyset \to dom(t')$ is such that $t' \circ (h \times h)_\restriction = h \circ t$.

(b) It is easy to see that $\mathcal{S}_0^p$ is a partial SUR-algebra. Let $\mathcal{S}'$ be a partial SUR-algebra and let $h : \{*\} \to S'$ be the unique function such that $h(*) = *' \in S'$, then clearly h is the unique Σ-structure morphism from $\mathcal{S}_0$ into $\mathcal{S}'$: since $(\emptyset, \emptyset) \in dom(t')$, note that $(h \times h)_\restriction : dom(t^p) := \{(\emptyset, \emptyset)\} \to dom(t')$ is such that $t' \circ (h \times h)_\restriction = h \circ t$.

□

3.4 Directed colimits of partial Surreal Algebras

One of the main general constructions in Mathematics is the colimit of an upward directed diagram. In the realm of partial SUR-algebras this turns out to be essential for the constructions of SUR-algebras and to obtain general results about them. We can recognize the utility of this process by the cumulative constructions of our main examples: No, SA, ST. Thus we will be concerned only with the colimit of **small** partial SUR-algebras, but over a possibly a *large* directed diagram.

This subsection is completely technical: we provide only some proofs, for the readers convenience. On the other hand, its consequences/applications are very interesting: see the entire Section 4.

Recall that:

• Given a regular "extended" cardinal κ (where a "$card(X) = \infty$" means that X is a proper class[23]), a partially ordered class $(I, \leq)$ will be κ-directed, if every subclass $I' \subseteq I$ such that $card(I') < \kappa$ admits an upper bound in I.

[23]Recall that in NBG, all the proper classes are in bijection, by the global form of the axiom of choice.

• $pSUR_s - alg$ denotes the full subcategory of $pSUR - alg$ determined by of all *small* partial SUR-algebras and its morphisms (then $SUR - alg \cap pSUR_s - alg = \emptyset$). Analogously, we will denote Σ_s-str the full subcategory of Σ-str determined by of all *small* partial Σ-structures and its morphisms.

59. The (first-order) directed colimit construction: Let $(I, \leq)$ is a ω-directed ordered class and consider $\mathcal{D} : (I, \leq) \to \Sigma_s - str$, $(i \leq j) \mapsto (\mathcal{S}_i \stackrel{h_{ij}}{\to} \mathcal{S}_j)$ be a diagram. Define:
• $S_\infty := (\sqcup_{i\in I} S_i)/\equiv$, the set-theoretical colimit, i.e. $\equiv$ is the least equivalence relation on the class $\sqcup_{i\in I} S_i$ such that $(a_i, i) \equiv (a_j, j)$ iff there is $k \geq i, j$ such that $h_{ik}(a_i) = h_{jk}(a_j) \in S_k$;
• $h_j : S_j \to S_\infty$, $a_j \mapsto [(a_j, j)]$;
• $* := [(*_i, i)]$ $(= [(*_j, j)], \forall i, j \in I)$;
• $-[(a_i, i)] = [(-_i a_i, i)]$;
• $[(a_i, i)] < [(a_j, j)]$ iff there is $k \geq i, j$ such that $h_{ik}(a_i) <_k h_{jk}(a_j) \in S_k$ □

With the construction above, it is straightforward to verify that $(S_\infty, <, -, *)$ is the colimit in the appropriate category of *first-order* (but possibly large) structures[24], with colimit co-cone $(h_j : S_j \to S_\infty)_{j\in I}$ and, if $\mathcal{D} : (I, \leq) \to pSUR_s - alg$, then the same (colimit) co-cone is in the "first-order part" of the category $pSUR - alg$, i.e., it satisfies the properties [pS1]–[pS4] presented in Definition 36. However, to "complete" the Σ-structure (respect. pSUR-algebra) we will need some extra conditions[25] as below:

Proposition 60. *Let $\mathcal{D} : (I, \leq) \to \Sigma_s - str$, $(i \leq j) \mapsto (\mathcal{S}_i \stackrel{h_{ij}}{\to} \mathcal{S}_j)$ be a diagram such that:*
(i) $(I, \leq)$ is a ω-directed ordered class and $h_{ij} : S_i \to S_j$ is a injective Σ-morphism, whenever $i \leq j$;
or;
(ii) $(I, \leq)$ is a ∞-directed ordered class (e.g. $(On, \leq)$),
then $S_\infty := (\sqcup_{i\in I} S_i)/\equiv$ is a (possibly large) partial Σ-structure and $(h_j : S_j \to S_\infty)_{j\in I}$ is a colimit cone in the category $\Sigma - str$. □

Proposition 61. *If $\mathcal{D} : (I, \leq) \to pSUR_s - alg$, $(i \leq j) \mapsto (\mathcal{S}_i \stackrel{h_{ij}}{\to} \mathcal{S}_j)$ is a diagram, where:*

[24] I.e., we drop the second-order part of the Σ-structure: the map $t : dom(t) \subseteq P_s(S_\infty) \times P_s(S_\infty) \to S_\infty$.

[25] These extra conditions are sufficient to obtain directed colimits in $pSUR - alg$ and are also the technical conditions that we will need in the sequel to establish some interesting results (see Theorem 69).

(i) $(I, \leq)$ is a ω-directed ordered class and $h_{ij} : S_i \to S_j$ is a injective pSUR-morphism, whenever $i \leq j$;
or;
(ii) $(I, \leq)$ is a ∞-directed ordered class (e.g. $(On, \leq)$)
then S_∞ is a (possibly large) partial SUR-algebra and $(h_j : S_j \to S_\infty)_{j \in I}$ is a colimit cone in the category $pSUR - alg$.

Proof. It is ease to describe a candidate for the pSUR-colimit over S_∞:

• $dom(t) := \bigcup_{i \in I}(h_i \times h_i)[dom(t_i)], h_i(t_i(A_i, B_i)) = t(h_i[A_i], h_i[B_i])$, i.e. $C_s^t(S_\infty) = \bigcup_{i \in I}\{(h_i[A_i], h_i[B_i]) : (A_i, B_i) \in C_s^{t_i}(S_i)\}$

• $t(h_i[A_i], h_i[B_i]) := h_i(t_i(A_i, B_i))$, for each $(A_i, B_i) \in C_s^{t_i}(S_i)$ and each $i \in I$.

Supposing that the last rule truly determines a function $t : C_s^t(S_\infty) \to S_\infty$, then is straightforward to check that: $(S_\infty, <, -, *, t)$ is a pSUR-algebra; $h_i : S_i \to S_\infty$ is a pSUR-algebra morphism and $(S_\infty, (h_j : S_j \to S_\infty)_{j \in I})$ is a colimit co-cone of the diagram $\mathcal{D} : (I, \leq) \to pSUR_s - alg$.

Thus, it remains only to verify that, under the hypothesis (i) or (ii), the assignment $t(h_i[A_i], h_i[B_i]) := h_i(t_i(A_i, B_i]))$ determines a function $t : C_s^t(S_\infty) \to S_\infty$. Suppose that $(h_i[A_i], h_i[B_i]) = (h_j[A_j], h_j[B_j])$ for some $i, j \in I$ and $(A_i, B_i) \in C_s^t(S_i)$, $(A_j, B_j) \in C_s^t(S_j)$. We have to show that $h_i(t_i(A_i, B_i)) = h_j(t_j(A_j, B_j))$.

Case (i):
Since $(I, \leq)$ is a ω-directed poset, select $k \in I$ such that $k \geq i, j$. Then $h_i = h_k \circ h_{ik}$ and $h_j = h_k \circ h_{jk}$ and $(h_k[h_{ik}[A_i], h_k[h_{ik}[B_i]]) = (h_k[h_{jk}[A_j], h_k[h_{jk}[B_j]])$. Since the transition morphisms in the diagram are injective, we have that the co-cone morphisms $h_l : S_l \to S_\infty$ are injective too, thus $(h_{ik}[A_i], h_{ik}[B_i]) = (h_{jk}[A_j], h_{jk}[B_j])$. Then: $h_{ik}(t_i(A_i, B_i)) = t_k(h_{ik}[A_i], h_{ik}[B_i]) = t_k(h_{jk}[A_j], h_{jk}[B_j]) = h_{jk}(t_j(A_j, B_j)$ and, composing with h_k, we obtain $h_i(t_i(A_i, B_i)) = h_j(t_j(A_j, B_j))$, as required.

Case (ii):
Since $(I, \leq)$ is ω-directed, for each $a \in A_i, a' \in A_j$ such that $h_i(a) = h_j(a')$, let $k_{aa'} \in I$ such that $k_{aa'} \geq i, j$ and $h_{ik_{aa'}}(a) = h_{jk_{aa'}}(a') \in S_{k_{aa'}}$. Since $(I, \leq)$ is ∞-directed and A_i, A_j are small, selected $k \geq k_{aa'}$ for each pair (a, a') as above. Then $h_{ik}(a) = h_{jk}(a')$ and $h_{ik}[A_i] = h_{jk}[A_j]$. An analogous reasoning with the sets B_i, B_j guarantees the existence of $l \in I$ such that $(h_{il}[A_i], h_{il}[B_i]) = (h_{jl}[A_j], h_{jl}[B_j])$. Then: $h_{il}(t_i(A_i, B_i)) = t_l(h_{il}[A_i], h_{il}[B_i]) = t_l(h_{jl}[A_j], h_{jl}[B_j]) = h_{jl}(t_j(A_j, B_j)$ and, composing with h_l, we obtain $h_i(t_i(A_i, B_i)) = h_j(t_j(A_j, B_j))$, as required. □

Proposition 62. *The subclass of morphisms $fpSUR \subseteq pSUR$ is* closed *under non-trivial directed colimits in the cases (i) and (ii) described in the Proposition above. More precisely: if $\mathcal{D} : (I, \leq) \to pSUR_s - alg$ is a ω-directed diagram where $(I, \leq)$ does not have maximum and satisfying (i) or/and (ii) above and such that*

$h_{ij} : S_i \to S_j$ is a fpSUR-morphism, whenever $i < j$, then the colimit co-cone $\forall j \in I$, $(h_j : S_j \to S_\infty)_{j \in I}$ is a formed by fpSUR-algebra morphisms. Moreover:
(a) If $(I, \leq)$ is ∞-directed and without maximum, then S_∞ is a SUR-algebra (thus it is a proper class);
(b) If the transition arrows $(h_{ij})_{i \leq j}$ are injective (respect. Σ − quasi-embedding, Σ-embedding), then the cocone arrows $(h_j)_{j \in I}$ are injective (respect. Σ − quasi-embedding, Σ-embedding);
(c) If $t_i : C_s^{t_i}(S_i) \to S_i$ is injective (respect. surjective/bijective), $\forall i \in I$, then $t^\infty : C_s^{t^\infty}(S^\infty) \to S^\infty$ is injective (respect. surjective/bijective).

Proof. If $\mathcal{D} : (I, \leq) \to pSUR_s - alg$ is a ω-directed diagram satisfying (i) or/and above and such that $h_{ij} : S_i \to S_j$ is a fpSUR-morphism, whenever $i < j$, then it is clear from the commutative condition in a co-cone that for each $j \in I \setminus \{max(I)\}$, $h_j : S_j \to S_\infty$ and for each $(A_j, B_j) \in C_s(S_j)$ then $(h_j[A_j], h_j[B_j]) \in C_s^t(S_\infty)$. Note that if there is $\top = max(I)$, then $h_\top : S_\top \overset{\cong}{\to} S_\infty$.

We will only provide a proof of item (a), the other items are left to the reader.

Let $(A, B) \in C_s(S_\infty)$, then for each $a \in A, b \in B$, we have $a < b$. Since $(I, \leq)$ is ω-directed, let $i_{ab} \in I$ such that $a = h_{i_{ab}}(a')$, $b = h_{i_{ab}}(b')$ where $a', b' \in S_{i_{ab}}$ and $a' <_{i_{ab}} b'$. Since A, B are small and $(I, \leq)$ is ∞-directed, there is a $k \in I$ such that $k \geq i_{ab}$, for each $(a, b) \in A \times B$ and thus, there is $a_k, b_k \in S_k$ such that $a = h_k(a_k)$, $b = h_k(b_k)$ and $a_k <_k b_k$. Forming subsets $A_k, B_k \subseteq S_k$ from the selection of the $a_k, b - k$ as above, then $(A_k, B_k) \in C_s(S_k)$. Since $(I, \leq)$ does not have a top, select $k' \in I$ such that $k' > k$ and then $(h_{kk'}[A_k], h_{kk'}[B_k]) \in C_s^t(S_{k'})$. Thus $(A, B) = (h_{k'}[h_{kk'}[A_k]], h_{k'}[h_{kk'}[B_k]]) \in C_s^t(S_\infty)$, finishing the proof. □

Example 63.

We have noted in Remark 46 that for each sequence of ordinal $\gamma < \beta < \alpha$, $i_{\gamma\beta} : SA_\gamma \hookrightarrow SA_\beta$ is fpSUR-algebra morphism. It is also a Σ-embedding. Then, for each $\alpha > 0$, $SA^{(\alpha)} \cong colim_{\gamma<\alpha} SA_\gamma$ as a pSUR-algebra and $i_\gamma^{(\alpha)} : SA_\gamma \hookrightarrow SA^{(\alpha)}$ determines a colimit co-cone of an ω-directed diagram[26] formed by fpSUR-algebras embeddings. Moreover $SA = SA_\infty \cong colim_{\gamma \in On} SA_\gamma$ and $i_\gamma^\infty : SA_\gamma \hookrightarrow SA$ determines a colimit co-cone a ∞-directed diagram over formed by fpSUR-algebras embeddings.

Analogous results holds for $ST^{(\alpha)} \cong colim_{\gamma<\alpha} ST_\gamma$, $\alpha > 0$, and
$ST \cong colim_{\gamma \in On} ST_\gamma$. □

[26]In fact it is κ directed diagram, where κ is any regular cardinal such that $\kappa \leq \alpha + \omega$.

4 Universal Surreal Algebras

In this section, we present some *categorical-theoretic* universal properties[27] concerning SUR-algebras and partial SUR-algebras. We will need notions, constructions and results developed in the previous sections to provide, for each small *partial* SUR-algebra I, a "best" SUR-algebra over I, $SA(I)$, (respect. a "best" *transitive* SUR-algebra over I, $ST(I)$). As a consequence of this result (and its proof) we will determine the SUR-algebras SA and ST in the category of SUR by universal properties that characterizes them uniquely up to unique isomorphisms: these will justify the adopted names "SA = the free surreal algebra" and "ST = the free transitive surreal algebra".

We start with the following

64. Main construction: Let $\mathcal{I} = (I, *, -, <, t)$ be a partial SUR-algebra. Consider:

(a) The set-theoretical pushout diagram over $(I \overset{t}{\leftarrow} C^t_s(I) \overset{incl}{\hookrightarrow} C_s(I))$:

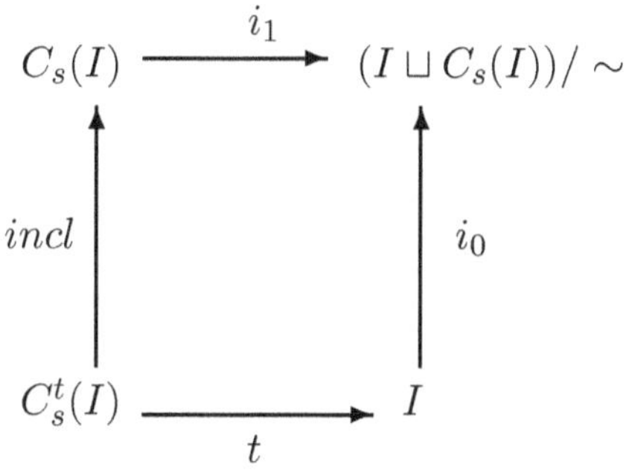

Note that:

• $(I \sqcup C_s(I))/\sim$, is the vertex of the set-theoretical pushout diagram, where $\sim$ is the least equivalence relation[28] on $I \sqcup C_s(I)$ such that $(x, 0) \sim ((A, B), 1)$ iff $(A, B) \in C^t_s(I)$ and $x = t(A, B)$.

• If I is small, then I^+ is small.

• $\forall (A, B), (C, D) \in C_s(I) \setminus C^t_s(I)$, $((A, B), 1) \sim ((C, D), 1)$ iff $(A, B) = (C, D)$ (by induction on the number of steps that witness the transitive closure).

• $\forall x, y \in I$, $(x, 0) \sim (y, 0)$ iff $x = y$ (by induction on the number of steps needed in the transitive closure).

[27] An analysis of model-theoretic universal properties of the "first-order part" of (partial) SUR-algebras, and its possible connections with categorial-theoretic universality presented here, will be theme of future research, see Section 5 for more details.

[28] Recall that the least equivalence relation on a set X that contains $R \subseteq X \times X$ is obtained from R adding the opposite relation R^{-1} and the diagonal relation Δ_X, and then taking the transitive closure $trcl(R \cup R^{-1} \cup \Delta_X) = R^{(eq,X)}$.

• Since $C_s^t(I) \hookrightarrow C_s(I)$ is injective function, then $i_0 : I \to (I \sqcup C_s(I))/\sim$, $x \mapsto [(x,0)]$ is an injective function (see above) and $(i_0)^+ : (P_s(I) \times P_s(I)) \mapsto (A,B) \mapsto (i_0[A], i_0[B])$ is an injective function.

(b) $\mathcal{I}^+ := (I^+, *^+, -^+, <^+, t^+)$ the Σ-structure defined below:
• $I^+ := (I \sqcup C_s(I))/\sim$.
• $*^+ := [(*,0)] = [((\emptyset,\emptyset),1)]$.
• $-^+[(x,0)] := [(-x,0)]$;
$-^+[((A,B),1)] := [((-B,-A),1)]$.
• Define $<^+$ by cases (only three):
$[(x,0)] <^+ [(y,0)]$ iff $x < y$;
$[(x,0)] <^+ [((A,B),1)]$ iff $x \in A$, whenever $(A,B) \in C_s(I) \setminus C_s^t(I)$;
$[((A,B),1)] <^+ [(y,0)]$ iff $y \in B$, whenever $(A,B) \in C_s(I) \setminus C_s^t(I)$.
Note that $C_s(i_0) = (i_0)^+{}_{\restriction} : C_s(I) \to C_s(I^+)$, $(A,B) \mapsto (i_0[A], i_0[B])$, is an injective function with adequate domain and codomain.
• Define $C_s^{t^+}(I^+) := range(C_s(i_0)) \subseteq C_s(I^+)$ (thus $C_s(I) \cong C_s^{t^+}(I^+)$) and $t^+ : C_s^{t^+}(I^+) \to I^+$, $(i_0[A], i_0[B]) \mapsto t^+(i_0[A], i_0[B]) := [((A,B),1)]$.
Note that $(A,B) \in dom(t)$ iff $t^+(i_0[A], i_0[B]) \in range(i_0)$.

Thus we obtain another set-theoretical pushout diagram that is isomorphic to the previous pushout diagram:

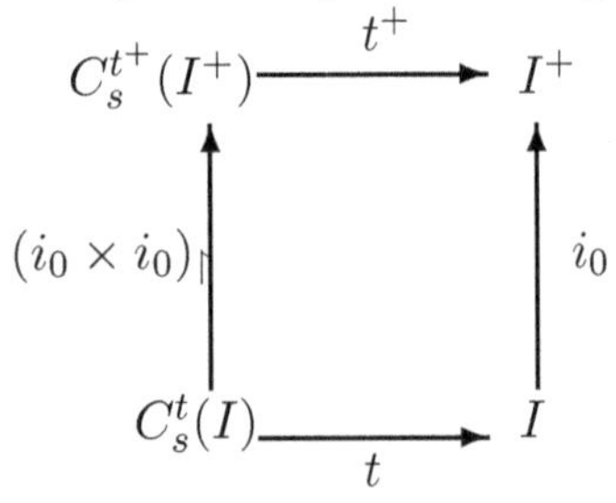

□

We describe below the main technical result in this section:

Lemma 65. *Let $\mathcal{I} = (I, *, -, <, t)$ be a (small) partial SUR-algebra and keep the notation in 64above. Then*

(a) $\mathcal{I}^+ = (I^+, *^+, -^+, <^+, t^+)$ is a (small) partial SUR-algebra.

(b) $i_0 : I \to I^+$ is a Σ-embedding and full morphism of partial SUR-algebras.

(c) If $t : C_s^t(I) \to I$ is injective (respect. surjective/bijective), then $t^+ : C_s^{t^+}(I^+) \to I^+$ is injective (respect. surjective/bijective).

(d) If $\mathcal{S}' = (S', *', -', <', t')$ is a partial SUR-algebra, then for each fpSUR-algebra morphism $f : \mathcal{I} \to \mathcal{S}'$ there is a unique pSUR-algebra morphism $f^+ : \mathcal{I}^+ \to \mathcal{S}'$

such that $f^+ \circ i_0 = f$. In particular, if $\mathcal{S}$ is a SUR-algebra, then f and f^+ are automatically fpSUR-algebras morphisms. Moreover:
• *If t' is injective, then f is a $\Sigma-$quasi-embedding iff f^+ is a $\Sigma-$quasi-embedding.*

Proof. Items (a), (b) and (c) are straightforward verifications. We will just sketch the proof of the universal property in item (d).

Candidate and uniqueness:
Suppose that there is a pSUR-algebra morphism $f^+ : \mathcal{I}^+ \to \mathcal{S}'$ such that $f^+ \circ i_0 = f$. Since $f : I \to S$ is a full partial SUR-algebra morphism, we have $(f \times f)_\restriction : C_s(I) \to C_s^{t'}(S')$. Then $(f^+ \times f^+)_\restriction : C_s^{t^+}(I^+) \to C_s^{t'}(S') : (\Gamma, \Delta) = (i_0[A], i_0[B]) \mapsto (f^+[\Gamma], f^+[\Delta]) = (f[A], f[B]) \in C_s^{t'}(S')$ and $f^+(t^+(\Gamma, \Delta)) = t'(f^+[\Gamma], f^+[\Delta]) = t'((f[A], f[B])) \in S'$. Since $range(i_0) \cup range(i_1) = S^+$, the function f^+ is determined by f:
• $f^+(z) = f(x) \in S'$, whenever $z = [(x, 0)] \in range(i_0)$;
• $f^+(z) = t'((f[A], f[B])) \in S'$, whenever $z = ([(A, B), 1)] \in range(i_1)$.
Existence:
Since $f : I \to S$ is a full partial SUR-algebra morphism, we have $(f \times f)_\restriction : C_s(I) \to C_s^{t'}(S')$, then the arrows

$$(C_s(I) \overset{t' \circ (f \times f)_\restriction}{\longrightarrow} S' \overset{f}{\longleftarrow} I)$$

yields a commutative co-cone over the diagram

$$(I \overset{t}{\leftarrow} C_s^t(I) \overset{incl}{\hookrightarrow} C_s(I)).$$

By the universal property of set-theoretical pushout, there is a unique function $f^+ : I^+ \to S'$ such that:
• $f^+ \circ i_0 = f$;
• $f^+ \circ i_1 = t' \circ (f \times f)_\restriction$.
Thus it remains only to check that $f' : I^+ \to S'$ is a pSUR-algebra morphism:
• $f^+(*^+) = f^+([*, 0]) = f(*) = *'$;
• $f^+(-^+[(x, 0)]) = f(-x) = -'f(x) = -'f^+([(x, 0)])$;
$f^+(-^+[((A, B), 1)]) = t'((f \times f)_\restriction(-B, -A)) = t'(f[-B], f[-A]) = -'t'(f[B], f[A]) = -'f^+([((A, B), 1)])$.
• If $[(x, 0)] <^+ [(y, 0)]$, then $x < y$ thus $f^+([(x, 0)]) = f(x) <' f(y) = f^+([(y, 0)])$;
If $(A, B) \in C_s(I) \setminus C_s^t(I)$:
- if $[(x, 0)] <^+ [((A, B), 1)]$, then $x \in A$ and $f(x) \in f[A]$. Since$(f[A], f[B]) \in C_s^{t'}(S')$, thus $f^+([(x, 0)]) = f(x) <' t'(f[A], f[B]) = f^+([((A, B), 1)])$;
- if $[((A, B), 1)] <^+ [(y, 0)]$, then $y \in B$ and $f(y) \in f[B]$. Since $(f[A], f[B]) \in C_s^{t'}(S')$, thus $f^+([((A, B), 1)]) = t'(f[A], f[B]) <' f(y) = f^+([(y, 0)])$.

• If $(\Gamma, \Delta) = (i_0[A], i_0[B]) \in C_s^{t^+}(I^+)$, then $(f^+[\Gamma], f^+[\Delta]) = (f[A], f[B]) \in C_s^{t'}(S')$ and
$f^+(t^+(\Gamma, \Delta)) = f^+(t^+((i_0[A], i_0[B])) = f^+([((A,B),1)]) = t' \circ (f \times f)_{\restriction}([((A,B),1)])$
$= t'(f[A], f[B]) = t'(f^+[i_0[A]], f^+[i_0[B]]) = t'(f^+[\Gamma], f^+[\Delta])$. □

Remark 66.

In the setting above, we can interpret the Conway's notions in a very natural way:
• $Old(I) := i_0[I] \cong I$;
• $Made(I) := I^+$;
• $New(I) := I^+ \setminus i_0[I] = New(I)$.

Note that if $t : C_s^t(I) \to I$ is surjective (e.g. $I = No^{(\alpha)}, SA^{(\alpha)}, ST^{(\alpha)}$, $\alpha \in On \setminus \{0\}$), then $t^+ : C_s^{t^+}(I^+) \to I^+$ and every "made member" is represented buy a Conway cut in of "old members". This representation is unique, whenever $t : C_s^t(I) \to I$ is bijective (e.g., $I = SA^{(\alpha)}, ST^{(\alpha)}$, $\alpha \in On \setminus \{0\}$).

When $I = SA^{(\alpha)}, \alpha \in On \setminus \{0\}$ and $C_s^t(I) = \{(A,B) \in C_s(SA^{(\alpha)}, <_{(\alpha)}) : t(A,B) = \langle A,B \rangle \in SA^{(\alpha)}$ $(t : C_s(I) \to I$ is bijective), then $t^+ : C_s^{t^+}(I^+) \to I^+$ can be identified with the (bijective) map $C_s(SA^{(\alpha)}, <_{(\alpha)}) \to SA_\alpha$. □

A slight modification in the construction of the Σ-structure presented in 64 above, just replacing $<^+$ by $<^+_{(tc)} := trcl(<^+)$, yields the following:

Lemma 67. *Let $\mathcal{I} = (I, *, -, <, t)$ be a (small) partial SUR-algebra and keep the notation in 64 above. Then*

(a) $\mathcal{I}^+_{(tc)} = (I^+, *^+, -^+, <^+_{(tc)}, t^+)$ is a (small) <u>transitive</u> partial SUR-algebra.

(b) $i_0 : I \to I^+_{(tc)}$ is a $\Sigma-$quasi-embedding (see Remark 52.(f)) and full morphism of partial SUR-algebras. Moreover, if $\mathcal{I}$ is a <u>transitive</u> SUR-algebra, then $i_0 : I \to I^+_{(tc)}$ is a Σ-embedding.

(c) If $t : C_s^t(I) \to I$ is injective (respect. surjective/bijective), then $t^+ : C_s^{t^+}(I^+) \to I^+$ is injective (respect. surjective/bijective).

(d) If $\mathcal{S}' = (S', *', -', <', t')$ is a partial <u>transitive</u> SUR-algebra, then for each fpSUR-algebra morphism $f : \mathcal{I} \to \mathcal{S}'$ there is a unique pSUR-algebra morphism $f^+ : \mathcal{I}^+ \to \mathcal{S}'$ such that $f^+ \circ i_0 = f$. In particular, if $\mathcal{S}$ is a transitive SUR-algebra, then f and f^+ are automatically fpSUR-algebras morphisms. □

When $I = ST^{(\alpha)}, \alpha \in On \setminus \{0\}$ and $C_s^t(I) = \{(A,B) \in C_s(ST^{(\alpha)}, <_{(\alpha)}) : t(A,B) = \langle A,B \rangle \in ST^{(\alpha)}$ $(t : C_s(I) \to I$ is bijective), then $t^+ : C_s^{t^+}(I^+_{(tc)}) \to I^+_{(tc)}$ can be identified with the (bijective) map $C_s(ST^{(\alpha)}, <_{(\alpha)}) \to ST_\alpha$.

Remark 68.

Note that applying the construction $(\ )^+$ to the SUR-algebra SA we obtain $(SA)^+ \cong C_s(SA) = SA$.

Applying both constructions $(\ )^+$ and $(\ )^+_{(tc)}$ to the SUR-algebra ST we obtain $(ST)^+ = (ST)^+_{(tc)} \cong C_s(ST) = ST$. □

Now we are ready to state and prove the main result of this section:

Theorem 69. *Let $\mathcal{I}$ be any small partial SUR-algebra. Then there exists SUR-algebras denoted by $SA(\mathcal{I})$ and $ST(\mathcal{I})$, and pSUR-morphisms $j_I^A : \mathcal{I} \to SA(\mathcal{I})$ and $j_I^T : \mathcal{I} \to ST(\mathcal{I})$ such that:*

(a)

(a1) j_I^A is a fpSUR-morphism and a Σ-embedding;

(a2) If $t : C_s^t(I) \to I$ is injective (respect. surjective/bijective), then $t^\infty : C_s^{t^\infty}(SA(I)) \to SA(I)$ is injective (respect. surjective/bijective);

(a3) $j_I^A : \mathcal{I} \to SA(\mathcal{I})$ satisfies the universal property: for each SUR-algebra $\mathcal{S}$ and each pSUR-morphism $h : \mathcal{I} \to \mathcal{S}$, there is a unique SUR-morphism $h_A : SA(\mathcal{I}) \to \mathcal{S}$ such that $h_A \circ j_I^A = h$. Moreover:

- *If t' is injective, then h is a $\Sigma-$quasi-embedding iff h_A is a $\Sigma-$quasi-embedding.*

(b)

(b1) j_I^T is a fpSUR-morphism and a $\Sigma-$ quasi-embedding, that is a Σ-embedding whenever $\mathcal{I}$ is transitive;

(b2) If $t : C_s^t(I) \to I$ is injective (respect. surjective/bijective), then $t^\infty : C_s^{t^\infty}(ST(I)) \to ST(I)$ is injective (respect. surjective/bijective);

(b3) $j_I^T : \mathcal{I} \to ST(\mathcal{I})$ satisfies the universal property: for each **transitive** *SUR-algebra $\mathcal{S}$ and each pSUR-morphism $h : \mathcal{I} \to \mathcal{S}$, there is a unique SUR-morphism $h_T : ST(\mathcal{I}) \to \mathcal{S}$ such that $h_T \circ j_I^T = h$.*

Proof.

Item (a): based on based on Lemma 65 and Proposition 62, we can define, by transfinite recursion a convenient *increasing* (compatible) family of diagrams $D_\alpha : [0, \alpha] \to pSUR-alg$, $\alpha \in On$, where:

(D0) $D_0(\{0\}) = I$;

(D1) For each $0 \leq \gamma < \beta < \alpha$, $D_\alpha(\gamma, \beta) = D_\beta(\gamma, \beta) : D_\beta(\gamma) \to D_\beta(\beta)$ is Σ-embedding and a fpSUR-morphism;

Just define $D_\alpha(\alpha) = (D_\alpha^{(\alpha)})^+$, where $D_\alpha^{(\alpha)} := colim_{\beta<\alpha} D_\alpha(\beta)$ and take, for $\beta < \alpha$, $D_\alpha(\beta, \alpha) = (h_\beta^\alpha)^+ : D_\alpha(\beta) \to (colim_{\beta<\alpha} D_\alpha(\beta))^+$ be the unique pSUR-morphism –that is automatically a fpSUR-morphism and a Σ-embedding, whenever h_β^α satisfies this conditions (see Lemma 65.(d))– such that $(h_\beta^\alpha)^+ \circ i_0 = h_\beta^\alpha$, where i_0 :

$(colim_{\beta<\alpha}D_\alpha(\beta)) \to (colim_{\beta<\alpha}D_\alpha(\beta))^+$ and where $h^\alpha_\beta : D_\alpha(\beta) \to (colim_{\beta<\alpha}D_\alpha(\beta))$ is the colimit co-cone arrow: by the recursive construction and by Proposition 62 h^α_β is a fpSUR-morphism and a Σ-embedding. This completes the recursion.

Gluing this increasing family of diagrams we obtain a diagram $D_\infty : On \to pSUR - alg$.

By simplicity we will just denote:

- $SA(I)_\alpha := D_\infty(\alpha)$, $\alpha \in On$;
- $SA(I)_\infty := colim_{\alpha\in On}SA(I)_\alpha$;
- $D_\alpha(\beta, \alpha) = j^A_{\beta,\alpha}$, for each $0 \leq \beta \leq \alpha \leq \infty$ (since the family $(D_\alpha)_\alpha$ is increasing, we just have introduce notation for "new arrows").

Then we set: $SA(I) := SA(I)_\infty$ and $j^A_I := j^A_{0,\infty}$.

The verification that $SA(I)$ is a SUR-algebra that satisfies the property in item (a2) and that j^A_I satisfies item (a1)[29], follows the recursive construction of the diagram and from a combination of Proposition 62 and Lemma 65.

By the same Lemma and Proposition combined, it can be checked by induction that for each $\alpha \in On$, there is a unique pSUR-morphism $h_\alpha : SA(\mathcal{I})_\alpha \to \mathcal{S}$ such that $h_\alpha \circ j^A_{0,\alpha} = h$ and such that h_α is injective (respect. $\Sigma - quasi$-embedding, Σ-embedding), whenever h is injective (respect. $\Sigma - quasi$-embedding, Σ-embedding). By applying one more time the colimit construction, we can guarantee that there is a unique pSUR-morphism $h^A := h_\infty : SA(\mathcal{I})_\infty \to \mathcal{S}$ such that $h^A \circ j^A_I = h$ and that it satisfies the additional conditions.

The proof of item (b) is analogous to the proof of item (a): basically we just have to replace to use of technical Lemma 65 by other technical Lemma 67. In general, we can on guarantee that $j^T_{\beta,\alpha}$ is a Σ-embedding and a fpSUR-morphism only for $0 < \beta < \alpha \leq \infty$. □

In particular, taking $I = \mathcal{S}_0$ as the *initial object* in pSUR-alg (see Proposition 58 in Subsection 3.3), we have that $SA \cong SA(I)$ and $ST \cong ST(I)$, and they satisfy corresponding universal properties:

Corollary 70.
(a) SA is universal (= initial object) over all SUR-algebras, i.e. for each SUR-algebra $\mathcal{S}$, there is a unique SUR-algebra morphism $f_\mathcal{S} : SA \to \mathcal{S}$.

(b) ST is universal (= initial object) over all **transitive** *SUR-algebras, i.e. for each* **transitive** *SUR-algebra $\mathcal{S}'$, there is a unique SUR-algebra morphism $h_{\mathcal{S}'} : ST \to \mathcal{S}'$.*

[29] In fact, $j^A_{\beta,\alpha}$ is Σ-embedding whenever $0 \leq \beta \leq \alpha \leq \infty$ and $j^A_{\beta,\alpha}$ is a fpSUR-morphism whenever $0 \leq \beta < \alpha \leq \infty$.

Proof. Item (a): Since for each each SUR-algebra $\mathcal{S}$ there is a unique pSUR-morphism $u_{\mathcal{S}} : \mathcal{S}_0 \to \mathcal{S}$ then, by Theorem 69.(a) above, $SA(\mathcal{S}_0)$ is a SUR-algebra that has the required universal property, thus we only have to guarantee that $SA \cong SA(\mathcal{S}_0)$. Taking into account the Remark 66 and the constructions performed in the proof of the item (a) in Theorem above, that we have a (lage) family of compatible pSUR-isomorphisms $ST_\alpha \cong SA(I)_\alpha, \forall \alpha \in On$. Thus $SA = \bigcup_{\alpha \in On} SA_\alpha \cong colim_{\alpha \in On} SA(\mathcal{S}_0)_\alpha = SA(\mathcal{S}_0)_\infty = SA(\mathcal{S}_0)$.

For item (b) the reasoning is similar: note that $I = \mathcal{S}_0 = \{*\}$ is a transitive partial SUR-algebra to conclude that $ST(\mathcal{S}_0)$ has the required universal property and note that by the proof of item (b) in Theorem 69 above, that $ST = \bigcup_{\alpha \in On} ST_\alpha \cong colim_{\alpha \in On} ST(\mathcal{S}_0)_\alpha = ST(\mathcal{S}_0)_\infty = ST(\mathcal{S}_0)$. □

This Corollary describes, in particular, that SA and ST are "rigid" as Σ-structures and :
- SA and $C_s(SA)$ are isomorphic SUR-algebras and the universal map $SA \to C_s(SA)$ is the unique iso from SA to $C_s(SA)$;
- ST and $C_s(ST)$ are isomorphic SUR-algebras and the universal map $ST \to C_s(ST)$ is the unique iso from ST to $C_s(ST)$.

We finish this section with an application of the Corollary 70 above: we obtain some non-existence results.

Corollary 71.
(i) Let $\mathcal{L}$ be a linear SUR-algebra, i.e., $<$ is a total relation (for instance take $\mathcal{L} = No$). Then there is no SUR-algebra morphism $h : \mathcal{L} \to ST$.
(ii) Let $\mathcal{T}$ be a transitive SUR-algebra, i.e., $<$ is a transitive relation (for instance take $\mathcal{T} = ST, No$). Then there is no SUR-algebra morphism $h : \mathcal{T} \to SA$.

Proof. (i) Suppose that there is a SUR-algebra morphism $h : \mathcal{L} \to ST$. Since the binary relation $<$ in L is acyclic and total, it is a strictly linear order, in particular it is transitive. Let $a, b \in L$, since L is linear, $a < b$ in $L \Leftrightarrow h(a) < h(b)$ in ST. Now, by the universal property of ST (see Theorem above) there is a unique SUR-algebra morphism $u : ST \to \mathcal{L}$ and then $h \circ u = id_{ST}$. Summing up, $h : (L, <) \to (ST, <)$ is an isomorphism of structures, thus $(ST, <)$ is a strictly ordered class, but the members of ST 0 and $\langle \{-1\}, \{1\} \rangle$ are not comparable by Fact 1.(c) in the Subsubsection 2.2.3, a contradiction.

(ii) Suppose that there is a SUR-algebra morphism $h : \mathcal{T} \to SA$. Since the binary relation $<$ in T is transitive, by the universal property of ST there is a (unique) SUR-algebra morphism $v : ST \to \mathcal{T}$, thus we get a SUR-algebra morphism

$g = h \circ v : ST \to SA$. By the universal property of SA there is a unique SUR-algebra morphism $u : SA \to ST$ (u is a inclusion) and then $g \circ u = id_{SA}$. Thus, for each $a, b \in SA$, $a < b$ in $SA \Leftrightarrow u(a) < u(b)$ in ST, but the members of SA denoted by -1 and 1 are not related (see Fact.(c) in the Subsubsection 2.2.2) and $u(-1) < u(1)$ in ST (by Fact 1.(c) in the Subsubsection 2.2.3), a contradiction. □

5 Final remarks and future works

The present work is essentially a collection of elementary results where we develop, from scratch, a new (we hope!) and complementary aspect of the Surreal Number Theory. In a continuation of the present work ([29]) we will establish links, in both directions, between SUR-algebras and ZF-algebras (the keystone of Algebraic Set Theory) and develop the first steps of a certain kind of set theory based (or ranked) on surreal numbers, that expands the relation between V and On.

In the sequel, we briefly present a (non-exhaustive) list of questions that have occurred to us during the elaboration of this work that we intend to address in the future.

Questions directly connect with the material presented in this work:

• We have described some general constructions in categories of partial SUR-algebra (with at least two kinds of morphisms): initial object, non-empty products, substructures and some kinds of directed inductive (co)limits. There are other general constructions available in these categories like quotients and coproducts? A preliminary analysis was made and indicates that the characterizations of the conditions where such constructions exists is a non trivial task.

• A specific construction like the (functor) cut surreal for SUR-algebras and its partial version turns out to be very useful to the development of the results of the (partial) SUR-algebra theory: the situation is, in some sense, parallel to the specific construction of rings of fractions construction in Commutative Algebra and Algebraic Geometry. There are other natural and nice specific constructions of (partial) SUR-algebra that, at least, provide new classes of examples?

• We have provided, by categorial methods, some universal results that characterizes the SUR-algebras SA and ST, and also some relative versions with base ("urelements") $SA(I)$, $ST(I')$ where I, I' are partial SUR-algebra satisfying a few constraints. There is an analog result satisfied by the SUR-algebra No? There are some natural expansions of No by convenient I'' are partial SUR-algebra, $No(I'')$, that also satisfies a universal property that characterizes it up to a unique isomorphism?

Future Works:

"Algebraic set theory" (AST) is a categorical approach to Set Theory (see [8]) based on universal properties satisfied by "relatively" free ZF-algebras: this is the primitive notion of the theory that allows to define (a posteriori) the notions of "be a set", "be an ordinal", or "be a member".

If V is the class of all sets, note that: $P_s(V)$ is a "large" small-complete lattice; $V = P_s(V)$; $u : V \to P(V)$ $x \mapsto \{x\}$ is an *endofunction*; $x \in y$ iff $u(x) \subseteq y$ in $(P_s(V), \subseteq)$.

A ZF-algebra is $(L, \bigvee, L \overset{s}{\to} L)$, L is a "large" small-complete lattice. Examples: $(V, \bigcup, u)$ is the free ZF-algebra; $(On, \bigcup, (-)^+)$ is the free ZF-algebra among the ones with increasing/inflationary "sucessor function".

There is a "derived" set theory: the theory of the free model (= free ZF-algebra) V, where $x \varepsilon y$ is **defined** as $s(x) \leq y$.

In [29], a sequel of present work, we have defined relations (in both directions) between certain class of equipped SUR-algebras (called *anchored* SUR-algebras) and certain class of ZF-algebras (called *standard* ZF-algebras), that "explains" and "expands" the relations $On \underset{b}{\overset{j}{\rightleftarrows}} No$.

a) The ZF-algebras V, On above described are *standard* ZF-algebras. In every standard ZF-algebra, the induced membership relation (as above) is well-founded.

b) Given a standard ZF-algebra and an operation on binary relations (e.g. "identity", "transitive closure", etc) we can build a corresponding Sur-algebra *space of signs*. (e.g. $Sig(V) \cong SA, Sig(On) \cong No$).

c) We say that a function $\beta : S \to \mathcal{C}$ from a SUR-algebra S onto an well-founded class $(\mathcal{C}, \prec)$ is an *anchor*, when it satisfies some convenient properties (SA, ST, No are naturally anchored; the birth function $b : No \to On$ is an anchor). This induces:
- an well-founded relation $\prec_\beta$ in S;
- a recursively-defined subclass of $HP_\beta(S) \subseteq S$ of "hereditary positive" members, with an induced ZF-algebra structure (e.g. $HP(SA) \cong V, HP(No) \cong On$).

d) Every SUR-algebra "space of signs" is naturally equipped with a anchor on its underlying standard ZF-algebra.

There are relationships, summarized by the diagram below, between the free SUR-algebra (SA), the class of all sets (V), the class of all ordinal numbers and the class of the surreal numbers:

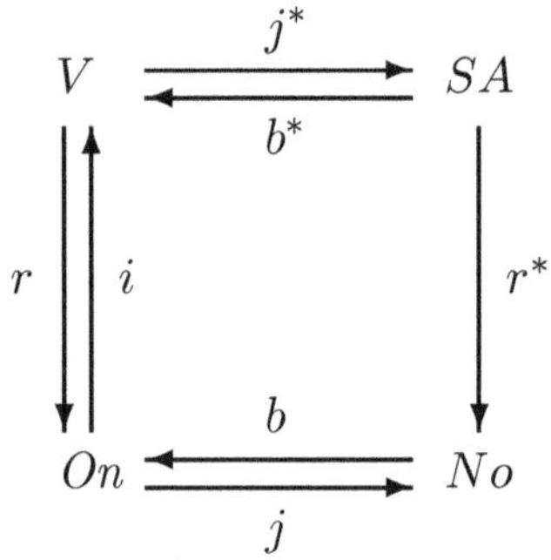

• The arrow r is the universal ZF-algebra morphism: the rank function.

• The arrow i is the inclusion $On \subseteq V$.

• The arrows b and j are as described in the Subsection 1.2; $b : No \to On$ is a anchor.

• r^* is universal SUR-algebra arrow.

• $j^*(x) = \langle j^*[x], \emptyset \rangle$ (defined by recursion)

• $b^*(\langle A, B \rangle) = b^*[A] \cup b^*[B]$ (defined by recursion); $b^* : SA \to V$ is an anchor.

We have the diagramatic relations (established by inductions):

• $r \circ i = id_{On}, \quad b \circ j = id_{No}, \quad b^* \circ j^* = id_V$

• $j = r^* \circ j^* \circ i, \quad r^* \circ j^* = j \circ r, \quad b \circ r^* \leq r \circ b^*$

Notation: $x, y, z \in V$, $u, v, w \in SA$. From the relation above we can derive:

• $u \prec v$ iff $b^*(u) \in b^*(v)$ (well founded relation)

• $u < v \;\Rightarrow\; u \prec v$ or $v \prec u$

• $x \in y \Rightarrow j^*(x) \prec j^*(y)$ and $j^*(x) < j^*(y)$.

• $x \in y \Leftrightarrow j^*(x) < j^*(y)$

These, in turn, suggests that SA, can encode an expansion of the class V, not only in a diagrammatic sense, but also according a set-theoretical viewpoint: In SA, an object $x = \langle A, B \rangle$ is "extensionally determined by simpler objects" since $A = \{y \in SA : y \prec x, y < x\}$ and $B = \{z \in SA : z \prec x, x < z\}$

Requirements for a theory "be" a set theory:

(a) Have the potential to define arrows (category) as entities of the theory, through some fundamental relation(s).

(b) Be the "internal theory" of a free object in a category (like in ZF-algebra setting).

(c) The derived internal category is "topos-like".

For instance $Set = cat(V)$, the topos of all sets and functions, is the category associated to the Zermelo-Fraenkel set theory (as usual), that, in turn, is the (classical) theory derived from the free ZF-algebra V.

In fact, in [29], we provide a first construction over SA that fulfills the three natural requirements above. However, trying to maintain determination by "binary" membership, we were limited to use only the subclass $Pos(SA) \subseteq SA$ of the "positive members" of SA and obtain a category $cat(Pos(SA))$ that turns out to be equivalent to $Set = cat(V)$.

Motivated by the comments above, we finish this work with some questions that will deserve future investigations:

• Can we obtain a wider (but still natural) expansion $cat(V) \rightarrow cat(SA)$ through construction on full SA by a *ternary* "incidence relation" $m(x, y, z)$ (i.e. $x < y < z$ and $x, z \prec y$)? This seems a natural question to pose about SA, since if $A < B$, then $a < \langle A, B\rangle < b$ and $a, b \prec \langle A, B\rangle$, for each $a \in A, b \in B$.

• We saw that the free/initial SUR-algebra SA is, in many senses, an expansion of the free/initial ZF-algebra V and its underlying set theory. Relatively constructions are available for SUR-algebras and for ZF-algebras (see [8]). In particular, it can be interesting examine possible natural expansions of set theories:

- with urelements B, $V(B)$, to some convenient relatively free SUR-algebra $SA(\hat{B})$;
- obtained from the free transitive SUR-algebra $ST \rightarrow No$

• A combination of the lines of research above mentioned can be a interesting ("second-order") task: it will be a line of development of general relative set theories that are genuinely "base independent".

References

[1] Peter Aczel, *Non-well-founded sets*, CSLI Lecture Notes, vol. 14, Stanford University, Center for the Study of Language and Information, Stanford, CA, 1988. With a foreword by Jon Barwise [K. Jon Barwise].

[2] Norman L. Alling, *Foundations of analysis over surreal number fields*, North-Holland Mathematics Studies, vol. 141, North-Holland Publishing Co., Amsterdam, 1987. Notas de Matemática [Mathematical Notes], 117.

[3] M. Aschenbrenner, L. van den Dries, and Joris van der Hoeven, *On numbers, germs, and transseries*, In Proc. Int. Cong. of Math. **1** (2018), no. 4, 1–24.

[4] ———, *The surreal numbers as a universal H- fields*, J. Eur. Math. Soc. **21** (2019), no. 4, 1179–1199.

[5] Vincent Bagayoko and Joris van der Hoeven, *Surreal substructures*, researchgate.net/publication/334277252 (2019), 61.

[6] J. H. Conway, *On numbers and games*, 2nd ed., A K Peters, Ltd., Natick, MA, 2001.

[7] Harry Gonshor, *An introduction to the theory of surreal numbers*, London Mathematical Society Lecture Note Series, vol. 110, Cambridge University Press, Cambridge, 1986.

[8] A. Joyal and I. Moerdijk, *Algebraic set theory*, London Mathematical Society Lecture Note Series, vol. 220, Cambridge University Press, Cambridge, 1995.

[9] Thomas Jech, *Set Theory*, Springer Monographs in Mathematics, Springer-Verlag, Berlin, 2003.

[10] Kenneth Kunen, *Set Theory*, Studies in Logic, vol. 34, College Publications, London, 2013.

[11] Norman L. Alling, *On the existence of real-closed fields that are η_α-sets of power $\aleph_\alpha$*, Trans. Amer. Math. Soc. **103** (1962), 341–352. MR0146089 (26 #3615)

[12] Norman L. Alling and Philip Ehrlich, *An abstract characterization of a full class of surreal numbers*, C. R. Math. Rep. Acad. Sci. Canada **8** (1986), no. 5, 303–308. MR859431 (87j:04008)

[13] ———, *An alternative construction of Conway's surreal numbers*, C. R. Math. Rep. Acad. Sci. Canada **8** (1986), no. 4, 241–246. MR850107 (87j:04007)

[14] Lou van den Dries and Philip Ehrlich, *Fields of surreal numbers and exponentiation*, Fund. Math. **167** (2001), no. 2, 173–188.

[15] Philip Ehrlich, *An alternative construction of Conway's ordered field* No, Algebra Universalis **25** (1988), no. 1, 7–16, DOI 10.1007/BF01229956. MR934997 (89d:04004a)

[16] ———, *All numbers great and small*, Real numbers, generalizations of the reals, and theories of continua, Synthese Lib., vol. 242, Kluwer Acad. Publ., Dordrecht, 1994, pp. 239–258. MR1340465

[17] ———, *Number systems with simplicity hierarchies: a generalization of Conway's theory of surreal numbers*, J. Symbolic Logic **66** (2001), no. 3, 1231–1258.

[18] ———, *Conway names, the simplicity hierarchy and the surreal number tree*, J. Log. Anal. **3** (2011), Paper 1, 26, DOI 10.4115/jla.2011.3.11. MR2769328 (2012i:06002)

[19] ———, *The absolute arithmetic continuum and the unification of all numbers great and small*, Bull. Symbolic Logic **18** (2012), no. 1, 1–45. MR2798267

[20] David Ellerman, *Adjoint Functors and Heteromorphisms*, ArXiv.org **0704.2207** (2007), 28.

[21] Rodrigo de Alvarenga Freire, *Grasping Sets Through Ordinals: On a Weak Form of the Constructibility Axiom*, South American Journal of Logic **2** (2016), no. 2, 347–359.

[22] Joel David Hamkins, *Every countable model of set theory embeds into its own constructible universe*, J. Math. Log. **13** (2013), no. 2, 1350006, 27.

[23] Joram Hirschfeld, *The Model Companion of ZF*, Proceedings of the American Mathematical Society **50** (1975), no. 1, 369–374.

[24] Salma Kuhlmann and Mickaël Matusinski, *The exponential-logarithmic equivalence classes of surreal numbers*, Order **32** (2015), no. 1, 53–68.

[25] E. G. K. López-Escobar, *Logic and mathematics: propositional calculus with only three primitive terms*, The many sides of logic, Stud. Log. (Lond.), vol. 21, Coll. Publ., London, 2009, pp. 153–170.

[26] Jacob Lurie, *The effective content of surreal algebra*, J. Symbolic Logic **63** (1998), no. 2, 337–371.

[27] Saunders MacLane, *The universality of formal power series fields*, Bull. Amer. Math. Soc. **45** (1939), 888–890.

[28] Dimi Rocha Rangel, *An algebraic framework to a theory of sets based on the surreal numbers*, PhD thesis, University of São Paulo, Brazil, 2018.

[29] Dimi Rocha Rangel and Hugo Luiz Mariano, *An algebraic (set) theory of surreal numbers, II*, preprint (2019), 33.

[30] Dana Scott, *On completing ordered fields*, Applications of Model Theory to Algebra, Analysis, and Probability (Internat. Sympos., Pasadena, Calif., 1967), Holt, Rinehart and Winston, New York, 1969, pp. 274–278.

Received 01 December 2019

Horn-Geometric Axioms for Faithfully Quadratic Rings

Hugo Rafael de Oliveira Ribeiro
Institute of Mathematics and Statistics, University of São Paulo, Brazil
hugor@ime.usp.br

Hugo Luiz Mariano
Institute of Mathematics and Statistics, University of São Paulo, Brazil
hugomar@ime.usp.br

Abstract

We present *explicit* Horn-geometric axiomatization for the theory of Faithfully Quadratic Rings, answering a question posed in [4].

Keywords: Quadratic Forms, Special Groups, Faithfully Quadratic Rings, Horn-Geometric sentences

Introduction

The classical theory of algebraic quadratic forms over fields of characteristic $\neq 2$ started in the late 1930's with E. Witt and was put foward in the late 1960's by A. Pfister. In the decade of 1980's, appeared the first abstract approach to quadratic forms theory as Marshall's abstract space of orderings ([5]). Only in early 1990's that arise a (finitary) first-order theory that generalizes, simultaneously, the reduced and non-reduced theory of quadratic forms called special groups ([2] and Definition 1).

In [4], the theory of Special Groups is employed to generalize results on quadratic forms over fields to a wide class of (commutative, unitary) rings, therein named faithfully quadratic rings, that includes rings with many units, reduced f-rings (herein rings of continuous real-valued functions) and strictly representable rings. If A is

The authors would like to thank the anonymous referee for his/her valuable suggestions that substantially improved our presentation.

a T-faithfully quadratic ring and $G_T(A)$ is its associated special group, then T-isometry and T-representation in the ring A are faithfully encoded by isometry and representation in $G_T(A)$. In fact, the theory can be extended to pairs (S,T) where T is a quadratic set and S is a T-subgroup (cf. Definition 6 below), which generalize Definition 2.2, p.10, of [4], and gives a unified treatment to the subject.

By a combination of results in [4], the first-order theory of T-faithfully quadra tic rings is *both* Horn and geometric. By Corollary 5.6, p.58, in [4], the theory of T-faithfully quadratic rings has a Horn axiomatization -which involves a non-constructive model-theoretic result of Keisler, Galvin and Shelah ([1], Theorem 6.2.5)- while Theorem 5.2, p. 54, in [4] shows that T-faithfully quadratic rings is a geometric theory. The question posed is if *explicit* Horn-geometric axioms could be provided for that theory (cf. paragraph right after the proof of Corollary 5.6, p. 58). It should be remarked that Horn-geometrical axioms for faithfully quadratic rings, in the case were T is the set of all squares of the ring, appears in Theorem 5.2.(a), p. 54, of [4].

In this work we provide a complete and explicit list of Horn-geometric axioms for T-quadratic faithfullness (Theorem 19). In the case where T is not necessarily the set of all squares of the ring, the contribution of our short work is the equivalence between T-isometry (Definition 2.17, p. 18, [4]) and the notion of T-congruence, introduced in Definition 12 of the paper, allowing the elimination of the disjunction appearing in underlined formula in p.56 of [4] and yielding a Horn-geometric axiomatic for the theory of Faithfully Quadratic Rings.

In all that follows: the word "ring" stands for a commutative unitary ring A, wherein 2 is a unit.

1 Preliminaries

Let A be a set and $\equiv$ a binary relation on A^2. We extend $\equiv$ to A^n, $n \geq 3$, by induction as follows: given $a_1, \ldots, a_n, b_1, \ldots, b_n \in A$, we say that $\langle a_1, \ldots, a_n \rangle \equiv_n \langle b_1, \ldots, b_n \rangle$ if there are $\alpha, \beta, z_3, \ldots, z_n \in A$ such that

$$\langle a_1, \alpha \rangle \equiv \langle b_1, \beta \rangle$$
$$\langle a_2, \ldots, a_n \rangle \equiv_{n-1} \langle \alpha, z_3, \ldots, z_n \rangle$$
$$\langle b_2, \ldots, b_n \rangle \equiv_{n-1} \langle \beta, z_3, \ldots, z_n \rangle.$$

For $n = 1$, we adopt the convention $\langle a \rangle \equiv_1 \langle b \rangle \Leftrightarrow a = b$. In general, the subscript of $\equiv_n$ is omitted and we use just $\equiv$. This extension of $\equiv$ comes from the inductive characterization of the quadratic forms isometry relation over fields of characteristic $\neq 2$.

Definition 1 (Def. 1.9, [4]). *A **proto-special group** (abbreviation: π-SG) is triple $(G, -1, \equiv_G)$ where G is a group of exponent 2, $-1 \in G$ (denote: $-x$ for $-1 \cdot x$, $\forall x \in G$) and $\equiv_G$ is a binary relation on $G \times G$ satisfying for all $a, b, c, d, x \in G$:*

SG-0 *$\equiv_G$ is an equivalence relation.*

SG-1 $\langle a, b\rangle \equiv_G \langle b, a\rangle$.

SG-2 $\langle a, -a\rangle \equiv_G \langle 1, -1\rangle$.

SG-3 $\langle a, b\rangle \equiv_G \langle c, d\rangle \Rightarrow ab = cd$.

SG-5 $\langle a, b\rangle \equiv_G \langle c, d\rangle \Rightarrow \langle xa, xb\rangle \equiv_G \langle xc, xd\rangle$.

*A π-SG is called **reduced (π-RSG)** if it satisfies* **red** *below and $-1 \neq 1$ (i.e. it is formally real).*

red $\langle a, a\rangle \equiv_G \langle 1, 1\rangle \Rightarrow a = 1$.

*Furthermore, a π-SG $(G, \equiv_G, -1)$ is called **pre-special group (pSG)** if for all $a, b, c, d \in G$*

SG-4 $\langle a, b\rangle \equiv_G \langle c, d\rangle \Rightarrow \langle a, -c\rangle \equiv_G \langle -b, d\rangle$

*and is a **special group (SG)** if satisfies $SG-4$ and*

SG-6 *The extension of $\equiv_G$ to G^3 is a transitive relation.*

Remark 1. *i) The axiom **SG-6** of Special Group definition seems to be very specific but it implies that for all $n \geq 1$, the extended relation $\equiv_n$ is an equivalence relation. See Theorem 1.23 of [2] for a proof of this fact.*

ii) A n-form over G is a tuple $\varphi = \langle a_1, \dots, a_n\rangle$, where $n \in \mathbb{N}$ and $a_1, \dots, a_n \in G$.

Definition 2 (Def. 1.3, [2]). *Let $(G, \equiv, -1)$ a π-SG and $\varphi = \langle a_1, \dots, a_n\rangle$ a form over G. The set of elements represented by φ is*

$$D_G(\varphi) = \{x \in G\colon \text{there are } b_2, \dots, b_n \in G \text{ such that } \langle x, b_2, \dots, b_n\rangle \equiv \varphi\}.$$

Example 3. • *Let K be a field of characteristic $\neq 2$. Consider $G(K) = K^\times/{K^\times}^2$, where $K^\times = K \setminus \{0\}$ is the set of all units of K, and the binary relation in $G(K)$ given by $\langle\overline{a}, \overline{b}\rangle \equiv \langle\overline{c}, \overline{d}\rangle$ iff $\overline{ab} = \overline{cd}$ in $G(K)$ and $a = cx^2 + dy^2$ for some $x, y \in K$. Then $(G(K), \equiv, \overline{-1})$ is a special group (Theorem 1.32 of [2]).*

- *The above construction can be generalized to a broad class of rings that satisfies the conditions of Definition 15 below: this includes the rings with many units, reduced f-rings (herein rings of continuous real-valued functions) and strictly representable rings. For more details, see [4].*

Definition 4 (Def. 1.4, [4]). *Let $\mathcal{L}$ a first-order language.*

*i) A formula φ is **positive-primitive (pp-formula)** if it is equivalent to $\exists \overline{v} \psi(\overline{v})$ where $\psi(\overline{v})$ is a conjunction of atomic formulas.*

*ii) A formula φ is called **Horn-geometrical** in the free variables $\overline{t}$ if it is the negation of an atomic formula or of the form $\forall \overline{v} \psi_1(\overline{v}, \overline{t}) \rightarrow \psi_2(\overline{y}, \overline{t})$ where ψ_1, ψ_2 are pp-formulas.*

Example 5.

- *The theory of commutative unitary rings is Horn-geometrical.*
- *The theory of (reduced) special groups is Horn-geometrical.*

The next definition sets the grounds to the theory of faithfull quadratic rings and slightly generalize the Definition 2.2, p.10, of [4], introducing the (unifying) notion of quadratic set.

Definition 6. *Let A be a ring.*

- *A subset $T \subseteq A$ is called **quadratic** if it is multiplicative (i.e., it closed under the ring product) and $A^2 \subseteq T$.*
- *Given a quadratic set $T \subseteq A$, a set $S \subseteq A^\times$ is called T**-subgroup** if it is multiplicative and $T^\times \cup \{-1\} \subseteq S$. When $T = A^2$, the A^2-subgroups are called q-subgroups.*

Let $T \subseteq A$ be a quadratic set and $S \subseteq A^\times$ be a T-subgroup. Given $a, b \in S$, the valued-represented mod T (p. 11, [4]) by $\langle a, b \rangle$ is

$$D^v_{S,T}(a, b) = \{c \in S \colon \text{there are } t, t' \in T \text{ such that } c = at + bt'\}.$$

Moreover, if $D^v_{S,T}(1, a)$ is multiplicative, for every $a \in S$, then (S, T) is called a **quadratic pair**.

Remark 2. *Let A be a ring.*

- *If $T = A^2$ or T is a preorder in A (i.e., if $A^2 \subseteq T, T \cdot T \subseteq T, T + T \subseteq T$), then T is a quadratic set in A.*

  Let T be a quadratic set in the ring A.

- *The smallest* T*-subgroup is* $T^\times \cup -T^\times$ *and the largest is* $A^\times$.
- *If* S *is a* T*-subgroup, then* S *is a subgroup of* $A^\times$.
- *Since* $0, 1 \in T$, $\{a, b\} \subseteq D^v_{S,T}(a, b)$.

Remark 3. *Let* A *be a ring,* T *be a quadratic set in* A *and* S *be a* T*-subgroup.*

- *If* (S, T) *is a quadratic pair, then* $D^v_{S,T}(1, a) \subseteq S$ *is a subgroup for all* $a \in S$. *In fact,* $1 \in D^v_{S,T}(1, a)$ *and given* $x \in D^v_{S,T}(1, a)$, *let* $u, v \in T$ *such that* $x = u + av$. *Then* $x^{-1} = x^{-2}x = ux^{-2} + avx^{-2} \in D^v_{S,T}(1, a)$. *Since* $D^v_{S,T}(1, a)$ *is multiplicative, it follows that* $D^v_{S,T}(1, a) \subseteq S$ *is a subgroup.*
- *If* (S, T) *is a quadratic pair and* $S' \subseteq S$ *is another* T*-subgroup, then* (S', T) *is also a quadratic pair. In particular, if* $(A^\times, T)$ *is a quadratic pair, then every* T*-subgroup determines a quadratic pair.*
- *A very interesting class of examples of quadratic pairs* (S, T) *is given in Theorem 2.3 in [3]: they are able to represent all reduced special groups i.e., for each special group* G, *there exists a quadratic pair* (S, T) *such that* $G \cong G_T(S)$. *In more details:* $A_G := C(X_G, \mathbb{R})$, *where* X_G *is the (boolean) space of all SG-morphisms* $G \to \{-1, 1\}$ *and* $T = A_G^2$.

In the field case, there is a useful criteria to determine a SG-subgroup (Definition 1.28 and Lemma 1.29, p.21, [2]). In the ring case, an analogous strategy works to obtain quadratic pairs.

Proposition 7. *Let* A *be a ring,* $T \subseteq A$ *quadratic set and* $S \subseteq A^\times$ *a* T*-subgroup. Suppose that for all* $p, q, u, v \in T$ *and* $a \in S$, *exists* $x \in A$ *such that*

$$(pua^2 + qv - xa) \in T \text{ and } (pv + qu + x) \in T. \qquad (*)$$

Then (S, T) *is a quadratic pair. In particular, if* $T = A^2$ *or* T *is a preorder, then* (S, T) *is a quadratic pair.*

Proof. Let $s, t \in D^v_{S,T}(1, a)$, $a \in A$. Take $p, q, u, v \in T$ with

$$s = pa + q \text{ and } t = ua + v.$$

Then $st = (pua^2 + qv) + (pv + qu)a$. By hypothesis, exists $x \in A$ such that $(pua^2 + qv - xa) \in T$ and $(pv + qu + x) \in T$. Then

$$st = (pua^2 + qv - xa) + (pv + qu + x)a \in D^v_{S,T}(1, a).$$

If T is closed under sums (preorder), then $(*)$ is satisfied with $x = 0$ for all $a \in A$. If $T = A^2$, then $p = p_1^2, q = q_1^2, u = u_1^2, v = v_1^2$; if we take $x = 2(p_1q_1u_1v_1)$, then $(pua^2 + qv - xa) = (p_1u_1a - q_1v_1)^2 \in T$ and $(pv + qu + x) = (p_1v_1 + q_1u_1)^2 \in T$. □

The next lemma is a version of Lemma 2.6, p.11, [4] for quadratic pairs and gives some properties for the valued-represented set of a 2-form. Note that the proofs follows without any major change.

Lemma 8. *Let A be a ring and (S,T) a quadratic pair. Let $x, y, u, v \in S$ and $t, w \in T^\times$.*

i) $uD^v_{S,T}(x,y) = D^v_{S,T}(ux,uy)$ *and* $D^v_{S,T}(x,y) = D^v_{S,T}(tx,wy)$.

ii) $u \in D^v_{S,T}(x,y) \Rightarrow ut \in D^v_{S,T}(x,y)$.

iii) $x \in D^v_{S,T}(1,y) \Rightarrow D^v_{S,T}(x,xy) = xD^v_{S,T}(1,y) = D^v_{S,T}(1,y)$.

iv) $u \in D^v_{S,T}(x,y) \Leftrightarrow D^v_{S,T}(u,uxy) = D^v_{S,T}(x,y)$.

v) Assuming $xy = uvt$, the following are equivalent:

(a) $D^v_{S,T}(x,y) = D^v_{S,T}(u,v)$;

(b) $D^v_{S,T}(x,y) \cap D^v_{S,T}(u,v) \neq \emptyset$

Proof. Items *i*) and *ii*) are straightforward. For the statement *iii*), the first equality is a direct consequence of *i*). For the last, since (S,T) is a quadratic pair, the set $D^v_{S,T}(1,a) \subseteq S$ is a subgroup (Remark 3) and thus

$$xD^v_{S,T}(1,a) \subseteq D^v_{S,T}(1,a) \text{ and } x^{-1}D^v_{S,T}(1,a) \subseteq D^v_{S,T}(1,a),$$

which establishes the equality.

iv) Since $u \in D^v_{S,T}(u,uxy)$, the direction $\Leftarrow$ is immediate. For the reverse, note that if $u \in D^v_{S,T}(x,y)$, then by item *i*) we have $ux \in D^v_{S,T}(x^2,xy) = D^v_{S,T}(1,xy)$ and thus by item *ii*) one can see that $D^v_{S,T}(ux,ux^2y) = D^v_{S,T}(1,xy)$. Multiplying this equation by x we obtain

$$D^v_{S,T}(u,uxy) = xD^v_{S,T}(ux,ux^2y) = xD^v_{S,T}(1,xy) = D^v_{S,T}(x,y).$$

v) The direction *a*) $\Rightarrow$ *b*) is immediate. For the other, assume that $z \in D^v_{S,T}(x,y) \cap D^v_{S,T}(u,v)$. Then by item *iv*) one has

$$D^v_{S,T}(x,y) = D^v_{S,T}(z,zxy) = D^v_{S,T}(z,zuvt) = D^v_{S,T}(z,zuv) = D^v_{S,T}(u,v).$$

□

Following the Chapter 2 of [4], fixed a quadratic pair (S,T), a candidate for special group is associated to the pair (S,T) in the following way:

- The group is given by $G_T(S) = S/T^{\times}$. Since $A^2 \subseteq T$, the group $G_T(S)$ has exponent 2. Furthermore, the distinguished element in $G_T(S)$ is $\overline{-1}$ (in general, denoted just by -1).
- For $\overline{a}, \overline{b}, \overline{c}, \overline{d} \in G_T(S)$, the 2-isometry is given by

  $$\langle \overline{a}, \overline{b} \rangle \equiv_{S,T} \langle \overline{c}, \overline{d} \rangle \Leftrightarrow \overline{ab} = \overline{cd} \text{ and } D^v_{S,T}(a,b) = D^v_{S,T}(c,d).$$

  Note that by Lemma 8.(i), the last equality of valued-represented sets it does not dependent of a particular choice of members in the equivalence classes.

The main properties of $G_T(S)$ are summarized below.

Proposition 9. *Let A be a ring and (S,T) a quadratic pair.*

i) The triple $(G_T(S), \equiv_{S,T}, -1)$ is a π-SG. Furthermore, for $x, y, z \in S$,

$$z \in D^v_{S,T}(x,y) \Leftrightarrow \overline{z} \in D_{G_T(S)}(\overline{x}, \overline{y}).$$

ii) If (S,T) is ***2-transversal****, that is, for all $x, y \in S$,*

$$D^v_{S,T}(x,y) = \{c \in S \colon \text{there are } u, v \in T^{\times} : c = ux + vy\},$$

then $G_T(S)$ is a pre-special group.

iii) $G_T(S)$ is reduced π-SG if, and only if, $-1 \notin T$ and $(T+T) \cap S \subseteq T$. In particular, if T is a proper preorder, then $G_T(S)$ is reduced.

Proof. The proof follows by very similar arguments of Lemma 2.7, p. 13, [4], which consists in an application of Lemma 8. □

2 Main Result

From now on, (S,T) will denote a quadratic pair in a ring A, unless explicitly stated otherwise. A tuple $\varphi = \langle a_1, ..., a_n \rangle$, $a_i \in S$, is called a n-form over S; also let $M(\varphi) = M((a_i)_{i \le n})$ be the $n \times n$ diagonal matrix whose non-zero elements are precisely the diagonal with $a_1, ..., a_n$, in this order. The notion of T-isometry appears in Definition 2.17, p. 18, [4]. The definition of T-congruence, to be shown equivalent to T-isometry, is introduced in order to yield a Horn-geometric axiomatization of the theory of T-faithfully quadratic rings.

Definition 10. *Let $\varphi = \langle a_1, ..., a_n\rangle$ and $\psi = \langle b_1, ..., b_n\rangle$ be two n-forms over S.*

- *The forms φ, ψ are **simply T-isometric**, written $\varphi \approx^*_{S,T} \psi$, if there are $t_1, ..., t_n \in T^\times$ such that $b_i = t_i a_i$ or $N \in GL_n(A)$ such that $NM(\varphi)N^t = M(\psi)$.*

- *φ and ψ are **simply T-congruent**, written $\varphi \cong^*_{S,T} \psi$, if there are $N \in GL_n(A)$ and $x_1, ..., x_n, y_1, ..., y_n \in T^\times$ such that*

$$NM((x_i)_{i\leq n})M(\varphi)N^t = M((y_i)_{i\leq n})M(\psi).$$

Proposition 11. *The relations $\approx^*_{S,T}$ (simply T-isometry) and $\cong^*_{S,T}$ (simply T-congruence) over the set of n-forms, $n \geq 1$, are reflexive and symmetric.*

Proof. Let $\varphi = \langle a_1, \ldots, a_n\rangle, \psi = \langle b_1, \ldots, b_n\rangle$ be forms over S.

- Reflexivity: Since $a_i = 1 \cdot a_i$ for all $i = 1, \ldots, n$, $\varphi \approx^*_{S,T} \varphi$.

  Similarly, we have $Id_n(A)Id_n(A)M(\varphi)Id_n(A)^t = Id_n(A)M(\varphi)$ and so $\varphi \cong^*_{S,T} \varphi$.

- Symmetry: Assume that $\varphi \approx^*_{S,T} \psi$. If there are $t_1, \ldots, t_n \in T^\times$ with $b_i = t_i a_i$, then $a_i = t_i^{-1} b_i$ and $t_i^{-1} = t_i^{-2} t_i \in T^\times$; so $\psi \approx^*_{S,T} \varphi$. If there is $N \in GL_n(A)$ such that $NM(\varphi)N^t = M(\psi)$, then $P = N^{-1}$ satisfies $PM(\psi)P^t = M(\varphi)$; so $\psi \approx^*_{S,T} \varphi$.

  Now suppose that $\varphi \cong^*_{S,T} \psi$. Then there are $x_i, y_i \in T^\times$, $i = 1, \ldots, n$, and $N \in GL_n(A)$ such that $NM((x_i)_{i\leq n})M(\varphi)N^t = M((y_i)_{i\leq n})M(\psi)$. Thus, $P = N^{-1}$ satisfies $PM((y_i)_{i\leq n})M(\psi)P^t = M((x_i)_{i\leq n})M(\varphi)$ and therefore $\psi \cong^*_{S,T} \varphi$.

□

Definition 12. *Let φ, ψ be two n-forms, $n \geq 1$, over S.*

- *The forms φ, ψ are said to be T-isometric, written $\varphi \approx_{S,T} \psi$, if there are n-forms $\varphi_0, \ldots, \varphi_k$ such that $\varphi_0 = \varphi, \varphi_k = \psi$ and $\varphi_i \approx^*_{S,T} \varphi_{i+1}$ for all $i = 0, \ldots, k-1$.*

- *The forms φ, ψ are T-congruent, written $\varphi \cong_{S,T} \psi$, if there are n-forms $\varphi_0, \ldots, \varphi_k$ such that $\varphi_0 = \varphi, \varphi_k = \psi$ and $\varphi_i \cong^*_{S,T} \varphi_{i+1}$ for all $i = 0, \ldots, k-1$.*

Note that the relations $\approx_{S,T}$ and $\cong_{S,T}$ are, respectively, the transitive closure of the relations $\approx^*_{S,T}$ and $\cong^*_{S,T}$ thus, by Proposition 11, they are equivalence relations on the set of n-forms over S.

The following (easily established) result is fundamental to obtain our main contribution.

Proposition 13. *Let $\varphi = \langle a_1, ..., a_n\rangle$ and $\psi = \langle b_1, ..., b_n\rangle$ be two n-forms over S. Then $\varphi \approx_{S,T} \psi$ iff $\varphi \cong_{S,T} \psi$.*

Proof. It's enough to show that 1) $\varphi \approx^*_{S,T} \psi \Rightarrow \varphi \cong_{S,T} \psi$ and 2) $\varphi \cong^*_{S,T} \psi \Rightarrow \varphi \approx_{S,T} \psi$.
1) Follows easily from definitions.
2) Assuming $\varphi \cong^*_{S,T} \psi$, take $x_1, ..., x_n, y_1, ..., y_n \in T^\times$ and $N \in GL_n(A)$ such that $NM((x_i)_{i\leq n})M(\varphi)N^t = M((y_i)_{i\leq n})M(\psi)$. Then we have

$$\varphi \approx^*_{S,T} \langle x_1a_1, ..., x_na_n\rangle \approx^*_{S,T} \langle y_1b_1, ..., y_nb_n\rangle \approx^*_{S,T} \psi.$$

□

Definition 14 (Def. 2.24, [4]). *Let A be a ring and (S, T) a quadratic pair. Let S be T-subgroup. Let $\varphi = \langle a_1, \ldots, a_n\rangle$ be a form over S and $\overline{\varphi} = \langle\overline{a_1}, \ldots, \overline{a_n}\rangle$ be the corresponding form in $G_T(S)$.*

a) $D_{S,T}(\varphi) = \{x \in S\colon$ there are $b_2, \ldots, b_n \in S$ such that $\overline{\varphi} \equiv_{G_T(S)} \langle\overline{x}, \overline{b_2}, \ldots, \overline{b_n}\rangle\}$ is the set of elements of S ***isometry-represented*** *mod T by φ in $G_T(S)$.*

b) $D^v_{S,T}(\varphi) = \{x \in S\colon$ there are $x_1, \ldots, x_n \in T$ such that $x = \sum_{i=1}^n x_ia_i\}$ is the set of elements of S ***value-represented*** *mod T by φ.*

c) $D^t_{S,T}(\varphi) = \{x \in S\colon$ there are $x_1, \ldots, x_n \in T^\times$ such that $x = \sum_{i=1}^n x_ia_i\}$ is the set of elements of S ***transversely represented*** *mod T by φ.*

d) The set $\mathcal{D}_{S,T}(\varphi)$ of elements of S ***inductively represented*** *mod T by φ is defined as follows:*

- *If $n = 1$, then $\mathcal{D}_{S,T}(\varphi) = \{a_1\}$;*
- *If $n = 2$, then $\mathcal{D}_{S,T}(\varphi) = D^v_{S,T}(\varphi)$;*
- *If $n \geq 3$, then*

$$\mathcal{D}_{S,T}(\varphi) = \bigcap_{k=1}^{n} \bigcup\{D^v_{S,T}(a_k, u)\colon u \in D^v_{S,T}(a_1, \ldots, \check{a_k}, \ldots, a_n)\}.$$

Remark 4. *Consider the notation from Definition 14. For all forms φ over S, the following is satisfied without any further assumptions:*

- $D_{S,T}(\varphi) \subseteq D^v_{S,T}(\varphi)$;
- $D^t_{S,T}(\varphi) \subseteq D^v_{S,T}(\varphi)$;
- $\mathcal{D}_{S,T}(\varphi) \subseteq D^v_{S,T}(\varphi)$;
- *If* $dim(\varphi) \leq 3$, *then* $\mathcal{D}_{S,T}(\varphi) \subseteq D_{S,T}(\varphi)$.

If $T = A^2$ or T is pre-order, Lemma 2.26 in [4], contains a proof of these statements. An analogous reasoning establishes the inclusions in the general case of quadratic pairs.

Definition 15 (Def. 3.1, [4])**.** *Let A be a ring and (S,T) a quadratic pair. The T-subgroup S is called T-**faithfully quadratic** if the following are satisfied:*

i) For all $a, b \in S$,

$$D^v_{S,T}(a,b) = D^t_{S,T}(a,b);$$

ii) For all form φ with dimension $n \geq 2$,

$$\mathcal{D}_{S,T}(\varphi) = D^v_{S,T}(\varphi);$$

iii) For all forms φ, ψ of same dimension and for all $a \in S$

$$\langle a \rangle \oplus \varphi \approx_{S,T} \langle a \rangle \oplus \psi \Rightarrow \varphi \approx_{S,T} \psi.$$

Definition 16 ([4], p. 27)**.** *Let A be a ring and $T \subseteq A$ a quadratic set. The ring A is called T-faithfully quadratic ring if*

- *The pair $(A^\times, T)$ is quadratic;*
- *The T-subgroup $A^\times$ is T-faithfully quadratic.*

In the case $T = A^2$, a A^2-faithfully quadratic is simply called faithfully quadratic ring.

The next result illustrates the potential of T-faithfully quadratic notion in provide a special group that faithfully represent the T-isometry and T-representation of a ring.

Theorem 17. *Let A be a ring and (S,T) a quadratic pair. Assume that S is T-faithfully quadratic. Let φ, ψ be forms over S of same dimension.*

i) $D^v_{S,T}(\varphi) = D^t_{S,T}(\varphi)$, *that is, an element in S is valued-represented iff it is transversally represented.*

ii) $D^v_{S,T}(\varphi) = D_{S,T}(\varphi)$, *that is, the set of valued-represented elements coincide with those isometry-represented in $G_T(S)$.*

iii) If $\varphi = \varphi_1 \oplus \cdots \oplus \varphi_k$, *then*

$$D^v_{S,T}(\varphi) = \bigcup \{D^v_{S,T}(u_1, \ldots, u_k) \colon u_i \in D^v_{S,T}(\varphi_i), i = 1, \ldots, k\}.$$

iv) $\varphi \approx_{S,T} \psi$ *if, and only if,* $\overline{\varphi} \equiv_{G_T(S)} \overline{\psi}$.

v) $(G_T(S), \equiv_{G_T(S)}, -1)$ *is a special group.*

Proof. The proof in the case $T = A^2$ or T is a preorder can be found in Theorems 3.3, 3.5 and 3.6 of [4] with emphasis in which axioms of faithfully quadratic pairs are needed to obtain each of above results. Very similar arguments entails the proof in for quadratic pairs. In 3.7, [4], there is a fruitful discussion about the reason and utility of faithfully quadratic rings to obtain a rich theory of quadratic forms over invertible coefficients. □

Proposition 18. *Let A be a ring and (S,T) a quadratic pair. Assume that S is T-faithfully quadratic. Given two forms φ and ψ of same dimension $n \geq 2$ such that $\varphi \cong_{S,T} \psi$, then it is possible to find $\varphi_0, ..., \varphi_{l(n)}$, $l(n) = 2^{n-1} - 1$, $\varphi_0 = \varphi, \varphi_{l(n)} = \psi$ such that $\varphi_i \cong^*_{S,T} \varphi_{i+1}$ for every $i < l(n)$.*

Proof. This can be obtained from a simple adaptation of the proof of Proposition 5.1, p.53, [4], combined with the Proposition 13. The proof of Lemma 2.21, p.20, [4], shows that $l(2) = 1$. The remaining of the proof follows essentially from Theorem 17.(*iv*) together with the inductive description of isometry in special groups (described before Definition 1). □

In Theorem 5.2, p.54, [4] is proved that the theory of faithfully quadratic rings is Horn-geometric and in Remark 5.3 of [4] is discussed how to adapt the proof of Theorem 5.2 to obtain a Horn-geometric axiomatization to faithfully quadratic q-subgroup. This result can be generalized to pre-orders and even for quadratic pairs. This is the content of next theorem.

Theorem 19. *The theory of quadratic pairs (S,T) such that S is T-faithfully quadratic is Horn-Geometrical in the language $L = \{+, \cdot, 0, 1, -1, S, T\}$, where S, T are unitary relational symbols.*

Proof. First of all, note that when (S,T) is a quadratic pair, $S \cap T = T^{\times}$. So we will use the formula $T^{\times}(a)$ as an abbreviation to the pp-formula $S(a) \wedge T(a)$. To simplify the formulas below, we should also note that $a \in A^{\times}$ is equivalent to $T^{\times}(a^2)$ and so $a \in A^{\times}$ is also a conjunction of atomic formulas. With this, for $n \geq 2$, we define:

- Let $(\phi_n)^v_{S,T}(a, b_1, ..., b_n)$ be the formula in $n+1$ free variables given by

$$\exists t_1, ..., t_n (\bigwedge_{i=1}^{n} (S(b_i) \wedge T(t_i)) \ \wedge \ S(a) \ \wedge \ a = \sum_{i=1}^{n} t_i b_i).$$

  Note that this formula is a pp-formula and it express $a \in D^v_{S,T}(b_1, ..., b_n)$.

- Let $(\phi_n)^t_{S,T}(a, b_1, ..., b_n)$ be the formula in $n+1$ free variables given by

$$\exists t_1, ..., t_n (\bigwedge_{i=1}^{n} (S(b_i) \wedge T^{\times}(t_i)) \ \wedge \ S(a) \ \wedge \ a = \sum_{i=1}^{n} t_i b_i).$$

  Note that this formula is a pp-formula and it encodes $a \in D^t_{S,T}(b_1, ..., b_n)$.

  Thus, fixed $n \geq 2$, let $l = l(n)$ from Proposition 18. Consider $(\psi_n)_{S,T}(\overline{a}, \overline{b})$ to be the formula in $2l$ variables given by

$$\exists \overline{c_0}, ..., \overline{c_l}, N_1, ..., N_l, \overline{x_1}, ..., \overline{x_l}, \overline{y_1}, ..., \overline{y_l}$$

$$(\bigwedge_{i=1}^{l} (\det(N_i) \in A^{\times 1}) \ \wedge \ \bigwedge_{i=1}^{l} (T^{\times}(\overline{x_i}) \wedge T^{\times}(\overline{y_i}) \wedge S(\overline{c_i})) \ \wedge \ \overline{c_0} = \overline{a} \ \wedge \ \overline{c_l} = \overline{b}$$

$$\wedge \ \bigwedge_{i=1}^{l} (N_i \cdot M((\overline{x_i})_{i \leq n}) \cdot M((\overline{c_{i-1}})_{i \leq n}) \cdot N_i^t = M((\overline{y_i})_{i \leq n}) \cdot M((\overline{c_i})_{i \leq n}))$$

  This formula is a pp-formula and it express $\langle a_1, ..., a_n \rangle \cong_{S,T} \langle b_1, ..., b_n \rangle$.

Now, a possible set of axioms to the theory of faithfully quadratic rings can be given by

[1] Note that the phrase "the determinant of N_i is an invertible element of A" can be encoded by a pp-formula in the language L.

i) axioms ensuring that $\{A, +, \cdot, 0, 1, -1\}$ is a commutative ring with unity with $2 \in A^{\times}$;

ii) $\forall a, b\ (T(a)\ \wedge\ T(b) \Rightarrow T(ab))$;

iii) $\forall a\ (T(a^2))$;

iv) $\forall a\ (S(a) \Rightarrow \exists u\ au = 1)$;

v) $\forall a, b\ (S(a)\ \wedge\ S(b) \Rightarrow S(ab))$;

vi) $\forall a, u\ (T(a)\ \wedge\ au = 1 \Rightarrow S(a))$;

vii) $S(-1)$;

viii) $\forall a, x, y\ ((\phi_2)^v_{S,T}(x, 1, a) \wedge (\phi_2)^v_{S,T}(y, 1, a) \Rightarrow (\phi_2)^v_{S,T}(xy, 1, a))$;

Until now the axioms basically express that (S, T) is a quadratic pair in the ring A with $2 \in A^{\times}$. Now we describe the axioms for S to be T-faithfully quadratic:

ix) $\forall a, b, x\ ((\phi_2)^v_{S,T}(x, a, b) \Rightarrow (\phi_2)^t_{S,T}(x, a, b))$;

For each $n \geq 2$,

x) $\forall a_1, ..., a_n, x\ ((\phi_n)^v_{S,T}(x, a_1, ..., a_n)\ \Rightarrow\ \bigwedge_{i=1}^n \exists u\ ((\phi_{n-1})^v_{S,T}(u, a_1, ..., \check{a}_i, ..., a_n)\ \wedge\ (\phi_2)^v_{S,T}(x, a_i, u)))$;

xi) $\forall a, a_1, ..., a_n, b_1, ..., b_n\ ((\psi_{n+1})_{S,T}(a, \overline{a}, a, \overline{b}) \Rightarrow (\psi_n)_{S,T}(\overline{a}, \overline{b}))$.

Let $\mathcal{T}$ be the theory determined by the sentences including in the itens $i) - xi)$ above. Note that $\mathcal{T}$ is a Horn-geometric theory. An argument analogous to that given in the end of the proof of Theorem 5.2, p. 56, in [4], will show that a $\langle A, S, T\rangle$ is a model of $\mathcal{T}$ iff A is a ring with $2 \in A^{\times}$, (S, T) is quadratic pair and S is T-faithfully quadratic. Furthermore, note that it is possible to require additionally that $\overline{1} \neq \overline{-1}$ in $G_T(S)$ by adding the axiom $\neg T(-1)$, which is yet a Horn-geometric formula. □

Remark 5. *In the above theorem, if the T-subgroup S can be described by a pp-sentence in the language $\{+, \cdot, 0, 1, -1, T\}$, then minor adaptations in the proof also yields a Horn-geometric theory without the symbol S in the language. This is the case when $S = A^{\times}$ or $S = T^{\times} \cup -T^{\times}$.*

References

[1] C.C. Chang, H.J. Keisler, **Model Theory**, North-Holland Publ. CO., Amsterdam, 1990.

[2] M. Dickmann, F. Miraglia, **Special Groups: Boolean-Theoretic Methods in the Theory of Quadratic Forms**, Memoirs of the AMS **689**, American Mathematical Society, Providence, USA, 2000.

[3] M. Dickmann, F. Miraglia, *Representation of reduced special groups in algebras of continuous functions*, Quadratic forms—algebra, arithmetic, and geometry, Contemp. Math., vol. 493, Amer. Math. Soc., Providence, RI, 2009, pp. 83–97.

[4] M. Dickmann, F. Miraglia, **Faithfully Quadratic Rings**, Memoirs of AMS **1128**, American Mathematical Society, Providence, USA, 2015.

[5] M. A. Marshall, **Spaces of Orderings and Abstract Real Spectra**, Lecture Notes in Mathematics **1636**, Springer-Verlag, Berlin, Germany, 1996.

 Received 20 December 2019

On superrings of polynomials and algebraically closed multifields

Kaique Matias de Andrade Roberto
Institute of Mathematics and Statistics, University of São Paulo, Brazil
kaique.roberto@usp.br

Hugo Luiz Mariano
Institute of Mathematics and Statistics, University of São Paulo, Brazil
hugomar@ime.usp.br

Abstract

The concept of multialgebraic structure – an "algebraic like" structure but endowed with multiple valued operations – has been studied since the 1930's; in particular, the concept of hyperrings was introduced by Krasner in the 1950's. Some general algebraic study has been made on multialgebras: see for instance [9] and [17]. More recently the notion of multiring have obtained more attention: a multiring is a lax hyperring, satisfying an weak distributive law, but hyperfields and multifields coincide. Multirings has been studied for applications in abstract quadratic forms theory ([12], [8]) and tropical geometry ([10]); a more detailed account of variants of concept of polynomials over hyperrings is even more recent ([10], [4]). In the present work we start a model-theoretic oriented analysis of multialgebras introducing the class of algebraically closed and providing variant proof of quantifier elimination flavor, based on new results on superring of polynomials ([4]).

Keywords: multialgebras; superring of polynomials; algebraically closed multifields.

We want to express our gratitude to the referees for their careful reading and valuable suggestions on the submitted version.

Introduction

The concept of multialgebraic structure – an "algebraic like" structure but endowed with multiple valued operations – has been studied since the 1930's; in particular, the concept of hyperrings was introduced by Krasner in the 1950's.

Some general algebraic study has been made on multialgebras: see for instance [9] and [17].

More recently the notion of multiring have obtained more attention: a multiring is a lax hyperring, satisfying an weak distributive law, but hyperfields and multifields coincide. Multirings has been studied for applications in abstract quadratic forms theory ([12], [8]) and tropical geometry ([10]); a more detailed account of variants of concept of polynomials over hyperrings is even more recent ([10], [4]).

In the present work we start a model-theoretic oriented analysis of multialgebras introducing the class of algebraically closed and providing variant proof of quantifier elimination flavor, based on new results on superring of polynomials as an Euclidean algorithm of division.

Overview of the work. In section 1 we develop the preliminaries results needed for the paper. Section 2 is devoted to a detailed account of the construction of supperring of polynomials ([4]) and to present some new results as the Euclidean algorithm of division for the superring of polynomials with coefficients over a hyperfield. Section 3 contains the main contributions of this paper: we introduce some concept of the algebraically closed hyperfield and give the first steps on a model theory of this class with a kind of quantifier elimination procedure. We finish the work in section 4 presenting some possible future developments.

1 Preliminaries

Our goals in this section are to develop the preliminary results needed for the work, and to provide a brief dictionary on multialgebras and hyperrings. We split it in two subsections, the first one contains general definitions and results on multialgebraic structures and the second one is focused specially on rings-like multi structures.

1.1 On Multialgebras

There are several definitions of multialgebra in the literature, considering that each multialgebra application in a specific area of Mathematics (mainly Algebra and Logic) requires a particular adaptation. Here, we adapt the notion of multialgebra used in [5]; the identity theory here presented is close to the exposed in [17].

Definition 1.1. *A **multialgebraic signature** is a sequence of parwise disjoint sets*

$$\Sigma = (\Sigma_n)_{n\in\mathbb{N}},$$

where $\Sigma_n = S_n \sqcup M_n$, which S_n is the set of strict multi-operation symbols and M_n is the set of multioperation symbols. In particular, $\Sigma_0 = S_0 \sqcup M_0$, F_0 is the set of symbols for constants and M_0 is the set of symbols for multi-constants. We also denote

$$\Sigma = ((S_n)_{n\geq 0}, (M_n)_{n\geq 0}).$$

Definition 1.2. *Let A be any set.*

i - A multi-operation of arity $n \in \mathbb{N}$ over a set A is a function

$$A^n \to \mathcal{P}^*(A) := \mathcal{P}(A) \setminus \{\emptyset\}.$$

ii - A multi-operation of arity $n \in \mathbb{N}$ over a set A, $A^n \to \mathcal{P}^*(A)$, is strict, *whenever it factors throuth the singleton function $s_A : A \rightarrowtail \mathcal{P}^*(A)$, $a \mapsto s_A(a) := \{a\}$. Thus it can be naturally identified with an ordinary n-ary operation $A^n \to A$.*

A 0-ary multi-operation (respectively *strict* multi-operation) on A can be identified with a non-empty subset of A (respectively a singleton subset of A).

Definition 1.3. *A **multialgebra** over a signature $\Sigma = ((S_n)_{n\geq 0}, (M_n)_{n\geq 0})$, is a set A endowed with a family of n-ary multioperations*

$$\sigma_n^A : A^n \to \mathcal{P}^*(A),\ \sigma_n \in S_n \sqcup M_n,\ n \in \mathbb{N},$$

such that: if $\sigma_n \in S_n$, then $\sigma_n^A : A^n \to \mathcal{P}^*(A)$ is a strict *n-ary multioperation.*

Remark 1.4.

i - Every algebraic signature $\Sigma = (F_n)_{n\in\mathbb{N}}$ is a multialgebraic signature where $M_n = \emptyset, \forall n \in \mathbb{N}$. Each algebra

$$(A, ((A^n \xrightarrow{f^A} A)_{f\in F_n})_{n\in\mathbb{N}})$$

over the algebraic signature Σ *can be naturally identified with a multi-algebra*

$$(A, ((A^n \stackrel{f^A}{\to} A \stackrel{s_A}{\rightarrowtail} \mathcal{P}^*(A))_{f \in F_n})_{n \in \mathbb{N}})$$

over the same signature.

ii - Every multialgebraic signature $\Sigma = ((S_n)_{n\in\mathbb{N}}, (M_n)_{n\in\mathbb{N}})$ *induces naturally a first-order language*

$$L(\Sigma) = ((F_n)_{n\in\mathbb{N}}, (R_{n+1})_{n\in\mathbb{N}})$$

where $F_n := S_n$ *is the set of n-ary operation symbols and* $R_{n+1} := M_n$ *is the set of (n+1)-ary relation symbols. In this way, multi-algebras*

$$(A, ((A^n \stackrel{\sigma^A}{\to} \mathcal{P}^*(A))_{\sigma \in S_n \sqcup M_n})_{n\in\mathbb{N}})$$

over a multialgebraic signature $\Sigma = (S_n \sqcup M_n)_{n\in\mathbb{N}}$ *can be naturally identified with the first-order structures over the language* $L(\Sigma)$ *that satisfies the* $L(\Sigma)$*-sentences:*

$$\forall x_0 \cdots \forall x_{n-1} \exists x_n (\sigma_n(x_0, \cdots, x_{n-1}, x_n)), \text{ for each } \sigma_n \in R_{n+1} = M_n, n \in \mathbb{N}.$$

Now we focus our attention into a more syntactic aspect of this multi-algebras theory. We start with a (recursive) definition of multi-terms:

Definition 1.5. *A* **(multi-)term** *on a multialgebra A of signature*

$$\Sigma = ((S_n)_{n\geq 0}, (M_n)_{n\geq 0})$$

is defined recursively as:

i - Variables $x_i, i \in \mathbb{N}$ *are terms.*

ii - If $t_0, \cdots, t_{n-1}$ *are terms and* $\sigma \in S_n \sqcup M_n$*, then* $\sigma(t_0, \cdots, t_{n-1})$ *is a term.*

We will call a multi-term t **strict**, *whenever it is composed only by combination of* **strict** *multi-operations and variables. The notion of* **occurrence** *of a variable in a term is as the usual. We will denote* $var(t)$ *as the (finite set of variables) that occurs in the term* t.

To define an interpretation for terms, we need a preliminary step. Given

$$\sigma \in S_n \sqcup M_n,$$

we "extend" $\sigma^A : A^n \to \mathcal{P}^*(A)$ to a n-ary operation in $\mathcal{P}^*(A)$,

$$\sigma^{\mathcal{P}^*(A)} : \mathcal{P}^*(A)^n \to \mathcal{P}^*(A),$$

by the rule:

$$\sigma^{\mathcal{P}^*(A)}(A_0, \cdots, A_{n-1}) := \bigcup_{a_0 \in A_0} \cdots \bigcup_{a_{n-1} \in A_{n-1}} \sigma^A(a_0, \cdots, a_{n-1}).$$

Definition 1.6. *The* ***interpretation of a term*** *t on a multialgebra A over a signature $\Sigma = ((S_n)_{n\geq 0}, (M_n)_{n\geq 0})$ is a function $t^A : A^{var(t)} \to \mathcal{P}^*(A)$ and is defined recursively as follows:*

i - The interpretation of a variable x_i, $x_i^A : A^{\{x_i\}} \to \mathcal{P}^*(A)$ is essentialy the singleton function of A:

$$x_i^A : A^{\{x_i\}} \cong A \rightarrowtail \mathcal{P}^*(A), \text{ is given by the rule } (\hat{a} : \{x_i\} \to A) \mapsto \{a\}.$$

ii - If $t = \sigma(t_0, \cdots, t_{n-1})$ is a term and $\sigma \in S_n \sqcup M_n$, denote $T = var(t)$ and $T_i = var(t_i)$. Then $T = \bigcup_{i<n} T_i$. Consider $t_{iT}^A : A^T \to \mathcal{P}^*(A)$ the composition

$$A^T \overset{proj^T_{T_i}}{\twoheadrightarrow} A^{T_i} \overset{t_i^A}{\to} \mathcal{P}^*(A),$$

where $proj^T_{T_i}$ is the canonical projection induced by the inclusion $T_i \hookrightarrow T$. Then $t^A : A^T \to \mathcal{P}^*(A)$ is the composition

$$A^T \overset{(t_{iT}^A)_{i<n}}{\longrightarrow} (\mathcal{P}^*(A))^n \overset{\sigma^{\mathcal{P}^*(A)}}{\longrightarrow} \mathcal{P}^*(A).$$

Definition 1.7. *Let A be a multialgebra A over a signature $\Sigma = ((S_n)_{n\geq 0}, (M_n)_{n\geq 0})$ and let t_1, t_2 be Σ-terms. We say that A realize that t_1 is* **contained in** *t_2, (notation: $A \models t_1 \sqsubseteq t_2$) whenever $t_1^A(\bar{a}) \subseteq t_2^A(\bar{a})$, for each tuple $\bar{a} : var(t_1) \cup var(t_2) \to A$.*

Apart from the notion of atomic formulas the definition of Σ-formulas for multialgebraic theories is similar to the (recursive) definition of first-order $L(\Sigma)$-formulas:

Definition 1.8. *The formulas of Σ are defined as follows:*

i- Atomic formulas are the formulas of type $t \sqsubseteq t'$, where t, t' are terms.

ii- If ϕ, ψ are formulas, then $\neg\phi$ and $\phi \vee \psi, \phi \wedge \psi, \phi \to \psi, \phi \leftrightarrow \psi$ are formulas.

iii- If ϕ is a formula and x_i is a variable, then $\forall x_i \phi$, $\exists x_i \phi$ are formulas.

The notion of occurrence *(respec. free occurrence) of a variable in a formula is as the usual. We will denote $fv(\phi)$ as the (finite) set of variables that occurs free in the formula ϕ.*

We use $t_1 =_s t_2$ to abbreviate the formula $(t_1 \sqsubseteq t_2) \wedge (t_2 \sqsubseteq t_1)$: this means that t_1 and t_2 are "strongly equal terms".

Definition 1.9. *The definition of interpretation of formulas $\phi(\bar{x})$ where*

$$fv(\phi) \subseteq \bar{x} \subseteq \{x_i : i \in \mathbb{N}\}$$

under a valuation of variables $v : \bar{x} \to A$ (or we will denote simply by $v = \bar{a}$) is:

i- $A \models_v t(\bar{x}) \sqsubseteq t'(\bar{x})$ iff $t^A(\bar{a}) \subseteq t'^A(\bar{a})$

ii- The case of complex formulas (given by the connectives $\neg$, $\vee$, $\wedge$, $\to$, $\leftrightarrow$, and quantifiers $\forall, \exists$) is as satisfaction of first-order $L(\Sigma)$-formulas in $L(\Sigma)$-structure on a valuation v.

Remark 1.10.

i- The theory of multi-algebras entails that for each term t, and each strict *term t',*

$$t \sqsubseteq t' \text{ iff } t =_s t'.$$

ii- In [17] contains a development of the identity theory for multialgebras, with another primitive notion: $t(\bar{x}) =_w t'(\bar{x})$; a Σ-multialgebra A satisfies the "weak identity" above iff there is some $\bar{a} \in A^{var(t) \cup var(t')}$ such that $t^A(\bar{a}) \cap t'^A(\bar{a}) \neq \emptyset$. This will not play any role in this work but is useful for applications of multi-algebraic semantics for complex logical systems ([9]).

There are many ways of define morphism for multialgebras. Follow below our choice:

Definition 1.11. *Let A and B be multialgebras of signature $\Sigma = ((S_n)_{n \geq 0}, (M_n)_{n \geq 0})$ and $\varphi : A \to B$ be a function.*

*i - φ is a **partial morphism** if for every $n \geq 0$, every $\sigma \in S_n$ and every $a_1, ..., a_n \in A$, we have*

$$\varphi(\sigma^A(a_1, ..., a_n)) \subseteq \sigma^B(\varphi(a_1), ..., \varphi(a_n)).$$

ii - *φ is a* ***morphism*** *if for every $n \geq 0$, every $\sigma \in S_n \sqcup M_n$ and every $a_1, ..., a_n \in A$, we have*

$$\varphi(\sigma^A(a_1, ..., a_n)) \subseteq \sigma^B(\varphi(a_1), ..., \varphi(a_n)).$$

iii - *φ is a* ***strong morphism*** *if for every $n \geq 0$, every $\sigma \in S_n \sqcup M_n$ and every $a_1, ..., a_n \in A$, we have*

$$\varphi(\sigma^A(a_1, ..., a_n)) = \sigma^B(\varphi(a_1), ..., \varphi(a_n)).$$

Remark 1.12.

i - *Let A, B be Σ-multialgebras. If B is a* strict multilagebra *(i.e. $\sigma_n^B(\bar{b})$ is unitary subset of B, for each $\sigma \in \Sigma$ and each tuple $\bar{b}$ in B), then the morphisms $A \to B$ coincide with the strong morphisms $A \to B$.*

ii - *There is a full and faithful concrete embedding of the category of ordinary algebraic structures over a signature Σ and homomorphisms into the category of Σ-multialgebras and (strong) morphisms: the image of this embedding is the class of strict multialgebras over Σ.*

iii - *The correspondence $\Sigma \mapsto L(\Sigma)$ induces a* concrete *isomorphism between the category of Σ-multialgebras and the category of $L(\Sigma)$- first order structures satisfying suitable $\forall\exists$ axioms. It is ease to see that this correspondence induces a bijection between injective strong embeddings of Σ-multialgebtras and $L(\Sigma)$-monomorphisms of first-order structures.*

We finish this subsection with two illustrative examples of multialgebras derived from an algebraic structure and from a first-order structure.

Example 1.13. *Let $(R, +, \cdot, 0, 1)$ be a commutative ring with $1 \neq 0$. Given $n \geq 1$, define an $(n+1)$-ary multioperation $*_n$ by the rule:*

$$d \in a_0 *_n a_1 *_n a_2 *_n ... *_n a_n \Leftrightarrow \text{ there is some } t \in R \text{ such that}$$
$$d = a_0 + a_1 t + a_2 t^2 + ... + a_n t^n.$$

The idea here, is that $a_0 *_n a_1 *_n a_2 *_n ... *_n a_n$ "analyze" the values taken in R by the polynomial $p(X) = a_0 + a_1 X + a_2 X^2 + ... + a_n X^n \in R[X]$. $*_n$ will be called ***The streching multialgebra of degree*** *n over R.*

Example 1.14. *Let $\mathcal{L} = \{0, 1, +, \cdot, \leq\}$ the language of ordered fields. Consider $\mathbb{R}$ as an ordered field. We can look at the ordering relation as a multioperation of arity 1. In agreement with our notation, we have*

$$\leq (a) := \{x \in \mathbb{R} : a \leq x\} = [a, +\infty).$$

From now on, all multi-algebras considered in this work will contain only operations of arities $0, 1, 2$. They will have strict constants and strict unary operations; the binary operations maybe strict or multivalued.

1.2 Multirings and Superrings

Now, we will get closer to the subject of our work.

Definition 1.15 (Adapted from definition 1.1 in [12]). *Let $\Sigma = ((S_n)_{n\geq 0}, (M_n)_{n\geq 0})$ be a multialgebraic signature where $S_0 = \{1\}$, $S_1 = \{r\}$, $S_n = \emptyset$ for all $n \neq 0, 1$ and $M_2 = \{\cdot\}$, $M_n = \emptyset$ for all $n \neq 2$. A multigroup is a Σ-structure $(G, \cdot, r, 1)$ where G is a (non-empty) set, 1 is an element of G, $r : G \to G$ is a function, $\cdot : G \times G \to \mathcal{P}^*(G)$, that satisfies the following formulas:*

i - $G \models 1 \cdot x =_s x$.

ii - $G \models x \cdot 1 =_s x$.

iii - $G \models [(x \cdot y) \cdot z] =_s [x \cdot (y \cdot z)]$.

iv - $G \models (z \sqsubseteq x \cdot y) \to [(x \sqsubseteq z + \cdot r(y)) \wedge (y \sqsubseteq r(x) \cdot z)]$.

A multimonoid is a multialgebra such that $S_1 = \{1\}$, $M_2 = \{\cdot\}$ and the other sets of symbols are empty, that satifies axioms (i), (ii), (iii) above.

A multimonoid/multigroup will said to be ***commutative (or abelian)*** *if satisfy:*

$$G \models x \cdot y =_s y \cdot x.$$

For multigroups, axiom (iii) can be replaced by the (apparently weaker) version:

$$G \models [(x \cdot y) \cdot z] \sqsubseteq [x \cdot (y \cdot z)]$$

In other words, an abelian multigroup is a first-order structure $(G, \cdot, r, 1)$ where G is a non-empty set, $r : G \to G$ is a function, 1 is an element of G, $\cdot \subseteq G \times G \times G$ is a ternary relation (that will play the role of binary multioperation, we denote $d \in a \cdot b$ for $(a, b, d) \in \cdot$) such that for all $a, b, c, d \in G$:

M1 - If $c \in a \cdot b$ then $a \in c \cdot (r(b)) \wedge b \in (r(a)) \cdot c$. We write $a - b$ to simplify $a + (-b)$.

M2 - $b \in a \cdot 1$ iff $a = b$.

M3 - If $\exists x(x \in a \cdot b \wedge t \in x \cdot c)$ then $\exists y(y \in b \cdot c \wedge t \in a \cdot y)$.

M4 - $c \in a \cdot b$ iff $c \in b \cdot a$.

Example 1.16.

a- *Suppose that* $(G, \cdot, (\ )^{-1}, 1)$ *is a ordinary group. Defining* $a * b = \{a \cdot b\}$ *and* $r(g) = g^{-1}$, *we have that* $(G, *, r, 1)$ *is a multigroup.*

b- *([17]) Let* $(G, \cdot, (\ )^{-1}, e)$ *be an ordinary group and let* $S \subseteq G$ *be a subset such that* $e \in S$ *and* $S^{-1} \subseteq S$, *define a binary relation* $a \sim_S b$ *iff* $b \cdot a^{-1} \in S$. *This is a reflexive and symmetric relation. Then take* $\sim_S^t$ *be the transitive closure of* $\sim_S$ *(note that if* S *is a subgroup of* G, *then* $\sim_S = \sim_S^t$*). Then* $G/\sim_S^t$ *with the inherit structure is a multigroup. In particular if* G *is a commutative group and* S *is a subgroup of* G, *then* $G/\sim_S^t$ *with the inherit structure is an ordinary abelian group.*

Definition 1.17 (Adapted from definition 2.1 in [12]). *A (commutative, unital) multiring is a multialgebraic structure* $(R, +, \cdot, -, 0, 1)$ *where* $(R, +, -, 0)$ *is a commutative multigroup,* $(R, \cdot, 1)$ *is a commutative (strict) monoid and that also satisfies the following axioms:*

- $R \models [x \cdot 0] =_s 0$ *(zero is absorving).*
- $R \models [z.(x + y)] \sqsubseteq [z.x + z.y]$ *(weak or semi distributive law).*

A multidomain is a non-trivial multiring without zero-divisors and a multifield is a non-trivial multiring such that every nonzero element is invertible.

A multiring is an hyperring if it satifies the full distributive law:

$$R \models [z(x + y)] =_s [zx + zy].$$

Of course, we extend this terminology for hyperdomains and hyperfields.

In other words, a multiring is a tuple $(R, +, \cdot, -, 0, 1)$ where R is a non-empty set, $\cdot : R \times R \to R$ and $- : R \to R$ are functions, 0 and 1 are elements of R, $+ \subseteq R \times R \times R$ is a relation. We denote $d \in a + b$ for $(a, b, d) \in +$. We require that $(R, +, -, 0)$ is a commutative multigroup and that all these satisfying the following properties for all $a, b, c, d \in R$:

M5 - $(a \cdot b) \cdot c = a \cdot (b \cdot c)$.

M6 - $a \cdot 1 = a$.

M7 - $a \cdot b = b \cdot a$.

M8 - $a \cdot 0 = 0$.

M9 - If $d \in a + b$ then $cd \in ca + cb$.(weak distributivity) $c.(a+b) \subseteq c.a + c.b$

Example 1.18.

a - Every ring, domain and field is gives rise naturally to a strict *multiring, multidomain and multifield, respectively. It is ease to see that the class of multifields and of hyperfields coincide.*

b - ([12]) $Q_2 = \{-1, 0, 1\}$ is multifield (of signals) with the usual product (in $\mathbb{Z}$) and the multivalued sum defined by relations

$$\begin{cases} 0 + x = x + 0 = x, \text{ for every } x \in Q_2 \\ 1 + 1 = 1, \ (-1) + (-1) = -1 \\ 1 + (-1) = (-1) + 1 = \{-1, 0, 1\} \end{cases}$$

This is a hyperfield of characteristic 0 (we will define the characteristic in 1.26).

c - ([10]) Let $K = \{0, 1\}$ with the usual product and the sum defined by relations $x + 0 = 0 + x = x$, $x \in K$ and $1 + 1 = \{0, 1\}$. This is a multifield called Krasner's multifield. Obviously, it has characteristic 2.

d - ([19])In the set $\mathbb{R}_+$ of positive real numbers, we define

$$a \bigtriangledown b = \{c \in \mathbb{R}_+ : |a - b| \leq c \leq a + b\}.$$

We have $\mathbb{R}_+$ with the usual product and $\bigtriangledown$ multivalued sum is a multifield, called triangle multifield. We denote this multifield by $\mathcal{T}\mathbb{R}_+$. Observe that $\mathcal{T}\mathbb{R}_+$ is not "double distributive":

$$(2 \bigtriangledown 1) \cdot (2 \bigtriangledown 1) = [1, 3] \cdot [1, 3] = [1, 9]$$

and

$$2 \cdot 2 \bigtriangledown 2 \cdot 1 \bigtriangledown 1 \cdot 2 \bigtriangledown 1 \cdot 1 = 4 \bigtriangledown 2 \bigtriangledown 2 \bigtriangledown 1 = [0, 9].$$

Example 1.19 (Kaleidoscope, Example 2.7 in [18]). *Let $n \in \mathbb{N}$ and define $X_n = \{-n, ..., 0, ..., n\}$. We define the n-**kaleidoscope multiring** by $(X_n, +, \cdot, 0, 1)$, where $+ : X_n \times X_n \to \mathbb{P}(X_n) \setminus \{\emptyset\}$ is given by the rules:*

$$a + b = \begin{cases} \{sgn(ab) \max\{|a|, |b|\}\} \text{ if } a, b \neq 0 \\ \{a\} \text{ if } b = 0 \\ \{b\} \text{ if } a = 0 \\ \{-a, ..., 0, ..., a\} \text{ if } b = -a \end{cases},$$

and $\cdot : X_n \times X_n \to \mathbb{P}(X_n) \setminus \{\emptyset\}$ is is given by the rules:

$$a \cdot b = \begin{cases} sgn(ab) \max\{|a|, |b|\} \text{ if } a, b \neq 0 \\ 0 \text{ if } a = 0 \text{ or } b = 0 \end{cases}.$$

In this sense, $X_0 = \{0\}$ and $X_1 = \{-1, 0, 1\} = Q_2$.

Example 1.20 (H-multifield, Example 2.8 in [18]). *Let $p \geq 1$ be a prime integer and $H_p := \{0, 1, ..., p-1\} \subseteq \mathbb{N}$. Now, define the binary multioperation and operation in H_p as follow:*

$$a + b = \begin{cases} H_p \text{ if } a = b,\ a, b \neq 0 \\ \{a, b\} \text{ if } a \neq b,\ a, b \neq 0 \\ \{a\} \text{ if } b = 0 \\ \{b\} \text{ if } a = 0 \end{cases}$$

$$a \cdot b = k \text{ where } 0 \leq k < p \text{ and } k \equiv ab \text{ mod } p.$$

$(H_p, +, \cdot, -, 0, 1)$ is a multifield such that for all $a \in H_p$, $-a = a$. For example, considering $H_3 = \{0, 1, 2\}$, using the above rules we obtain these tables

+	0	1	2
0	{0}	{1}	{2}
1	{1}	{0, 1, 2}	{1, 2}
2	{2}	{1, 2}	{0, 1, 2}

·	0	1	2
0	0	0	0
1	0	1	2
2	0	2	1

In fact, these H_p is a kind of generalization of K, in the sense that $H_2 = K$.

Here is a lemma stating the basic properties concerning multirings:

Lemma 1.21. *Any multiring R satisfies the formulas:*

a - $-(0) =_s 0$.

b - $-(-(x)) =_s x$.

c - $z \sqsubseteq x + y \leftrightarrow -(y) \sqsubseteq (-(x)) + (-(z))$.

d - $-(xy) =_s (-x)y =_s x(-y)$.

The general definition of the concepts of morphisms and strong morphisms for multialgebraic structures take the following form in the case of multirings:

Definition 1.22 (Definition 2.9 in [18]). *Let A and B multirings. A map $f : A \to B$ is a morphism if for all $a, b, c \in A$:*

i - $c \in a + b \Rightarrow f(c) \in f(a) + f(b)$;

ii - $f(-a) = -f(a)$;

iii - $f(0) = 0$;

iv - $f(ab) = f(a)f(b)$;

v - $f(1) = 1$.

f is ***a strong morphism*** *if is a morphism and for all $a, b \in A$, $f(a + b) = f(a) + f(b)$.*

For multirings, there are types of "substructure" that can be considered. Let $f\colon A \to B$ a multiring morphism. If f is injective and a strong morphism, we say that A is **strongly embedded** in B. If f is injective, strong morphism and for all $a, b \in A$ and $c \in B$ if $c \in f(a) + f(b)$, then $c \in \mathrm{Im}(f)$, then A is a **submultiring** of B. Note that in the rings case, all these notions coincide.

To the best of our knowledge, the concept of superring first appears in ([4]). There are many important advances and results in hyperring theory, and for instance, we recommend for example, the following papers: ([1]), ([3]), ([4]), ([2]), ([13]), ([16]), ([15]), ([14]).

Definition 1.23 (Definition 5 in [4]). *A superring is a structure* $(S, +, \cdot, -, 0, 1)$ *such that:*

i - $(S, +, -, 0)$ *is a commutative multigroup.*

ii - $(S, \cdot, 1)$ *is a commutative multimonoid.*

iii - 0 *is an absorbing element:* $a \cdot 0 = \{0\} = 0 \cdot a$*, for all* $a \in S$.

iv - *The weak/semi distributive law holds: if* $d \in c.(a + b)$ *then* $d \in (ca + cb)$*, for all* $a, b, c, d \in S$.

v - *The rule of signals holds:* $-(ab) = (-a)b = a(-b)$*, for all* $a, b \in S$.

A superdomain is a non-trivial superring without zero-divisors in this new context, i.e. whenever

$$0 \in a \cdot b \text{ iff } a = 0 \text{ or } b = 0$$

A superfield is a non-trivial superring such that every nonzero element is invertible in this new context, i.e. whenever

$$\text{For all } a \neq 0 \text{ exists } b \text{ such that } 1 \in a \cdot b.$$

A superring is strong if for all $a, b, c, d \in S$, $d \in c \cdot (a + b)$ *iff* $d \in ca + cb$.

Definition 1.24. *Let A and B superrings. A map* $f : A \to B$ *is a morphism if for all* $a, b, c \in A$*:*

i - $c \in a + b \Rightarrow f(c) \in f(a) + f(b)$;

ii - $c \in a \cdot b \Rightarrow f(c) \in f(a) \cdot f(b)$;

iii - $f(-a) = -f(a)$;

iv - $f(0) = 0$;

v - $f(1) = 1$.

f *is* ***a strong morphism*** *if is a morphism and for all* $a, b \in A$, $f(a+b) = f(a)+f(b)$ *and* $f(a \cdot b) = f(a) + f(b)$.

The reader interested in Logic but not familiar with multialgebras may have some troubles with the terminology "multi, hyper, super" used in the multialgebra context. For their benefit, we propose the following dictionary that, in particular emphasize the number of multioperations in the structure at sight:

Definition 1.25 (Dictionary).

i - ${}_0Ring$ *will be denote the (traditional) category of commutative rings with unit; its objects will be called* ***0-rings****.*

ii - ${}_1Ring$ *will be denote the category of commutative multirings; its objects will be called* ***1-rings****.* ${}_1FRing$ *will be denote the category of commutative hyperrings; Its objects will be called* ***full 1-rings****.*

iii - ${}_2Ring$ *will be denote the category of commutative superrings; its objects will be called* ***2-rings****.* ${}_2FRing$ *will be denote the category of strong commutative superrings; its objects will be called* ***full 2-rings****.*

In this sense, ${}_0Ring = {}_0FRing$*. These definitions can (and will be) carried to subcategories: for example,* ${}_1Field$ *is the category of multifields (and we have that* ${}_1Field = {}_1FField$*).*

From now on, we will use the conventions just above.

Let $(R, +, \cdot, -, 0, 1)$ be a 2-ring, $p \in \mathbb{N}$ and a p-tuple $(a_0, a_1, ..., a_{p-1})$.

We define the finite sum by:

$$x \in \sum_{i<0} a_i \text{ iff } x = 0,$$

$$x \in \sum_{i<p} a_i \text{ iff } x \in y + a_{p-1} \text{ for some } y \in \sum_{i<p-1} a_i, \text{if } p \geq 1.$$

The finite product is given by:

$$x \in \prod_{i<0} a_i \text{ iff } x = 1,$$

$$x \in \prod_{i<p} a_i \text{ iff } x \in y \cdot a_{p-1} \text{ for some } y \in \prod_{i<p-1} a_i, \text{if } p \geq 1.$$

Thus, if $(\vec{a}_0, \vec{a}_1, ..., \vec{a}_{p-1})$ is a p-tuple of tuples $\vec{a}_i = (a_{i0}, a_{i1}, ..., a_{im_i})$, then we have the finite sum of finite products:

$$x \in \sum_{i<0} \prod_{j<m_i} a_{ij} \text{ iff } x = 0,$$

$$x \in \sum_{i<p} \prod_{j<m_i} a_{ij} \text{ iff } x \in y + z \text{ for some } y \in \sum_{i<p-1} \prod_{j<m_i} a_{ij}$$

$$\text{and } z \in \prod_{j<m_{p-1}} a_{p-1,j},\ p \geq 1.$$

Now, we translate some basic facts that holds in rings (0-rings) to 2-rings. Before, we need some terminology:

Definition 1.26.

i - *An **ideal** of a 2-ring A is a non-empty subset $\mathfrak{a}$ of A such that $\mathfrak{a}+\mathfrak{a}\subseteq\mathfrak{a}$ and $A\mathfrak{a}\subseteq\mathfrak{a}$. An ideal $\mathfrak{p}$ of A is said to be prime if $1\notin\mathfrak{p}$ and $ab\subseteq\mathfrak{p}\Rightarrow a\in\mathfrak{p}$ or $b\in\mathfrak{p}$. An ideal $\mathfrak{m}$ is maximal if it is proper and for all ideals $\mathfrak{a}$ with $\mathfrak{m}\subseteq\mathfrak{a}\subseteq A$, then $\mathfrak{a}=\mathfrak{m}$ or $\mathfrak{a}=A$. We will denote $Spec(A)=\{\mathfrak{p}\subseteq A:\mathfrak{p}$ is a prime ideal$\}$.*

ii - *The **characteristic** of a 2-ring is the smaller integer $n\geq 1$ such that*

$$0\in\sum_{i<n}1,$$

otherwise the characteristic is zero. For full 2-domains, this is equivalent to say that n is the smaller integer such that

$$\textit{For all } a,\ 0\in\sum_{i<n}a.$$

iii - *A **polynomial expression** in the variables x_{ij}, is a multiterm of the form*

$$\sum_{i<p}\prod_{j<m_i}x_{ij}.$$

iv - *Let S be a subset of a 2-ring A. We define the **ideal generated by** S as $\langle S\rangle:=\bigcap\{\mathfrak{a}\subseteq A$ ideal$: S\subseteq\mathfrak{a}\}$. If $S=\{a_1,...,a_n\}$, we easily check that*

$$\langle a_1,...,a_n\rangle=\sum Aa_1+...+\sum Aa_n,\ \textit{where}\ \sum Aa=\bigcup_{n\geq 1}\{\underbrace{Aa+...+Aa}_{n\ times}\}.$$

Note that if A is a full 2-ring, then $\sum Aa=Aa$.

Lemma 1.27. *Let A be a 2-ring.*

i - *For all $n\in\mathbb{N}$ and all $a_0,...,a_{n-1}\in A$, the sum $a_0+...+a_{n-1}$ and product $a_0\cdot...\cdot a_{n-1}$ does not depends on the order of the entries.*

ii - *For every term $t(y_1,...,y_n)$ on the 2-ring language, exists variables x_{ij} such that A satisfies the formula*

$$t(y_1,...,y_n)\sqsubseteq\sum_{i<p}\prod_{j<m_i}x_{ij}.$$

Moreover, if A is a full *2-ring, it satisfies the formula*

$$t(y_1,...,y_n)=_s\sum_{i<p}\prod_{j<m_i}x_{ij}.$$

Proof.

i - There is nothing to prove if $n = 0, 1$. If $n = 2$ this is just the commutativity of $+$ and $\cdot$. For $n \geq 3$, the result follows by induction, using the commutativity and the associativity of $+$ and $\cdot$.

ii - This follows by induction on the complexity of the term by the repeated use of associativity of $+$ and $\cdot$ and the weak/semi distributivity law. If A is a full 2-ring, the proof use the full distributivity law instead of its weak version.

□

2 Multipolynomials

This section is devoted to a detailed account of the construction of supperring of polynomials.

Even if the rings-like multi-algebraic structure have been studied for more than 70 years, the idea of considering notions of polynomial in the rings-like multialgebraic structure seems to have considered only in the present century: for instance in [10] some notion of multi polynomials is introduced to obtain some applications to algebraic and tropical geometry, in [4] a more detailed account of variants of concept of multipolynomials over hyperrings is applied to get a form of Hilbert's Basissatz.

Our main result in this section is the Theorem 2.6 that provides a Euclidean division algorithm for 2-rings of multipolynomials in one variable with coefficients in a 1-field.

Here we will stay close to [4] perspective: let $(R, +, -, \cdot, 0, 1)$ be a 2-ring and set

$$R[X] := \{(a_n)_{n\in\omega} : \exists t \forall n(n \geq t \to a_n = 0)\}.$$

Of course, we define the **degree** of $(a_n)_{n\in\omega}$ to be the smallest t such that $n \geq t \to a_n = 0$. Now define the binary multioperations $+, \cdot : R[X] \times R[X] \to \mathcal{P}^*(R[X])$, a unary operation $- : R[X] \to R[X]$ and elements $0, 1 \in R[X]$ by

$$(c_n)_{n\in\omega} \in (a_n)_{n\in\omega} + (b_n)_{n\in\omega} \text{ iff } \forall n(c_n \in a_n + b_n)$$
$$(c_n)_{n\in\omega} \in ((a_n)_{n\in\omega} \cdot (b_n)_{n\in\omega} \text{ iff } \forall n(c_n \in a_0 \cdot b_n + a_1 \cdot b_{n-1} + ... + a_n \cdot b_0)$$
$$-(a_n)_{n\in\omega} = (-a_n)_{n\in\omega}$$
$$0 := (0)_{n\in\omega}$$
$$1 := (1, 0, ..., 0, ...)$$

For convenience, we denote elements of $R[X]$ by $\boldsymbol{a} = (a_n)_{n\in\omega}$. Beside this, we denote

$$\begin{aligned} 1 &:= (1, 0, 0, ...), \\ X &:= (0, 1, 0, ...), \\ X^2 &:= (0, 0, 1, 0, ...) \end{aligned}$$

etc. In this sense, our "monomial" a_iX^i is denoted by $(0, ...0, a_i, 0, ...)$, where a_i is in the i-th position; in particular, we will denote $\underline{b} = (b, 0, 0, ...)$ and we frequently identify $b \in R \leftrightsquigarrow \underline{b} \in R[X]$.

The properties stated in the lemma below it immediately follows from the definitions involving $R[X]$:

Lemma 2.1. *Let R be a 2-ring and $R[X]$ as above and $n, m \in \mathbb{N}$.*

a - $\{X^{n+m}\} = X^n \cdot X^m$.

b - *For all $a \in R$, $\{aX^n\} = \underline{a} \cdot X^n$.*

c - *Given $\boldsymbol{a} = (a_0, a_1, ..., a_n, 0, 0, ...) \in R[X]$, with with $\deg \boldsymbol{a} \leq n$ and $m \geq 1$, we have*

$$\boldsymbol{a}X^m = (0, 0, ..., 0, a_0, a_1, ..., a_n, 0, 0, ...) = a_0X^m + a_1X^{m+1} + ... + a_nX^{m+n}.$$

d - *For $\boldsymbol{a} = (a_n)_{n\in\omega} \in R[X]$, with $\deg \boldsymbol{a} = t$,*

$$\{\boldsymbol{a}\} = a_0 \cdot 1 + a_1 \cdot X + ... + a_t \cdot X^t = a_0 + X(a_1 + a_2X + ... + a_nX^{t-1}).$$

e - $cX^k.(\boldsymbol{a} + \boldsymbol{b}) = cX^k.\boldsymbol{a} + cX^k.\boldsymbol{b}$.

Fact 2.2.

i - *$R[X]$ is a 2-ring.*

ii - *The map $a \in R \mapsto \underline{a} = (a, 0, \cdots, 0, \cdots)$ defines a strong injective 2-ring homomorphism $R \rightarrowtail R[X]$.*

iii - *For an ordinary ring R (identified with a strict suppering), the 2-ring $R[X]$ is naturally isomorphic to (the 2-ring associated to) the ordinary ring of polynomials in one variable over R.*

Remark 2.3. *If R is a full 2-ring, does not hold in general that $R[X]$ is also a full 2-ring. In fact, even if R is a 1-field, there are examples, e.g. $R = K, Q_2$, such that $R[X]$ is not a full 2-ring (see [4]).*

Definition 2.4. *$R[X]$ will be called the* ***2-ring of polynomials*** *with one variable over R. The elements of $R[X]$ will be called (multi)polynomials. We denote $R[X_1, ..., X_n] := (R[X_1, ..., X_{n-1}])[X_n]$.*

Theorem 2.5. *$R[X]$ is a 2-domain iff R is a 2-domain.*

Proof. We just need to prove that $R[X]$ is a 2-domain iff R is a 2-domain, since the rest is consequence of this. ($\Leftarrow$) Let $(a_n)_{n\in\omega}, (b_n)_{n\in\omega} \in R[X]$ such that $\underline{0} \in (a_n)_{n\in\omega} \cdot (b_n)_{n\in\omega}$. Suppose $a_0 \neq 0$. Since R is a superdomain, we have $b_0 = 0$. Now, we have $0 \in a_0b_1 + a_1b_0$, and since $b_0 = 0$, we conclude $0 \in a_0b_1 + 0$, and so $0 \in a_0b_1$. Since $a_0 \neq 0$, we have $b_1 = 0$. Repeating this process t steps, when t is the maximum of degrees involved we have that $(b_n)_{n\in\omega} = \underline{0}$.

($\Rightarrow$) Immediate, since $R \rightarrowtail R[x]$ is an injective strong 2-ring homorphism. □

Now we are read to state and prove the main result in this section.

Theorem 2.6 (Euclid's Division Algorithm)**.** *Let K be a 1-field. Given polynomials $\boldsymbol{a}, \boldsymbol{b} \in K[X]$ with $\boldsymbol{b} \neq 0$, there exists $\boldsymbol{q}, \boldsymbol{r} \in K[X]$ such that $\boldsymbol{a} \in \boldsymbol{qb} + \boldsymbol{r}$, with $\deg \boldsymbol{r} < \deg \boldsymbol{b}$ or $\boldsymbol{r} = 0$.*

Proof. Let $n = \deg \boldsymbol{a}$ and $m = \deg \boldsymbol{b}$. We proceed by induction on n. Note that if $m \geq n$, then is sufficient take $\boldsymbol{q} = 0$ and $\boldsymbol{r} = \boldsymbol{a}$., so we can suppose $m \leq n$. If $m = n = 0$, then $\boldsymbol{a} = (a_0, 0, ...0, ...)$ and $\boldsymbol{b} = (b_0, 0, ..., 0, ...)$ are both non zero constants, so is sufficient take $\boldsymbol{q} = (a_0/b_0, 0, 0, ..., 0, ...)$ and $\boldsymbol{r} = 0$.

Now, suppose $n \geq 1$. Write $\boldsymbol{a} = a_0 + X(a_1 + ... + a_nX^{n-1}) = a_0 + X\boldsymbol{a}'$ and $\boldsymbol{b} = b_0 + X(b_1 + ... + b_mX^{m-1}) = b_0 + X\boldsymbol{b}'$, with $a_n, b_m \neq 0$ and $\deg \boldsymbol{a}' < n$, $\deg \boldsymbol{b}' < m$. Then

$$\begin{aligned} \boldsymbol{a} - a_0 \cdot 1 &\in X\boldsymbol{a}' \\ \boldsymbol{b} - b_0 \cdot 1 &\in X\boldsymbol{b}'. \end{aligned}$$

Then $(\boldsymbol{a} - a_0 \cdot 1) - (\boldsymbol{b} - b_0 \cdot 1) \subseteq X\boldsymbol{a}' - X\boldsymbol{b}' = X(\boldsymbol{a}' - \boldsymbol{b}')$. Since all polynomials in $\boldsymbol{a}' - \boldsymbol{b}'$ have degree $< n$, by induction we can write

$$\boldsymbol{a}' - \boldsymbol{b}' \subseteq \boldsymbol{qb}' + \boldsymbol{r}$$

with $\deg \boldsymbol{r} < \deg \boldsymbol{b}' < m-1$ or $\boldsymbol{r} = 0$. Substuting we obtain

$$\begin{aligned}
(\boldsymbol{a} - a_0 \cdot 1) - (b - b_0 \cdot 1) &\subseteq X(\boldsymbol{q}\boldsymbol{b}' + \boldsymbol{r}) \Rightarrow \\
(\boldsymbol{a} - a_0 \cdot 1) - (b - b_0 \cdot 1) + (b - b_0 \cdot 1) &\subseteq X(\boldsymbol{q}\boldsymbol{b}' + \boldsymbol{r}) + (b - b_0 \cdot 1) \Rightarrow \\
(\boldsymbol{a} - a_0 \cdot 1) &\subseteq X(\boldsymbol{q}\boldsymbol{b}' + \boldsymbol{r}) + (b - b_0 \cdot 1) \Rightarrow \\
\boldsymbol{a} + (a_0 \cdot 1 - a_0 \cdot 1) &\subseteq X(\boldsymbol{q}\boldsymbol{b}' + \boldsymbol{r}) + (b - b_0 \cdot 1 + a_0 \cdot 1) \Rightarrow \\
\boldsymbol{a} &\subseteq X(\boldsymbol{q}\boldsymbol{b}' + \boldsymbol{r}) + (b - b_0 \cdot 1 + a_0 \cdot 1).
\end{aligned}$$

On the other hand,

$$\begin{aligned}
X(\boldsymbol{q}\boldsymbol{b}' + \boldsymbol{r}) + (b - b_0 \cdot 1 + a_0 \cdot 1) &\subseteq X(\boldsymbol{q}\boldsymbol{b}' + \boldsymbol{r}) + (b - b_0 \cdot 1 + a_0 \cdot 1) + \boldsymbol{q} \cdot b_0 - \boldsymbol{q} \cdot b_0 \\
&= \boldsymbol{q}(X\boldsymbol{b}' + b_0) + (X\boldsymbol{r} - \boldsymbol{q} \cdot b_0 + a_0 \cdot 1 - b_0 \cdot 1) \\
&= \boldsymbol{q}\boldsymbol{b} + (X\boldsymbol{r} - \boldsymbol{q} \cdot b_0 + a_0 \cdot 1 - b_0 \cdot 1).
\end{aligned}$$

So $\boldsymbol{a} \subseteq \boldsymbol{q}\boldsymbol{b} + (X\boldsymbol{r} - \boldsymbol{q} \cdot b_0 + a_0 \cdot 1 - b_0 \cdot 1)$ with $\deg(X\boldsymbol{r} - \boldsymbol{q} \cdot b_0 + a_0 \cdot 1 - b_0 \cdot 1) < \deg \boldsymbol{b}$, as desired. □

Remark 2.7.

i - *Note that the polynomials q and r of Theorem 2.6 are not unique in general: if $\boldsymbol{a} \in \boldsymbol{b}\boldsymbol{q} + \boldsymbol{r}$, then $\boldsymbol{a} \in \boldsymbol{b}(\boldsymbol{q} + 1 - 1) + \boldsymbol{r}$ and $\boldsymbol{a} \in \boldsymbol{b}\boldsymbol{q} + (\boldsymbol{r} + 1 - 1)$, then, if $\{0\} \neq 1 - 1$, we have many q's and r's.*

However, if R is a ring (or 0-ring, in agreement with our notation), then Theorem 2.6 provide the usual Euclid Algorithm, with the uniqueness of the quotient and remainder.

ii - *The Theorem 2.6 above gives immediately another proof of Theorem 6 in [4]. i.e. every ideal in $K[X]$ is a principal ideal: for a non zero ideal $I \subseteq K[X]$ select a nonzero polynomial $\boldsymbol{b}(x) \in I$ with minimal degree, then $I = F[X].\boldsymbol{b}(x)$.*

3 Beginning the model theory of algebraically closed multifields

This section contains the main contributions of this paper: we introduce some concept of the algebraically closed 1-field and give the first steps on a model theory of this class with a kind of quantifier elimination procedure.

3.1 On algebraically closed multifields

Let R, S be 2-rings and $h : R \to S$ be a morphism. Then h extends naturally to the 2-rings multipolynomials $h^X : R[X] \to S[X]$:

$$(a_n)_{n\in\mathbb{N}} \in R[X] \;\mapsto\; (h(a_n))_{n\in\mathbb{N}} \in S[X]$$

Now let $s \in S$ we have the **h-evaluation** of s at $\boldsymbol{a} \in R[X]$, $degree(\boldsymbol{a}) \leq n$ by

$$\boldsymbol{a}^h(s) = ev^h(s, \boldsymbol{a}) = \{s' \in S : s' \in h(a_0) + h(a_1).s + h(a_2).s^2 + ... + h(a_n).s^n\}.$$

In particular if $T \supseteq R$ is a 2-ring extension and $\alpha \in T$, we have the **evaluation** of α at $\boldsymbol{a} \in R[X]$ by

$$\boldsymbol{a}(\alpha) = ev(\alpha, \boldsymbol{a}) = \{b \in T : b \in a_0 + a_1\alpha + a_2\alpha^2 + ... + a_n\alpha^n\}.$$

A **root** of $\boldsymbol{a}$ in T is an element $\alpha \in T$ such that $0 \in ev(\alpha, \boldsymbol{a})$. A 2-ring R is **algebraically closed** if every non constant polynomial in $R[X]$ has a root in R.

Observe that, if F is a field, the evaluation of $F[X]$ as a 1-ring coincide with the usual evaluation. Therefore, if F is algebraically closed as 1-field and 2-field, then will be algebraically closed in the usual sense.

Unfortunately, in dealing with multipolynomials, strange situations appears:

Example 3.1 (Finite Algebraically Closed 1-Field). *The 1-field $K = \{0, 1\}$ is algebraically closed. In fact, if $\mathbf{p} = a_0 + a_1X + a_2X^2 + ... + a_nX^n \in K[X]$, with $a_n \neq 0$, then $\mathbf{p}(1) = K$, since $1 + 1 = \{0, 1\}$.*

3.2 A quantifier elimination procedure

Instead of these "anomalies", we have a quantifier elimination procedure for any *infinite* algebraically closed 1-fields. We will describe this (that is a variation of Theorem 9.2.1 in [7]) after the following technical lemma:

Lemma 3.2 (Reduction Lemma). *Let A be a 2-ring, $t_1(\bar{x}), t_2(\bar{x})$ be terms on the full 2-ring language and let $v = \bar{a} : \bar{x} \to A$*

i - $t_1^A(\bar{a}) \subseteq t_2^A(\bar{a})$ iff $0 \in (t_2 - t_1)^A(\bar{a})$.

ii - Given any atomic formula, $t_1(\bar{x}) \sqsubseteq t_2(\bar{x})$, there is a polynomial term $p(\bar{x}) \in R[\bar{x}]$ such that

$$A \models_v (t_1(\bar{x}) \sqsubseteq t_2(\bar{x})) \;\leftrightarrow\; (0 \sqsubseteq p(\bar{x})).$$

Proof. Item (i) follows immediately from the axiom **M1** of superrings. Item (ii) follows from item (i) above, the item (ii) of Lemma 1.27 and by a repeated use of Theorem 2.5. □

Let $\mathcal{L}$ be the language of 1-rings. For each 1-ring R, let $\mathcal{L}(R)$ be the language extending $\mathcal{L}$ by adding all elements of R as *strict* constant symbols. Let Γ' be the 1-ring axioms. Let extend Γ' by (in)equalities and relations of the form

$$a_0 \neq b_0;\ c_1 = a_1.b_1;\ c_2 \in a_2 + b_2;\ a_i, b_i, c_i \in R$$

that are true in R ("the diagram of R"). Denote the set of formulas obtained by $\Gamma'(R)$. A model of $\Gamma'(R)$ is a 1-ring that contains a subset $\overline{R} = \{\overline{a} : a \in R\}$ and $\overline{R}$ is an isomomorphic copy of R inside this model.

If $R = K$ is a 1-field and Γ is the 1-field axioms, then a model of $\Gamma(K)$ is a 1-field that contains a subset $\overline{K} = \{\overline{a} : a \in K\}$ and $\overline{K}$ is a 1-field isomorphic to K. Then a model of $\Gamma(K)$ is (up to a isomorphism) a 1-field containing K.

Now, we extend $\Gamma(K)$ to a new set of axioms $\tilde{\Gamma}(K)$ adding

$$\forall z_0 ... \forall z_n \exists x [0 \in z_0 + z_1 x + ... + z_{n-1} x^{n-1} + x^n],\ n \geq 1. \qquad \text{(AC)}$$

and because the counter Example 3.1, we add also the family of axioms

$$\exists z_0 ... \exists z_{n-1} \bigvee_{i<j<n} [z_i \neq z_j],\ n \geq 2.$$

A model F of $\Gamma(K)$ is also a model of $\tilde{\Gamma}(K)$ iff F is infinite and algebraically closed. Our aim is to describe a quantifier elimination procedure for $\tilde{\Gamma}(F)$. By the reduction Lemma 3.2, F regards every atomic formula as equivalent modulo $\Gamma(K)$ to a polynomial "equation" $0 \in f(X_1, ..., X_n)$.

Since $K[\bar{X}]$ is a 2-domain (by an iteration of Theorem 2.5), a conjunction of inequations

$$\bigwedge_{i=1}^{m} [0 \neq g_i(\bar{X})]$$

is equivalent to the "inequation" $0 \notin g_1(\bar{X}) ... g_n(\bar{X})$. Then, to obtain a quantifier elimination for $\tilde{\Gamma}(K)$ is sufficient eliminate Y from the formula

$$\exists Y [0 \in f_1(\bar{X}, Y) \wedge ... \wedge 0 \in f_m(\bar{X}, Y) \wedge 0 \notin g(\bar{X}, Y)] \qquad (1)$$

with $f_1, ..., f_m, g \in R[X_1, ..., X_m, Y]$.

Theorem 3.3 (Quantifier Elimination Procedure). *Let K be an infinite 1-field and $\varphi(X_1, ..., X_n, Y)$ the formula in 1. Then $\varphi(X_1, ..., X_n, Y)$ is equivalent modulo $\tilde{\Gamma}(R)$ to a boolean combination of atomic formulas $\psi(X_1, ..., X_r)$, $r \geq n$.*

Proof. The proof consists in three parts:

A - Reduction to the case that only one of $f_1, ..., f_m$ involves Y.

Move each conjunction that appears in (1) and that does not involve Y to the left of $\exists Y$ according to the rule "$\exists Y[\varphi \wedge \psi] \equiv \varphi \wedge \exists Y[\psi]$ if Y does not appear in φ". Thus we assume $\deg_Y(f_i(\bar{X}, Y)) \geq 1$, $i = 1, ..., m$ and $m \geq 2$.

We now perform an induction on $\sum \deg_Y(f_i(\bar{X}, Y))$:

Let $p(\bar{X}, Y)$ and $q(\bar{X}, Y)$ be multipolynomials with coefficients in R such that $0 \leq \deg_Y p(\bar{X}, Y) \leq \deg_Y q(\bar{X}, Y) = d$. Write $p(\bar{X}, Y)$ in the form

$$p(\bar{X}, Y) = a_k(\bar{X})Y^k + a_{k-1}(\bar{X})Y^{k-1} + ... + a_0(\bar{X}) \tag{2}$$

with $a_j \in R[\bar{X}]$. For each j with $0 \leq j \leq k$ let

$$p_j(\bar{X}, Y) = a_j(\bar{X})Y^j + a_{j-1}(\bar{X})Y^{j-1} + ... + a_0(\bar{X})$$

If $0 \notin a_j(\bar{X})$, division of $q(\bar{X}, Y)$ by $p_j(\bar{X}, Y)$ produces $q_j(\bar{X}, Y)$ and $r_j(\bar{X}, Y)$ in $R[\bar{X}, Y]$ for which

$$a_j(\bar{X})^d q(\bar{X}, Y) \subseteq q_j(\bar{X}, Y)p_j(\bar{X}, Y) + r_j(\bar{X}, Y), \tag{3}$$

and $\deg_Y(r_j) < \deg_Y(p_j) \leq d$.

Let F be a model of $\Gamma(K)$. If $x_1, ..., x_n, y$ are elements of F such that $0 \in a_l(\bar{x})$ for $l = j+1, ..., k$ and $0 \notin a_j(\bar{x})$, then $[0 \in p(\bar{x}, y) \wedge 0 \in q(\bar{x}, y)]$ is equivalent in F to $[0 \in p_j(\bar{x}, y) \wedge 0 \in r_j(\bar{x}, y)]$. Therefore, the formula $[0 \in p(\bar{X}, Y) \wedge 0 \in q(\bar{X}, Y)]$ is equivalent modulo $\Gamma(K)$ to the formula

$$\left(\bigvee_{j=0}^{k} [0 \in a_k(\bar{X}) \wedge ... \wedge 0 \in a_{j+1}(\bar{X}) \wedge 0 \notin a_j(\bar{X}) \wedge 0 \in p_j(\bar{X}, Y) \wedge 0 \in r_j(\bar{X}, Y)]\right)$$
$$\vee [0 \in a_k(\bar{X}) \wedge ... \wedge 0 \in a_0(\bar{X}) \wedge 0 \in q(\bar{X}, Y)]. \tag{4}$$

Apply the outcome of (4) to $f_1(\bar{X}, Y)$ and $f_m(\bar{X}, Y)$ (of 1). With the rule "$\exists Y[\varphi \vee \psi] \equiv \exists Y\varphi \vee \exists Y\psi$" we have replaced (1) by disjunction of statements of form (1) in each which the sum corresponding to $\sum \deg_Y(f_i(\bar{X}, Y))$ is smaller. Using the induction assumption, we conclude that m may be taken to be at most 1.

B - Reduction to the case that $m = 0$.

Continue the notation of part A which left us at the point of considering how to eliminate Y from $p(\bar{X}, Y)$ in

$$\exists Y[0 \in p(\bar{X}, Y) \wedge 0 \notin g(\bar{X}, Y)]. \tag{5}$$

Consider a model F of $\tilde{\Gamma}(K)$ and elements $x_1, ..., x_n \in F$. If $0 \notin p(\bar{x}, Y)$ then (since F is algebraically closed) the statement

$$F \models \exists Y[0 \in p(\bar{x}, Y) \wedge 0 \notin g(\bar{x}, Y)]$$

is equivalent to the statement

$$p(\bar{x}, Y) \text{ does not divide } g(\bar{x}, Y)^k \text{ in } F[X].$$

Therefore, with $q(\bar{X}, Y) = g(\bar{X}, Y)^k$ and in the notation of (2) and (3), formula (5) is equivalent modulo $\tilde{\Gamma}(K)$ to the formula

$$\left(\bigvee_{j=0}^{k} [0 \in a_k(\bar{X}) \wedge ... \wedge 0 \in a_{j+1}(\bar{X}) \wedge 0 \notin a_j(\bar{X}) \wedge \exists Y[\in r_j(\bar{X}, Y)]]\right)$$
$$\vee[0 \in a_k(\bar{X}) \wedge ... \wedge 0 \in a_0(\bar{X}) \wedge \exists Y[0 \in g(\bar{X}, Y)]]$$

a disjunction of statements of form (1) with $m = 0$.

C - Completion of the proof.

By part B we are in the point of removing Y from a statement of the form

$$\exists Y[0 \notin a_l(\bar{X})Y^l + a_{l-1}(\bar{X})Y^{l-1} + ... + a_0(\bar{X})].$$

Since models of $\tilde{\Gamma}(K)$ are infinite 1-fields, this formula is equivalent modulo $\tilde{\Gamma}(K)$ to

$$0 \notin a_l(\bar{X}) \vee ... \vee 0 \notin a_0(\bar{X}),$$

completing the quantifier elimination procedure. □

Remark 3.4.

i- The previous result subsumes the usual one, i.e., if K is a ordinary algebraically closed field then it is an infinite algebraically closed (strict) 1-field and Theorem 3.3 is just the usual quantifier elimination result.

ii- In general, when the translate the Theorem 3.3 to a result on first-order relational structures (see Remark 1.12) we get that the theory of algebraically closed 1-fields perceives that some formulas are always equivalent to some $\forall\exists$-formulas: but this results seems not so ease to have a previous and direct intuition (and consequent proof), i.e. without the use of the language of multialgebras and its results.

4 Final remarks and future works

We finish the work presenting here some possible future developments.

- It could be interesting describe and explore an alternative notion of algebraically closed multifield based on an alternative notion of of root of a polynomial, taking in account factorizations, for example, if $p(x) \in (x-b)q(x)$ for some $q(x)$, then b can be seem as a root of $p(x)$: by Theorem 7 in [4], this in fact *coincide* with the other notion of root of a polynomial $p(x) \in F[x]$ whenever F is a hyperfield.
- It is true that any full 1-field has a kind of algebraic closure?
- In what sense the theory of algebraically closed 1-fields could be considered a model completion of the 1-field theory?
- We gave a first step in model theory of 1-fields. It will be interesting consider, in the same line, the model theory of 1-fields endowed with some extra structure: orderings ([12], valuations ([11], etc.
- The Theorem 3.3 could be adapted for real closed and henselian 1-fields?
- In [17] was started the development of a identity theory and a universal algebra like theory for multi structures. However, a full model theory of multi structures, in the vein of chapter 1 of [6], should be an object of interest (as the present work suggests) and it is seems to be unknown.

References

[1] Madeline Al Tahan, Sarka Hoskova-Mayerova, and Bijan Davvaz. Some results on (generalized) fuzzy multi-hv-ideals of hv-rings. *Symmetry*, 11(11):1376, 2019.

[2] R Ameri, M Eyvazi, and S Hoskova-Mayerova. Advanced results in enumeration of hyperfields. *Aims Mathematics*, 5(6):6552–6579, 2020.

[3] R Ameri, A Kordi, and S Hoskova-Mayerova. Multiplicative hyperring of fractions and coprime hyperideals. *Analele Universitatii" Ovidius" Constanta-Seria Matematica*, 25(1):5–23, 2017.

[4] Reza Ameri, Mansour Eyvazi, and Sarka Hoskova-Mayerova. Superring of polynomials over a hyperring. *Mathematics*, 7(10):902, 2019.

[5] Marcelo E Coniglio, Aldo Figallo-Orellano, and Ana C Golzio. Non-deterministic algebraization of logics by swap structures. *arXiv preprint arXiv:1708.08499*, 2017.

[6] Razvan Diaconescu. *Institution-independent model theory*. Springer Science & Business Media, 2008.

[7] Michael D. Fried and Moshe Moshe Jarden. *Field Arithmetic*. Springer, 2008.

[8] Pawel Gladki and Krzysztof Worytkiewicz. Witt rings of quadratically presentable fields. *Categories and General Algebraic Structures*, 12(1):1–23, 2020.

[9] Ana Claudia Golzio. A brief historical survey on hyperstructures in algebra and logic. *South American Journal of Logic*, 2018.

[10] Jaiung Jun. Algebraic geometry over hyperrings. *Advances in Mathematics*, 323:142–192, 2018.

[11] Jaiung Jun. Valuations of semirings. *Journal of Pure and Applied Algebra*, 222(8):2063–2088, 2018.

[12] Murray Marshall. Real reduced multirings and multifields. *Journal of Pure and Applied Algebra*, 205(2):452–468, 2006.

[13] Ch G Massouros. Theory of hyperrings and hyperfields. *Algebra and Logic*, 24(6):477–485, 1985.

[14] Christos G Massouros and Gerasimos G Massouros. On join hyperrings. In *Proceedings of the 10th International Congress on Algebraic Hyperstructures and Applications, Brno, Czech Republic*, pages 203–215, 2009.

[15] Geronimos G Massouros and Christos G Massouros. Homomorphic relation on hyperingoinds and join hyperrings. *Ratio Mathematica*, 13(1):61–70, 1999.

[16] Anastase Nakassis. Recent results in hyperring and hyperfield theory. *International Journal of Mathematics and Mathematical Sciences*, 11, 1988.

[17] Cosmin Pelea and Ioan Purdea. Multialgebras, universal algebras and identities. *Journal of the Australian Mathematical Society*, 81(1):121–140, 2006.

[18] Hugo Rafael Ribeiro, Kaique Matias de Andrade Roberto, and Hugo Luiz Mariano. Functorial relationship between multirings and the various abstract theories of quadratic forms. *to appear in São Paulo Journal of Mathematical Sciences, https://doi.org/10.1007/s40863-020-00185-1*, 2020.

[19] Oleg Viro. Hyperfields for tropical geometry I. Hyperfields and dequantization. *arXiv preprint arXiv:1006.3034*, 2010.

 Received 29 December 2020

Connecting abstract logics and adjunctions in the theory of (π-)institutions: some theoretical remarks and applications

Gabriel Bittencourt Rios
Institute of Mathematics and Statistics, University of São Paulo, Brazil
gabriel.bit@usp.br

Daniel de Almeida Souza
Institute of Mathematics and Statistics, University of São Paulo, Brazil
daniel.almeida.souza@usp.br

Darllan Conceição Pinto
Institute of Mathematics, Federal University of Bahia, Brazil

This author was funded by FAPESB, Grant APP0072/2016.

darllan@ufba.br

Hugo Luiz Mariano
Institute of Mathematics and Statistics, University of São Paulo, Brazil
hugomar@ime.usp.br

Abstract

In the present work, a natural sequel to [17], we further discuss the existence of adjunctions between categories of institutions and of π-institutions. This is done at both a foundational and an applied level. Firstly, we reformulate and conceptually clarify such adjunctions in terms of the 2-categorical data involved in the construction of categories of institution-like structures. More precisely, we remark that the process used for passing from rooms to institutions ([10]) can be extended, due to its 2-functoriality, to more general room-like and institution-like structures in such a way that the aforementioned adjunctions are all seen to arise from simpler adjunctions at the room-like level. Secondly, and mostly independently, we provide some applications of such adjunctions to abstract logics, mainly to the setting of propositional logics and filter pairs ([2]); we also generalize the process of skolemization, a classical device from predicate logic, to the institutional setting.

Keywords: (π-)institutions, abstract logics, adjunctions

Introduction

The concept of *institution* was introduced by J. A. Goguen and R. M. Burstall (see [13]) in order to present a unified mathematical formalism for the notion of a formal logical system, i.e. it provides a *"...categorical abstract model theory which formalizes the intuitive notion of logical system, including syntax, semantic, and satisfaction relation between them..."* ([10]). This means that it encompasses the abstract concept of universal model theory for a logic: it contains a satisfaction relation between models and sentences that is "stable under change of notation". The are several natural examples of institutions, and a systematic study of abstract model theory based on the general notion of institution is presented in Diaconescu's book [10].

A proof-theoretical variation of the notion of institution, the concept of π-*institution*, was introduced by Fiadeiro and Sernadas in [12]: it formalizes the notion of a deductive system and *"...replace the notion of model and satisfaction by a primitive consequence operator (à la Tarski)"*. Categories of propositional logics endowed with natural notions of translation morphisms provide examples of π-institutions. Voutsadakis has developed an intensive study of abstract algebraic logic based on the concept of π-institution, see for instance [21].

Certain relations between institutions and π-institutions were established in [12] and [21]. On the other hand, it seems that the explicit functorial connections between the category of institutions (with comorphisms) and that of π-institutions (with comorphisms) first appeared in [17]: indeed, the category of π-institutions is isomorphic to a full coreflective subcategory of the category of institutions. In the present (ongoing) work, we expand the study initiated in [17] by establishing new adjunctions concerning categories of institution-like structures and sketching new connections between these and abstract logics. Thus the goal of the article is twofold: firstly, a categorical analysis in the setting of the abstract theory of models (respectively, theory of proof) given by institution theory (respectively, π-institution theory); secondly, applications to presentations of propositional logics (abstract logics, filter pairs) and abstract predicate logic devices (skolemization).

Overview of the paper:

In **Section 1** we recall, for the reader's convenience, the definitions of institution and of π-institution, as well as their respective notions of (co)morphism. In **Section 2** we expand the work in [17] by presenting new adjunctions involving categories of categories, diagrams, institutions, and π-institutions. **Section 3** is devoted to extending the construction of the category of rooms – as presented in [10] – in a way that applies to more general categories of institution-like structures. This is done by applying classical 2-categorical machinery (such as the 2-Yoneda embedding and the Grothendieck construction) and, although being relatively straightforward from a technical point of view, its 2-functoriality allows us to provide a crucial conceptual simplification of the aforementioned adjunctions between categories of institution-like structures: they are seen to arise as images (under a 2-functor of *institutional realization*) of adjunctions between their generating categories of room-like structures. In **Section 4**, we present some institutions and π-institutions of abstract propositional logics, not only the ones obtained by the former adjunctions, useful for establishing an abstract Glivenko's theorem for algebraizable logics regardless of their signatures associated ([19]). We have also defined a institution for each filter pair -general and finitary version (see [2])- in fact, we provide a functor from the category of filter pairs to the category of institutions that can be restricted to a functor from the category of propositional logics to the category of institutions and, moreover, that can be extended to a functor from the "multialgebraic" setting (logics and filter pairs), useful to deal with complex logics, as Logics of Formal Inconsistency (LFIs) ([7]), thought non-deterministic semantics of matrices ([4]). **Section 5** introduces a new institutional device: skolemization; which is applied to get, by borrowing from FOL, a form of downward Löwenheim-Skolem for the setting of multialgebras. **Section 6** finishes the paper presenting some remarks and perspectives of future developments.

1 Preliminaries: categories of institutions and π-institutions

In this first section we recall, for the reader's convenience, the definition of institution and π-institution with their respective notions of morphisms and comorphisms, consequently defining their categories. We also add a subsection recalling the main results in [17]: the adjunction between the categories of institutions and π-institutions endowed with its *comorphisms*.

1.1 Categories of institutions

Definition 1.1. *An institution $I = (\mathbb{S}ig, Sen, Mod, \models)$ consists of*

$$\begin{array}{ccccc} & & \mathbb{S}ig & & \\ & \swarrow^{Mod} & & \searrow^{Sen} & \\ (\mathbb{C}at)^{op} & & \models & & \mathbb{S}et \end{array}$$

1. *a category $\mathbb{S}ig$, whose the objects are called* signature*,*

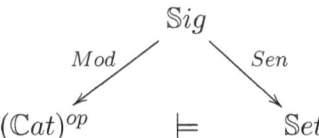

2. *a functor* $Sen : \mathbb{S}ig \to \mathbb{S}et$*, for each signature a set whose elements are called* sentence *over the signature*

3. *a functor* $Mod : (\mathbb{S}ig)^{op} \to \mathbb{C}at$*, for each signature a category whose the objects are called* model,

4. *a relation* $\models_\Sigma \subseteq |Mod(\Sigma)| \times Sen(\Sigma)$ *for each* $\Sigma \in |\mathbb{S}ig|$*, called* Σ-satisfaction, *such that for each morphism* $h : \Sigma \to \Sigma'$*, the compatibility condition*

$$M' \models_{\Sigma'} Sen(h)(\phi) \ if \ and \ only \ if \ Mod(h)(M') \models_\Sigma \phi$$

holds for each $M' \in |Mod(\Sigma')|$ *and* $\phi \in Sen(\Sigma)$

Example 1.2. *Let Lang denote the category of languages* $L = ((F_n)_{n\in\mathbb{N}}, (R_n)_{n\in\mathbb{N}})$, *– where* F_n *is a set of symbols of n-ary function symbols and* R_n *is a set of symbols of n-ary relation symbols,* $n \geqslant 0$ *– and language morphisms*[1]. *For each pair of cardinals* $\aleph_0 \leqslant \kappa, \lambda \leqslant \infty$*, the category Lang endowed with the usual notion of* $L_{\kappa,\lambda}$*-sentences (=* $L_{\kappa,\lambda}$*-formulas with no free variable), with the usual association of category of structures and with the usual (tarskian) notion of satisfaction, gives rise to an institution* $I(\kappa, \lambda)$.

Definition 1.3. *Let* I *and* I' *be institutions.*

(a) An institution **morphism** $h = (\Phi, \alpha, \beta) : I \to I'$ *consists of:*

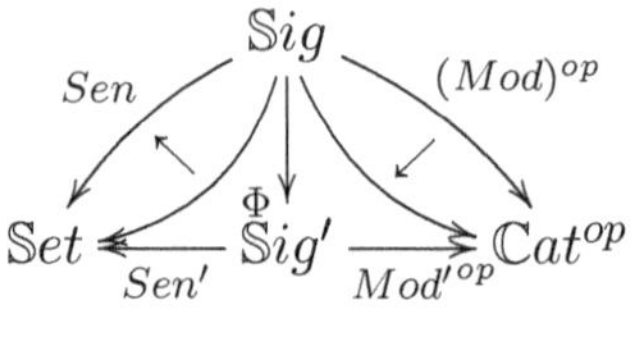

- *a functor* $\Phi : \mathbb{S}ig \to \mathbb{S}ig'$
- *a natural transformation* $\alpha : Sen' \circ \Phi \Rightarrow Sen$
- *a natural transformation* $\beta : Mod \Rightarrow Mod' \circ \Phi^{op}$

Such that the following compatibility condition *holds:*

$$m \models_\Sigma \alpha_\Sigma(\varphi') \quad iff \quad \beta_\Sigma(m) \models'_{\Phi(\Sigma)} \varphi'$$

For any $\Sigma \in \mathbb{S}ig$*, any* Σ*-model* m *and any* $\Phi(\Sigma)$*-sentence* φ'.

[1] That can be chosen "strict" (i.e., $F_n \mapsto F'_n$, $R_n \mapsto R'_n$) or chosen be "flexible" (i.e., $F_n \mapsto \{n - ary - terms(L')\}$, $R_n \mapsto \{n - ary - atomic - formulas(L')\}$).

(b) *A triple* $f = \langle \phi, \alpha, \beta \rangle : I \to I'$ *is a* **comorphism** *between the given institutions if the following conditions hold:*

- $\phi : \mathbb{S}ig \to \mathbb{S}ig'$ *is a functor.*
- *natural transformations* $\alpha : Sen \Rightarrow Sen' \circ \phi$ *and* $\beta : Mod' \circ \phi^{op} \Rightarrow Mod$ *satisfying:*

$$m' \models'_{\phi(\Sigma)} \alpha_\Sigma(\varphi) \ iff \ \beta_\Sigma(m') \models_\Sigma \varphi$$

For any $\Sigma \in \mathbb{S}ig$, $m' \in Mod'(\phi(\Sigma))$ *and* $\varphi \in Sen(\Sigma)$.

Given comorphisms $f : I \to I'$ and $f' : I \to I''$, notice that $f' \bullet f := \langle \phi' \circ \phi, \alpha' \bullet \alpha, \beta' \bullet \beta \rangle$ defines a comorphism $f' \bullet f : I \to I''$, where $(\alpha' \bullet \alpha)_\Sigma = \alpha'_{\phi(\Sigma)} \circ \alpha_\Sigma$ and $(\beta' \bullet \beta)_\Sigma = \beta_\Sigma \circ \beta'_{\phi(\Sigma)}$. Let $Id_I := \langle Id_{\mathbb{S}ig}, Id, Id \rangle : I \to I$. It is straightforward to check that these data determines a category[2]. We will denote by $\mathbf{Ins}_{co}$ this category of institution comorphisms. Of course, using analagous methods one can also define $\mathbf{Ins}_{mor}$—the category of institution morphisms.

Example 1.4. *Given two pairs of cardinals* (κ_i, λ_i), *with* $\aleph_0 \leqslant \kappa_i, \lambda_i \leqslant \infty$, $i = 0, 1$, *such that* $\kappa_0 \leqslant \kappa_1$ *and* $\lambda_0 \leqslant \lambda_1$, *then it is induced a morphism and a comorphism of institutions* $(\Phi, \alpha, \beta) : I(\kappa_0, \lambda_0) \to I(\kappa_1, \lambda_1)$, *given by the* same data: $\mathbb{S}ig_0 = Lang = \mathbb{S}ig_1$, $Mod_0 = Mod_1 : (Lang)^{op} \to \mathbb{C}at$, $Sen_i = L_{\kappa_i, \lambda_i}$, $i = 0, 1$, $\Phi = Id_{Lang} : \mathbb{S}ig_0 \to \mathbb{S}ig_1$, $\beta := Id : Mod_i \Rightarrow Mod_{1-i}$, $\alpha := inclusion : Sen_0 \Rightarrow Sen_1$.

1.2 Categories of π-institutions

Definition 1.5. *A* π*-institution* $J = \langle \mathbb{S}ig, Sen, \{C_\Sigma\}_{\Sigma \in |\mathbb{S}ig|} \rangle$ *is a triple with its first two components exactly the same as the first two components of an institution and, for every* $\Sigma \in |\mathbb{S}ig|$, *a closure operator* $C_\Sigma : \mathcal{P}(Sen(\Sigma)) \to \mathcal{P}(Sen(\Sigma))$, *such that, for every* $f : \Sigma_1 \to \Sigma_2 \in Mor(\mathbb{S}ig)$, *the following holds:*

$$Sen(f)(C_{\Sigma_1}(\Gamma)) \subseteq C_{\Sigma_2}(Sen(f)(\Gamma)), \ for \ all \ \Gamma \subseteq Sen(\Sigma_1).$$

Definition 1.6. *Let* J *and* J' *be* π*-institutions.*

(a) *A* **morphism** *between* J *and* J' *is a pair* $\langle \Phi, \alpha \rangle$ *such that:*

- $\Phi : \mathbb{S}ig \to \mathbb{S}ig'$ *is a functor*

[2] As usual in category theory, the set theoretical size issues on such global constructions of categories can be addressed by the use of at least two Grothendieck universes.

- $\alpha : Sen'\Phi \Rightarrow Sen$ *is a natural transformation*

And, for all $\Gamma \cup \{\varphi\} \subseteq Sen'(\Phi\Sigma)$, *the following holds:*

$$\varphi \in C_{\Phi\Sigma}(\Gamma) \Rightarrow \alpha_\Sigma(\varphi) \in C_\Sigma(\alpha_\Sigma(\Gamma))$$

(b) $\langle \Phi, \alpha \rangle : J \to J'$ *is a* ***comorphism*** *between* π*-institution if:*

- $\Phi : \mathbb{S}ig \to \mathbb{S}ig'$ *is a functor*
- $\alpha : Sen \Rightarrow Sen'\Phi$ *is a natural transformation*

Such that, for all $\Gamma \cup \{\varphi\} \subseteq Sen(\Sigma)$, *we have:*

$$\varphi \in C_\Sigma(\Gamma) \Rightarrow \alpha_\Sigma(\varphi) \in C_{\Phi\Sigma}(\alpha_\Sigma(\Gamma))$$

Given π-institution morphisms (respec. comorphisms) $\langle F, \alpha \rangle : J \to J'$ and $\langle G, \beta \rangle : J' \to J''$, $g \cdot f$ is defined as $\langle GF, \alpha \cdot \beta F \rangle$ (respec. $\langle GF, \beta F \cdot \alpha \rangle$), routine calculations show the composition is well defined. The identity morphism and comorphism are both given by $\langle 1_{\mathbb{S}ig}, 1_{Sen} \rangle$. These remarks lead us to define $\pi\mathbf{Ins}_{mor}$ and $\pi\mathbf{Ins}_{co}$ the categories of, respectively, institution morphisms and comorphisms.

Remark 1.7. *It is easy to see that* π*-institution can be equivalently described by a triple* $\langle \mathbb{S}ig, Sen, \{\vdash_\Sigma\}_{\Sigma \in |\mathbb{S}ig|} \rangle$ *where the first two components are simply the ones used for* π*-institutions and the third component is a family, indexed by* $\Sigma \in |\mathbb{S}ig|$, *of* tarskian consequence relations $\vdash_\Sigma \subseteq \mathcal{P}(Sen(\Sigma)) \times Sen(\Sigma)$ *such that for every arrow* $f : \Sigma_1 \to \Sigma_2$ *in* $\mathbb{S}ig$ *the induced function* $Sen(f) : Sen(\Sigma_1) \to Sen(\Sigma_2) \in Mor(\mathbb{S}et)$ *is a* logical translation, *i.e. for each* $\Gamma \cup \{\varphi\} \subseteq Sen(\Sigma_1)$

$$\Gamma \vdash_{\Sigma_1} \varphi \;\Rightarrow\; Sen(f)[\Gamma] \vdash_{\Sigma_2} Sen(f)(\varphi)$$

1.3 An adjunction between $\mathbf{Ins}_{co}$ and $\pi\mathbf{Ins}_{co}$

For the reader's convenience, we recall here the adjunction between $\mathbf{Ins}_{co}$ and $\pi\mathbf{Ins}_{co}$ established in [17]; thus all the proofs will be omitted.

Let $I = \langle \mathbb{S}ig, Sen, Mod, \models \rangle$ be an institution. Given $\Sigma \in |\mathbb{S}ig|$, consider

$$\Gamma^\star = \{m \in Mod(\Sigma);\ m \models_\Sigma \varphi \ \mathit{for\ all}\ \varphi \in \Gamma\} \ \text{ and}$$

$$M^\star = \{\varphi \in Sen(\Sigma);\ m \models_\Sigma \varphi \ \mathit{for\ all}\ m \in M\}$$

for any $\Gamma \subseteq Sen(\Sigma)$ and $M \subseteq Mod(\Sigma)$. Notoriously, these mappings establish a Galois connection. Thus $C^I_\Sigma(\Gamma) := \Gamma^{\star\star}$ defines a closure operator for any $\Sigma \in |\mathbb{S}ig|$ ([21]). We can now define the first part of our adjunction:

$$\mathbf{Ins}_{co} \xrightarrow{\quad F \quad} \pi\mathbf{Ins}_{co}$$

$$\begin{array}{ccc}
I & \longmapsto & \langle \mathbb{S}ig^I, Sen^I, \{C^I_\Sigma\}_{\Sigma\in|\mathbb{S}ig|}\rangle \\
{\scriptstyle\langle\phi,\alpha,\beta\rangle}\downarrow & \longmapsto & \downarrow{\scriptstyle\langle\phi,\alpha\rangle} \\
J & \longmapsto & \langle \mathbb{S}ig^J, Sen^J, \{C^J_\Sigma\}_{\Sigma\in|\mathbb{S}ig|}\rangle
\end{array}$$

For the other side of the adjunction consider the application:

$$\pi\mathbf{Ins}_{co} \xrightarrow{\quad G \quad} \mathbf{Ins}_{co}$$

$$\begin{array}{ccc}
J & \longmapsto & \langle \mathbb{S}ig^J, Sen^J, Mod^J, \models^J\rangle \\
{\scriptstyle\langle\phi,\alpha\rangle}\downarrow & \longmapsto & \downarrow{\scriptstyle\langle\phi,\alpha,\alpha^{-1}\rangle} \\
J' & \longmapsto & \langle \mathbb{S}ig^{J'}, Sen^{J'}, Mod^{J'}, \models^{J'}\rangle
\end{array}$$

Where:

- Mod^J is taken as:

$$\mathbb{S}ig^{op} \xrightarrow{\quad Mod^J \quad} \mathbb{C}at$$

$$\begin{array}{ccc}
\Sigma & \longmapsto & \{C_\Sigma(\Gamma) : \Gamma \subseteq Sen(\Sigma)\} \\
{\scriptstyle f}\downarrow & \longmapsto & \uparrow{\scriptstyle Sen(f)^{-1}} \\
\Sigma' & \longmapsto & \{C_{\Sigma'}(\Gamma) : \Gamma \subseteq Sen(\Sigma')\}
\end{array}$$

  With $Mod^J(\Sigma)$ being viewed as a "co-discrete category"[3].

- For each Σ we let $\models^J_\Sigma \subseteq |Mod(\Sigma)| \times Sen(\Sigma)$ as the relation:

$$m \models^J_\Sigma \varphi \quad iff \quad \varphi \in m$$

  For any $m \in Mod(\Sigma)$ and $\varphi \in Sen(\Sigma)$

[3] I.e., a class of objects C endowed with the trivial groupoid structure of all ordered pairs, $C \times C$.

Theorem 1.8. *The functors $F : \mathbf{Ins}_{co} \to \pi\mathbf{Ins}_{co}$ and $G : \pi\mathbf{Ins}_{co} \to \mathbf{Ins}_{co}$ defined above establish an adjunction $G \dashv F$ between the categories $\mathbf{Ins}_{co}$ and $\pi\mathbf{Ins}_{co}$. Moreover, $F \circ G = Id_{\pi\mathbf{Ins}_{co}}$ and the unity of this adjunction, the natural transformation $\eta : Id_{\pi\mathbf{Ins}_{co}} \to F \circ G$, is the identity. Thus the category $\pi\mathbf{Ins}_{co}$ can be seen to be a full coreflective subcategory of $\mathbf{Ins}_{co}$.*

2 Adjunctions between Inst, π-Inst, Cat, Diag

In this section we continue and expand the analysis of categorical relations between categories whose objects are categories endowed with some extra structure like categories of (π-)institutions, categories of categories and categories of Set-based diagrams.

2.1 An adjunction between $\mathbf{Ins}_{mor}$ and $\pi\mathbf{Ins}_{mor}$

It is natural to ask whether we could achieve a similar adjunction considering morphisms instead of comorphisms, that is, taking $\mathbf{Ins}_{mor}$ and $\pi\mathbf{Ins}_{mor}$ instead of $\mathbf{Ins}_{co}$ and $\pi\mathbf{Ins}_{co}$. In this subsection, we sketch a proof that the category of π-institutions and morphisms is isomorphic to a full coreflective subcategory of the category of institutions and morphisms: this is a natural variant of the results in [17] which were recalled in subsection 1.3.

Let $I = \langle \mathbb{S}ig, Sen, Mod, \models \rangle$ be an institution. Given $\Sigma \in |\mathbb{S}ig|$ let:

$$\Gamma^* := \{m \in Mod(\Sigma) : m \models_\Sigma \varphi \ for\ all\ \varphi \in \Gamma\}$$

and

$$M^* := \{\varphi \in Sen(\Sigma) : m \models_\Sigma \varphi \ for\ all\ m \in M\}$$

for any $\Gamma \subseteq Sen(\Sigma)$ and $M \subseteq |Mod(\Sigma)|$. These mappings cleary define a Galois connection between $\mathcal{P}(Sen(\Sigma))$ and $\mathcal{P}(|Mod(\Sigma)|)$. Therefore, $Con^I_\Sigma(\Gamma) := \Gamma^{**}$ defines a closure operator on $\mathcal{P}(Sen(\Sigma))$ for any $\Sigma \in |\mathbb{S}ig|$.

Lemma 2.1. *Let $\langle \phi, \alpha, \beta \rangle : I \to I'$ be an arrow in $\mathbf{Ins}_{mor}$ and $\sigma \in |\mathbb{S}ig|$. Given $\Gamma \subseteq Sen(\Sigma)$ and $M \subseteq |Mod(\Sigma)|$ the following holds:*

- $\beta_\Sigma[(\alpha_\Sigma[\Gamma])^*] \subseteq \Gamma^*$
- $\alpha_\Sigma[(\beta_\Sigma[M])^*] \subseteq M^*$

Proof: The proof is similar to the one of **Lemma 2.8** in [17]

□

Consider now the following functor:

$$F : \mathbf{Ins}_{mor} \to \pi\mathbf{Ins}_{mor}$$
$$I \mapsto \langle \mathbb{S}ig, Sen, \{Con^{I}_{\Sigma}\}_{\Sigma \in |\mathbb{S}ig|} \rangle$$

The proof that F is well defined on objects can be found on [17]. The action on morphisms is defined as follows:

$$I \xrightarrow{\langle \phi, \alpha, \beta \rangle} I'$$
$$F(I) \xrightarrow{\langle \phi, \alpha \rangle} F(I')$$

Consider now the following application,

$$G : \pi\mathbf{Ins}_{mor} \to \mathbf{Ins}_{mor}$$
$$J \to \langle \mathbb{S}ig, Sen, Mod^{J}, \models^{J} \rangle$$

Where:

- $Mod^{J} : \mathbb{S}ig^{op} \to \mathbb{C}at$ is defined as:

$$\Sigma \xrightarrow{f} \Sigma' \mapsto \{C_{\Sigma'}(\Gamma') : \Gamma' \subseteq Sen(\Sigma')\} \xrightarrow{Sen(f)^{-1}} \{C_{\Sigma}(\Gamma) : \Gamma \subseteq Sen(\Sigma)\}$$

- For each $\Sigma \in |\mathbb{S}ig|$, $\models^{J}_{\Sigma} \subseteq |Mod^{J}(\sigma)| \times Sen(\Sigma)$ is defined such that, give $m \in |Mod(\Sigma)|$ and $\varphi \in Sen(\sigma)$, $m \models^{J}_{\Sigma} \varphi$ iff $\varphi \in m$.
  The proof that Mod^{J} is well defined and that $G(J)$ satisfies the compatibility condition and is indeed an institution can be found in [17]

Given a morphism $f = \langle \phi, \alpha \rangle : J \to J'$ in $\pi\mathbf{Ins}_{mor}$ define, for $\Sigma \in |\mathbb{S}ig|$ and $m \in |Mod^{J}(\Sigma)|$, $\beta_{\Sigma}(m) := \alpha^{-1}_{\Sigma}(m)$. Let us prove that $\beta_{\Sigma} : Mod^{J}(\Sigma) \to Mod^{J'}(\phi(\Sigma))$.

$$\begin{array}{ccc}
\mathcal{P}(Sen(\Sigma)) & \xleftarrow{\alpha^{-1}_{\Sigma}} & \mathcal{P}(Sen'(\phi(\Sigma))) \\
{\scriptstyle Sen(f)^{-1}} \uparrow & & \uparrow {\scriptstyle Sen('\phi(f))^{-1}} \\
\mathcal{P}(Sen(\Sigma')) & \xleftarrow[\alpha^{-1}_{\Sigma'}]{} & \mathcal{P}(Sen'(\phi(\Sigma')))
\end{array}$$

Let us register prove the compatibility condition for morphisms. Given $\Sigma \in |\mathbb{S}ig|$, $m \in Mod^J(\Sigma)$ and $\varphi \in Sen(\phi(\Sigma))$ we have:

$$\begin{aligned} m \models^J_\Sigma \alpha_\Sigma(\varphi) &\iff \alpha_\Sigma(\varphi) \in m \\ &\iff \varphi \in \alpha_\Sigma^{-1}(m) \\ &\iff \varphi \in \beta_\Sigma(m) \\ &\iff \beta_\Sigma(m) \models^{J'}_{\phi(\Sigma)} \varphi \end{aligned}$$

It follows that $G(f) = \langle \phi, \alpha, \beta \rangle$ is a morphism of institutions. To prove G a functor simply notice that, given $f = \langle \phi, \alpha \rangle : J \to J'$ and $f' = \langle \phi', \alpha' \rangle : J' \to J''$ in $\pi\mathbf{Ins}_{mor}$, $G(f' \cdot f) = \langle \phi' \cdot \phi, \alpha' \cdot \alpha\phi, (\alpha' \cdot \alpha\phi)^{-1} \rangle = \langle \phi' \cdot \phi, \alpha' \cdot \alpha\phi, \alpha^{-1}\phi \cdot \alpha'^{-1} \rangle = G(f') \cdot G(f)$ and, for any π-institution J, routine calculations show $G(1_J) = 1_{G(I)}$.

In fact, as in [17], we have the following:

Theorem 2.2. *The functors* $\mathbf{Ins}_{mor} \underset{G}{\overset{F}{\rightleftarrows}} \pi\mathbf{Ins}_{mor}$ *establish and adjunction* $G \dashv F$. *Moreover, since* $F \circ G = Id_{\pi\mathbf{Ins}_{mor}}$ *and the unity of this adjunction, the natural transformation* $\eta : Id_{\pi\mathbf{Ins}_{mor}} \to F \circ G$, *is the identity. Thus the category* $\pi\mathbf{Ins}_{mor}$ *can be seen as a full coreflective subcategory of* $\mathbf{Ins}_{mor}$.

2.2 Adjunctions between CAT and $\pi\mathbf{Ins}_{co}$

In this section we detail left and right adjoints for the forgetful functor from $\pi\mathbf{Ins}_{co}$ to **CAT**. Something of notice here is the similarity between the functors shown here and the adjoints to the forgetful functor from **Top** to $\mathbb{S}et$. Indeed, we describe a left adjoint that associates categories to their "discrete" π-institution, where every set is closed, and a right adjoint that maps to their "codiscrete" π-institution, where the only closed sets are the empty set and the entire set of formulas. The place of these two constructions in the theory of π-institutions is then similar to the place of the "(co)discrete" topology in point set topology. That is to say, as illustrative examples of pathologies.

Let us commence by the right adjoint. We begin by defining an action on the objects of **CAT**; given a category $\mathcal{A}$ let $\top\mathcal{A} := \langle \mathcal{A}, *, \{Con_c\}_{a\in|\mathcal{A}|} \rangle$ where $* : \mathcal{A} \to \mathbb{S}et$ is the constant functor to the singleton set and, for each object a in $\mathcal{A}$ and $\Gamma \subseteq \{*\}$, we define $Con_a(\Gamma) = \{*\}$. It is clear that Con_a is closure operator on $\{*\}$. Moreover, for any arrow $a \xrightarrow{f} a'$ in $\mathcal{A}$ and $\Gamma \subseteq \{*\}$, we have that $*f(Con_a(\Gamma)) = Con_{a'}(*f(\Gamma))$ and thus $\top\mathcal{A}$ is a π-institution.

We can now extend $\top$ to morphisms. Given some functor $F : \mathcal{A} \to \mathcal{B}$, we see that there is a unique $! : * \Rightarrow *F$; furthermore, routine calculations show $\varphi \in Con_a(\Gamma) \Rightarrow$

$!_a(\varphi) \in Con_{Fa}(!_a(\Gamma))$ for $\{\varphi\} \cup \Gamma \subseteq \{*\}$. Define then $\top F = \langle F, ! \rangle$ the remarks above showing it a comorphism between $\top\mathcal{A}$ and $\top\mathcal{B}$.

To prove that $\top$ behaves functorially notice, firstly, that the lone arrow $* \Rightarrow *$ is 1_* so $\top(1_{\mathcal{A}}) = \langle 1_{\mathcal{A}}, 1_* \rangle = 1_{\top\mathcal{A}}$. Finally, the below diagram guarantees that the composition is well behaved.

$$\begin{array}{ccccc} *c & \dashrightarrow & *Fc & \dashrightarrow & *GFc \\ \downarrow & & \downarrow & & \downarrow \\ *c' & \dashrightarrow & *Fc' & \dashrightarrow & *GFc' \end{array}$$

Theorem 2.3. *Let* $U : \pi\mathbf{Ins}_{co} \to \mathbf{CAT}$ *the forgetful functor, taking each* π-*institution to its signature category and each comorphism to its first coordinate. The functors* $\top : \mathbf{CAT} \to \pi\mathbf{Ins}_{co}$ *and* $U : \pi\mathbf{Ins}_{co} \to \mathbf{CAT}$ *establish an adjunction* $\top \vdash U$ *with counit* $\eta_{\mathcal{A}} = 1_{\mathcal{A}}$.

Proof: Given some a π-institution J and a functor $F : \mathbb{S}ig^J \to \mathcal{A}$, consider the below diagram:

$$\begin{array}{ccccc} \mathcal{A} & \xleftarrow{1_{\mathcal{A}}} & U\top\mathcal{A} & & \top\mathcal{A} \\ & \nwarrow^{F} & \uparrow F & \longleftarrow & \uparrow \langle F,\alpha \rangle \\ & & \mathbb{S}ig^J & & J \end{array}$$

Where α is the single arrow $Sen \Rightarrow *F$. Given $\{\varphi\} \cup \Gamma \subseteq Sen(\Sigma)$ we have that $\varphi \in C_\Sigma(\Gamma) \Rightarrow \alpha_\Sigma(\varphi) = *$. As $Con_{F\Sigma}(\alpha_\Sigma(\Gamma)) = \{*\}$ it follows that $\varphi \in C_\Sigma(\Gamma) \Rightarrow \alpha_\Sigma(\varphi) \in Con_{F\Sigma}(\alpha_\Sigma(\Gamma))$ and thus $\langle F, \alpha \rangle$ is indeed a comorphism between J and $D\mathcal{A}$. As $\langle F, \alpha \rangle$ is clearly the only arrow that makes the diagram commute, the result follows. □

We can now describe the left adjoint. Consider the following functor:

$$\bot : \mathbf{CAT} \longrightarrow \pi\mathbf{Ins}_{co}$$

$$\begin{array}{ccc} \mathcal{A} & \longmapsto & \langle \mathcal{A}, \varnothing, (Con_a)_{a \in |\mathcal{A}|} \rangle \\ F \downarrow & \longmapsto & \downarrow \langle F, ! \rangle \\ \mathcal{B} & \longmapsto & \langle \mathcal{B}, \varnothing, (Con_b)_{b \in |\mathcal{B}|} \rangle \end{array}$$

Where $\varnothing$ is the constant functor to the empty set, Con_a is the single closure operator on the empty set and ! is the unique natural transformation $\varnothing \Rightarrow \varnothing F$. By vacuity, $\langle F, ! \rangle$ satisfies the comorphism condition. Proving that $\bot$ is indeed a functor uses similar arguments to the ones given above.

Theorem 2.4. *Let U as above. The functors $\bot$ and U establish an adjunction $\bot \dashv U$ with unit $\epsilon_{\mathcal{A}} = 1_{\mathcal{A}}$.*

Proof: Given some a π-institution J and a functor $F : \mathcal{A} \to \mathbb{S}ig^J$, consider the below diagram:

$$\begin{array}{ccccc} \mathcal{A} & \xrightarrow{1_{\mathcal{A}}} & U\bot\mathcal{A} & & \bot\mathcal{A} \\ & \searrow^{F} & \downarrow F & \longleftarrow & \downarrow \langle F,\alpha\rangle \\ & & \mathbb{S}ig^J & & J \end{array}$$

Where α is the only natural transformation $\varnothing \Rightarrow Sen^J F$. We argue by vacuity to show that $\langle F, \alpha\rangle$ is a comorphism. Since $\langle F, \alpha\rangle$ it is clearly the only arrow that makes the diagram commute, the result follows.

□

Remark 2.5. *It is easy to see how one would go on defining the $\pi\mathbf{Ins}_{mor}$ versions of the functors $\top$ and $\bot$. This, of course, prompt us to question if these functors still define an adjunction. Routine calculations show that the directions would be reversed, that is, in the $\pi\mathbf{Ins}_{mor}$ case we have: $\bot \vdash U \vdash \top$*

Remark 2.6. *Let us consider a generalization of $\pi\mathbf{Ins}_{co}$ for a moment. Given a concrete category $\mathcal{C}$, i.e. a faithful functor $|-| : \mathcal{C} \to Set$, a $\mathcal{C}-\pi$–institution is a triple of the form $\langle \mathbb{S}ig, Sen : \mathbb{S}ig \to \mathcal{C}, (C_\Sigma : \mathcal{P}|Sen(\Sigma)| \to \mathcal{P}|Sen(\Sigma)|)_{\Sigma \in |\mathcal{C}|}\rangle$ where $\mathbb{S}ig$ is a category, Sen a functor and C_Σ a closure operator on $\mathcal{P}|Sen(\Sigma)|$ satisfying structurality; furthermore, one can easily generalize a version of comorphisms for $\mathcal{C}-\pi$–institutions. Consider then $\mathcal{C}-\pi\mathbf{Ins}_{co}$— the category of $\mathcal{C}-\pi$–institution comorphisms.*
Let 1 a terminal object in the concrete category $\mathcal{C}$. We can now define a functor $\top_{\mathcal{C}} : \mathbf{CAT} \to \mathcal{C}-\pi\mathbf{Ins}_{co}$ as

$$\mathcal{A} \xrightarrow{F} \mathcal{B} \mapsto \langle \mathcal{A}, 1, (Con_a)_{a \in Ob(\mathcal{A})}\rangle \xrightarrow{\langle F,\alpha\rangle} \langle \mathcal{B}, 1, (Con_b)_{b \in Ob(\mathcal{B})}\rangle$$

Where 1 is the constant functor to the terminal object, $Con_a(\Gamma) = |Sen(a)|$ for each $a \in Ob(\mathcal{A})$ and $\Gamma \subseteq |Sen(a)|$ and α is the unique $1 \Rightarrow 1F$. Using the methods analogous we see that $\top_{\mathcal{C}} \vdash forgetful$. Suppose now that $\mathcal{C}$ had a initial object 0, one can easily see how to define $\bot_{\mathcal{C}}$ — the left adjoint to the forgetful — mimicking $\bot$.
It is common, specially when dealing with propositional logics, to define the syntax as an algebraic structure instead of a set. This remark could be of use in that scenario.

2.3 Adjunctions $Diag_{co} \leftrightarrows \pi\mathbf{Ins}_{co}$

We begin this section by describing $Diag_{co}(C)$ and $Diag_{mor}(C)$, the categories of diagrams for a given category C. Diagrams for **Set** can be initially seen as π-institutions minus the consequence relation and the 2-categorially minded will recognize diagrams for C as the Grothendieck construction for $\mathbf{CAT}(-, C)$. After this introduction, we proceed to obtain right and left adjoints to the the forgetful $Diag_{co}(\mathbf{Set}) \to \pi\mathbf{Ins_{co}}$. Finally, we further this result to categories adjoint to **Set**. In this sense the purpose of this section is twofold:

- Firstly, it may serve as a path to the theory of "generalized" π-institutions, that is, π-institutions having sentence functors over any arbitrary category, not only **Set**. This practice of taking sentences in categories different of **Set** is common in logic, a notorious example being that of propostional logic where sentences are taken as free algebras.

- Secondly, it introduces, albeit tacitly, the 2-categorial ideas which will be used in the next section. Indeed, the idea of diagrams will be explored again in section 3.2.

Let C be a category. Denote $Diag_{co}(C)$ the category whose objects are pair (A, F), where $F : A \to C$ is a covariant functor and such that $Hom((A, F), (A', F'))$ is the (meta)class of all pairs (T, α) where $T : A \to A'$ is a functor and $\alpha : F \to F' \circ T$ is a natural transformation. Let $id_{(A,F)} := (id_A, id_F)$ and if $(T', \alpha') \in Hom((A', F'), (A'', F''))$, then $(T', \alpha') \bullet (T, \alpha) := (T' \circ T, \alpha'_T \circ \alpha)$. $Diag_{mor}(C)$ denotes the category with the same objects as $Diag_{co}(C)$ and, for arrows, $(T, \alpha) \in Hom((A, F), (A', F'))$ iff $T : A \to A'$ is a functor and $\alpha : F' \circ T \to F$ is a natural transformation; identities are the same as in $Diag_{co}(C)$ and compositions are adapted accordingly: $(T', \alpha') \bullet (T, \alpha) := (T' \circ T, \alpha \circ \alpha'_T)$.

Now consider the category $\pi\mathbf{Ins}_{co}$ and the obvious forgetful functor $U : \pi\mathbf{Ins}_{co} \to Diag_{co}(Set)$ given by:

$$\pi\mathbf{Ins}_{co} \xrightarrow{\quad U \quad} Diag_{co}(Set)$$

$$\begin{array}{ccc}
\langle \mathbb{S}ig, Sen, (C_\Sigma)_{\Sigma \in |\mathbb{S}ig|} \rangle & \longmapsto & \langle \mathbb{S}ig, Sen \rangle \\
{\scriptstyle \langle F,\alpha \rangle}\big\downarrow & \longmapsto & \big\downarrow{\scriptstyle \langle F,\alpha \rangle} \\
\langle \mathbb{S}ig', Sen', (C'_\Sigma)_{\Sigma \in |\mathbb{S}ig|} \rangle & \longmapsto & \langle \mathbb{S}ig', Sen' \rangle
\end{array}$$

The main result of this subsection is that U has a left adjoint $L : Diag_{co}(Set) \to \pi\mathbf{Ins}_{co}$ and a right adjoint $R : Diag_{co}(Set) \to \pi\mathbf{Ins}_{co}$. Thus $U : \pi\mathbf{Ins}_{co} \to Diag_{co}(Set)$ preserves all limits and all colimits.

We will provide just the definitions of the functors, since the proof of the universal properties are straightforward.

$L : Diag_{co}(Set) \to \pi\mathbf{Ins}_{co}$ is given by: $L(A, F) := (A, F, (C_a^{min})_{a\in|A|})$, where $C_a^{min} : P(F(a)) \to P(F(a))$ is such that:

$$\Gamma \in P(F(a)) \mapsto C_a^{min}(\Gamma) := \Gamma$$

It is ease to see that $L(A, F)$ satisfies the coherence condition in the definition of π-institution.

The action of L on morphisms is very simple:

$$L(((A, F) \overset{(T,\alpha)}{\to} (A', F')) = (A, F, (C_a^{min})_{a\in|A|}) \overset{(T,\alpha)}{\to} (A', F', (C'^{min}_{a'})_{a'\in|A'|});$$

this clearly determines a morphism of π-institutions.

For each $(A, F) \in |Diag_{co}(Set)|$, we have the identity arrow $id_{(A,F)} : (A, F) \to U(L(A, F))$ and this is a initial object in the comma category $(A, F) \downarrow U$. Thus L is left adjoint to U and we have just described the component (A, F) of the unity of this adjunction.

Similarly, we have a functor $R : Diag_{co}(Set) \to \pi\mathbf{Ins}_{co}$ with action $R(A, F) := (A, F, (C_a^{max})_{a\in|A|})$, where $C_a^{max} : P(F(a)) \to P(F(a))$ is such that:

$$\Gamma \in P(F(a)) \mapsto C_a^{max}(\Gamma) := F(a)$$

With the obvious action on arrows, R becomes the right adjoint to U.

Remark 2.7. *Given category C and a functor $C \xrightarrow{E} \mathbf{Set}$ with left adjoint $\mathbf{Set} \xrightarrow{\mathcal{L}} C$ (respec. right adjoint $\mathbf{Set} \xrightarrow{\mathcal{R}} C$) we can form $Diag_{co}(C) \xrightarrow{\tilde{E}} Diag_{co}(\mathbf{Set})$ and $Diag_{co}(\mathbf{Set}) \xrightarrow{\tilde{L}} Diag_{co}(C)$ by composing:*

$$\tilde{E}((T, \alpha) : (A, F) \to (A', F')) = (T, E\alpha)$$

and likewise for $\tilde{\mathcal{L}}$ (respec. $\tilde{\mathcal{R}}$). It is straightforward that $\tilde{E}$ has as left adjoint $\tilde{\mathcal{L}}$ (respec. right adjoint $\tilde{\mathcal{R}}$). We can then compose this adjunction with the one obtained above to obtain $\pi\mathbf{Ins}_{co} \underset{L\circ\tilde{\mathcal{R}}}{\overset{\tilde{E}\circ U}{\underset{\top}{\rightleftarrows}}} Diag_{co}(C)$. (respec. $Diag_{co}(C) \underset{\tilde{\mathcal{L}}\circ U}{\overset{R\circ\tilde{E}}{\underset{\top}{\rightleftarrows}}} \pi\mathbf{Ins}_{co}$).

We summarize below the adjunctions previously presented. It can be described an analogous diagram for "morphisms" instead of "co-morphism".

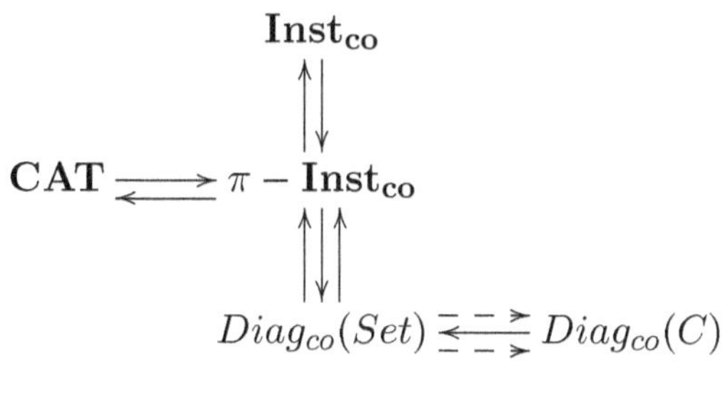

3 Adjunctions at the level of room-like structures

Accordding to [10], page 47,

"The presentation of institutions as functors was given already in [13] and the 2-categorical structure of the category of institutions has been studied in [9] ."

This section aims at describing how a standard construction from 2-category theory – the Grothendieck construction, which associates a fibration to a pseudofunctor in a 2-functorial way – allows us to reformulate the above adjunctions between categories of institution-like structures in a way which is general and systematic, and which provides conceptually clearer equivalent descriptions of the same phenomena. This is done in two main steps:

1. We borrow from [10] the definition of the *category of rooms*, denoted by $\mathbb{R}oom$ – which can be used to provide a concise description of the category of institutions[4] – and generalize it in a straightforward way (to categories of room-like structures), so as to obtain analogous reconstructions of categories of institution-like structures.

2. By using the (non-trivial) facts that (a) the process of associating fibrations to pseudofunctors defines a 2-categorical equivalence, and (b) the 2-categorical Yoneda embedding is 2-fully faithful, we are able to conclude that the 2-functorial procedure (described below) which sends categories of room-like structures to categories of institution-like structures is also 2-fully faithful. As a corollary, any 2-categorical connections between categories of institution-like objects can be "pulled-back" to a corresponding construction at the level of room-like structures. For the purposes of this paper, we shall only be con-

[4]In [10], this description is used to show that $\mathbf{Ins}_{mor}$ is a complete category.

cerned with the particular case of recovering instutition-level adjunctions in terms of much simpler room-level adjunctions.

The definition of categories of room-like objects is illustrated in terms of three archetypal examples: for institutions (as usual), for π-institutions (a direct analogous), and for the category of small categories (which turns out to be an extremal example).

It should also be remarked that the aforementioned procedure comes naturally, and quite generally, in two variants: one suitable for describing *morphisms* between institution-like structures, and one suitable for describing *comorphisms* between them.

Before introducing the actual definitions, we outline as follows the background to be considered: as described in [10], the category of institutions and morphisms can be obtained by means of a standard categorical construction often referred to as the *Grothendieck construction.* There, a central role is played by the so-called *category of rooms*, denoted by $\mathbb{R}oom$: individually, an institution having $\mathbb{S}ig$ as its category of signatures corresponds to a functor $\mathbb{S}ig \longrightarrow \mathbb{R}oom$; on the other hand, (co)morphisms of institutions should also take into account base-change functors between different categories of signatures. The Grothendieck construction provides an adequate framework for studying this kind of phenomena. More precisely, given a 1-category $\mathcal{C}$ (regarded as a strict 2-category with trivial 2-cells), the Grothendieck construction, which we shall denote by $-^\sharp$, associates to each pseudofunctor $F : \mathcal{C} \longrightarrow \mathbf{CAT}$ a 1-category $F^\sharp$ together with a structure (*projection*) functor $F^\sharp \longrightarrow \mathcal{C}$ onto the base category. Most importantly, it constitutes a pseudofunctor

$$-^\sharp : [\mathcal{C}, \mathbf{CAT}] \longrightarrow \mathbf{CAT}/\mathcal{C},$$

where:

- $[\mathcal{C}, \mathbf{CAT}]$ denotes the 2-category of pseudofunctors $\mathcal{C} \longrightarrow \mathbf{CAT}$, pseudonatural transformations, and modifications.
- $\mathbf{CAT}/\mathcal{C}$ denotes the slice 2-category defined in the obvious way.

Our main interest will be the case where $\mathcal{C}$ is $\mathbb{C}at$, the 1-category of categories. We shall also need to consider the 2-categorical Yoneda (pseudo)functor

$$\begin{aligned} Y : \mathbf{C} &\longrightarrow [\mathbf{C}^{op}, \mathbf{CAT}] \\ c &\longmapsto \mathbf{C}(-, c) \end{aligned}$$

associated to a (possibly weak) 2-category $\mathbf{C}$, and variations thereof. A pseudofunctor equivalent to one of the form $\mathbf{C}(-, c)$ is called a *representable 2-presheaf*. We will be concerned with (restrictions to $\mathbf{CAT}$ of) 2-presheaves on a (suitably large) 2-category of categories which are represented by variations of $\mathbb{R}oom$. For instance, $\mathbf{Ins}_{co}$ is described in [10] as the Grothendieck construction $\mathbf{CAT}(-^{op}, \mathbb{R}oom)^\sharp$ of the Yoneda-like 2-presheaf $\mathbf{CAT}(-^{op}, \mathbb{R}oom)$ on $\mathbf{CAT}$. Our goal in this section will be to provide an alternative description of the above adjunctions between categories of institution-like structures (such as institutions and π-institutions), by noticing that (i) it is easy to describe $\mathbb{R}oom$-like categories from which other categories of institution-like structures can be obtained through a similar Yoneda-followed-by-Grothendieck procedure, and (ii) the notion of adjunction is available for any 2-category, and adjunctions in this sense are preserved by pseudofunctors.

As for categorical prerequisites, we restrict ourselves to providing quick (and mostly ad-hoc) descriptions of some of the necessary constructions from 2-category theory, including the Grothendieck construction; hence the reader is strongly encouraged to have a prior basic knowledge on these topics. For that purpose, we refer to [10] and [20] for a brief introduction, and to [14] for a more detailed discussion.

The present section does not aim at completeness; instead, it consists in a brief introduction, including basic constructions a few functioning examples, to the idea of canonically producing new (resp. recovering well-known) 2-categorical information on categories of institution-like structures in terms of their simpler counterparts: categories of room-like structures.

3.1 2-categorical preliminaries

We start by fixing some notations and defining the 2-categorical constructions alluded to above. The basic language of 2-category theory will be freely used. Unless otherwise specified, by a 2-category we mean a *strict* 2-category. If $\mathcal{C}$ is a 1-category, we regard it as a 2-category whenever necessary. We denote by $\mathbf{CAT}$ the 2-category of categories, functors, and natural transformations, and by $\mathbb{C}at$ the 1-category of categories and functors. Given 2-categories $\mathbf{C}$ and $\mathbf{D}$, we denote by $[\mathbf{C}, \mathbf{D}]$ the corresponding category of pseudofunctors, pseudonatural transformations, and modifications. If $\mathbf{C}$ is a 2-category, we denote by $\mathbf{C}^{op}$ (resp. $\mathbf{C}^{co}$, $\mathbf{C}^{coop}$) the 2-category obtained by reversing the 1-cells (resp. 2-cells, both 1-cells and 2-cells). By a contravariant pseudofunctor from $\mathbf{C}$ to $\mathbf{D}$ we mean a pseudofunctor $\mathbf{C}^{op} \longrightarrow \mathbf{D}$. By a 2-presheaf (resp. category of 2-presheaves) we mean a pseudofunctor $\mathbf{C}^{op} \longrightarrow \mathbf{CAT}$ (resp. a 2-category $[\mathbf{C}^{op}, \mathbf{CAT}]$).

3.1.1 The Grothendieck construction

The Grothendieck construction can be defined in two similar versions: taking as input either a contravariant **CAT**-valued pseudofunctor (i.e. a 2-presheaf), or a covariant one.

Definition 1. *(Grothendieck construction for contravariant pseudofunctors)*

Let $\mathcal{C}$ be a 1-category. Given a pseudofunctor $F : \mathcal{C}^{op} \longrightarrow \mathbf{CAT}$, we define its *Grothendieck construction* or *Grothendieck category*, denoted by $F^\sharp$, as the 1-category given by the following data:

- Its objects are pairs (c, x), where $c \in Ob(\mathcal{C})$ and $x \in Ob(F(c))$.
- An arrow $(c, x) \longrightarrow (d, y)$ is a pair (f, ϕ), where $f \in \mathcal{C}(c, d)$ and $\phi \in F(c)(x, Ff(y))$.
- The composite of morphisms $(f, \phi) : (c, x) \longrightarrow (d, y)$ and $(g, \psi) : (d, y) \longrightarrow (e, z)$ is defined as

$$(g \circ f \ , \ \alpha_z^{f,g} \circ F(f)(\psi) \circ \phi),$$

  where $\alpha^{f,g}$ is the natural isomorphism (associated to F by the definition of a pseudofunctor) $F(f) \circ F(g) \implies F(g \circ f)$. See

$$x \xrightarrow{\ \phi\ } Ff(y) \xrightarrow{Ff(\psi)} Ff(Fg(z)) = (Ff \circ Fg)(z) \xrightarrow{\alpha_z^{f,g}} F(g \circ f)(z).$$

The reader will be able to check that composition is associative and that each object possesses an identity arrow (by using the natural isomorphisms $\alpha^c : 1_{F(c)} \implies F(id_c)$). The category $F^\sharp$ is canonically endowed with a (*projection*) functor $F^\sharp \longrightarrow \mathcal{C}$ given by $(c, x) \longmapsto c$ and $(f, \phi) \longmapsto f$.

Now, suppose given a 1-cell in $[\mathcal{C}^{op}, \mathbf{CAT}]$, i.e. a pseudonatural transformation $\eta : F \implies G$. We define a functor $\eta^\sharp : F^\sharp \longrightarrow G^\sharp$ as follows:

- $\eta^\sharp((c, x)) = (c, \eta_c(x))$ for each $(c, x) \in Ob(F^\sharp)$.
- For each $(f, \phi) : (c, x) \longrightarrow (d, y)$ in $F^\sharp$, we define $\eta^\sharp((f, \phi)) : (c, \eta_c(x)) \longrightarrow (d, \eta_d(y))$ as

$$(f \ , \ \gamma_y^f \circ \eta_c(\phi)),$$

where γ^f is the natural isomorphism (associated to η by the definition of a pseudonatural transformation) as in

$$\begin{array}{ccc} F(d) & \xrightarrow{\eta_d} & G(d) \\ {\scriptstyle F(f)}\downarrow & \overset{\gamma^f}{\nearrow\!\!\!\nearrow} & \downarrow{\scriptstyle G(f)} \\ F(c) & \xrightarrow[\eta_c]{} & G(c). \end{array}$$

See

$$\eta_c(x) \xrightarrow{\eta_c(\phi)} \eta_c(F(f)(y)) \xrightarrow{\gamma^f_y} G(f)(\eta_d(y)).$$

The reader will be able to check that $\eta^\sharp$ is indeed a functor. Also, it is clear that it is compatible with the projections $F^\sharp \longrightarrow \mathcal{C}$ and $G^\sharp \longrightarrow \mathcal{C}$, so that we can regard $\eta^\sharp$ as a 1-cell in the slice 2-category $\mathbf{CAT}/\mathcal{C}$.

Finally, suppose given a 2-cell in $[\mathcal{C}^{op}, \mathbf{CAT}]$, i.e. a modification $\mu : \eta \Rrightarrow \chi$ between pseudonatural transformations $\eta,\ \chi : F \Longrightarrow G$. We define a natural transformation $\mu^\sharp : \eta^\sharp \Longrightarrow \chi^\sharp$ as follows: for each $(c, x) \in Ob(F^\sharp)$, we take

$$\mu^\sharp_{(c,x)} : \eta^\sharp((c, x)) = (c, \eta_c(x)) \longrightarrow \chi^\sharp((c, x)) = (c, \chi_c(x))$$

to be $(id_c, \beta^c_{\chi_c(x)} \circ (\mu_c)_x)$, where β^c is the natural isomorphism (associated to G by the definition of a pseudofunctor) $1_{G(c)} \Longrightarrow G(id_c)$. See

$$\eta_c(x) \xrightarrow{(\mu_c)_x} \chi_c(x) \xrightarrow{\beta^c_{\chi_c(x)}} G(id_c)(\chi_c(x)).$$

The reader will be able to check that $\mu^\sharp$ is indeed a natural transformation. Furthermore, it can be verified that by sending a pseudofunctor F to a category $F^\sharp$, a pseudonatural transformation $\eta : F \Longrightarrow G$ to a functor $\eta^\sharp : F^\sharp \longrightarrow G^\sharp$, and a modification $\mu : \eta \Rrightarrow \chi$ to a natural transformation $\mu^\sharp : \eta^\sharp \Longrightarrow \chi^\sharp$, we have defined a pseudofunctor

$$-^\sharp : [\mathcal{C}^{op}, \mathbf{CAT}] \longrightarrow \mathbf{CAT}/\mathcal{C}.$$

Definition 2. *(Grothendieck construction for covariant pseudofunctors)*

Let $\mathcal{C}$ be a 1-category. Given some pseudofunctor $F : \mathcal{C} \longrightarrow \mathbf{CAT}$, we define its *Grothendieck construction* or *Grothendieck category*, denoted by $F_\sharp$, as the 1-category given by the following data:

- Its objects are pairs (c, x), where $c \in Ob(\mathcal{C})$ and $x \in Ob(F(c))$.
- An arrow $(c, x) \longrightarrow (d, y)$ is a pair (f, ϕ), where $f \in \mathcal{C}(c, d)$ and $\phi \in F(d)(Ff(x), y)$.
- The composite of morphisms $(f, \phi) : (c, x) \longrightarrow (d, y)$ and $(g, \psi) : (d, y) \longrightarrow (e, z)$ is defined as

$$(g \circ f \ , \ \psi \circ F(g)(\phi) \circ (\alpha_x^{f,g})^{-1}),$$

where $\alpha^{f,g}$ is the natural isomorphism (associated to F by the definition of a pseudofunctor) $F(f) \circ F(g) \implies F(g \circ f)$. See

$$F(g \circ f)(x) \xrightarrow{(\alpha_x^{f,g})^{-1}} (Fg \circ Ff)(x) = Fg(Ff(x)) \xrightarrow{Fg(\phi)} Fg(y) \xrightarrow{\psi} z.$$

The reader will be able to check that composition is associative and that each object possesses an identity arrow (by using the natural isomorphisms $\alpha^c : 1_{F(c)} \implies F(id_c)$). As in the previous definition, $F_\sharp$ has a canonical projection functor $F_\sharp \longrightarrow \mathcal{C}$ given by $(c, x) \longmapsto c$ and $(f, \phi) \longmapsto f$. (Here, the reader might recognize it as what is called in the literature an *opfibration*, or that it realizes $F_\sharp$ as an *opfibered category* over $\mathcal{C}$).

Suppose given a 1-cell in $[\mathcal{C}, \mathbf{CAT}]$, i.e. a pseudonatural transformation $\eta : F \implies G$. We define a functor $\eta_\sharp : F_\sharp \longrightarrow G_\sharp$ as follows:

- $\eta_\sharp((c, x)) = (c, \eta_c(x))$ for each $(c, x) \in Ob(F_\sharp)$.
- For each $(f, \phi) : (c, x) \longrightarrow (d, y)$ in $F_\sharp$, we define $\eta_\sharp((f, \phi)) : (c, \eta_c(x)) \longrightarrow (d, \eta_d(y))$ as

$$(f \ , \ \eta_d(\phi) \circ (\gamma_x^f)^{-1}),$$

where γ^f is the natural isomorphism (associated to η by the definition of a pseudonatural transformation) as in

$$\begin{array}{ccc} F(d) & \xrightarrow{\eta_d} & G(d) \\ {\scriptstyle F(f)}\uparrow & \overset{\gamma^f}{\searrow\!\!\searrow} & \uparrow{\scriptstyle G(f)} \\ F(c) & \xrightarrow[\eta_c]{} & G(c). \end{array}$$

See

$$G(f)(\eta_c(x)) \xrightarrow{(\gamma^f_x)^{-1}} \eta_d(F(f)(x)) \xrightarrow{\eta_d(\phi)} \eta_d(y).$$

The reader will be able to check that $\eta_\sharp$ is indeed a functor. Again, it is clearly compatible with the projections $F_\sharp \longrightarrow \mathcal{C}$ and $G_\sharp \longrightarrow \mathcal{C}$, so that we can regard $\eta_\sharp$ as a 1-cell in the slice 2-category $\mathbf{CAT}/\mathcal{C}$.

Suppose given a 2-cell in $[\mathcal{C}, \mathbf{CAT}]$, i.e. a modification $\mu : \eta \Rrightarrow \chi$ between pseudonatural transformations η, $\chi : F \Longrightarrow G$. We define a natural transformation $\mu_\sharp : \eta_\sharp \Longrightarrow \chi_\sharp$ as follows: for each $(c, x) \in Ob(F_\sharp)$, we take

$$(\mu_\sharp)_{(c,x)} : \eta_\sharp((c,x)) = (c, \eta_c(x)) \longrightarrow \chi_\sharp((c,x)) = (c, \chi_c(x))$$

to be $(id_c, (\mu_c)_x \circ (\beta^c_{\eta_c(x)})^{-1})$, where β^c is the natural isomorphism (associated to G by the definition of a pseudofunctor) $1_{G(c)} \Longrightarrow G(id_c)$. See

$$G(id_c)(\eta_c(x)) \xrightarrow{(\beta^c_{\eta_c(x)})^{-1}} \eta_c(x) \xrightarrow{(\mu_c)_x} \chi_c(x).$$

The reader will be able to check that $\mu_\sharp$ is indeed a natural transformation. As before, it can be verified that by sending a pseudofunctor F to $F_\sharp$, a pseudonatural transformation $\eta : F \Longrightarrow G$ to $\eta_\sharp : F_\sharp \longrightarrow G_\sharp$, and a modification $\mu : \eta \Rrightarrow \chi$ to $\mu_\sharp : \eta_\sharp \Longrightarrow \chi_\sharp$, we have defined a pseudofunctor

$$-_\sharp : [\mathcal{C}, \mathbf{CAT}] \longrightarrow \mathbf{CAT}/\mathcal{C}.$$

3.1.2 Representable pseudofunctors

Let $\mathbf{C}$ be a 2-category. For each $c \in Ob(\mathbf{C})$, we define a pseudofunctor (in fact, a strict 2-functor) $\mathbf{C}(-, c) : \mathbf{C}^{op} \longrightarrow \mathbf{CAT}$ as follows:

- Each $d \in Ob(\mathbf{C})$ is sent to the hom-category $\mathbf{C}(d, c)$.
- Each 1-cell $f : d \longrightarrow e$ in $\mathbf{C}$ is sent to the functor $\mathbf{C}(f, c) : \mathbf{C}(e, c) \longrightarrow \mathbf{C}(d, c)$ given by precomposition of both 1-cells and 2-cells with f.

- Each 2-cell $\eta : f \Longrightarrow g$ between 1-cells f, $g : d \longrightarrow e$ is sent to the natural transformation

$$\mathbf{C}(\eta, c) : \mathbf{C}(f, c) \Longrightarrow \mathbf{C}(g, c)$$

given by precomposition with η, that is, by associating to each 1-cell $h : e \longrightarrow c$ (i.e. object of $\mathbf{C}(e, c)$) the 2-cell (i.e. morphism of $\mathbf{C}(d, c)$)

$$\mathbf{C}(\eta, c)_h = h \circ \eta : h \circ f \longrightarrow h \circ g.$$

Next, given a 1-cell $p : c \longrightarrow c'$ in $\mathbf{C}$, we define a pseudonatural transformation (in fact, a strict 2-natural transformation) $\mathbf{C}(-, p) : \mathbf{C}(-, c) \Longrightarrow \mathbf{C}(-, c')$ as follows:

- To each $d \in Ob(\mathbf{C})$ we associate the functor (i.e. 1-cell in **CAT**) $\mathbf{C}(d, p) : \mathbf{C}(d, c) \longrightarrow \mathbf{C}(d, c')$ given by postcomposition of both 1-cells and 2-cells with f.
- As we are only dealing with strict 2-categories, composition of 1-cells in $\mathbf{C}$ is strictly associative, hence we can fill the square diagrams thus obtained with identity natural transformations.

Given a 2-cell $\eta : p \Longrightarrow p'$ between p, $p' : c \longrightarrow c'$, we define a modification $\mathbf{C}(-, \eta) : \mathbf{C}(-, p) \Rrightarrow \mathbf{C}(-, p')$ by associating to each $d \in Ob(\mathbf{C})$ the natural transformation $\mathbf{C}(d, \eta) : \mathbf{C}(d, p) \Longrightarrow \mathbf{C}(d, p')$ given on each $f \in Ob(\mathbf{C}(d, c))$ by $\mathbf{C}(d, \eta)_f = \eta \circ f : p \circ f \longrightarrow p' \circ f$.

Routine diagram chasing shows that the above constructions define a strict 2-functor $\mathbf{C} \longrightarrow [\mathbf{C}^{op}, \mathbf{CAT}]$, which we denote by $\mathcal{Y}_{\mathbf{C}}$ and call the *Yoneda embedding* associated to $\mathbf{C}$.

Remark 3.1. *The above constructions can be adapted to produce a Yoneda embedding for any weak 2-category $\mathbf{C}$. In this case, $\mathcal{Y}_{\mathbf{C}}$ will in general only be a (non-strict) pseudofunctor. Also, the term embedding used here may be misleading in that the 2-categorical statement analogous to the Yoneda lemma, although true, is not nearly immediate from the above discussion. An elementary but not-so-short proof is given in [5].*

3.1.3 Adjunctions in a 2-category

Definition 3. *Let $\boldsymbol{C}$ be a 2-category. An adjunction in $\boldsymbol{C}$ is a quadruple $(f, g, \eta, \varepsilon)$, where:*

- *f and g are 1-cells in $\boldsymbol{C}$ of the form $f : c \longrightarrow d$, $g : d \longrightarrow c$.*
- *η and ε are 2-cells of the form $\eta : id_c \implies g \circ f$, $\varepsilon : f \circ g \implies id_d$.*
- *These satisfy the identities $(\varepsilon f) \circ (f\eta) = 1_f$ and $(g\varepsilon) \circ (\eta g) = 1_g$.*

We denote the existence of such an adjunction by $f \dashv g$.

For our purposes, the crucial property of adjunctions in 2-categories is that they are (up to isomorphism) preserved by any pseudofunctor:

Lemma 4. *Let $F : \boldsymbol{C} \longrightarrow \boldsymbol{D}$ be a pseudofunctor, and $(f, g, \eta, \varepsilon)$ an adjunction in $\boldsymbol{C}$. Then F induces an adjunction $(F(f), F(g), \bar{\eta}, \bar{\varepsilon})$ in $\boldsymbol{D}$.*

Proof. Let $f : c \longrightarrow d$, $g : d \longrightarrow c$. Take $\bar{\eta} : id_{F(c)} \implies F(g) \circ F(f)$ to be the composite

$$id_{F(c)} \overset{\alpha^c}{\implies} F(id_c) \overset{F(\eta)}{\implies} F(g \circ f) \overset{(\alpha^{g,f})^{-1}}{\implies} F(g) \circ F(f),$$

where α^c and $\alpha^{g,f}$ are the 2-cells associated to F as a pseudofunctor. Analogously, take $\bar{\varepsilon} : F(f) \circ F(g) \implies id_{F(d)}$ to be the composite

$$F(f) \circ F(g) \overset{\alpha^{f,g}}{\implies} F(f \circ g) \overset{F(\varepsilon)}{\implies} F(id_d) \overset{(\alpha^d)^{-1}}{\implies} id_{F(d)}.$$

Now, notice that

$$(\bar{\varepsilon}F(f)) \circ (F(f) \circ \bar{\eta}) = (((\alpha^d)^{-1}F(\varepsilon)\alpha^{f,g})F(f)) \circ (F(f)((\alpha^{g,f})^{-1}F(\eta)\alpha^c))$$

is given by the following composite of 2-cells:

$$F(f) \overset{F(f)\alpha^c}{\implies} F(f) \circ F(id_c) \overset{F(f)F(\eta)}{\implies} F(f) \circ F(g \circ f) \overset{F(f)(\alpha^{g,f})^{-1}}{\implies} F(f) \circ F(g) \circ F(f) \implies$$

$$\overset{\alpha^{f,g}F(f)}{\implies} F(f \circ g) \circ F(f) \overset{F(\varepsilon)F(f)}{\implies} F(id_d) \circ F(f) \overset{(\alpha^d)^{-1}F(f)}{\implies} F(f).$$

On the other hand, the equality $(\varepsilon f) \circ (f\eta) = 1_f$ implies (by functoriality of $\mathbf{C}(c,d) \longrightarrow \mathbf{D}(F(c), F(d)))$ $F(\varepsilon f) \circ F(f\eta) = 1_{F(f)}$. The left-hand side equals the composite of 2-cells

$$F(f) \overset{F(f\eta)}{\Longrightarrow} F(f \circ g \circ f) \overset{F(\varepsilon f)}{\Longrightarrow} F(f),$$

which (by expanding $id_{F(f \circ g \circ f)}$ through the coherence laws of F as a pseudo-functor) can be rewritten as

$$F(f) \overset{F(f\eta)}{\Longrightarrow} F(f \circ g \circ f) \overset{(\alpha^{f,g\circ f})^{-1}}{\Longrightarrow} F(f) \circ F(g \circ f) \overset{F(f)(\alpha^{g,f})^{-1}}{\Longrightarrow} F(f) \circ F(g) \circ F(f) \Longrightarrow$$

$$\overset{\alpha^{f,g}F(f)}{\Longrightarrow} F(f \circ g) \circ F(f) \overset{\alpha^{f\circ g,f}}{\Longrightarrow} F(f \circ g \circ f) \overset{F(\varepsilon f)}{\Longrightarrow} F(f).$$

Again by using the coherence laws of F, it can be shown (as the reader will be able to do in detail) that the following equalities hold:

$$(F(f)F(\eta)) \circ (F(f)\alpha^c) = (\alpha^{f,g\circ f})^{-1} \circ F(f\eta) : F(f) \Longrightarrow F(f) \circ F(g \circ f),$$

$$((\alpha^d)^{-1}F(f)) \circ (F(\varepsilon)F(f)) = F(\varepsilon f) \circ \alpha^{f\circ g,f} : F(f \circ g) \circ F(f) \Longrightarrow F(f).$$

It follows that the two composites of 2-cells above are equal, so that $(\bar{\varepsilon}F(f)) \circ (F(f) \circ \bar{\eta}) = 1_{F(f)}$, which is the first desired identity. The second one can be shown analogously. □

3.2 Categories of institutions as Grothendieck categories

[10] describes a procedure to recover $\mathbf{Ins}_{mor}$ as a Grothendieck category. It is done by introducing the so-called *category of rooms*, denoted by $\mathbb{R}oom$ (see below), so that $\mathbf{Ins}_{mor}$ is canonically equivalent (isomorphic, in fact) to $\mathbf{CAT}((-)^{op}, \mathbb{R}oom)^\sharp$. Before recalling this construction, it will be convenient to define (or better, to fix notation for) a general notion of $\mathbb{R}oom$-like category which can be applied to produce other categories of institution-like objects.

Definition 5.

Let $\mathcal{C}$ be a 1-category. We say that a 1-category R is a *category of rooms* for $\mathcal{C}$ if there exists an equivalence of categories $\mathcal{C} \simeq \mathbf{CAT}(-^{op}, R)^{\sharp}$, where the right-hand side denotes the category obtained as in

$$\begin{array}{ccccccc}
\mathbf{CAT} & & [\mathbf{CAT}^{op}, \mathbf{CAT}'] & & [\mathbb{C}at^{op}, \mathbf{CAT}'] & & \mathbf{CAT}'/\mathbb{C}at \\
\rotatebox{90}{$\in$} & & \rotatebox{90}{$\in$} & & \rotatebox{90}{$\in$} & & \rotatebox{90}{$\in$} \\
R & \longmapsto & \mathbf{CAT}(-^{op}, R) & \longmapsto & \mathbf{CAT}(-^{op}, R) & \longmapsto & \mathbf{CAT}(-^{op}, R),^{\sharp}
\end{array}$$

where we denote by $\mathbf{CAT}'$ a 2-category of categories defined in a Grothendieck universe larger than that of $\mathbf{CAT}$. As discussed in the previous subsection, both the Yoneda embedding for 2-categories and the Grothendieck construction are pseudofunctorial. It is then immediate that the above construction gives rise to a pseudofunctor (in fact, a strict 2-functor)

$$\begin{aligned}
\mathbf{CAT} &\longrightarrow \mathbf{CAT}'/\mathbb{C}at \\
R &\longmapsto \mathbf{CAT}(-^{op}, R)^{\sharp}.
\end{aligned}$$

It will be denoted by **ins** and called *institutional realization.*

It often happens that the right Grothendieck construction to be used is that from Definition 2, for covariant pseudofunctors. We say that R is a *category of co-rooms* for $\mathcal{C}$ if there exists an equivalence of categories $\mathcal{C} \simeq (\mathbf{CAT}(-^{op}, R)_{\sharp})^{op}$. See

$$\begin{array}{ccccccccc}
\mathbf{CAT} & & [\mathbf{CAT}^{op}, \mathbf{CAT}'] & & [\mathbb{C}at^{op}, \mathbf{CAT}'] & & \mathbf{CAT}'/\mathbb{C}at^{op} & & \mathbf{CAT}'^{co}/\mathbb{C}at \\
\rotatebox{90}{$\in$} & & \rotatebox{90}{$\in$} & & \rotatebox{90}{$\in$} & & \rotatebox{90}{$\in$} & & \rotatebox{90}{$\in$} \\
R & \mapsto & \mathbf{CAT}(-^{op}, R) & \mapsto & \mathbf{CAT}(-^{op}, R) & \mapsto & \mathbf{CAT}(-^{op}, R)_{\sharp} & \mapsto & (\mathbf{CAT}(-^{op}, R)_{\sharp}).^{op}
\end{array}$$

Once again, we obtain a pseudofunctor (in fact, a strict 2-functor)

$$\begin{aligned}
\mathbf{CAT} &\longrightarrow \mathbf{CAT}'^{co}/\mathbb{C}at \\
R &\longmapsto (\mathbf{CAT}(-^{op}, R)_{\sharp})^{op},
\end{aligned}$$

which we denote by **coins** and call *institutional co-realization.*

Remark 3.2. *It is clear that* **CAT** *plays no distinguished role in this construction besides being a 2-category. The inner op as in* $\mathbf{CAT}(-^{op}, R)$ *and* $(\mathbf{CAT}(-^{op}, R)_{\sharp})^{op}$ *corresponds (see Example 6) to the fact that we wish the functors sending signatures to categories of models to be contravariant. The outer op as in* $(\mathbf{CAT}(-^{op}, R)_{\sharp})^{op}$ *(as well as its absence from* $\mathbf{CAT}(-^{op}, R)$*) corresponds to the fact that we wish any morphism between institution-like objects to have the same direction as its corresponding functor between signature categories. The co as in* $\mathbf{CAT}'^{co}/\mathbb{C}at$ *is due to the fact that the pseudofunctor taking a category to its opposite reverses the direction of natural transformations, but not of functors. Since left-right adjunctions in* $\mathbf{CAT}'$ *correspond to right-left adjunctions in* $\mathbf{CAT}'^{co}$*, Lemma 4 implies that* ***coins*** *sends left-right adjunctions in* **CAT** *to right-left adjunctions in* $\mathbf{CAT}'^{co}/\mathbb{C}at$*.*

We list below some examples of room categories for some categories of institution-like objects. Proofs will not be given, but the reader will be able to provide them without difficulty.

Example 6. *(*$\mathbb{R}oom$*, a room category for* $\mathbf{Ins}_{mor}$ *and* $\mathbf{Ins}_{co}$*)*

Define a category $\mathbb{R}oom$ as follows:

- Its objects are triples $\langle S, M, (R_m)_{m \in Ob(M)} \rangle$, where S is a set, M is a category, and, for each $m \in Ob(M)$, $R_m : S \to 2 = \{0, 1\}$ is a function.
- A morphism $\langle S, M, (R_m)_{m \in Ob(M)} \rangle \xrightarrow{(\sigma,\mu)} \langle S', M', (R'_{m'})_{m' \in Ob(M')} \rangle$ consists of a function $\sigma : S' \to S$ and a functor $\mu : M \to M'$ such that $R'_{\mu m}(s) = R_m \sigma(s)$ for every $m \in Ob(M)$ and $s \in Ob(S)$.
- Composition is given by $(\sigma', \mu') \circ (\sigma, \mu) = (\sigma \circ \sigma', \mu' \circ \mu)$.

It is clear that $\mathbb{R}oom$ is indeed a category. Then, in the terminology introduced above, we have

$$\mathbf{Ins}_{mor} \cong \mathbf{ins}(\mathbb{R}oom),$$

$$\mathbf{Ins}_{co} \cong \mathbf{coins}(\mathbb{R}oom).$$

Both projections $\mathbf{ins}(\mathbb{R}oom) \longrightarrow \mathbb{C}at$ and $\mathbf{coins}(\mathbb{R}oom) \longrightarrow \mathbb{C}at$ recover the underlying category of signatures of an institution. For more on this example, we refer the reader to [10].

Example 7. *(*$\pi\mathbb{R}oom$*, a room category for* $\pi\mathbf{Ins}_{mor}$ *and* $\pi\mathbf{Ins}_{co}$*)*

Define a category $\pi\mathbb{R}oom$ as follows:

- Its objects are pairs $\langle S, C\rangle$, where S is a set and $C : 2^S \longrightarrow 2^S$ is a closure operator (we give $2^S \cong \mathscr{P}(S)$ the canonical ordering).
- A morphism $\langle S, C\rangle \xrightarrow{\sigma} \langle S', C'\rangle$ consists of a function $\sigma : S' \longrightarrow S$ such that $\sigma^* \circ C = C' \circ \sigma^*$, where $\sigma^* : 2^S \longrightarrow 2^{S'}$ is the function given by pulling back along σ (or by taking preimages).
- Composition is given by $\sigma' \circ_{\pi\mathbb{R}oom} \sigma = \sigma \circ_{Set} \sigma'$.

It is clear that $\pi\mathbb{R}oom$ is indeed a category. It is easily shown that

$$\pi\mathbf{Ins}_{mor} \cong \mathbf{ins}(\pi\mathbb{R}oom),$$

$$\pi\mathbf{Ins}_{co} \cong \mathbf{coins}(\pi\mathbb{R}oom).$$

Both projections $\mathbf{ins}(\pi\mathbb{R}oom) \longrightarrow \mathbb{C}at$ and $\mathbf{coins}(\pi\mathbb{R}oom) \longrightarrow \mathbb{C}at$ recover the underlying category of signatures of a π-institution.

Example 8. *(The terminal category, a room category for $\mathbb{C}at$)*

Let $1 = \{*\}$ denote the terminal category. It is immediate that both $\mathbf{ins}(1)$ and $\mathbf{coins}(1)$ are canonically isomorphic to $\mathbb{C}at$ via the projections provided by the Grothendieck construction.

Example 9. *(Institution-like structures versus diagrams)*

ins and **coins** are essentially the same, respectively, as the constructions of categories of diagrams $Diac_{mor}$ and $Diag_{co}$ given (in an ad hoc way) in Section 2. Indeed, for any category $\mathcal{C}$ there are canonical isomorphisms of categories

$$\mathbf{ins}(\mathcal{C}) \cong Diag_{mor}(\mathcal{C}^{op}),$$

$$\mathbf{coins}(\mathcal{C}) \cong Diag_{co}(\mathcal{C}^{op}),$$

both given on objects by sending a pair $(\mathcal{A}, F : \mathcal{A} \to \mathcal{C})$ to $(\mathcal{A}, F^{op} : \mathcal{A} \to \mathcal{C}^{op})$.

Moreover, for each $\mathcal{C}$ we have an isomorphism

$$Diag_{mor}(\mathcal{C}) \cong Diag_{co}(\mathcal{C}^{op})$$

also given by sending a pair $(\mathcal{A}, F : \mathcal{A} \to \mathcal{C})$ to $(\mathcal{A}, F^{op} : \mathcal{A} \to \mathcal{C}^{op})$. It then follows that for each $\mathcal{C}$ we have a sequence of isomorphisms

$$\mathbf{ins}(\mathcal{C}) \cong Diag_{mor}(\mathcal{C}^{op}) \cong Diag_{co}(\mathcal{C}) \cong \mathbf{coins}(\mathcal{C}^{op}).$$

An immediate corollary of this is:

- $\mathbb{R}oom^{op}$ (resp. $\pi\mathbb{R}oom^{op}$) is a category of rooms for $\mathbf{Ins}_{co}$ (resp. $\pi\mathbf{Ins}_{co}$).
- $\mathbb{R}oom^{op}$ (resp. $\pi\mathbb{R}oom^{op}$) is a category of co-rooms for $\mathbf{Ins}_{mor}$ (resp. $\pi\mathbf{Ins}_{mor}$).

Although the constructions of categories of diagrams and of institutional realizations are equally expressive, **ins** and **coins** fit better into the institutional framework, while $Diag_{mor}$ and $Diac_{co}$ would be more natural from a general categorical point of view.

3.3 Recovering adjunctions between categories of (π-)institutions

Lemma 4 ensures us that **ins** preserves adjunctions, and that **coins** reverses adjunctions. As a result, the adjunctions between categories of institution-like objects described in the previous sections can be given a simple and uniform treatment as images under **ins** or **coins** of certain adjunctions between the room categories attributed to them in the previous subsection.

Example 10. *($\mathbf{Ins}_{mor}$ and $\pi\mathbf{Ins}_{mor}$)*

Define functors $\mathscr{F} : \mathbb{R}oom \longrightarrow \pi\mathbb{R}oom$ and $\mathscr{G} : \pi\mathbb{R}oom \longrightarrow \mathbb{R}oom$ as follows:

- For each object $r = \langle S, M, (R_m)_{m\in Ob(M)}\rangle$ of $\mathbb{R}oom$, we define $\mathscr{F}(r)$ as $\langle S, C^r\rangle$, where $C^r : \mathscr{P}(S) \longrightarrow \mathscr{P}(S)$ is given by sending each $S' \subset S$ to

  $$\{s \in S \text{ such that } R_m(s) = 1 \text{ for every } m \in Ob(M) \text{ such that } R_m(S') = \{1\}\}.$$

  A morphism $\langle S, M, (R_m)_{m\in Ob(M)}\rangle \xrightarrow{(\sigma,\mu)} \langle S', M', (R'_{m'})_{m'\in Ob(M')}\rangle$ is sent to σ.

- For each object $r = \langle S, C\rangle$ of $\pi\mathbb{R}oom$, we define $\mathscr{G}(r)$ as

  $\langle S, \mathscr{P}(S), (\chi_m)_{m\in Ob(\mathscr{P}(S))}\rangle$, where $\mathscr{P}(S)$ is given the structure of a co-discrete category, and for each $m \subset S$, $\chi_m : S \longrightarrow 2$ is the characteristic function of m.

  A morphism $\langle S, C\rangle \xrightarrow{\sigma} \langle S', C'\rangle$ is sent to (σ, σ^*), where $\sigma^* : \mathscr{P}(S) \longrightarrow \mathscr{P}(S')$ is the functor between co-discrete categories given on objects by taking preimages.

One can then easily describe an adjunction $\mathscr{G} \dashv \mathscr{F}$ and show that $\mathscr{G}$ is fully faithful (hence it realizes $\pi\mathbb{R}oom$ as a coreflective subcategory of $\mathbb{R}oom$). It follows from Lemma 4, and from the fact that pseudofunctors preserve isomorphisms between 1-cells, that the functors

$$\mathbf{ins}(\mathscr{F}) : \mathbf{ins}(\mathbb{R}oom) \cong \mathbf{Ins}_{mor} \longrightarrow \mathbf{ins}(\pi\mathbb{R}oom) \cong \pi\mathbf{Ins}_{mor},$$
$$\mathbf{ins}(\mathscr{G}) : \mathbf{ins}(\pi\mathbb{R}oom) \cong \pi\mathbf{Ins}_{mor} \longrightarrow \mathbf{ins}(\mathbb{R}oom) \cong \mathbf{Ins}_{mor}$$

satisfy $\mathbf{ins}(\mathscr{G}) \dashv \mathbf{ins}(\mathscr{F})$, and that $\mathbf{ins}(\mathscr{G})$ realizes $\mathbf{ins}(\pi\mathbb{R}oom)$ (resp. $\pi\mathbf{Ins}_{mor}$) as a coreflective subcategory of $\mathbf{ins}(\mathbb{R}oom)$ (resp. $\mathbf{Ins}_{mor}$).

Example 11. *($\mathbf{Ins}_{co}$ and $\pi\mathbf{Ins}_{co}$)*

Let $\mathscr{F}$ and $\mathscr{G}$ be as in the previous example. The same argument shows that the functors

$$\mathbf{coins}(\mathscr{F}) : \mathbf{coins}(\mathbb{R}oom) \cong \mathbf{Ins}_{co} \longrightarrow \mathbf{coins}(\pi\mathbb{R}oom) \cong \pi\mathbf{Ins}_{co},$$
$$\mathbf{coins}(\mathscr{G}) : \mathbf{coins}(\pi\mathbb{R}oom) \cong \pi\mathbf{Ins}_{co} \longrightarrow \mathbf{coins}(\mathbb{R}oom) \cong \mathbf{Ins}_{co}$$

satisfy $\mathbf{coins}(\mathscr{F}) \dashv \mathbf{coins}(\mathscr{G})$, and that $\mathbf{coins}(\mathscr{G})$ realizes $\mathbf{coins}(\pi\mathbb{R}oom)$ (resp. $\pi\mathbf{Ins}_{co}$) as a reflective subcategory of $\mathbf{coins}(\mathbb{R}oom)$ (resp. $\mathbf{Ins}_{co}$).

Example 12. *(Categories of (π-)institutions and $\mathbb{C}at$)*

We leave to the reader the exercise of defining adjoints (left, right, or both) to the terminal functors $\mathbb{R}oom \to 1$ and $\pi\mathbb{R}oom \to 1$ using the methods described here, in order to produce several canonical adjunctions between $\mathbb{C}at$ and categories of (π-)institutions.

Example 13. *(Categories of (π-)institutions and categories of diagrams)*

Any adjunction of the form

$$\mathbb{R}oom \underset{R}{\overset{L}{\rightleftarrows}} \mathcal{C}^{op} \quad (L \dashv R)$$

induces two adjunctions: one between $\mathbf{Ins}_{mor} = \mathbf{ins}(\mathbb{R}oom)$ and $\mathbf{ins}(\mathcal{C}^{op}) \cong Diag_{mor}(\mathcal{C})$, and one between $\mathbf{Ins}_{co} = \mathbf{coins}(\mathbb{R}oom)$ and $\mathbf{coins}(\mathcal{C}^{op}) \cong Diag_{co}(\mathcal{C})$. Analogously, an adjunction of the form

$$\mathbb{R}oom \underset{R}{\overset{L}{\rightleftarrows}} \mathcal{C} \quad (L \dashv R)$$

induces an adjunction between $\mathbf{Ins}_{mor} = \mathbf{ins}(\mathbb{R}oom)$ and $\mathbf{ins}(\mathcal{C}) \cong Diag_{co}(\mathcal{C})$, and another one between $\mathbf{Ins}_{co} = \mathbf{coins}(\mathbb{R}oom)$ and $\mathbf{coins}(\mathcal{C}) \cong Diag_{mor}(\mathcal{C})$. Analogously for $\pi\mathbb{R}oom$ (or any category whatsoever) in place of $\mathbb{R}oom$, and for $R \dashv L$ in place of $L \dashv R$.

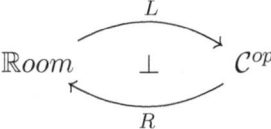

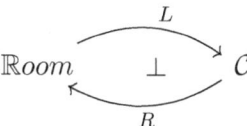

4 Propositional logics and (π-)institutions

In this section, we present several different ways of connecting abstract propositional logics to institutions and π-institutions.

In subsection 4.1 we have described the π-institutions associated to categories of abstract propositional logics and some forms of translation morphisms, as developed in [17]. This naturally lead us to search an analogous "model-theoretical" version of it that is different from the canonical one i.e., that obtained by applying the functor $G : \pi\mathbf{Ins}_{co} \to \mathbf{Ins}_{co}$ (see subsections 1.3 and 2.1). This is achieved in section 4.2, based on the development made in the section 3.1 of [19]: we provide (another) *institutions* for each category of propositional logics, through the use of the notion of a *matrix* for a propositional logic. It should be mentioned that the use of institutional-theoretic devices are useful for establishing an abstract Glivenko's theorem for algebraizable logics regardless of their particular signatures associated (see [19]).

In [2] was introduced the concept of (finitary) filter pair, that can be seem as a categorial presentation of a propositional logic, in fact the category of logics is isomorphic to a coreflective subcategory of the category of filter pairs. In the subsection 4.3 we present a functor $\mathcal{F}i \to \mathbf{Ins}_{mor}$, from the category of filter pairs, $\mathcal{F}_i$, to the category of all institutions and morphisms, $\mathbf{Ins}_{mor}$. This is qualitatively different connection from the obtained in subsections 4.1 and 4.2 between propositional logic and (π-)institution. From the adjunctions between the categories of logics and of filter pairs, $\mathcal{L} \leftrightarrows \mathcal{F}i$, and the adjunction between the categories of institutions and of π-institutions, $\pi - \mathbf{Ins}_{mor} \leftrightarrows \mathbf{Ins}_{mor}$, we obtain directly functors: $\mathcal{F}i \to \pi - \mathbf{Ins}_{mor}$, $\mathcal{L} \to \mathbf{Ins}_{mor}$, $\mathcal{L} \to \pi - \mathbf{Ins}_{mor}$. We finish this section with some remarks, indicating some generalizations concerning the use of multialgebras (a concept that will appear again in Section 5) in the setting of abstract propositional logic, including a natural generalization of the notion of filter pairs.

4.1 A π-institution for the abstract propositional logics

Here we describe the π-institutions associated to categories of abstract propositional logics and some forms of translation morphisms, as developed in [17].

In [1], [11] and [16] are considered some categories of propositional logics, namely $\mathcal{L}_s$ and $\mathcal{L}_f$, where:

- the objects are of the form $l = (\Sigma, \vdash)$, where $\Sigma = (\Sigma_n)_{n \in \mathbb{N}}$ is finitary signature, $Form(\Sigma) = Fm_\Sigma(X)$ is the absolutely free Σ-algebra of formulas on a fixed enumerable set of variables X and $\vdash \subseteq P(Form(\Sigma)) \times Form(\Sigma)$ is a tarskian consequence operator;

- the morphisms $f : (\Sigma, \vdash) \to (\Sigma', \vdash')$ are of the form $f : \Sigma \to \Sigma'$ with the former category having "strict" (n-ary symbol to n-ary symbol) morphisms and the latter "flexible" (n-ary symbol to n-ary term) morphisms.

To the category $\mathcal{L}_f$ is associated an π-institution J_f in the following way:

- $\mathbb{S}ig_f := \mathcal{L}_f$;
- $Sen_f : \mathbb{S}ig_f \to \mathbb{S}et$ is given by $(g : (\Sigma, \vdash) \to (\Sigma', \vdash)) \mapsto (\hat{g} : Form(\Sigma) \to Form(\Sigma'))$, where $\hat{g}$ is the usual expansion to formulas;
- For each $l = (\Sigma, \vdash) \in |\mathbb{S}ig_f|$ and $\Gamma \subseteq Form(\Sigma)$, we define $C_l(\Gamma) := \{\phi \in Form(\Sigma) : \Gamma \vdash_l \phi\}$.

An analogous process is used to form J_s from $\mathcal{L}_s$.

In [16], the "inclusion" functor $(+)_L : \mathcal{L}_s \to \mathcal{L}_f$ induces a comorphism (and also a morphism) on the associated π-institutions $(+) := ((+)_L, \alpha^+) : J_s \to J_f$, where, for each $l = (\Sigma, \vdash) \in \mathbb{S}ig_s = \mathcal{L}_s$, $\alpha^+(l) = Id_{Form(\Sigma)} : Form(\Sigma) \to Form(\Sigma)$. The paper also presents a right adjoint $(-)_L : \mathcal{L}_f \to \mathcal{L}_s$ to the "inclusion" functor. Essentially this fuctor sends a signature Σ to its derived one $(-)_L\Sigma := (Form(\Sigma)[n])_{n\in\mathcal{N}}$. We have also a comorphism of π-institutions associated to this functor. Notice that given some logic $l = (\Sigma, \vdash)$, we have $Sen_s(-)_L(l) = Form((-)_L\Sigma) = Form(\Sigma)$. So the fuctor $(-)_L$ induces a comorphism $((-)_L, \alpha^-)$ where α^- is the identity between formulas. It will be interesting understand the role of these adjoint pair of functors between the logical categories $(\mathcal{L}_f, \mathcal{L}_s)$ at the π-institutional level (J_f, J_s).

4.2 An institution for the abstract propositional logics

We now present an alternative institutionalization of propositional logic. This assignment is used in [19] to establish an abstract Glivenko's theorem for algebraizable logics.

Let $l = (\Sigma, \vdash)$ be a logic and $M \in \Sigma - Str$. A subset F of M is a l-filter is for every $\Gamma \cup \{\varphi\} \subseteq Form(\Sigma)$ such that $\Gamma \vdash \varphi$ and every valuation $v : Form(\Sigma) \to A$, if $v[\Gamma] \subseteq F$ then $v(\varphi) \in F$. The pair $\langle M, F\rangle$ is then said to be a matrix model of l. The class of all matrix model of l is denoted by $Matr_l$.This class is the class of objects of a category, also denoted by $Matr_l$: a morphism $h : \langle M, F\rangle \to \langle M', F'\rangle$ is a Σ-homomorphism $h : M \to M'$ such that $h^{-1}[F'] = F$; composition and identities are inherited from $\Sigma - Str$.

From to the category of logics $\mathcal{L}_f$ (also to $\mathcal{L}_s$), we define:

- $\mathbb{S}ig := \mathcal{L}_f$, the category of propositional logics $l = (\Sigma, \vdash)$ and flexible morphisms.
- $Sen : \mathbb{S}ig \to \mathbb{S}et$ where $Sen(l) = \mathcal{P}(Form(\Sigma)) \times Form(\Sigma)$ and given $f \in Mor_{\mathbb{S}ig}(l_1, l_2)$ then $Sen(f) : Sen(l_1) \to Sen(l_2)$ is such that $Sen(f)(\langle \Gamma, \varphi \rangle) = \langle f[\Gamma], f(\varphi) \rangle$. It is easy to see that Sen is a functor.
- $Mod : \mathbb{S}ig \to \mathbb{C}at^{op}$ where $Mod(l) = Matr_l$ and given $f \in Mor_{\mathbb{S}ig}(l_1, l_2)$, $Mod(f) : Matr_{l_2} \to Matr_{l_1}$ such that $Mod(f)(\langle M', F' \rangle) = \langle f^\star(M'), F' \rangle$. Here $f^\star : \Sigma' - str \to \Sigma - str$ is a functor that "commutes over Set" induced by the morphism f where the interpretation of connectives are: $c_n^{f^\star M'} := f(c_n)^{M'}$ for all $c_n \in \Sigma$ (more details in [19]).
- Given $l = (\Sigma, \vdash) \in |\mathbb{S}ig|$, $\langle M, F \rangle \in |Mod(l)|$ and $\langle \Gamma, \varphi \rangle \in Sen(l)$ define the relation $\models_l \subseteq |Mod(l)| \times Sen(l)$ as:

$$\langle M, F \rangle \models_l \langle \Gamma, \varphi \rangle \ iff \ for \ all \ v{:}Form(\Sigma) \to M, \ if \ v[\Gamma] \subseteq F, \ then \ v(\varphi) \in F.$$

In [19], section 3.1, it is proven that this construction defines indeed an institution.

It should be noted that this institution and the π-institution described in the previous subsection, shares the same $\mathbb{S}ig$ $(= \mathcal{L}_f)$, but *are not* connected by the canonical relation (adjunction) between institutions and π-institutions.

4.3 Filter pairs as institutions

The notion of (finitary) filter pair, introduced in [2], can be seem as a categorical presentation of a propositional logic. Here we recall the precise definition of this notion and associate an institution to the category of all filter pairs.

Definition 4.1. *Let Σ be a signature. A* **Filter Pair** *over Σ is a pair (F, i), consisting of a contravariant functor $F : \Sigma - str^{op} \to$* **CLat**, *from Σ-structures to complete lattices, and a collection of maps $i = (i_M : F(M) \to (\mathcal{P}(M), \subseteq))_{M \in \Sigma - str}$ such that is a natural transformation.*

$$\begin{array}{ccccc} M & & F(M) & \xrightarrow{i_M^F} & (\mathcal{P}(M);\subseteq) \\ {\scriptstyle f}\downarrow & & {\scriptstyle F(f)}\uparrow & & \uparrow{\scriptstyle f^{-1}} \\ N & & F(N) & \xrightarrow[i_N^F]{} & (\mathcal{P}(N);\subseteq) \end{array}$$

Remark 4.2. *Let (F, i) be a filter pair and X be a set. The relation $\vdash \subseteq \mathcal{P}(Fm_\Sigma(X)) \times Fm_\Sigma(X)$ such that for any $\Gamma \cup \{\varphi\} \subseteq Fm_\Sigma(X)$, $\Gamma \vdash \varphi$ iff for any $a \in F(Fm_\Sigma(X))$ if $\Gamma \subseteq i_{Fm_\Sigma(X)}(a)$ then $\varphi \in i_{Fm_\Sigma(X)}(a)$ is a tarskian consequence relation. Then we have a propositional logic associated with the filter pair (F, i) such that the set of variables is X.*

Below is the definition of a finitary filter pair so that its associated propositional logic is finitary.

Definition 4.3. *Let Σ be a signature. A* **finitary filter pair** *over Σ is a filter pair (F, i) which F is a functor from Σ-structures to algebraic lattices such that for any $M \in \Sigma-str$, i_M preserves arbitrary infima (in particular $i_M(\top) = M$) and directed suprema.*

Definition 4.4. The category of Filter Pairs: *Consider the category $\mathcal{F}i$ defined in the following manner:*

- ***Objects:*** *Filters pairs (F, i^F).*
- ***Morphisms:*** *Let (F, i^F) be a filter pair over a signature Σ and $(F', i^{F'})$ be a filter pair over a signature Σ'. A morphism $(F, i^F) \to (F', i^{F'})$ is a pair (H, j) such that $H : \Sigma'-str \to \Sigma-str$ is a* signature functor *and $j : F' \Rightarrow F \circ H$ is a natural transformation such that given $M' \in Obj(\Sigma'-str)$,*

$$i^F_{H(M')} \circ j_{M'} = i^{F'}_{M'}.$$

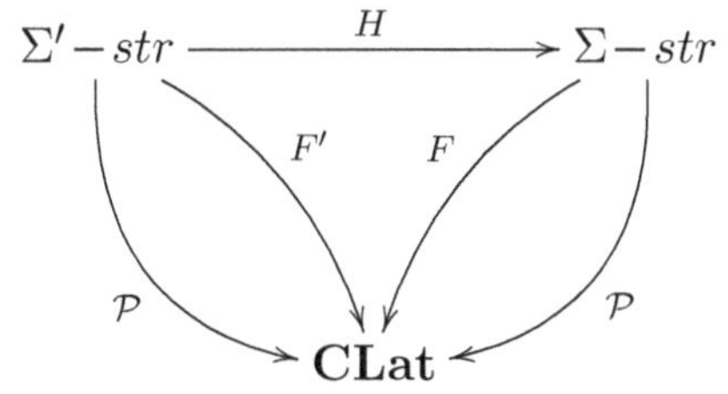

- **Identities:** *For each signature Σ and each filter pair (F, i^F) over Σ, $Id_{(F,i^F)} := (Id_{\Sigma-str}, Id_F)$.*
- **Composition:** *Given morphisms $(H, j), (H', j')$ in $\mathcal{F}i$.*

$$(H', j') \bullet (H, j) = (H \circ H', j \bullet j')$$

Where $(j \bullet j')_{M''} := j_{H'(M'')} \circ j'_{M''}$.

Observe that

$$i^F_{H\circ H'(M'')} \circ ((j \bullet j')_{M''}) = i^{F''}_{M''}$$

Indeed:

$$\begin{aligned} i^F_{H\circ H'(M'')} \circ ((j \bullet j')_{M''}) &= i^F_{H\circ H'(M'')} \circ (j_{H'(M'')} \circ j'_{M''}) \\ &= (i^F_{H\circ H'(M'')} \circ j_{H'(M'')}) \circ j'_{M''} \\ &= i^{F'}_{H'(M'')} \circ j'_{M''} \\ &= i^{F''}_{M''} \end{aligned}$$

It is straightforward to check that the composition is associative and that identity laws hold.

In [2] was defined a category of finitary filter pairs and presented it as functorial encoding of the category of all (finitary, propositional) logics: in fact the category of propositional logics and flexible morphisms can be represented as a coreflective full subcategory of the category of filter pairs.

Fact 4.5.

- *For any signature functor $H : \Sigma' - Str \to \Sigma - Str$, there is a signature morphism $m_H : \Sigma \to \Sigma'$, such that $m_H(c_n) = \eta_H(X)(c_n(x_0, ..., x_{n-1}))$, where $\eta_H(X) : Form_\Sigma(X) \to H(Form_{\Sigma'}(X))$ (see Lemma 3.17 of [2]). We consider the functor*

$$\begin{array}{llcl} \mathbb{L}: & \mathcal{F}i & \to & \mathcal{L}_f \\ & (G, i^G) & & l_G \\ & \downarrow (H, j) & \mapsto & \downarrow m_H \\ & (G', i^{G'}) & & l_{G'} \end{array}$$

- *The functor $\mathbb{F} : \mathcal{L} \to \mathcal{F}i$*

$$\begin{array}{llcl} \mathbb{F}: & \mathcal{L}_f & \to & \mathcal{F}i \\ & l & & (Fi_l, \iota) \\ & h\downarrow & \mapsto & \mathbb{F}(h)\downarrow \\ & l' & & (Fi_{l'}, \iota') \end{array}$$

where $\mathbb{F}(h) = (h^\star, j^\star)$ and the natural transformation $j^\star : Fi_{l'} \Rightarrow Fi_l \circ h^\star$ is given by a family of inclusions, i.e., let $M' \in \Sigma' - str$ and $F' \in Fi_{l'}(M')$, then $j^\star_{M'}(F') := F'$.

- *The functor $\mathbb{F} : \mathcal{L} \to \mathcal{F}i$ is full, faithful, injective on the objects and is left adjoint to the functor $\mathbb{L}$. By a well known result of category theory, the unity of this adjunction is an isomorphism. Moreover it is easy to see that the components of the natural transformation that is the unity of this adjunction is given, for each logic $l \in Obj(\mathcal{L}_f)$, by the identity $id_l : l \to \mathbb{L} \circ \mathbb{F}(l) = l$.*

  The components of the counit of this adjunction is given by, for each signature Σ and each filter pair (G, i^G) over Σ:

$$(Id_{\Sigma-Str}, j^G) : (Fi_{l_G}, \iota) \to (G, i^G)$$

  where $j^G_M : G(M) \to \mathcal{F}i_{l_G}(M)$ is the unique factorization of $i^G_M : G(M) \to \wp(M)$ through $\iota_M : \mathcal{F}i_{l_G}(M) \hookrightarrow P(M)$. Thus for each logic l', j^G induces by composition a (natural) bijection:

$$\mathcal{F}i(\mathbb{F}(l'), (G, i^G)) \;\cong\; \mathcal{L}_f(l', \mathbb{L}(G, i^G)).$$

- *The same constructions of the above functors provide a more general adjunction relating the category of filter pairs and propositional logics which are non-finitary.*

Proposition 4.6. *Every filter pair (F, i) over a signature Σ determines an institution $I_{(F,i)}$ where:*

- $Sig_I = \Sigma - str$;
- $(Sig_I \xrightarrow{Sen_I} \mathbb{S}et) = (\Sigma - str \xrightarrow{forgetful} \mathbb{S}et)$;
- $(Sig_I^{op} \xrightarrow{Mod_I} \mathbf{CAT}) = (\Sigma - str^{op} \xrightarrow{F} \mathbf{CLat} \rightarrowtail \mathbf{CAT})$;
- *for each $M \in Ob(Sig_I) = Ob(\Sigma - str)$, define*
  $\models_M \subseteq Ob(Mod_I(M)) \times Sen_I(M) = F(M) \times |M|$ *as:*

$$t \models_M m \quad iff \quad m \in i_M(t)$$

Moreover, when i_M preserves arbitrary infima, the π-institution $P_{(F,i)}$ cannonically associated to $I_{(F,i)}$ is such that for each $M \in Ob(Sig_I) = Ob(\Sigma - str)$, $\mathcal{C}_M : P(Sen_I) \to P(Sen_I)$ is given by

$$(X \subseteq |M|) \;\mapsto\; i_M(t_X),$$

where $t_X := \bigwedge\{t \in F(M) : X \subseteq i_M(t)\}$

Proof: Sig_I, Sen_I and Mod_I associated with a filter pair (F, i) are well defined. It remains to prove the compatibility condition. Let $h : M \to M'$ be a morphism in $Sig_I = \Sigma - str$ and $a \in F(M')$ such that $a \models_{M'} h(m)$. So $h(m) \in i_{M'}(a)$ and since i is a natural transformation we have $m \in h^{-1} \circ i_{M'}(a) = i_M \circ F(h)(a)$. Then $F(h)(a) \models_M m$.

The associated π-instituion takes $X \subseteq P(U(M))$ into $i_M(T_X) = i_M(\bigwedge\{T \in F(M) : X \subseteq i_M(T)\} = \bigcap\{i_M(T) : X \subseteq i_M(T)\}$

□

Proposition 4.7. (Every morphism of filter pair induces a institution morphism.) *Given morphism* $(F, i) \xrightarrow{(H,j)} (F', i')$ *then* $I_{(F,i)} \xleftarrow{(H,Id,j)} I_{(F',i')}$ *is a institution morphism.*

Proof: We just need to prove that (H, Id, j) satisifies the compatibility condition. Let $M' \in \Sigma' - str$, $m' \in F'(M')$ and $\varphi \in H(M')$.

$$\begin{aligned} m' \models_{M'} Id_{M'}\varphi &\iff \varphi \in i'_{M'}(m') \\ &\iff \varphi \in i_{H(M')} \circ j_{M'}(m') \\ &\iff j_{M'}(m') \models_{H(M')} \varphi \end{aligned}$$

The result follows

□

Using propositions 4.6 and 4.7 we can now define the (contravariant) functor:

$$\mathcal{F}i \xrightarrow{D} \mathbf{Ins}_{mor}$$

$$\begin{array}{ccc} (F, i) & \longmapsto & I_{(F,i)} \\ {\scriptstyle (H,j)}\downarrow & \longmapsto & \uparrow{\scriptstyle (H,Id,j)} \\ (F', i') & \longmapsto & I_{(F',i')} \end{array}$$

Verifying functoriality is straightforward.

Remark 4.8.
- *From the adjunction* $\mathbf{Ins}_{mor} \rightleftarrows \pi - \mathbf{Ins}_{mor}$ *described in section 2.1, we obtain directly a functor* $\mathcal{F}i \to \pi - \mathbf{Ins}_{mor}$*.*
- *From the adjuction* $\mathcal{L}_f \rightleftarrows \mathcal{F}i$*, recalled in Fact 4.5, we obtain functors* $\mathcal{L}_f \to \mathbf{Ins}_{mor}$ *and* $\mathcal{L}_f \to \pi - \mathbf{Ins}_{mor}$*.*

4.4 Generalizations

In this final subsection we provide a kind of generalization of the previous subsections: we explore the extension of the category of propositional logics by the category of filter pairs to "extend" the (π-)institution of logics to a (π-)institution of filter pairs; we extend the concept of filter pairs allowing multialgebras as the domain of a filter pair and thus we extend the functor from filter pairs to the category of institutions to a funtor from the category of multifilter pairs to institutions.

Remark 4.9. *The institution (respec. π-institution) associated to the abstract propositional logics as described in subsection 4.2 (respec. 4.1) can be "extended", through the adjunction $(\mathbb{F}, \mathbb{L}) : \mathcal{L}_f \rightleftarrows \mathcal{F}i$ (see Fact 4.5) to a institution (respec. π-institution) for the filter pairs (apart from size issues):*

- * $\mathbb{S}ig' = \mathcal{F}i$;

  * $Sen' : \mathbb{S}ig \to \mathbb{S}et$ *is given by* $((H, j) : (G, i^G) \to (G', i^{G'})) \mapsto (\eta_H(X) : Fm_\Sigma(X) \to H(Fm_{\Sigma'}(X)))$, *where* $G : \Sigma - Str^{op} \to \mathbf{CLat}$ *and* $G' : \Sigma' - Str^{op} \to \mathbf{CLat}$;

  * *For each* $(G, i^G) \in |\mathbb{S}ig|$ *and* $\Gamma \subseteq Fm_\Sigma(X)$, *we define* $C'_{(G,i^G)}(\Gamma) := \{\phi \in Fm_\Sigma(X) : \Gamma \vdash_{\mathbb{L}(G,i^G)} \phi\}$.

  Denoting $(\mathbb{S}ig, Sen, (C_\bullet))$ *the π-institution of propositional logics (subsection 4.1), note that:*

  * $Sen' \circ \mathbb{F} = Sen$.

  * *For each* $(\Sigma, \vdash) \in |\mathcal{L}_f|$, $C'_{\mathbb{F}(\Sigma,\vdash)} = C_{(\Sigma,\vdash)}$.

  Thus $(\mathbb{F}, id_{Sen})$ *is, simultaneously, a morphism and a comorphism of π-institutions* $(\mathbb{S}ig, Sen, (C_\bullet)) \to (\mathbb{S}ig', Sen', (C'_\bullet))$.

- * $\mathbb{S}ig' = \mathcal{F}i$;

  * $Sen' : \mathbb{S}ig' \to \mathbb{S}et$ *where* $Sen'(G, i^G) = \mathcal{P}(Fm_\Sigma(X)) \times Fm_\Sigma(X)$ *and given* $(H, j) \in Mor_{\mathbb{S}ig'}((G, i^G), (G', i^{G'}))$ *then* $Sen(H, j) : Sen(G, i^G) \to Sen(G', i^{G'})$ *is such that* $Sen(H, j)(\langle\Gamma, \varphi\rangle) = \langle\eta_H(X)[\Gamma], \eta_H(X)(\varphi)\rangle$.

  * $Mod' : \mathbb{S}ig' \to \mathbb{C}at^{op}$ *where* $Mod'(G, i^G) = Matr_{\mathbb{L}(G,i^G)}$ *and given* $(H, j) \in Mor_{\mathbb{S}ig'}((G, i^G), (G', i^{G'}))$, $Mod'(H, j) : Matr_{\mathbb{L}(G',i^{G'})} \to Matr_{\mathbb{L}(G,i^G)}$ *such that* $Mod'(H, j)(\langle M', F'\rangle) = \langle H(M'), F'\rangle$.

  * *Given* $(G, i^G)) \in |\mathbb{S}ig'|$, $\langle M, F\rangle \in |Mod'(G, i^G)|$ *and* $\langle\Gamma, \varphi\rangle \in Sen'(G, i^G)$ *define the relation* $\models'_{(G,i^G)} \subseteq |Mod'(G, i^G)| \times Sen'(G, i^G)$ *as:*

$$\langle M, F\rangle \models'_{(G,i^G)} \langle \Gamma, \varphi\rangle \; iff \; for \; all \; Fm_\Sigma(X) \xrightarrow{v} M, \; v(\varphi) \in F \; for \; v[\Gamma] \subseteq F$$

Denoting $(\mathbb{S}ig, Sen, Mod, (\models_\bullet))$ *the institution of propositional logics (subsection 4.2), note that:*

* $Sen' \circ \mathbb{F} = Sen$.

* $Mod' \circ \mathbb{F} = Mod$

* *For each* $l = (\Sigma, \vdash) \in |\mathcal{L}_f|$, *each* $\langle \Gamma, \varphi\rangle \in Sen(l)$ *and each* $\langle M, F\rangle \in |Mod(l)|$

$$\langle M, F\rangle \models'_{\mathbb{F}(l)} \langle \Gamma, \varphi\rangle \; iff \; \langle M, F\rangle \models_l \langle \Gamma, \varphi\rangle.$$

Thus $(\mathbb{F}, id_{Sen}, id_{Mod})$ *is, simultaneously, a morphism and a comorphism of institutions* $(\mathbb{S}ig, Sen, Mod, (\models_\bullet)) \to (\mathbb{S}ig', Sen', Mod', (\models'_\bullet))$.

The institution obtained above can be extended to the case of multialgebras and that this also extends the institution for N-matrix semantics to propositional logic ([4]) allowing us to use the institution theory in order to analyze logical properties of non-algebraizable logics. Moreover, another work in progress, we are trying, using filter pairs, to establish a multialgebraic semantics for propositional logics that are not algebraizable, for example Logic of Formal Inconsistency (LFI's) ([7]), and possibly to obtain a kind of transfer theorem between metalogical and multialgebraic properties.

Remark 4.10 (Multialgebras).

- *A n-ary multioperation on a set A is a function* $F : A^n \to \mathcal{P}^*(A)$, *where* $\mathcal{P}^*(A) = A \backslash \{\emptyset\}$. *To each ordinary n-ary multioperation on A,* $f : A^n \to A$ *is associated a (strict) n-ary operation on* $A : F : A^n \to \mathcal{P}^*(A)$ *given by* $F := s_A \circ f$, *where* $s_A : A \to \mathcal{P}^*(A), x \mapsto s_A(x) = \{x\}$.

- *A multialgebraic signature is a sequence of pairwise disjoint sets* $\Sigma = (\Sigma_n)_{n\in\mathbb{N}}$, *where* $\Sigma_n = S_n \sqcup M_n$, *where* S_n *is the set of strict multioperation symbols and* M_n *is the set of multioperation symbols. In particular,* $\Sigma_0 = S_0 \sqcup M_0$, F_0 *is the set of symbols for constants and* M_0 *is the set of symbols for multiconstants. We also denote* $\Sigma = ((S_n)_{n\geqslant 0}, (M_n)_{n\geqslant 0})$.

- *A multialgebra over a signature* $\Sigma = ((S_n)_{n\geqslant 0}, (M_n)_{n\geqslant 0})$, *is a set A endowed with a family of n-ary multioperations*

$$\sigma_n^A : A^n \to \mathcal{P}^*(A), \; \sigma_n \in S_n \sqcup M_n, \; n \in \mathbb{N},$$

such that: if $\sigma_n \in S_n$, *then* $\sigma_n^A : A^n \to \mathcal{P}^*(A)$ *is a* strict *n-ary multioperation.*

- *If A and B are Σ-multialgebras, then a Σ-morphism from A to B is a function $h : A \to B$ such that for each $n \in \mathbb{N}$, each $\sigma_n \in S_n \sqcup M_n$ and each $a_0, \cdots, a_{n-1} \in A$*
$$h[\sigma^A(a_0, \cdots, a_{n-1})] \subseteq \sigma^B(h(a_0), \cdots, h(a_{n-1})).$$

- *Σ-morphisms between Σ-multialgebras can be composed in a natural way and they form a category Σ-Malg. It is clear that Σ-alg, the category of ordinary Σ-algebras is isomorphic to the a full subcategory of strict Σ-multialgebras. $s : \Sigma - Alg \hookrightarrow \Sigma - Malg$.*

- *Every algebraic signature $\Sigma = (F_n)_{n\in\mathbb{N}}$ is a multialgebraic signature where $M_n = \emptyset, \forall n \in \mathbb{N}$. Each algebra $(A, ((A^n \stackrel{f^A}{\to} A)_{f\in F_n})_{n\in\mathbb{N}})$ over the algebraic signature Σ can be naturally identified with a multialgebra $(A, ((A^n \stackrel{f^A}{\to} A \stackrel{s_A}{\rightarrowtail} \mathcal{P}^*(A))_{f\in F_n})_{n\in\mathbb{N}})$ over the same signature.*

- *Every multialgebraic signature $\Sigma = ((S_n)_{n\in\mathbb{N}}, (M_n)_{n\in\mathbb{N}})$ induces naturally a first-order language $L(\Sigma) = ((F_n)_{n\in\mathbb{N}}, (R_{n+1})_{n\in\mathbb{N}})$ where $F_n := S_n$ is the set of n-ary operation symbols and $R_{n+1} := M_n$ is the set of (n+1)-ary relation symbols. In this way, multialgebras $(A, ((A^n \stackrel{\sigma^A}{\to} \mathcal{P}^*(A))_{\sigma\in S_n\sqcup M_n})_{n\in\mathbb{N}})$ over a multialgebraic signature $\Sigma = (S_n \sqcup M_n)_{n\in\mathbb{N}}$ can be naturally identified with the first-order structures over the language $L(\Sigma)$ that satisfies the $L(\Sigma)$-sentences: $\forall x_0 \cdots \forall x_{n-1} \exists x_n(\sigma_n(x_0, \cdots, x_{n-1}, x_n))$, for each $\sigma_n \in R_{n+1} = M_n, n \in \mathbb{N}$.* [5]

- *Now we focus our attention into a more syntactic aspect of this multialgebras theory. We start with a (recursive) definition of (multi)terms: variables $x_i, i \in \mathbb{N}$ are terms; if $t_0, \cdots, t_{n-1}$ are terms and $\sigma \in S_n \sqcup M_n$, then $\sigma(t_0, \cdots, t_{n-1})$ is a term.*

- *To define an interpretation for terms, we need a preliminary step. Given $\sigma \in S_n \sqcup M_n$, we "extend" $\sigma^A : A^n \to \mathcal{P}^*(A)$ to a n-ary operation in $\mathcal{P}^*(A)$, $\sigma^{\mathcal{P}^*(A)} : \mathcal{P}^*(A)^n \to \mathcal{P}^*(A)$, by the rule:*
$$\sigma^{\mathcal{P}^*(A)}(A_0, \cdots, A_{n-1}) := \bigcup_{a_0\in A_0} \cdots \bigcup_{a_{n-1}\in A_{n-1}} \sigma^A(a_0, \cdots, a_{n-1}).$$
In this way, $\mathcal{P}^*(A)$ is an ordinary Σ-algebra. Moreover
$$\sigma^{\mathcal{P}^*(A)}(\{a_0\}, \cdots, \{a_{n-1}\}) = \sigma^A(a_0, \cdots, a_{n-1}).$$

[5]We will address this correspondence in Example 5.2.

- *The association above determines a functor* $p : \Sigma - Malg \to \Sigma - alg$ *and, the family of singleton maps* $s_A : A \to (s \circ p)(A)$, $A \in |\Sigma - Malg|$, *is a natural transformation.*

Remark 4.11 (Multifilter pairs and institutions).

- *It is straightfoward to extend the notion of filter pair* (G, i^G), *where the domain of the functor* G *is the category* $\Sigma - alg$ *to the concept of* multifilter pair, *where the domain of the functor* G *is the category* $\Sigma - Malg$. *With a natural notion of morphism of mult-filter pair we obtain a category* $m\mathcal{F}i$ *of multifilter pairs.*

- *The previously described functors* $s : \Sigma - alg \to \Sigma - Malg$ *and* $p : \Sigma - Malg \to \Sigma - alg$ *provide a pair of functors* $\mathcal{F}i \rightleftarrows m\mathcal{F}i$.

- *The functor* $\mathcal{F}i \to \mathbf{Ins}_{mor}$ *can be extended to a funtor* $m\mathcal{F}i \to \mathbf{Ins}_{mor}$.

We summarize below some of the functors previously presented.

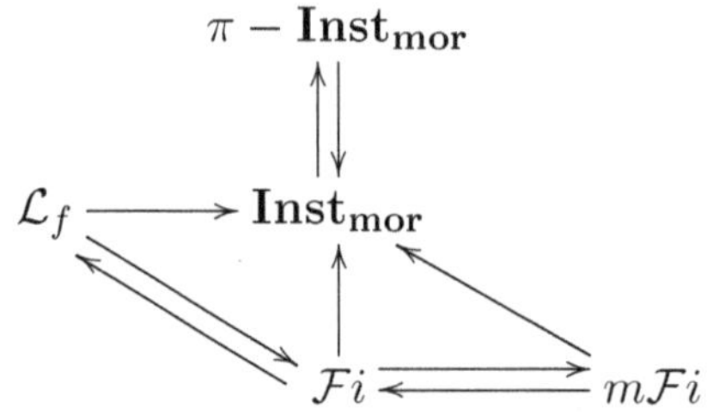

5 Skolemization, a new institutional device

Skolemization is an important tool of classical model theory, this section seeks to develop it in the context of institutions. We also prove a borrowing theorem and apply it to obtain a form of downward Löwenheim-Skolem for the setting of multialgebras.

Given an institution I, we say that $\langle I, S, (\mathcal{I}_\Sigma)_{\Sigma \in |\mathbb{S}ig|}, (\tau_\Sigma)_{\Sigma \in |\mathbb{S}ig|} \rangle$ is an skolemization for I iff:

- S is a functor of the form

$$(Mod)^\sharp \xrightarrow{\ S\ } (Mod^{\mathbb{P}res})^\sharp$$

$$\begin{array}{ccc} \langle \Sigma, M\rangle & \longmapsto & \langle (\Sigma_S, S_\Sigma), M_{S\Sigma}\rangle \\ {\scriptstyle\langle f,u\rangle}\downarrow & \longmapsto & \downarrow{\scriptstyle\langle g,v\rangle} \\ \langle \Sigma', N\rangle & \longmapsto & \langle (\Sigma'_S, S_{\Sigma'}), N_{S\Sigma'}\rangle \end{array}$$

  Where $^\sharp$ denotes the Grothendieck construction. We refer to S as the skolem functor.

- For each $\Sigma \in |\mathbb{S}ig|$, $\Sigma \xrightarrow{\tau_\Sigma} \Sigma_S$ is an arrow in $\mathbb{S}ig$ satisfying $M_{S\Sigma} \restriction_{\tau_\Sigma} = M$ for all $M \in |Mod(\Sigma)|$. Given $M \in Mod(\Sigma)$ we say that $M' \in Mod(\Sigma_S)$ is a skolemization of M if $M' \restriction_{\tau_\Sigma} = M$ and $M' \models_{\Sigma_S} S_\Sigma$

- For each signature Σ, $\mathcal{I}_\Sigma$ is an inclusion system in $Mod(\Sigma_S)$ such that, if the Σ_S-models M' and N' are skolemizations of M and N respectively and $M' \hookrightarrow N'$ then $M^\star = N^\star$. [6]

Example 5.1. FOL[1]
Let **FOL**[1] *stand for the institution of unsorted first order logic and consider the functor:*

$$(Mod)^\sharp \xrightarrow{\ Skolem\ } (Mod^{\mathbb{P}res})^\sharp$$

$$\begin{array}{ccc} \langle \Sigma, M\rangle & \longmapsto & \langle (\Sigma_S, S_\Sigma), M_{S\Sigma}\rangle \\ {\scriptstyle\langle f,u\rangle}\downarrow & \longmapsto & \downarrow{\scriptstyle\langle f',u\rangle} \\ \langle \Sigma', N\rangle & \longmapsto & \langle (\Sigma'_S, S_{\Sigma'}), N_{S\Sigma'}\rangle \end{array}$$

Where Σ_S and S_Σ are, respectively, the skolem expansion and theory of Σ and $M_{S\Sigma}$ is any skolemization of M with the same underlying set. Let F^Σ_ψ be the skolem function of the Σ-formula ψ and define f' as follows: if $x \in \Sigma$ simply let $f'(x) = f(x)$, else we have $x = F^\Sigma_\psi$ for some ψ in $Sen(\Sigma)$ and then we let $f'(x) = F^{\Sigma'}_{Sen\, f(\psi)}$.

For each first order signature Σ, let $\mathcal{I}_\Sigma$ be the usual inclusion system on $Mod^{\mathbf{FOL^1}}(\Sigma)$ and define $\tau_\Sigma : \Sigma \to \Sigma_S$ as $\tau_\Sigma(x) = x$. It is easy to see that

$$\langle \mathbf{FOL^1}, Skolem, (\mathcal{I}_\Sigma)_{\Sigma \in |\mathbb{S}ig^{\mathbf{FOL^1}}|}, (\tau_\Sigma)_{\Sigma \in |\mathbb{S}ig^{\mathbf{FOL^1}}|}\rangle$$

[6]Given $M \in |Mod(\Sigma)|$, define $M^\star := \{\varphi \in Sen(\Sigma) : M \models_\Sigma \varphi\}$

is a skolemization for $\mathbf{FOL}^1$.

Theorem 14. *Let* I *institution with skolemization* $\langle I, S, (\mathcal{I}_\Sigma)_{\Sigma\in|\mathbb{S}ig^I|}, (\tau_\Sigma)_{\Sigma\in|\mathbb{S}ig^I|}\rangle$. *Given an institution* J *and a morphism* $\langle \phi, \alpha, \beta\rangle : J \to I$ *if:*

- ϕ *is fully faithful,*
- *For each* $\Sigma_i \in |\mathbb{S}ig^I|$ *there is some* $\Sigma_j \in |\mathbb{S}ig^J|$ *such that* $\phi(\Sigma_j) \cong (\phi\Sigma_i)_S$ *in* $\mathbb{S}ig^I$. *Let* $i_{\Sigma_i} : (\Sigma_j) \to (\Sigma_i)_S$ *denote the isomorphism arrow,*
- *Each* β_Σ *is an isomorphism, and*
- *Each* α_Σ *is semantically surjective, that is, for every* $\varphi \in Sen^J(\Sigma)$ *there is some* $\psi \in \alpha_\Sigma[Sen^I(\phi\Sigma)]$ *such that* $\varphi^\star = \psi^\star$.

Then $\langle J, S', (\mathcal{I}'_\Sigma)_{\Sigma\in|\mathbb{S}ig^J|}, (\tau'_\Sigma)_{\Sigma\in|\mathbb{S}ig^J|}\rangle$ *has a skolemization where*

- *If* $\mathcal{I}_{\phi\Sigma} = \langle I, E\rangle$ *then* $\mathcal{I}'_\Sigma = \langle I', E'\rangle$ *where* I' *and* E' *are the images of* $\beta_\Sigma^{-1} Mod^I i_{\phi\Sigma}$ *restricted to* I *and* E *respectively,*
- *For each* Σ, τ'_Σ *is the unique arrow satisfying* $\phi(\tau'_\Sigma) = i_{\phi\Sigma}^{-1} \cdot \tau_\phi\Sigma$.

Proof: Consider the application

$$m : (Mod^J)^\sharp \longrightarrow (Mod^I\phi)^\sharp$$

$$\begin{array}{ccc} \langle \Sigma, M\rangle & \longmapsto & \langle \phi(\Sigma), \beta_\Sigma(M)\rangle \\ \downarrow{\scriptstyle\langle f,u\rangle} & \longmapsto & \downarrow{\scriptstyle\langle \phi(f), \beta_\Sigma(u)\rangle} \\ \langle \Sigma', N\rangle & \longmapsto & \langle \phi(\Sigma'), \beta_{\Sigma'}(N)\rangle \end{array}$$

Let us prove that m is a functor. Given arrows $\langle \Sigma, M\rangle \xrightarrow{\langle f,u\rangle} \langle \Sigma', N\rangle \xrightarrow{\langle g,v\rangle} \langle \Sigma'', W\rangle$ in $(Mod^J)^\sharp$ we have:

$$\begin{aligned} m(\langle g, v\rangle \cdot \langle f, u\rangle) &= m(\langle gf, Mod^J fv \cdot u\rangle) \\ &= \langle \phi(gf), \beta_\Sigma(Mod^J fv \cdot u)\rangle \\ &= \langle \phi(g) \cdot \phi(f), (\beta_\Sigma Mod^J f)(v) \cdot \beta_\Sigma(u)\rangle \\ &= \langle \phi(g) \cdot \phi(f), (Mod^I\phi(f)\beta_{\Sigma'})(v) \cdot \beta_\Sigma(u)\rangle \\ m(\langle g, v\rangle) \cdot m(\langle f, u\rangle) &= \langle \phi(g), \beta_{\Sigma'}(v)\rangle \cdot \langle \phi(f), \beta_\Sigma(u)\rangle \end{aligned}$$

As m clearly satisfies the identity laws we have that m is well defined.

Consider now the functors $(Mod^I\phi)^\sharp \xrightarrow{\mathcal{J}} (Mod^I)^\sharp \xrightarrow{S} (Mod^{\mathbb{P}res^I})^\sharp$. Composing:

$$
\begin{array}{ccc}
(Mod^J)^\sharp & \xrightarrow{\quad S\mathcal{J}m \quad} & (Mod^{\mathbb{P}res^I})^\sharp \\
 & & \\
\langle \Sigma, M \rangle & \longmapsto & \langle ((\phi\Sigma)_S, S_{\phi\Sigma}), (\beta_\Sigma(M))_{S\phi\Sigma} \rangle \\
\big\downarrow \langle f,u \rangle & \longmapsto & \big\downarrow \langle \psi, v \rangle \\
\langle \Sigma', N \rangle & \longmapsto & \langle ((\phi\Sigma')_S, S_{\phi\Sigma'}), (\beta_{\Sigma'}(N))_{S_{\phi\Sigma}} \rangle
\end{array}
$$

We now have what we need to define a functor $S' : (Mod^J)^\sharp \to (Mod^{\mathbb{P}res^J})^\sharp$. Given $\langle \Sigma, M \rangle \in |(Mod^J)^\sharp|$, let $S'(\langle \Sigma, M \rangle) := \langle (\breve{\Sigma}, S_{\breve{\Sigma}}), M_{\breve{\Sigma}} \rangle$ where:

- $\breve{\Sigma}$ is an object in $\mathbb{S}ig^J$ such that there is an isomorphism $i_{\phi\Sigma} : \phi(\breve{\Sigma}) \xrightarrow{\sim} (\phi\Sigma)_S$ in $\mathbb{S}ig^I$
- $S_{\breve{\Sigma}} := \alpha_{\breve{\Sigma}}(Sen^I i_{\phi\Sigma}^{-1}(S_{\phi\Sigma}))$
- $\breve{M} := \beta_{\breve{\Sigma}}^{-1} Mod^I i_{\phi\Sigma}((\beta_\Sigma)_{S\phi\Sigma})$

And, given an arrow $\langle f, u \rangle$ in $(Mod^J)^\sharp$, let $S'(\langle f, u \rangle) := \langle \breve{\psi}, \breve{v} \rangle$, where:

- $\phi(\breve{\psi})$ is the lone arrow that makes the below square commute

$$
\begin{array}{ccc}
(\phi\Sigma)_S & \xrightarrow{\ \psi\ } & (\phi\Sigma')_S \\
\cong \big\downarrow & & \big\downarrow \cong \\
\phi(\breve{\Sigma}) & \underset{\phi(\breve{\psi})}{\dashrightarrow} & \phi(\breve{\Sigma}')
\end{array}
$$

- $\breve{v} := \beta_{\breve{\Sigma}}^{-1}(Mod^I i_{\phi\Sigma}(v))$

First, let us prove that $S'(\langle f, u \rangle)$ is a morphism in $(Mod^{\mathbb{P}res^J})^\sharp$.

$$Sen^I\psi(S_{\phi\Sigma}) \subseteq S_{\phi\Sigma'}$$
$$\alpha_{\breve{\Sigma}'}(Sen^I i_{\phi\Sigma'}^{-1}(Sen^I\psi(S_{\phi\Sigma})) \subseteq \alpha_{\breve{\Sigma}'}(Sen^I i_{\phi\Sigma'}^{-1}(S_{\phi\Sigma'}))$$

As $\alpha_{\breve{\Sigma'}} \cdot Sen^I i^{-1}_{\phi\Sigma'} \cdot Sen^I \psi = \alpha_{\breve{\Sigma'}} \cdot Sen^I \phi\breve{\psi} \cdot Sen^I i^{-1}_{\phi\Sigma} = Sen^J \breve{\psi} \cdot \alpha_{\breve{\Sigma}} \cdot Sen^I i^{-1}_{\phi\Sigma}$ it follows that $Sen^J \breve{\psi}(S_{\breve{\Sigma}}) \subseteq S_{\breve{\Sigma'}}$.

Now, we prove that S' is functorial. It is clear that $S'\langle 1_\Sigma, 1_M\rangle = \langle 1_{\breve{\Sigma}}, 1_{\breve{M}}\rangle = 1_{S'\langle\Sigma,M\rangle}$ and, given a pair of arrows

$$\langle((\phi\Sigma)_S, S_{\phi\Sigma}), (\beta_\Sigma M)_{S\phi\Sigma}\rangle \xrightarrow{\langle\psi_1, w\rangle} \langle((\phi(\Sigma'))_S, S_{\phi(\Sigma')}), (\beta'_\Sigma N)_{S\phi(\Sigma')}\rangle$$

$$and$$

$$\langle((\phi(\Sigma'))_S, S_{\phi(\Sigma')}), (\beta'_\Sigma N)_{S\phi(\Sigma')}\rangle \xrightarrow{\langle\psi_2, y\rangle} \langle((\phi(\Sigma''))_S, S_{\phi(\Sigma'')}), (\beta''_\Sigma W)_{S\phi(\Sigma'')}\rangle$$

We have:

$$\begin{array}{ccccc} (\phi\Sigma)_S & \xrightarrow{\psi_1} & (\phi\Sigma')_S & \xrightarrow{\psi_2} & (\phi\Sigma'')_S \\ \cong\downarrow & & \downarrow\cong & & \downarrow\cong \\ \phi(\breve{\Sigma}) & \underset{\phi(\breve{\psi_1})}{\dashrightarrow} & \phi(\breve{\Sigma'}) & \underset{\phi(\breve{\psi_2})}{\dashrightarrow} & \phi(\breve{\Sigma''}) \end{array}$$

Notice that, by definition, $\phi(\breve{\psi_2 \cdot \psi_1})$ is the unique arrow that makes the outer rectangle commute. It follows that $\phi(\breve{\psi_2 \cdot \psi_1}) = \phi(\breve{\psi_2}) \cdot \phi(\breve{\psi_1})$ and so, by faithfulness, $\breve{\psi_2 \cdot \psi_1} = \breve{\psi_2} \cdot \breve{\psi_1}$.

Moreover, let $\bullet$ and $\circ$ stand for the composition of the second coordinate in, respectively, $(Mod^J)^\sharp$ and $(Mod^{\mathbb{P}res^J})^\sharp$. We then have:

$$\begin{aligned} \breve{w} \circ \breve{y} &= Mod^J \breve{\psi_1} \beta^{-1}_{\breve{\Sigma'}} Mod^I i_{\phi\Sigma}(w) \cdot \beta^{-1}_{\breve{\Sigma}} Mod^I i_{\phi\Sigma}(y) \\ &= \beta^{-1}_{\breve{\Sigma}} Mod^I \phi\breve{\psi_1} Mod^I i_{\phi\Sigma'}(w) \cdot \beta^{-1}_{\breve{\Sigma}} Mod^I i_{\phi\Sigma}(y) \\ &= \beta^{-1}_{\breve{\Sigma}} Mod^I i_{\phi\Sigma} Mod^J \psi_1(w) \cdot \beta^{-1}_{\breve{\Sigma}} Mod^I i_{\phi\Sigma}(y) \\ \breve{w \bullet y} &= \beta^{-1}_{\breve{\Sigma}} Mod^I i_{\phi\Sigma}(Mod^J \psi_1(w) \cdot y) \end{aligned}$$

We now have a functor $S' : (Mod^J)^\sharp \to (Mod^{\mathbb{P}res^J})^\sharp$. Finally, let us prove that S' indeed forms a skolemization.

First, notice that $i^{-1}_{\phi\breve{\Sigma}} \cdot \tau_{\phi\Sigma} \in Sig^I(\phi\Sigma, \phi\breve{\Sigma})$. Define then τ'_Σ as the arrow in $Sig^J(\Sigma, \breve{\Sigma})$ satisfying $\phi(\breve{\tau}) = i^{-1}_{\phi\breve{\Sigma}} \cdot \tau$. Given some $M \in |Mod^J\Sigma|$ we have:

$$\begin{aligned} \breve{M} \restriction_{\breve{\tau}} &= Mod^J \breve{\tau} \cdot \beta^{-1}_{\breve{\Sigma}} Mod^I i_{\phi\Sigma}((\beta_\Sigma(M))_{S\phi\Sigma}) \\ &= \beta^{-1}_\Sigma (Mod^I \phi\breve{\tau} Mod^I i_{\phi\Sigma}((\beta_\Sigma(M))_{S\phi\Sigma})) \\ &= \beta^{-1}_\Sigma (Mod^I \tau((\beta_\Sigma(M))_{S\phi\Sigma}) \\ M &= \beta^{-1}_\Sigma(\beta_\Sigma(M)) \end{aligned}$$

Now given $\mathcal{I}_{\phi\Sigma} = \langle \mathcal{U}, E \rangle$ we define $\mathcal{I}'_\Sigma = \langle \mathcal{U}', E' \rangle$ as:

- For any object i in $\mathcal{U}$, $\beta_{\check{\Sigma}}^{-1} Mod^I i_{\phi\Sigma}(i)$ is an object of $\mathcal{U}'$
  For any arrow a in $\mathcal{U}$, $\beta_{\check{\Sigma}}^{-1} Mod^I i_{\phi\Sigma}(a)$ is an arrow of $\mathcal{U}'$
- For any object e in E, $\beta_{\check{\Sigma}}^{-1} Mod^I i_{\phi\Sigma}(e)$ is an object of E'
  For any arrow b in E, $\beta_{\check{\Sigma}}^{-1} Mod^I i_{\phi\Sigma}(b)$ is an arrow of E'

Routine calculations show $\mathcal{I}'_\Sigma$ is an inclusion system in $Mod^J\check{\Sigma}$.

Finally, suppose that the $\check{\Sigma}$-models M' and N' are skolemizations of, respectively, the Σ-models M and N and that $M' \hookrightarrow N'$. Clearly then $(\beta_{\check{\Sigma}}(M')) \restriction_{i^{-1}_{\phi\Sigma}} \hookrightarrow (\beta_{\check{\Sigma}}(N')) \restriction_{i^{-1}_{\phi\Sigma}}$. Moreover, using structurality and the morphism compatibility condition we have that:

$$M' \models_{\check{\Sigma}} S_{\check{\Sigma}} \iff M' \models \alpha_{\check{\Sigma}}(Sen^I i^{-1}_{\phi\Sigma}(S_{\phi\Sigma})) \iff Mod^I i^{-1}_{\phi\Sigma} \beta_{\check{\Sigma}}(M') \models_{(\phi\Sigma)_S} S_{\phi\Sigma}$$

It follows then that

$$((\beta_{\check{\Sigma}}(M')) \restriction_{i^{-1}_{\phi\Sigma} \cdot \tau})^\star = ((\beta_{\check{\Sigma}}(N')) \restriction_{i^{-1}_{\phi\Sigma} \cdot \tau})^\star$$

Or equivalently,

$$((\beta_{\check{\Sigma}}(M')) \restriction_{i^{-1}_{\phi\check{\tau}}})^\star = ((\beta_{\check{\Sigma}}(N')) \restriction_{i^{-1}_{\phi\check{\tau}}})^\star$$

By naturality,

$$(\beta_\Sigma(Mod^I \check{\tau}(M')))^\star = (\beta_\Sigma(Mod^I \check{\tau}(N')))^\star$$

Since M' and N' are skolemizations, we have that $M' \restriction_{\check{\tau}} = M$ and $N' \restriction_{\check{\tau}} = N$. Now notice that

$$M \models \alpha_\Sigma(\varphi) \iff \beta_\Sigma(M) \models \varphi \iff \beta_\Sigma(N) \models \varphi \iff N \models \alpha_\Sigma(\varphi)$$

As α_Σ is semantically surjective the result follows.

□

As an illustration of the previous theorem we present the following:

Example 5.2. ***(Multialgebras have the Downward Löwenheim-Skolem property)***
We now describe **MA**—*the institution of (unsorted) multialgebras*[7]*. As signatures we simply use (unsorted) first order signatures. The intuition here is that function symbols are to be interpreted as functions and relations as multioperations.*
Let us describe the syntax. The terms are built in a first order manner with the caveat that relation symbols can too be used to form terms, that is, functions are allowed to take relations as arguments and we can compose relations. For the formulas, we have two atoms: $t \succ t'$*, interpreted as set inclusion, and* $t \doteq t'$*, interpreted as (deterministic) equality. The full set of formulas is built by using quantification and Boolean connectives, the sentences being the formulas without free variables. For the semantics we let the category of models of given signature be the category of multialgebras of that signature. A more detailed characterization of this institution can be found in [15].*
We can now describe a morphism $\mathbf{MA} \xrightarrow{\langle \phi, \alpha, \beta \rangle} \mathbf{FOL}^1$*:*

- *We start by defining the functor*

$$\phi : \mathbb{S}ig^{\mathbf{MA}} \longrightarrow \mathbb{S}ig^{\mathbf{FOL}^1}$$

$$\begin{array}{ccc} \langle (\mathcal{F}_i)_{i<\omega}, (\mathcal{M}_i)_{i<\omega} \rangle & \longmapsto & \langle (\mathcal{F}_i)_{i<\omega}, (\mathcal{R}_i)_{i<\omega} \rangle \\ f \downarrow & \longmapsto & \downarrow f \\ \langle (\mathcal{F}'_i)_{i<\omega}, (\mathcal{M}'_i)_{i<\omega} \rangle & \longmapsto & \langle (\mathcal{F}'_i)_{i<\omega}, (\mathcal{R}'_i)_{i<\omega} \rangle \end{array}$$

  Where $\mathcal{R}_{i+1} := \{r_m : m \in M_i\}$*. It is easy to see that* ϕ *is well defined and fully faithful. Moreover, we have that the functor is essentially surjective.*

- *Given* $\Sigma \in |\mathbb{S}ig^{\mathbf{MA}}|$ *we define* $\alpha_\Sigma : Sen^{\mathbf{FOL}^1}(\phi\Sigma) \to Sen^{\mathbf{MA}}(\Sigma)$ *recursively:*

$$\begin{aligned} \alpha_\Sigma(x_i) &= x_i \\ \alpha_\Sigma(f(t_1 \cdots t_n)) &= f(\alpha_\Sigma(t_1) \cdots \alpha_\Sigma(t_n)) \\ \alpha_\Sigma(t \approx t') &= \alpha_\Sigma(t) \doteq \alpha_\Sigma(t') \\ \alpha_\Sigma(r_m(t_1 \cdots t_{n+1})) &= m(\alpha_\Sigma(t_1) \cdots \alpha_\Sigma(t_n)) \succ t_{n+1} \end{aligned}$$

$$\alpha(A \wedge B) = \alpha_\Sigma(A) \wedge \alpha_\Sigma(B); \quad \alpha_\Sigma(\neg A) = \neg \alpha_\Sigma(A); \quad \alpha_\Sigma(\exists x_i(A)) = \exists x_i(\alpha_\Sigma(A))$$

[7]Here we consider a wide sense of n-ary multioperation on a set A: this is just a function $F : A^n \to \mathcal{P}(A)$, allowing $\varnothing$ in the range.

Elementary induction shows that α is indeed a natural transformation. Notice that the set $\alpha_\Sigma[Sen^{\mathbf{FOL}^1}(\phi\Sigma)]$ consists of formulas built of terms where there is no composition with multioperations. The idea we use to show that α_Σ is semantically surjective is simple: suppose we have the formula

$f(x_1 \cdots m(y_1 \cdots y_k) \cdots x_n) \doteq x_{n+1}$ where $m(y_1 \cdots y_k)$ happens in the j-th place, we simply introduce a new variable and restrict its domain, i.e., we consider the formula $\forall x_j(m(y_1 \cdots y_k) \succ x_j \wedge f(x_1 \cdots x_j \cdots x_n)) \doteq x_{n+1}$. Using a similar technique for inclusion[8] and proceeding by induction on nested formulas the proof follows.[9]

- *Given some signature Σ consider the functor*

$$\beta_\Sigma : Mod^{\mathbf{MA}}(\Sigma) \longrightarrow Mod^{\mathbf{FOL}^1}(\phi\Sigma)$$

$$\begin{array}{ccc}
\langle W, (F_i)_{i<\omega}, (M_i)_{i<\omega}\rangle & \longmapsto & \langle W, (F_i)_{i<\omega}, (R_i)_{i<\omega}\rangle \\
h\big\downarrow & \longmapsto & \big\downarrow h \\
\langle W', (F_i')_{i<\omega}, (M_i')_{i<\omega}\rangle & \longmapsto & \langle W', (F_i')_{i<\omega}, (R_i')_{i<\omega}\rangle
\end{array}$$

Where $r_m = \{x_1 x_2 \cdots x_{i+1} \in M^{i+1} : x_{i+1} \in m(x_1 \cdots x_i)\}$ and $R_{i+1} := \bigcup_{m \in M_i} r_m$. It is easy to see that β_Σ is well defined and that $(\beta_\Sigma)_{\Sigma \in |\mathbb{S}ig^{\mathbf{MA}}|}$ ensemble into a natural transformation. Furthermore simple arguments show that $\langle \phi, \alpha, \beta \rangle$ indeed forms an institution morphism. Finally, we define an inverse for β_Σ

$$Mod^{\mathbf{MA}}(\Sigma) \longleftarrow Mod^{\mathbf{FOL}^1}(\phi\Sigma) : \beta_\Sigma^{-1}$$

$$\begin{array}{ccc}
\langle W, (F_i)_{i<\omega}, (M_i)_{i<\omega}\rangle & \longleftarrow & \langle W, (F_i)_{i<\omega}, (R_i)_{i<\omega}\rangle \\
h\big\downarrow & \longleftarrow & \big\downarrow h \\
\langle W', (F_i')_{i<\omega}, (M_i')_{i<\omega}\rangle & \longleftarrow & \langle W', (F_i')_{i<\omega}, (R_i')_{i<\omega}\rangle
\end{array}$$

Where $m_r(x_1 \cdots x_i) := \{x_{i+1} \in W : r(x_1 \cdots x_{i+1})\}$ and $M_i := \bigcup_{r \in R_{i+1}} m_r$.

[8] For example, if f and g are function symbols and m is a multioperation, then the formula $f(m(x)) \succ g(y)$ is equivalent to $\exists z((m(x) \succ z) \wedge (f(z) \doteq g(y)))$

[9] Note that the full proof would have to address equalities between multioperations and inclusions between functions. The former being equivalent to $\perp$ and the latter to an equality, for instance, $f(x) \succ g(y)$ and $f(x) \doteq g(y)$

This proves that **MA** *has a skolemization. Observe that the inclusion system of this skolemization is the standard one, that is, an inclusion simply means a subalgebra. Using this fact and a similar technique to skolem hulls one can now easily prove a downward Löwenheim-Skolem result for multialgebras.*

6 Final remarks and future works

We finish the present work presenting some perspectives of future developments.

Remark 6.1. *The adjunctions obtained in Section 2 lead us to research about the relationship between the types of representations of propositional logics and their institutions and π-institution developed in Section 4:*

1. *The result of these analyzes may provide us with a way to study metalogical properties of abstract propositional logics and their algebraic or categorical properties, for instance, the relation between Craig's interpolation in an abstract logics and the amalgamation properties of its algebraic or categorical semantic. In particular, it could be interesting examine the possibility of generalize the work in [3], describing a Craig interpolation property for institutions associated to multialgebras: this is a natural (non-deterministic) matrix semantics for complex logics as the* **LFI's***, the logics of formal inconsistencies (see [8]).*

2. *By a convenient modification of this matrix institution, is presented in section 3.2 of [19] an institution for each "equivalence class" of algebraizable logic: this furnished technical means to apply notions and results from the theory of institutions in the propositional logic setting and to derive, from the introduction of the notion of "Glivenko's context", a strong and general form of Glivenko's Theorem relating two "well-behaved" logics.*

Remark 6.2. *Another interesting discussion – already suggested in [10] – which can be posed is how to repeat the whole discussion of Section 3 with a version of the Grothendieck construction for indexed 2-categories in order to directly produce the 2-category of institutions, as well as related 2-categories of institution-like structures. The technical categorical devices necessary for developing this idea are presented in [6], for example.*

Remark 6.3. *The borowing result presented in section 5 leads us to question which institutions have the skolemization property in a non-trivial way. Furthermore, in predicate logic skolemization is deeply related to the idea of indiscernibles, which* *leads the authors to question if an institution-independent formalization of this idea is possible. Another question is if whether skolemization of an institution I implies the skolemization of* $\mathbb{P}res^I$*; if so, then in any skolemizable institution every theory would admit some expansion to a model-complete theory.*

References

[1] P. Arndt, R. A. Freire, O. O. Luciano, and H. L. Mariano, *A global glance on categories in logic.* Logica Universalis, **1** (2007), 3–39.

[2] P. Arndt, H. L. Mariano and D. C. Pinto, *Finitary Filter Pairs and Propositional Logics,* South American Journal of Logic, **4(2)** (2018), 257–280.

[3] P. Arndt, H. L. Mariano and D. C. Pinto. *Horn filter pairs and Craig interpolation property,* in preparation.

[4] A. Avron, A. Zamansky, *Non-Deterministic Semantics for Logical Systems.*, In: Gabbay D., Guenthner F. (eds) Handbook of Philosophical Logic. Handbook of Philosophical Logic, vol 16, 2005.

[5] I. Bakovic, *Bicategorical Yoneda lemma,* preprint, https://www2.irb.hr/korisnici/ibakovic/yoneda.pdf.

[6] I. Bakovic . *Fibrations of bicategories,* preprint, https://www2.irb.hr/korisnici/ibakovic/groth2fib.pdf.

[7] W. Carnielli, M. E. Coniglio, João Marcos, *Logics of Formal Inconsistency.* In: Gabbay D., Guenthner F. (eds) Handbook of Philosophical Logic. Handbook of Philosophical Logic, vol 14. Springer, Dordrecht, 2007.

[8] M. E. Coniglio, A. Figallo-Orellano, A. C. Golzio, *Non-deterministic algebraization of logics by swap structures,* Logic Journal of the IGPL, to appear. First published online: November 29, 2018. DOI: 10.1093/jigpal/jzy072.

[9] R. Diaconescu, *Grothendieck Institutions,* Applied Categorical Structures **10** (2002).

[10] R. Diaconescu, *Institution-independent Model Theory,* Birkhauser Basel - Boston - Berlin, 2008.

[11] V.L. Fernandez and M.E. Coniglio, *Fibring algebraizable consequence system.* Proceedings of CombLog 04 - Workshop on Combination of Logics: Theory and Application (2004), 93–98.

[12] J. Fiadeiro and A. Sernadas, *Structuring theories on consequence,* D. Sannella and A. Tarlecki (eds.), Recent Trends in Data Type Specification, Lecture Notes in Comput. Sci. **332** (1988), 44–72.

[13] J. A. Goguen and R. M. Burstall, *Institutions: abstract model theory for specification and programming,* Journal of the ACM (JACM), **39(1)** (1992), 95–146.

[14] P. Johnstone, *Sketches of an elephant: A topos theory compendium,* Oxford University Press, 2002.

[15] Y. Lamo, *The Institution of Multialgebras - a general framework for algebraic software development.*, PhD thesis, University of Bergen, (2002).

[16] H. L. Mariano and C. A. Mendes, *Towards a good notion of categories of logics.* arXiv preprint, http://arxiv.org/abs/1404.3780, (2014).

[17] H. L. Mariano and D. C. Pinto, *Remarks on Propositional Logics and the Categorial Relationship Between Institutions and π-Institutions*, South American Journal of Logic, **3(1)** (2017), 111–121.

[18] H. L. Mariano and D. C. Pinto, *Algebraizable Logics and a functorial encoding of its morphisms*, Logic Journal of the IGPL **25(4)** (2017), 524–561.

[19] H. L. Mariano and D. C. Pinto, *An abstract approach to Glivenko's theorem*, arXiv preprint (2016), https://arxiv.org/pdf/1612.03410.pdf.

[20] nLab, https://ncatlab.org/nlab/show/HomePage.

[21] G. Voutsadakis, *Categorical abstract algebraic logic: algebraizable institutions*, Applied Categorical Structures, **10** (2002), 531–568.

Received 31 December 2019

Some Classical Modal Logics with a Necessity/Impossibility Operator

Cezar A. Mortari
Department of Philosophy, Federal University of Santa Catarina
c.mortari@ufsc.br

Abstract

In this paper we examine modal logics in which the modal operator $\Box$ can be read as necessity, or impossibility, or both. Consider classical modal logics, for example; that is, the logics closed under the following rule of inference: $A \leftrightarrow B \;/\; \Box A \leftrightarrow \Box B$. Here $\Box$ usually represents necessity. But it also can be read as possibility, impossibility, contingency, non-necessity, and even negation. On the other hand, in the rule $A \to B \;/\; \Box B \to \Box A$ the $\Box$ operator can no longer be read as necessity, or possibility — but it makes sense to read it as impossibility or negation: if A implies B and B is impossible, so is A.

In this paper we will deal with only the necessity/impossibility readings of $\Box$. We consider several modal formulas and, using neighborhood semantics, identify, on frames, conditions corresponding to them. We consider several logics obtained by adding one or more of these formulas as axioms, and prove determination theorems for them. Besides the preliminary results presented in this paper, we conclude indicating some topics for further research.

Keywords: classical modal logics; neighborhood semantics; impossibility operator.

1 Introduction

This work was inspired by an old paper of Richard Sylvan's ([6]) on relational semantics for some strict classical modal logics. In that paper, Sylvan intended to present relational semantics for all well-known systems of modal logic of Lewis, Lemmon and Fey's, considering relational frames that included, in addition to normal worlds, one or more sets of non-normal ones. The non-normal worlds considered in that paper had the following conditions for formulas with the necessity operator (where x is some world):

Thanks to the anonymous referees for many helpful corrections and suggestions. Thanks also to the audience at the Brazilian Logic Conference.

* opposite: $\Box A$ is true at x iff A is false at every world accessible to x;
* contrary: $\Box A$ is true at x iff A is false at some world accessible to x;
* perverse: $\Box A$ is true at x iff A is true at some world accessible to x;
* rafferty: $\Box A$ is arbitrarily true or false at x.

As we know, Sylvan's proposed semantics did not work for some of the systems he had in mind, but the logics characterized by frames with non-normal worlds remain interesting in their own right. Being non-normal, they can find applications in the areas of epistemic and deontic logic (where, for instance, usual inference rules such as necessitation are deemed too strong).

In [5] we examined the logics of relational frames with one or more kinds of non-normal worlds. Among the logics there considered, the smallest was one in which the interpretation of $\Box$ was neutral with regard to necessity, impossibility, possibility or non-necessity. So we had a set of formulas that were valid no matter how $\Box$ was interpreted.

In this paper, we intend to examine, from the point of view of neighborhood semantics, some logics in which the interpretation of $\Box$ can be either necessity or impossibility (further work will deal with other readings). By a necessity interpretation (or reading) of $\Box$ we mean the following: $\Box A$ is true at a world x if and only if A is true at every world accessible to x. An impossibility interpretation (or reading) of $\Box$ means: $\Box A$ is true at a world x if and only if A is false at every world accessible to x.[1] Using relational frames for the semantics, this would give us the case where the set of worlds in a frame is split into only normal and opposite worlds. However, weaker systems can be devised resorting to neighborhood semantics and considering classical modal logics as a starting point.

A modal logic is *classical* if it is closed under the following rule of inference:

RE. $A \leftrightarrow B \ / \ \Box A \leftrightarrow \Box B$.

Now $\Box$ usually represents a necessity operator, but it can have several interpretations, like possibility, but also non-necessity, impossibility, contingency, non-contingency, and even negation. This means that we can read $\Box$ in many different ways, and only additional requirements (syntactic and/or semantic) will fix some interpretation as necessity, or impossibility, and so on.

Thus, in this paper we will consider classical modal logics in which the interpretation of $\Box$ is either necessity or impossibility.

[1] Of course, this still leaves room for different interpretations of necessity and impossibility: logical, metaphysical, epistemic, deontic, temporal and so on.

2 Preliminaries

We will work in a basic modal language consisting of a countable set Φ of propositional variables (for which we will use p, q etc.), the propositional constant $\bot$, and the primitive operators $\neg$, $\wedge$ and $\Box$. The standard operators $\vee$, $\to$, $\leftrightarrow$, $\Diamond$, and the constant $\top$ are defined in the usual way. We will also use A, B etc. as meta-variables for formulas.

All logics considered in this paper will be extensions of classical propositional logic, so they include the set PL of all tautologies, as well as being closed under modus ponens (MP) and uniform substitution of variables (US). Thus, for the purposes of this paper, a *logic* can be seen as a set of formulas that includes all tautologies of classical propositional logic and is closed under modus ponens (MP) and uniform substitution (US). A logic is closed under some rule of inference iff it contains the conclusion of the rule whenever it contains the rule's hypotheses. If all instances of some schema belong to a logic, or if the logic is closed under some rule of inference, we say that it *has* or *provides* that schema or rule.

Logics can also be presented axiomatically, and by a *theorem* of a logic we understand a formula which is provable in an axiom system for that logic. Inference rules are then to be understood in the folowing sense: if the premisses of a rule are theorems, so is its conclusion.

As said above, a modal logic is classical if it is closed under the rule RE:

RE. $A \leftrightarrow B \;/\; \Box A \leftrightarrow \Box B$.

The smallest classical modal logic, E, can be axiomatized by

PL. A, if A is an instance of a tautology of classical propositional logic,

and the inference rules MP (modus ponens), US (uniform substitution), and RE. As we know, E has no theorems of the form $\Box A$; as examples of E-theorems we have $\Box p \leftrightarrow \neg\neg\Box p$ (which is an instance of a tautology) and $\Box\Box p \leftrightarrow \Box\neg\neg\Box p$ (which is obtained from the previous theorem by applying RE).

E can be extended by several different axioms schemes. Two well-known ones are the following:

M. $\Box(A \wedge B) \to (\Box A \wedge \Box B)$

C. $(\Box A \wedge \Box B) \to \Box(A \wedge B)$

If, as usual, we read $\Box$ as a necessity operator, both formulas are acceptable: a conjunction is necessary if and only if both of its elements are. M is also acceptable if we read $\Box$ as possibility (if $A \wedge B$ is possible, so are A and B). Schema C, on the other hand, turns out to be false under this reading: A and B can be individually possible, but if one excludes the other, their conjunction is impossible. However, C is again acceptable if we read $\Box$ as *impossibility*: if A is impossible, and so is B, then the conjunction $A \wedge B$ is also impossible. Actually, if either A or B is impossible, their conjunction is impossible, too, what we could express using this formula (which we will call W′ for reasons below):

$$\text{W}'. \quad (\Box A \vee \Box B) \rightarrow \Box(A \wedge B)$$

Since we are only considering extensions of E by means of formulas which are acceptable when $\Box$ is read as necessity or impossibility, we will start with the following three groups of axiom schemes (some other schemas will be discussed later on).

(I)

$$\text{M}. \quad \Box(A \wedge B) \rightarrow (\Box A \wedge \Box B)$$

$$\text{M}'. \quad (\Box A \vee \Box B) \rightarrow \Box(A \vee B)$$

$$\text{Y}. \quad \Box(A \wedge B) \rightarrow (\Box A \vee \Box B)$$

$$\text{N}. \quad \Box\top$$

(II)

$$\text{W}. \quad \Box(A \vee B) \rightarrow (\Box A \wedge \Box B)$$

$$\text{W}'. \quad (\Box A \vee \Box B) \rightarrow \Box(A \wedge B)$$

$$\text{Z}. \quad \Box(A \vee B) \rightarrow (\Box A \vee \Box B)$$

$$\text{O}. \quad \Box\bot$$

(III)

$$\text{C}. \quad (\Box A \wedge \Box B) \rightarrow \Box(A \wedge B)$$

$$\text{V}. \quad (\Box A \wedge \Box B) \rightarrow \Box(A \vee B)$$

Formulas in the first group are acceptable under a necessity reading, but not impossibility. That is, if the truth conditions for $\Box A$ are the usual ones ($\Box A$ is true at x iff A is true at every accessible world x'), all these formulas come out intuitively true. But not impossibility: if $\Box A$ is true at x iff A is *false* at every accessible world x', then obviously $\Box\top$ will come out false as well. And the same goes for M, M′ and Y. Notice, by the way, that Y is a weaker form of M — it follows from it by propositional logic alone. And we will prove later that M and M′ are deductively equivalent with regard to E.

On the second group, we have it the other way round: these formulas are acceptable under an impossibility reading ($\Box$ meaning 'false at every accessible world'), but not necessity. Z is a weaker form of W, and W and W′ will be proven equivalent.

Finally, formulas on the third group are neutral, so to speak: they admit both readings.

The role played by the schemes listed above can also be played by some inference rules. Consider, for instance, the inference rule RM:

RM. $A \to B \;/\; \Box A \to \Box B$

A well-known result (see [2]) is that a logic L has RM iff it has RE and M. We will show, with regard to our candidate axioms, that the following inference rules, divided into three groups as well, can be employed in place of them to generate a logic.

(I) RM. $A \to B \;/\; \Box A \to \Box B$

RM′. $(A \vee B) \leftrightarrow C \;/\; (\Box A \vee \Box B) \to \Box C$

RY. $A \leftrightarrow (B \wedge C) \;/\; \Box A \to (\Box B \vee \Box C)$

RN. $A \;/\; \Box A$

(II) RW. $A \to B \;/\; \Box B \to \Box A$

RW′. $A \to (B \wedge C) \;/\; (\Box B \vee \Box C) \to \Box A$

RZ. $(A \vee B) \leftrightarrow C \;/\; \Box C \to (\Box A \vee \Box B)$

RO. $\neg A \;/\; \Box A$

(III) RC. $(A \wedge B) \leftrightarrow C \;/\; (\Box A \wedge \Box B) \to \Box C$

RV. $(A \vee B) \leftrightarrow C \;/\; (\Box A \wedge \Box B) \to \Box C$

Let us now establish some facts about the connections between these axioms and rules. First, as said before, M and M′, and W and W′, are equivalent: a logic has one of them iff it has the other. This is also the case for RM and RM′, and RW and RW′.

Theorem 2.1. *Let L be a logic. Then:*

(a) *L has* RE *and* M *iff it has* RM *iff it has* RM′ *iff it has* RE *and* M′*;*

(b) *L has* RE *and* W *iff it has* RW *iff it has* RW′ *iff it has* RE *and* W′.

Proof. We prove case (b); the first part of (a) is a known result, and the demonstration of the rest follows similar lines. In what follows, we use PL to indicate that something follows by classic propositional logic.

First, suppose that L has RE and W, and that $A \to B$ is a thesis. Then so are $B \leftrightarrow (A \vee B)$ by PL, and $\Box B \leftrightarrow \Box(A \vee B)$ by RE. Using W, that is, $\Box(A \vee B) \to (\Box A \wedge \Box B)$, we get $\Box B \to \Box A$ again by propositional logic, so L has RW.

Suppose now that it has RW, and that $A \leftrightarrow B$ is a thesis. Then so are $\Box A \rightarrow \Box B$ and $\Box B \rightarrow \Box A$. Two applications of RW and propositional logic give us $\Box A \leftrightarrow \Box B$, so L has RE. For W, since $A \rightarrow (A \vee B)$ and $B \rightarrow (A \vee B)$ are tautologies, RW gives us $\Box(A \vee B) \rightarrow A$, $\Box(A \vee B) \rightarrow B$, and W follows immediately.

To prove that L has RW$'$, if it has RE and W$'$, suppose that $A \rightarrow (B \wedge C)$ is a thesis. Then so are $A \rightarrow B$ and $A \rightarrow C$, from what $A \leftrightarrow (A \wedge B)$ and $A \leftrightarrow (A \wedge C)$ follow by PL. Using RE we obtain $\Box A \leftrightarrow \Box(A \wedge B)$ and $\Box A \leftrightarrow \Box(A \wedge C)$, and it follows that $\Box(A \wedge B) \leftrightarrow \Box(A \wedge C)$ is also a thesis. Consider now these two instances of W$'$: $(\Box A \vee \Box B) \rightarrow \Box(A \wedge B)$ and $(\Box A \vee \Box C) \rightarrow \Box(A \wedge C)$. By PL we obtain $(\Box B \vee \Box C) \rightarrow \Box A$, so L has RW$'$.

Suppose now that L has RW$'$. Since $(A \wedge B) \rightarrow (A \wedge B)$ is a tautology, RW$'$ gives us immediately $(\Box A \vee \Box B) \rightarrow \Box(A \wedge B)$, which is W$'$. That L also has RE is easily shown.

We prove now that L has RW iff it has RW$'$. So suppose L has RW and that $A \rightarrow (B \wedge C)$ is a thesis. Then so are $A \rightarrow B$ and $A \rightarrow C$. Using RW we obtain $\Box B \rightarrow \Box A$ and $\Box C \rightarrow \Box A$, from what we get, by PL, $(\Box B \vee \Box C) \rightarrow \Box A$, so L has RW$'$. For the other direction, suppose that L has RW$'$, and that $A \rightarrow B$ is a thesis. Then so are $A \rightarrow (B \wedge B)$ (by PL) and $(\Box B \vee \Box B) \rightarrow \Box A$, using RW$'$. Now $\Box B \rightarrow \Box A$ follows by PL, so L has RW. □

Corollary 2.2. *A classical logic has* M *iff it has* M$'$, *and it has* W *iff it has* W$'$.

Accordingly, we will only consider M and W in building our logics. As for other axioms and rules, we can prove the following equivalences.

Theorem 2.3. *Let L be a logic. Then:*

(a) *L has* RY *iff it has* RE *and* Y*;*

(b) *L has* RZ *iff it has* RE *and* Z*;*

(c) *L has* RC *iff it has* RE *and* C*;*

(d) *L has* RV *iff it has* RE *and* V.

Proof. Proofs of above results are not difficult; we show (c) as an illustration. So suppose L has RC. Since $(A \wedge B) \leftrightarrow (A \wedge B)$ is a tautology and so a theorem, one application of RC gives us $(\Box A \wedge \Box B) \rightarrow \Box(A \wedge B)$, so we have C. To show that L has RE, consider:

1.	$A \leftrightarrow B$	hypothesis
2.	$(A \wedge A) \leftrightarrow B$	1 PL
3.	$(\Box A \wedge \Box A) \rightarrow \Box B$	2 RC
4.	$\Box A \rightarrow \Box B$	3 PL
5.	$(B \wedge B) \leftrightarrow A$	1 PL
6.	$(\Box B \wedge \Box B) \rightarrow \Box A$	5 RC
7.	$\Box B \rightarrow \Box A$	6 PL
8.	$\Box A \leftrightarrow \Box B$	4,7 PL

For the other direction, suppose L has RE and C, and that $(A \wedge B) \leftrightarrow C$ is a theorem. Applying RE we obtain $\Box(A \wedge B) \leftrightarrow \Box C$ From this and C, that is, $(\Box A \wedge \Box B) \rightarrow \Box(A \wedge B)$, we otain $(\Box A \wedge \Box B) \rightarrow \Box C$ by propositional logic. □

Some further results are:

Theorem 2.4. *Let L be a classical logic. Then:*

(a) *if L has* M*, then it has* Y *and* V*;*

(b) *if L has* W*, then it has* Z *and* C*;*

(c) *L has* RN *iff it has* N*, and has* RO *iff it has* O*.*

Proof. So suppose L is a classical logic; we prove cases (a) and (b) as examples.

(a) Suppose L has M. Since $(\Box A \wedge \Box B) \rightarrow (\Box A \vee \Box B)$ is a tautology, L has Y. For V, since $A \rightarrow (A \vee B)$ and $B \rightarrow (A \vee B)$ are tautologies, using RM (which L has since it has M) we get both $\Box A \rightarrow \Box(A \vee B)$ and $\Box B \rightarrow \Box(A \vee B)$, and V, that is, $(\Box A \wedge \Box B) \rightarrow \Box(A \vee B)$, follows immediately by propositional logic.

(b) Now suppose L has W. Then Z follows from W by propositonal logic alone. Now both $(A \wedge B) \rightarrow A$ and $(A \wedge B) \rightarrow B$ are tautologies; using RW we get $\Box A \rightarrow \Box(A \wedge B)$ and $\Box B \rightarrow \Box(A \wedge B)$, from which C follows by propositional logic. □

3 Extending E

Having now listed some candidate axiom schemas and rules, let us see which logics we get combining them. We will start with axioms of group III only, that is, the core logics where □ admits, indifferently, a necessity or an impossibility reading. Our candidate axioms (from this group) are C and V (and of course the corresponding rules). Besides the minimal classical logic E, we obtain EV, EC, and EVC. That they are all different from each other will be shown later, after we introduce frames and models. So, initially, EVC is the strongest logic where □ still can be read either as

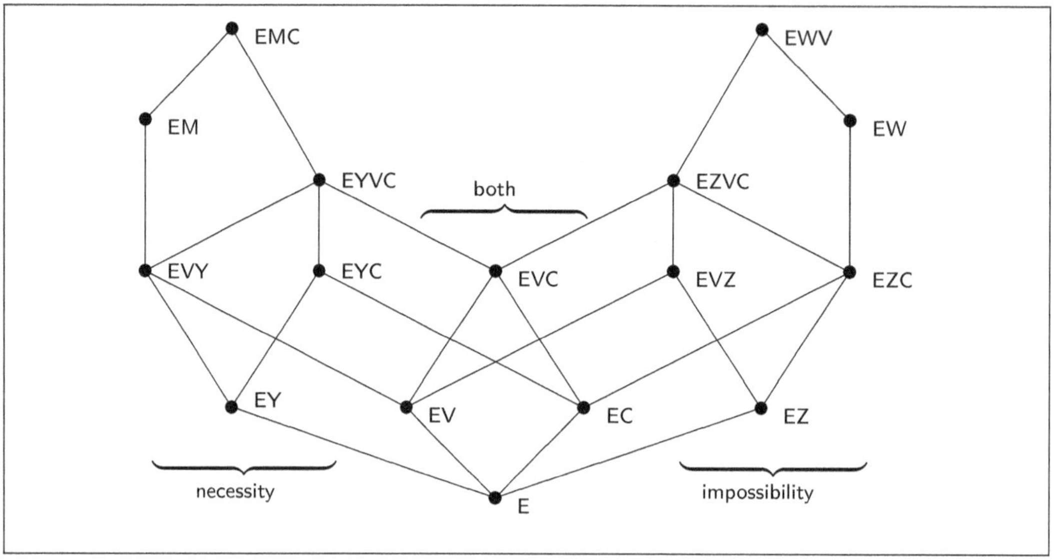

Figure 1: Some logics

necessity or impossibility. If we now add axioms from the two remaining groups, I and II, we will force the interpretation of □ in one or the other direction.

Combinations of M, C, and N yield the usual eight classical modal logics presented in Brian Chellas's book [2]. Analogously, if we take combinations of W, V, and O we get eight systems (E being one of them). Combinations of most of these schemes (with the exception of N and O) give us the logics depicted in Figure 1, where, at the center, we have E and three of its extensions in which □ can be read either as necessity or impossibility.

For the other logics, let us start with the necessity reading. Adding Y to the four center logics gives us four other systems. Now, if a logic has M, then it has Y and V, so adding M to E gives us an extension of EVY. Finally, if we add C to EM, we get them all, so EMC extends all other logics in the diagram which have a necessity reading of □. And naturally we can add N to these logics, getting known classical logics like EN or EMN. EMCN, in particular, is another name for the smallest normal logic, K.

Going in the other direction, let us take the core logics and extend them with axioms (and/or rules) from group II. We arrive at a similar picture on the right of Figure 1. First, we have four Z extensions of the four center logics. And since a logic has Z and C if it has W, we have EW and finally EWV, the strongest logic in the diagram in which □ means impossibility. We can also extend these logics with O; in this case, EWVO will be the smallest normal logic with □ meaning impossibility.

The logics in the diagram are all distinct, but to show that, we need first to talk about semantics.

4 Frames and models

Definition 4.1. A *frame* is a structure $\mathfrak{F} = \langle U, S \rangle$, where U is a nonempty set, the universe of the frame, and S is a function that associates to each $x \in U$ a set of subsets of U (that is, $S(x) \subseteq \mathcal{P}(U)$).

A *valuation* V on U is a function from the set Φ of propositional variables to $\mathcal{P}(U)$.

Definition 4.2. Let $\mathfrak{F} = \langle U, S \rangle$ be a frame and V a valuation in U. A *model* is a pair $\mathfrak{M} = \langle \mathfrak{F}, V \rangle$; we say that $\mathfrak{M}$ is *based on* $\mathfrak{F}$.

Definition 4.3. Let $\mathfrak{M} = \langle U, S, V \rangle$ be a model and x an element of U. A formula A is *true* at the point x, what is denoted by $\mathfrak{M}, x \Vdash A$, when:

$\mathfrak{M}, x \Vdash \mathbf{p}$ iff $x \in V(\mathbf{p})$, for $\mathbf{p} \in \Phi$;

$\mathfrak{M}, x \nVdash \bot$;

$\mathfrak{M}, x \Vdash \neg A$ iff $\mathfrak{M}, x \nVdash A$;

$\mathfrak{M}, x \Vdash A \wedge B$ iff $\mathfrak{M}, x \Vdash A$ and $\mathfrak{M}, x \Vdash B$;

$\mathfrak{M}, x \Vdash \Box A$ iff $\|A\|^{\mathfrak{M}} \in S(x)$, with $\|A\|^{\mathfrak{M}} = \{y \in U : \mathfrak{M}, y \Vdash A\}$.

The set $\|A\|^{\mathfrak{M}}$ from the above definition is called the *truth set* of A in $\mathfrak{M}$. When there is no risk of confusion, we will drop the superscript and write simply $\|A\|$.

Definition 4.4. A formula A is *true in a model* $\mathfrak{M} = \langle U, S, V \rangle$, what we denote by $\mathfrak{M} \Vdash A$, if it is true at every $x \in U$. A formula A is *valid at a point* x in a frame $\mathfrak{F}$, what we denote by $\mathfrak{F}, x \Vdash A$, if it is true at x in every model based on $\mathfrak{F}$. A formula A is *valid in a frame* $\mathfrak{F}$ if it is valid at every point of the frame, and *valid in a class* C of frames, what we denote by $\mathsf{C} \Vdash A$, if it is valid in every frame of the class.

We say that a formula B is a *tautological consequence* of a set $A_1, \ldots, A_m$ of formulas, for $(m \geq 0)$, iff $(A_1 \wedge \ldots \wedge A_m) \to B$ is an instance of a tautology of classical propositional logic. (Of course, if $m = 0$, $(A_1 \wedge \ldots \wedge A_m) \to B$ is just B.)

Lemma 4.5. *Let* C *be any class of frames. Then:*

(i) *if* B *is a tautological consequence of* $A_1, \ldots, A_m$ *, for* $(m \geq 0)$*, and* $\mathsf{C} \Vdash A_1$*,* $\ldots$*,* $\mathsf{C} \Vdash A_m$*, then* $\mathsf{C} \Vdash B$*;*

(ii) *if* $\mathsf{C} \Vdash A \leftrightarrow B$*, then* $\mathsf{C} \Vdash \Box A \leftrightarrow \Box B$*.*

Proof. (i) The proof is straightforward, owing to the fact that every tautology is valid in every state of every frame, and that modus ponens preserves validity.

(ii) Suppose $\mathsf{C} \Vdash A \leftrightarrow B$. Thus $\mathsf{C} \Vdash A$ iff $\mathsf{C} \Vdash B$, and it follows that, for every model $\mathfrak{M} = \langle U, S, V \rangle$ based on a frame in C, $\|A\|^{\mathfrak{M}} = \|B\|^{\mathfrak{M}}$. Hence, for every $x \in U$, $\|A\|^{\mathfrak{M}} \in S(x)$ iff $\|B\|^{\mathfrak{M}} \in S(x)$. So $\mathfrak{M}, x \Vdash \Box A$ iff $\mathfrak{M}, x \Vdash \Box B$, from what it follows that $\mathsf{C} \Vdash \Box A \leftrightarrow \Box B$. □

5 Soundness

The logic of a class C of frames is the set of all formulas valid in every frame of that class. With no restrictions imposed on the frames, the logic of the class of all frames is the minimal classical modal logic knwon as E.

To obtain determination (that is, soundness and completeness) theorems for the other logics we need to define the corresponding property of each of the schemas M, Y, N, C, V, W, Z, and O, which, as can be easily shown, are not valid in the class of all frames. For some of them the properties are well-known in the literature (see [2]).

Let $\mathfrak{M} = \langle U, S \rangle$ be a frame, x an element of U, and X and Y subsets of $\mathcal{P}(U)$. We consider the following natural conditions:

(I) (m) if $X \cap Y \in S(x)$, then $X \in S(x)$ and $Y \in S(x)$;

(y) if $X \cap Y \in S(x)$, then $X \in S(x)$ or $Y \in S(x)$;

(n) $U \in S(x)$;

(II) (w) if $X \cup Y \in S(x)$, then $X \in S(x)$ and $Y \in S(x)$;

(z) if $X \cup Y \in S(x)$, then $X \in S(x)$ or $Y \in S(x)$;

(o) $\emptyset \in S(x)$;

(III) (c) if $X \in S(x)$ and $Y \in S(x)$, then $X \cap Y \in S(x)$;

(v) if $X \in S(x)$ and $Y \in S(x)$, then $X \cup Y \in S(x)$.

For M′ and W′, the conditions would be:

(m′) if $X \in S(x)$ or $Y \in S(x)$, then $X \cup Y \in S(x)$;

(w′) if $X \in S(x)$ or $Y \in S(x)$, then $X \cap Y \in S(x)$.

But we will not use them, since, as shown before, in classical logics M′ and W′ are equivalent to M and W, respectively.

A known result is that condition (m) is equivalent to *closure under supersets*, that is:

(m*) if $X \in S(x)$ and $X \subseteq Y$, then $Y \in S(x)$.

That is, if a set X is in $S(x)$, for some $x \in U$, then all its supersets are, too. Analogously, we can show that (w) is equivalent to *closure under subsets*, that is:

(w*) if $X \in S(x)$ and $Y \subseteq X$, then $Y \in S(x)$.

That is, is a set X belongs to $S(x)$, then all its subsets belong, too.

The following lemma establishes a correspondence between axiom schemas and their corresponding condition.

Lemma 5.1. *The schemas* M, Y, N, C, V, W, Z *and* O *are valid in a frame* $\mathfrak{F}$ *iff* $\mathfrak{F}$ *satisfies the conditions* (m), (y), (n), (c), (v), (w), (z), *and* (o), *respectively.*

Proof. The results for M, C, and N are already known (see [3] for instance).

We show W and Z as examples. Let $\mathfrak{F} = \langle U, S \rangle$ be a frame in which (w) holds, $\mathfrak{M}$ a model based on $\mathfrak{F}$, A and B formulas, and x an element of U such that $x \Vdash \Box(A \vee B)$. It follows that $\|A \vee B\| \in S(x)$ and, since (what can easily be shown) $\|A \vee B\| = \|A\| \cup \|B\|$, that $\|A\| \cup \|B\| \in S(x)$. Given that (w) holds in $\mathfrak{M}$, it follows that $S(x)$ contains $\|A\|$ and $\|B\|$. But then $x \Vdash \Box A$ and $x \Vdash \Box B$, from what it follows that W is valid in a frame satisfying (w).

Let now $\mathfrak{F}$ be a frame in which (w) does not hold. That is, there is a $u \in U$ and subsets X and Y of U such that $X \in S(u)$, $Y \subseteq X$, but $Y \notin S(u)$. Let $\mathcal{M}$ be a model based on $\mathfrak{F}$ such that $V(p) = X$, $V(q) = Y$. Clearly $X \cup Y = X$, so $u \Vdash \Box(p \vee q)$, since $\|p \vee q\| = X$ and $X \in S(u)$. We also have that $u \Vdash \Box p$. However, since $Y \notin S(u)$ and $\|q\| = Y$, $u \nVdash \Box q$. Now this models falsifies $\Box(p \vee q) \to \Box p \wedge \Box q$, an instance of W.

For Z, let $\mathfrak{F} = \langle U, S \rangle$ be a frame in which (z) holds, $\mathfrak{M}$ a model based on $\mathfrak{F}$, A and B formulas and x an element of U such that $x \Vdash \Box(A \vee B)$. It follows that $\|A \vee B\| \in S(x)$ and, since $\|A \vee B\| = \|A\| \cup \|B\|$, that $\|A\| \cup \|B\| \in S(x)$. Given that (z) holds in $\mathfrak{M}$, it follows that $S(x)$ contains either $\|A\|$ or $\|B\|$. But then either $x \Vdash \Box A$ or $x \Vdash \Box B$, so $x \Vdash \Box A \vee \Box B$, from what it follows that Z is valid in in a frame satisfying (z).

For the other direction, let now $\mathfrak{F}$ be a frame in which (z) does not hold. That is, there is a $u \in U$ and subsets X and Y of U such that $X \cup Y \in S(u)$, but neither

X nor Y are in $S(u)$. Let $\mathcal{M}$ be a model based on $\mathfrak{F}$ such that $V(p) = X$, $V(q) = Y$. Clearly $X \cup Y = X$, so $u \Vdash \Box(p \vee q)$, since $\|p \vee q\| = X \cup Y$ and $X \cup Y \in S(u)$. However, we also have that $u \nVdash \Box p$ and $u \nVdash \Box q$, since neither $\|p\| = V(p)$ nor $\|q\| = V(q)$ are in $S(u)$. Hence, this models falsifies $\Box(p \vee q) \rightarrow \Box p \vee \Box q$, an instance of Z. □

The following general soundness result, proven in [2], holds for our logics in Figure 1.

Theorem 5.2 (Soundness). *Let $S_1, \ldots, S_n$ be schemas valid respectively in classes of frames C_i satisfying the corresponding condition (s_i), for $1 \leq i \leq n$. Then the logic is sound with respect to the class $\mathsf{C}_1 \cap \ldots \cap \mathsf{C}_n$.*

We can now show some results to the effect that some axiom schema does not follow from certain others. The strategy is the usual one: find a model for the strongest logic without some specific schema, and show that the schema fails in that model. For instance, consider C, $(\Box A \wedge \Box B) \rightarrow \Box(A \wedge B)$, and take a look at the diagram in Figure 1. It is known that C is not a thesis of EM (see [2]). We need to show that it is not a thesis of EVZ, too.

In fact, we can prove the following theorem, establishing the distinctness of all logics in Figure 1.

Theorem 5.3. *The logics depicted in Figure 1, as well as their extensions with* N *or* O, *are all distinct.*

Proof. Let us consider each of our axiom schemas.

[N and O]. Consider the model where $U = \{1\}$, $S(1) = \emptyset$, and $V(p) = \emptyset$ for all $p \in \Phi$. Conditions (m), (y), (c), (v), (w) and (z) hold trivially, so this is both an EMC and an EWV model, what makes it a model of every logic below EMC and EWV in the diagram of Figure 1. However, since neither $\emptyset$ nor U are in $S(1)$, both $\Box\top$ and $\Box\bot$ are false in 1. Thus, all systems on the diagram are distinct from their N and O extensions.

[C.] The two strongest logics without C are EM and EVZ. That C is not a thesis of EM is a known result (see [2]). Now consider the model where $U = \{1, 2\}$, $S(1) = S(2) = \{\{1\}, \{2\}, \{1, 2\}\}$, and $V(p) = \{1\}$, $V(q) = \{2\}$. Condition (v) is easily verified, since the union of any two sets in $S(1)$ belongs to $S(1)$. (m) and (z) also hold, so this is both an EM-model and an EVZ-model, as well as a model for all their sublogics. However, $1 \Vdash \Box p$, $1 \Vdash \Box q$, but $1 \nVdash \Box(p \wedge q)$, since $\|p \wedge q\| = \emptyset$, and $\emptyset \notin S(1)$. So C fails. Thus EM, EVZ and all their sublogics are distinct from their C-extensions.

[Z.] EMC is the strongest logic without Z. Consider the model where $U = \{1,2\}$, $S(1) = S(2) = \{\{1,2\}\}$, $V(p) = \{1\}$, $V(q) = \{2\}$. Conditions (m) and (c) are verified. Now $1 \Vdash \Box(p \vee q)$, since $\|p \vee q\| = \{1,2\}$ and $\{1,2\} \in S(1)$, but $1 \nVdash \Box p$ and $1 \nVdash \Box q$, so this model falsifies an instance of Z. Recall that W entails Z, and notice that this model falsifies W as well. Hence EMC and its sublogics are distinct from EZ and its extensions up to EWV.

[Y.] EWV is the strongest logic without Y. Consider the model where $U = \{1,2\}$, $S(1) = S(2) = \{\emptyset\}$, $V(p) = \{1\}$, $V(q) = \{2\}$. Conditions (w) and (v) are satisfied (two sets belong to $S(1)$ iff their union belongs, too). However, $1 \Vdash \Box(p \wedge q)$, since $\|p \wedge q\| = \emptyset$, but $1 \nVdash \Box p$ and $1 \nVdash \Box q$, falsifying an instance of Y. Since M entails Y, this model falsifies M as well. Hence EWV and its sublogics are distinct from EY and its extensions up to EMC.

[W.] EMC and EZVC are the strongest logics without W. We have shown above that Z is not a theorem of EMC and, since W entails Z, neither is W. Now consider the model where $U = \{1,2\}$, $S(1) = S(2) = \{\{1\}\}$, $V(p) = \{1\}$, $V(q) = \emptyset$. Notice that conditions (c), (v), and (z) are all satisfied, so this is an EZVC-model, as well as a model for all its sublogics. However, $1 \Vdash \Box(p \vee q)$, since $\|p \vee q\| = \{1\}$, but $1 \nVdash \Box q$, since $\emptyset \notin S(1)$. Hence W fails, and EW is distinct from EMC, EZVC, and all their sublogics.

[M.] EWV and EYVC are the strongest logics without M. We have shown above that Y is not a theorem of EWV and, since M entails Y, neither is M. Consider now the model where $U = \{1,2\}$, $S(1) = S(2) = \{\{1\}\}$, $V(p) = \{1\}$, $V(q) = \emptyset$. Notice that conditions (y), (c), and (v) are all satisfied, so this is an EVCY-model, as well as a model for all its sublogics. However, $1 \Vdash \Box(p \wedge q)$, since $\|p \wedge q\| = \{1\}$, but $1 \nVdash \Box q$, so M fails. Hence EM is distinct from EWV, EYVC, and all their sublogics.

[V.] EW and EYC are the strongest logics without V. Consider first the model $U = \{1,2\}$, $S(1) = S(2) = \{\emptyset, \{1\}, \{2\}\}$, $V(p) = \{1\}$, $V(q) = \{2\}$. (w) is satisfied, so this is an EW-model. Now $1 \vDash \Box p$ and $1 \Vdash \Box q$, but $1 \nVdash \Box(p \vee q)$, since $\|p \vee q\| = U$ and $U \notin S(1)$. Hence V is falsified in an EW-model. Consider now the model where $U = \{1,2,3\}$, $S(1) = S(2) = S(3) = \{\{1,2\}, \{2\}, \{2,3\}\}$. Conditions (y) and (c) are satisfied, so we have a model for EYC. We see that $1 \vDash \Box p$, $1 \Vdash \Box q$, but $1 \nVdash \Box(p \vee q)$, since $\|p \vee q\| = U$ and $U \notin S(1)$. Hence V is falsified in an EYC-model. Thus EW, EYC and all their sublogics are distinct from their V-extensions. □

6 Completeness

Being fond of canonical models, we will use them to demonstrate completeness theorems for several of the logics considered here. We start with a few definitions and lemmas (where L is any of our logics).

Definition 6.1. A set Γ of formulas is L-inconsistent if there is a finite subset of formulas $\{\alpha_1, \ldots, \alpha_n\} \subseteq \Gamma$ such that $\vdash_L \neg(\alpha_1 \wedge \ldots \wedge \alpha_n)$; otherwise Γ is L-consistent.

If Γ is finite, i.e., $\Gamma = \{\gamma_1, \ldots, \gamma_n\}$, Γ is L-consistent if and only if $\nvdash_L \neg(\gamma_1 \wedge \ldots \wedge \gamma_n)$. And a singleton $\{\alpha\}$ is L-consistent, of course, if and only if $\nvdash_L \neg\alpha$.

Definition 6.2. Let Γ be a set of formulas and α a formula. We say that $\Gamma \vdash_L \alpha$ if there is a finite subset of formulas $\{\alpha_1, \ldots, \alpha_n\} \subseteq \Gamma$ such that $\vdash_L (\alpha_1 \wedge \ldots \wedge \alpha_n) \to \alpha$.

Definition 6.3. A set Γ of formulas is *maximal* if, for every formula α, either $\alpha \in \Gamma$ or $\neg\alpha \in \Gamma$. Γ is a *maximal consistent set* (MCS) if it is maximal and consistent.

The proofs of the following lemmas are standard, so we will omit them.

Lemma 6.4. *Let Δ be an MCS, and α and β any formulas. Then:*

(i) $\bot \notin \Delta$

(ii) $\alpha \in \Delta$ *iff* $\neg\alpha \notin \Delta$;

(iii) $\alpha \wedge \beta \in \Delta$ *iff* $\alpha \in \Delta$ *and* $\beta \in \Delta$;

(iv) $\alpha \vee \beta \in \Delta$ *iff* $\alpha \in \Delta$ *or* $\beta \in \Delta$;

(v) $\alpha \to \beta \in \Delta$ *iff* $\alpha \notin \Delta$ *or* $\beta \in \Delta$;

(vi) $\alpha \leftrightarrow \beta \in \Delta$ *iff* $\alpha \in \Delta$ *and* $\beta \in \Delta$, *or* $\alpha \notin \Delta$ *and* $\beta \notin \Delta$;

(vii) *if* $\alpha \in \Delta$ *and* $\alpha \to \beta \in \Delta$ *then* $\beta \in \Delta$.

Lemma 6.5 (Lindenbaum). *Let Γ be a consistent set of formulas. Then there is an MCS Δ such that $\Gamma \subseteq \Delta$.*

Lemma 6.6. $\vdash_L \alpha$ *iff for every L-MCS* Γ, $\alpha \in \Gamma$.

Where L is a logic, let $\mathfrak{S}_L$ be the set of all maximal consistent sets (MCSs) of formulas in L. Let $|A|_L = \{\Gamma \in \mathfrak{S}_L : A \in \Gamma\}$. The set $|A|_L$ is called the *proof set* of A in L. A subset X of $\mathfrak{S}_L$ a called a *proof set* if $X = |A|_L$ for some A. Notice that there are subsets of $\mathfrak{S}_L$ which are not proof sets of any formula; such sets will be called *non-proof sets*.

Definition 6.7. Let L be a classical modal logic. We say that $\mathfrak{M}_L = \langle U_L, S_L, V_L \rangle$ is a *canonical model for L* iff it satisfies the following conditions:

(i) $U_L = \mathfrak{S}_L$;

(ii) $|A|_L \in S_L(\Gamma)$ iff $\Box A \in \Gamma$, for all $\Gamma \in U_L$ and every A;

(iii) $V_L(\mathbf{p}) = |\mathbf{p}|_L$, for every $\mathbf{p} \in \Phi$.

Now a logic L does not have just one canonical model, even though the universe of a canonical model always is the set $\mathfrak{S}_L$ of all maximal L-consistent sets. The reason is the following: a proof set $|A|_L$ belongs to $S(\Gamma)$, for some $\Gamma \in U_L$, if and oly if $\Box A \in \Gamma$. However, $S(\Gamma)$ can also contain any number of *non-proof* sets. We can, thus, have several different canonical model constructions for a logic, from the *smallest* canonical model ($S(\Gamma)$ contains only the proof sets of formulas A such that $\Box A \in \Gamma$) to the *largest* one ($S(\Gamma)$ contains in addition all non-proof sets).

Lemma 6.8. *Let $\mathfrak{M}$ be a canonical model for a logic L. Then, for every wff A and every $\Gamma \in U_L$, $\mathfrak{M}, \Gamma \Vdash A$ iff $A \in \Gamma$.*

Proof. By induction on formulas. Let Γ be some element of U_L.
(a) $A = \mathbf{p}$, for some $\mathbf{p} \in \Phi$. By definition, $\Gamma \Vdash \mathbf{p}$ iff $\Gamma \in V_L(\mathbf{p})$ iff $\Gamma \in |\mathbf{p}|_L$. By construction of $|\mathbf{p}|_L$, Γ is a set in $|\mathbf{p}|_L$ iff $\mathbf{p} \in \Gamma$.
(b) $A = \bot$. By definition, $\Gamma \nVdash \bot$. And since every element of U is a consistent set, $\bot \notin \Gamma$.
(c) $A = \neg B$. By definition, $\Gamma \Vdash \neg B$ iff $\Gamma \nVdash B$. By the inductive hypothesis, $\Gamma \Vdash B$ iff $B \in \Gamma$, so $\Gamma \nVdash B$ iff $B \notin \Gamma$. Now $B \notin \Gamma$ iff $\neg B \in \Gamma$. Thus $\Gamma \Vdash \neg B$ iff $\neg B \in \Gamma$.
(d) $A = B \wedge C$. By definition, $\Gamma \Vdash B \wedge C$ iff $\Gamma \Vdash B$ and $\Gamma \Vdash C$. By the inductive hypothesis, $\Gamma \Vdash B$ iff $B \in \Gamma$, and $\Gamma \Vdash C$ iff $C \in \Gamma$. Now $B \in \Gamma$ and $C \in \Gamma$ iff $B \wedge C \in \Gamma$. Thus $\Gamma \Vdash B \wedge C$ iff $B \wedge C \in \Gamma$.
(e) $A = \Box B$. Suppose that $\Gamma \in U_L$. By definition, $\Gamma \Vdash \Box B$ iff $\|B\|^{\mathfrak{M}} \in S_L(\Gamma)$. By the inductive hypothesis, for every $\Delta \in U$ we have that $\Delta \Vdash B$ iff $B \in \Delta$; that is, $\|B\|^{\mathfrak{M}} = |B|_L$. So $\|B\|^{\mathfrak{M}} \in S_L(\Gamma)$ iff $|B|_L \in S_L(\Gamma)$. Now, by definition of S_L, $|B|_L \in S_L(\Gamma)$ iff $\Box B \in \Gamma$. Hence, $\Gamma \Vdash \Box B$ iff $\Box B \in \Gamma$. □

From this lemma it follows immediately that:

Theorem 6.9 (Completeness for E). *Let Γ be an E-consistent set of formulas. Then Γ has a model.*

Proof. Suppose that Γ is E-consistent. By a standard Lindenbaum argument we can show that there exists an E-MCS Δ such that $\Gamma \subseteq \Delta$. Since Δ is an E-MCS, Δ is a state in a canonical model $\mathfrak{M}_{\mathsf{E}}$ for E. By the previous lemma, $\mathfrak{M}_{\mathsf{E}}, \Delta \Vdash A$, for every $A \in \Delta$. Since $\Gamma \subseteq \Delta$, Γ has a model. □

As a consequence, E is determined by the class of all frames. With regard to the other logics, we need to show that they have canonical models satisfying the needed conditions, because not every canonical model does. As an example, take EM. To show that this logic is determined by the class of all frames satisfying condition (m), we need to show that there is a canonical model whose frame belongs to that class. However, a well-known result is that the *smallest* EM-canonical model does not satisfies (m), and this is why: let $|A|$ the proof set of some atomic formula A, Γ some point in the smallest canonical model, and suppose $|A| \in S(\Gamma)$. Let now X be any non-proof set such that $|A| \subseteq X$. If condition (m) were satisfied, X should belong to $S(\Gamma)$, but it doesn't, because, in the smallest canonical model, only proof sets belong to $S(\Gamma)$.

Let us begin with logics regarding which the smallest canonical model is all we need.

Proposition 6.10. *Let L be any logic that has N or O. Then the smallest L-canonical model satisfies the corresponding (n) or (o) condition.*

Proof. We show (o) as an example. Let $\mathfrak{M}_L = \langle U_L, S_L, V_L \rangle$ be a canonical model for a logic that has O. Then, since $\Box\bot \in \Gamma$ for every Γ in U_L, $|\bot|_L \in S_L(\Gamma)$. Since $|\bot| = U_L$, $U_L \in S_L(\Gamma)$. □

Proposition 6.11. *Let L be one of the logics* E, EC, EV, EVC, *or their extensions with N or O. Then the smallest L-canonical model satisfies the corresponding conditions.*

Proof. For E, EC, EN and ECN the result is known. As an example, consider the smallest canonical model for EV. Let X and Y be subsets of U such that $X \in S(\Gamma)$ and $Y \in S(\Gamma)$. But then there are formulas A and B such that $X = |A|$ and $Y = |B|$. By the definition of a canonical model, $\Box A \in \Gamma$ and $\Box B \in \Gamma$, and it follows by V that $\Box(A \wedge B) \in \Gamma$. So $|A \wedge B| \in S(\Gamma)$. And since, what can easily be shown, $|A \wedge B| = |A| \cap |B|$, $X \cap Y \in S(\Gamma)$ and condition (v) is verified. □

For some other logics, the *largest* canonical model will do the job.

Proposition 6.12. *Let L be a logic.*

(a) *If L is EY, EZ, or their extensions with N or O, the largest L-canonical model satisfies the corresponding conditions.*

Proof. Consider the largest EY-canonical model. Let X and Y be subsets of U such that $X \cap Y \in S(\Gamma)$, for some $\Gamma \in U$. If either X or Y is a nonproof set, then trivially X or Y is in $S(\Gamma)$. So suppose both X and Y are proof sets. For some A and B, $X = |A|$ and $Y = |B|$. Then $X \cap Y = |A| \cap |B| = |A \wedge B|$. But then $\Box(A \wedge B) \in \Gamma$, from what it follows that $\Box A$ or $\Box B$ are in Γ, so either X or Y belong to $S(\Gamma)$.

For the other logics, the proof is analogous. □

How do we proceed, then, to obtain the desired canonical models for the other logics? The trick consists in taking the smallest model and adding enough non-proof sets to $S_L(\Gamma)$, for every $\Gamma \in U_L$, so the model will satisfy the desired condition—and still be a canonical model for the logic in question. For EM and EMC there is a well-known solution: supplementation (see, for instance, [2]).

Here the details. Let $M = \langle U, S, V\rangle$ be the smallest canonical model for a logic L. The *supplementation* of $\mathfrak{M}$ is the model $\mathfrak{M}^+ = \langle U_L, S^+, V_L\rangle$, such that, for every $\Gamma \in U_L$ and every $X \subseteq U_L$,

$$X \in S_L^+(\Gamma) \quad \text{iff} \quad Y \subseteq X \text{ for some } Y \in S_L(\Gamma).$$

That is, $S_L^+(\Gamma) = \{X \subseteq U_L : X \subseteq |A|_L \text{ for some } \Box A \in \Gamma\}$. Obviously, for every $\Gamma \in U_L$, $S_L(\Gamma) \subseteq S_L^+(\Gamma)$.

Of course, we need to show that the supplemention is indeed a canonical model. This, and the proof of the next proposition, are done in [2], ch. 9, but see the proof of Proposition 6.14 below, which is analogous.

Proposition 6.13. *Let L be one of the logics* EM, EMC, EMN, EMCN. *Then the supplementaion of the smallest L-canonical model satisfies the corresponding conditions.*

For EW, EWV and their extensions with O, we can use a similar construction. Let $M = \langle U, S, V\rangle$ be the smallest canonical model for a logic L. The *anti-supplementation* of $\mathfrak{M}$ is the model $\mathfrak{M}_L^- = \langle U_L, S_L^-, V_L\rangle$ be the model where, for every $\Gamma \in U_L$ and every $X \subseteq U_L$,

$$X \in S_L^-(\Gamma) \quad \text{iff} \quad X \subseteq Y \text{ for some } Y \in S_L(\Gamma).$$

That is, $S_L^-(\Gamma) = \{X \subseteq U_L : X \subseteq |A|_L \text{ for some } \Box A \in \Gamma\}$. Obviously, for every $\Gamma \in U_L$, $S_L(\Gamma) \subseteq S_L^-(\Gamma)$.

Proposition 6.14. *Let L be one of the logics* EW, EWV, EWO, EWVO. *Then the anti-supplementation of the smallest L-canonical model satisfies the corresponding conditions.*

Proof. Suppose W is a theorem of L. We have to prove that the anti-supplementation $\mathfrak{M}_L^-$ is a canonical model for L. To do is, it is enough to show that condition (ii) of the definition is satisfied, that is, for every A and every $\Gamma \in U_L$,

$$|A|_L \in S_L^-(\Gamma) \quad \text{iff} \quad \Box A \in \Gamma.$$

If $\Box A \in \Gamma$ then $|A|_L \in S_L(\Gamma)$, since $\mathfrak{M}_L$ is canonical for L, and so $|A|_L \in S_L^-(\Gamma)$. For the other direction, suppose that $|A|_L \in S_L^-(\Gamma)$. Thus, for some $Y \supseteq |A|_L$,

$Y \in S_L(\Gamma)$. Since $\mathfrak{M}_L$ is the smallest canonical model, this means that, for some B, $Y = |B|_L$. It follows that $|A|_L \subseteq |B|_L$, and $\Box B \in \Gamma$. Now this means that $\vdash_L A \to B$, and it follows by RW (which L has, if it has W) that $\vdash_L \Box B \to \Box A$. Hence, $\Box B \in \Gamma$.

So $\mathfrak{M}_L^-$ is a canonical model for L. We show that is satisfies the required conditions. (That (o) is verified was already shown above.)

[EW.] Let Γ be an element of U_L, and X and Y be subsets of U_L such that $X \cup Y \in S_L^-(\Gamma)$. By construction, there must be some formula A such that $|A|_L \in S(\Gamma)$ and $X \cup Y \subseteq |A|_L$. But obviously $X \subseteq X \cup Y$ and $Y \subseteq X \cup Y$, so $X \subseteq |A|_L$, $Y \subseteq |A|_L$ and, again by construction, $X \in S_L^-(\Gamma)$ and $Y \in S_L^-(\Gamma)$.

[EWV.] Let Γ be an element of U_L, and X and Y be subsets of U_L such that both X and Y are in $S_L^-(\Gamma)$. Then there are formulas A and B such that $X \subseteq |A|$, $Y \subseteq |B|$, and both $|A|$ and $|B|$ are in $S(\Gamma)$. Hence $\Box A$ and $\Box B$ are in Γ, from what it follows by V that $\Box(A \vee B) \in \Gamma$, and hence that $|A \vee B| \in S(\Gamma)$. Given that $X \subseteq |A|$ and $Y \subseteq |B|$, $X \cup Y \subseteq |A \vee B|$. Thus, $X \cup Y \in S_L^-(\Gamma)$.

[EWO and EWVO.] Since $\Box\bot$ is a theorem of these logics, $\Box\bot$ belongs to every MCS Γ in the canonical model, so $|\bot| \in S(\Gamma)$. □

This leaves us with logics having Y or Z, and also in these cases the smallest canonical model will not work.

For logics having Y, we can define the following construction, a kind of *weak supplementation.* Consider the smallest L-canonical model $\mathfrak{M}_L = \langle U_L, S_L, V_L\rangle$, and let $\mathfrak{M}_L^* = \langle U_L, S_L^*, V_L\rangle$ be the model where, for every $\Gamma \in U_L$ and every $X \subseteq U_L$,

$$X \in S_L^*(\Gamma) \quad \text{iff} \quad X \in S_L(\Gamma) \text{ or } X \text{ is a non-proof set and, for some } Y \in S_L(\Gamma),\ Y \subseteq X.$$

Since only non-proof sets are added to $S_L(\Gamma)$, $\mathfrak{M}_L^*$ is a canonical model for L.

For logics having Z, we can define a similar construction, a kind of *weak anti-supplementation.* Consider the smallest L-canonical model $\mathfrak{M}_L = \langle U_L, S_L, V_L\rangle$, and let $\mathfrak{M}_L^\star = \langle U_L, S_L^\star, V_L\rangle$ be the model where, for every $\Gamma \in U_L$ and every $X \subseteq U_L$,

$$X \in S_L^\star(\Gamma) \quad \text{iff} \quad X \in S_L(\Gamma) \text{ or } X \text{ is a non-proof set and, for some } Y \in S_L(\Gamma),\ X \subseteq Y.$$

Since only non-proof sets are added to $S_L(\Gamma)$, $\mathfrak{M}_L^\star$ is a canonical model for L.

Proposition 6.15. *Let L be a logic.*

(a) *If L is* EY, EYC, EVY,EYVC, *or their extensions with N or O, the weak supplementation of the smallest L-canonical model satisfies the corresponding conditions.*

(b) *If L is* EZ, EZC, EVZ,EZVC, *or their extensions with N or O, the weak anti-supplementation of the smallest L-canonical model satisfies the corresponding conditions.*

Proof. We show case (b); (a) is analogous.

Suppose that Z is a theorem of L, and let X and Y be any subsets of U_L such that $X \cup Y \in S^{\star}_L(\Gamma)$. If X and Y are proof sets then, for some A and B, $X = |A|$ and $Y = |B|$. But then $X \cup Y = |A| \cup |B| = |A \vee B|$, so $\Box(A \vee B) \in \Gamma$. Given that Z is a theorem, then $\Box A \vee \Box B \in \Gamma$, and then either $|A| \in S^{\star}_L(\Gamma)$ or $|B| \in S^{\star}_L(\Gamma)$, so (z) holds.

Suppose now that X and Y are non-proof sets, and that $X \cup Y$ is a non-proof set, too. Then, for some A such that $|A| \in S_L(\Gamma)$, $X \cup Y \subseteq |A|$. Now, since $X \subseteq X \cup Y$, we have that $X \subseteq |A|$, so by construction $X \in S^{\star}_L(\Gamma)$ and (z) holds. If $X \cup Y$ is a proof set, again $X \in S^{\star}_L(\Gamma)$ and (z) holds.

If now one of X, Y is a proof set and the other not, the argument is analogous.

We can also show that, if C or V are theorems, that conditions (c) or (v) holds. □

From all of the above results, determination theorems follow for all logics in Figure 1 and their extensions with N or O. (Decidability should follow using filtrations, but this would be a subject for another paper.)

7 More on neutral readings

In the previous sections, we discussed only four logics (E, EV, EC, and EVC) in which □ can be read either as necessity or impossibility. But there are more.

Consider the following schemes:

K. $\Box(A \to B) \to (\Box A \to \Box B)$

X. $(\Box(A \to B) \wedge \Box(B \to C)) \to \Box(A \to C)$

X_1. $(\Box(A \to B) \wedge \Box(B \to C)) \to \Box\top$

K is a standard axiom in normal modal logics, and X and X_1 are discussed in [3], a paper by B. Chellas and K. Segerberg where prenormal modal logics were introduced.

The reader familiar with [3] may wonder why we are listing both X and X_1 as candidate axioms. They are equivalent in *prenormal* logics, but this does not in general. A logic is said *prenormal* if it provides the following schema:

nK. $\Box\top \to (\Box(A \to B) \to (\Box A \to \Box B))$.

Now if a logic is prenormal, it has X iff it has qX, which is the following:

qX. $\Diamond\bot \to ((\Box(A \to B) \land \Box(B \to C)) \to \Box(A \to C))$.

In [3], Chellas and Segerberg show that qX is equivalent to X_1, and also that a prenormal logic has X iff it has qX. This result does not carry to classical modal logics which are not prenormal. It is easily shown that if a classical logic has X, then it has qX (or X_1 for that matter), but we can have (and will exhibit later) models in which X_1 is true but X is false.

The role of K and X_1 can also be played by the following inference rules.

RK′. $A \leftrightarrow (B \to C)$ / $\Box A \to (\Box B \to \Box C)$

RX_1. $A \lor B$ / $(\Box A \land \Box B) \to \Box\top$

A word about nomenclature. RX_1 is called RqX in [3]. And for consistency, RK′ should be named simply RK. However, RK is already standard for the following, much stronger inference rule:

RK. $(A_1 \land \ldots \land A_n) \to B$ / $(\Box A_1 \land \ldots \land \Box A_n) \to \Box B$, for $n \geq 0$.

In fact, a logic closed under RK is a *normal modal logic*, having M, C, X and N as theorems.

Let us now establish some facts about the connections between these axioms and rules.

Theorem 7.1. *Let L be a classical logic. Then:*

(a) *L has* RK′ *iff it has* K*;*

(b) *L has* RX_1 *iff it has* X_1.

Some further results are:

Theorem 7.2. *Let L be a classical logic. Then:*

(a) *if L has* V*, then it has* X_1*;*

(b) *if L has* V *and* K*, then it has* X*;*

(c) *if L has* M*, then it has* C *iff it has* K *iff it has* X*;*

(d) *if L has* W*, then it has* K*;*

(e) *if L has* W *and* X_1*, then it has* X*;*

(f) *if L has* X*,* C*, and* Y*, then it has* V*;*

(g) *L has* X *iff it has* X_1 *and* K.

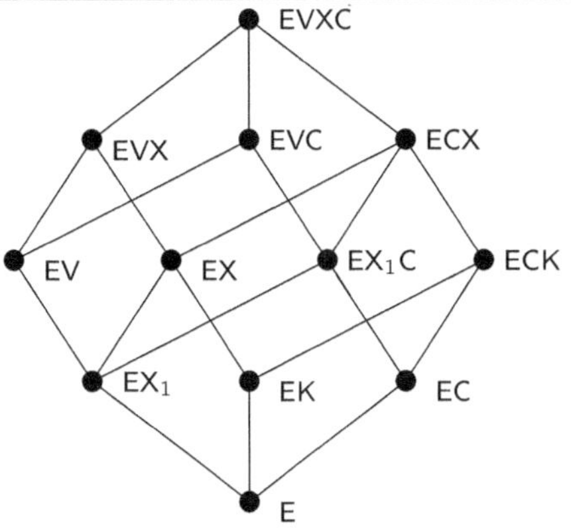

Figure 2: More Group III logics

Proofs of these results are not difficult, sometimes pretty much straightforward, so we will omit them here. As a first example, we show that if a classical logic has V and K, then it has X. If L has V, then it has RV. Now, since $(A \to B) \vee (B \to C)$ is a tautology, so is $((A \to B) \vee (B \to C)) \leftrightarrow \top$. By RV, we get $\Box(A \to B) \wedge \Box(B \to C) \to \Box\top$, that is, X_1 which is equivalent to Chellas and Segerberg's qX. If now L also has K, then it has nK and is prenormal, in which case it has qX iff it has X. Thus, if L has V and K, it has X. As a second example, if a classical logic L has X, then it follows easily that it has X_1 and K. For the other direction, if L has K then it is prenormal and, having X_1, also has X.

So if we take K, X_1, and X in consideration, we obtain the twelve logics (shown in Figure 2) in which $\Box$ can be read either as necessity or impossibility, instead of the four we started with.

We can show that they are all distinct. For this, we need the natural conditions corresponding to axiom schemas K, X_1 and X. They are the following:

(k) if $-X \cup Y \in S(x)$ and $X \in S(x)$, then $Y \in S(x)$;

(x_1) if $X \in S(x)$, $Y \in S(x)$, and $X \cup Y = U$, then $U \in S(x)$;

(x) if $-X \cup Y \in S(x)$ and $-Y \cup Z \in S(x)$, then $-X \cup Z \in S(x)$;

A note on the conditions (x_1) and (x). In [3], Chellas and Segerberg present the following condition for qX, an axiom equivalent to X_1:

(qx) if $X \in S(x)$ and $Y \in S(x)$, then $X \cup Y \neq U$.

And they state (p. 22, fn. 13) that (qx) and (x) are equivalent conditions. Now this is not in general the case, but (qx) and (x_1) are equivalent in [3] because Chellas and

Segerberg work with frames having normal and queer points, having an acessibility relation (for truth conditions in normal points) and a neighborhood function (for queer points). They also require, of their frames, for every queer point $x \in U$, that $U \notin S(x)$. It is also worth noticing that, for their (prenormal) frames, (qx) and (x_1) are equivalent. Since we are *not* requiring that $U \notin S(x)$, we can show that these conditions are not equivalent. Of course, if a frame satisfies (x), then it also satisfies (x_1), and (qx). For suppose that two subsets X and Y of U are such that that $X \in S(x)$, $Y \in S(x)$, If we also have that $X \cup Y = U$, then $Y = -X \cup Y$, and $X = -Y \cup X$. Assuming that (x) holds, if $-X \cup Y$ and $-Y \cup X$ are in $S(x)$, the so is $-X \cup X$ — but this is just U.

The following is an example of a model in which (x_1) holds, but (x) fails.

Let $\mathfrak{M} = \langle U, S, V\rangle$ be such that $U = \{1, 2, 3, 4\}$ and S such that for all $x \in U$, $S(x) = \{\{2, 3, 4\}, \{1, 4\}, U\}$, V such that $V(p) = \{1, 2\}$, $V(q) = \{2, 3\}$, and $V(r) = \{4\}$. First, $\|p \to q\| = -\|p\| \cup \|q\| = \{2, 3, 4\}$, and $\|q \to r\| = -\|q\| \cup \|r\| = \{1, 4\}$. So, for instance, $1 \Vdash \Box(p \to q)$ and $1 \Vdash \Box(q \to r)$. However, $1 \nVdash \Box(p \to r)$, since $\|p \to r\| = -\|p\| \cup \|r\| = \{3, 4\}$, and $\{3, 4\} \notin S(1)$. On the other hand, (x_1) holds: for every $x \in U$, $\{2, 3, 4\} \in S(x)$, $\{1, 4\} \in S(x)$, $\{2, 3, 4\} \cup \{1, 4\} = U$, and $U \in S(x)$. (Incidentally, this is also an EV-model, so X is not a theorem of EV.)

The correspondence between axiom schemas and their natural conditions is show below.

Lemma 7.3. *The schemas* K, X_1, *and* X *are valid in a frame* $\mathfrak{F}$ *iff* $\mathfrak{F}$ *satisfies the conditions* (k), (x_1), *and* (x), *respectively.*

Soundness results follow easily from this.

Theorem 7.4. *The twelve logics shown in Figure 2 are all distinct.*

Proof. Some of the results were already proven in Theorem 5.3; others (like EC and EK being distinct logics) are known in the literature (see [2]). We need to show that K (hence X) fails in an EVC-model, that X_1 (hence X) fail in an ECK-model, and that V fails in an ECX-model.

[For K.] Consider the model where $U = \{1, 2, 3\}$, $S(1) = \{\{1\}, \{1, 2\}, U\}$, $V(p) = \{1\}$, $V(q) = \{1, 3\}$. Conditions (c), and (v) are satisfied, so this is an EVC-model. However, K is falsified. We see that $1 \Vdash \Box p$ and also that $1 \vdash \Box(p \to q)$ (since $\|p \to q\| = -\|p\| \cup \|q\| = U$), but $1 \nVdash \Box q$.

[For X_1.] Consider the model where $U = \{1, 2\}$, $S(1) = \{\emptyset, \{1\}, \{2\}\}$, $V(p) = \{1\}$. This model is closed under intersections, so (c) is verfied. Condition (k) also holds. So this is an ECK-model. However, $1 \Vdash \Box p$ and $1 \Vdash \Box\neg p$. Since $p \vee \neg p$ is a tautology, $(\Box p \wedge \Box\neg p) \to \Box\top$ should be true at 1, but it is not, since $\{1\} \cup \{2\} = U$, but $U \notin S(1)$, so X_1 (and hence X) fails.

[For V.] Consider now the model where $U = \{1,2,3\}$, $S(1) = S(2) = S(3) = \{\{1\},\{2\},\emptyset\}$, $V(p) = \{1\}$, $V(q) = \{2\}$. This model is closed under intersections, so (c) is verified. Since no union of sets belonging to $S(1)$ is equal to U, (x) is also vacuosly verified. So this is an ECX-model. However, V is falsified, since $1 \Vdash \Box p$, $1 \Vdash \Box q$, but $1 \nVdash \Box(p \vee q)$, given that $\|p \vee q\| \notin S(1)$. □

For completeness, we need to find a canonical model for a logic satisfying the required conditions.

Proposition 7.5. *For* EX_1 *and* $\mathsf{EX}_1\mathsf{C}$, *the smallest canonical model satisfies the corresponding conditions.*

Proof. We already know that (c) is verified in the smallest canonical model of a logic having C. So consider X_1, some Γ in the smallest canonical model, and suppose X and Y are sets in $S(\Gamma)$ such that $X \cup Y = U$. But then there are formulas A and B such that $X = |A|$ and $Y = |B|$. This implies that $|A| \cup |B| = |\top|$ and $A \vee B$ is a theorem, from what we obtain $\Box A \wedge \Box B \to \Box\top$. Now we have that $\Box A$ and $\Box B$ are in Γ, so $|\top| = U$ is in $S(\Gamma)$, and condition (x_1) is verified. □

For the remaining logics we will need some constructions introduced by Roy Benton in an (unfortunately still) unpublished paper of 1975 (see [1]).[2] Completeness for EK and ECK was proven in [1] using another canonical model construction, *overlay canonical models*. An overlay canonical model is a model where

> $X \in S_L(\Gamma)$ iff there are formulas A and B such that: $|A|_L \subseteq X \subseteq |B|_L$ and, for every C, if $|A|_L \subseteq |C|_L \subseteq |B|_L$ then $\Box C \in \Gamma$.

Proposition 7.6. *For* EK, ECK, EX, ECX, EVX, *and* EVXC, *overlay canonical models satisfy the corresponding conditions.*

Proof. Since a logic has X iff it has X_1 and K, it is enough to show that overlay canonical models verify (x_1) and (k). We know ([1]) that they satisfy both (k) and (c). To see that (x_1) is verified, suppose there are sets X and Y belonging to $S(\Gamma)$, for some $\Gamma \in U_L$, such that $X \cup Y = U$. Then there are formulas A, B, C and D such that $|A| \subseteq X \subseteq |B|$ and $|C| \subseteq X \subseteq |D|$. Obviously $|B| \cup |D| = U$, so $B \vee D$ is a theorem and, since by definition $\Box B$ and $\Box D$ are in Γ, $\Box\top \in \Gamma$ and $U \in S(\Gamma)$.

So an overlay canonical model satisfies conditions (k) and (x_1).

[2]Thanks to Prof. B. Chellas for putting me in contact with Prof. Roy Benton, and many thanks to Prof. Benton for kindly providing me a copy of his paper.

(v) Suppose there are setx X and Y belonging to $S(\Gamma)$, for some $\Gamma \in U_L$, such that $X \cup Y = U$. Then there are formulas A, B, C and D such that $|A| \subseteq X \subseteq |B|$ and $|C| \subseteq X \subseteq |D|$. Two uses of V give us that $|A \vee B|$ and $|C \vee D|$ are in $S(\Gamma)$. Thus we have $|A \vee B| \subseteq X \cup Y \subseteq |C \vee D|$. Let E be a formula such that $|A \vee B| \subseteq |E| \subseteq |C \vee D|$. We need to show that $\Box E \in \Gamma$. Now clearly $|E| = |E| \cap (|B| \cup |D|)$, that is, $|E| = (|E| \cap |B|) \cup (|E| \cap |D|)$. But since $|A| \subseteq (|E| \cap |B|) \subseteq |B|$ and $|C| \subseteq (|E| \cap |D|) \subseteq |D|$, we have that $\Box(E \wedge B)$ and $\Box(E \wedge D)$ are in Γ, so $\Box E \in \Gamma$ and we are done. $\Box$

This is not the end of the story, however. In [3], Chellas and Segerberg present an hierarchy of *Cresswell logics*, that is, logics having the Cresswell rule:

RC_n. $A_0 \vee \ldots \vee A_{n-1} \;/\; (\Box A_0 \wedge \ldots \wedge \Box A_{n-1}) \to \Box\top$, for $n \geq 2$.

Notice that in all these logics $\Box$ still can be read as impossibility. And a good question is how far can we still go adding more axioms without having to commit to one of the readings of $\Box$. In (*previous work*), using relational frames with normal and opposite worlds, and no conditions on the accessibility relation, we presented a logic which extends the strongest Cresswell logic PX_ω. Extensions of this logic obtained by imposing conditions (like reflexivity) on the accessibility relation remain to be investigated.

Finally, logics in which $\Box$ can be read either as necessity or impossibility are natural candidates for (weak?) *noncontingency logics*, a topic also worth investigating.

8 Final remarks

In this final section, we discuss some further results and point directions to additional research on the topic of this paper.

8.1 Mixing the readings

If we look again at the diagram on Figure 1, we see that, on the left side, we have EMC and its sublogics, down to E, logics in which $\Box$ can (or must, if the logic contains M) be read as necessity. We also have extensions of these logics adding N ($\Box\top$) as a thesis. On the right side, we have EWV and its sublogics, down to E, logics in which $\Box$ can (or must, if the logic contains W) be read as impossibility. We also have the extensions of these logics by adding O ($\Box\bot$) as a thesis. Now, what would happen if we mix things up? For instance, suppose we add O to EMC, or N to EWV?

Take, for instance, EM plus O, and consider a model for this logic. Conditions (m) and (o) will have to hold. So, for every $x \in U$, $\emptyset \in S(x)$ — this is (o). Now

(m) is equivalent to closure under supersets, so, for every $x \in U$, and every $X \subseteq U$, $X \in S(x)$. As a result, for every formula A, $\Box A$ turns out to be valid. For similar reasons, adding N to EW will give us the same result. So EMO, EMCO, EWN and EWVN are all the same logic — the normal system called Ver in [4], which we obtain adding, to the smallest normal logic K, $\Box A$ as an axiom.

A somewhat weaker logic is what we get combining EMC and EWV. Since W entails C in E, and M entails V, this logic is just EMW. Having conditions (m) and (w) on our models gives us the following picture: for every $x \in U$, either $S(x) = \emptyset$, or $S(x) = \mathcal{P}(U)$, where $\mathcal{P}(U)$ is the power set of U. So either nothing is necessary/impossible, or everything is.

Weaker logics can be obtained, say, by adding Z to EY or EM, and so on.

8.2 Some thoughts on negation

Consider again the schemes below:

Y. $\Box(A \wedge B) \to (\Box A \vee \Box B)$

W. $\Box(A \vee B) \to (\Box A \wedge \Box B)$

W′. $(\Box A \vee \Box B) \to \Box(A \wedge B)$

V. $(\Box A \wedge \Box B) \to \Box(A \vee B)$

Replacing $\Box$ with $\neg$ we get the four cases of De Morgan's Laws:

$\mathrm{Y}_\neg$. $\neg(A \wedge B) \to (\neg A \vee \neg B)$

$\mathrm{W}_\neg$. $\neg(A \vee B) \to (\neg A \wedge \neg B)$

$\mathrm{W}'_\neg$. $(\neg A \vee \neg B) \to \neg(A \wedge B)$

$\mathrm{V}_\neg$. $(\neg A \wedge \neg B) \to \neg(A \vee B)$

Now this suggests that we could add one or more of the axioms above to E to obtain logics with an additional, non-classical negation represented by $\Box$ — for instance, a paraconsistent one. As we proved before, W and W′ are equivalent in classical modal logics, so this leaves us with only Y, W and V. Considering now that Z and C are theses, if W is, the only schemes that do not admit a negation reading are M and N: $\neg(A \wedge B) \to (\neg A \wedge \neg B)$ and $\neg\top$ are intuitively false. As a result, the only two logics in the diagram of Figure 1 that do not admit a negation reading of $\Box$ are EM and EMC (as well as extensions of other logics with N). The remaining logics can be seen as encoding different properties of a non-classical negation.

What happens now to standard formulas involving negation like $\neg\neg A \to A$, $A \to \neg\neg A$ and $A \vee \neg A$? Neither $\Box\Box A \to A$, $A \to \Box\Box A$, nor $A \vee \Box A$ are theses of E, as simple semantical arguments will show.

Take excluded middle, for instance: $A \vee \Box A$ (remember we are reading $\Box$ here as negation). Consider now a model $\langle U, S, V \rangle$ such that $U = \{1\}$, $S(1) = \emptyset$, $V(p) = \emptyset$. Since $1 \notin V(p)$, $1 \nVdash p$. Now $\emptyset \notin S(1)$, so $\Box p$ is false at 1. Hence 1 falsifies $p \vee \Box p$.

If we wish to add $A \vee \Box A$ as a thesis, we will have to consider models based on frames satisfying the condition (for $x \in U$ and $X \subseteq U$):

(3e) $x \in X$ or $X \in S(x)$.

And we can show that:

Proposition 8.1. $\mathsf{E} + A \vee \Box A$ *is determined by the class of frames in which (3e) holds.*

Proof. Let x be an element of U, and suppose $x \nVdash A \vee \Box A$. Then $x \nVdash A$ and $x \nVdash \Box A$. So $x \notin \|A\|$ and, since (3e) holds, $\|A\| \in S(x)$. But then $x \Vdash \Box A$, a contradiction.

To show completeness, we take the largest canonical model, that is, the canonical model in which, for every $\Gamma \in U_L$, $S(\Gamma)$ also contains, in addition to proof sets, all $X \subseteq U$ which are non-proof sets. We show that (3e) holds in the largest canonical model for $\mathsf{E} + A \vee \Box A$. So let $\Gamma \in U_L$ and $X \subseteq U_L$. If $\Gamma \in X$, we are done. So suppose $\Gamma \notin X$. Now, if X is a proof set, then, for some A, $X = |A|$. Since $\Gamma \notin |A|$, $\Gamma \in -|A| = |\neg A|$, so $\neg A \in \Gamma$, from what it follows that $\Box A \in \Gamma$ and $|A| \in S(\Gamma)$. And if X is not a proof set, then $X \in S(\Gamma)$, since we are in the largest canonical model. Hence (3e) holds. $\Box$

Further work can be done investigating the logics obtained by adding $A \to \Box\Box A$, $\Box\Box A \to A$ and other schemes involving negation (like contraposition principles) to the logics in Figure 1 which admit a negation reading of $\Box$.

To sum it up, we presented some preliminary results about logics in which $\Box$ can be read as necessity, impossibility, or both. Further work is being done, also regarding other interpretations of $\Box$: contingency, non-necessity, and non-classical negations. We hope to deal with these topics in further work.

References

[1] Benton, Roy A. "Strong modal completeness with respect to neighborhood semantics." Unpublished manuscript, Department of Philosophy, The University of Michigan, 1975.

[2] Chellas, B. F. *Modal Logic: an introduction.* Cambridge: Cambridge University Press, 1980.

[3] Chellas, B. F. & Segerberg, K. "Modal Logics in the Vicinity of S1." *Notre Dame Journal of Formal Logic* **37**(1): 1–25, 1996.

[4] Hughes, G. E. & Cresswell, M. J. *A New Introduction to Modal Logic.* London, New York: Routledge, 1996.

[5] Mortari, C. A. "Revisiting Sylvan's Semantics for S0.6° and related systems." *South American Journal of Logic* **2**(1): 1–33, 2016.

[6] Sylvan, R. "Relational semantics for all Lewis, Lemmon and Feys' modal logics, most notably for the systems between S0.3° and S1." *The Journal of Non-Classical Logic* **6**(2): 19–40, 1989.

Received 30 December 2019

Logicism in The Eyes of The Author of *Tractatus Logico-Philosophicus* (and of Philosophical Remarks)

Pedro Noguez
Federal University of Rio Grande do Sul (UFRGS) , Graduate Program - Philosophy Department, Porto Alegre (RS), Brazil
pedro.noguez@ufrgs.br

Abstract

Along his philosophical development, Wittgenstein struggled in different ways to reconcile two guiding principles: (i) letting psychological investigations aside when asking for the nature of meaning, truth, and number; (ii) not going beyond what was in each time regarded as "the limits of language", i.e., not speaking nonsense. As he inherited principle (i) from Frege and Russell, his continuous critique to logicism can come in aid of an evaluation of whether principle (ii)'s being launched against that doctrine was or wasn't in conflict with (i). In this paper, I argue that even though psychological *investigations* played no role in Wittgenstein's early account of propositional content, propositional content was construed by him on traditional assumptions about how meaningful thought can be conveyed by us through the intentional use of symbols. These assumptions about how thoughts get their meaningfulness are the grounds for principle (ii). Since Frege's and Russell's theses about the reduction of arithmetic to Logic made appeal to no such assumptions, I conclude that Wittgenstein's critique of logicism fails to cope with (i) by virtue of stronger, psychologistic committment with (ii).

An initial draft of this paper was presented in May, 2019, at the Brazilian Logic Conference, to whose organization and participants I am deeply indebted. I am especially grateful to my former advisor, Gisele Secco (UFSM), for having so patiently introduced me to the study of *Tractatus*'s aphorisms for three years as I was an undergraduate. I also thank Professor Paulo Faria (UFRGS), who first led my attention to Goldfarb's defense of logicism against Poincaré; Professor Anderson Nakano, for encouragement, colleagues Rodrigo Ferreira (UFRGS) and Valquíria Machado (UFRGS), for valuable conversations and advise, and Professor Eros Carvalho (UFRGS), for having shown me how contingent some traditional assumptions about the nature of mentality are. I would like also to express my gratitude to the anonymous referees of IFCoLog Journal of Logic and its Applications. This work was funded and thus made possible by the Brazilian governmental agency CAPES.

It has already been argued that two main aims guided Wittgenstein's philosophical development [1].One aim was to try to avoid introducing psychological investigations into his speculations on the nature of meaning, truth, and arithmetic. The second one was to try to avoid speaking what he himself regarded as nonsense in each phase of his thought. Since rejection of psychologism in the philosophy of logic is a position that Wittgenstein directly inherited from both Frege and Russell, then, since Wittgenstein rejects logicism as well, one should expect to find some connection between what he regarded as nonsense to speak, on the one hand, and what he thought the logicists were trying to do. There are some suspicions, however, that Wittgenstein's thoughts on what would be nonsense to speak reveal, in spite of his second-order beliefs about psychologism, that he was in fact subordinating the laws of *Wahrsein* or, of *being* true, to the laws of *Fürwahrhalten* or of *to be held as* true. In this paper, I argue that such suspicions are sound. [2]

Wittgenstein started to think about the essence of descriptions when he was thinking the thoughts expressed in the *Tractatus Logico-Philosophicus.* It would not be an exaggeration, I think, to claim that all there was to his doctorate thesis was an act of defiance toward Frege's and (mainly) Russell's claim that mathematical content is but Logical content in disguise [3] . The great German master and the English friend of Wittgenstein's had quite different views on how this reduction was to be carried out and demonstrated. In any case, for both of them the fundamental subject-matter of mathematics (including elementary number theory, geometry [4], and higher analysis), the natural numbers, were to be construed as sets, of equinumerous sets, of things which in turn are truly said to be thus and so. "To be thus and so" can take as multiple forms as there are concepts, for Frege, or propositional functions, for Russell: both concepts and propositional functions articulate truths (and falsities) with some individuals, (ordered) pairs of individuals, and so on. And so each concept or propositional function determines an extension, and Frege and Russell demonstrated that, with a few assumptions concerning extensions in general, numbers can be understood as sets of equinumerous extensions, to the effect that, thus construed, they satisfy Peano's axioms for the series of natural numbers.

Furthermore, the logicist construal of numbers was a way of complying with two philosophical intuitions. One of them, already stated, was that applying mathematics in other branches of science, as well as in ordinary discourse (part of what an

[1]See, for instance, [5]

[2]I refer the reader to [8], p.35, to see how else, besides the way I put it, the author of the Tractatus might be deemed as unsuccessful in guarding himself against psychologism.

[3]I like to use the capital 'L' to distinguish what Frege and Russell regarded as that to which arithmetic was a part, "the laws of Truth" from what we today call 'logic'.

[4]Frege actually fell short of defending that geometrical truths were analytic (FA, §89).

explanation of the nature of number is supposed to address) consists of verifying properties of that which is commonly true about each element of some set of things, as Frege was already trying to show in his *Grundlagen der Arithmetik*. The second philosophical intuition was that whether a given number belongs to some "to be thus and so" is so much independent from any feature of our minds as the facts that such and such things are thus and so. The latter *desideratum* is what anti-psychologism is all about: not drawing conclusions about what the content of a proposition (from mathematics or otherwise) is, from considerations about the nature of mind.

In §63 of Foundations of Arithmetic, Frege introduced the cardinality operator for concepts which, nowadays we read as

Nx:Fx

The definition of this operator was based on Hume's Principle, according to which the cardinalities of two concepts Fx and Gx are identical if and only if there is a one-one correlation between the elements of the extensions of F and G, or,

$$F \approx G \leftrightarrow Nx : Fx = Nx : Gx.$$

Hume's Principle provides the criterion for either distinguishing or identifying attributable cardinalities to any two given concepts, preserving symmetry, transitivity and reflexivity to the relation of identity. Fx and Gx are equinumerous if and only if there is a relation R such that every object of the extension of Fx holds R to one and only one object of the extension of Gx, and every object of the extension of Gx holds R to one and only one object of the extension of Fx.

In order to express same-cardinality relations between concepts from Hume's Principle we need the notion of unicity,

$$\exists ! x Hx = def \exists x (Hx \wedge \forall y (Hy \rightarrow (y = x)),$$

so that the complete definition of equinumerosity reads

$$F \approx G = def \exists R ((\forall x (Fx \rightarrow \exists ! y (Gy \wedge Rxy)) \wedge (\forall x (Gx \rightarrow \exists ! y (Fy \wedge Rxy))).$$

And thus Frege defines, in FA's §73, the notion of number of a concept as "class of equinumerous classes":

$$Nx : Fx = def G : G \approx F$$

But Hume's Principle is a biconditional. So being, it could indicate either that numbers are fundamentally classes of equinumerous classes or that classes of equinumerous classes are fundamentally numbers. As it happens, though, as much for Frege as for Russell, it is not possible to make reference to numbers without making reference to classes of objects (and, thus, to objects); at least not once Hume's Principle is assumed alongside the thesis that every natural number is the number of a concept (or propositional function). On the other hand, given both assumptions, it is possible to make reference to objects, concepts and propositional functions without making reference to numbers. For one-one correlations among extensions, which ground the concept of number, are by their turn grounded on the relation of subsumption that holds between objects and functions, and subsumption is grounded on facts of reality, not on a concept's extension. As the serial notion of number emerges abstractly from the differences in size of extensions, the objects of study of arithmetic ground themselves on the objects of study of Logic broadly understood, not the opposite.

Frege and Russell claimed that Logic was a Science [5]. For it assumed not only the rules that governed the construction of well-formed formulas, plus axioms and postulates that allow one to derive new formulas afresh. Logic, as a Science, dealt also with sets of things, and memberhood in any of the sets of things dealt with by Logic – and therefore the identity of these sets – should be conceived of as delimited by nothing less than all possible shapes that truth itself can take: the only sets admissible by Logic are extensions of predicates, of some "to be thus and so". Thus Logic does not oppose form to propositional content by excluding the latter (as most today might be inclined to characterize logic), but rather purports to deal with such content through the knowledge of its form: the context of logicism's foundational enterprise is that of grounds of justification alone, no care for how we discover it [6]. It concerns not the legitimacy of attributing a number to a concept, but rather the would-be grounds for the legitimacy of any numerical attribution, inside or outside of pure mathematics. As Russell states in the Introduction to *Principia Mathematica*,

> Most mathematical investigation is concerned not with the analysis of the complete process of reasoning, but with the presentation of such an abstract of the proof as is sufficient to convince a properly instructed mind. For such investigations the detailed presentation of the steps in reasoning is of course unnecessary, provided that the detail is carried far enough to guard against error. In this connection it may be remembered

[5]For Frege's expression of this viewpoint, see Basic Laws of Arithmetic, vol. I, Foreword, p. xv. For Russell's, see Introduction to Mathematical Philosophy, Chap. XVI, p. 169.

[6]See [3], p. 65

> that the investigations of Weierstrass and others of the same school have shown that, even in the common topics of mathematical thought, much more detail is necessary than previous generations of mathematicians had anticipated. (*PM*, Introduction, p. 03)

This much is sufficient for us to understand how arithmetic was, to the logicist view, a science that investigated a subset of the laws concerning all that which is either true or false. Also, we can see how such topic would not be subordinated to psychological laws or metaphysical assumptions about the mind, according to which the mind relates to it. For the classes that constitute the fundamental subject-matter of arithmetic are extensions, and these are determined independently of any trace of the human mind. It is correct, to a certain extent, to identify what Logic thus construed deals with, with semantics [7] : the only qualification needed, not to misguide oneself here is that symbols, or language for that purpose, played a merely auxiliary and heuristic role: namely, for the logicist derivations to be shown and surveyed. Thoughts, propositions, and their derivative notions of propositional function, concept, argument, and classes, were not essentially linked to symbols as far as Frege and Russell were concerned with them.

The science of Logic as a human enterprise would have been a very special one, because of the absolute generality of its concerns; but that would not be enough to qualify it as a completely different kind of science: as any science, it made assumptions that it could not prove. For instance, that there are at least as many logical objects (things that fill argument-places in first-order propositional functions) as there are natural numbers. For suppose it is true that every number is a class of equinumerical extensions of propositional functions. In that case, what if any general state of the world was determined by the distribution and mutual relations of some 10^{20} objects? Then we would have to admit that, for our inconvenience, an equation such as

$$10^{20} + 1 = 10^{20}.10$$

is correct, because any number above 10^{20} would denote a class of classes which ones would contain a proper subset with the necessary amount of members to be on-one correlated with any class denoted by 10^{20} , while also containing one or more remaining members, which is impossible if there are only 10^{20} objects. So all classes that were defined as exceeding the class-members of 10^{20} by any number would be identical to the null class, and, therefore, for any numbers n, m such that

[7]7 See [3].

$10^{20} < n < m,$

we would have

$n = m,$

which is more than just an inconvenient result, but rather a contradiction. Hence, if natural numbers are to be regarded as classes of extensions at all, there must be extensions to each of all magnitudes. And, of course, they must be extensions, i.e., there must be at least one propositional function defining the classes that will be encompassed by each and all natural numbers.

This much of unprovable assumptions was seen as a deep mistake of logicism in the eyes of the author of the *Tractatus*: to claim that *a priori* formal sciences received their content from such matters that could only be *decided* by experience (if at all) would be to claim that we can in some sense anticipate experience through calculus (TLP, 5.552-5.5541). In Wittgenstein's view, there was no sense in speaking of meaning and truth while ignoring the way we potentially come up with meaningful, sometimes truthful *acts* by mastering the use of symbols. It is we who make ourselves pictures of facts (TLP, 2.1), and such pictures must share the form, the possibilities of their elements' arrangements with one another, with the elements of facts. Taking seriously what saying something must be if we do make such pictures, and keeping silent on the actual, unthinkable multiplicity of the elements that are respective to worldly, thought-like, and language-like states of affairs, Wittgenstein purported otherwise to explain the very notions with which Frege and Russell had concerned themselves.

As Juliet Floyd says, "[Wittgenstein's] denial that numerals are names, that numbers are objects or (second-order) properties, is best read as recasting the whole idea of what the drawing of categorial or logical distinctions can accomplish." ([7], p.312) The drawing of categorial distinctions, such as those of function and argument, would, according to Wittgenstein, be relevant for us to understand the nature of whatever bears truth or falsity, but only at the level of its possibility: for that alone is what can be anticipated with *a priori* certainty. We have all rights to speculation at this level because there can be no illogical thoughts (TLP, 3.02-03, 5.4731). The actual, deep, structure of thought, however, cannot be anticipated lest through logical analysis, i.e., through the application of the symbolic normativity of mathematical logic (as a formal system) to language, that would render explicit what is logically essential to the expression of thoughts, leaving aside the superficial clothing that thoughts ordinarily wear (TLP, 4.002). In his view, all that can be accomplished by the complete regimentation of the way we express our thoughts is this: a regimentation of our descriptive practices, such that all that we mean in thinking will become clearest *am Symbol allein* and will not request for further elucidation, nor for philosophical theses of any kind (TLP, 6.52, 6.521). If this is all we can do through the application of formal systems, however, then what right do we have, to say anything determinate about how many "logical atoms" there must be in the world? None, for sure. This is Wittgenstein's negative stand against logicism, insofar as the latter thesis relies on assumptions about (i) the logical multiplicity of the world, and (ii) which kinds of concepts or propositional functions grounding the existence of extensions there are. For logicism to get off the ground it is necessary, first, that (i) "the logical multiplicity of the world" (i.e., of the set of individuals of which something might be true) is at least aleph-zero. Second, (ii) that the range of second order variables admits a standard interpretation, so that it is equal to the power set of the set of individuals, thence there being no class without a corresponding concept or propositional function [8]. Yet neither (i) nor (ii) are subject-matters about which any form of calculus alone can reveal to us how things actually are.

> TLP 5.551: Our fundamental principle is that every question that can be decided at all by logic can be decided off-hand.
>
> (And if we get into a situation where we need to answer such a problem by looking at the world, this shows that we are on a fundamentally wrong track.)

The next step is to see that anything that can be thought to exist can also be said to exist. As far as the possibility of truth and falsity of propositions, and of existence or non-existence of states of affairs, is concerned, there is a triad of language, thought, and the world, which *defines* the only logically admissible form of possibility. Whereas whatever can be said or thought to be the case can be either true or false, this is matched by the world by the transcendentally established space of possibilities, of existence or non-existence, of each state of affairs (TLP, 4.023, 4.032, 6.13). Each proposition, a possible representation by any means, will be either true or false as soon as it is actually presented, i.e. said, depending on the existence or non-existence, at such time, of the state of affairs it represents (TLP,

[8] A well-known alternative to the predicative "restrictions" on the range of second-order variables is the admission of Ramsey's propositional functions in extension, which we will not have time to discuss here (see [11]). Although Ramsey claimed to be vindicating the logicist project, he was just moving away from it in a direction that was the Platonist opposite from that of Wittgenstein's. After all, Ramsey's numbers would not be anything that resembled properties of concepts: classes and propositional functions in extension would exist on their own.

4.2). Finally, we have a picture of what the sense of a proposition is: it is the state of affairs it represents, regardless of whether the latter exists or not (TLP, 4.1). If the state of affairs the proposition represents does exist when the proposition is said, the proposition will be true; if it does not, the proposition will be false. Further, by just inverting this relation of truth of a proposition to existence of the state of affairs it represents, we know also what the sense of the negation of a proposition is (TLP, 4.06-4.062).

The way propositions (symbols, thoughts) manage to represent states of affairs *regardless* of whether the latter exist or not is because a proposition must in essence be an articulation of symbols that are *not* themselves articulated, but that rather stand for those parts of reality which by transcendental necessity exist. And so we have *names* on the side of language and thought, and *objects*, the substance of the world, on the latter's side (TLP, 2.021, 4.21-4.221). Since an articulation of names implies comparability with an articulation of objects, and thus the former's truth or falsity, if the elements that constitute the ultimate analysis of propositional sense were themselves articulate, then the sense of a proposition would ultimately rely on the truth of another,and the study of propositional sense could not, on its turn, rely solely on that which can be anticipated *a priori*. One doesn't learn anything by knowing that objects exist, because there could be no thought if they didn't: objects are the nods of the structure of any fact (therefore, of anything we can think of); when a state of affairs doesn't exist, it is because its nods are in fact composing another structure.

But here is where Wittgenstein arrives to a surprising problem to which his assumptions led him, and to which only he has thought of a solution (a situation before which I think any philosopher should at least pause): the problem of saying that a situation is possible. If saying that a situation s is possible is itself possible, then there must be a proposition, true or false as far as we can tell from our armchairs, that represents the state of affairs of s's being possible. Call it "P(s)". The problem is that, if P(s) were false, then s itself would have been impossible. However, as we saw above, a situation's being possible amounts to the existence of the objects that would constitute it in accord (i.e., accordingly articulated) if the situation obtained. And this, by its turn, is regarded by Wittgenstein as sufficient grounds for holding that, however unprecisely distinguished *am Symbol allein*, there are names within the essential repertoire of thought corresponding to each of those objects, whereas one of the possible articulations of these names is the one which will turn out true if s obtains. Since this one possible articulation of the right names is what the sense of the proposition p, expressing the obtainment of s, is, then, if P(s) were false, p, the proposition that says that s exists, would have no sense. And P(s) itself would be nonsense, since it would be saying of nothing that it is possible.

The great solution to the problem, How do we say that a situation is possible?, is in part to declare that it is not possible to say it, but also that it doesn't matter, because it *shows* itself that a situation s is possible by there being a proposition (some picture one can make) which is able to depict s (TLP, 4.1212, 4.022). There can be no novelties (nor relevant questions) concerning what can exist in reality, because the whole space of worldly possibilities are met through their correspondents right in the head (TLP, 4.024, 4.116).

Any proposition p can be negated, resulting in a new proposition, $\sim p$, that will be a truth- function of p. That is to say, instead of rendering $\sim p$ true, the existence of the state of affairs depicted by p will render it false, and vice-versa. The appearance of conflict with the demand that the sense of a proposition cannot depend on the truth of another is dissolved by attention to the fact that it is still a comparison of p to the world that will determine whether $\sim p$ is true. Hence any proposition which is the negation of another has no sense of its own, but only the sense of the negated proposition with inverted "truth-poles", i.e., an inverted relation of dependence of its (inessential) presentation's truth-value to the existence of the state of affairs depicted by the negated proposition. In other words, if the logical space were initially composed of only three propositions p, q, r, then their respective negations, $\sim p, \sim q, \sim r$, would not be further possible thoughts, representing further possible situations. Truth and falsity, as well as existence and non-existence come in pairs within the space of logical possibilities (TLP, 4.06-4.064).

Hence from our brief interlude into the distinction between what can be said and what shows itself, we arrive at another distinction, that of what is essential to the sense of a proposition to what is accidental to it. We already saw that the essence of the sense of $\sim p$ is the sense of p. It follows that it would be misleading to treat the sign '$\sim$ ' as a propositional function, because, whereas the fulfillment of the argument-places of first-order propositional functions such as f(x) yield propositions that add up to the general sum of the logical space, the fulfillment of $\sim (x)$, with x ranging over propositions, yields no such additions, whatever proposition x is substituted for. Instead of treating negation as a propositional function, therefore, Wittgenstein labeled it a "truth-operation" (TLP, 5.2341, 5.254), that of symbolically expressing the inversion of truth-poles of a given proposition, resulting in a new propositional sign, but not in a new propositional sense – the formal difference between the old and the new propositions is the fruit of our choice in expressing ourselves, and is no part of the essence of propositions; only, the possibility of thus operating on symbolic expression is (TLP, 4.51). As for those propositions whose symbols are not obtained out of truth-operations, and that therefore directly express only the essence of a propositional sense, we shall call them "elementary propositions", which are truth-functions of themselves (TLP, 5).

Because any of the other logical connectives, '$\wedge$', '$\vee$', '$\rightarrow$', can be reduced to different ways of applying the operation of joint negation (TLP, 5.502), represented by N'(ζ) – where ζ can stand for any sum of elementary propositions – the conclusion is that all propositions are truth-functions of elementary propositions (TLP, 5.3). Or, what amounts to the same thing, all propositions result from a finite number of applications of truth-operations on elementary propositions (TLP, 5.5- 5.502). As a special case, there will be propositions of the kind of T = N'(N'((N'p), p)) – same as p or $\sim p$ – which are true on the sole basis of how the truth-operations were applied to p, whatever proposition p is. Propositions such as T are tautologies. Since no situations' obtaining is accountable for its truth, a tautology is a proposition whose presentation says nothing; it has no sense (TLP, 5.142). Still, and precisely because of that, a tautology is necessarily true. A further application of negation on a tautology yields a contradiction, which again says nothing, but this time is necessarily false. The presentation of tautologies, though able to reveal interesting symbolic properties, can deliver no further insights on the essence of propositions, much less about the extension of any concept. The essence of propositions, their general form as truth-functions of elementary symbolic arrangements, is already the essence of the world (TLP, 5.47-472). Finally we have, at the level that precedes all empirical or contingent concerns, the notions of possibility, necessity, and impossibility. The final touch of the Tractarian view is to say that *these* modal notions are the only ones with which we are entitled to deal (TLP, 5.132-5.1361, 6.37).

The question to be made is: is the meaning of what we think or say a thing that must be grasped in its entirety in order to be grasped at all? If it is, then the facts that, first, we indeed think propositions, and, second, that we understand them, will be crucial ones for us to understand what meanings are, just as Wittgenstein seems to have thought. For we cannot hold at the same time, both that nothing is left undetermined by meaning *as we grasp it*, and that some empirical discovery will reveal what we meant in the past. Hence,

> TLP, 4.1121 Psychology is no nearer related to philosophy than is any other natural science.
>
> The theory of knowledge is the philosophy of psychology.
>
> Does not my study of symbolism correspond to the study of thought processes which philosophers held to be so essential to the philosophy of logic? Only they became entangled more often than not in inessential psychological investigations, and there is an analogous danger for my method.

At this point, if not already earlier, we can appreciate how Wittgenstein's early reflections about the essence of descriptions did involve a psychologistic conception of that which bears truth or falsity. Thoughts' logical compositionality – or, in other words, the construal of thought and psychological processes in general as being constitutively inferential, and of psychological states as true or false representations – is to this day a largely shared philosophical (and scientific) point of departure in investigating the nature of mind [9]. It is nevertheless (even if disregarding all but assertive states of mind) a claim about acts through which we depict to ourselves, pictures of facts, and not about the contents of possible judgments *per se*. If Wittgenstein equates these things, then that's precisely where he lets psychological considerations bear on his account of propositional contents. The fact that no psychological *investigations* were allowed within the course of Wittgenstein's reflections about propositions (no doubt about that) offers no support to the supposition that he shared Frege's and Russell's stance against psychologism – again, in spite of what Wittgenstein himself might have thought he was doing. For *a priori* speculations about the relation that our minds bear to propositional contents are just as irrelevant for determining what are propositional contents as empirical results about the workings of our minds.

As the premises that allowed the logicist thesis to be established went beyond the assumptions that could be made solely on *a priori* grounds, Wittgenstein imbued himself with the task of offering a new account for the formal sciences. Such account should preserve their intimate connection with all the sciences but, at the same time, should present Logic, in particular, as a "totally different kind than another science" ([20], p.10): its generality would not come from its interest in the laws of *Wahrsein*, but rather from its interest on the (transcendental) laws of *Fürwahrhalten*, that is, of saying, thinking, which indeed pervades every scientific enterprise. Since his principle was that we must sharply distinguish what can be anticipated through pure reason from that which cannot (TLP, 5.551), he tried to develop a way to draw all relevant logical distinctions (such as function, argument, and the logical connectives) that were needed to account for any form of inference without, nevertheless, assuming that inference-permissions among statements should be grounded on extensions of concepts. As we saw, the problem for him was not our inability to tell which are the extensions (this is too obviously outside the scope of Logic), but rather that we are unable to tell *a priori* even what kinds of extensions there are (hence, we are unable to tell even whether arithmetic *could* be grounded on Logic).

In the 1930's, shortly after Wittgenstein's return to philosophy, he continued to criticize logicism, this time using arguments that appeared more ostensibly than

[9] See [19], and also [16] for evaluating the pervasiveness of the notion of mental representation within the domain of cognitive science, as well as its predicaments and alternatives. See also [2], who links such paradigm to its Kantian roots.

ever to to inculcate psychological constrains on the investigation of the nature of number. Wittgenstein argued that the mere possibility of there being a one-one correlation between elements of two extensions – e.g., the set of spoons and the set of cups to be arranged on a table – presupposed the determination of each concept's number, implying that numbers cannot be defined through correlations among concepts' extensions. Explaining to Waismann and Schlick what he meant by saying that there are as many spoons as cups, Wittgenstein said:

> What I mean is obviously this: I can allot the spoons to the cups because there is the right number of spoons. But to explain this I must presuppose the concept of number. It is not the case that a correlation defines number; rather, number makes a correlation possible. This is why you cannot explain number by means of correlation (equinumerosity). You must not explain number by means of correlation; you can explain it by means of possible correlation, and this precisely presupposes number. [10]

According to [12], Wittgenstein's argument above consisted in claiming that the logicist definition would be an epistemologically defective one, because it would explain number by one-one correlation, whereas we cannot know *that* there is a one-one correlation without knowing *which* number is the number of each correlated extension. So Wittgenstein's argument above would in sum be that the content of a numerical attribution to some set (say, to the set of spoons) is irreducible to that of a statement of possible correlation with another set (like the set of cups) because, in order to know that the content of the latter statement is true, one must first of all know that the content of the former is. As Goldfarb states in his defence of logicism against the attacks of Poincaré, however, one has to remember that the logicist thesis presupposes a clear distinction between "the mental or physical conditions under which a person comes to understand, appreciate, or believe a proposition and the ultimate rational basis of the proposition" ([6], p. 67). Was Wittgenstein so carelessly allowing psychological investigations to interfere with his thoughts about the relation of Logic to mathematics?

A better defense of Wittgenstein's argument quoted above is provided by Anderson Nakano ([15]), who supposes that in the 1930's Wittgenstein still held the Tractarian thesis according to which the possibility of a situation is shown by the existence of a proposition depicting it, i.e., by the possibility of thinking the situation. A one-one correlation between two classes can be stated in two ways: either by giving each one *in extenso*, as in

[10]Cited from [12], p. 64. See also [21], §118

$$\{a, b\} \approx \{c, d\}$$

,

or by intentionally defining them as the extensions of concepts F and G, thus, as we put it before,

$$F \approx G$$

In the first case, the existence of a correlation would be expressed as

$$\exists R((aRc \wedge bRd) \vee (aRd \wedge bRc)),$$

and, were it not for the sameness of number of classes a, b and c, d, a disjunction like the one above could not even be written, whereas we know in advance that if they do have the same number of elements, then the expression of their correlation will contain n! disjuncts. The possibility of depicting such a correlation (i.e., the existence of a Tractarian proposition that represents it) presupposes, therefore, sameness of number of the two classes.

The second case, however, is the one that should most interest us, since it was the logicist thesis that numbers are classes of *intentionally* defined equinumerous extensions that Wittgenstein had in mind when speaking to his Viennese retinue. The Tractarian line of reasoning Nakano ([15]) attributes to Wittgenstein is the following. The existence of a one-one correlation between the elements of the extensions of F and G would mean that there is a proposition,

$$\exists R((\forall x(Fx \rightarrow \exists! y(Gy \wedge Rxy)) \wedge (\forall x(Gx \rightarrow \exists! y(Fy \wedge Rxy))),$$

whose truth would imply that F and G have the same amount of elements in their respective extensions. However, as Nakano points out, "the possibility of an one-one correlation, shown by the existence of [the proposition that represents the correlation] cannot be used to define sameness of number, since sameness of number is said by [the above's indicated logical consequence] and *what can be shown, cannot be said* (4.1212)." (Nakano, [15], p. 14, my emphasis)

That is to say, when Wittgenstein said that "You must not explain number by means of correlation; you can explain it by means of possible correlation, and this precisely presupposes number", he meant that, either way, the possibility of there being a possible correlation should be understood as the existence of a proposition that depicts it, and there isn't any such proposition if there is no sameness of number. However, should there be such a proposition, sameness of number between conceptual extensions would show itself in it; hence, it could not be said. The reason I think this is a more charitable reading of Wittgenstein's argument than that of Marion and Okada ([12]) is that, were Wittgenstein to hold that the definition of number served the purpose of finding out what one needs to know, prior to obtaining the knowledge of a correct numerical attribution, Wittgenstein's misunderstanding of the logicist thesis would be far more profound than if he were merely equating the existence of a (possibly thought, written, or spoken) proposition with the possibility of the situation it depicts.

Even though there is considerable doubt as to how much of Tractarian ideas were still held by Wittgenstein when he wrote Philosophical Remarks and had his conversations with members of the Vienna Circle, probably the relation between possibilities and existence of respective propositions was not yet thrown away, as is shown by this passage:

> What sort of an impossibility is the impossibility, e.g., of a 1-1 correlation between 3 circles and 2 crosses? (...) That a 1-1 correlation is possible is shown in that a significant proposition, true or false, asserts that it obtains. And that the correlation discussed above is not possible is shown by the fact that we cannot describe it. ([21], §119)

As we saw earlier, the dependence of the sense of a proposition on the truth of another would mean that if the latter proposition were false, the former would yield no sense. Possibilities show themselves by our being able to think them, i.e., by our being able to say they are true (which is not to say that the possibility of a situation requires someone to *actually* think them) [11] . As a consequence, the only pair of coupled presuppositions of logic, the ascription of sense to propositions and reference to names (TLP, 6.124), had to be determined once and for all, and all there would be to "Logic" to investigate was what already lied in front of us, that showed itself. This, that was the sense of propositions, was earlier characterized as what would be the case if the proposition were true; in the PR, not only what would be the case, but also the phenomenological frame it would have to be discovered to be true in, mattered. But the determination of sense before truth was still imperative, and for the same old reason that all that is necessary for truth to be possible must enter into the constitution of some of our doings, i.e., sayings.

> PR, §28 Expecting is connected with looking for: looking for something presupposes that I know what I am looking for, without what I am

[11] I am grateful to one of the anonymous reviewers of this paper for having me warned against this possible misinterpretation.

> looking for having to exist.
> Earlier I would have put this by saying that searching presupposes the elements of the complex, but not *the* combination that I was looking for. And that isn't a bad image: for, in the case of language, that would be expressed by saying that the sense of a proposition only presupposes the grammatically correct use of certain words.

Following the Tractarian lead, every proposition is a possible symbolic manipulation inside a system (language) whose logical multiplicity, however occluded by lack of perspicuity, matches that of the world (TLP, 5.555). What is said through symbols is how things are, and both how things are and how things can be is outside the reach of our control (TLP, 6.124). Nevertheless, we manage to express it. The idea that "If you exclude the element of intention from language, its whole function then collapses" (PR, §20) is not stated in Wittgenstein's early aphorisms; even so, the idea that a proposition is the result of an activity, like making ourselves pictures of facts (TLP, 2.1) or using perceptible signs as a projection of a state of affairs (TLP, 3.11), is – and surely saying is no unintentional activity.

Wittgenstein did not mistake thoughts for mental images, nor did he ground the necessity of tautologies on empirically discovered psychological laws. But some of our intentional actions were, for him (at least until the early 1930's), intrinsically contentful, in such a way that all knowledge concerning contents of possible judgments lied within the reach of one's armchair. Known without observation [12], the descriptions under which some actions are propositions (are that which one says) would describe the only rightful bearers of truth or falsity, and the supposition that any form of calculus (either arithmetical or logical) bore some form of content should be dismissed, if only because such supposition would get in conflict with the determinateness of what is *a priori* knowable. Therefore, Wittgenstein's decisive refusal of Frege's and Russell's views on the content of arithmetical theorems (the claim that the very question about it would be *nonsense*) is directly drawn from substantial assumptions about our mental powers, which concerned *how we know what we mean when we think.*

> TLP, 4.411 It seems plausible even at first sight that the introduction of elementary propositions is fundamental for the understanding of other kinds of propositions. Indeed, the understanding of general propositions depends *paupably* on that of elementary propositions.

The "postulate of the determinateness of sense", same as "the postulate of the possibility of simple signs" (TLP, 3.24) is not so much about sense, but about our

[12] I'm relying on [1] account of the concept of intention.

minds. It is the postulate of the *a priori* knowability of sense. Assumption of the mind's fit to formal treatment, in order to elucidate its own thoughtful activities was an essential part to Wittgenstein's argument against logicism, for it was a premise to his conclusions about what the content of a proposition is. The non-conceptual basis of elementary propositions, from which all ordinary concepts would be constructible through further and further applications of the N operation (primarily) upon sets of such elementary basis, is a requirement of the world-mirroring thesis on the mind's workings. It is also the ground on which rests the distinction between what can be said and what shows itself in it, which as we saw is the best line of defense we can attribute (as far as I know) to Wittgenstein's rehearsal of critiques to logicism in the middle of his career. Postulating the possibility of simple sings in line with TLP, 3.24 is no dirtying one's hands with psychological research, but this is it for Wittgenstein's anti- psychologism, (under the best interpretative efforts) both in the *Tractatus Logico-Philosophicus* and in the Philosophical Remarks. For assuming a thesis about the nature of mind – even if not through empirical research – in order to carry out conclusions about what the sense of a proposition is, is no better compliance with anti-psychologism than doing research. For both Frege and Russell, such would not be surprising grounds to object to their theories, as they shew (for those who had eyes to see it), respectively, in the *Vorwort* to the *Grundgesetze der Arithmetik* and in the Preface to *Principia Mathematica*:

> As to the question, why and with what right we acknowledge a logical law to be true, logic can respond only by reducing it to other logical laws. Where this is not possible, it can give no answer. Stepping outside logic, one can say: our nature and external circumstances force us to judge, and when we judge we cannot discard this – law of identity, for example – but have to acknowledge it if we do not want to lead our thinking into confusion and in the end abandon judgement altogether. I neither want to dispute nor to endorse this opinion, but merely note that what we have here is not a logical conclusion. What is offered here is not a ground of being true but of our taking to be true. ([10], p. xvii)

> In mathematics, the greatest degree of self-evidence is usually not to be found quite at the beginning, but at some later point; hence the early deductions, until they reach this point, give reasons rather for believing the premises because true consequences follow from them, than for believing the consequences because they follow from the premises. ([18], p. v)

References

[1] Anscombe, G.E.M (1963) Intention. US: Harvard University Press

[2] Brook, A. (2004) Kant, Cognitive Science, and Contemporary Neo-Kantianism. Journal of Consciousness Studies 11(10-11):1-25

[3] Coffa, A., *The semantic tradition from Kant to Carnap: to the Vienna Station*, US: Cambridge University Press, 1991.

[4] Cuter, J.V.G., Operations and Truth Operations in The Tractatus, *Philosophical Investigations*, 28(1):63-75, 2005.

[5] Engelmann, M.L., *Wittgenstein's Philosophical Development: Phenomenology, Grammar, Method, and The Anthropological View*, UK: Palgrave Macmillan, 2013.

[6] Goldfarb, W. Poincaré agains the logicists. *History and Philosophy of Modern Mathematics*, pp. 61-81, US: University of Minnesota Press, 1988.

[7] Floyd, J., Number and Ascriptions of Number in Wittgenstein's Tractatus. *From Frege to Wittgenstein: Perspectives on Early Analytic Philosophy*, pp. 308-352, UK: Oxford University Press, 2002.

[8] Frascolla, P., *Wittgenstein's Philosophy of Mathematics*, US: Routledge, 1994.

[9] Frege, G., Sobre a Justificação de uma Conceitografia e Os Fundamentos da Aritmética, BR: Abril S.A, 1983.

[10] Frege, G., *Basic Laws of Arithmetic*, UK: Oxford University Press, 2013.

[11] Marion, M., *Wittgenstein, Finitism, and the Foundations of Mathematics*, UK: Clarendon Press, 1998.

[12] Marion, M. and Okada, M., Wittgenstein on equinumerosity and surveyability, *Grazer Philosophische Studien* 89(1)61-78, 2014.

[13] Narboux, J.P., How Showing takes care of itself, *Philosophical Topics*, 42(2)201-262, 2014.

[14] Nakano, A., Numbers in elementary propositions: Some remarks on writings before and after some remarks on logical form, *Nordic Wittgenstein Review*, 6(1):85-103, 2017.

[15] Nakano, A., On Ramsey's reason to amend Principia Mathematica's logicism and Wittgenstein's reaction, *Synthese* (forthcoming).

[16] Ramsey, W., *Representation Reconsidered.* US: Cambridge University Press, 2007.

[17] Russell, B., *Introduction to Mathematical Philosophy*, RU: George Allen & Unwin, 1919,

[18] Russel, B. and Whitehead, A.N., *Principia Mathematica*, UK: Cambridge University Press, 2nd, 1927. ed.

[19] Wilson, R.A. and Keil, F.C., *The MIT Encyclopedia of the Cognitive Sciences*, UK: MIT Press, 1999.

[20] Wittgeinstein, L., *Letters to Russell*, Keynes and Moore, ed. G. H . von Wright. Ithaca: Cornell University Press, 1974.

[21] Wittgenstein, L., *Philosophical Remarks.* UK: Basil Blackwell, 1975.

[22] Wittgenstein, L., *Tractatus Logico-Philosophicus*, ed. Marc Joseph, translated by Frank

Ramsey and C.K. Odgen, UK: Broadview Press, 2014.

 Received 30 December 2019

Proof-search, analytic tableaux, models and counter-models, in Hypo constructive semantics for Minimal and Intuitionistic Propositional Logic

Wagner Sanz
Faculdade de Filosofia, Universidade Federal de Goiás
wsanz@ufg.br

Abstract

The aim is to show how to obtain models and counter-models in Hypo semantics for Intuitionistic Propositional Logic (IPL). The case of Minimal Propositional Logic (MPL) is a byproduct. Hypo is an alternative constructive proof-theoretical semantics where the concept of hypothetical consequence assumes a protagonist role in place of the usual semantical concept of constructive proof or of any other semantical value. The paper starts by examining proof-search in a handy LJ propositional sequent calculus ($hLJp$) and it goes on to show how the rules orienting this search are a natural consequence of Hypo semantics. Then, a refutation procedure for the so-called antisequents is presented. Finally, counter-models for the validity of a sentence in IPL and MPL are briefly exemplified and compared to the usual Kripke counter-models.

1 Introduction

From a philosophical point of view, a semantical clause for a logical constant is supposed to contain a clarification of meaning in terms of its constituents via a basic primitive predicate/relation. A model theory is then obtained by adding stipulations of how this basic predicate/relation applies in the case of atomic sentences.

Hypo semantics [10] and [11] has been proposed as a new alternative constructive semantics for Intuitionistic Propositional Logic (IPL) and Minimal Propositional Logic (MPL). The objective is to show how to obtain models and counter-models

We wish to express our gratitude to two anonymous referees that made valuable suggestions for the present article.

within it. Its models and counter-models are distinct of those presented in [7], since Hypo has distinct clauses for disjunction and implication. Nonetheless, the counter-model search has some similarities in both semantics.

Hypo models are entirely motivated on proof-theoretical grounds, deduction being conceived under a hypothetical paradigm as discussed by [3] and [12]. On the other hand, the traditional paradigm conceives validity of inferences to be a matter of preserving truth or, in the constructive case, preserving provability. In Hypo, the relation of consequence assumes the basic semantical role. Truth preservation becomes a derived property. The starting point for the exposition below is the problem of proof-search in a sequent calculus for IPL and MPL under the hypothetical paradigm.

Gentzen's *Untersuchungen* paper [5] is the original source of the sequent calculus. It contains a proof of equivalence between sequent calculus, the axiomatic calculus and the natural deduction calculus, for both Intuitionistic and Classical Logic. The *hauptsatz*, or main theorem, just proves the cut elimination property for sequents. The paper brings, as a corollary to this theorem, the procedure of proof-search for the Sequent System of Intuitionistic Propositional Logic (LJp). And, since minimal logic is embedded in intuitionistic logic, the same procedure works for the Sequent System of Minimal Propositional Logic (LMp). The novelty in the procedures to be presented below lies in its relation to Hypo clauses.

Analytic tableaux are well known methods for searching the refutation of a sentence. They have a close connection with proof-search. They were investigated by some of the leading researchers in logic as in [1] and [13]. For the particular case of IPL the tableau method is intimately connected with Kripke semantics [7] for this logic. A full exposition can be found in [4].

The exposition starts with a definition of a handy sequent system for IPL and MPL. The handy system enjoys a kind of cut elimination property which is proved in a very short and direct way. As it contains neither contraction nor thinning, this system reveals itself as a nice ground for proof-search.

Proof-search is illuminated by Hypo semantical clauses. Actually, the search corresponds to an analytic reading of its clauses. *Antisequents* are the constructive negation of sequents, and a refutation procedure for antisequents is going to be obtained through an analytic reading of the antisequent clauses derived from Hypo clauses. Together, these procedures constitute the ground for obtaining models and counter-models in Hypo semantics.

Besides [7] and [4], [8] is another important antecedent concerning the construction of counter-models for non-theorems of IPL in the setting of proof-theoretic semantics, although they use the Kripke model framework. These authors proposed a calculus $CRIP$ which they use as a basis in the "proof-search" for antisequents offering then a non-looping method for building Kripke trees refuting non-theorems of IPL. However, as we see it, once Hypo semantics is well understood, it becomes unnecessary to define any antisequent system, if counter-models are the objective. Also, some simplification in the resulting counter-models can be achieved.

More recently [9] also presented a sequent system for the implicational fragment of MPL over which they define how to prove sequents and how to obtain counter-models. They employ new sequent structures containing "bags" of formulas, remaining in the framework of Kripke counter-models.

Finally, we briefly compare Kripke counter-models and the new Hypo way of establishing counter-models. One of the intuitions behind Hypo models is adapted from the work of Pinto and Dyckhoff, the definition of a Hypo basis.

The important point to be highlighted is that, different from Kripke semantics, Hypo is a truly motivated proof-theoretical semantics.

From now on we use the device of giving a second reading in a context by using brackets, like in the heading below. The expression in brackets are supposed to substitute the expression coming before.

2 A Handy Intuitionistic [Minimal] Sequent Calculus

Assume a recursively defined sentential language $\mathfrak{L}$ containing one distinguished atomic sentence for the absurd "$\bot$" and no sentential parameters or variables.[1] We use capital Latin letters C, D, etc., to represent sentences of $\mathfrak{L}$ and small Latin letters c, d, etc., to represent sentences belonging to $\mathfrak{at}\mathfrak{L}$, the subset of atomic sentences of $\mathfrak{L}$. The Latin letters are metaparameters for sentences. Multisets are sets admitting multiple copies of an element. We use capital Greek letters Γ and Δ to represent finite multisets of $\mathfrak{L}$, including the empty one. Small Greek letters γ and δ are used to represent finite multisets of $\mathfrak{at}\mathfrak{L}$, including the empty one. Greek letters are metaparameters for finite multisets of sentences.

"$\Gamma \vdash C$" represents sequents. The left side of the sequent contains a multiset of hypotheses indicated by Γ, the right side contains only one sentence C, the *consequentia*. Negation is defined as: $\neg C := C \to \bot$. The symbols "$\Leftarrow$", "$\Rightarrow$" and "$\Leftrightarrow$" are metalinguistic constants representing inference from right to left, inference from left to right and inference in both directions, respectively, all of them should be constructively interpreted. The linear presentation of rules is just a shorter way of giving the usual vertical presentation of inferences in sequent calculus.

[1] Sentence and proposition are roughly equivalent concepts if one does not care for philosophical distinctions. In any event, it should be reiterated that, from a constructive point of view, truth-values are not acceptable as the interpretation of propositions.

Definition 2.1. *The handy LJ Sequent System for propositional logic (hLJp) consists of the following initial sequents and sequent rules, hLMp is equal to hLJp without the (absurd) initial sequent:*

Initial sequents

$(basic) : \Gamma, C \vdash C$

$(absurd) : \Gamma, \bot \vdash C$

Inference rules

Structural[2] $(MP) : (\vdash C \; and \vdash C \to D) \Rightarrow \vdash D$

Left Side (LS)

$(\wedge) : \Gamma, C \wedge D \vdash E \Leftarrow \Gamma, C, D \vdash E$

$(\vee) : \Gamma, C \vee D \vdash E \Leftarrow (\Gamma, C \vdash E \; and \; \Gamma, D \vdash E)$

$(\to) : \Gamma, C \to D \vdash E \Leftarrow (\Gamma, D \vdash E \; and \; \Gamma \vdash C)$

Right Side (RS)

$(\wedge) : \Gamma \vdash C \wedge D \Leftarrow (\Gamma \vdash C \; and \; \Gamma \vdash D)$

$(\vee) : \Gamma \vdash C \vee D \Leftarrow (\Gamma \vdash C \; or \; \Gamma \vdash D)^3$

$(\to) : \Gamma \vdash C \to D \Leftarrow \Gamma, C \vdash D$

A system having some resemblance with $hLJp$ and designed for proof-search is $LJT*$ by [8, p. 2]. $LJT*$ contains neither thinning nor contraction like $hLJp$, but it contains four left introduction implication rules. [9] when dealing with proof-search defined their LMp system with a left implication introduction rule having a stronger condition than the one in $hLJp$.

Lemma 2.2. *The axioms of IPL [MPL] are provable in hLJp [hLMp], that is:*[4] *(i)* $\vdash C \to (C \wedge C)$; *(ii)* $\vdash (C \wedge D) \to (D \wedge C)$; *(iii)* $\vdash (C \to D) \to ((C \wedge E) \to (D \wedge E))$; *(iv)* $\vdash ((C \to D) \wedge (D \to E)) \to (C \to E)$; *(v)* $\vdash C \to (D \to C)$; *(vi)* $\vdash (C \wedge (C \to D)) \to D$; *(vii)* $\vdash C \to (C \vee D)$; *(viii)* $\vdash (C \vee D) \to (D \vee C)$; *(ix)* $\vdash ((C \to E) \wedge (D \to E)) \to ((C \vee D) \to E)$; *(x)* $\vdash (C \to \bot) \to (C \to D)$ *(only in IPL); (xi)* $\vdash ((C \to D) \wedge (C \to (D \to \bot))) \to (C \to \bot)$.

Proof. Notice that the axioms can be proved without using (MP). We give the proof of axiom (iv) $\vdash ((C \to D) \wedge (D \to E)) \to (C \to E)$ in linear fashion, the others are not more difficult. $C \vdash C$ and $C, D \vdash D$ by $(basic)$. $C, C \to D \vdash D$, by

[2] The usual role of the cut rule is fulfilled by (MP)

[3] This is a shorter way of writing: $(\Gamma \vdash C \vee D \Leftarrow \Gamma \vdash C)$ and $(\Gamma \vdash C \vee D \Leftarrow \Gamma \vdash D)$.

[4] See [6, pp. 105–106]

$(\rightarrow LS)$. $C, C \rightarrow D, E \vdash E$ by (*basic*). By $(\rightarrow LS)$, $C, C \rightarrow D, D \rightarrow E \vdash E$. By $(\rightarrow RS)$, $C \rightarrow D, D \rightarrow E \vdash C \rightarrow E$. By $(\wedge LS)$, $(C \rightarrow D) \wedge (D \rightarrow E) \vdash C \rightarrow E$. Finally, $\vdash ((C \rightarrow D) \wedge (D \rightarrow E)) \rightarrow (C \rightarrow E)$ by $(\rightarrow RS)$. □

Theorem 2.3. *All theorems of IPL [MPL] are provable in hLJp [hLMp].*

Proof. The *modus ponens* rule of IPL $[MPL]$ holds in $hLJp$ $[hLMp]$ as (MP). Therefore, any axiomatic proof in IPL $[MPL]$ can be translated into a proof in $hLJp$ $[hLMp]$, since, by Lemma 2.2, all axioms of IPL $[MPL]$ are provable in $hLJp$ $[hLMp]$. □

We assume acquaintance with LJ $[LM]$ sequent calculus for propositional logic as presented in [5], it is called in our context as LJp $[LMp]$.[5] Gentzen had already proved the equivalence of IPL $[MPL]$ in axiomatic presentation with LJp $[LMp]$.[6] We notice that the following rules belong to LJp $[LMp]$, but they are absent in $hLJp$ $[hLMp]$ as also in Pinto and Dyckhoff's system mentioned above:

$(thinning) : \Gamma \vdash C \Rightarrow \Gamma, D \vdash C$

$(contraction) : \Gamma, D, D \vdash C \Rightarrow \Gamma, D \vdash C$

$(cut) : (\Gamma \vdash C \text{ and } C, \Delta \vdash D \Rightarrow \Gamma \vdash D$

$$\begin{array}{ccc}
 & 1.C \rightarrow D \vdash \neg\mathbf{C} \vee \mathbf{D}^{\checkmark} & \\
\swarrow & \downarrow & \searrow R2 * (1) \\
2.C \rightarrow D \vdash \neg\mathbf{C} & 3.\mathbf{C} \rightarrow \mathbf{D} \vdash D & 4.\mathbf{C} \rightarrow \mathbf{D}^{\checkmark} \vdash \neg C \vee D^{\checkmark} \\
\downarrow R3(2) & \downarrow L3 * (3) & \downarrow L3 * (4) \\
5.C \rightarrow D, C \vdash \bot & 6.D \vdash D \text{ (ini)} & 9.D \vdash \neg C \vee D \\
 & 7. \vdash C \otimes & 10. \vdash C \otimes \\
\downarrow & & \\
12.D, C \vdash \bot \otimes & & \\
13.C \vdash C \text{ (ini)} & &
\end{array}$$

Proof-search asks which rule might have been applied last in a putative proof of a sequent. $hLJp$ proofs contain neither contractions nor thinnings. We cal it the *the-last-sequent-introduced* question. Also, as will be shown below, (MP) can be

[5]Gentzen does not employ the absurd constant "$\bot$", he uses negation "$\neg$". Hypotheses are given as finite lists, which would then require the interchange rule–$\Gamma, E, F, \Delta \vdash C \Rightarrow \Gamma, F, E, \Delta \vdash C$. See [5, pp. 83–85]

[6]See [5, pp. 103–105].

disregarded. Hence, if every provable sequent can be proved without using thinnings, contractions or cuts, then both: (i) no sentence occurrence disappears in a proof; (ii) no occurrence is multiplied inside a sequent in the process of proof-search. The search will then only examine subsentences of the occurrences in the sequent. Therefore, $hLJp$ offers a suitable ground for effecting proof-search by means of a case analysis depending on which side of the sequent the logical constant is placed.

Theorem 2.4. *All inference rules of $hLJp$ [$hLMp$] are derivable in Gentzen's LJp [LMp] including the basic and the absurd sequents.*

Proof. This is immediate for each (LS) and (RS) rule of $hLJp$ [$hLMp$], as also for (*basic*) and (*absurd*) sequents. Finally, (MP) is derivable in $hLJp$ [$hLMp$] by using the (*cut*) rule in LJp [LMp], given that $C, C \to D \vdash D$ is provable in LJp [LMp]. □

Corollary 2.5. *If a sequent $\Gamma \vdash C$ is provable in $hLJp$ [$hLMp$], then it is provable in LJp [LMp].*

Proof. Immediate from Theorem 2.4. □

Theorem 2.6. *$hLJp$ [$hLMp$] and LJp [$hLMp$] are equivalent.*

Proof. Every sequent provable in $hLJp$ [$hLMp$] is provable in LJp [LMp] according to Corollary 2.5. Every theorem in axiomatic IPL [MPL] is provable in $hLJp$ [$hLMp$] according to Theorem 2.3. Since LJp [LMp] and IPL [MPL] are equivalent, as Gentzen proved in his *hauptsatz* paper [5] section V, then $hLJp$ [$hLMp$] and LJp [LMp] are equivalent. □

Lemma 2.7. *If a sequent $\Gamma \vdash C$ is provable in $hLJp$ and Γ is nonempty, then the last rule used in the proof of $\Gamma \vdash C$ is not an instance of (MP).*

Proof. Immediate by observing that the form of (MP) in $hLJp$ does not admit a conclusion containing hypotheses. □

A sentence occurrence in a sequent inside a proof in $hLJp$ [$hLMp$] is *a main occurrence* with respect to an inference if the sequent is either the conclusion of a (LS) or (RS) rule and the occurrence is introduced by this inference, or if the sequent is a premise of a (LS) or (RS) rule and the occurrence is the explicit condition for applying it. All occurrences of a (MP) rule are *main* occurrences. All other occurrences are called *side occurrences* with respect to the inference, that is, they belong to the context.

Lemma 2.8. *If, for an atomic sentence e not occurring in either Γ or C, a sequent $e, \Gamma \vdash C$ is provable in hLJp [hLMp], then no rule in the proof is (MP).*

Proof. By induction on the number of proof steps. The basis is trivial. Suppose that the proof has $n+1$ steps. Suppose that for proofs containing n steps or less the assertion holds. It is clear that the $n+1$ step must be either an (LS) rule or an (RS) rule, since (MP) does not admit hypotheses, according to Lemma 2.7. Additionally, the hypothesis e was not introduced in the $n+1$ step since it is atomic. Hence it is a side occurrence in the conclusion. Also, it has to be a side occurrence in the premises. Remind that it occurs neither in Γ nor C. The result follows by the induction hypotheses. □

Theorem 2.9. *For every proof of $\Gamma \vdash C$ in hLJp [hLMp], there is a (MP)-free proof of it.*[7]

Proof. According to Corollary 2.5, if a sequent $\Gamma \vdash C$ is provable in *hLJp* [*hLMp*], then it is also provable in *LJp* [*LMp*]. Let e be an atomic sentence not occurring in $\Gamma \vdash C$. By thinning the sequent $e, \Gamma \vdash C$ is also provable in *LJp* [*LMp*]. By Theorem 2.6, $e, \Gamma \vdash C$ is provable in *hLJp* [*hLMp*]. By Lemma 2.8, the proof of $e, \Gamma \vdash C$ is (MP)-free. Since e is atomic, it occurs only as a side occurrence in the hypotheses of any sequent in the proof, so it can be erased in all of them. □

This seems to be one of the shortest proofs of a property that is connected to Gentzen's *hauptsatz*.[8] It means that rule (MP) can be disregarded in proof-search inside *hLJp* [*hLMp*]. Theorem 2.9 has two immediate interesting consequences. From now on, "$\Gamma \nvdash C$" means that $\Gamma \vdash C$ is not provable in a constructivist sense. It is called *antisequent* in [8]. In our article it is equivalent to say that: $(\Gamma \vdash C) \Rightarrow \perp$, where $\perp$ is the absurd constant in the metalanguage.

Corollary 2.10. *(i) hLJp [hLMp] is consistent, that is, $\nvdash \perp$; (ii) $\vdash C \vee D \Rightarrow (\vdash C$ or $\vdash D)$.*

Proof. By considering what would be the last inference rule in a (MP)-free proof according to Theorem 2.9. □

3 Metaprinciples for LJp $[LMp]$

The following metaequivalences make explicit the meaning of the logical constants either as a hypothesis or as a *consequentia* in a sequent inside *LJp* [*LMp*].

[7]It is contraction-free and cut-free.

[8]That (MP) is related to the (cut) rule means, at least, that both express a transitivity property.

Theorem 3.1. *It holds for the logical constants of LJp [LMp] that:*

Left side, as a hypothesis

$L1 : \Gamma, C \wedge D \vdash E \Leftrightarrow \Gamma, C, D \vdash E$

$L2 : \Gamma, C \vee D \vdash E \Leftrightarrow \Gamma, C \vdash E \textit{ and } \Gamma, D \vdash E$

$L3 : \Gamma, C \to D \vdash E \Leftrightarrow \textit{for any } \Delta \supseteq \Gamma, (\Delta, C \vdash D \Rightarrow \Delta \vdash E)$

Right side, as consequentia*:*

$R1 : \Gamma \vdash C \wedge D \Leftrightarrow \Gamma \vdash C \textit{ and } \Gamma \vdash D$

$R2 : \Gamma \vdash C \vee D \Leftrightarrow \textit{for any } E, ((\Gamma, C \vdash E \textit{ and } \Gamma, D \vdash E) \Rightarrow \Gamma \vdash E)$

$R3 : \Gamma \vdash C \to D \Leftrightarrow \Gamma, C \vdash D$

Proof. $L1, L2, R1$ and $R3$ can be proved straightforwardly in a first step. We consider the cases of $L3$ and $R2$ and their proofs in a second step. $L3$ is proved as follows. Direction $\Rightarrow$. Suppose $\Gamma, C \to D \vdash E$. Suppose that $\Delta, C \vdash D$ for $\Delta \supseteq \Gamma$. Clearly, $\Delta \vdash C \to D$ by $R3$. Thus, $\Gamma, \Delta \vdash E$ by (*cut*). By (*contractions*), $\Delta \vdash E$. Direction $\Leftarrow$. Suppose that, for any $\Delta \supseteq \Gamma$, $(\Delta, C \vdash D \Rightarrow \Delta \vdash E)$. By instantiation, $(\Gamma, C \to D, C \vdash D) \Rightarrow (\Gamma, C \to D \vdash E)$. $C \to D \vdash C \to D$ is a basic sequent. By $R3$, $C \to D, C \vdash D$. Therefore, $\Gamma, C \to D \vdash E$. Now we consider $R2$, which is proved as follows. Direction $\Rightarrow$. Suppose $\Gamma \vdash C \vee D$. Suppose $\Gamma, C \vdash E$ and $\Gamma, D \vdash E$ for an E whatever. By $L2$, $\Gamma, C \vee D \vdash E$. By (*cut*), $\Gamma \vdash E$. Direction $\Leftarrow$. Suppose that, for any E, $((\Gamma, C \vdash E \text{ and } \Gamma, D \vdash E) \Rightarrow \Gamma \vdash E)$. By instantiation, $(\Gamma, C \vdash C \vee D \text{ and } \Gamma, D \vdash C \vee D) \Rightarrow \Gamma \vdash C \vee D$. $\Gamma, C \vee D \vdash C \vee D$ is basic. By $L2$, $\Gamma, C \vdash C \vee D$ and $\Gamma, D \vdash C \vee D$. Therefore, $\Gamma \vdash C \vee D$. □

Double inference rules were anticipated by Došen in a series of writings, from [2] to [3]. For him, the meaning of each logical constant can be described in a structural way by double inferences, as those presented in Theorem 3.1. This is clearly the case when they do not involve quantification. But, in two cases, this meaning explanation is one sided. Disjunction on the right side and implication on the left side cannot be presented as regular inference rules because of the second order quantification in them. Hence, only the use of the constant in the other side follows the pattern of double inferences. The explicitation of meaning in those two cases require a quantification that cannot be circumvented. If it can be completely described in structural terms or not is then a question of defining what should we understand by the concept of a structural description.

In our perspective, the meaning explanation for logical constants should pay attention to two distinct behaviors of logical constants: the behavior as a hypothesis and the behavior as a *consequentia*.

As can be seen, most of $hLJp$ [$hLMp$] inferences can then be read bidirectionally. Only (MP), ($\to LS$) and ($\vee RS$) cannot. Proof-search development employs the necessary condition inference attached to each logical constant. That is, it employs the reading from left to right in $L1$ to $R3$. This is the *analytic* or *decompositional* reading of the rule and it presents the strict necessary conditions attached to the elimination of the logical constant in the respective side. From right to left we have the *synthetic* reading, and it presents the strict sufficient conditions for the introduction of the logical constant in the respective side.

The left to right reading of ($\to LS$) and ($\vee RS$) cannot be used owing to the fact that the condition in these rules is not strictly necessary and sufficient.

Theorem 3.2. *($\to LS$) and ($\vee RS$) are consequences of the synthetic reading of $L3$ and $R2$ above in LJp [LMp], respectively.*

Proof. First ($\to LS$). Suppose that $\Gamma \vdash C$ and $\Gamma, D \vdash E$. Suppose that, for a $\Delta \supseteq \Gamma$, $\Delta, C \vdash D$. By (*thinnings*), $\Gamma, \Delta, C \vdash D$ and $\Gamma, \Delta, D \vdash E$. Hence, $\Gamma, \Delta, C \vdash E$, by (*cut*) and (*contractions*). By (*thinnings*), $\Gamma, \Delta \vdash C$. Thus, again, by (*cut*) and some (*contractions*) $\Gamma, \Delta \vdash E$. Therefore, $\Gamma, C \to D \vdash E$ by $L3$. Second, ($\vee RS$). Suppose $\Gamma \vdash A$. Suppose that $\Gamma, A \vdash E$ and $\Gamma, B \vdash E$. By (*cut*) and some (*contractions*), $\Gamma \vdash E$. The other case is similar. $\square$

4 Proof-search in $hLJp$ [$hLMp$]

As already pointed, proof-search is done through an analysis of complex sentences, through the so called decompositional reading. The analysis goes on until an initial sequent is reached or until all sentences in the sequent are atomic. In two cases the analytic readings involve quantification and, thus, a dependence on an endless number of sequents, according to $L3$ and $R2$. Thus, a guiding heuristics has to be employed. It consists in asking the *the-last-sequent-introduced* question.

When searching for a proof, if either an implication in the left side or a disjunction in the right side comes under analysis, then either ($\to LS$) or ($\vee RS$) had to be employed in case the occurrence was the last one introduced. But, it might also be the case that it was not the last. Then, another complex occurrence should be the last. Hence, the analysis should take another alternative. The occurrence discarded as being the last is marked and it should not undergo analysis again, as in the following implications:

$$\Gamma, C \to D \vdash E \Rightarrow ((\Gamma, D \vdash E \text{ and } \Gamma \vdash C) \text{ or } \Gamma, C \to D^{\checkmark} \vdash E)$$

$$\Gamma \vdash C \vee D \Rightarrow ((\Gamma \vdash \text{ or } \Gamma \vdash D) \text{ or } \Gamma \vdash C \vee D^{\checkmark})$$

This left to right quasi-analytic reading is fortunate because the following equivalences hold, as can be easily verified.

Theorem 4.1. *It holds for the logical constants of LJp [LMp] that:*[9]

$L3*: \Gamma, C \to D \vdash E \Leftrightarrow ((\Gamma, D \vdash E \; and \; \Gamma \vdash C) \; or \; \Gamma, C \to D \vdash E)$

$R2*: \Gamma \vdash C \vee D \Leftrightarrow ((\Gamma \vdash C \; or \; \Gamma \vdash D) \; or \; \Gamma \vdash C \vee D)$

Proof. Immediate. □

Proof search is going to be based on $L3*$ and $R2*$ together with $L1, L2$, $R1$ and $R3$ under a controlling strategy for preventing the procedure from going into a loop. It might be the case that the respective occurrences–implication in the left side or disjunction in the right side–were not the last introduced in the sequent. However, since at least one of the occurrences has to be the last in case there is a proof of the sequent, only a finite amount of analysis must be effected over the occurrences inside that sequent in the worst case scenario.[10] If, after examining each occurrence, no proof could be found, then the sequent is not provable. Actually, things are better than this scenario since in other cases we can employ the analytical reading. The development can be assumed to be definitive because provability, as well as refutability, will be preserved. Observe, additionally, that there is no alternative branching in $L1$, $L2$, $R1$ and $R3$, only in $L3*$ and $R2*$.

For showing a *sequent to be provable* it is enough to exhibit at least *one correct finished path in a search-tree.* If all sequents in a path were analyzed, then the *path is finished.* An initial sequent is considered immediately analyzed. A non-initial *sequent* which contains no more than atomic sentences is considered analyzed and *closed.* It is marked with "⊗". A *path is closed* if it contains at least one closed sequent. A finished non-closed path contains a proof. For abreviating the proof-search procedure, we can assume, when convenient, that a closed sequent turns finished the path to which it belongs. Other sequents are analyzed as follows.

Definition 4.2. *STRATEGY for proof-search – It consists in developing all non-finished paths going through a non-analyzed sequent picked at will by choosing one of its non-marked non-atomic sentence occurrences at will and applying the analytic reading of the rules in Theorem 4.1 or the other rules in Theorem 3.1, depending on which side of the sequent the sentence occurs, such that the development must take care of marked formulas as described below:*

---

[9]Of course, now the reading from right to left is no longer the statement of a strict sufficient condition for the logical constant in question.

[10]The worst case scenario is that where all the hypotheses are implications and the *consequentia* is a disjunction: $C_1 \to D_1, \ldots, C_{n-1} \to D_{n-1} \vdash C_n \vee D_n$.

	Left Rules	Right rules	
$L1$	$\Gamma, C \wedge D \vdash E$ $\vdots$ $\downarrow$ $\Gamma^*, C, D \vdash E^*$	$\Gamma \vdash C \wedge D$ $\vdots$ $\downarrow$ $\Gamma^* \vdash C$ $\Gamma^* \vdash D$	$R1$
$L2$	$\Gamma, C \vee D \vdash E$ $\vdots$ $\downarrow$ $\Gamma^*, C \vdash E^*$ $\Gamma^*, D \vdash E^*$	$\Gamma \vdash C \vee D^{\checkmark}$ $\vdots$ $\swarrow$ $\downarrow$ $\searrow$ $\Gamma^* \vdash C$ $\quad$ $\Gamma^* \vdash D$ $\quad$ $\Gamma^* \vdash C \vee D^{\checkmark}$	$R2*$
$L3*$	$\Gamma, C \supset D^{\checkmark} \vdash E$ $\vdots$ $\downarrow$ $\searrow$ $\Gamma^*, D \vdash E^*$ $\quad$ $\Gamma, C \supset D^{\checkmark} \vdash E$ $\Gamma^* \vdash C$	$\Gamma \vdash C \supset D$ $\vdots$ $\downarrow$ $\Gamma^*, C \vdash D$	$R3$

Γ^* and E^* are the result of erasing any marks "$\checkmark$" that might occur in Γ and E. If there is none, then $\Gamma^* = \Gamma$ and $E^* = E$.[11] *The rightmost branch in the rules $L3*$ and $R2*$ with a repetition is going to be written down only if there is left either a right implication or a right disjunction unmarked in the sequent developed noticing that the formula occurrence being developed has been marked "$\checkmark$" and it cannot again be the object of analysis in this rightmost path of $R2*$ or $L3*$. Any developed sequent is considered analyzed. END of STRATEGY*

When a sequent is developed by rules $L1, L2, R1$ or $R3$, the conjunction of the sequents written down in the path is equivalent to the sequent developed–according to the decompositional reading. If it is developed by $L3*$ or by $R2*$, there will be alternative paths. The disjunction of the conjunction of sequents in each path is equivalent to the analyzed sequent. Therefore, no information is lost through developments.

[11] This is going to be the case only if the sequent developed belongs to the rightmost path in $L3*$ and $R2*$. Any other development erases the marks.

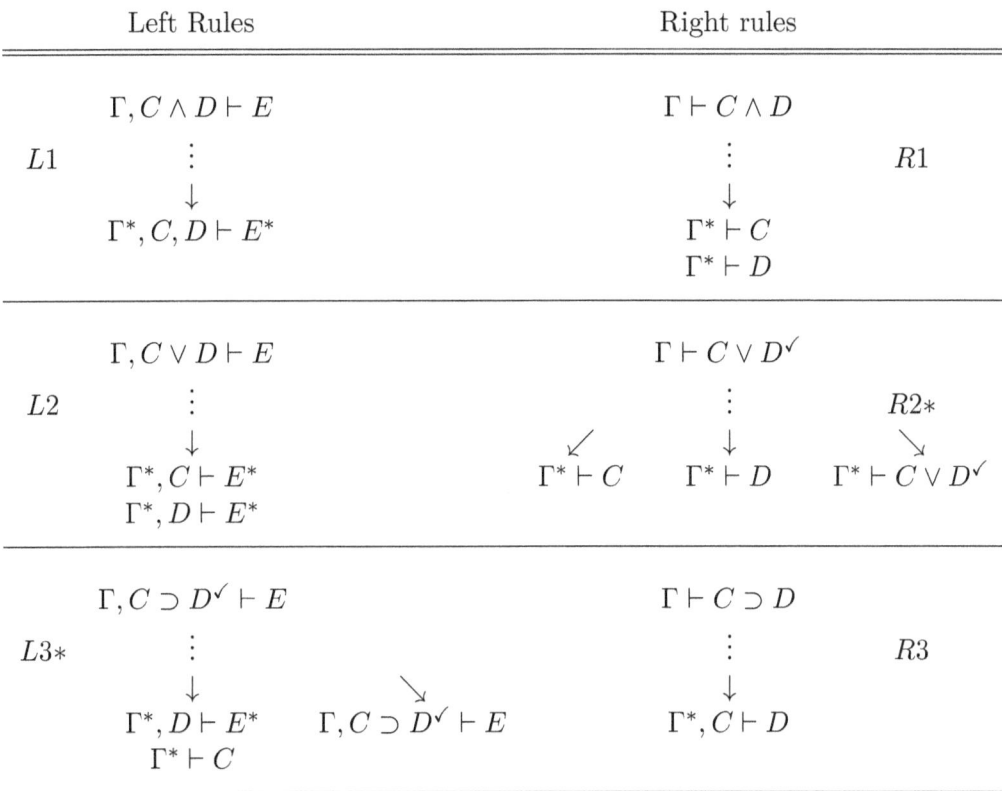

The rightmost branch in $L3*$ and $R2*$ develops the supposition of being the last for every occurrence of either a left implication or a right disjunction in a given sequent.

Example 4.3 illustrates a successful proof-search. The path has a proof when read bottom-up. It is finished and open, that is, not-closed:

$$
\begin{array}{cc}
1.\neg C \vee D \vdash \mathbf{C} \to \mathbf{D} & \\
\downarrow & R3(1) \\
2.\neg \mathbf{C} \vee \mathbf{D}, C \vdash D & \\
\downarrow & L2(2) \\
3.\neg \mathbf{C}^{\checkmark}, C \vdash D & \\
4.D, C \vdash D \text{ (initial)} & \\
\downarrow & L3*(3) \\
5.\bot, C \vdash D \text{ (initial)} & \\
6.C \vdash C \text{ (initial)} &
\end{array}
$$

Example 4.3.

Example 4.4 below shows the case of an unsuccessful proof-search. Every path is closed. Although sequent 8 could still be developed, the procedure stops since the path contains the closed atomic sequent 9.

$$
\begin{array}{ccc}
 & 1.C \to D \vdash \neg \mathbf{C} \vee \mathbf{D}^{\checkmark} & \\
\swarrow & \downarrow & \searrow R2*(1) \\
2.C \to D \vdash \neg \mathbf{C} & 3.\mathbf{C} \to \mathbf{D}^{\checkmark} \vdash D & 4.\mathbf{C} \to \mathbf{D}^{\checkmark} \vdash \neg C \vee D^{\checkmark} \\
\downarrow R3(2) & \downarrow L3*(3) & \downarrow L3*(4) \\
5.\mathbf{C} \to \mathbf{D}^{\checkmark}, C \vdash \bot & 6.D \vdash D \text{ (ini)} & 8.D \vdash \neg C \vee D \\
 & 7. \vdash C \otimes & 9. \vdash C \otimes \\
\downarrow L3*(5) & & \\
10.D, C \vdash \bot \otimes & & \\
11.C \vdash C \text{ (ini)} & &
\end{array}
$$

Example 4.4.

Theorem 4.5. *STRATEGY is a finite procedure.*

Proof. Each sequent $\Gamma \vdash C$ in a path has a degree that is the sum of the degrees of each sentence occurring in the sequent–its number of connectives–, noticing that the absurd is atomic and has degree 0. All occurrences in the new sequents are subformulas that descend from occurrences in the developed sequent. The number of sequents resulting from a development through $L1, L2, R1$ or $R3$ is finite and their degree is less than the degree of the sequent developed. Sequents resulting from a development by $L3*$ and $R2*$ are finite in number and of a lesser degree than the sequent developed, with the exception sometimes of one third rightmost path in which the developed sequent is repeated. However, if repeated, it is repeated only a finite number of times until all left implications and right disjunction occurrences are marked. Hence, the procedure is finite. It is impossible to go on endlessly. □

All rules $L1, L2, L3*, R1, R2*, R3$, are such that if one side of the equivalence is provable [refutable], then the other side must also be provable [refutable] . Therefore, when looking for a proof of a sequent inside $hLJp$ [$hLMp$], if the sequent undergoes a transformation according to one of the rules, then at least in one path all sequents written down in a development are provable if, and only if, the former sequent was provable.

Theorem 4.6. *There is a proof of* $\Gamma \vdash C$ *in* $hLJp$ *[*$hLMp$*] if, and only if, there is a finished open path in the proof-search* $STRATEGY$ *for* $\Gamma \vdash C$*.*

Proof. (i) Left to right. Suppose we have a proof of $\Gamma \vdash C$ in $hLJp$ [$hLMp$]. Therefore by Theorem 2.9 there is a proof of it that does not contain (MP). This proof corresponds to an open finished path of STRATEGY read bottom up, since each (LS) or (RS) introduction rule corresponds to a development by analysis through $L1, L2, L3*, R1, R2*$ and/or $R3$. (ii) Right to left. Suppose the proof-search STRATEGY gives us a finished open path for $\Gamma \vdash C$. All sequents in this path are analyzed, and those that contain no more than atomic occurrences are initial. They are immediately provable in $hLJp$ [$hLMp$]. Any non-initial sequent in the path is the consequence of the synthetic readings of $L1, L2, L3*, R1, R2*$ and/or $R3$, depending on which was the development effected. Now, all synthetic readings are (LS) or (RS) inference rules of $hLJp$ [$hLMp$], with the exception of two cases. Those two cases are either $\Gamma, C \to D \vdash E \Leftarrow \Gamma, C \to D^{\checkmark} \vdash E$ using rule $L3*$ or $\Gamma \vdash C \vee D \Leftarrow \Gamma \vdash C \vee D^{\checkmark}$ using rule $R2*$. In both cases, the inference can be erased since its premise and conclusion are identical. □

Corollary 4.7. *Decidability -* $\Gamma \vdash C$ *is decidable in* $hLJp$ *[*$hLMp$*].*

Proof. By Theorems 4.6 and 4.5, we have a decision procedure for $hLJp$ [$hLMp$]. □

Corollary 4.7 is not yet sufficient to guarantee a strong *tertium non datur* in which negation is understood in a constructive sense. Only after defining what are bases the conditions for such assertion are fullfilled, in Theorem 5.9.

5 Hypo semantics

The proof-search procedure was established without major considerations concerning the semantics of IPL or MPL with exception of the explanation clauses in Theorem 3.1. They are the basis of Hypo semantics in [10] and [11]. Rules $L1$ to $R3$ state the necessary and sufficient conditions for the use of a logical constant, either as a hypothesis or as a *consequentia* inside a sequent.

5.1 Semantical clauses

To the above clauses in Theorem 3.1 some others must be added in order to obtain a full semantics.

5.1.1 Complementing Hypo semantical principles for intuitionistic [minimal] propositional logic

Let's use $\overline{\top}$ as an abreviation for representing any *constructive tautology* in the metalanguage.

Theorem 5.1. *Equivalences for* $\perp$ *in* LJp *(but not in* LMp*):*

$L4 : \Gamma, \perp \vdash c \Leftrightarrow \overline{\top}$ *(c is atomic)*

$R4 : \Gamma \vdash \perp \Leftrightarrow$ *for any atomic* e, $\Gamma \vdash e$

Proof. $L4$ is easy. Now $R4$. Direction $\Leftarrow$ first. From the supposition that, for any atomic e, $\Gamma \vdash e$, by instantiation, we have $\Gamma \vdash \perp$ since $\perp$ is atomic. Direction $\Rightarrow$, suppose $\Gamma \vdash \perp$. However, $\perp \vdash e$ for any atomic sentence e in LJp. By (cut), for any atomic e, $\Gamma \vdash e$. □

Besides clauses for logical constants, there must be principles laying out how to operate with hypotheses. Actually, we have been using their syntactical counterpart in the proofs of equivalence above. Although sequent structural rules are usually read as syntactical principles, we claim that they have semantical content. This must be the case in Hypo semantics since consequence is its fundamental relation. Below, small Latin letter c represents an atomic sentence.

Definition 5.2. *Hypo structural principles for* LJp *[*LMp*]:*

$(atomicIdempotence) : c \vdash c$

$(Load) : \Gamma \vdash C \Rightarrow \Gamma, D \vdash C$

$(atomicDrop) : (\Gamma \vdash c \text{ and } c, \Gamma \vdash E) \Rightarrow \Gamma \vdash E$

Theorem 5.3. *The full (Drop) principle and the full (Idempotence) principle hold in Hypo, that is:*[12]
$(Drop) : (\Gamma \vdash C \text{ and } C, \Gamma \vdash E) \Rightarrow \Gamma \vdash E$
$(Idempotence) : C \vdash C.$

Proof. (*Drop*) is proved by induction on the degree of C. Degree 0 (zero) is immediate from (*atomicDrop*). Induction step, suppose that C is of degree n. By the induction hypothesis (IH), full (*Drop*) holds for degrees less than n. There are three cases: (a), (b) and (c). (a) $C = F \wedge G$. Suppose $\Gamma \vdash F \wedge G$. By $R1$, $\Gamma \vdash F$ and $\Gamma \vdash G$. By (*Load*), $\Gamma, G \vdash F$. Suppose $F \wedge G, \Gamma \vdash E$. By $L1$, $F, G, \Gamma \vdash E$. By (IH), (*Drop*) holds for F, hence $\Gamma, G \vdash E$. And again, it holds for G, thus $\Gamma \vdash E$. (b) $C = F \vee G$. Suppose $\Gamma \vdash F \vee G$. By $R2$, for any E, $((\Gamma, F \vdash E$ and $\Gamma, G \vdash E) \Rightarrow \Gamma \vdash E)$. Suppose $F \vee G, \Gamma \vdash E$. By $L2$, $F, \Gamma \vdash E$ and $G, \Gamma \vdash E$. By *meta modus ponens*, $\Gamma \vdash E$. (c) $C = F \to G$. Suppose $\Gamma \vdash F \to G$. By $R3$, $\Gamma, F \vdash G$. Suppose $F \to G, \Gamma \vdash E$. By $L3$, for any $\Delta \supseteq \Gamma$, $(\Delta, C \vdash D \Rightarrow \Delta \vdash E)$. By instantiation, $\Gamma, F \vdash G \Rightarrow \Gamma \vdash E$. By *meta modus ponens*, $\Gamma \vdash E$. (*Idempotence*) is also proved by induction on the degree of C. □

5.2 Atomic bases and Hypo models

The traditional semantical values of constructive provability or of classical truth have no role in Hypo semantics. The consequence relation, which the sequents are supposed to express, constitutes the basic semantical brick. Now, a basis consists in establishing this relation among basic hypotheses and atomic *consequentia*.

Let the set of positive atoms be defined as $\mathfrak{A}^+ = \{d | d \in \mathfrak{at}\mathfrak{L} - \{\bot\}\}$, and negative atoms as the set $\mathfrak{A}^- = \{d \to \bot | d \in \mathfrak{at}\mathfrak{L} - \{\bot\}\}$. In both cases we say that the atomic sentence d is the core of the atom. Basic hypotheses are of two kinds: positive and negative atoms. In *LMp bases*, negative atoms are omitted and $\bot$ is irrelevant in the language, so it can be disregraded. In *LJp bases*, both kinds of atoms are needed. Let $\mathfrak{M} = \{\omega \cup \delta | \omega$ is a multiset $\subseteq \mathfrak{A}^+$ and δ a multiset $\subseteq \mathfrak{A}^-\}$. We say that $\gamma \in \mathfrak{M}$ is *normal* iff no negative atom in γ has the same core of a positive atom in γ. Let $\mathfrak{H} = \{\gamma \in \mathfrak{M} | \gamma$ is finite and normal$\}$ be the *set of basic hypotheses*. Let $\mathfrak{P} = \{< \gamma, c > | \gamma \in \mathfrak{H}$ and $c \in \mathfrak{at}\mathfrak{L}\}$ be the set of *basic pairs* of $\mathfrak{L}$. Let $\mathfrak{R_B}^{\oplus}$ and $\mathfrak{R_B}^{\ominus}$ be both defined constructively such that $\mathfrak{R_B}^{\oplus} \subseteq \mathfrak{P}$ and $\mathfrak{R_B}^{\ominus} \subseteq \mathfrak{P} - \mathfrak{R_B}^{\oplus}$. The *atomic consequence relation on basis* $\mathfrak{B}$ *over* $\mathfrak{L}$ is defined as follows:

$$\gamma \vdash^{\mathfrak{B}} c \text{ if } < \gamma, c > \in \mathfrak{R_B}^{\oplus} \text{ and } \gamma \nvdash^{\mathfrak{B}} c \text{ if } < \gamma, c > \in \mathfrak{R_B}^{\ominus}$$

[12]The (*Cut*) rule is easily provable from (*Drop*) and (*Idempotence*).

The reason for considering sentences of the form $e \to \bot$ as basic (negative) hypotheses for *LJp* is that not all suppositions are of the form "suppose that e were the case". They can also be of the form "suppose that e were not the case". These should have space in a proof-theoretic semantics under the hypothetical paradigm, since we are taking hypotheses to be a primitive notion in our environment.

We follow the usual constructive practice of representing that e is not the case as $e \to \bot$. Constructively speaking, it might occur that $\nvdash^{\mathfrak{B}} e$ for some atomic sentence e and, at the same time, $e \to \bot \vdash^{\mathfrak{B}} \bot$, from which it follows that $\vdash^{\mathfrak{B}} \neg\neg e$ given the usual definitions of negation and the clauses for implication. This relation $e \to \bot \vdash^{\mathfrak{B}} \bot$ has then to be established at a basic level, in a basis $\mathfrak{B}$ for intuitionistic logic.

An atomic basis $\mathfrak{B}'$ is an *extension of the atomic basis* $\mathfrak{B}$ whenever, for all $< \gamma, c > \in \mathfrak{P}$, $(\gamma \vdash^{\mathfrak{B}} c \Rightarrow \gamma \vdash^{\mathfrak{B}'} c)$.

Decidable atomic bases $\mathfrak{B}$ are those in which $\mathfrak{R}_{\mathfrak{B}}{}^{\oplus} \cup \mathfrak{R}_{\mathfrak{B}}{}^{\ominus} = \mathfrak{P}$.

The *empty basis* ($\varnothing$) is the atomic basis over $\mathfrak{P}$ such that, for every $< \gamma, c >$: $\gamma \vdash^{\varnothing} c \Leftrightarrow c \in \gamma$. This basis is the *ground basis* for *LMp* and it can be extended to other bases. The *atomic sequents valid in the empty basis* are all initial sequents of *LMp*. The extension *absurd-empty basis* ($abs\varnothing$) is such that $\gamma \vdash^{abs\varnothing} c \Leftrightarrow (c \in \gamma$ or $\bot \in \gamma)$. It is the *ground basis* for *LJp* and it can also be extended. The *atomic sequents valid* in $abs\varnothing$ are exactly those that are initial in *LJp*. We are going to refer to both bases as the empty basis when there is no risk of confusion.

Given an atomic basis $\mathfrak{B}$, an *intuitionist* [*minimal*] *model* $\mathfrak{M}_{\mathfrak{B}}$ is the closure of $\mathfrak{B}$ by the sequent rules of *LJp* [*LMp*] if $\mathfrak{B}$ extends $abs\varnothing$ [$\varnothing$]. It contains all *valid sequents* in basis $\mathfrak{B}$.

Theorem 5.4. *Any sequent valid in the empty basis is also valid in any extension of it. Conversely, any sequent invalid in an extension of the empty basis is invalid in the empty basis.*

Proof. According to the definitions of empty basis and of model. □

Theorem 5.5. $\vdash^{[abs]\varnothing} C \vee D \Leftrightarrow (\vdash^{[abs]\varnothing} C \text{ } or \vdash^{[abs]\varnothing} D)$.

Proof. By rule ($\vee RS$) the right to left direction is proved. If there is a proof of the disjunction in the empty basis, given that the basis validates only the atomic initial sequents of *hLMp* [*hLJp*], then by Corollary 2.10 (ii) there is a proof of one of the disjuncts, which then proves the left to right direction component. □

The semantical principles of Hypo are now complete. The (*atomicDrop*) principle establishes the condition under which an atomic hypothesis can be disregarded.

Notice that there is no semantical principle corresponding to contraction, hypotheses are being considered as given through multisets.

Atomic bases are the ground for presenting models and counter-models for constructive logical constants.

Sets of hypotheses can be contradictory.[13] We say that a multiset Γ of sentences is inconsistent[14] if, and only if, $\Gamma \vdash^{\varnothing} \bot$; and a multiset Γ is trivial if, and only if, for all atomic e, $\Gamma \vdash^{\varnothing} e$. The definition of the absurd above states that it follows from a set of hypotheses when this set is trivial in intuitionistic logic.

Now, there is a procedure for obtaining models in which a given set of sentences holds. Different paths of a tree generated by STRATEGY give us a basis that serves as a model.

Definition 5.6. Model-search - *Given a proof-search tree according to definition 4.2 for $\Gamma \vdash C$ where all sequents are fully analyzed, a model for $\Gamma \vdash C$ is given either by: (i) the empty basis in case the tree contains a non-closed finished path (i.e., a proof); or (ii) the basis obtained by extending the empty basis with all non-initial atomic sequents belonging to a given closed path.*

$$
\begin{array}{cc}
1. \vdash (\mathbf{C} \to \mathbf{D}) \to \mathbf{D} & \\
\downarrow & R3(1) \\
2. \mathbf{C} \to \mathbf{D}^{\checkmark} \vdash D & \\
\downarrow & L3*(2) \\
3. D \vdash D \text{ (initial)} & \\
4. \vdash C \otimes &
\end{array}
$$

Example 5.7.

Concerning example 5.7, a basis proving $\vdash C$ is enough for proving $\vdash (C \to D) \to D$ according to the path in example 5.7. If C and D are atomic, then we obtained a definition for a basis, that is, for the basis $\mathfrak{B} = \{\vdash C\}$ extending the empty basis. It holds for it that $\vdash^{\mathfrak{B}} (C \to D) \to C$. If they are not, then $\vdash C$ has to be further analyzed by means of STRATEGY.[15]

The same procedure for finding a basis can be used for the case of a finite set of sequents, all then being put at the beginning of the path starting the tree.

Now, a basis can be trivial, in the sense that it proves any atomic sentence. For example, any intuitionistic model whose basis contain rules $\vdash a$ and $a \vdash \bot$. Hence,

[13] That is, it contains a sentence and its contradictory negation.

[14] Another concept that can be defined this way is the concept of an *impossible* multiset.

[15] We remember that, according to the observation already made above, it might be the case that $e \to \bot \vdash^{\mathfrak{B}} \bot$ but $\nvdash^{\mathfrak{B}} e$ for a certain basis $\mathfrak{B}$.

Hypo models are not assumed to be consistent by definition, they must be proved to be so.

Theorem 5.8. *The empty bases are not trivial:* $\nvdash^{[abs]\varnothing} \perp$.

Proof. Immediate from their definitions. □

Theorem 5.9. Tertium non-datur *for consequence in the empty basis - either* $\Gamma \vdash^{[abs]\varnothing} C$ *or* $\Gamma \nvdash^{[abs]\varnothing} C$ *in hLMp [hLJp].*

Proof. According to 4.7 there is a procedure for determining for any sequent if it is provable or not by the STRATEGY procedure. In the empty basis, any atomic sequent is decidable according to the definition of a basis. STRATEGY develops sequents according to one of the rules $L1, L2, L3*, R1, R2*$ or $R3$ of Theorems 4.1 and 3.1. Each development is an equivalence. This equivalences have the form of a disjunction of conjunctions. Thus, after completely analyzing all sequents in a path through STRATEGY, if all paths are closed, that is, if none is a proof, then there is in each conjunctive expression (sentences in the same path) of a disjunction (alternative paths) one atomic sequent that is not provable. Any atomic non-provable sequent is in fact refutable inside the empty basis. Therefore, the initial sequent of the tree is also refutable, since in each path there is at least one refutable atomic sequent. □

6 Antisequents and non-consequence

Next, negations of *hLJp* [*hLMp*] sequents, that is, antisequents representing unprovability in the empty basis and its extensions are examined. Since proof-search has already been established for *hLJp*, a refutation procedure for antisequents is investigated without formulating any new formal system. From a certain perspective, this is dispensable if the objective is to decide when an antisequent holds or not.

6.1 Antisequents and analytic tableaux

"$\underline{\perp}$" represents the absurd in the metalanguage.

Theorem 6.1. *It holds for the empty basis with LMp [LJp] that:*

Left side, as a hypothesis

$(\overline{L1}) : \Gamma, C \wedge D \nvdash^{[abs]\varnothing} E \Leftrightarrow \Gamma, C, D \nvdash^{[abs]\varnothing} E$

$(\overline{L2}) : \Gamma, C \vee D \nvdash^{[abs]\varnothing} E \Leftrightarrow (\Gamma, C \nvdash^{[abs]\varnothing} E \text{ or } \Gamma, D \nvdash^{[abs]\varnothing} E)$

$(\overline{L3}) : \Gamma, C \rightarrow D \nvdash^{[abs]\varnothing} E \Leftrightarrow (\text{for any } \Delta \supseteq \Gamma : (\Delta, C \vdash^{[abs]\varnothing} D \Rightarrow \Delta \vdash^{[abs]\varnothing} E)) \Rightarrow \underline{\perp}$

Right side, as consequentia*:*

$(\overline{R1}) : \Gamma \nvdash^{[abs]\varnothing} C \wedge D \Leftrightarrow (\Gamma \nvdash^{[abs]\varnothing} C \text{ } or \text{ } \Gamma \nvdash^{[abs]\varnothing} D)$

$(\overline{R2}) : \Gamma \nvdash^{[abs]\varnothing} C \vee D \Leftrightarrow (for\ any\ E((\Gamma, C \vdash^{[abs]\varnothing} E \text{ } and \text{ } \Gamma, D \vdash E) \Rightarrow \Gamma \vdash^{[abs]\varnothing} E)) \Rightarrow \perp.$

$(\overline{R3}) : \Gamma \nvdash^{[abs]\varnothing} C \rightarrow D \Leftrightarrow \Gamma, C \nvdash^{[abs]\varnothing} D$

Proof. By using Theorems 3.1 and 5.9, remaining constructive in the metatheory, recalling that $\Gamma \nvdash^{[abs]\varnothing} C$ means that $(\Gamma \vdash^{[abs]\varnothing} C) \Rightarrow \perp$. □

Similar to $L3$ and $R2$, $\overline{L3}$ and $\overline{R2}$ involve an endless number of cases. One possible solution is to consider the contraposition of rules $(LS \rightarrow)$ and $(RS\vee)$:

$$\Gamma, C \rightarrow D \nvdash^{[abs]\varnothing} E \Rightarrow (\Gamma \nvdash^{[abs]\varnothing} C \text{ or } \Gamma, D \nvdash^{[abs]\varnothing} E)$$

$$\Gamma \nvdash^{[abs]\varnothing} C \vee D \Rightarrow (\Gamma \nvdash^{[abs]\varnothing} C \text{ and } \Gamma \nvdash^{[abs]\varnothing} D)$$

They are enough for effecting the analysis of the antisequents. In fact, the following equivalences hold:

Theorem 6.2.

$(\overline{L3*}) : \Gamma, C \rightarrow D \nvdash^{[abs]\varnothing} E \Leftrightarrow ((\Gamma \nvdash^{[abs]\varnothing} C \text{ } or \text{ } \Gamma, D \nvdash^{[abs]\varnothing} E) \text{ } and \text{ } \Gamma, C \rightarrow D \nvdash^{[abs]\varnothing} E)$

$(\overline{R2*}) : \Gamma \nvdash^{[abs]\varnothing} C \vee D \Leftrightarrow ((\Gamma \nvdash^{[abs]\varnothing} C \text{ } and \text{ } \Gamma \nvdash^{[abs]\varnothing} D) \text{ } and \text{ } \Gamma \nvdash^{[abs]\varnothing} C \vee D)$

The extra condition in the right to left reading contains too much to serve as an explanation of the logical constant. And yet, without them, the right to left inference would be incorrect, as illustrated by the following examples: (i) $c \rightarrow d, (c \rightarrow d) \rightarrow e \nvdash^{[abs]\varnothing} e \Leftarrow ((c \rightarrow d) \rightarrow e \nvdash^{[abs]\varnothing} c$ or $(c \rightarrow d) \rightarrow e, d \nvdash^{[abs]\varnothing} e)$; (ii) $e \wedge (c \vee d) \nvdash^{[abs]\varnothing} c \vee d \Leftarrow (e, (c \vee d) \nvdash^{[abs]\varnothing} c$ and $e, (c \vee d) \nvdash d)$. Together with the above antisequent principles, the following also holds.

Theorem 6.3.

$(antiLoad) : \Gamma, D \nvdash C \Rightarrow \Gamma \nvdash C$

$(antiDrop) : \Gamma \nvdash E \Rightarrow (\Gamma \nvdash C \text{ } or \text{ } C, \Gamma \nvdash E)$

Proof. (*antiLoad*) and (*antiDrop*) are the contrapositions of (*Load*) and (*Drop*) respectively. □

Theorem 6.4. *If* $C \rightarrow \perp \nvdash^{abs\varnothing} E$, *then* $C \rightarrow D \nvdash^{abs\varnothing} E$.

Proof. Suppose $C \to \bot \nvdash^{abs\varnothing} E$. By (*antiDrop*), either $C \to \bot \nvdash^{abs\varnothing} C \to D$ or $C \to D, C \to \bot \nvdash^{abs\varnothing} E$. Clearly, $C \to \bot \vdash^{abs\varnothing} C \to D$. Hence $C \to D, C \to \bot \nvdash^{abs\varnothing} E$. By (*antiLoad*), $C \to D \nvdash^{abs\varnothing} E$. □

Corollary 6.5. *For any atomic c, d and e:* $c \to d \nvdash^{abs\varnothing} e$.

Proof. By Theorem 6.4 since $c \to \bot \nvdash^{abs\varnothing} e$, given that $< c \to \bot, e >$ is a basic pair. □

Analytic *tableaux* can be seen as the result of an analysis searching to establish $\nvdash^{[abs]\varnothing} C$ for a sentence C. In case all different paths in the *tableau* are shown to be impossible, i.e., closed, then it is impossible to establish $\nvdash^{[abs]\varnothing} C$. That is, the supposition that $\nvdash^{[abs]\varnothing} C$ holds is absurd. Given that $\Gamma \vdash^{[abs]\varnothing} C$ or $\Gamma \nvdash^{[abs]\varnothing} C$ for LMp $[LJp]$ according to Theorem 5.9, then $\vdash^{[abs]\varnothing} C$. Otherwise, if at least one path is finished and open, then $\vdash^{[abs]\varnothing} C$ is unprovable.

In order to improve visual accuracy, from now on we avoid the superscript for bases. The reader can adequately fulfill it in the right places.

The searching procedure for antisequents to be presented below is inspired in STRATEGY. In STRATEGY, any attempt at a development considers the sentence picked as if it were the last occurrence introduced in a putative proof of the sequent, this guiding heuristics keeps orienting the search. Adapted for the context, it is going to be assumed that the occurrence chosen for development in the given antisequent could not be the last sentence occurrence introduced in a proof.[16] Most of the cases are immediately analyzed. Exceptions are the left implication and right disjunction cases, which oblige us to consider other alternatives when picking an occurrence.

The search procedure takes the complementary action of STRATEGY in each situation. As before, if all antisequents in a path were analyzed, then the *path is finished.* An antisequent of form $\Gamma, C \nvdash C$ [or form $\Gamma, \bot \nvdash C$], that is *the negation of an initial sequent*, in the empty basis of LMp $[LJp]$, is considered *immediately analyzed.*[17] If a *path* contains such kind of sequents, then it is *closed, otherwise* it is *open.* Any path in which such an antisequent occurs contains an impossibility. Any antisequent that is a basic pair is considered *immediately analyzed.* An open finished path indicates how the antisequent under examination is obtainable from antisequents for basic pairs in the empty basis.

Definition 6.6. *ANTISTRATEGY – Take any non marked occurrence of any non fully analyzed antisequent and develop it in all open paths going through the antisequent according to the following rules:*

[16] If it were, then the antisequent would not be correct.

[17] In order to spare search steps, a path containing such a sequent can be considered finished even if not all antisequents in the path were analyzed.

	Left Rules	Right rules	
$\overline{L1}$	$\Gamma, C \wedge D \nvdash E$ $\vdots$ $\downarrow$ $\Gamma^*, C, D \nvdash E^*$	$\Gamma \nvdash C \wedge D$ $\vdots$ $\downarrow \qquad \searrow$ $\Gamma^* \nvdash C \qquad \Gamma^* \nvdash D$	$\overline{R1}$
$\overline{L2}$	$\Gamma, C \vee D \nvdash E$ $\vdots$ $\downarrow \qquad \searrow$ $\Gamma^*, C \nvdash E^* \qquad \Gamma^*, D \nvdash E^*$	$\Gamma \nvdash C \vee D^{\checkmark}$ $\vdots$ $\downarrow$ $\Gamma^* \nvdash C$ $\Gamma^* \nvdash D$	$\overline{R2*}$
$\overline{L3*}$	$\Gamma, C \supset D^{\checkmark} \nvdash E$ $\vdots$ $\downarrow \qquad \searrow$ $\Gamma^*, D \nvdash E^* \qquad \Gamma \nvdash C$	$\Gamma \nvdash C \supset D$ $\vdots$ $\downarrow$ $\Gamma^*, C \nvdash D$	$\overline{R3}$

When the analysis is done through $\overline{R2*}$ or $\overline{L3*}$, the occurrence picked receives a mark "✓", but the sequent developed is not regarded as (fully) analyzed, unless all of its left implications and its right disjunction occurrences (in case there is one) become all marked. When the analysis is made through the other rules, the sequent is always considered fully analyzed. All new sequents written down contain only unmarked occurrences. The procedure goes on until all paths are finished. END of STRATEGY

Now, consider example 6.7, for atomic sentences c and d:[18]

[18]From now on we omit the superscript for the empty basis in order to improve visual accuracy.

$1. c \to d^{\checkmark} \nvdash \neg c \vee d^{\checkmark}$

$\downarrow$ $\quad \overline{R2*}(1)$

$2. c \to d \nvdash \neg c$

$3. c \to d \nvdash d$

$\downarrow$ $\quad \overline{R3}(2)$

$4. c \to d^{\checkmark}, c \nvdash \bot$

$\swarrow \qquad \searrow$ $\quad \overline{L3*}(4)$

$5. d, c \nvdash \bot \qquad 6. c \nvdash c \otimes$

$\downarrow \qquad \searrow$ $\quad \overline{L3*}(1)$

$7. d \nvdash \neg c \vee d \qquad 8. \nvdash c$

$\downarrow$ $\quad \overline{R2*}(7)$

$9. d \nvdash \neg c$

$10. d \nvdash d \otimes$

Example 6.7.

The antisequent 1 has been analyzed into two separated steps, one is given in antisequents 2 and 3, the other in antisequents 7 and 8. Each development picks one occurrence which becomes then marked. The antisequent becomes completely analyzed in the second development. Antisequent 3 does not require further analysis given what is stated in Corollary 6.5. Antisequents 6 and 10 close the path (no need to develop antisequent 9). There is only one open path finishing in antisequent 8. Thus, the starting antisequent holds; that is, it is not refutable. The development of antisequent 4 shows that $c \to d$ could not be introduced for proving $c \to d, c \vdash \bot$. It further means that the first development of antisequent 1 cannot be proved, otherwise the paths starting in such development would all be closed, which they are not. Then the procedure goes back to antisequent 1 in order to effect the second development.

Theorem 6.8. *ANTISTRATEGY is finite.*

Proof. In the worst scenario, each new antisequent written down after a development is of a lower degree. The degree of an antisequent is the sum of the degrees of all occurrences in the antisequent, assuming that the absurd constant is atomic and has degree 0. With two exceptions, the antisequent developed is considered analyzed. In antisequents developed through $\overline{L3*}$ or $\overline{R2*}$ at least one formerly non-marked formula becomes marked and, since the number of complex occurrences in the antisequent is finite, after a finite number of developments it has to become analyzed, either because the rule applied leaves it fully analyzed or because all complex sentences are marked. Hence, the process cannot go on indefinitely. □

Now, if an antisequent holds, ANTISTRATEGY returns at least one non-closed finished path.

Consider example 6.9, for the atomic sentence c:

$$
\begin{array}{ll}
1. \nvdash c \vee \neg c^{\checkmark} & \\
\downarrow & \overline{R2} * (1) \\
2. \nvdash c & \\
3. \nvdash \neg c & \\
\downarrow & \overline{R3}(3) \\
4. c \nvdash \bot &
\end{array}
$$

Example 6.9.

That is, both, c and $\neg c$ for atomic c, do not hold in the empty basis, and this explains why the third middle excluded is not valid in LJp even if this logic is decidable. Actually, it is not even necessary to appeal to decidability in order to see that the search has returned a counter-model. $c \nvdash^{abs\varnothing} \bot$ is a basic or ground fact in the empty basis. This is going to be examined in the next section.[19]

Now, in order to realize that ANTISTRATEGY generates closed search trees only for provable sequents of $hLJp$ $[hLMp]$ it is enough to realize that the rules in ANTISTRATEGY are complementary to the rules of STRATEGY.

Theorem 6.10. *STRATEGY contains an open finished branch for* $\Gamma \vdash C$ *in hLJp [hLMp] if, and only if, all finished paths obtained through ANTISTRATEGY for* $\Gamma \nvdash C$ *are closed.*

Proof. From left to right. Suppose STRATEGY contains an open finished branch for $\Gamma \vdash C$ in $hLJp$ $[hLMp]$. By Theorem 4.6, $\Gamma \vdash C$ is provable in $hLJp$ $[hLMp]$. The ANTISTRATEGY tree for $\Gamma \nvdash C$ is finite according to Theorem 6.8, all its paths are finished. Suppose there is a non closed path for $\Gamma \nvdash C$ obtained through ANTISTRATEGY. Then, no antisequent in this path is of the form $\Gamma, D \nvdash D$ or also of form $\Gamma, \bot \nvdash D$ in the case of $hLJp$. Since $\Gamma \vdash C$ is provable, then $\Gamma \nvdash C$ is refutable, because $hLJp$ $[hLMp]$ is consistent according to Corollary 2.10. And every

[19]When asking if $\vdash C \vee \neg C$ holds, this has to be read as: it is the case that, for any C, $\vdash C \vee \neg C$. And, if the instance obtained by substituting an atomic sentence in place of C were not a constructive tautology, then the universal closure cannot be a constructive tautology at all, which explains why the example we are considering uses an atomic c whatever. From Kripke semantics perspective, the sentence does not hold in general if there is a world α in a model in which $\alpha \nvDash c \vee \neg c$. Then, according to the disjunction clause in this semantics, $\alpha \nvDash c$ and $\alpha \nvDash \neg c$. But, in order to have $\alpha \nvDash \neg c$ there must be another world β different from α, but accessible from α, such that $\beta \vDash c$ and $\beta \nvDash \bot$.

sequent in the path has also to be refutable according to the barred rules $\overline{L1}$ to $\overline{R3}$, since they are equivalences. Hence, there is at least one atomic refutable antisequent in the path. It is atomic because otherwise the path would not be finished. As it has to be refutable, it is of the form $\gamma, e \nvdash e$ or also of form $\gamma, \bot \nvdash e$ in case of $hLJp$. However, this contradicts the assumption that there was no antisequent of forms $\Gamma, D \nvdash D$ or also of form $\Gamma, \bot \nvdash D$. Now, from right to left. Suppose all finished paths for $\Gamma \nvdash C$ are closed. It means that each one contains a sequent of the form $\Gamma, C \nvdash C$ in the case of a search in $hLMp$ or also of form $\Gamma, \bot \nvdash C$ in the case of a search in $hLJp$. Since the set of alternative paths constitutes the whole set of alternative developments for $\Gamma \nvdash C$, and since each sequent in a path is implied by $\Gamma \nvdash C$, then $\Gamma \nvdash C$ is absurd, refutable. According to Theorem 5.9, $\Gamma \vdash C$. By Theorem 4.6 again, there is an open finished path for $\Gamma \vdash C$. □

As pointed out in the introduction, an antisequent system for IPL had already been proposed in [8]. It is called $CRIP$. As we saw above, there are two cases where the rules for introducing logical constants in antisequents involve a secnd order quantification: implication in the left side and disjunction in the right side. $CRIP$ uses sequents with multiple conclusions for dealing with the case of disjunction in the right side. It is the following (we use their original numbers):

(6) $\Gamma \nvdash C, D, \Delta \Rightarrow \Gamma \nvdash C \vee D, \Delta$

Multiple conclusions at the right render the calculus and the search procedure more complex. For dealing with left implication, the authors use now four distinct rules:

(7) $\Gamma, p, B \nvdash \Delta \Rightarrow \Gamma, p, p \to B \nvdash \Delta$

(8) $\Gamma, C \to B, D \to B \nvdash \Delta \Rightarrow \Gamma, (C \vee D) \to B \nvdash \Delta$

(9) $\Gamma, C \to (D \to B) \nvdash \Delta \Rightarrow \Gamma, (C \wedge D) \to B \nvdash \Delta$

(10) $\Gamma, B \nvdash \Delta \Rightarrow \Gamma, (C \to D) \to B \nvdash \Delta$.

In a proof search, those four $CRIP$ rules above provide an analysis that simplifies the implication in the left. However, it is not guaranteed that the analysis can go further when the antecedent is atomic. The problem is then managed in the axiom of this system in Figure 2 of [8, p. 227]. Naturally, $CRIP$'s axiom constitutes the starting point of a derivation for any antisequent. This axiom has offered us a good clue for the formulation of Hypo bases, in particular the notion of a basic pair. The antisequents for basic pairs in Hypo are just a little bit simpler than their axiom, in view of Theorem 6.4 and Corollary 6.5.

$CRIP's$ antisequents admitt in the right side sets of sentences. Thus, for dealing with implication in the right side, the authors are obliged to offer the complex rule (11) in Figure 2 of [8, p. 227]. When the right side allows only single conclusions, the rule for right introduction of ANTISTRATEGY –$\Gamma, C \nvdash D \Rightarrow \Gamma \nvdash C \to D$–is simpler.

Now, two conclusions are to be extracted from what was just said, concerning the search in ANTISTRATEGY. Maybe it could be simplified in view of those four $CRIP$ rules for left implication. In contrast, we achieved some simplification and systematization with rules $L3/L3*$ and $R3$ with respect to $CRIP's$ axiom and rule (11).

6.2 How to obtain Hypo models and counter-models?

Models for sequents and for antisequents can be extracted, respectively, from trees obtained by STRATEGY and ANTISTRATEGY. A *model for an antisequent* is obtained by collecting all basic pairs belonging to a fully analyzed non-closed path of ANTI-STRATEGY. A model for an antisequent is a *counter-model for the* associated *sequent*, like example 5.

Let us examine some cases of counter-model production in Hypo and in Kripke semantics. In what follows, antisequents and sequents continue to be used even if the subject matter is semantics.

Example 6.9 contains a model for the antisequent $\nvdash c \vee \neg c$. The empty basis is the model since: $\nvdash^{abs\emptyset} c$ and $c \nvdash^{abs\emptyset} \bot$. That is, the empty basis gives a counter-model for the validity of $c \vee \neg c$. ANTISTRATEGY generates a sequence of sets of hypotheses that is similar to a counter-model for $c \vee \neg c$ in Kripke semantics.

Examining ANTISTRATEGY from the perspective of possible worlds, the search in example 5 starts by considering what happens in "worlds" of hypotheses. Antisequent 1 is being considered in the world α of no hypotheses. Next, because of $\overline{R2*}$, antisequents 2 and 3 must also be the case in α. The clause for implication in Kripke semantics requires that all further accessible worlds be such that, if the antecedent is forced on it, then the consequent has also to be forced. Then, since negation is defined via implication, this definition has to be employed for the case of antisequent 3. Hypo semantics also requires the consequent to hold in every world of hypotheses in which the antecedent holds. Worlds in Hypo are just finite sets of hypotheses. Translating the example 5 into the language of worlds, antisequent 4 is the case in a world β where the atomic hypothesis c is forced but the absurd is not, that is, in any world where c is forced but absurd is not forced.

In Kripke semantics no world forces the absurd, all of them are *ab initio* consistent worlds. In Hypo semantics, sets of hypotheses can be inconsistent. In each world of a Kripke model either an atomic sentence is forced or it is not. In each "world of hypotheses" in Hypo either a sentence is derivable from a set of hypotheses or it is not, according to Corollary 4.7 if the basis is decidable. Bases in Hypo semantics can be undecidable. The empty basis is decidable.

A major difference between Hypo and Kripke semantics concerns disjunction. The following example illustrates it. We put side by side a Hypo structure in the empty basis, based on ANTISTRATEGY, with the corresponding Kripke counter-model. What happens when looking for counter-models for $(c \to (d \vee e)) \to ((c \to d) \vee (c \to e))$?

Examples 6.11 and 6.12 illustrates what happens in a Kripke and a Hypo model respectively.

Kripke model

$9.1.\delta \models d \vee e$	$9.2.\gamma \models d \vee e$
$8.1.\delta \models d$	$8.2.\gamma \not\models d$
$7.1.\delta \not\models e$	$7.2.\gamma \models e$
$6.1.\delta \models c$	$6.2.\gamma \models c$
$\nwarrow$	$\nearrow$

$$5.\beta \not\models c \to d$$
$$4.\beta \not\models c \to e$$
$$3.\beta \not\models (c \to d) \vee (c \to e)$$
$$2.\beta \models c \to (d \vee e)$$
$$\uparrow$$
$$1.\alpha \not\models (c \to (d \vee e)) \to ((c \to d) \vee (c \to e))$$

Example 6.11.

Example 6.12. *See example at the end of the paper.*

Observe that the formula is not intuitionistically valid.[20] Does 1 hold in the no hypotheses set of the Hypo model (Hm) in the empty basis? Does 1 hold in an α world of a Kripke model (Km)? It holds in the empty set of hypotheses of Hm only if in a successor set of hypotheses 3 holds. In Km, 1 holds in α if, on the supposition that 2 holds in β, 3 is the case in β. In Hm the sequent corresponding to step 2 of Km $c \to (d \vee e) \vdash c \to (d \vee e)$ trivially holds. Now according to $\overline{R2*}$, first alternative, 3 holds in Hm only if both 4 and 5 hold in Hm, which corresponds to world β in

[20] Same numbers indicate similar steps in each model of example 6.

Km. In Km, 4 requires us to look for a world δ where 7.1 is the case when 6.1 holds. And 5 requires a world γ where 8.2 is the case when 6.2 holds. Notice that in 6.1 and 6.2 the atomic sentence being considered is the same! Disjunction clauses in Kripke and Hypo semantics are different. Kripke semantics requires that at least one of the disjuncts hold in the same world in which the disjunction holds. This is not the case in Hypo, just consider that 7 and 8 hold in Hm. Observe that 8.1 and 7.2 are required in Km, otherwise 2 would not be the case.

The construction of a Kripke counter-model stops at δ and γ. Hypo needs some few worlds more, but none of the steps 2, 6.1, 6.2 or 9 is required.[21] A world, in Kripke semantics, is defined by a special kind of "valuation" for atomic sentences: either it is forced or it is not forced in that world, *tertium non datur*. As a consequence, because of the semantical clauses, any sentence is either forced or not forced. Hence, worlds δ and γ force hypotheses c and $c \rightarrow (d \vee e)$. Kripke strategy cannot be used in general for obtaining a Hypo counter-model. Any Kripke "valuation" forcing these two hypotheses should force c and force either d or e.[22] This is not the case in Hypo model in the example and it is not the case of Hypo in general.

Developing ANTISTRATEGY for the antisequent $\nvdash (c \rightarrow (d \vee e)) \rightarrow ((c \rightarrow d) \vee (c \rightarrow e))$, we discovered that it holds in a basis where the following antisequents for basic pairs hold: $10 : \nvdash c$; $16 : c, d \nvdash e$; $17 : c, e \nvdash d$. But, of course, those three antisequents hold in the empty basis. Therefore, ANTISTRATEGY gives the empty basis as a counter-model for the sequent $\vdash (c \rightarrow (d \vee e)) \rightarrow ((c \rightarrow d) \vee (c \rightarrow e))$. That is, the sentence is not logically valid in intuitionistic or minimal logic.

[8] obtain counter-models similar to Kripke counter-mo-dels by using the antisequent system $CRIP$. They were close to discover a new semantics for intuitionistic logic, but missed it by a few steps. The idea of taking worlds as finite sets of hypotheses is already in their work. However, since their antisequents have multiple conclusions, the corresponding Kripke models became complex trees, one for each conclusion in the multiset.

The previous example was of an admissible non-valid formula in intuitionistic logic. Let's consider now a non-valid and non-admissible formula in example 6.13:

Example 6.13.

According to ANTISTRATEGY, any basis in which $c \nvdash d$ holds gives a counter-example to Peirce axiom. This is the case again of the empty basis. Also, in this basis, for the empty set of hypotheses: $\nvdash d$. Thus, d would only be derived if some

[21] Thirteen lines in Km against eleven in Hs.

[22] The Kripke counter-model is successful only if no contradiction arises in the forcing.

Hypo model	Kripke model
$7. c \nvdash d$ from 4	$7. \gamma \nvDash d$ from 5
	$6. \gamma \vDash c$ from 5
	$\uparrow$
	$5. \beta \nvDash c \to d$ from 2 and 3
$3. (c \to d) \to d \nvdash d$ from 1	$3. \beta \nvDash d$ from 1
	$2. \beta \vDash (c \to d) \to d$ from 1
$\nwarrow$ $\nearrow$	$\uparrow$
$1. \nvdash ((c \to d) \to d) \to d$	$1. \alpha \nvDash ((c \to d) \to d) \to d$
$4. \nvdash c \to d$ from 3	

extra hypothesis were called in. However, $(c \to d) \to d \nvdash d$ since $\nvdash c \to d$, given that $c \nvdash d$.

Summing up, Hypo counter-models for a sequent are obtained by using ANTI-STRATEGY over the corresponding antisequent. The models involve a basis, usually the empty basis, and a structure of worlds of hypotheses. These structures are not identical to Kripke models, although they have some similarities. Models for a sequent can be directly read from the tree developed by STRATEGY.

The major divergences among Kripke and Hypo semantics are in the implication and the disjunction clauses. We proved that Hypo clauses for disjunction and implication are metaproperties of LJp. Kripke clauses for disjunction and implication are not or, at least, its adequacy is proved by completness using classical principles. This is an important reason, among others, for asserting that Hypo is a truly proof-theoretical semantics.

7 Conclusion

In the above we presented proof-search for sequents and refutation-search for antisequents in intuitionistic logic and minimal logic. They were based on a handy sequent system for both logics. This system has neither contraction nor thinning rules. Additionally, we proved that rule (MP) is eliminable from proofs in this system–like (cut) eliminability–what makes the proof-search direct from sentences to subsentences. The proof showing the (MP) elimination property is remarkably concise.

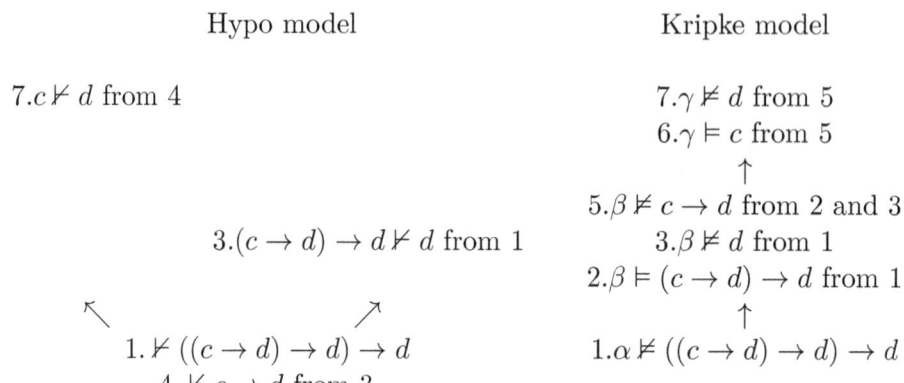

The rules of the handy system, with two exceptions, correspond to the synthetic reading of the clauses in Hypo semantics. The clauses of the semantics have been presented and proved as metaproperties of sequents in Gentzen's LJp [LMp]. The analytical readings of the clauses are the basis for searching a proof of sequents in the handy system via STRATEGY. The contraposition of the clauses is the basis on which refutation of antisequents via ANTISTRATEGY is built.

Finally, Hypo bases and models are introduced and briefly discussed. Models are directly read from STRATEGY. On the other hand, counter-models are obtained by using ANTISTRATEGY. The structure of sets of hypotheses in a Hypo counter-model has been compared with Kripke counter-models. There are some similarities among them, but they are not equal. A counter-model for the provability of a sequent in Hypo is given by the empty basis model. Above we gave reasons why we claim Hypo semantics to be a proof-theoretical semantics, and different from Kripke semantics. We also noticed that once the search procedure for antisequents is defined there is no need of characterizing a system for antisequents like [8] *CRIP*.

References

[1] Evert Beth. *The foundations of mathematics - a study in the philosophy of science.* North-Holland, Amsterdam, 1959.

[2] Kosta Došen. *Logical constants: an essay in proof theory.* PhD thesis, University of Oxford, 1980.

[3] Kosta Došen. Inferential semantics. In Thomas Piecha and Peter Schroeder-Heister, editors, *Advances in proof-theoretic semantics*, volume 43 of *Trends in Logic*, pages 147–162. Springer, Dordrecht, 2016.

[4] Melvin Fitting. *Intuitionistic logic model theory and forcing.* North-Holland, Amsterdam, 1969.

[5] Gerhard Gentzen. Investigations into logical deduction. In M.Szabo, editor, *The collected papers of Gerhard Gentzen*, pages 68–131. North-Holland, 1969.

[6] Arendt Heyting. Intuitionism: an introduction. In L.Brouwer *et alli*, editor, *Studies in logic and the foundations of mathematics.* North-Holland, Amsterdam, 1956.

[7] Saul Kripke. Semantical analysis of intuitionistic logic i. In J.N. Crossley and M.A.E. Dummett, editors, *Formal systems and recursive functions*, volume 40 of *Studies in logic and the foundations of mathematics*, pages 92–130. Elsevier, 1965.

[8] Luis Pinto and Roy Dyckhoff. Loop-free construction of counter-models in intuitionistic propositional logic. In M Behara *et alli*, editor, *Conference A: mathematics and theoretical physics I*, pages 225–231. De Gruyter, 1995.

[9] Jefferson Santos, Bruno Vieira, and Edward Haeusler. A unified procedure for provability and counter-model generation in minimal implicational logic. *Electronic notes in theoretical computer science*, 324:165–179, 2016.

[10] Wagner de Campos Sanz. Hypo: a simple constructive semantics for intuitionistic sentential logic, soundness and completeness. In Enrique Alonzo *et alli*, editor, *Aventuras en el mundo de la lógica: ensayos en honor a Maria Manzano*, pages 377–410. College, London, 2019.

[11] Wagner de Campos Sanz. Hypo: a simple constructive semantics for intuitionistic sentential logic; soundness and completeness. In *Proof-theoretic semantics: assessment and future perspectives. Proceedings of the third Tübingen conference on proof-theoretic semantics, 27–30 March 2019*, pages 153–178, 2019.

[12] Peter Schroeder-Heister. Proof-theoretical semantics. In Edward Zalta *et alli*, editor, *Stanford encyclopedia of philosophy.* Metaphysics research lab center for the study of language and information stanford university, 2012.

[13] Raymond Smullyan. *First order logic.* Dover, New York, 1995.

Example 6.12

Hypo model (over the empty basis)

$$8. c, c \to (d \lor e) \nvdash d \ \overline{R3}/5$$
$$7. c, c \to (d \lor e) \nvdash e \ \overline{R3}/4$$

$$\uparrow$$

$$17. c, e \nvdash d \ \overline{L2}/15 \qquad 16. c, d \nvdash e \ \overline{L2}/14 \qquad 15. c, d \lor e \nvdash d \ \overline{L3*}/8 \qquad 5. c \to (d \lor e) \nvdash c \to d \ \overline{R2*}/3$$
$$14. c, d \lor e \nvdash e \ \overline{L3*}/7 \qquad 4. c \to (d \lor e) \nvdash c \to e \ \overline{R2*}/3$$
$$3. c \to (d \lor e) \nvdash (c \to d) \lor (c \to e) \ \overline{R3}/1$$

$$\nwarrow \qquad \uparrow \qquad \uparrow \qquad \nearrow$$

$$1. \nvdash (c \to (d \lor e)) \to ((c \to d) \lor (c \to e))$$
$$10. \nvdash c \ \overline{L3*}/3$$

Received 16 January 2020

A NOTE ON TARSKI'S REMARKS ABOUT THE NON-ADMISSIBILITY OF A GENERAL THEORY OF SEMANTICS

GARIBALDI SARMENTO
Departamento de Filosofia, Universidade Federal da Paraiba
garibaldi.sarmento@academico.ufpb.br

Abstract

Taking Tarski's semantic theory as a starting point, we can formulate a hypothesis about the construction of a generalized semantics theory for recursive arithmetic, so that it is possible to prove Tarski's conjecture (i.e., there is *no* general semantics for *all* transfinite languages for recursive arithmetic).

Key-words: general semantics, recursive arithmetic, recursive ω-rule.

1 Introduction

Leibniz and other philosophers had cogitated on what would be the strict construction for the *universal* language of mathematics, a *mathesis universalis*, or a uniquely logical and mathematical language.

In the lecture "Remarks before the Princeton Bicentennial Conference on Problems in Mathematics" Gödel observes that: "Tarski has stressed in his lecture (and I think justly) the great importance of the concept of general recursiveness (or Turing's computability). It seems to me that this importance is largely due to the fact that with this concept one has for the first time succeeded in giving an *absolute* definition of an interesting epistemological notion, i.e., one *not* depending on the formalism chosen." [My italics] [1]. Likewise, I note that Tarski's metatheorem (see **Theorem 2.3.1**)—by expressing that 'the notion of arithmetical truth is not arithmetically definable in its own consistent arithmetical language'—also has an absolute character in the sense of Gödel. Since the argument in Tarski's metatheorem applies to

[1]See p. 84 in [1].

any formal language suitable for arithmetic. This theorem can be employed as an epistemological principle for demarcating the construction of a universal language for mathematics.

In this essay I take into account some remarks of Tarski [7] concerning the *non-admissibility* of a generalized semantics.

According to Tarski [7]:

> In particular it would be incorrect to suppose that the *relativization* of the concept of truth— ... —would open the way to some *general theory* of this concept which would embrace all possible or at least *all* formalized languages. ... The *language* of the general theory of truth would then contain a *contradiction* for exactly the same reason as does colloquial language. [My italics].

To explain this statement, it will be helpful to consider the *main* problem below, which we shall call Tarski's *Conjecture* (*T.C.*).

Accordingly with Tarski's definition [see [7]] the concept of truth *is* relative to a given formal language $\mathbb{L}$ (the concept of 'truth-in-$\mathbb{L}$') in which $\mathbb{L}$ *is* constant. A general semantics metatheory consists of a theory for which the formal language $\mathbb{L}$ (with respect to the 'trans-linguistic' concept of 'truth-in-$\mathbb{L}$') *is* variable, by ranging through the formalized languages of a class of linguistic structures. It is arguable whether, according to Tarski's theory, a generalized semantics theory is admissible or not.

A general semantics is *admissible* if, and only if, the class of (formalized) languages $\mathcal{C}_{\mathbb{L}}$ does *not* contain *all* possible languages with respect to this general theory.

Now we can formulate the following (semantic-)metamathematical problem.

Main problem. *Would a general semantics metatheory for* **RA** *be admissible? More precisely, a general theory in which it is allowable to express and prove all true arithmetic sentences in a transfinite higher system, so that the concept of truth for such a theory is definable in a transfinite language for recursive arithmetic?*

Suppose that there *exists* a *transfinite* sequence of transfinite sequences of axiomatic systems for recursive arithmetic expressed of the following form:

$$\left\langle \mathbb{A}_j^{(i)} \right\rangle_{i,j} = \bigcup_{i \leq \varepsilon_0} \bigcup_{j < \tau} \mathbb{A}_j^{(i)},$$

with $\tau \leq \omega^{\omega^{\omega}}$ and $\varepsilon_0 = \lim_{n \to \infty} \omega_n$, by defining: $\omega_0 = \omega$ and $\omega_{n+1} = \omega^{\omega_n}$.

In $\left\langle \mathbb{A}_{\mathsf{j}}^{(\mathsf{i})} \right\rangle_{\mathsf{i,j}}$ the subscript symbol i denotes the *order* of the language $\mathbb{L}^{(\mathsf{i})}$ for the axiomatic system $\mathbb{A}_{\mathsf{j}}$, with $\mathsf{j} < \tau$, such that:

1. $\mathbb{A}_{0}^{(\mathsf{i})} = \mathbf{RA}^{\mathbf{i}}$, i.e., recursive arithmetic of ith-order;

2. $\mathbb{A}_{\mathsf{k}+1}^{(\mathsf{i})}$ consists of $\mathbb{A}_{\mathsf{k}}^{(\mathsf{i})}$ added to all sentences $\mathrm{Pr}_{\mathbb{A}_{\mathsf{k}}^{(\mathsf{i})}}(\overline{\varphi}) \rightarrow \varphi$ where $\mathrm{Pr}_{\mathbb{A}_{\mathsf{k}}^{(\mathsf{i})}}(\overline{\varphi})$ denotes an (ith-order) arithmetical sentence expressing that φ *is* provable *in* $\mathbb{A}_{\mathsf{k}}^{(\mathsf{i})}$;

3. $\mathbb{A}_{\mathsf{k}}^{(\mathsf{i})} = \bigcup_{\mathsf{j}<\mathsf{k}} \mathbb{A}_{\mathsf{j}}^{(\mathsf{i})}$, if k is a limit ordinal.

Now we can state the following *hypothesis* for a general metatheory of (ϑ-order) recursive arithmetic:

Hypothesis H: let $\mathcal{A}$ be the *general metatheory* for **RA** such that:

$$\mathcal{A} := \left\langle \mathbb{A}_{j}^{(i)} \right\rangle_{i,j} \bigcup \{ \text{ generalized version of Tarski's } \omega - \text{rule } \}^{2}$$

representable in the *language* of transfinite order $\mathbb{L}^{(\varepsilon_0)}$.

Admitting this hypothesis, it is possible to show that, in fact, there is *not* a *unique* maximal language for a transfinite higher order system containing all proofs for every true (ϑ-order) arithmetic sentence.

We can symbolize the sketch of the argument for that solution as follows: if $H\colon \mathcal{A} + \Phi$, where Φ stands for the principle of transfinite *induction* in ϑth-order recursive arithmetic $\mathbf{RA}^{(\vartheta)}$ ($\vartheta \leq \varepsilon_0$), then $H \implies$ *T. C.* via **Theorem 2.3.1**.

The structure of the argumentation is given as follows.

Section 2 deals with the main results that constitute the metamathematical basis of the statement *H*.

Section 3 contains the solution to the main problem, i.e., that *H* implies Tarski's Conjecture.

Section 4 contains: a brief discussion of the philosophical consequences of Tarski's Conjecture for the foundations of arithmetic and a (possible) non-immediate metamathematical consequence with respect to Hilbert's Program.

[2]See **Definition 3.1.1** of the generalized version of Tarski's ω-rule.

2 Metamathematical results for establishing hypothesis *H*

In this section we consider some remarks and semantic assumptions for defining a general truth theory for recursive arithmetic in a transfinite hierarchy of higher-order languages.

The *Hypothesis H* is implied by the following results:

1. Gentzen's proof [3] of the *consistency* of first-order recursive arithmetic $\mathbf{RA}^1$ [via transfinite induction];
2. theorem of Shoenfield-Feferman [2]: all true sentences of elementary number theory are provable in the recursive progression $\mathbb{A}_k$ based on the *reflection principle*: $\mathbb{A}_{k+1}$ consists of $\mathbb{A}_k$ together with all sentences of the form $(\forall \mathbf{x})\mathsf{Pr}_{\mathbb{A}_k}(\overline{\phi}(\mathsf{nm}_\mathbf{x})) \to (\forall \mathbf{x})\phi(\mathbf{x})$; in which $\mathsf{Pr}_{\mathbb{A}_k}(\overline{\phi}(\mathsf{nm}_\mathbf{x}))$ expresses that the result of substituting the $(\mathbf{x}+\mathbf{1})$-st numeral in ϕ is provable in $\mathbb{A}_k$ [via *restricted* (or recursive) ω-rule];
3. Tarski's *undefinability theorem* for formalized languages of *infinite* order.

2.1 Remarks about Gentzen's consistency proof of arithmetic

According to Mostowski [6]:

> An adequate formulation of the principle of transfinite induction in its full generality is possible only *in* set theory. Gentzen ... *used a much more restricted principle which can be expressed in purely arithmetical terms.* His principle has the form
>
> $$(*) \qquad \wedge_y [\, \wedge_x (x \prec y \to A(x)) \to A(y)] \to \wedge_x A(x)$$
>
> where A is any arithmetical formula and $\prec$ an arithmetically definable well-ordering of integers. As compared with the set-theoretical transfinite induction this principle *is* limited in a twofold way: First, we do *not* speak of sets and we use the principle *only to* show that all integers satisfy an arithmetical formula A. Secondly, we do *not* formulate the principle *for any* well-ordering (which would require a certain amount of set theory) but *only for* the very special well-orderings that can be defined arithmetically for integers.

> For many formulae defining well-orderings of integers the formulae $(*)$ *is* provable in Peano's arithmetic; hence by Gödel's second undecidability theorem such special cases of $(*)$ *cannot* yield the consistency proof. Gentzen's discovery was that *there is* a formula $\mathsf{x} \prec \mathsf{y}$ which defines a well-ordering of integers of the type $\varepsilon_0 = \omega + \omega^{\omega} + \omega^{\omega^{\omega}} + \ldots$ and which has the property that the induction principle $(*)$ for this well-ordering allows us to prove the consistency of arithmetic. It follows that for such a well-ordering $\prec$ the principle $(*)$ is *not* provable in arithmetic.
>
> ... The general idea [of Gentzen's proof] is that to each formal proof there is defined a transfinite number $\alpha \prec \varepsilon_0$ called the height of the proof; it is shown that if this proof would have as its end formula $0 \neq 0$, then so would also a proof with a lesser height. Hence the existence of a formal inconsistency would violate the induction principle. [My italics][3].

As observed in Mostowski's excerpt Gentzen's theory obtained by adding (quantifier-free) transfinite induction principle $(*)$ to (first-order) primitive recursive arithmetic $\mathbf{RA^1}$ demonstrates the consistency of (first-order) Peano arithmetic ($\mathbf{PA^1}$). Nonetheless, $\mathbf{RA^1}$ does *not* contain $\mathbf{PA^1}$ because of *all* instances of induction schema are axioms of $\mathbf{PA^1}$. By another hand, Gentzen's theory is *not* contained in $\mathbf{PA^1}$, since by Gödel's [4] incompleteness result the well-ordering of integers of the type ε_0, i.e., the limit of the sequence: ω, ω^{ω}, $\omega^{\omega^{\omega}}$, ..., is *not* provable in $\mathbf{PA^1}$.

It should be observed that the formal system $\mathcal{A}$ (see hypothesis H above) contrasts sharply with Gentzen's theory, for we apply the set-theoretical transfinite induction for ordinal numbers and for any well-ordering via definability in terms of ordinals in a higher order system.

2.2 On Feferman's result

Suppose we firstly construct the axiom system $\mathsf{A_k}$ where k is an ordinal number and such that the set of numbers of formulae *deducible* from $\mathsf{A_k}$ let be given by the formula $\mathsf{D_k}$.[4] According to Feferman we can define a formula $\Gamma^{\varphi}_{\mathsf{k}}$, for every $\varphi(\mathsf{x})$ (with one free variable) by expressing : *if* $\varphi(\overline{\mathsf{n}})$ *is demonstrable from* $\mathsf{A_k}$ *for all symbols (term-names)* $\overline{\mathsf{n}}$ *of natural numbers, then* $(\forall\mathsf{x})\varphi(\mathsf{x})$ *is true.*[5].

Thus, it is possible to define the following transfinite recursive (countable) sequence of axiomatic systems:

---

[3]See pp.49-50.

[4]Feferman [2] shows how $\mathsf{D_k}$ is defined for an appropriate (countable) sequence of ordinals with limit $\mathsf{k} \leq \omega^{\omega^{\omega}}$.

[5]Feferman [2] provides a recursive procedure for defining $\Gamma^{\varphi}_{\mathsf{k}}$ explicitly.

$$A_{k+1} = A_k \bigcup \{\Gamma_k^{\varphi} : \text{for all } \varphi\}$$

$$A_\lambda = \bigcup_{k<\lambda} A_k, \text{ if } \lambda \text{ is a limit ordinal.}$$

Now we can state *Feferman's Theorem.*

Theorem 2.2.1. *All true formulae in* $\mathbf{RA^1}$ *are deducible from* $\bigcup_{\mathsf{k}<\tau} \mathsf{A_k}$, *with* $\tau \leq \omega^{\omega^\omega}$.

Proof: See [2].

Remark. Feferman [2] points out how this theorem can be extended to recursive arithmetic formalized in higher order systems.

2.3 Tarski's metatheorem

We conclude Section 2 by stating Tarski's metatheorem for infinite order languages. Let $\mathbb{L}$ be an *infinite* order language and let $\mathcal{T}_{\mathbb{L}}$ be a *truth* theory in $\mathbb{L}$, i.e., $\mathcal{T}_{\mathbb{L}}$ is constituted by metatheory axioms in addition to the every instance of *convention* **T** [or schema **T**] for any sentence of $\mathbb{L}$. It should be remarked that, by presupposing the consistency of the metatheory, such extension is consistent [see Tarski [7], theorem III, §5, pp. 256-257].

According to Tarski we can introduce in $\mathcal{T}_{\mathbb{L}}$ a *rule* of infinite induction [ω-rule].

Definition 2.3.1. *Let be given* $\varphi_1, \ldots, \varphi_n, \ldots$ *a list of all sentences of* $\mathbb{L}$. *If each one of the following formulae* $\mathbb{F}(\varphi_1), \ldots, \mathbb{F}(\varphi_n), \ldots$ *will be provable in* $\mathcal{T}_{\mathbb{L}}$, *where* $\mathbb{F}(\zeta)$ *denotes a metalinguistic expression whatsoever, then*

$$\mathcal{T}_{\mathbb{L}} \vdash \wedge_{i\in\omega}\mathbb{F}(\varphi_i).$$

Remark. Let $\mathcal{T}_{\mathbb{L}}^{\omega} := \mathcal{T}_{\mathbb{L}} \cup \{\omega - \text{rule}\}$. Tarski has proved that $\mathcal{T}_{\mathbb{L}}^{\omega}$ *is* categorical, but the $\mathrm{Con}_{\mathcal{T}_{\mathbb{L}}^{\omega}}$ is an open problem.

According to Tarski:

> The metalanguage then becomes a language of higher order and thus one which is essentially richer in grammatical forms than the language we are investigating . ... But now we are in a position to define the concept of truth for any language of finite or infinite order, ... It *is*

> impossible to give an adequate *definition of truth* for a language in which the *arithmetic of the natural numbers can be constructed*, if the order of the metalanguage in which the investigations are carried out does *not* exceed the order of the language investigated. [My italics][6]

We can establish the following version of Tarski's metatheorem of *truth undefinability* for infinite order languages:

Theorem 2.3.1. ***Tarski's Metatheorem.*** *Assuming the consistency of the metatheory, then the concept of truth in any infinite order language is undefinable from primitive concepts of the metatheory.*

Proof: See [7].

3 Hypothesis *H* implies Tarski's Conjecture

Considering the *transfinite* hierarchy of formal systems $\mathcal{A}$ we have a correlative transfinite hierarchy of concepts of truth *definability* according Tarski's Theory.

3.1 Completeness of $\mathcal{A}$

Now by replacing respectively the *truth* theory $\mathcal{T}_{\mathbb{L}}$ for $\mathcal{A}$ and the metalinguistic formula $\mathbb{F}$ for the principle of transfinite *induction* Φ in ϑth-order recursive arithmetic $\mathbf{RA}^{(\vartheta)}$, $\vartheta \leq \varepsilon_0$, in which the following statement *is* true in $\mathbb{L}^{(\vartheta+1)}$:

$$\Phi(\varphi^{(\vartheta)}) : \bigwedge_{k} (\bigwedge_{\lambda} (\lambda \prec k \rightarrow \varphi^{(\vartheta)}(\lambda)) \rightarrow \varphi^{(\vartheta)}(k)) \rightarrow \bigwedge_{k} \varphi^{(\vartheta)}(k)$$

where $\lambda \prec \mathsf{k}$ ($\lambda, \mathsf{k} \in \mathrm{On}$) is a schema of sentences which defines a well-ordering of ordinal numbers into type ε_0, and $\varphi^{(\vartheta)}$ denotes a ϑth-order arithmetic formula (quantifier-free) in $\mathbf{RA}^{(\vartheta)}$.

Therefore, we can state the following generalized version of Tarski's ω-rule *in* $\mathcal{A}$.

Definition 3.1.1. *If* $\Phi(\varphi_{\mathsf{i}}^{(\vartheta)})$ *is provable in* $\mathcal{A}$ *for any sentence* $\varphi_{\mathsf{i}}^{(\vartheta)} \in \mathbb{L}^{(\vartheta)}$, *then* $\mathcal{A} \vdash \bigwedge_{\mathsf{i}\in\omega} \Phi(\varphi_{\mathsf{i}}^{(\vartheta)})$ *with* $\mathsf{i} = 1, 2, \ldots$

[6] See [7], p.272.

We shall call this generalized ω-rule of *transfinite* ϑ-rule, with $\vartheta \leq \varepsilon_0$.

Remark. Note that for $\vartheta = 1$, this transfinite ϑ-rule is *reducible* to Schoenfield's recursive [7]ω-rule. Moreover, $\Phi(\varphi_i^{(1)})$, $i = 1, 2, \ldots$, *is* the *first-order* (transfinite) induction principle.

Suppose that the principle of transfinite induction Φ *is* true in the language $\mathbb{L}^{(\varepsilon_0+1)}$ (Φ *is* provable *in* $\mathbf{RA}^{(\varepsilon_0)}$), then any proof for a $\mathbf{RA}^{(\vartheta)}$ theorem, $\vartheta \leq \varepsilon_0$, in the higher order language $\mathbb{L}^{(\vartheta+1)}$ is *replaceable* by a proof from such principle Φ via transfinite ϑ-rule.

Hence, for such concept of demonstrability, we can formulate the following 'strong' *completeness theorem.*

Theorem 3.1.1. ***Completeness Metatheorem.*** *Every arithmetic sentence $\varphi^{(\vartheta)}$ of $\mathbf{RA}^{(\vartheta)}$ ($\vartheta \leq \varepsilon_0$) is decidable from $\Phi(\varphi^{(\vartheta)})$ in the general metatheory $\mathcal{A}$.*

Proof. It follows of the definition of $\mathcal{A}$ and transfinite ϑ-rule applied to $\Phi(\varphi^{(\vartheta)})$ for any $\varphi^{(\vartheta)} \in \mathbf{RA}^{(\vartheta)}$ with $\vartheta \leq \varepsilon_0$.

3.2 *H* $\Longrightarrow$ *T. C.*

Now we can establish the main result of this study. We are going to show how, by means of Tarski's semantic, may be proved that $\mathcal{A}$ has the necessary closure condition by introducing the concept of truth for the transfinite language $\mathbb{L}^{(\vartheta)}$, $\vartheta \leq \varepsilon_0$.

Suppose that $\mathrm{Con}_{\mathcal{A}}$, i.e., the consistency of the metatheory $\mathcal{A}$, let $\mathcal{C}_{\mathbb{L}}^{\mathcal{A}}$ be defined as follows:

$$\mathcal{C}_{\mathbb{L}}^{\mathcal{A}} := \bigcup_{\vartheta \leq \varepsilon_0} \mathbb{L}^{(\vartheta)}$$

where $\mathcal{C}_{\mathbb{L}}^{\mathcal{A}}$ denotes the class of *all* transfinite languages of ϑ-order for recursive arithmetic ($1 \leq \vartheta \leq \varepsilon_0$) such that $\mathcal{A}$ is the *general metatheory.*

Thus, there exists (at least) a transfinite language $\mathbb{L}^{(\xi)}$, with $\varepsilon_0 < \xi$, in which the concept of truth for $\mathcal{A}$ *is* definable and such that $\mathbb{L}^{(\xi)} \notin \mathcal{C}_{\mathbb{L}}^{\mathcal{A}}$ by Tarski's metatheorem. We can take, without lost of generality, $\xi = \varepsilon_0 + 1$ because the semantic closure of

[7]Given an axiomatic set A we shall define the set $\mathsf{Rs}^{(\omega)}(\mathsf{A})$ as the *least* set of sentences Γ (closed under logic deduction) in which if there exists a recursive function suppling a proof *in* Γ for each n of every sentence $\varphi_i(\bar{n})$, then it contains $(\forall x)\varphi_i(x)$.

$\mathbb{L}^{(\varepsilon_0)}$ can *not* be contained in $\mathbb{L}^{(\varepsilon_0)}$, otherwise the *truth* of $\Phi(\varphi^{(\xi)})$ should be derived *in* $\mathbb{L}^{(\varepsilon_0)}$, contradiction.[8]

4 Concluding Remarks

In this section we consider some philosophical consequences of Tarski's Conjecture in the framework of mathematical foundations. If we accept the following criterion for a foundational approach:

($\mathcal{F}$) The foundation should be expressible like as a formal system;

then, the epistemological analysis of *T. C.* entails that there is a metamathematical *limit* to formalize a basis for arithmetic that satisfies criterion $\mathcal{F}$.

Thus, *no* formalized *universal* language could contain the whole theory of numbers and, therefore, the arithmetic presupposes a *limitedness* transfinite sequence of higher order languages.

Nonetheless, if it is possible to *translate* system $\mathcal{A}$ into the axiomatic system developed by Gödel [5] whose smallest model consists of primitive recursive functionals with the strong schema of induction $\Phi(\varphi^{(\vartheta)})$ [which has the same strength as $\mathbf{PA^1}$], then the *completeness* metatheorem [see **Theorem 3.1.1**] allows to take into account a *new* approach for a metamathematical program analogous to Hilbert's Program.

References

[1] Davis. M., *The Undecidable: Basic Papers on Undecidable Propositions, Unsolvable Problems and Computable Functions.* Hewlett, N.Y., Raven Press, (1965).

[2] Feferman, S., Transfinite Recursive Progression of Axiomatic Theories, *The Journal Symbolic Logic*, vol.27, pp. 256-316, (1962).

[3] Gentzen, G., Die Widerspruchsfreiheit der reinen Zahlentheorie, *Math. Ann.*, vol.112, pp. 493-565, (1936).

[4] Gödel, K., Über formal unentscheidbare Sätze der Principia Mathematica und verwandter Systeme I., *Monatsh. Math. Phys.*, 38: pp.173-198, (1931).

[5] Gödel, K., Über eine bisher noch nicht benützte Erweiterung des finiten Standpunktes, *Logica (Studia Paul Bernays dedicata).* Ed. Griffon, Neuchatel, pp. 76*83, (1959).

[8]This means that, if $\mathcal{A}$ is consistent, then by setting $\lfloor\Phi(\varphi^{(\xi)})\rfloor$ for the code in $\mathbb{L}^{(\varepsilon_0)}$ of $\Phi(\varphi^{(\xi)})$, there is *no* formula $W(x)$ of $\mathbb{L}^{(\varepsilon_0)}$ such that $\vdash \Phi(\varphi^{(\xi)}) \leftrightarrow W(\lfloor\Phi(\varphi^{(\xi)})\rfloor)$ for every sentence $\varphi^{(\xi)} \in \mathbb{L}^{(\xi)}$

[6] Mostowski, A., *Thirty Years of Foundational Studies*, Lectures on the development of mathematical logic and the study of the foundations of mathematics in 1930-1964, Oxford, Basil Blackwell, (1966).

[7] Tarski, A., *Logic, Semantics, Metamathematics*, Ed. Hackett Publishing Company (second edition) J. Corcoran ed., (1983).

Received 29 February 2019

www.ingramcontent.com/pod-product-compliance
Lightning Source LLC
Chambersburg PA
CBHW081342230426
43667CB00017B/2698
* 9 7 8 1 8 4 8 9 0 3 8 9 0 *